Student's Solutions Manual

for use with

Calculus

Second Edition

Robert T. Smith
Millersville University

Roland B. Minton
Roanoke College

Prepared by
Frank Purcell and Elka Block
Twin Prime Editorial

Mc
Graw
Hill

Boston   Burr Ridge, IL   Dubuque, IA   Madison, WI   New York   San Francisco   St. Louis
Bangkok   Bogotá   Caracas   Kuala Lumpur   Lisbon   London   Madrid   Mexico City
Milan   Montreal   New Delhi   Santiago   Seoul   Singapore   Sydney   Taipei   Toronto

McGraw-Hill Higher Education

*A Division of The **McGraw-Hill** Companies*

Student's Solutions Manual for use with
CALCULUS, SECOND EDITION
ROBERT T. SMITH AND ROLAND B. MINTON

Published by McGraw-Hill Higher Education, an imprint of The McGraw-Hill Companies, Inc.,
1221 Avenue of the Americas, New York, NY 10020. Copyright © The McGraw-Hill Companies,
Inc., 2002, 2000. All rights reserved.

This book is printed on acid-free paper.

1 2 3 4 5 6 7 8 9 0 QPD QPD 0 3 2 1

ISBN 0-07-239859-0

www.mhhe.com

Table of Contents

Chapter 0

Section 0.1

5. $3x + 2 < 11$
$3x < 9$
$x < 3$
$(-\infty, 3)$

7. $2x - 3 < -7$
$2x < -4$
$x < -2$
$(-\infty, -2)$

9. $4 - 3x < 6$
$-3x < 2$
$x > -\dfrac{2}{3}$
$\left(-\dfrac{2}{3}, \infty\right)$

11. $4 \le x + 1 < 7$
$3 \le x < 6$
$[3, 6)$

13. $-2 < 2 - 2x < 3$
$-4 < -2x < 1$
$2 > x > -\dfrac{1}{2}$
$-\dfrac{1}{2} < x < 2$
$\left(-\dfrac{1}{2}, 2\right)$

15. $x^2 + 3x - 4 > 0$
$(x + 4)(x - 1) > 0$

$x < -4$ or $x > 1$
$(-\infty, -4) \cup (1, \infty)$

17. $x^2 - x - 6 < 0$
$(x - 3)(x + 2) < 0$

$(-2, 3)$

19. $3x^2 + 4 > 0$
$3x^2 > -4$
$x^2 > -\dfrac{4}{3}$
This is always true.
$(-\infty, \infty)$

21. $|x - 3| < 4$
$-4 < x - 3 < 4$
$-1 < x < 7$
$(-1, 7)$

23. $|3 - x| < 1$
$-1 < 3 - x < 1$
$-4 < -x < -2$
$4 > x > 2$
$2 < x < 4$
$(2, 4)$

25. $|2x + 1| > 2$
$2x + 1 < -2$ or $2x + 1 > 2$
$2x < -3$ $2x > 1$
$x < -\dfrac{3}{2}$ or $x > \dfrac{1}{2}$
$\left(-\infty, -\dfrac{3}{2}\right) \cup \left(\dfrac{1}{2}, \infty\right)$

27. $\dfrac{x + 2}{x - 2} > 0$

$x < -2$ or $x > 2$
$(-\infty, -2) \cup (2, \infty)$

29. $\dfrac{x^2-x-2}{(x+4)^2} > 0$

$\dfrac{(x-2)(x+1)}{(x+4)^2} > 0$

$$
\begin{array}{l}
\underset{2}{\overset{\;-\quad\quad 0\;+}{\rule{4cm}{0.4pt}}} \!\!\!\to (x-2)\\[2mm]
\underset{-1}{\overset{-\quad 0\quad +}{\rule{4cm}{0.4pt}}} \!\!\!\to (x+1)\\[2mm]
\underset{-4}{\overset{+\;\;0\quad\quad +}{\rule{4cm}{0.4pt}}} \!\!\!\to (x+4)^2\\[2mm]
\underset{-4\quad -1\quad\;\; 2}{\overset{+\;\;x\;+\;\;0\;\;-\;\;0\;+}{\rule{5cm}{0.4pt}}} \!\!\!\to \dfrac{(x-2)(x+1)}{(x+4)^2}
\end{array}
$$

$x < -4$ or $-4 < x < -1$ or $x > 2$

$(-\infty,\,-4) \cup (-4,\,-1) \cup (2,\,\infty)$

31. $\dfrac{-8x}{(x+1)^3} < 0$

$$
\begin{array}{l}
\underset{0}{\overset{+\quad\quad 0\quad -}{\rule{4cm}{0.4pt}}} \!\!\!\to (-8x)\\[2mm]
\underset{-1}{\overset{-\quad 0\quad +}{\rule{4cm}{0.4pt}}} \!\!\!\to (x+1)^3\\[2mm]
\underset{-1\quad 0}{\overset{+\;\;x+0\quad -}{\rule{4.5cm}{0.4pt}}} \!\!\!\to \dfrac{-8x}{(x+1)^3}
\end{array}
$$

$x < -1$ or $x > 0$

$(-\infty,\,-1) \cup (0,\,\infty)$

33. $d\{(2,1),(4,4)\} = \sqrt{(4-2)^2+(4-1)^2}$

$\qquad\qquad\qquad = \sqrt{4+9}$

$\qquad\qquad\qquad = \sqrt{13}$

35. $d\{(-1,-2),(3,-2)\}$

$\quad = \sqrt{[3-(-1)]^2+[-2-(-2)]^2}$

$\quad = \sqrt{16+0}$

$\quad = 4$

37. $d\{(0,2),(-2,6)\} = \sqrt{(-2-0)^2+(6-2)^2}$

$\qquad\qquad\qquad = \sqrt{4+16}$

$\qquad\qquad\qquad = \sqrt{20}$

39. $d\{(1,1),(3,4)\} = \sqrt{(3-1)^2+(4-1)^2}$

$\qquad\qquad\qquad = \sqrt{4+9}$

$\qquad\qquad\qquad = \sqrt{13}$

$d\{(1,1),(0,6)\} = \sqrt{(0-1)^2+(6-1)^2}$

$\qquad\qquad\qquad = \sqrt{1+25} = \sqrt{26}$

$d\{(3,4),(0,6)\} = \sqrt{(0-3)^2+(6-4)^2}$

$\qquad\qquad\qquad = \sqrt{9+4} = \sqrt{13}$

$(\sqrt{13})^2 + (\sqrt{13})^2 = (\sqrt{26})^2$

This statement is true, so yes, this is a right triangle.

41. $d\{(-2,3),(2,9)\} = \sqrt{[2-(-2]^2+(9-3)^2}$

$\qquad\qquad\qquad = \sqrt{16+36}$

$\qquad\qquad\qquad = \sqrt{52}$

$d\{(-2,3),(-4,13)\} = \sqrt{[-4-(-2)]^2+(13-3)^2}$

$\qquad\qquad\qquad\quad = \sqrt{4+100} = \sqrt{104}$

$d\{(2,9),(-4,13)\} = \sqrt{(-4-2)^2+(13-9)^2}$

$\qquad\qquad\qquad\quad = \sqrt{36+16} = \sqrt{52}$

$(\sqrt{52})^2 + (\sqrt{52})^2 = (\sqrt{104})^2$

This statement is true, so yes, this is a right triangle.

43.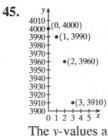

The y-values are increasing by
550, 650, 750, …
The next population is $3200 + 850 = 4050$.

45.

The y-values are decreasing by 10, 30, 50, …
The next population is $3910 - 70 = 3840$.

47. When a calculation is done using a finite number of digits, the result can only contain a finite number of digits. Therefore, the result can be written as a quotient of two integers (where the denominator is a power of 10), so the result must be rational.

49. $\dfrac{\left(\frac{1}{2}\right)^7 - \left(\frac{2}{3}\right)^{12}}{\left(\frac{1}{2}\right)^7} \approx .013$ or 1.3%

51.

P	win %
.551	.568
.587	.593
.404	.414
.538	.556
.605	.615

Section 0.2

5. (1, 2) and (3, 6):

$$m = \frac{6-2}{3-1} = \frac{4}{2} = 2$$

(3, 6) and (0, 0):

$$m = \frac{0-6}{0-3} = \frac{-6}{-3} = 2$$

Since the slopes are the same, the points must be colinear.

7. (2, 1) and (0, 2):

$$m = \frac{2-1}{0-2} = \frac{1}{-2} = -\frac{1}{2}$$

(0, 2) and (4, 0):

$$m = \frac{0-2}{4-0} = \frac{-2}{4} = -\frac{1}{2}$$

Since the slopes are the same, the points must be colinear.

9. (3, 1) and (4, 4):

$$m = \frac{4-1}{4-3} = \frac{3}{1} = 3$$

(4, 4) and (5, 8):

$$m = \frac{8-4}{5-4} = \frac{4}{1} = 4$$

Since the slopes are not equal, the points are not colinear.

11. (4, 1) and (3, 2):

$$m = \frac{2-1}{3-4} = \frac{1}{-1} = -1$$

(3, 2) and (1, 3):

$$m = \frac{3-2}{1-3} = \frac{1}{-2} = -\frac{1}{2}$$

Since the slopes are not equal, the points are not colinear.

13. $m = \dfrac{6-2}{3-1} = \dfrac{4}{2} = 2$

15. $m = \dfrac{3-2}{3-1} = \dfrac{1}{2}$

17. $m = \dfrac{-1-(-6)}{1-3} = \dfrac{5}{-2} = -\dfrac{5}{2}$

19. $m = \dfrac{-0.4-(-1.4)}{-1.1-0.3} = \dfrac{1.0}{-1.4} = -\dfrac{5}{7}$

21. $P_2 = (2, 5)$

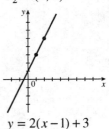

$y = 2(x-1)+3$

23. $P_2 = (2, 2)$

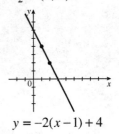

$y = -2(x-1)+4$

25. $P_2 = (0, 1)$

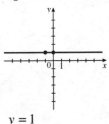

$y = 1$

27. $P_2 = (4, 2)$

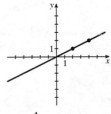

$y = \dfrac{1}{2}(x-2)+1$

29. $P_2 = (3.3, 2.3)$

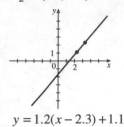

$y = 1.2(x - 2.3) + 1.1$

31. parallel—both have slope 3

33. perpendicular—slopes are -2 and $\dfrac{1}{2}$

35. neither—slopes are 2 and -2

37. neither—slopes are 4 and $\dfrac{1}{4}$

39. $y = -\dfrac{1}{2}x + \dfrac{1}{2}$ and $y = -\dfrac{1}{2}x + \dfrac{3}{4}$

parallel—both have slope $-\dfrac{1}{2}$

41. a. $y = 2(x - 2) + 1$

    b. $y = -\dfrac{1}{2}(x - 2) + 1$

43. a. $y = -1$

    b. $x = 0$

45. a. $y = 2(x - 3) + 1$

    b. $y = -\dfrac{1}{2}(x - 3) + 1$

47. $P_1(1, 1), P_2 = (2, 3)$

$m = \dfrac{3-1}{2-1} = \dfrac{2}{1} = 2$

$y = 2(x - 1) + 1$

At $x = 4$: $y = 2(4 - 1) + 1 = 7$

49. $P_1 = (0.5, 4), P_2 = (1, 3)$

$M = \dfrac{3-4}{1-0.5} = \dfrac{-1}{0.5} = -2$

$y = -2(x - 1) + 3$

At $x = 4$: $y = -2(4 - 1) + 3 = -3$

51. $P_1 = (-2, 1), P_2(-1, 1.5)$

$m = \dfrac{1.5 - 1}{-1 - (-2)} = \dfrac{0.5}{1} = 0.5$

$y = 0.5(x + 2) + 1$

At $x = 4$: $y = 0.5(4 + 2) + 1 = 4$

53. function

55. not a function

57. function

59. both

Note: It is a rational function because it can be

written as $f(x) = \dfrac{x^3 - 4x + 1}{1}$.

61. rational

63. neither

65. both

Note: It is a rational function because it can be

written as $f(x) = \dfrac{3 - 2x + x^4}{1}$.

67. $x + 2 \geq 0$

$x \geq -2$

$[-2, \infty)$

69. $(-\infty, \infty)$

71. Determine where the denominator is 0.

$x^2 - 1 = 0$

$x^2 = 1$

$x = \pm 1$

Domain: $x \neq \pm 1$

$(-\infty, -1) \cup (-1, 1) \cup (1, \infty)$

73. $x^2 + 1 \geq 0$

$x^2 \geq -1$

This is always true.

$(-\infty, \infty)$

75. $f(0) = 0^2 - 0 - 1 = -1$

$f(2) = 2^2 - 2 - 1 = 1$

$f(-3) = (-3)^2 - (-3) - 1 = 11$

$f\left(\dfrac{1}{2}\right) = \left(\dfrac{1}{2}\right)^2 - \dfrac{1}{2} - 1 = -\dfrac{5}{4}$

77. $f(0) = \sqrt{0+1} = 1$

$f(3) = \sqrt{3+1} = 2$

$f(-1) = \sqrt{-1+1} = 0$

$f\left(\dfrac{1}{2}\right) = \sqrt{\dfrac{1}{2}+1} = \sqrt{\dfrac{3}{2}} = \dfrac{\sqrt{6}}{2}$

79. Possible answer: 50 ft $\le x \le$ width of lot

81. Possible answer: $0 \le x \le$ number made

83. This is not a function because you could receive different scores on tests for which you studied the same number of hours. There are many possible y-values for one value of x.

85. This is not a function because a person's weight is not determined by the amount of exercise. There are many possible y-values for one value of x.

87. Possible answer: days 1 and 9

89. Possible answer: flat—constant speed; up—speed increasing (going downhill); down—speed decreasing (going uphill)

91. Possible answer: moderate usage during day, increasing in evening, low usage at night

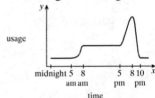

93. Possible answer: highest during good songs, lowest during bad songs, moderate during commercials

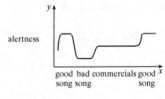

95. $98.6 = -1.8h + 212$

$1.8h = 113.4$

$h = 63$

63,000 feet

97. $P_1 = (120, 9100),\ P_2 = (60, 10{,}000)$

$m = \dfrac{10{,}000 - 9100}{60 - 120} = -15$

$y = -15(x - 120) + 9100$

When $x = 90$,

$y = -15(90 - 120) + 9100 = 9550$ rpm

Section 0.3

5.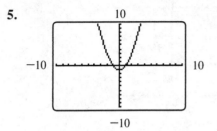

7.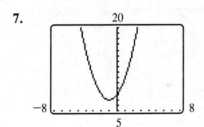

9.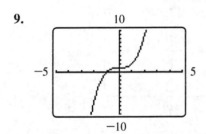

11.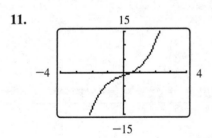

13.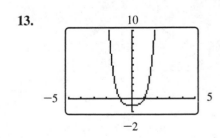

15.

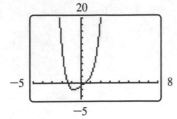

17.

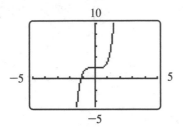

19.

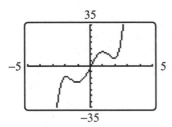

21.

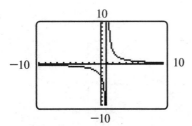

23.

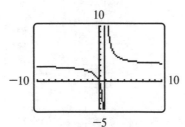

25.

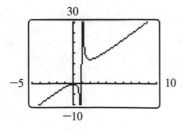

27.

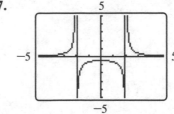

29.

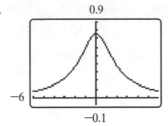

31.

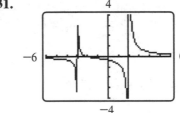

33.

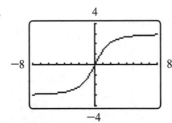

35.

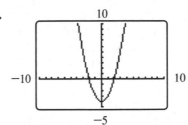

x-intercepts: $x = -2$, $x = 2$
y-intercept: $y = -4$

37.

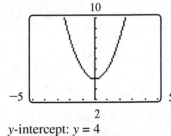

y-intercept: $y = 4$

39.

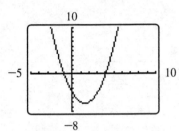

x-intercepts: $x = -1$, $x = 4$
y-intercept: $y = -4$

41.

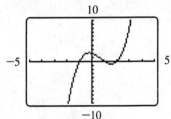

x-intercepts: $x = -1$, $x = 1$, $x = 2$
y-intercept: $y = 2$

43.

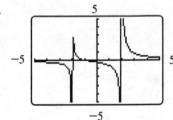

x-intercept: $x = -1$

y-intercept: $y = -\dfrac{1}{4}$

45. $x^2 - 4 = 0$
    $x^2 = 4$
     $x = \pm 2$

47. $x^2 + 3x - 10 = 0$
    $(x + 5)(x - 2) = 0$
    $x = -5$, $x = 2$

49. $x^2 + 4 = 0$
     $x^2 = -4$
    No vertical asymptotes

51. $x^3 + 3x^2 + 2x = 0$
    $x(x^2 + 3x + 2) = 0$
    $x(x + 2)(x + 1) = 0$
    $x = 0$, $x = -2$, $x = -1$

53. $x^2 - 9 = 0$
     $x^2 = 9$
      $x = \pm 3$

55.

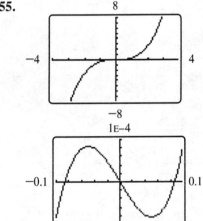

57.

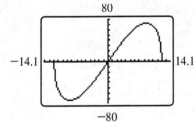

59.

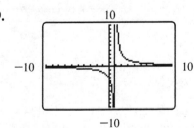

Horizontal asymptote: $y = 0$

7

61.

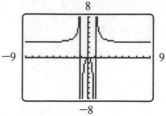

Horizontal asymptote: $y = 3$

63.

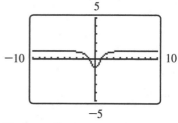

Horizontal asymptote: $y = 1$

65.

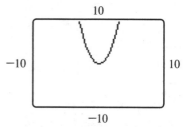

The graph of $y = x^2$ on $-10 \le x \le 10$, $-10 \le y \le 10$, is identical to the graph of $y = 2(x-1)^2 + 3$ on certain graphing windows centered at the point $(1, 3)$. One possibility is $-9 \le x \le 11, -17 \le y \le 23$; another is $-4 \le x \le 6, -2 \le y \le 8$.

67. $\sqrt{y^2}$ is the distance from the x-axis.

$\sqrt{x^2 + (y-2)^2}$ is the distance from the point $(0, 2)$.

$y^2 = x^2 + (y-2)^2$

$y^2 = x^2 + y^2 - 4y + 4$

$4y = x^2 + 4$

$y = \dfrac{1}{4}x^2 + 1$ parabola

69. The terminal velocity for the lead ball will be $\sqrt{2}$ times that of the wood ball.

Section 0.4

5. $x^2 - 4x + 3 = 0$

$(x-3)(x-1) = 0$

$x = 3, \ x = 1$

7. $x^2 - 2x - 15 = 0$

$(x-5)(x+3) = 0$

$x = 5, \ x = -3$

9. $x^2 - 4x + 2 = 0$

$x = \dfrac{4 \pm \sqrt{(-4)^2 - 4(1)(2)}}{2(1)}$

$= \dfrac{4 \pm \sqrt{8}}{2}$

$= \dfrac{4 \pm 2\sqrt{2}}{2}$

$x = 2 \pm \sqrt{2}$

11. $2x^2 + 4x - 1 = 0$

$x = \dfrac{-4 \pm \sqrt{4^2 - 4(2)(-1)}}{2(2)}$

$= \dfrac{-4 \pm \sqrt{24}}{4}$

$= \dfrac{-4 \pm 2\sqrt{6}}{4}$

$x = -1 \pm \dfrac{\sqrt{6}}{2}$

13. $x^3 - 3x^2 + 2x = 0$

$x(x^2 - 3x + 2) = 0$

$x(x-2)(x-1) = 0$

$x = 0, \ x = 2, \ x = 1$

15. $x^3 + x^2 - 4x - 4 = 0$

$x^2(x+1) - 4(x+1) = 0$

$(x+1)(x^2 - 4) = 0$

$x = -1, \ x = \pm 2$

17. $x^4 - 1 = 0$

$(x^2 + 1)(x^2 - 1) = 0$

$x = \pm 1$

19. $x^6 + x^3 - 2 = 0$

$(x^3 + 2)(x^3 - 1) = 0$

$x = -\sqrt[3]{2},\ x = 1$

21.

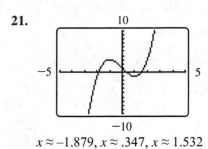

$x \approx -1.879,\ x \approx .347,\ x \approx 1.532$

23.

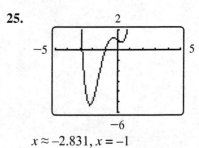

$x \approx -1.325$

25.

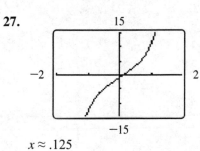

$x \approx -2.831,\ x = -1$

27.

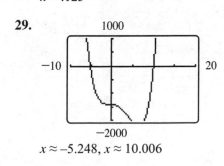

$x \approx .125$

29.

$x \approx -5.248,\ x \approx 10.006$

31.

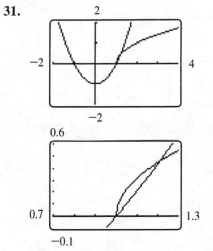

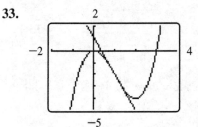

There are 2 solutions: $x = 1$, $x \approx 1.206$.

33.

The graph does not clearly show the number of intersection points. Solve algebraically.

$$x^3 - 3x^2 = 1 - 3x$$
$$x^3 - 3x^2 + 3x - 1 = 0$$
$$(x-1)^3 = 0$$
$$x = 1$$

There is 1 solution: $x = 1$.

35.

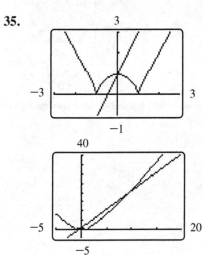

After zooming out, the graph shows that there are 2 solutions: $x = 0$, $x \approx 9.534$.

37.

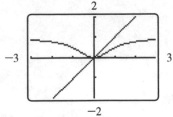

The graph shows that there is 1 solution.
Solve algebraically.

$$\frac{x^2}{x^2+1} = x$$
$$x^2 = x(x^2+1)$$
$$0 = x(x^2+1) - x^2$$
$$0 = x^3 - x^2 + x$$
$$0 = x(x^2 - x + 1)$$

$$x = 0 \text{ or } x = \frac{1 \pm \sqrt{(-1)^2 - 4(1)(1)}}{2} = \frac{1}{2} \pm \frac{\sqrt{3}}{2}i$$

There is 1 real solution: $x = 0$.

39.

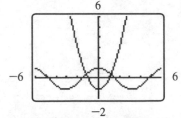

The graph shows that there are 2 solutions:
$x \approx \pm 1.177$.

41. One way to show these values is to solve the equation by completing the square.

$$ax^2 + bx + c = 0$$
$$a\left(x^2 + \frac{b}{a}x + \left(\frac{b}{2a}\right)^2\right) + c - \frac{b^2}{4a} = 0$$
$$a\left(x + \frac{b}{2a}\right)^2 + c - \frac{b^2}{4a} = 0$$
$$4a^2\left(x + \frac{b}{2a}\right)^2 + 4ac - b^2 = 0$$
$$4a^2\left(x + \frac{b}{2a}\right)^2 = b^2 - 4ac$$
$$\left(x + \frac{b}{2a}\right)^2 = \frac{b^2 - 4ac}{4a^2}$$
$$x + \frac{b}{2a} = \pm\sqrt{\frac{b^2 - 4ac}{4a^2}}$$
$$x + \frac{b}{2a} = \pm\frac{\sqrt{b^2 - 4ac}}{2|a|}$$
$$x + \frac{b}{2a} = \frac{\pm\sqrt{b^2 - 4ac}}{2a}$$
$$x = \frac{-b \pm \sqrt{b^2 - 4ac}}{2a}$$

43.

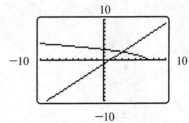

45. $\qquad x - 2 = \sqrt{7 - 2x}$

$$x^2 - 4x + 4 = 7 - 2x$$
$$x^2 - 2x - 3 = 0$$
$$(x - 3)(x + 1) = 0$$

$x = 3$ true solution
$x = -1$ extraneous

47. $\quad x^2 + 2x + 5 = 0$

$$x = \frac{-2 \pm \sqrt{2^2 - 4(1)(5)}}{2(1)}$$
$$= \frac{2 \pm \sqrt{-16}}{2}$$
$$= \frac{-2 \pm 4i}{2}$$
$$x = -1 \pm 2i$$

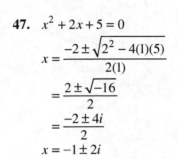

no real zeros, no x-intercepts

49. $\quad x^2 + 2x - 5 = 0$

$$x = \frac{-2 \pm \sqrt{2^2 - 4(1)(-5)}}{2(1)}$$
$$= \frac{-2 \pm \sqrt{24}}{2}$$
$$= \frac{-2 \pm 2\sqrt{6}}{2}$$
$$x = -1 \pm \sqrt{6}$$

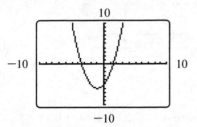

2 real zeros, 2 x-intercepts

51. $\qquad x^3 - 1 = 0$

$$(x - 1)(x^2 + x + 1) = 0$$

$x = 1 \quad$ or $\quad x = \dfrac{-1 \pm \sqrt{1^2 - 4(1)(1)}}{2}$

$x = 1 \quad$ or $\quad x = \dfrac{-1 \pm i\sqrt{3}}{2}$

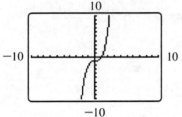

1 real zero, 1 x-intercept

53. $\qquad x^2 - 1 = x$

$$x^2 - x - 1 = 0$$
$$x = \frac{1 \pm \sqrt{(-1)^2 - 4(1)(-1)}}{2(1)}$$
$$x = \frac{1 \pm \sqrt{5}}{2}$$

55. $\qquad x^3 = x$

$$x^3 - x = 0$$
$$x(x^2 - 1) = 0$$
$$x(x + 1)(x - 1) = 0$$
$$x = 0, \; x = \pm 1$$

11

57. $x^2 + c = x$

$x^2 - x + c = 0$

$x = \dfrac{1 \pm \sqrt{(-1)^2 - 4(1)(c)}}{2(1)}$

$= \dfrac{1 \pm \sqrt{1 - 4c}}{2}$

There are two real fixed points if
$1 - 4c > 0$

$-4c > -1$

$c < \dfrac{1}{4}$

There are no real fixed points if
$1 - 4c < 0$

$c > \dfrac{1}{4}$

59. $5 = 5 + 0.14(130) - \dfrac{16.3(130)^2}{v^2}$

$18.2 = \dfrac{275,470}{v^2}$

$v^2 = \dfrac{275,470}{18.2}$

$v = \sqrt{\dfrac{275,470}{18.2}} \approx 123.03 \text{ ft/s}$

$(123.03)\left(\dfrac{3600}{5280}\right) \approx 83.9 \text{ mph}$

61. $.5\sqrt{h} = 4.5$

$\sqrt{h} = 9$

$h = 81 \text{ feet}$

Section 0.5

5. a. $\left(\dfrac{\pi}{4}\right)\left(\dfrac{180°}{\pi}\right) = 45°$

b. $\left(\dfrac{\pi}{3}\right)\left(\dfrac{180°}{\pi}\right) = 60°$

c. $\left(\dfrac{\pi}{6}\right)\left(\dfrac{180°}{\pi}\right) = 30°$

d. $\left(\dfrac{4\pi}{3}\right)\left(\dfrac{180°}{\pi}\right) = 240°$

7. a. $(180)\left(\dfrac{\pi}{180}\right) = \pi$

b. $(270)\left(\dfrac{\pi}{180}\right) = \dfrac{3\pi}{2}$

c. $(120)\left(\dfrac{\pi}{180}\right) = \dfrac{2\pi}{3}$

d. $(30)\left(\dfrac{\pi}{180}\right) = \dfrac{\pi}{6}$

9.

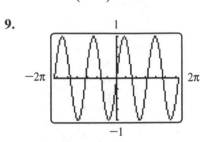

11.

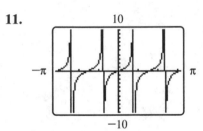

13.

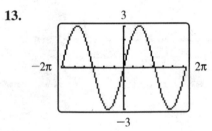

15.

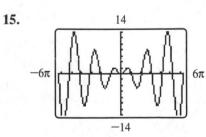

17.

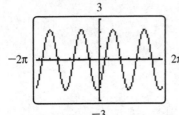

19.

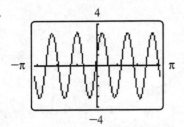

21.

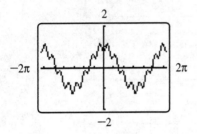

23.

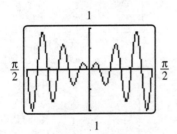

25.

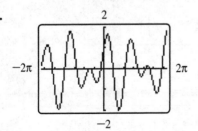

27. Domain: all real numbers except odd multiples of $\dfrac{\pi}{2}$: range: all real numbers

29. Domain: all real numbers except integer multiples of π; range: all real numbers

31. amplitude = 3

period = $\dfrac{2\pi}{2} = \pi$

frequency = $\dfrac{1}{\pi}$

33. amplitude = 5

period = $\dfrac{2\pi}{3}$

frequency = $\dfrac{3}{2\pi}$

35. amplitude = 3

period = $\dfrac{2\pi}{2} = \pi$

frequency = $\dfrac{1}{\pi}$

37. amplitude = 4

period = 2π

frequency = $\dfrac{1}{2\pi}$

39. $\sin(\alpha - \beta) = \sin[\alpha + (-\beta)]$
$$= \sin\alpha\cos(-\beta) + \sin(-\beta)\cos\alpha$$
$$= \sin\alpha\cos\beta - \sin\beta\cos\alpha$$

41. $\tan^2\theta + 1 = \dfrac{\sin^2\theta}{\cos^2\theta} + 1$

$$= \dfrac{\sin^2\theta + \cos^2\theta}{\cos^2\theta}$$

$$= \dfrac{1}{\cos^2\theta}$$

$$= \sec^2\theta$$

43. $5\cos(x + \beta) = 5\cos x \cos\beta - 5\sin x \sin\beta$
$$= 4\cos x - 3\sin x$$

$5\cos\beta = 4$ and $5\sin\beta = 3$

$\cos\beta = \dfrac{4}{5}$ and $\sin\beta = \dfrac{3}{5}$

This is possible since

$\left(\dfrac{4}{5}\right)^2 + \left(\dfrac{3}{5}\right)^2 = \dfrac{16}{25} + \dfrac{9}{25} = 1.$

Estimate β by trial-and-error.

$\beta \approx 0.6435$ radians

45. $\cos 2x$ has period $\dfrac{2\pi}{2} = \pi$, so it repeats at intervals of width π, 2π, 3π, etc.

$3 \sin \pi x$ has period $\dfrac{2\pi}{\pi} = 2$, so it repeats at intervals of width 2, 4, 6, etc.

Since one list consists of irrational numbers and the other of rational numbers, there can't be any numbers on both lists. $f(x)$ is not periodic.

47. $\sin 2x$ has period $\dfrac{2\pi}{2} = \pi$, so it repeats at intervals of width π, 2π, 3π, etc.

$\cos 5x$ has period $\dfrac{2\pi}{5}$, so it repeats at intervals of width $\dfrac{2\pi}{5}, \dfrac{4\pi}{5}, \dfrac{6\pi}{5}$, etc.

Since 2π is the first number to appear on both lists, $f(x)$ is periodic with period 2π.

49. $2\cos x - 1 = 0$

$\cos x = \dfrac{1}{2}$

$x = \pm\dfrac{\pi}{3} + 2n\pi$

51. $\cos^2 x + \cos x = 0$
$(\cos x)(\cos x + 1) = 0$
$\cos x = 0 \text{ or } \cos x = -1$
$x = \dfrac{\pi}{2} + n\pi, \ x = \pi + 2n\pi$

53. $\sin^2 x - 4\sin x + 3 = 0$
$(\sin x - 1)(\sin x - 3) = 0$
$\sin x = 1 \text{ or } \sin x = 3$
There are no solutions for $\sin x = 3$, so the solutions are $x = \dfrac{\pi}{2} + 2n\pi$.

55. $\sin^2 x + \cos x - 1 = 0$
$(1 - \cos^2 x) + \cos x - 1 = 0$
$(\cos x)(\cos x - 1) = 0$
$\cos x = 0 \text{ or } \cos x = 1$
$x = \pm\dfrac{\pi}{2} + 2n\pi, \ x = 2n\pi$

57. $\cos^2 \theta = 1 - \left(\dfrac{1}{3}\right)^2 = 1 - \dfrac{1}{9} = \dfrac{8}{9}$

$\cos \theta = \sqrt{\dfrac{8}{9}} = \dfrac{\sqrt{8}}{3} \text{ or } \dfrac{2\sqrt{2}}{3}$

59. $\cos^2 \theta = 1 - \left(\dfrac{1}{2}\right)^2 = 1 - \dfrac{1}{4} = \dfrac{3}{4}$

$\cos \theta = -\sqrt{\dfrac{3}{4}} = -\dfrac{\sqrt{3}}{2}$

61.

3 solutions: $x = 0, \ x \approx 1.109, \ x \approx 3.698$

63.

2 solutions $x \approx -1.455, \ x \approx 1.455$

65.

1 solution: $x \approx 1.249$

67.

2 solutions: $x = 0, \ x \approx .877$

69.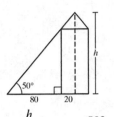

$\dfrac{h}{2} = \tan 20°$

$h = 2\tan 20° \approx 0.73$ miles

71.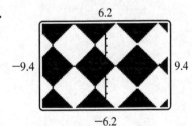

$\dfrac{h}{80 + 20} = \tan 50°$

$h = 100\tan 50° \approx 119.2$ feet

73. $v(t) = 170 \sin 2\pi ft$

$\dfrac{2\pi}{2\pi f} = \dfrac{\pi}{30}$

$f = \dfrac{30}{\pi}$

meter voltage $= \dfrac{170}{\sqrt{2}} = 85\sqrt{2} \approx 120.2$ volts

75. The fluctuation could be caused by seasonal variations. Assuming sales are slowest in January, $t = 0$ corresponds to April. Since $s(12) - s(0) = 24$, sales are increasing by $24,000 per year.

77.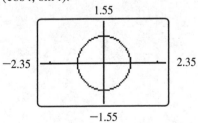

79. The graph is a circle—namely, the unit circle—because coordinates on the unit circle are given by $(\cos t, \sin t)$.

Section 0.6

5. $2^{-3} = \dfrac{1}{2^3} = \dfrac{1}{8}$

7. $3^{1/2} = \sqrt{3}$

9. $5^{2/3} = \sqrt[3]{5^2} = \sqrt[3]{25}$

11. $4^{-2/3} = \dfrac{1}{4^{2/3}} = \dfrac{1}{\sqrt[3]{4^2}} = \dfrac{1}{\sqrt[3]{16}}$

13. $\dfrac{1}{x^2} = x^{-2}$

15. $\sqrt{x} = x^{1/2}$

17. $\dfrac{2}{x^3} = 2x^{-3}$

19. $\dfrac{1}{2\sqrt{x}} = \dfrac{1}{2}x^{-1/2}$

21. $4^{3/2} = (\sqrt{4})^3 = 2^3 = 8$

23. $\dfrac{\sqrt{8}}{2^{1/2}} = \dfrac{\sqrt{8}}{\sqrt{2}} = \sqrt{4} = 2$

25. $e^2 \approx 7.389$

27. $2e^{-1/2} \approx 1.213$

29. $\dfrac{12}{e} \approx 4.415$

31. $10e^{-.01} \approx 9.900$

33.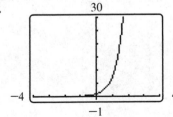

15

35.

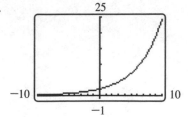

47.

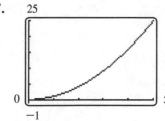

37.

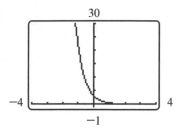

49.

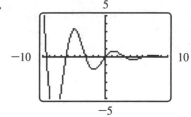

39.

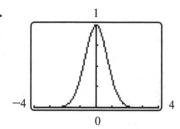

51. $e^{2x} = 2$

$$\ln e^{2x} = \ln 2$$
$$2x = \ln 2$$
$$x = \frac{\ln 2}{2}$$

41.

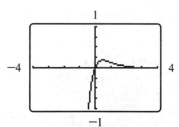

53. $2e^{-2x} = 1$

$$e^{-2x} = \frac{1}{2}$$
$$\ln e^{-2x} = \ln\left(\frac{1}{2}\right)$$
$$-2x = \ln 1 - \ln 2 = -\ln 2$$
$$x = \frac{\ln 2}{2}$$

55. $\ln 2x = 4$

$$2x = e^4$$
$$x = \frac{e^4}{2}$$

43.

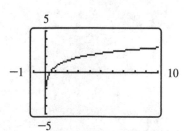

57. $2\ln 4x - 1 = 6$

$$2\ln 4x = 7$$
$$\ln 4x = \frac{7}{2}$$
$$4x = e^{7/2}$$
$$x = \frac{e^{7/2}}{4}$$

45.

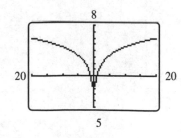

59. $e^{2\ln x} = 4$

$e^{\ln x^2} = 4$

$x^2 = 4$

$x = 2, -2$

Must have $x > 0$

$x = 2$

61. a. $\log_3 7 = \dfrac{\ln 7}{\ln 3} \approx 1.771$

b. $\log_4 60 = \dfrac{\ln 60}{\ln 4} \approx 2.953$

c. $\log_3 \dfrac{1}{24} = \dfrac{\ln(1/24)}{\ln 3} \approx -2.893$

63. a. Since $3^2 = 9$, $\log_3 9 = 2$

b. Since $4^3 = 64$, $\log_4 64 = 3$

c. Since $3^{-3} = \dfrac{1}{27}$, $\log_3 \dfrac{1}{27} = -3$

65. $\log 3 - \ln 4 = \ln\left(\dfrac{3}{4}\right)$

67. $\dfrac{1}{2}\ln 4 - \ln 2 = \ln 4^{1/2} - \ln 2 = \ln 2 - \ln 2 = 0$

69. $\ln \dfrac{3}{4} + 4\ln 2 = \ln \dfrac{3}{4} + \ln 16$

$= \ln\left(\dfrac{3}{4} \cdot 16\right)$

$= \ln 12$

71. Using $f(0) = 2$:

$2 = ae^{b(0)} = a$

$f(x) = 2e^{bx}$

Using $f(2) = 6$:

$6 = 2e^{2b}$

$3 = e^{2b}$

$\ln 3 = \ln e^{2b} = 2b$

$b = \dfrac{\ln 3}{2}$

$f(x) = 2e^{(\frac{1}{2}\ln 3)x}$

73. Using $f(0) = 4$:

$4 = ae^{b(0)} = a$

$f(x) = 4e^{bx}$

Using $f(2) = 2$:

$2 = 4e^{2b}$

$\dfrac{1}{2} = e^{2b}$

$\ln\left(\dfrac{1}{2}\right) = \ln e^{2b} = 2b$

$b = \dfrac{\ln\left(\frac{1}{2}\right)}{2} = -\dfrac{\ln 2}{2}$

$f(x) = 4e^{-\left(\frac{1}{2}\ln 2\right)x}$

75. Using $f(0) = -2$:

$-2 = ae^{b(0)} = a$

$f(x) = -2e^{bx}$

Using $f(1) = -3$:

$-3 = -2e^{b}$

$\dfrac{3}{2} = e^{b}$

$\ln\left(\dfrac{3}{2}\right) = \ln e^{b} = b$

$f(x) = -2e^{\left[\ln\left(\frac{3}{2}\right)\right]x}$

77. $1 - \left(\dfrac{9}{10}\right)^{10} \approx 0.651$

79. $1 - \dfrac{1}{e} \approx .632$

81.

$u = \ln x$	.4055	.6931	.9163	1.0986	1.2528	1.3863
$v = \ln y$	2.0719	2.7619	3.2977	3.7350	4.1051	4.4257

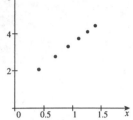

$$m \approx \frac{4.4257 - 2.0719}{1.3863 - 0.4055} = 2.4;\ v = 2.4u + b$$

$$2.0719 \approx 2.4(0.4055) + b$$

$$b \approx 1.1$$

Using exercise 80,

$$b = \ln a$$

$$a = e^b = e^{1.1} \approx 3$$

83. Let x be the number of decades after 1780.

$u = \ln x$	0	.693	1.099	1.386	1.609	1.792	1.946	2.079
$v = \ln$ (Population)	15.184	15.485	15.795	16.081	16.370	16.653	16.959	17.264

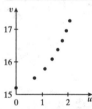

The plot does not look linear; exponential function

85. a. $\quad 7 = -\log [H^+]$

$\qquad [H^+] = 10^{-7}$

b. $\quad [H^+] = 10^{-8}$

c. $\quad [H^+] = 10^{-9}$

For each increase in pH of one, $[H^+]$ changes by a factor of $\frac{1}{10}$.

87. a. $\quad \log E = 4.4 + 1.5(4) = 10.4$

$\qquad\qquad E = 10^{10.4}$

b. $\quad \log E = 4.4 + 1.5(5) = 11.9$

$\qquad\qquad E = 10^{11.9}$

c. $\quad \log E = 4.4 + 1.5(6) = 13.4$

$\qquad\qquad E = 10^{13.4}$

For each increase in M of one, E changes by a factor of $10^{1.5} \approx 31.6$.

89. a. $80 = 10\log\left(\dfrac{I}{10^{-12}}\right)$

$8 = \log\left(\dfrac{I}{10^{-12}}\right)$

$10^8 = \dfrac{I}{10^{-12}}$

$I = 10^8 \cdot 10^{-12} = 10^{-4}$

b. $I = 10^{-3}$

c. $I = 10^{-2}$

For each increase in dB of ten, I changes by a factor of 10.

91.

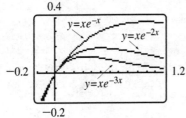

$y = xe^{-x}$: maximum at $x = 1$

$y = xe^{-2x}$: maximum at $x = \dfrac{1}{2}$

$y = xe^{-3x}$: maximum at $x = \dfrac{1}{3}$

$y = xe^{-kx}$: maximum at $x = \dfrac{1}{k}$

Section 0.7

5. $(f \circ g)(x) = f(g(x))$

$= f(\sqrt{x-3})$

$= \sqrt{x-3} + 1$

$D : \{x \mid x \ge 3\}$

$(g \circ f)(x) = g(f(x))$

$= g(x+1)$

$= \sqrt{x+1-3}$

$= \sqrt{x-2}$

$D : \{x \mid x \ge 2\}$

7. $(f \circ g)(x) = f(g(x))$

$= f(\ln x)$

$= e^{\ln x}$

$= x$

$D : \{x \mid x > 0\}$

$(g \circ f)(x) = g(f(x))$

$= g(e^x)$

$= \ln e^x$

$= x$

$D:$ ℝ (all real numbers)

9. $(f \circ g)(x) = f(g(x))$

$= f(\sin x)$

$= \sin^2 x + 1$

$D:$ ℝ (all real numbers)

$(g \circ f)(x) = g(f(x))$

$= g(x^2 + 1)$

$= \sin(x^2 + 1)$

$D:$ ℝ (all real numbers)

11.   Possible answer:

$f(x) = \sqrt{x}; \ g(x) = x^4 + 1$

13.   Possible answer:

$f(x) = \dfrac{1}{x}; \ g(x) = x^2 + 1$

15.   Possible answer:

$f(x) = x^2 + 3; \ g(x) = 4x + 1$

17.   Possible answer:

$f(x) = x^3; \ g(x) = \sin x$

19.   Possible answer:

$f(x) = \cos x; \ g(x) = 4x$

21.   Possible answer:

$f(x) = e^x; \ g(x) = x^2 + 1$

23.   Possible answer:

$f(x) = \sqrt{x}; \ g(x) = e^x + 1$

25.   Possible answer:

$f(x) = \ln x; \ g(x) = 3x - 5$

27. Shift the graph down 3 units.

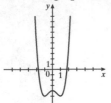

29. Shift the graph right 3 units.

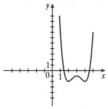

31. Multiply the scale on the x-axis by $\frac{1}{2}$.

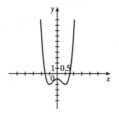

33. Multiply the scale on the y-axis by 4 and then shift the graph down 1 unit.

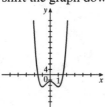

35. Shift the graph right 4 units.

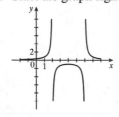

37. Multiply the scale on the x-axis by $\frac{1}{2}$.

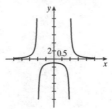

39. Shift the graph left 3 units and then multiply the scale on the x-axis by $\frac{1}{3}$.

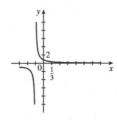

41. Multiply the scale on the y-axis by 2 and then shift the graph down 4 units.

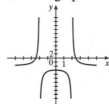

43. $f(x) = x^2 + 2x + 1$
$$= (x^2 + 2x + 1) + 1 - 1$$
$$f(x) = (x + 1)^2$$
Shift $y = x^2$ to the left 1 unit.

45. $f(x) = x^2 + 2x + 4$
$$= (x^2 + 2x + 1) + 4 - 1$$
$$f(x) = (x + 1)^2 + 3$$
Shift $y = x^2$ to the left 1 unit and up 3 units.

47. $f(x) = 2x^2 + 4x + 4$
$$= 2(x^2 + 2x + 1) + 4 - 2$$
$$f(x) = 2(x + 1)^2 + 2$$
Shift $y = x^2$ to the left 1 unit, then multiply the scale on the y-axis by 2, then shift up 2 units. Alternately, using $f(x) = 2[(x + 1)^2 + 1]$, shift left 1 unit and up 1 unit, then double the vertical scale.

49.

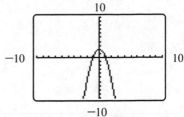

Graph is reflected across the *x*-axis and the scale on the *y*-axis is multiplied by 2.

51.

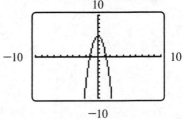

Graph is reflected across the *x*-axis, the scale on the *y*-axis is multiplied by 3, and the graph is shifted up 2 units.

53.

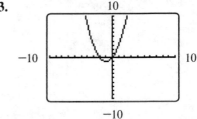

Graph is reflected across the *y*-axis.

55.

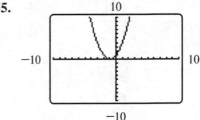

Graph is reflected across the *y*-axis and shifted up 1 unit.

57. The graph is reflected across the *x*-axis and the scale on the *y*-axis is multiplied by $|c|$.

59. The passenger feels the car bouncing up and down. The bounces gradually subside.

61.
$$x_4 \approx \cos .65 \approx .796$$
$$x_5 \approx \cos .796 \approx .70$$
$$x_6 \approx \cos .70 \approx .765$$
$$x_7 \approx \cos .765 \approx .721$$
$$x_8 \approx \cos .721 \approx .751$$
$$x_9 \approx \cos .751 \approx .731$$
$$x_{10} \approx \cos .731 \approx .744$$
$$x_{11} \approx \cos .744 \approx .735$$
$$x_{12} \approx \cos .735 \approx .742$$
$$x_{13} \approx \cos .742 \approx .737$$
$$x_{14} \approx \cos .737 \approx .740$$
$$x_{15} \approx \cos .740 \approx .738$$
$$x_{16} \approx \cos .738 \approx .73959$$
$$x_{17} \approx \cos .73959 \approx .73874$$
$$x_{18} \approx \cos .73874 \approx .73931$$
$$x_{19} \approx \cos .73931 \approx .73893$$
$$x_{20} \approx \cos .73893 \approx .73919$$

63. They converge to 0.

65. Any fixed point must be a solution of the equation $f(x) = x$ because $x_{n+1} = x_n$ and $x_{n+1} = f(x_n)$ implies $f(x_n) = x_n$.

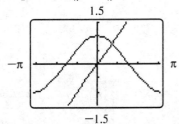

$\cos x = x$
Using a graph, the only solution is (.739085, .739085). This agrees with the results of exercise 61.

Section 0.8

3. First point (1, 2)

Second point	m_{sec}	Second point	m_{sec}
(2, 5)	3	(0, 1)	1
(1.1, 2.21)	2.1	(0.9, 1.81)	1.9
(1.01, 2.0201)	2.01	(0.99, 1.9801)	1.99

slope ≈ 2

5. First point (0, 1)

Second point	m_{sec}	Second point	m_{sec}
(1, cos 1)	−.4597	(−1, cos (−1))	.4597
(0.1, cos .1)	−.0500	(−.1, cos (−.1))	.0500
(.01, cos .01)	−.0050	(−.01, cos (−.01))	.0050

slope ≈ 0

7. First point (0, 2)

Second point	m_{sec}	Second point	m_{sec}
(1, 3)	1	(−1, 1)	1
(.1, 2.001)	.01	(−.1, 1.999)	.01
(.01, 2.000001)	.0001	(−.01, 1.999999)	.0001

slope ≈ 0

9. First point (0, 1)

Second point	m_{sec}	Second point	m_{sec}
$(1, \sqrt{2})$	.4142	(−1, 0)	1
$(.1, \sqrt{1.1})$	.4881	$(-.1, \sqrt{.9})$	.5132
$(.01, \sqrt{1.01})$	.4988	$(-.01, \sqrt{.99})$	.5013

slope $\approx .5$

11. First point (0, 1)

Second point	m_{sec}	Second point	m_{sec}
$(1, e)$	1.7183	$(-1, e^{-1})$	.6321
$(.1, e^{.1})$	1.0517	$(-.1, e^{-.1})$	.9516
$(.01, e^{.01})$	1.0050	$(-.01, e^{-.01})$	.9950

slope ≈ 1

13. First point $(1, 0)$

Second point	m_{sec}		Second point	m_{sec}
$(2, \ln 2)$	.6931		$(.5, \ln .5)$	1.3863
$(1.1, \ln 1.1)$	.9531		$(.9, \ln .9)$	1.0536
$(1.01, \ln 1.01)$	.9950		$(.99, \ln .99)$	1.0050

slope ≈ 1

15. a. $d_4 = d\{(0, 1), (.5, 1.25)\} + d\{(.5, 1.25), (1, 2)\} + d\{(1, 2), (1.5, 3.25)\} + d\{(1.5, 3.25), (2, 5)\}$

$$= \sqrt{(.5-0)^2 + (1.25-1)^2} + \sqrt{(1-.5)^2 + (2-1.25)^2} + \sqrt{(1.5-1)^2 + (3.25-2)^2}$$
$$+ \sqrt{(2-1.5)^2 + (5-3.25)^2}$$
$$\approx 4.6267$$

b. $d_8 = d\{(0, 1), (.25, 1.0625)\} + d\{(.25, 1.0625), (.5, 1.25)\} + d\{(.5, 1.25), (.75, 1.5625)\}$

$$+ d\{(.75, 1.5625), (1.2)\} + d\{(1, 2), (1.25, 2.5625)\} + d\{(1.25, 2.5625), (1.5, 3.25)\}$$
$$+ d\{(1.5, 3.25), (1.75, 4.0625)\} + d\{(1.75, 4.0625), (2, 5)\}$$
$$\approx 4.6417$$

c. 4.64678

17. a. $d_4 = d\left\{(0, 1), \left(\dfrac{\pi}{8}, \cos\dfrac{\pi}{8}\right)\right\} + d\left\{\left(\dfrac{\pi}{8}, \cos\dfrac{\pi}{8}\right), \left(\dfrac{\pi}{4}, \cos\dfrac{\pi}{4}\right)\right\} + d\left\{\left(\dfrac{\pi}{4}, \cos\dfrac{\pi}{4}\right), \left(\dfrac{3\pi}{8}, \cos\dfrac{3\pi}{8}\right)\right\}$

$$+ d\left\{\left(\dfrac{3\pi}{8}, \cos\dfrac{3\pi}{8}\right), \left(\dfrac{\pi}{2}, 0\right)\right\}$$
$$\approx 1.9063$$

b. $d_8 = d\left\{(0, 1), \left(\dfrac{\pi}{16}, \cos\dfrac{\pi}{16}\right)\right\} + d\left\{\left(\dfrac{\pi}{16}, \cos\dfrac{\pi}{16}\right), \left(\dfrac{\pi}{8}, \cos\dfrac{\pi}{8}\right)\right\}$

$$+ d\left\{\left(\dfrac{\pi}{8}, \cos\dfrac{\pi}{8}\right), \left(\dfrac{3\pi}{16}, \cos\dfrac{3\pi}{16}\right)\right\} + d\left\{\left(\dfrac{3\pi}{16}, \cos\dfrac{3\pi}{16}\right), \left(\dfrac{\pi}{4}, \cos\dfrac{\pi}{4}\right)\right\}$$

$$+ d\left\{\left(\dfrac{\pi}{4}, \cos\dfrac{\pi}{4}\right), \left(\dfrac{5\pi}{16}, \cos\dfrac{5\pi}{16}\right)\right\} + d\left\{\left(\dfrac{5\pi}{16}, \cos\dfrac{5\pi}{16}\right), \left(\dfrac{3\pi}{8}, \cos\dfrac{3\pi}{8}\right)\right\}$$

$$+ d\left\{\left(\dfrac{3\pi}{8}, \cos\dfrac{3\pi}{8}\right), \left(\dfrac{7\pi}{16}, \cos\dfrac{7\pi}{16}\right)\right\} + d\left\{\left(\dfrac{7\pi}{16}, \cos\dfrac{7\pi}{16}\right), \left(\dfrac{\pi}{2}, 0\right)\right\}$$

$$\approx 1.9091$$

c. 1.91010

19. a. $d_4 = d\left\{(0,1), \left(\frac{3}{4}, \sqrt{\frac{7}{4}}\right)\right\} + d\left\{\left(\frac{3}{4}, \sqrt{\frac{7}{4}}\right), \left(\frac{3}{2}, \sqrt{\frac{5}{2}}\right)\right\} + d\left\{\left(\frac{3}{2}, \sqrt{\frac{5}{2}}\right), \left(\frac{9}{4}, \sqrt{\frac{13}{8}}\right)\right\} + d\left\{\left(\frac{9}{4}, \sqrt{\frac{13}{4}}\right), (3, 2)\right\}$

≈ 3.1673

b. $d_8 = d\left\{(0, 1), \left(\frac{3}{8}, \sqrt{\frac{11}{8}}\right)\right\} + d\left\{\left(\frac{3}{8}, \sqrt{\frac{11}{8}}\right), \left(\frac{3}{4}, \sqrt{\frac{7}{4}}\right)\right\} + d\left\{\left(\frac{3}{4}, \sqrt{\frac{7}{4}}\right), \left(\frac{9}{8}, \sqrt{\frac{17}{8}}\right)\right\}$

$+ d\left\{\left(\frac{9}{8}, \sqrt{\frac{17}{8}}\right), \left(\frac{3}{2}, \sqrt{\frac{5}{2}}\right)\right\} + d\left\{\left(\frac{3}{2}, \sqrt{\frac{5}{2}}\right), \left(\frac{15}{8}, \sqrt{\frac{23}{8}}\right)\right\}$

$+ d\left\{\left(\frac{15}{8}, \sqrt{\frac{23}{8}}\right), \left(\frac{9}{4}, \sqrt{\frac{13}{4}}\right)\right\} + d\left\{\left(\frac{9}{4}, \sqrt{\frac{13}{4}}\right), \left(\frac{21}{8}, \sqrt{\frac{29}{8}}\right)\right\}$

$+ d\left\{\left(\frac{21}{8}, \sqrt{\frac{29}{8}}\right), (3, 2)\right\}$

≈ 3.1677

c. 3.16784

21. a. $d_4 = d\{(-2, 5), (-1, 2)\} + d\{(-1, 2), (0, 1)\} + d\{(0, 1), (1, 2)\} + d\{(1, 2), (2, 5)\}$

≈ 9.1530

b. $d_8 = d\left\{(-2, 5), \left(-\frac{3}{2}, \frac{13}{4}\right)\right\} + d\left\{\left(-\frac{3}{2}, \frac{13}{4}\right), (-1, 2)\right\} + d\left\{(-1, 2), \left(-\frac{1}{2}, \frac{5}{4}\right)\right\}$

$+ d\left\{\left(-\frac{1}{2}, \frac{5}{4}\right), (0, 1)\right\} + d\left\{(0, 1), \left(\frac{1}{2}, \frac{5}{4}\right)\right\} + d\left\{\left(\frac{1}{2}, \frac{5}{4}\right), (1, 2)\right\}$

$+ d\left\{(1, 2), \left(\frac{3}{2}, \frac{13}{4}\right)\right\} + d\left\{\left(\frac{3}{2}, \frac{13}{4}\right), (2, 5)\right\}$

≈ 9.2534

c. 9.29357

23. $a_0 = \frac{1}{1} = 1; \; a_1 = \frac{3+1}{2+1} = \frac{4}{3};$

$a_2 = \frac{6+1}{4+1} = \frac{7}{5}; \; a_3 = \frac{9+1}{6+1} = \frac{10}{7}$

$\lim\limits_{n\to\infty} a_n = \lim\limits_{n\to\infty} \frac{3+\frac{1}{n}}{2+\frac{1}{n}} = \frac{3}{2}$

25. $a_0 = 0; \; a_1 = \frac{1}{2}; \; a_2 = \frac{2}{3}; \; a_3 = \frac{3}{4}$

$\lim\limits_{n\to\infty} a_n = \lim\limits_{n\to\infty} \frac{1}{1+\frac{1}{n}} = 1$

27. $a_0 = \frac{0+2}{0+1} = 2; \; a_1 = \frac{1+2}{1+1} = \frac{3}{2}; \; a_2 = \frac{2+2}{4+1} = \frac{4}{5};$

$a_3 = \frac{3+2}{9+1} = \frac{1}{2}$

$\lim\limits_{n\to\infty} a_n = \lim\limits_{n\to\infty} \frac{1+\frac{2}{n}}{n+\frac{1}{n}} = \lim\limits_{n\to\infty} \frac{1}{n} = 0$

29. $u_0 = 0; \; u_1 = \frac{2}{1+4} = \frac{2}{5}; \; a_2 = \frac{8}{4+4} = 1;$

$a_3 = \frac{18}{9+4} = \frac{18}{13}$

$\lim\limits_{n\to\infty} a_n = \lim\limits_{n\to\infty} \frac{2}{1+\frac{4}{n^2}} = \frac{2}{1} = 2$

31. $a_0 = \dfrac{2}{0+1} = 2;\ a_1 = \dfrac{2}{1+1} = 1;\ a_2 = \dfrac{2}{2+1} = \dfrac{2}{3};$

$a_3 = \dfrac{2}{3+1} = \dfrac{1}{2}$

$\lim\limits_{n\to\infty} a_n = \lim\limits_{n\to\infty} \dfrac{2}{n+1} = 0$

33. $a_1 = 1 - \dfrac{1}{1} = 0;\ a_2 = 1 - \dfrac{1}{2} = \dfrac{1}{2};\ a_3 = 1 - \dfrac{1}{3} = \dfrac{2}{3};$

$a_4 = 1 - \dfrac{1}{4} = \dfrac{3}{4}$

$\lim\limits_{n\to\infty} a_n = \lim\limits_{n\to\infty}\left(1 - \dfrac{1}{n}\right) = 1.$

35. $a_1 = e^{-1},\ a_2 = e^{-2};\ a_3 = e^{-3};\ a^4 = e^{-4}$

$\lim\limits_{n\to\infty} a_n = \lim\limits_{n\to\infty} e^{-n} = \lim\limits_{n\to\infty} \dfrac{1}{e^n} = 0$

37. $a_1 = \cos 1;\ a_2 = \cos\dfrac{1}{2};\ a_3 = \cos\dfrac{1}{3};\ a_4 = \cos\dfrac{1}{4}$

$\lim\limits_{n\to\infty} a_n = \lim\limits_{n\to\infty} \cos\left(\dfrac{1}{n}\right) = \cos 0 = 1$

39. $u_0 = 0;\ a_1 = \dfrac{1}{\sqrt{2}};\ a_2 = \dfrac{2}{\sqrt{5}};\ a_3 = \dfrac{3}{\sqrt{10}}$

$\lim\limits_{n\to\infty} a_n = \lim\limits_{n\to\infty} \dfrac{n}{\sqrt{n^2\left(1+\frac{1}{n^2}\right)}}$

$= \lim\limits_{n\to\infty} \dfrac{1}{\sqrt{\left(1+\frac{1}{n^2}\right)}}$

$= \dfrac{1}{\sqrt{1}}$

$= 1$

41. $a_1 = 4 + 1 - 2 = 3;\ a_2 = 4 + \dfrac{1}{2} - \dfrac{2}{4} = 4;$

$a_3 = 4 + \dfrac{1}{3} - \dfrac{2}{9} = \dfrac{37}{9} = 4\dfrac{1}{9};$

$a_4 = 4 + \dfrac{1}{4} - \dfrac{2}{16} = \dfrac{33}{8} = 4\dfrac{1}{8}$

$\lim\limits_{n\to\infty} a_n = \lim\limits_{n\to\infty}\left(4 + \dfrac{1}{n} - \dfrac{2}{n^2}\right) = 4$

43. $a_n = \dfrac{1}{10^n}$, so $\lim\limits_{n\to\infty} a_n = 0.$ Since

$\lim\limits_{n\to\infty} a_n = 0,\ 1 - .99\overline{9} = 0,$ so $.99\overline{9} = 1.$

Chapter 0 Review

1. See Example 1.1.

$3x + 1 > 4$

$3x > 3$

$x > 1$

$(1, \infty)$

3. See Example 1.4.

$x^2 - 2x - 8 > 0$

$(x-4)(x+2) > 0$

$x < -2$ or $x > 4$

$(-\infty, -2) \cup (4, \infty)$

5. See Example 1.7.

$|1 - 2x| < 3$

$-3 < 1 - 2x < 3$

$-4 < -2x < 2$

$2 > x > -1$

$-1 < x < 2$

$(-1, 2)$

7. See Theorem 1.2.

$d\{(0, 1), (2, 5)\} = \sqrt{(2-0)^2 + (5-1)^2}$

$= \sqrt{2^2 + 4^2}$

$= \sqrt{4 + 16}$

$= \sqrt{20}$

9. See Example 2.1.

$m = \dfrac{7-3}{0-2} = \dfrac{4}{-2} = -2$

11. See Theorem 2.1.

parallel; both have slope $= 3$

13. See Example 1.9.

$$d\{(1, 2), (2, 4)\} = \sqrt{(2-1)^2 + (4-2)^2}$$
$$= \sqrt{1+4}$$
$$= \sqrt{5}$$

$$d\{(2, 4), (0, 6)\} = \sqrt{(0-2)^2 + (6-4)^2}$$
$$= \sqrt{4+4}$$
$$= \sqrt{8}$$

$$d\{(1, 2), (0, 6)\} = \sqrt{(0-1)^2 + (6-2)^2}$$
$$= \sqrt{1+16}$$
$$= \sqrt{17}$$

For a right triangle, must have

$$(\sqrt{5})^2 + (\sqrt{8})^2 = (\sqrt{17})^2$$

Since this statement is not true, the triangle is not a right triangle.

15. See Example 2.5.
Line goes through (1, 1) and (3, 2)

$$m = \frac{2-1}{3-1} = \frac{1}{2}$$

$$y = \frac{1}{2}(x-1) + 1$$

when $x = 4$,

$$y = \frac{1}{2}(4-1) + 1 = \frac{5}{2}.$$

17. See Example 2.4.

$$y = -\frac{1}{3}[x - (-1)] - 1$$

$$y = -\frac{1}{3}(x+1) - 1$$

19. See Example 2.9.
The graph passes the vertical line test, so it is a function.

21. See Example 2.12.

$$4 - x^2 \geq 0$$
$$4 \geq x^2$$
$$-2 \leq x \leq 2$$
$$[-2, 2]$$

For exercises 23–36, see Example 3.2.

23.

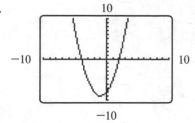

25.

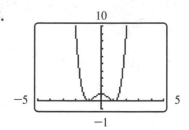

27.

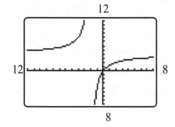

29.

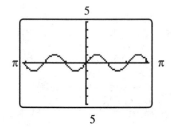

31.

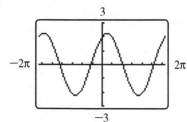

33.

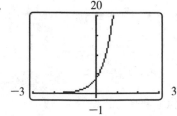

35.

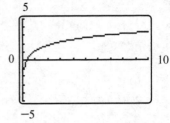

37. See Examples 3.2 and 4.1.
$$x^2 + 2x - 8 = 0$$
$$(x + 4)(x - 2) = 0$$
$$x = -4, \ x = 2$$
x-intercepts: $x = -4, x = 2$
y-intercept: $y = -8$

39. See Example 3.4.
Vertical asymptote: $x = -2$
Horizontal asymptote: $y = 4$

41. See Example 4.1.
$$x^2 - 3x - 10 = 0$$
$$(x - 5)(x + 2) = 0$$
$$x = 5, \ x = -2$$

43. See Example 4.3.

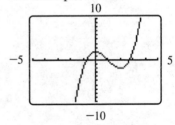

$$x^3 - 3x^2 + 2 = 0$$
$$(x - 1)(x^2 - 2x - 2 = 0)$$
$$x = 1 \text{ or } x = \frac{2 \pm \sqrt{2^2 - 4(1)(-2)}}{2} = 1 \pm \sqrt{3}$$
$$x = 1, \ x = 1 - \sqrt{3} \approx -.732, \ x = 1 + \sqrt{3} \approx 2.732$$

45. See Example 4.6.

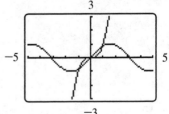

There are 3 solutions.

47. See Example 5.7.
$\cos 2x$ has period $\dfrac{2\pi}{2} = \pi$ and repeats at intervals of width $\pi, 2\pi, 3\pi$, etc.

$2 \sin 4x$ has period $\dfrac{2\pi}{4} = \dfrac{\pi}{2}$ and repeats at intervals of width $\dfrac{\pi}{2}, \pi, \dfrac{3\pi}{2}$, etc.

The function is periodic with period π.

49. See Example 5.8.

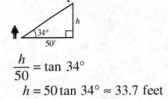

$$\frac{h}{50} = \tan 34°$$
$$h = 50 \tan 34° \approx 33.7 \text{ feet}$$

51. See page 51.

a. $5^{-1/2} = \dfrac{1}{5^{1/2}} = \dfrac{1}{\sqrt{5}}$

b. $3^{-2} = \dfrac{1}{3^2} = \dfrac{1}{9}$

53. See Example 6.7.
$$\ln 8 - 2\ln 2 = \ln 8 - \ln 2^2$$
$$= \ln 8 - \ln 4$$
$$= \ln\left(\frac{8}{4}\right)$$
$$= \ln 2$$

55. See Example 6.8.
$$3e^{2x} = 8$$
$$e^{2x} = \frac{8}{3}$$
$$\ln e^{2x} = \ln\left(\frac{8}{3}\right)$$
$$2x = \ln\left(\frac{8}{3}\right)$$
$$x = \frac{1}{2}\ln\frac{8}{3}$$

57. See Example 7.2.

$$(f \circ g)(x) = f(g(x)) = f(\sqrt{x-1})$$
$$= (\sqrt{x-1})^2$$
$$= x-1$$

$D : \{x \mid x \geq 1\}$

$$(g \circ f)(x) = g(f(x)) = g(x^2)$$
$$= \sqrt{x^2 - 1}$$

$D : \{x \mid x \leq -1 \text{ or } x \geq 1\}$

59. See Example 7.3.

Possible answer: $f(x) = e^x$; $g(x) = 3x^2 + 2$

61. See Example 7.9.

$$f(x) = x^2 - 4x + 1$$
$$= x^2 - 4x + 4 + 1 - 4$$
$$f(x) = (x-2)^2 - 3$$

Shift the graph of $y = x^2$ to the right 2 units and down 3 units.

63. See Example 8.1.

First point: (2, 0)

Second point	m_{sec}	Second point	m_{sec}
(3, 3)	3	(1, −1)	1
(2.1, .21)	2.1	(1.9, −.19)	1.9
(2.01, .0201)	2.01	(1.99, −.0199)	1.99

slope ≈ 2

65. See Example 8.2.

a. $d_4 = d\left\{(0, 0), \left(\dfrac{\pi}{16}, \sin\dfrac{\pi}{16}\right)\right\} + d\left\{\left(\dfrac{\pi}{16}, \sin\dfrac{\pi}{16}\right), \left(\dfrac{\pi}{8}, \sin\dfrac{\pi}{8}\right)\right\}$

$\qquad + d\left\{\left(\dfrac{\pi}{8}, \sin\dfrac{\pi}{8}\right), \left(\dfrac{3\pi}{16}, \sin\dfrac{3\pi}{16}\right)\right\} + d\left\{\left(\dfrac{3\pi}{16}, \sin\dfrac{3\pi}{16}\right), \left(\dfrac{\pi}{4}, \sin\dfrac{\pi}{4}\right)\right\}$

$\qquad = \sqrt{\left(\dfrac{\pi}{16}\right)^2 + \left(\sin\dfrac{\pi}{16}\right)^2} + \sqrt{\left(\dfrac{\pi}{8} - \dfrac{\pi}{16}\right)^2 + \left(\sin\dfrac{\pi}{8} - \sin\dfrac{\pi}{16}\right)^2}$

$\qquad + \sqrt{\left(\dfrac{3\pi}{16} - \dfrac{\pi}{8}\right)^2 + \left(\sin\dfrac{3\pi}{16} - \sin\dfrac{\pi}{8}\right)^2} + \sqrt{\left(\dfrac{\pi}{4} - \dfrac{3\pi}{16}\right)^2 + \left(\sin\dfrac{\pi}{4} - \sin\dfrac{3\pi}{16}\right)^2}$

$\qquad \approx 1.0580$

b. $d_8 = d\left\{(0, 0), \left(\dfrac{\pi}{32}, \sin\dfrac{\pi}{32}\right)\right\} + d\left\{\left(\dfrac{\pi}{32}, \sin\dfrac{\pi}{32}\right), \left(\dfrac{\pi}{16}, \sin\dfrac{\pi}{16}\right)\right\} + d\left\{\left(\dfrac{\pi}{16}, \sin\dfrac{\pi}{16}\right), \left(\dfrac{3\pi}{32}, \sin\dfrac{3\pi}{32}\right)\right\}$

$\qquad + d\left\{\left(\dfrac{3\pi}{32}, \sin\dfrac{3\pi}{32}\right), \left(\dfrac{\pi}{8}, \sin\dfrac{\pi}{8}\right)\right\} + d\left\{\left(\dfrac{\pi}{8}, \sin\dfrac{\pi}{8}\right), \left(\dfrac{5\pi}{32}, \sin\dfrac{5\pi}{32}\right)\right\}$

$\qquad + d\left\{\left(\dfrac{5\pi}{32}, \sin\dfrac{5\pi}{32}\right), \left(\dfrac{3\pi}{16}, \sin\dfrac{3\pi}{16}\right)\right\} + d\left\{\left(\dfrac{3\pi}{16}, \sin\dfrac{3\pi}{16}\right), \left(\dfrac{7\pi}{32}, \sin\dfrac{7\pi}{32}\right)\right\}$

$\qquad + d\left\{\left(\dfrac{7\pi}{32}, \sin\dfrac{7\pi}{32}\right), \left(\dfrac{\pi}{4}, \sin\dfrac{\pi}{4}\right)\right\}$

$\qquad \approx 1.0581$

67. See Example 8.3.

$$a_2 = \frac{3}{4-3} = 3; \quad a_3 = \frac{3}{9-3} = \frac{1}{2}; \quad a_4 = \frac{3}{16-3} = \frac{3}{13}; \quad a_5 = \frac{3}{25-3} = \frac{3}{22}$$

$$\lim_{n \to \infty} a_n = \lim_{n \to \infty}\left(\frac{3}{n^2 - 3}\right) = 0$$

Chapter 1

Section 1.1

5. a. −2   **b.** 2

   c. does not exist   **d.** 1

   e. $\dfrac{1}{2}$ **f.** −1

   g. 3   **h.** does not exist

   i. 2   **j.** 2

7.

x	$f(x)$	x	$f(x)$
1.5	2.22	0.5	1.71
1.1	2.05	0.9	1.95
1.01	2.00	0.99	1.99
1.001	2.00	0.999	2.00

$\lim\limits_{x\to 1^{+}} f(x) = 2$ $\lim\limits_{x\to 1^{-}} f(x) = 2$

$\lim\limits_{x\to 1} f(x) = 2$

9.

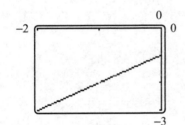

on [−2, 0] by [−3, 0]

x	$\dfrac{x^2-1}{x+1}$	x	$\dfrac{x^2-1}{x+1}$
−0.9	−1.9	−1.1	−2.1
−0.99	−1.99	−1.01	−2.01
−0.999	−1.999	−1.001	−2.001

$\lim\limits_{x\to -1} \dfrac{x^2-1}{x+1} = -2$

11.

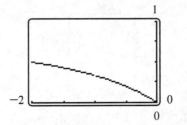

on [−2, 0] by [0, 1]

x	$\dfrac{x^2+x}{x^2-x-2}$	x	$\dfrac{x^2+x}{x^2-x-2}$
−0.9	0.31	−1.1	0.35
−0.99	0.33	−1.01	0.34
−0.999	0.33	−1.001	0.33

$\lim\limits_{x\to -1} \dfrac{x^2+x}{x^2-x-2} = \dfrac{1}{3}$

13.

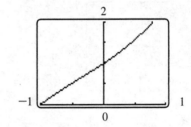

on [−1, 1] by [0, 2]

x	$\dfrac{x^2+x}{\sin x}$	x	$\dfrac{x^2+x}{\sin x}$
0.1	1.10	−0.1	0.90
0.01	1.01	−0.01	0.99
0.001	1.00	−0.001	1.00

$\lim\limits_{x\to 0} \dfrac{x^2+x}{\sin x} = 1$

15.

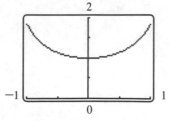

on [–1, 1] by [0, 2]

x	$\dfrac{\tan x}{\sin x}$	x	$\dfrac{\tan x}{\sin x}$
0.1	1.005	–0.1	1.005
0.01	1.000	–0.01	1.000
0.001	1.000	–0.001	1.000

$$\lim_{x \to 0} \frac{\tan x}{\sin x} = 1$$

17.

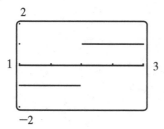

on [1, 3] by [–2, 2]

x	$\dfrac{x-2}{|x-2|}$	x	$\dfrac{x-2}{|x-2|}$
2.1	1	1.9	–1
2.01	1	1.99	–1
2.001	1	1.999	–1

$$\lim_{x \to 2} \frac{x-2}{|x-2|} \text{ does not exist.}$$

There is a break in the graph at $x = 2$.

19.

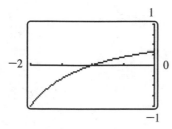

on [–2, 0] by [–1, 1]

x	$\dfrac{x^2-1}{x^2+2x-3}$	x	$\dfrac{x^2-1}{x^2+2x-3}$
–0.9	0.05	–1.1	–0.05
–0.99	0.005	–1.01	–0.005
–0.999	0.00	–1.001	0.00

$$\lim_{x \to -1} \frac{x^2-1}{x^2+2x-3} = 0$$

21.

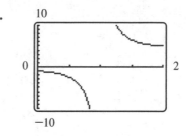

on [0, 2] by [–10, 10]

x	$\dfrac{x^2+1}{x-1}$	x	$\dfrac{x^2+1}{x-1}$
1.1	22.1	0.9	–18.1
1.01	202.01	0.99	–198.01
1.001	2002	0.999	–1998

$$\lim_{x \to 1} \frac{x^2+1}{x-1} \text{ does not exist.}$$

There is a break in the graph at $x = 1$.

23.

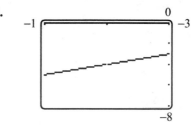

on [–3, –1] by [–8, 0]

x	$\dfrac{x^2-4}{x+2}$	x	$\dfrac{x^2-4}{x+2}$
–1.9	–3.9	–2.1	–4.1
–1.99	–3.99	–2.01	–4.01
–1.999	–3.999	–2.001	–4.001

$$\lim_{x \to -2} \frac{x^2-4}{x+2} = -4$$

25.

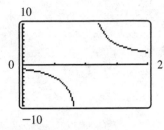

on [0, 2] by [−10, 10]

x	$\dfrac{x^2-1}{x^2-2x+1}$	x	$\dfrac{x^2-1}{x^2-2x+1}$
1.1	21	0.9	−19
1.01	201	0.99	−199
1.001	2001	0.999	−1999

$\lim\limits_{x\to 1}\dfrac{x^2-1}{x^2-2x+1}$ does not exist.

There is a break in the graph at $x = 1$.

27. If $g(a) = 0$ and $f(a) \neq 0$, $\lim\limits_{x\to a}\dfrac{f(x)}{g(x)}$ does not exist.

29. One possibility is

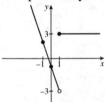

31. One possibility is

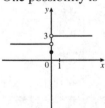

33. The first argument gives the correct value; it is better than the second argument since it doesn't depend on specific values of x.

35.

x	$(1+x)^{1/x}$	x	$(1+x)^{1/x}$
0.1	2.59	−0.1	2.87
0.01	2.70	−0.01	2.73
0.001	2.7169	−0.001	2.7196

$\lim\limits_{x\to 0}(1+x)^{1/x} \approx 2.7182818$

37.

x	$x^{\sec x}$
0.1	0.099
0.01	0.010
0.001	0.001

$\lim\limits_{x\to 0^+} x^{\sec x} = 0$

For negative x the values of $x^{\sec x}$ are complex numbers. Both the real and the imaginary part of $x^{\sec x}$ approach 0 as $x \to 0^-$, so $\lim\limits_{x\to 0^-} x^{\sec x} = 0$.

39. A possible answer is

$$f(x) = \frac{x^2}{x}$$

$$g(x) = \begin{cases} 1 \text{ if } x \leq 0 \\ -1 \text{ if } x > 0 \end{cases}$$

41. As x gets arbitrarily close to a, $f(x)$ gets arbitrarily close to L.

Section 1.2

5. $\lim\limits_{x\to 0}(x^2 - 3x + 1) = 0^2 - 3(0) + 1 = 1$

7. $\lim\limits_{x\to 1}\sqrt{x^2 + 2x + 4} = \sqrt{1^2 + 2(1) + 4} = \sqrt{7}$

9. $\lim\limits_{x\to 2}\dfrac{x-5}{x^2+4} = \dfrac{2-5}{2^2+4} = -\dfrac{3}{8}$

11. $\lim\limits_{x\to 3}\dfrac{x^2-x-6}{x-3} = \lim\limits_{x\to 3}\dfrac{(x-3)(x+2)}{x-3}$
$= \lim\limits_{x\to 3}(x+2)$
$= 3 + 2$
$= 5$

13. $\lim\limits_{x\to 2}\dfrac{x^2-x-2}{x^2-4} = \lim\limits_{x\to 2}\dfrac{(x-2)(x+1)}{(x+2)(x-2)}$
$= \lim\limits_{x\to 2}\dfrac{x+1}{x+2}$
$= \dfrac{2+1}{2+2}$
$= \dfrac{3}{4}$

15. $\lim\limits_{x\to1}\dfrac{x^3-1}{x^2+2x-3}=\lim\limits_{x\to1}\dfrac{(x-1)(x^2+x+1)}{(x+3)(x-1)}$

$=\lim\limits_{x\to1}\dfrac{x^2+x+1}{x+3}$

$=\dfrac{1^2+1+1}{1+3}$

$=\dfrac{3}{4}$

17. $\lim\limits_{x\to0}\dfrac{\sin x}{\tan x}=\lim\limits_{x\to0}\dfrac{\sin x}{\frac{\sin x}{\cos x}}$

$=\lim\limits_{x\to0}\cos x$

$=\cos 0$

$=1$

19. $\lim\limits_{x\to0}\dfrac{xe^{-2x+1}}{x^2+1}=\dfrac{0(e^{-2(0)+1})}{0^2+1}=\dfrac{0}{1}=0$

21. $\lim\limits_{x\to0}\dfrac{\tan 2x}{x}=\lim\limits_{x\to0}\dfrac{\sin 2x}{x\cos 2x}$

$=\lim\limits_{x\to0}\dfrac{2\sin x\cos x}{x\cos 2x}$

$=\left[\lim\limits_{x\to0}\dfrac{\sin x}{x}\right]\cdot\left[\lim\limits_{x\to0}\dfrac{2\cos x}{\cos 2x}\right]$

$=(1)\left(\dfrac{2\cos 0}{\cos(2\cdot0)}\right)$

$=2$

23. $\lim\limits_{x\to1}\dfrac{x-1}{\sqrt{x}-1}=\lim\limits_{x\to1}\dfrac{(\sqrt{x}+1)(\sqrt{x}-1)}{\sqrt{x}-1}$

$=\lim\limits_{x\to1}(\sqrt{x}+1)$

$=\sqrt{1}+1$

$=2$

25. $\lim\limits_{x\to2^-}f(x)=\lim\limits_{x\to2^-}(3x^2-2x+1)$

$=3(2)^2-2(2)+1$

$=9$

$\lim\limits_{x\to2^+}f(x)=\lim\limits_{x\to2^+}(x^3+1)=2^3+1=9$

$\lim\limits_{x\to2}f(x)=9$

27. $\lim\limits_{x\to2^-}f(x)=\lim\limits_{x\to2^-}2x$

$=2(2)$

$=4$

$\lim\limits_{x\to2^+}f(x)=\lim\limits_{x\to2^+}x^2=2^2=4$

$\lim\limits_{x\to2}f(x)=4$

29. $\lim\limits_{x\to0}f(x)=\lim\limits_{x\to0}(3x+1)$

$=3(0)+1$

$=1$

31. $\lim\limits_{x\to-1^-}f(x)=\lim\limits_{x\to-1^-}(2x+1)$

$=2(-1)+1$

$=-1$

$\lim\limits_{x\to-1^+}f(x)=\lim\limits_{x\to-1^+}3=3$

$\lim\limits_{x\to-1}f(x)=$ does not exist

33.

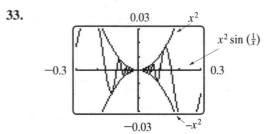

on [–0.3, 0.3] by [–0.03, 0.03]

x	$x^2\sin\left(\frac{1}{x}\right)$		x	$x^2\sin\left(\frac{1}{x}\right)$
–0.1	0.0054		0.1	–0.005
–0.01	5×10^{-5}		0.01	-5×10^{-5}
–0.001	-8×10^{-7}		0.001	8×10^{-7}

Conjecture: $\lim\limits_{x\to0}x^2\sin\left(\dfrac{1}{x}\right)=0$

Let $f(x)=-x^2$, $h(x)=x^2$.

Then $f(x)\le x^2\sin\left(\dfrac{1}{x}\right)\le h(x)$.

$\lim\limits_{x\to0}(-x^2)=0$, $\lim\limits_{x\to0}(x^2)=0$

Therefore, by the Squeeze Theorem,

$\lim\limits_{x\to0}x^2\sin\left(\dfrac{1}{x}\right)=0$.

35. Let $f(x) = 0,\ h(x) = \sqrt{x}$.

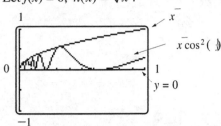

on $[0, 1]$ by $[-1, 1]$

From the graph we see that

$$f(x) \le \sqrt{x}\cos^2(1/x) \le h(x).$$

$$\lim_{x\to 0^+} 0 = 0, \quad \lim_{x\to 0^+} \sqrt{x} = 0$$

Therefore, by the Squeeze Theorem,

$$\lim_{x\to 0^+} \sqrt{x}\cos^2\left(\frac{1}{x}\right) = 0.$$

37. $\lim_{x\to 4^+} \sqrt{16 - x^2}$ does not exist because the domain

of the function is $[-4, 4]$.

39. $\lim_{x\to -2^+} \sqrt{x^2 + 3x + 2}$ does not exist because the

domain of the function is $(-\infty, -2) \cup (-1, \infty)$.

41. $\lim_{x\to 0} \dfrac{\sqrt{1 - \cos x}}{x} = \sqrt{\dfrac{1}{2}} = \dfrac{\sqrt{2}}{2}$

43. $\lim_{x\to a^-} f(x) = \lim_{x\to a^-} g(x) = g(a)$ because $g(x)$ is a

polynomial. Similarly,

$$\lim_{x\to a^+} f(x) = \lim_{x\to a^+} h(x) = h(a).$$

45. a. $\lim_{x\to 2}(x^2 - 3x + 1)$

$= 2^2 - 3(2) + 1$ (Theorem 2.4)

$= -1$

b. $\lim_{x\to 0} \dfrac{x - 2}{x^2 + 1}$

$= \dfrac{\lim_{x\to 0}(x - 2)}{\lim_{x\to 0}(x^2 + 1)}$ (Theorem 2.3(iv))

$= \dfrac{\lim_{x\to 0} x - \lim_{x\to 0} 2}{\lim_{x\to 0} x^2 + \lim_{x\to 0} 1}$ (Theorem 2.3(ii))

$= \dfrac{0 - 2}{0 + 1}$ (Theorem 2.1 and Corollary 2.2)

$= -2$

47. $\lim_{h\to 0} \dfrac{f(2 + h) - f(2)}{h}$

$= \lim_{h\to 0} \dfrac{(2 + h)^2 + 2 - (2^2 + 2)}{h}$

$= \lim_{h\to 0} \dfrac{4h + h^2}{h}$

$= \lim_{h\to 0} 4 + h$

$= 4$

49. $\lim_{h\to 0} \dfrac{f(0 + h) - f(0)}{h}$

$= \lim_{h\to 0} \dfrac{(0 + h)^3 - (0)^3}{h}$

$= \lim_{h\to 0} \dfrac{h^3}{h}$

$= \lim_{h\to 0} h^2$

$= 0$

51. $m = \lim_{h\to 0} \dfrac{\sqrt{1 + h} - 1}{h} \cdot \dfrac{\sqrt{1 + h} + 1}{\sqrt{1 + h} + 1}$

$= \lim_{h\to 0} \dfrac{1 + h - 1}{h(\sqrt{1 + h} + 1)}$

$= \lim_{h\to 0} \dfrac{h}{h(\sqrt{1 + h} + 1)}$

$= \lim_{h\to 0} \dfrac{1}{\sqrt{1 + h} + 1}$

$= \dfrac{1}{\sqrt{1 + 0} + 1}$

$= \dfrac{1}{2}$

53. $\lim_{x\to 0^+} (1 + x)^{1/x} = e \approx 2.71828$

55. $\lim_{x\to 0^+} x^{-x^2} = 1$

57. $\lim\limits_{x \to a}[f(x)]^3 = \left[\lim\limits_{x \to a} f(x)\right]\left[\lim\limits_{x \to a} f(x)\right]\left[\lim\limits_{x \to a} f(x)\right]$ (Theorem 2.3 (iii))

$\qquad\qquad = L \cdot L \cdot L$

$\qquad\qquad = L^3$

$\lim\limits_{x \to a}[f(x)]^4 = \left[\lim\limits_{x \to a} f(x)\right]\left[\lim\limits_{x \to a}[f(x)]^3\right] = L \cdot L = L^4$

59. We can't split the limit of a product into a product of limits unless we know that both limits exist; the limit of the product of a term tending toward 0 and a term with an unknown limit is not 0 but instead is unknown.

61. One possibility is $f(x) = \dfrac{1}{x}$, $g(x) = -\dfrac{1}{x}$.

63. Yes. If it were true that $\lim\limits_{x \to a}[f(x) + g(x)]$ exists,

then by Theorem 2.3 (ii), it would also be true

that $\lim\limits_{x \to a} g(x) = \lim\limits_{x \to a}\big[[f(x) + g(x)] - [f(x)]\big]$

$\qquad\qquad = \lim\limits_{x \to a}[f(x) + g(x)] - \lim\limits_{x \to a} f(x)$

exists. This is clearly a contradiction.

65. $\lim\limits_{x \to 0^+} T(x) = \lim\limits_{x \to 0^+}(0.14x) = 0$

If you have no income, you have no tax liability.

$\lim\limits_{x \to 10,000^-} T(x) = 0.14(10,000) = 1400$

$\lim\limits_{x \to 10,000^+} T(x) = 1500 + 0.21(10,000) = 3600$

$\lim\limits_{x \to 10,000} T(x)$ does not exist. If you have $10,000

of taxable income, your tax liability makes a huge increase.

67. $\lim\limits_{x \to 3^-}[x] = 2$; $\lim\limits_{x \to 3^+}[x] = 3$

Therefore $\lim\limits_{x \to 3}[x]$ does not exist.

Section 1.3

5. $x = -2, x = 2$

7. $x = -2, x = 1, x = 4$

9. $x = -2, x = 2, x = 4$

11. $f(1)$ is not defined and $\lim\limits_{x \to 1} f(x)$ does not exist.

13. $f(0)$ is not defined and $\lim\limits_{x \to 0} f(x)$ does not exist.

15. $\lim\limits_{x \to 2^-} f(x) = \lim\limits_{x \to 2^-}(x^2) = 4$

$\lim\limits_{x \to 2^+} f(x) = \lim\limits_{x \to 2^+}(3x - 2) = 4$

$\lim\limits_{x \to 2} f(x) = 4$; $f(2) = 3$

$\lim\limits_{x \to 2} f(x) \ne f(2)$

17. $f(x) = \dfrac{x - 1}{(x + 1)(x - 1)}$

removable discontinuity at $x = 1$, removed by

$g(x) = \dfrac{1}{x + 1}$

non-removable discontinuity at $x = -1$

19. No discontinuities

21. $f(x) = \dfrac{x^2 \sin x}{\cos x}$

non-removable discontinuities at $x = \dfrac{\pi}{2} + n\pi$, $n =$

$\ldots, -3, -2, -1, 0, 1, 2, 3, \ldots$

23. $f(x) = \dfrac{x \cos x}{\sin x}$

non-removable discontinuities at $x = n\pi$, n integer, $n \ne 0$

removable discontinuity at $x = 0$, removed by

$g(x) = \begin{cases} \cos x, & x = 0 \\ \dfrac{x \cos x}{\sin x}, & x \ne 0 \end{cases}$

25. non-removable discontinuity at $x = 1$

27. $\lim\limits_{x\to-1^-} f(x) = \lim\limits_{x\to-1^-} (3x-1) = -4$

$\lim\limits_{x\to-1^+} f(x) = \lim\limits_{x\to-1^+} (x^2+5x) = -4$

$\lim\limits_{x\to1^-} f(x) = \lim\limits_{x\to1^-} (x^2+5x) = 6$

$\lim\limits_{x\to1^+} f(x) = \lim\limits_{x\to1^+} (3x^3) = 3$

non-removable discontinuity at $x = 1$

29. $x+3 \geq 0$

$x \geq -3$

$[-3, \infty)$

31. $(-\infty, \infty)$

33. $(-\infty, \infty)$

35. $x+1 > 0$

$x > -1$

$(-1, \infty)$

37. $\lim\limits_{x\to10000^-} T(x) = \lim\limits_{x\to10000^-} 0.14x$

$= 0.14(10,000)$

$= 1400$

$\lim\limits_{x\to10000^+} T(x) = \lim\limits_{x\to10000^+} (c+0.21x)$

$= c+0.21(10,000)$

$= c+2100$

$c+2100 = 1400$

$c = -700$

39. a. For taxable amounts over \$128,100 but not over \$278,450, your tax liability is \$(a) + 36% of the amount over \$128,100. Thus $T(x) = a + .36(x - 128,100)$ when $128,100 < x \leq 278,450$. From example 3.8, $a = 34,573.50$, so $T(x) = 34,573.50 + .36(x - 128,100)$.

b. For taxable amounts over \$278,450, your tax liability is \$(b) + 39.6% of the amount over \$278,450. Thus $T(x) = b + .396(x - 278,450)$ when $x > 278,450$. From example 3.8, $b = 88,699.50$, so $T(x) = 88,699.50 + .396(x - 278,450)$.

41. $f(2) = -3, f(3) = 2$

Since $f(2) < 0$ and $f(3) > 0$, the Intermediate Value Theorem says there is a zero in the interval $[2, 3]$.

a	b	f(a)	f(b)	midpoint	f(midpoint)
2	3	-3	2	2.5	-0.75
2.5	3	-0.75	2	2.75	0.5625
2.5	2.75	-0.75	0.5625	2.625	-0.109375
2.625	2.75	-0.109375	0.5625	2.6875	0.223
2.625	2.6875	-0.109375	0.223	2.65625	0.557

The interval $[2.625, 2.65625]$ contains the zero.

43. $f(-1) = 1, f(0) = -2$

Since $f(-1) > 0$ and $f(0) < 0$, the Intermediate Value Theorem says there is a zero in the interval $[-1, 0]$.

a	b	f(a)	f(b)	midpoint	f(midpoint)
-1	0	1	-2	-0.5	-0.125
-1	-0.5	1	-0.125	-0.75	0.578
-0.75	-0.5	0.578	-0.125	-0.625	0.256
-0.625	-0.5	0.256	-0.125	-0.5625	0.072
-0.5625	-0.5	0.072	-0.125	-0.53125	-0.025

The interval $[-0.5625, -0.53125]$ contains the zero.

45. $f(0) = 1, f(1) = \cos 1 - 1 \approx -0.46$
Since $f(0) > 0$ and $f(1) < 0$, the Intermediate Value Theorem says there is a zero in the interval $[0, 1]$.

a	b	$f(a)$	$f(b)$	midpoint	f(midpoint)
0	1	1	–0.46	0.5	0.378
0.5	1	0.378	–0.46	0.75	–0.018
0.5	0.75	0.378	–0.018	0.625	0.186
0.625	0.75	0.186	–0.018	0.6875	0.085
0.6875	0.75	0.085	–0.018	0.71875	0.034

The interval $[0.71875, 0.75]$ contains the zero.

47. $\lim_{x \to 2^+} f(x) = \lim_{x \to 2^+} (3x - 1) = 5$
$f(2) = 3(2) - 1 = 5$
$f(x)$ is continuous from the right at $x = 2$.

49. $\lim_{x \to 2^+} f(x) = \lim_{x \to 2^+} (3x - 3) = 3$
$f(2) = 2^2 = 4$
$f(x)$ is not continuous from the right at $x = 2$.

51. A function is continuous from the left at $x = a$ if
$\lim_{x \to a^-} f(x) = f(a)$.

a. $\lim_{x \to 2^-} f(x) = \lim_{x \to 2^-} x^2 = 4$
$f(2) = 5$
$f(x)$ is not continuous from the left at $x = 2$.

b. $\lim_{x \to 2^-} f(x) = \lim_{x \to 2^-} x^2 = 4$
$f(2) = 3$
$f(x)$ is not continuous from the left at $x = 2$.

c. $\lim_{x \to 2^-} f(x) = \lim_{x \to 2^-} x^2 = 4$
$f(2) = 4$
$f(x)$ is continuous from the left at $x = 2$.

d. $f(x)$ is not continuous from the left at $x = 2$ because $f(2)$ is undefined.

53. Need $g(30) = 100$ and $g(34) = 0$.
$m = \dfrac{0 - 100}{34 - 30} = -25$
$y = -25(x - 34)$
$g(T) = -25(T - 34)$

55.

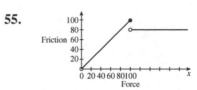

The graph is discontinuous at $x = 100$. This is when the box starts moving.

57. Let $f(t)$ be her distance from home as a function of time on Monday. Let $g(t)$ be her distance from home as a function of time on Tuesday. Let t be given in minutes, with $t = 0$ corresponding to 7:13 a.m. Then she leaves for home at $t = 4$ and returns home at $t = 406$. Let $h(t) = f(t) - g(t)$. If $h(t) = 0$ for some t, then the saleswoman was at exactly the same place at the same time on both Monday and Tuesday.
$h(4) = f(4) - g(4) < 0$ because $f(4) < g(4)$.
$h(406) = f(406) - g(406) = f(406) > 0$.
By the Intermediate Value Theorem, there is a t in the interval $[4, 406]$ such that $h(t) = 0$.

59.

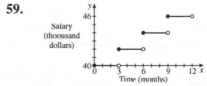

The function $f(t)$ is continuous.

61. We already know $f(x) \neq 0$ for $a < x < b$.
Suppose $f(d) < 0$ for some d, $a < d < b$.
Then by the Intermediate Value Theorem, there is an e in the interval $[c, d]$ such that $f(e) = 0$. But this e would also be between a and b, which is impossible. Thus, $f(x) > 0$ for all $a < x < b$.

Section 1.4

5. $\lim\limits_{x \to 1^-} \dfrac{1-2x}{x^2-1} = \infty$

7. $\lim\limits_{x \to 1} \dfrac{1-2x}{x^2-1}$ does not exist.

9. $\lim\limits_{x \to -1^+} \dfrac{1-2x}{x^2-1} = -\infty$

11. $\lim\limits_{x \to 2^-} \dfrac{4-x}{(x-2)^2} = \infty$

13. $\lim\limits_{x \to 2} \dfrac{4-x}{(x-2)^2} = \infty$

15. $\lim\limits_{x \to 2^-} \dfrac{-x}{\sqrt{4-x^2}} = -\infty$

17. $\lim\limits_{x \to -1} (x^2 - 2x - 3)^{-2/3} = \infty$

19. $\lim\limits_{x \to -\infty} \dfrac{-x}{\sqrt{4+x^2}} = \lim\limits_{x \to -\infty} \dfrac{-x}{-x\sqrt{\frac{4}{x^2}+1}}$

$= \lim\limits_{x \to -\infty} \dfrac{1}{\sqrt{\frac{4}{x^2}+1}}$

$= \dfrac{1}{\sqrt{1}}$

$= 1$

21. $\lim\limits_{x \to -\infty} \dfrac{x^3 - 2x + 1}{3x^3 + 4x^2 - 1} = \lim\limits_{x \to -\infty} \dfrac{x^3\left(1 - \frac{2}{x^2} + \frac{1}{x^3}\right)}{x^3\left(3 + \frac{4}{x} - \frac{1}{x^3}\right)}$

$= \lim\limits_{x \to -\infty} \dfrac{1 - \frac{2}{x^2} + \frac{1}{x^3}}{3 + \frac{4}{x} - \frac{1}{x^3}}$

$= \dfrac{1}{3}$

23. $\lim\limits_{x \to \infty} \dfrac{x^3 - 2x + 5}{3x^2 + 4x - 1} = \lim\limits_{x \to \infty} \dfrac{x^3\left(1 - \frac{2}{x^2} + \frac{5}{x^3}\right)}{x^2\left(3 + \frac{4}{x} - \frac{1}{x^2}\right)}$

$= \lim\limits_{x \to \infty} \dfrac{x\left(1 - \frac{2}{x^2} + \frac{5}{x^3}\right)}{3 + \frac{4}{x} - \frac{1}{x^2}}$

$= \infty$

25. $\lim\limits_{x \to \infty} \dfrac{x^2 - \sin x}{x^2 + 4x - 1} = \lim\limits_{x \to \infty} \dfrac{x^2\left(1 - \frac{\sin x}{x^2}\right)}{x^2\left(1 + \frac{4}{x} - \frac{1}{x^2}\right)}$

$= \lim\limits_{x \to \infty} \dfrac{1 - \frac{\sin x}{x^2}}{1 + \frac{4}{x} - \frac{1}{x^2}}$

$= \dfrac{1}{1}$

$= 1$

27. $\lim\limits_{x \to \infty} \dfrac{3x^3 - x + 5}{4x^3 + 4x^2 - 1} = \lim\limits_{x \to \infty} \dfrac{x^3\left(3 - \frac{1}{x^2} + \frac{5}{x^3}\right)}{x^3\left(4 + \frac{4}{x} + \frac{1}{x^3}\right)}$

$= \lim\limits_{x \to \infty} \dfrac{3 - \frac{1}{x} + \frac{5}{x^3}}{4 + \frac{4}{x} - \frac{1}{x^3}}$

$= \dfrac{3}{4}$

29. $\lim\limits_{x \to \infty} \left[\sqrt{x^2 + 3} - x\right] = 0$

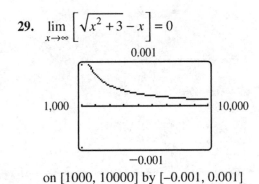

0.001

1,000 | 10,000

−0.001

on [1000, 10000] by [−0.001, 0.001]

31. $\lim\limits_{x \to \infty} e^{2x} = \infty$

33. $\lim\limits_{x \to \infty} \sin 2x$ does not exist.

35. $\lim\limits_{x \to \infty} (e^{-3x} \cos 2x) = 0$

37. $\lim\limits_{x \to \infty} \ln 2x = \infty$

39. $\lim\limits_{x\to 0^+}(x\ln 2x)=0$

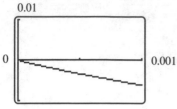

on $[0, 0.001]$ by $[-0.01, 0.01]$

41. $4+x^2=0$ Never
No vertical asymptotes

$$\lim_{x\to\infty}\frac{x}{\sqrt{4+x^2}}=\lim_{x\to\infty}\frac{x}{x\sqrt{\frac{4}{x^2}+1}}$$

$$=\lim_{x\to\infty}\frac{1}{\sqrt{\frac{4}{x^2}+1}}$$

$$=\frac{1}{\sqrt{1}}$$

$$=1$$

$$\lim_{x\to-\infty}\frac{x}{\sqrt{4+x^2}}=\lim_{x\to-\infty}\frac{x}{-x\sqrt{\frac{4}{x^2}+1}}$$

$$=\lim_{x\to-\infty}\frac{-1}{\sqrt{\frac{4}{x^2}+1}}$$

$$=\frac{-1}{\sqrt{1}}$$

$$=-1$$

HA: $y=1$ and $y=-1$.

43. $4-x^2=0$
$\qquad 4=x^2$
VA: $x=\pm 2$

$$\lim_{x\to\pm\infty}\frac{x}{4-x^2}=\lim_{x\to\pm\infty}\frac{x}{x^2\left(\frac{4}{x^2}-1\right)}$$

$$=\lim_{x\to\pm\infty}\frac{1}{x\left(\frac{4}{x^2}-1\right)}$$

$$=0$$

HA: $y=0$

45. $4-x^2=0$
$\qquad 4=x^2$
VA: $x=\pm 2$

$$\lim_{x\to\pm\infty}\frac{x^3}{4-x^2}=\lim_{x\to\pm\infty}\frac{x^3}{x^2\left(\frac{4}{x^2}-1\right)}$$

$$=\lim_{x\to\pm\infty}\frac{x}{\left(\frac{4}{x^2}-1\right)}$$

$$=\mp\infty$$

No horizontal asymptote.

$$-x^2+4\overline{)\begin{array}{l}-x\\x^3\\-(+x^3-4x)\\4x\end{array}}$$

SA: $y=-x$

47. $4+x^2=0$ Never
No vertical asymptotes.

$$\lim_{x\to\pm\infty}\frac{x^3}{4+x^2}=\lim_{x\to\pm\infty}\frac{x^3}{x^2\left(\frac{4}{x^2}+1\right)}$$

$$=\lim_{x\to\pm\infty}\frac{x}{\frac{4}{x^2}+1}$$

$$=\pm\infty$$

No horizontal asymptote.

$$x^2+4\overline{)\begin{array}{l}x\\x^3\\-(x^3+4x)\\-4x\end{array}}$$

SA: $y=x$

49. $x^2-x+1=0$

$$x=\frac{1\pm\sqrt{1-4}}{2}\quad\text{Not real}$$

No vertical asymptotes

$$\lim_{x\to\pm\infty}\frac{2x^3}{x^2-x+1}=\lim_{x\to\pm\infty}\frac{2x^3}{x^2\left(1-\frac{1}{x}+\frac{1}{x^2}\right)}$$

$$=\lim_{x\to\pm\infty}\frac{2x}{1-\frac{1}{x}+\frac{1}{x^2}}$$

$$=\pm\infty$$

No horizontal asymptote.

$$x^2-x+1\overline{)\begin{array}{l}2x+2\\2x^3\\-(2x^3-2x^2+2x)\\2x^2-2x\\-(2x^2-2x+2)\\-2\end{array}}$$

SA: $y=2x+2$

51.

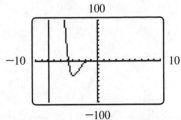

on [–10, 10] by [–100, 100]

HA: $y = 0$ approached only as $x \to \infty$.

The graph crosses the horizontal asymptote an infinite number of times.

53. In Exercise 52, we found that

$$\lim_{x \to \infty} f(x) = \lim_{x \to 0^+} f\left(\frac{1}{x}\right) \text{ and }$$

$$\lim_{x \to -\infty} f(x) = \lim_{x \to 0^-} f\left(\frac{1}{x}\right).$$

Suppose $f(x) = (1+x)^{1/x}$.

Then $f\left(\dfrac{1}{x}\right) = f\left(1 + \dfrac{1}{x}\right)^x$.

Therefore, since $\lim\limits_{x \to 0^+} (1+x)^{1/x} = \lim\limits_{x \to 0^-} (1+x)^{1/x}$,

we can say $\lim\limits_{x \to \infty} \left(1 + \dfrac{1}{x}\right)^x = \lim\limits_{x \to -\infty} \left(1 + \dfrac{1}{x}\right)^x$.

55. $h(0) = \dfrac{300}{1 + 9(.8^0)} = \dfrac{300}{10} = 30$ mm

$$\lim_{t \to \infty} \frac{300}{1 + 9(.8^t)} = 300 \text{ mm}$$

57. $\lim\limits_{x \to 0^+} \dfrac{80x^{-.3} + 60}{9x^{-.3} + 5} \left(\dfrac{x^{.3}}{x^{.3}}\right) = \lim\limits_{x \to 0^+} \dfrac{80 + 60x^{.3}}{9 + 5x^{.3}}$

$$= \frac{80}{9}$$

$$\approx 8.89 \text{ mm}$$

$$\lim_{x \to \infty} \frac{80x^{-.3} + 60}{9x^{-.3} + 5} = \frac{60}{5} = 12 \text{ mm}$$

59. $f(x) = \dfrac{80x^{-0.3} + 60}{10x^{-0.3} + 30}$

61. $\lim\limits_{t \to \infty} v_N = \lim\limits_{t \to \infty} \dfrac{Ft}{m} = \infty$

$$\lim_{t \to \infty} v_E = \lim_{t \to \infty} \frac{Fct}{\sqrt{m^2 c^2 + F^2 t^2}}$$

$$= \lim_{t \to \infty} \frac{Fct}{t\sqrt{\dfrac{m^2 c^2}{t^2} + F^2}}$$

$$= \lim_{t \to \infty} \frac{Fc}{\sqrt{\dfrac{m^2 c^2}{t^2} + F^2}}$$

$$= \frac{Fc}{\sqrt{F^2}}$$

$$= c$$

63. $\lim\limits_{t \to \infty} (e^{-at} \sin t) = 0$ because as $t \to \infty$, $e^{-at} \to 0$

while $\sin t$ oscillates between –1 and 1.

A: about 5 seconds, B: about 18 seconds

65. $\lim\limits_{x \to -\infty} p_n(x) = \lim\limits_{x \to -\infty} (a_n x^n + a_{n-1} x^{n-1} + \cdots + a_0)$

$$= \lim_{x \to -\infty} \left[x^n \left(a_n + \frac{a_{n-1}}{x} + \cdots + \frac{a_0}{x} \right) \right]$$

When the degree n is odd, if a_n is positive, the limit as $x \to -\infty$ is $-\infty$, and if a_n is negative, the limit as $x \to -\infty$ is $+\infty$.

67. One example is $f(x) = \dfrac{x^2 - 1}{x + 1}$.

69. $\lim\limits_{x \to 0^+} x^{1/(\ln x)} = e \approx 2.71828$

71. $\lim\limits_{x \to \infty} \left(1 - \dfrac{1}{x}\right)^x = \dfrac{1}{e} \approx 0.36788$

73. $\lim\limits_{x \to \infty} x^{1/x} = 1$

39

Section 1.5

5. $\left|x^2 + 1 - 1\right| < \dfrac{1}{10}$

$$\left|x^2\right| < \dfrac{1}{10}$$

$$|x| < \dfrac{\sqrt{10}}{10} \approx .32 = \delta$$

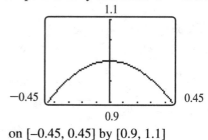

on [−0.32, 0.32] by [0.9, 1.1]

7. $\left|\cos x - 1\right| < \dfrac{1}{10}$

$$-\dfrac{1}{10} < \cos x - 1 < \dfrac{1}{10}$$

$$0.9 < \cos x < 1.1$$

Experimentally determine $\delta = 0.45$.

on [−0.45, 0.45] by [0.9, 1.1]

9. $\left|\sqrt{x+3} - 2\right| < \dfrac{1}{10}$

$$-\dfrac{1}{10} < \sqrt{x+3} - 2 < \dfrac{1}{10}$$

$$1.9 < \sqrt{x+3} < 2.1$$

$$3.61 < x + 3 < 4.41$$

$$-0.39 < x - 1 < 0.41$$

Let $\delta = 0.39$.

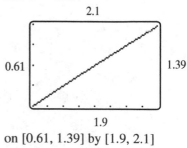

on [0.61, 1.39] by [1.9, 2.1]

11. $\left|\dfrac{x+2}{x^2} - 3\right| < \dfrac{1}{10}$

$$-\dfrac{1}{10} < \dfrac{x+2}{x^2} - 3 < \dfrac{1}{10}$$

$$2.9 < \dfrac{x+2}{x^2} < 3.1$$

Experimentally determine $\delta = 0.02$

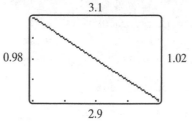

on [0.98, 1.02] by [2.9, 3.1]

13. $\left|3x - 0\right| < \varepsilon$

$$-\varepsilon < 3x < \varepsilon$$

$$-\dfrac{\varepsilon}{3} < x < \dfrac{\varepsilon}{3}$$

$$\delta = \dfrac{\varepsilon}{3}$$

15. $\left|(3x + 2) - 8\right| < \varepsilon$

$$8 - \varepsilon < 3x + 2 < 8 + \varepsilon$$

$$6 - \varepsilon < 3x < 6 + \varepsilon$$

$$2 - \dfrac{\varepsilon}{3} < x < 2 + \dfrac{\varepsilon}{3}$$

$$-\dfrac{\varepsilon}{3} < x - 2 < \dfrac{\varepsilon}{3}$$

$$\delta = \dfrac{\varepsilon}{3}$$

17. $\left|(3 - 4x) - (-1)\right| < \varepsilon$

$$-\varepsilon < 4 - 4x < \varepsilon$$

$$-\dfrac{\varepsilon}{4} < 1 - x < \dfrac{\varepsilon}{4}$$

$$|x - 1| < \dfrac{\varepsilon}{4}$$

$$\delta = \dfrac{\varepsilon}{4}$$

19. $\left|\left(1 + \dfrac{x}{2}\right) - \dfrac{1}{2}\right| < \varepsilon$

$$-\varepsilon < \dfrac{1}{2} + \dfrac{x}{2} < \varepsilon$$

$$-2\varepsilon < 1 + x < 2\varepsilon$$

$$-2\varepsilon < x - (-1) < 2\varepsilon$$

$$\delta = 2\varepsilon$$

21. $\left| (x^2 - 1) - 0 \right| < \varepsilon$

$\qquad \left| x^2 - 1 \right| < \varepsilon$

$\qquad \left| x + 1 \right| \left| x - 1 \right| < \varepsilon$

Consider $x \in [0, 2] \, (\delta = 1)$.

Then $\left| x + 1 \right| \le 3$, so require

$3 \left| x - 1 \right| < \varepsilon$

$\qquad \left| x - 1 \right| < \dfrac{\varepsilon}{3}$

$\qquad \delta = \min \left\{ 1, \dfrac{\varepsilon}{3} \right\}$

23. $\left| (x^2 - 1) - 3 \right| < \varepsilon$

$\qquad \left| x^2 - 4 \right| < \varepsilon$

$\qquad \left| x + 2 \right| \left| x - 2 \right| < \varepsilon$

Consider $x \in [1, 3] \, (\delta = 1)$.

Then $\left| x + 2 \right| \le 5$, so require

$5 \left| x - 2 \right| < \varepsilon$

$\qquad \left| x - 2 \right| < \dfrac{\varepsilon}{5}$

$\qquad \delta = \min \left\{ 1, \dfrac{\varepsilon}{5} \right\}$

25. $\delta = \dfrac{\varepsilon}{\left| m \right|}$

The formula does not depend on a because a line has constant slope.

27. $\displaystyle \lim_{x \to a^-} f(x) = L$ if given any number $\varepsilon > 0$, there is another number $\delta > 0$, such that $\left| f(x) - L \right| < \varepsilon$ whenever $0 < a - x < \delta$.

$\displaystyle \lim_{x \to a^+} f(x) = L$ if given any number $\varepsilon > 0$, there is another number $\delta > 0$, such that $\left| f(x) - L \right| < \varepsilon$ whenever $0 < x - a < \delta$.

29. $\dfrac{2}{x - 1} > M = 100$

$\dfrac{x - 1}{2} < \dfrac{1}{100}$

$x - 1 < \dfrac{1}{50}$

$\delta = \dfrac{1}{50}$

31. $\cot x > 100$

$\dfrac{\cos x}{\sin x} > 100$

Experimentally determine $\delta = 0.0095$.

33. $\dfrac{2}{\sqrt{4 - x^2}} > 100$

Experimentally determine $\delta = 0.0001$.

35. $\left| \dfrac{x^2 - 2}{x^2 + x + 1} - 1 \right| < 0.1$

Experimentally determine $M = 12$.

37. $\left| \dfrac{x^2 + 3}{4x^2 - 4} - 0.25 \right| < 0.1$

Experimentally determine $N = -10$.

39. $\left| e^{-2x} \right| < 0.1$

$\dfrac{1}{e^{2x}} < 0.1$

$10 < e^{2x}$

$\ln 10 < 2x$

$\dfrac{\ln 10}{2} < x$

$M = \dfrac{\ln 10}{2} \approx 1.15$

41. $\left| \dfrac{2}{x^3} - 0 \right| < \varepsilon$

$\dfrac{2}{x^3} = \left| \dfrac{2}{x^3} \right| < \varepsilon$

$\dfrac{2}{\varepsilon} < x^3$

$\sqrt[3]{\dfrac{2}{\varepsilon}} < x$

Choose $M = \sqrt[3]{\dfrac{2}{\varepsilon}}$.

43.

$$\left|\frac{1}{x^k} - 0\right| < \varepsilon$$

$$\frac{1}{x^k} = \left|\frac{1}{x^k}\right| < \varepsilon$$

$$\frac{1}{\varepsilon} < x^k$$

$$\frac{1}{\sqrt[k]{\varepsilon}} < x$$

Choose $M = \dfrac{1}{\sqrt[k]{\varepsilon}}$.

45.

$$\left|\left(\frac{1}{x^2+2} - 3\right) - (-3)\right| < \varepsilon$$

$$\left|\frac{1}{x^2+2}\right| < \varepsilon$$

$$\frac{1}{x^2+2} < \varepsilon$$

$$\frac{1}{\varepsilon} < x^2 + 2$$

$$\frac{1}{\varepsilon} - 2 < x^2$$

$$-\sqrt{\frac{1}{\varepsilon} - 2} > x$$

Choose $N = -\sqrt{\dfrac{1}{\varepsilon}}$.

47.

$$-\frac{2}{(x+3)^4} < N$$

$$-\frac{2}{N} > (x+3)^4$$

$$\sqrt[4]{-\frac{2}{N}} > |x+3|$$

Choose $\delta = \sqrt[4]{-\dfrac{2}{N}}$.

49.

$$\frac{4}{(x-5)^2} > M$$

$$\frac{4}{M} > (x-5)^2$$

$$\frac{2}{\sqrt{M}} > |x-5|$$

Choose $\delta = \dfrac{2}{\sqrt{M}}$.

51. $\lim\limits_{x\to 1^-} f(x) = 2$

$\lim\limits_{x\to 1^+} f(x) = 4$

Choose $\varepsilon = 1$.

53. $\lim\limits_{x\to 1^-} f(x) = 2$

$\lim\limits_{x\to 1^+} f(x) = 4$

Choose $\varepsilon = 1$.

55.

$$\left|2r^2 - 8\right| < \varepsilon$$

$$\left|2(r^2 - 4)\right| < \varepsilon$$

$$|r+2||r-2| < \frac{\varepsilon}{2}$$

Take $r \in [1, 3]$.

Then, $|r+2| \le 5$

$$5|r-2| < \frac{\varepsilon}{2}$$

$$|r-2| < \frac{\varepsilon}{10}$$

$$\delta = \min\left\{1, \frac{\varepsilon}{10}\right\}$$

Section 1.6

3. As $h \to 0, f(a+h) \to f(a)$, so there could be a loss-of-significance error in the numerator.

5. $\lim\limits_{x\to\infty} x\left(\sqrt{4x^2+1} - 2x\right) = 0.25$

$$\left|\frac{e^x + x}{e^x - x^2} - 1\right| < 0.1$$

on [0, 1000000] by [0.15, 0.35]
Error begins near $x = 200,000$.

$$x(\sqrt{4x^2+1} - 2x) \cdot \frac{\sqrt{4x^2+1} + 2x}{\sqrt{4x^2+1} + 2x}$$

$$= \frac{x(4x^2+1-4x^2)}{\sqrt{4x^2+1} + 2x}$$

$$= \frac{x}{\sqrt{4x^2+1} + 2x}$$

$$= f(x)$$

7. $\displaystyle\lim_{x\to\infty}\sqrt{x}\left(\sqrt{x+4}-\sqrt{x+2}\right)=1$

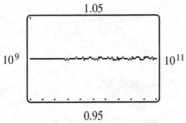

on $[10^9,10^{11}]$ by $[0.95, 1.05]$

Error begins near $x=10^{10}$.

$$\sqrt{x}(\sqrt{x+4}-\sqrt{x+2})\cdot\frac{\sqrt{x+4}+\sqrt{x+2}}{\sqrt{x+4}+\sqrt{x+2}}$$

$$=\frac{\sqrt{x}[(x+4)-(x+2)]}{\sqrt{x+4}+\sqrt{x+2}}$$

$$=\frac{2\sqrt{x}}{\sqrt{x+4}+\sqrt{x+2}}=f(x)$$

9. $\displaystyle\lim_{x\to\infty}x\left(\sqrt{x^2+4}-\sqrt{x^2+2}\right)=1$

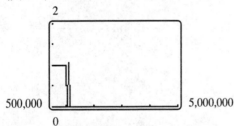

on $[500000, 5000000]$ by $[0, 2]$
Error begins near $x=900,000$.

$$x(\sqrt{x^2+4}-\sqrt{x^2+2})\cdot\frac{\sqrt{x^2+4}+\sqrt{x^2+2}}{\sqrt{x^2+4}+\sqrt{x^2+2}}$$

$$=\frac{x[x^2+4-(x^2+2)]}{\sqrt{x^2+4}+\sqrt{x^2+2}}$$

$$=\frac{2x}{\sqrt{x^2+4}+\sqrt{x^2+2}}=f(x)$$

11. $\displaystyle\lim_{x\to0}\frac{1-\cos 2x}{12x^2}=\frac{1}{6}$

on $[-10^{-5},10^{-5}]$ by $[0, 0.3]$

Error begins near $x=9\times10^{-7}$.

$$\frac{1-\cos 2x}{12x^2}\cdot\frac{1+\cos 2x}{1+\cos 2x}=\frac{\sin^2 2x}{12x^2(1+\cos 2x)}$$

$$=f(x)$$

13. $\displaystyle\lim_{x\to1}\frac{x^2+x-2}{x-1}=\lim_{x\to1}\frac{(x+2)(x-1)}{x-1}$

$$=\lim_{x\to1}(x+2)$$

$$=3$$

$\displaystyle\lim_{x\to1}\frac{x^2+x-2.01}{x-1}$ does not exist.

15. $f(1)=0;\ g(1)=0.00159265$
$f(10)=0;\ g(10)=-0.0159259$
$f(100)=0;\ g(100)=-0.158593$
$f(1000)=0;\ g(1000)=-0.999761$

17. $(1.000003-1.000001)\times10^7=20$

On a computer with a 6-digit mantissa, the calculation would be

$(1.00000-1.00000)\times10^7=0$.

Chapter 1 Review

1. See Example 1.5.

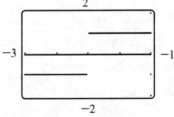

on $[-3, -1]$ by $[-2, 2]$

x	$\frac{x+2}{|x+2|}$		x	$\frac{x+2}{|x+2|}$
-1.9	1		-2.1	-1
-1.99	1		-2.01	-1
-1.999	1		-2.001	-1

$\displaystyle\lim_{x\to-2}\frac{x+2}{|x+2|}$ does not exist.

3. See Example 4.6.

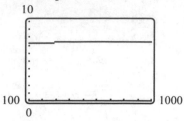

on [100, 1000] by [0, 10]

x	$\left(1+\dfrac{2}{x}\right)^x$
10	6.1917
100	7.2446
1000	7.3743
10,000	7.3876

$$\lim_{x\to\infty}\left(1+\frac{2}{x}\right)^x = e^2 \approx 7.4$$

5. See p. 87.

a. 1   **b.** −2

c. does not exist   **d.** 0

7. See Definition 3.1.
$x=-1,\ x=1$

9. See Example 2.4.
$$\lim_{x\to2}\frac{x-2}{x^2-4}=\lim_{x\to2}\frac{x-2}{(x+2)(x-2)}$$
$$=\lim_{x\to2}\frac{1}{x+2}$$
$$=\frac{1}{2+2}=\frac{1}{4}$$

11. See p. 84.

$\displaystyle\lim_{x\to-2}\frac{x+2}{\sqrt{x^2-4}}$ does not exist because the function

is not defined to the right of $x=-2$.

13. See Example 2.7.
$$\lim_{x\to0}\frac{\tan x}{\sec x}=\frac{\tan 0}{\sec 0}=\frac{0}{1}=0$$

15. See Example 2.10.
$$\lim_{x\to1^-}f(x)=\lim_{x\to1^-}(2x+1)$$
$$=3$$
$$\lim_{x\to1^+}f(x)=\lim_{x\to1^+}(x^2+1)$$
$$=2$$
$\displaystyle\lim_{x\to1}f(x)$ does not exist.

17. See Example 2.6.
$$\lim_{x\to1^+}\sqrt{x^2+1}=\sqrt{2}$$

19. See Examples 4.2, 4.3.
$$\lim_{x\to1^+}\frac{x+1}{x^2-1}=\lim_{x\to1^+}\frac{x+1}{(x+1)(x-1)}$$
$$=\lim_{x\to1^+}\frac{1}{\underset{+}{x-1}}$$
$$=\infty$$

21. See Example 4.7.
$$\lim_{x\to\infty}\frac{x^2-4}{3x^2+x+1}=\lim_{x\to\infty}\frac{x^2\left(1-\frac{4}{x^2}\right)}{x^2\left(3+\frac{1}{x}+\frac{1}{x^2}\right)}$$
$$=\lim_{x\to\infty}\frac{1-\frac{4}{x^2}}{3+\frac{1}{x}+\frac{1}{x^2}}$$
$$=\frac{1}{3}$$

23. See Section 1.4.
$$\lim_{x\to\infty}e^{-3x}=0$$

25. See Section 1.4.
$$\lim_{x\to\infty}\ln 2x=\infty$$

27. See Example 4.7.
$$\lim_{x\to-\infty}\frac{2x}{x^2+3x-5}=\lim_{x\to-\infty}\frac{2x}{x^2\left(1+\frac{3}{x}+\frac{5}{x^2}\right)}$$
$$=\lim_{x\to-\infty}\frac{2}{x\left(1+\frac{3}{x}+\frac{5}{x^2}\right)}$$
$$=0$$

29. See Example 2.9.

$$0 \le \frac{x^2}{x^2+1} < 1$$

$$0 \le \frac{2x^3}{x^2+1} < 2x$$

$$\lim_{x\to 0} 0 = 0; \lim_{x\to 0} 2x = 0$$

By the Squeeze Theorem,

$$\lim_{x\to 0} \frac{2x^3}{x^2+1} = 0.$$

31. See Example 3.4.

$$f(x) = \frac{x-1}{x^2+2x-3} = \frac{x-1}{(x+3)(x-1)}$$

f has a non-removable discontinuity at $x = -3$ and a removable discontinuity at $x = 1$.

33. See Examples 3.2, 3.3.

$$\lim_{x\to 0^-} f(x) = \lim_{x\to 0^-} \sin x = 0$$

$$\lim_{x\to 0^+} f(x) = \lim_{x\to 0^+} x^2 = 0$$

$$\lim_{x\to 2^-} f(x) = \lim_{x\to 2^-} x^2 = 4$$

$$\lim_{x\to 2^+} f(x) = \lim_{x\to 2^+} (4x-3) = 5$$

f has a non-removable discontinuity at $x = 2$.

35. See Examples 3.4, 3.6.

$$f(x) = \frac{x+2}{x^2-x-6} = \frac{x+2}{(x-3)(x+2)}$$

$(-\infty, -2), (-2, 3), (3, \infty)$

37. See Example 3.6.

$$f(x) = \sin(1+e^x)$$

$(-\infty, \infty)$

39. See Examples 4.7, 4.8.

$$f(x) = \frac{x+1}{(x-2)(x-1)}$$

VA: $x = 1$, $x = 2$

$$\lim_{x\to \pm\infty} \frac{x+1}{x^2-3x+2} = \lim_{x\to \pm\infty} \frac{x\left(1+\frac{1}{x}\right)}{x^2\left(1-\frac{3}{x}+\frac{2}{x^2}\right)}$$

$$= \lim_{x\to \pm\infty} \frac{1+\frac{1}{x}}{x\left(1-\frac{3}{x}+\frac{2}{x^2}\right)}$$

$$= 0$$

HA: $y = 0$

41. See Examples 4.7, 4.8.

$$f(x) = \frac{x^2}{x^2-1} = \frac{x^2}{(x+1)(x-1)}$$

VA: $x = -1$, $x = 1$

$$\lim_{x\to \pm\infty} \frac{x^2}{x^2-1} = \lim_{x\to \pm\infty} \frac{x^2}{x^2\left(1-\frac{1}{x^2}\right)}$$

$$= \lim_{x\to \pm\infty} \frac{1}{1-\frac{1}{x^2}}$$

$$= \frac{1}{1}$$

$$= 1$$

HA: $y = 1$

43. See Example 5.4.

$$\left|\sin x - 1\right| < \frac{1}{10}$$

$$-\frac{1}{10} < \sin x - 1 < \frac{1}{10}$$

$$0.9 < \sin x < 1.1$$

Experimentally determine $\delta = 0.4$.

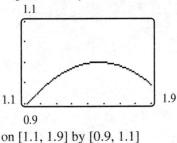

1.1

1.1   1.9

0.9

on [1.1, 1.9] by [0.9, 1.1]

45. See Example 5.2.

$$\left|(2x+1)-1\right| < \varepsilon$$

$$\left|2x\right| < \varepsilon$$

$$\left|x\right| < \frac{\varepsilon}{2}$$

Choose $\delta = \frac{\varepsilon}{2}$.

47. See Example 5.7.

$$\left|\frac{x^2}{x^2+1} - 1\right| < 0.1$$

Experimentally determine $M = 3$.

49. (a) See Example 6.6.

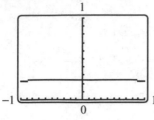

on $[-1, 1]$ by $[0, 1]$

x	$\dfrac{1-\cos x}{2x^2}$		x	$\dfrac{1-\cos x}{2x^2}$
0.1	0.2498		−0.1	0.2498
0.01	0.2500		−0.01	0.2500
0.00	0.2500		−0.001	0.2500

$$\lim_{x \to 0} \frac{1-\cos x}{2x^2} = \frac{1}{4}$$

(b) Use window $[-1\times10^{-5}, 1\times10^{-5}]$ by $[0,1]$.

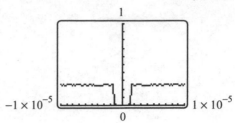

(c) $\dfrac{\sin^2 x}{2x^2(1+\cos x)}$

51. The limit of θ' as x approaches 0 is 66 radians per second, far faster than the player can maintian focus. From about 12 feet on in to the plate the player can't keep her eye on the ball.

46

Chapter 2

Section 2.1

5.

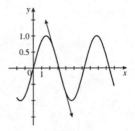

7.

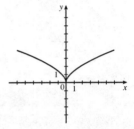

9. At $x = 1$ the slope of the tangent line is approximately -1.

11. C, B, A, D

13. $f(x) = x^3 - x$

a. $m_{sec} = \dfrac{f(2) - f(1)}{2 - 1}$

$= \dfrac{6 - 0}{1}$

$= 6$

b. $m_{sec} = \dfrac{f(3) - f(2)}{3 - 2}$

$= \dfrac{24 - 6}{1}$

$= 18$

c. $m_{sec} = \dfrac{f(2) - f(1.5)}{2 - 1.5}$

$= \dfrac{6 - 1.875}{.5}$

$= 8.25$

d. $m_{sec} = \dfrac{f(2.5) - f(2)}{2.5 - 2}$

$= \dfrac{13.125 - 6}{.5}$

$= 14.25$

e. $m_{sec} = \dfrac{f(2) - f(1.9)}{2 - 1.9}$

$= \dfrac{6 - 4.959}{.1}$

$= 10.41$

f. $m_{sec} = \dfrac{f(2.1) - f(2)}{2.1 - 2}$

$= \dfrac{7.161 - 6}{.1}$

$= 11.61$

g. Tangent line slope estimate: 11

15. $f(x) = \cos x^2$

a. $m_{sec} = \dfrac{f(2) - f(1)}{2 - 1}$

$= \dfrac{-.65 - .54}{1}$

$= -1.19$

b. $m_{sec} = \dfrac{f(3) - f(2)}{3 - 2}$

$= \dfrac{-.91 - (-.65)}{1}$

$= -.26$

c. $m_{sec} = \dfrac{f(2) - f(1.5)}{2 - 1.5}$

$= \dfrac{-.65 - (-.63)}{.5}$

$= -.04$

d. $m_{sec} = \dfrac{f(2.5) - f(2)}{2.5 - 2}$

$= \dfrac{1.00 - (-.65)}{.5}$

$= 3.3$

e. $m_{sec} = \dfrac{f(2) - f(1.9)}{2 - 1.9}$

$= \dfrac{-.65 - (-.89)}{.1}$

$= 2.4$

f. $m_{sec} = \dfrac{f(2.1) - f(2)}{2.1 - 2}$

$= \dfrac{-.30 - (-.65)}{.1}$

$= 3.5$

g. Tangent line slope estimate: 3

17.

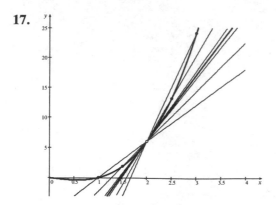

19. The answer is an animation.

21. $f(x) = x^2 - 2$, $a = 1$

$$m_{\tan} = \lim_{h \to 0} \frac{f(1+h) - f(1)}{h}$$

$$= \lim_{h \to 0} \frac{(1+h)^2 - 2 - (1^2 - 2)}{h}$$

$$= \lim_{h \to 0} \frac{1 + 2h + h^2 - 2 + 1}{h}$$

$$= \lim_{h \to 0} \frac{2h + h^2}{h}$$

$$= \lim_{h \to 0} 2 + h$$

$$= 2$$

$f(1) = 1^2 - 2 = -1$

So the equation of the tangent line is
$y = 2(x - 1) + (-1)$ or $y = 2x - 3$

23. $f(x) = x^2 - 3x$, $a = -2$.

$m_{\tan}$

$$= \lim_{h \to 0} \frac{f(-2+h) - f(-2)}{h}$$

$$= \lim_{h \to 0} \frac{(-2+h)^2 - 3(-2+h) - ((-2)^2 - 3(-2))}{h}$$

$$= \lim_{h \to 0} \frac{4 - 4h + h^2 + 6 - 3h - 10}{h}$$

$$= \lim_{h \to 0} \frac{-7h + h^2}{h}$$

$$= \lim_{h \to 0} -7 + h$$

$$= -7$$

$f(-2) = (-2)^2 - 3(-2) = 10$

So the equation of the tangent line is
$y = -7(x + 2) + 10$ or $y = -7x - 4$

25. $f(x) = \dfrac{2}{x+1}$, $a = 1$

$$m_{\tan} = \lim_{h \to 0} \frac{f(1+h) - f(1)}{h}$$

$$= \lim_{h \to 0} \frac{\frac{2}{(1+h)+1} - \frac{2}{1+1}}{h}$$

$$= \lim_{h \to 0} \frac{\frac{2}{2+h} - 1}{h}$$

$$= \lim_{h \to 0} \frac{\left(\frac{2-(2+h)}{2+h}\right)}{h}$$

$$= \lim_{h \to 0} \frac{\left(\frac{-h}{2+h}\right)}{h}$$

$$= \lim_{h \to 0} \frac{-1}{2+h}$$

$$= -\frac{1}{2}$$

$f(1) = \dfrac{2}{1+1} = 1$

So the equation of the tangent line is

$y = -\dfrac{1}{2}(x - 1) + 1$ or $y = -\dfrac{1}{2}x + \dfrac{3}{2}$

27. $f(x) = \sqrt{x+3}$, $a = -2$

$$m_{\tan} = \lim_{h \to 0} \frac{f(-2+h) - f(-2)}{h}$$

$$= \lim_{h \to 0} \frac{\sqrt{(-2+h)+3} - \sqrt{-2+3}}{h}$$

$$= \lim_{h \to 0} \frac{\sqrt{h+1} - 1}{h}$$

$$= \lim_{h \to 0} \frac{\sqrt{h+1} - 1}{h} \cdot \frac{\sqrt{h+1} + 1}{\sqrt{h+1} + 1}$$

$$= \lim_{h \to 0} \frac{h + 1 - 1}{h(\sqrt{h+1} + 1)}$$

$$= \lim_{h \to 0} \frac{1}{\sqrt{h+1} + 1}$$

$$= \frac{1}{2}$$

$f(-2) = \sqrt{-2+3} = \sqrt{1} = 1$

So the equation of the tangent line is

$y = \dfrac{1}{2}(x + 2) + 1$ or $y = \dfrac{1}{2}x + 2$

29. $f(x) = |x - 1|$ $a = 1$

$$\lim_{h \to 0^+} \frac{f(1+h) - f(1)}{h} = \lim_{h \to 0^+} \frac{|1+h-1| - |1-1|}{h}$$

$$= \lim_{h \to 0^+} \frac{|h| - 0}{h}$$

$$= \lim_{h \to 0^+} \frac{h}{h}$$

$$= 1$$

$$\lim_{h \to 0^-} \frac{f(1+h) - f(1)}{h} = \lim_{h \to 0^-} \frac{|1+h-1| - |1-1|}{h}$$

$$= \lim_{h \to 0^-} \frac{|h| - 0}{h}$$

$$= \lim_{h \to 0^-} \frac{-h}{h}$$

$$= -1$$

Since the one sided limits are different,

$\lim_{h \to 0} \dfrac{f(1+h) - f(1)}{h}$ does not exist and thus the tangent line does not exist.

31. $f(x) = \begin{cases} -2x^2 & \text{if } x < 0 \\ x^3 & \text{if } x \geq 0 \end{cases}$ at $a = 0$

$$\lim_{h \to 0^+} \frac{f(0+h) - f(0)}{h} = \lim_{h \to 0^+} \frac{h^3 - 0^3}{h}$$

$$= \lim_{h \to 0^+} h^2$$

$$= 0$$

$$\lim_{h \to 0^-} \frac{f(0+h) - f(0)}{h} = \lim_{h \to 0^-} \frac{-2h^2 - 0}{h}$$

$$= \lim_{h \to 0^-} -2h$$

$$= 0$$

So the slope of the tangent line is 0.

33. $f(x) = \begin{cases} x^2 - 1 & \text{if } x < 0 \\ x^3 + 1 & \text{if } x \geq 0 \end{cases}$ at $a = 0$

$$\lim_{x \to 0^+} f(x) = \lim_{x \to 0^+} x^3 + 1 = 1$$

$$\lim_{x \to 0^-} f(x) = \lim_{x \to 0^-} x^2 - 1 = -1$$

Since the function is not continuous at $a = 0$, the tangent line does not exist.

35. $f(t) = 16t^2 + 10$

a. $v_{avg} = \dfrac{f(2) - f(0)}{2 - 0}$

$$= \frac{16(2)^2 + 10 - (16(0)^2 + 10)}{2 - 0}$$

$$= \frac{64}{2} = 32$$

b. $v_{avg} = \dfrac{f(2) - f(1)}{2 - 1}$

$$= \frac{16(2)^2 + 10 - (16(1)^2 + 10)}{2 - 1}$$

$$= 48$$

c. $v_{avg} = \dfrac{f(2) - f(1.9)}{2 - 1.9}$

$$= \frac{16(2)^2 + 10 - (16(1.9)^2 + 10)}{2 - 1.9}$$

$$= \frac{6.24}{.1}$$

$$= 62.4$$

d. $v_{avg} = \dfrac{f(2) - f(1.99)}{2 - 1.99}$

$$= \frac{16(2)^2 + 10 - (16(1.99)^2 + 10)}{2 - 1.99}$$

$$= \frac{.6384}{.01}$$

$$= 63.84$$

e. $v(2) = \lim_{h \to 0} \dfrac{f(2+h) - f(2)}{h}$

$$= \lim_{h \to 0} \frac{16(2+h)^2 + 10 - (16(2^2) + 10)}{h}$$

$$= \lim_{h \to 0} \frac{16(4 + 4h + h^2) + 10 - 64 - 10}{h}$$

$$= \lim_{h \to 0} \frac{64h + 16h^2}{h}$$

$$= \lim_{h \to 0} 64 + 16h$$

$$= 64$$

37. $f(t) = \sqrt{t^2 + 8t}$

a. $v_{avg} = \dfrac{f(2) - f(0)}{2 - 0}$

$= \dfrac{\sqrt{4 + 16} - \sqrt{0}}{2}$

$= \dfrac{\sqrt{20}}{2}$

$= \sqrt{5}$

$= 2.236068$

b. $v_{avg} = \dfrac{f(2) - f(1)}{2 - 1}$

$= \dfrac{\sqrt{20} - \sqrt{1 + 8}}{1}$

$= \sqrt{20} - 3$

$= 1.472136$

c. $v_{avg} = \dfrac{f(2) - f(1.9)}{2 - 1.9}$

$= \dfrac{\sqrt{20} - \sqrt{1.9^2 + 8(1.9)}}{.1}$

$= \dfrac{\sqrt{20} - \sqrt{18.8}}{.1}$

$= \dfrac{.1350862}{.1}$

$= 1.3508627$

d. $v_{avg} = \dfrac{f(2) - f(1.99)}{2 - 1.99}$

$= \dfrac{\sqrt{20} - \sqrt{(1.99)^2 + 8(1.99)}}{2 - 1.99}$

$= \dfrac{\sqrt{20} - \sqrt{19.8801}}{.01}$

$= \dfrac{.0134253}{.01}$

$= 1.3425375$

e. $v(2) = \lim\limits_{h \to 0} \dfrac{f(2 + h) - f(2)}{h}$

$= \lim\limits_{h \to 0} \dfrac{\sqrt{(2 + h)^2 + 8(2 + h)} - \sqrt{20}}{h}$

$= \lim\limits_{h \to 0} \dfrac{\sqrt{h^2 + 12h + 20} - \sqrt{20}}{h}$

$= \lim\limits_{h \to 0} \dfrac{\sqrt{h^2 + 12h + 20} - \sqrt{20}}{h} \cdot \dfrac{\sqrt{h^2 + 12h + 20} + \sqrt{20}}{\sqrt{h^2 + 12h + 20} + \sqrt{20}}$

$= \lim\limits_{h \to 0} \dfrac{h^2 + 12h + 20 - 20}{h(\sqrt{h^2 + 12h + 20} + \sqrt{20})}$

$= \lim\limits_{h \to 0} \dfrac{h + 12}{\sqrt{h^2 + 12h + 20} + \sqrt{20}}$

$= \dfrac{12}{2\sqrt{20}} = \dfrac{6}{\sqrt{20}} = 1.3416408$

39. $f(t) = -16t^2 + 5,\ a = 1$

$v(1) = \lim\limits_{h \to 0} \dfrac{f(1 + h) - f(1)}{h}$

$= \lim\limits_{h \to 0} \dfrac{-16(1 + h)^2 + 5 - (-16(1)^2 + 5)}{h}$

$= \lim\limits_{h \to 0} \dfrac{-16 - 32h - 16h^2 + 5 + 11}{h}$

$= \lim\limits_{h \to 0} \dfrac{-32h - 16h^2}{h}$

$= \lim\limits_{h \to 0} -32 - 16h = -32$

41. $f(t) = \sqrt{t + 16},\ a = 0$

$v(0) = \lim\limits_{h \to 0} \dfrac{f(0 + h) - f(0)}{h}$

$= \lim\limits_{h \to 0} \dfrac{\sqrt{h + 16} - \sqrt{16}}{h}$

$= \lim\limits_{h \to 0} \dfrac{\sqrt{h + 16} - 4}{h} \cdot \dfrac{\sqrt{h + 16} + 4}{\sqrt{h + 16} + 4}$

$= \lim\limits_{h \to 0} \dfrac{h + 16 - 16}{h(\sqrt{h + 16} + 4)}$

$= \lim\limits_{h \to 0} \dfrac{1}{\sqrt{h + 16} + 4} = \dfrac{1}{\sqrt{16} + 4} = \dfrac{1}{8}$

43. The hiker reached the top at the highest point on the graph. (about 1.75 hours) The hiker was going the fastest on the way up at the point whose tangent line has the greatest slope. (about 1.5 hours) The hiker was going the fastest on the way down at the point where the tangent line has the least slope (about 4 hours). Where the graph is level, the hiker was resting, or walking on flat ground.

45.

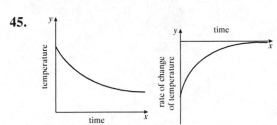

47. $\theta(t) = 0.4t^2$

3 rotations corresponds to $t = 3(2\pi) = 6\pi$

$$\lim_{h \to 0} \frac{\theta(6\pi + h) - \theta(6\pi)}{h}$$

$$= \lim_{h \to 0} \frac{.4(6\pi + h)^2 - .4(6\pi)^2}{h}$$

$$= \lim_{h \to 0} \frac{.4(36\pi^2 + 12\pi h + h^2 - 36\pi^2)}{h}$$

$$= \lim_{h \to 0} \frac{4.8\pi h + .4h^2}{h}$$

$$= \lim_{h \to 0} 4.8\pi + .4h$$

$$= 4.8\pi$$

So the angular velocity at the time of release is $= 4.8\pi \approx 15$ rad/s.

49.
$$v_{avg} = \frac{f(s) - f(r)}{s - r}$$

$$= \frac{as^2 + bs + c - (ar^2 + br + c)}{s - r}$$

$$= \frac{a(s^2 - r^2) + b(s - r)}{s - r}$$

$$= \frac{a(s + r)(s - r) + b(s - r)}{s - r}$$

$$= a(s + r) + b$$

$$v(r) = \lim_{h \to 0} \frac{f(r + h) - f(r)}{h}$$

$$= \lim_{h \to 0} \frac{a(r + h)^2 + b(r + h) + c - (ar^2 + br + c)}{h}$$

$$= \lim_{h \to 0} \frac{a(r^2 + 2rh + h^2 - r^2) + b(r + h - r)}{h}$$

$$= \lim_{h \to 0} \frac{ah(2r + h) + bh}{h}$$

$$= \lim_{h \to 0} a(2r + h) + b$$

$$= 2ar + b$$

Similarly, $v(s) = 2as + b$.

$$a(s + r) + b = \frac{2}{2}(a(s + r) + b)$$

$$= \frac{2(a(s + r) + b)}{2}$$

$$= \frac{2as + 2ar + 2b}{2}$$

$$= \frac{(2ar + b) + (2as + b)}{2}$$

51. $f(t) = \sqrt{100 + 8t}$

At $t = 3$,

$$m_{\tan} = \lim_{h \to 0} \frac{f(3 + h) - f(3)}{h}$$

$$= \lim_{h \to 0} \frac{\sqrt{100 + 8(3 + h)} - \sqrt{100 + 8(3)}}{h}$$

$$= \lim_{h \to 0} \frac{\sqrt{124 + 8h} - \sqrt{124}}{h}$$

$$= \lim_{h \to 0} \frac{\sqrt{124 + 8h} - \sqrt{124}}{h} \cdot \frac{\sqrt{124 + 8h} + \sqrt{124}}{\sqrt{124 + 8h} + \sqrt{124}}$$

$$= \lim_{h \to 0} \frac{124 + 8h - 124}{h(\sqrt{124 + 8h} + \sqrt{124})}$$

$$= \lim_{h \to 0} \frac{8}{\sqrt{124 + 8h} + \sqrt{124}} = \frac{8}{2\sqrt{124}}$$

$$\approx .3592106$$

At $t = 2$, we have from example 1.5,

$$m_{\tan} = \frac{4}{\sqrt{116}} \approx .3713906.$$

So the rate of change at $t = 3$ is less than the rate of change at $t = 2$.

53. $\lim\limits_{h\to 0}\dfrac{f(a+h)-f(a)}{h}$

Let $h = x - a$.

Then $\lim\limits_{h\to 0}\dfrac{f(a+h)-f(a)}{h}$

$= \lim\limits_{x-a\to 0}\dfrac{f(a+x-a)-f(a)}{x-a}$

$= \lim\limits_{x\to a}\dfrac{f(x)-f(a)}{x-a}$

55. Compute the slope of the tangent to the path at

$x = \dfrac{1}{2}$:

$m_{\tan}$

$= \lim\limits_{h\to 0}\dfrac{f\left(\dfrac{1}{2}+h\right)-f\left(\dfrac{1}{2}\right)}{h}$

$= \lim\limits_{h\to 0}\dfrac{\dfrac{1}{4}+h+h^2-\dfrac{1}{4}}{h}$

$= \lim\limits_{h\to 0}h+1$

$= 1$

Since the line joining $\left(\dfrac{1}{2},\dfrac{1}{4}\right)$ to $\left(1,\dfrac{3}{4}\right)$ has slope

1, the car will hit the tree.

Section 2.2

5. $f(x) = 3x + 1,\ a = 1$

$f'(1) = \lim\limits_{h\to 0}\dfrac{f(1+h)-f(1)}{h}$

$= \lim\limits_{h\to 0}\dfrac{3(1+h)+1-(3+1)}{h}$

$= \lim\limits_{h\to 0}\dfrac{3+3h+1-4}{h}$

$= \lim\limits_{h\to 0}\dfrac{3h}{h}$

$= \lim\limits_{h\to 0}3$

$= 3$

$f'(1) = \lim\limits_{b\to 1}\dfrac{f(b)-f(1)}{b-1}$

$= \lim\limits_{b\to 1}\dfrac{3b+1-(3+1)}{b-1}$

$= \lim\limits_{b\to 1}\dfrac{3b-3}{b-1}$

$= \lim\limits_{b\to 1}\dfrac{3(b-1)}{b-1}$

$= \lim\limits_{b\to 1}3$

$= 3$

7. $f(x) = \sqrt{3x+1},\ a = 1$

$f'(1) = \lim\limits_{h\to 0}\dfrac{f(1+h)-f(1)}{h}$

$= \lim\limits_{h\to 0}\dfrac{\sqrt{3(1+h)+1}-\sqrt{3(1)+1}}{h}$

$= \lim\limits_{h\to 0}\dfrac{\sqrt{4+3h}-2}{h}\cdot\dfrac{\sqrt{4+3h}+2}{\sqrt{4+3h}+2}$

$= \lim\limits_{h\to 0}\dfrac{4+3h-4}{h(\sqrt{4+3h}+2)}$

$= \lim\limits_{h\to 0}\dfrac{3}{\sqrt{4+3h}+2}$

$= \dfrac{3}{\sqrt{4+3(0)}+2}$

$= \dfrac{3}{4}$

$f'(1) = \lim\limits_{b\to 1}\dfrac{f(b)-f(1)}{b-1}$

$= \lim\limits_{b\to 1}\dfrac{\sqrt{3b+1}-\sqrt{3(1)+1}}{b-1}$

$= \lim\limits_{b\to 1}\dfrac{\sqrt{3b+1}-2}{b-1}\cdot\dfrac{3}{3}$

$= \lim\limits_{b\to 1}\dfrac{3(\sqrt{3b+1}-2)}{(\sqrt{3b+1}-2)(\sqrt{3b+1}+2)}$

$= \lim\limits_{b\to 1}\dfrac{3}{\sqrt{3b+1}+2}$

$= \dfrac{3}{\sqrt{3(1)+1}+2}$

$= \dfrac{3}{4}$

9. $f(x) = x^2 + 2x,\ a = 0$

$f'(0) = \lim\limits_{h\to 0}\dfrac{f(0+h)-f(0)}{h}$

$= \lim\limits_{h\to 0}\dfrac{h^2+2h-(0^2+2\cdot 0)}{h}$

$= \lim\limits_{h\to 0}\dfrac{h^2+2h}{h}$

$= \lim\limits_{h\to 0}h+2$

$= 2$

11. $f(x) = x^3 + 4, \ a = -1$

$$\begin{aligned}
f'(-1) &= \lim_{b \to -1} \frac{f(b) - f(-1)}{b - (-1)} \\
&= \lim_{b \to -1} \frac{b^3 + 4 - ((-1)^3 + 4)}{b + 1} \\
&= \lim_{b \to -1} \frac{b^3 + 1}{b + 1} \\
&= \lim_{b \to -1} \frac{(b^2 - b + 1)(b + 1)}{b + 1} \\
&= \lim_{b \to -1} b^2 - b + 1 \\
&= 3
\end{aligned}$$

13. $f(x) = 3x^2 + 1$

$$\begin{aligned}
f'(x) &= \lim_{h \to 0} \frac{f(x + h) - f(x)}{h} \\
&= \lim_{h \to 0} \frac{3(x + h)^2 + 1 - (3(x)^2 + 1)}{h} \\
&= \lim_{h \to 0} \frac{3x^2 + 6xh + 3h^2 + 1 - (3x^2 + 1)}{h} \\
&= \lim_{h \to 0} \frac{6xh + 3h^2}{h} \\
&= \lim_{h \to 0} 6x + 3h \\
&= 6x
\end{aligned}$$

15. $f(x) = \dfrac{3}{x + 1}$

$$\begin{aligned}
f'(x) &= \lim_{b \to x} \frac{f(b) - f(x)}{b - x} \\
&= \lim_{b \to x} \frac{\dfrac{3}{b + 1} - \dfrac{3}{x + 1}}{b - x} \\
&= \lim_{b \to x} \frac{\dfrac{3(x + 1) - 3(b + 1)}{(b + 1)(x + 1)}}{b - x} \\
&= \lim_{b \to x} \frac{-3(b - x)}{(b + 1)(x + 1)(b - x)} \\
&= \lim_{b \to x} \frac{-3}{(b + 1)(x + 1)} \\
&= \frac{-3}{(x + 1)^2}
\end{aligned}$$

17. $f(x) = \sqrt{3x + 1}$

$$\begin{aligned}
f'(1) &= \lim_{h \to 0} \frac{f(x + h) - f(x)}{h} \\
&= \lim_{h \to 0} \frac{\sqrt{3(x + h) + 1} - \sqrt{3(x) + 1}}{h} \\
&= \lim_{h \to 0} \frac{\sqrt{3x + 3h - 1} - \sqrt{3x + 1}}{h} \cdot \frac{\sqrt{3x + 3h - 1} + \sqrt{3x + 1}}{\sqrt{3x + 3h - 1} + \sqrt{3x + 1}} \\
&= \lim_{h \to 0} \frac{3x + 3h - 1 - (3x + 1)}{h(\sqrt{3x + 3h - 1} + \sqrt{3x + 1})} \\
&= \lim_{h \to 0} \frac{3h}{h(\sqrt{3x + 3h - 1} + \sqrt{3x + 1})} \\
&= \lim_{h \to 0} \frac{3}{(\sqrt{3x + 3h - 1} + \sqrt{3x + 1})} \\
&= \frac{3}{2\sqrt{3x + 1)}}
\end{aligned}$$

19. $f(x) = x^3 + 2x - 1$

$$\begin{aligned}
f'(x) &= \lim_{b \to x} \frac{f(b) - f(x)}{b - x} \\
&= \lim_{b \to x} \frac{b^3 + 2b - 1 - (x^3 + 2x - 1)}{b - x} \\
&= \lim_{b \to x} \frac{b^3 - x^3 + 2(b - x)}{b - x} \\
&= \lim_{b \to x} \frac{(b - x)(b^2 + 2bx + x^2 + 2)}{b - x} \\
&= \lim_{b \to x} b^2 + 2bx + x^2 + 2 \\
&= 3x^2 + 2
\end{aligned}$$

21. c

23. a

25. b

27.

29.

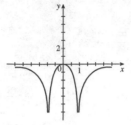

31.

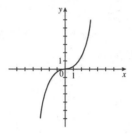

33. The function is not differentiable at $x = -3$, $x = -1$, $x = 3$

35.

x-interval	slope of secant line
(0.2, 1.0)	4.25
(0.4, 1.0)	4.5
(0.6, 1.0)	4.75
(0.8, 1.0)	5
(1.0, 1.2)	5
(1.0, 1.4)	4.75
(1.0, 1.6)	4.5
(1.0, 1.8)	4.25

The slope of the tangent line at $x = 1$ is approximately 5.

37.

Time Interval	Average Velocity
(1.6, 2.0)	8.5
(1.7, 2.0)	9
(1.8, 2.0)	9.5
(1.9, 2.0)	10
(2.0, 2.1)	10
(2.0, 2.2)	9.5
(2.0, 2.3)	9
(2.0, 2.4)	8.5

The velocity at $t = 2$ is approximately 10.

39. $f(x) = x^{2/3}$

From the graph, we see that $f(x)$ is continuous at $x = 0$. Also, at $x = 0, f(0) = 0^{2/3} = 0$, and
$$\lim_{x \to 0} f(x) = \lim_{x \to 0} x^{2/3} = 0 = f(0).$$
So $f(x)$ is continuous at $x = 0$.

$$
\begin{aligned}
f'(0) &= \lim_{h \to 0} \frac{f(0+h) - f(0)}{h} \\
&= \lim_{h \to 0} \frac{(0+h)^{2/3} - 0^{2/3}}{h} \\
&= \lim_{h \to 0} \frac{h^{2/3}}{h} \\
&= \lim_{h \to 0} \frac{1}{h^{1/3}} \\
&= \frac{1}{0} \text{ undefined}
\end{aligned}
$$

So $f'(0)$ does not exist.

41. $1 + \ln 1 = 1$

43. $f(x) = \begin{cases} 2x + 1 \text{ if } x < 0 \\ 3x + 1 \text{ if } x \geq 0 \end{cases}$

$$
\begin{aligned}
D_+ f(0) &= \lim_{h \to 0^+} \frac{f(h) - f(0)}{h} \\
&= \lim_{h \to 0^+} \frac{3h + 1 - 1}{h} \\
&= 3
\end{aligned}
$$

$$
\begin{aligned}
D_- f(0) &= \lim_{h \to 0^-} \frac{f(h) - f(0)}{h} \\
&= \lim_{h \to 0^-} \frac{2h + 1 - 1}{h} \\
&= 2
\end{aligned}
$$

45. $f(x) = \begin{cases} g(x) & \text{if } x < 0 \\ k(x) & \text{if } x \geq 0 \end{cases}$

$f(x)$ is continuous at $x = 0$, and $g(x)$ and $k(x)$ are differentiable at $x = 0$.

$$D_+ f(0) = \lim_{h \to 0^+} \frac{f(h) - f(0)}{h}$$

$$= \lim_{h \to 0} \frac{k(h) - k(0)}{h}$$

$$= k'(0).$$

$$D_- f(0) = \lim_{h \to 0^-} \frac{f(h) - f(0)}{h}$$

$$= \lim_{h \to 0} \frac{g(h) - g(0)}{h}$$

$$= g'(0)$$

If $f(x)$ has a jump discontinuity at $x = 0$,
$D_- f(0) \neq g'(0)$.

47. If $f'(x) > 0$ for all x, then the tangent line to $f(x)$ at each value of x has positive slope, so the function must be an increasing function.

49.

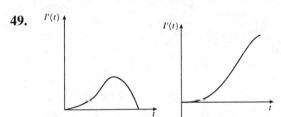

51. Let $f(x) = -1 - x^2$; then for all x, we have $f(x) \leq x$. But if $x = -1$,

$$f'(-1) = \lim_{h \to 0} \frac{f(-1 + h) - f(-1)}{h}$$

$$= \lim_{h \to 0} \frac{2(-1 + h) - 3 - (2(-1) - 3)}{h}$$

$$= \lim_{h \to 0} \frac{-2 + 2h - 3 + 2 + 3}{h}$$

$$= \lim_{h \to 0} \frac{2h}{h}$$

$$= 2$$

So $f'(x)$ is not always less than 1.

53. meters per second

55. items per dollar

57. $\lceil$ rotates to $\rfloor$ in 180°
$\neg$ rotates to $\rfloor$ in 90°

So it takes $\dfrac{180°}{\frac{11°}{.001 \text{ second}}} \approx .016$ seconds to

identify $\lceil$

And $\dfrac{90°}{\frac{11°}{.001 \text{ second}}} = .008$ seconds to identify $\neg$.

Section 2.3

5. $f(x) = x^3 - 2x + 1$

$$f'(x) = \frac{d}{dx}(x^3) - \frac{d}{dx}(2x) + \frac{d}{dx}(1)$$

$$= 3x^2 - 2\frac{d}{dx}(x) + 0$$

$$= 3x^2 - 2(1)$$

$$= 3x^2 - 2$$

7. $f(x) = 3x^2 - 4$

$$f'(x) = \frac{d}{dx}(3x^2) - \frac{d}{dx}(4)$$

$$= 3\frac{d}{dx}(x^2) - 0$$

$$= 3(2x)$$

$$= 6x$$

9. $g(x) = 4$

$$g'(x) = \frac{d}{dx}(4) = 0$$

11. $f(t) = 3t^3 - 2\sqrt{t}$

$$f'(t) = \frac{d}{dt}(3t^3) - \frac{d}{dt}(2\sqrt{t})$$

$$= 3\frac{d}{dt}(t^3) - 2\frac{d}{dt}(t^{1/2})$$

$$= 3(3t^2) - 2\left(\frac{1}{2}t^{-1/2}\right)$$

$$= 9t^2 - \frac{1}{\sqrt{t}}$$

13. $f(x) = \dfrac{3}{x} - 8x + 1$

$$f'(x) = \frac{d}{dx}\left(\frac{3}{x}\right) - \frac{d}{dx}(8x) + \frac{d}{dx}(1)$$

$$= 3\frac{d}{dx}(x^{-1}) - 8\frac{d}{dx}(x) + 0$$

$$= 3(-x^{-2}) - 8(1)$$

$$= -\frac{3}{x^2} - 8$$

15. $h(x) = \dfrac{10}{\sqrt{x}} - 2x$

$$h'(x) = \frac{d}{dx}\left(\frac{10}{\sqrt{x}}\right) - \frac{d}{dx}(2x)$$

$$= 10\frac{d}{dx}\left(x^{-1/2}\right) - 2\frac{d}{dx}(x)$$

$$= 10\left(-\frac{1}{2}x^{-3/2}\right) - 2(1)$$

$$= -5x^{-3/2} - 2$$

17. $f(s) = 2s^{3/2} - 3s^{-1/3}$

$$f'(s) = \frac{d}{ds}\left(2s^{3/2}\right) - \frac{d}{ds}\left(3s^{-1/3}\right)$$

$$= 2\frac{d}{ds}\left(s^{3/2}\right) - 3\frac{d}{ds}\left(s^{-1/3}\right)$$

$$= 2\left(\frac{3}{2}s^{1/2}\right) - 3\left(-\frac{1}{3}s^{-4/3}\right)$$

$$= 3\sqrt{s} + s^{-4/3}$$

19. $f(x) = 2\sqrt[3]{x} + 3$

$$f'(x) = \frac{d}{dx}\left(2\sqrt[3]{x}\right) + \frac{d}{dx}(3)$$

$$= 2\frac{d}{dx}\left(x^{1/3}\right) + 0$$

$$= 2\left(\frac{1}{3}x^{-2/3}\right)$$

$$= \frac{2}{3}x^{-2/3}$$

21. $f(x) = x(3x^2 - \sqrt{x}) = 3x^3 - x\sqrt{x} = 3x^3 - x^{3/2}$

$$f'(x) = \frac{d}{dx}(3x^3) - \frac{d}{dx}\left(x^{3/2}\right)$$

$$= 3\frac{d}{dx}(x^3) - \frac{d}{dx}(x^{3/2})$$

$$= 3(3x^2) - \left(\frac{3}{2}x^{1/2}\right)$$

$$= 9x^2 - \frac{3}{2}\sqrt{x}$$

23. $f(x) = \dfrac{3x^2 - 3x + 1}{2x}$

$$= \frac{3x^2}{2x} - \frac{3x}{2x} + \frac{1}{2x}$$

$$= \frac{3}{2}x - \frac{3}{2} + \frac{1}{2}x^{-1}$$

$$f'(x) = \frac{d}{dx}\left(\frac{3}{2}x\right) - \frac{d}{dx}\left(\frac{3}{2}\right) + \frac{d}{dx}\left(\frac{1}{2}x^{-1}\right)$$

$$= \frac{3}{2}\frac{d}{dx}(x) - 0 + \frac{1}{2}\frac{d}{dx}(x^{-1})$$

$$= \frac{3}{2}(1) + \frac{1}{2}(-1x^{-2})$$

$$= \frac{3}{2} - \frac{1}{2x^2}$$

25. $f(x) = x^4 + 3x^2 - 2$

$$f'(x) = \frac{d}{dx}(x^4 + 3x^2 - 2) = 4x^3 + 6x$$

$$f''(x) = \frac{d}{dx}(4x^3 + 6x) = 12x^2 + 6$$

27. $f(x) = x^6 - \sqrt{x} = x^6 - x^{1/2}$

$$\frac{df}{dx} = \frac{d}{dx}\left(x^6 - x^{1/2}\right)$$

$$= 6x^5 - \frac{1}{2}x^{-1/2}$$

$$\frac{d^2 f}{dx^2} = \frac{d}{dx}\left(6x^5 - \frac{1}{2}x^{-1/2}\right)$$

$$= 30x^4 - \frac{1}{2}\left(-\frac{1}{2}x^{-3/2}\right)$$

$$= 30x^4 + \frac{1}{4}x^{-3/2}$$

29. $f(t) = 4t^2 - 12 + \dfrac{4}{t^2} = 4t^2 - 12 + 4t^{-2}$

$$f'(t) = \frac{d}{dt}(4t^2 - 12 + 4t^{-2})$$

$$= 8t^2 - 0 + 4(-2t^{-3})$$

$$= 8t^2 - 8t^{-3}$$

$$f''(t) = \frac{d}{dt}(8t - 8t^{-3})$$

$$= 8 - 8(-3t^{-4})$$

$$= 8 + 24t^{-4}$$

$$f'''(t) = \frac{d}{dt}(8 + 24t^{-4})$$

$$= 0 + 24(-4t^{-5})$$

$$= -96t^{-5}$$

31. $f(x) = x^4 + 3x^2 - 2$

$f'(x) = \dfrac{d}{dx}(x^4 + 3x^2 - 2) = 4x^3 + 6x$

$f''(x) = \dfrac{d}{dx}(4x^3 + 6x) = 12x^2 + 6$

$f'''(x) = \dfrac{d}{dx}(12x^2 + 6) = 24x$

$f^{(4)}(x) = \dfrac{d}{dx}(24x) = 24$

33. $f(x) = \dfrac{x^2 - x + 1}{\sqrt{x}} = x^{3/2} - x^{1/2} + x^{-1/2}$

$f'(x) = \dfrac{d}{dx}\left(x^{3/2} - x^{1/2} + x^{-1/2}\right)$

$= \dfrac{3}{2}x^{1/2} - \dfrac{1}{2}x^{-1/2} - \dfrac{1}{2}x^{-3/2}$

$f''(x) = \dfrac{d}{dx}\left(\dfrac{3}{2}x^{1/2} - \dfrac{1}{2}x^{-1/2} - \dfrac{1}{2}x^{-3/2}\right)$

$= \dfrac{3}{4}x^{-1/2} + \dfrac{1}{4}x^{-3/2} + \dfrac{3}{4}x^{-5/2}$

$f'''(x) = \dfrac{d}{dx}\left(\dfrac{3}{4}x^{-1/2} + \dfrac{1}{4}x^{-3/2} + \dfrac{3}{4}x^{-5/2}\right)$

$= -\dfrac{3}{8}x^{-3/2} - \dfrac{3}{8}x^{-5/2} - \dfrac{15}{8}x^{-7/2}$

35. $s(t) = -16t^2 + 40t + 10$

$v(t) = s'(t) = -32t + 40$

$a(t) = v'(t) = -32$

37. $s(t) = \sqrt{t} + 2t = t^{1/2} + 2t$

$v(t) = s'(t) = \dfrac{1}{2}t^{-1/2} + 2$

$a(t) = v'(t) = -\dfrac{1}{4}t^{-3/2}$

39. $h(t) = -16t^2 + 40t + 5, \ a = 1$

$v(t) = h'(t) = -32t + 40$

$v(1) = -32(1) + 40 = 8$

The object is going up.

$a(t) = v'(t) = -32$

$a(1) = -32$

The object is slowing down.

41. $h(t) = 10t^2 - 24t, \ a = 2$

$v(t) = h'(t) = 20t - 24$

$v(2) = 20(2) - 24 = 16$

The object is going up.

$a(t) = v'(t) = 20$

$a(2) = 20$

The object is speeding up.

43. $f(x) = 4\sqrt{x} - 2x, \ a = 4$

$f'(x) = \dfrac{d}{dx}\left(4x^{1/2} - 2x\right) = 2x^{-1/2} - 2$

$f'(4) = 2(4)^{-1/2} - 2 = -1$

$f(4) = 4\sqrt{4} - 2(4) = 0$

The tangent line is $y = -1(x - 4) + 0$ or

$y = -x + 4.$

45. $f(x) = x^2 - 2, \ a = 2$

$f'(x) = 2x$

$f'(2) = 4$

$f(2) = 2^2 - 2 = 2$

So the tangent line is $y = 4(x - 2) + 2$ or

$y = 4x - 6.$

47. $f(x) = x^3 - 3x + 1$

$f'(x) = 3x^2 - 3$

The tangent line to $y = f(x)$ is horizontal when

$f'(x) = 0.$

$f'(x) = 3x^2 - 3 = 0$

$3(x^2 - 1) = 0$

$3(x + 1)(x - 1) = 0$

$x = -1$ or $x = 1$

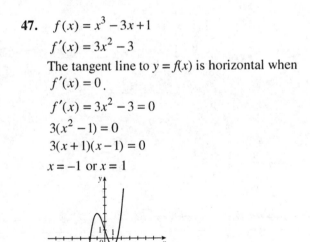

49. $f(x) = x^{2/3}$

$f'(x) = \frac{2}{3}x^{-1/3} = \frac{2}{3\sqrt[3]{x}}$

The slope of the tangent line to $y = f(x)$ does not exist when $f'(x)$ is not defined. That is,

when $x = 0$, $\frac{2}{3\sqrt[3]{0}}$ is not defined.

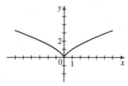

51. a. $f(x)$

   b. $f''(x)$

   c. $f'(x)$

53. $f(x) = \sqrt{x} = x^{1/2}$

$f'(x) = \frac{1}{2}x^{-1/2}$

$f''(x) = \frac{1}{2}\left(-\frac{1}{2}\right)x^{-3/2}$

$f'''(x) = \left(\frac{1}{2}\right)\left(\frac{-1}{2}\right)\left(\frac{-3}{2}\right)x^{-5/2}$

$f^{(n)}(x) = \left(\frac{1}{2}\right)\left(\frac{-1}{2}\right)\left(\frac{-3}{2}\right)\cdots\left(\frac{-(2n-3)}{2}\right)x^{-(2n-1)/2}$

$= (-1)^{n-1}\frac{1\cdot3\cdots\cdots(2n-3)}{2^n}x^{-(2n-1)/2}$

55. $f(x) = ax^2 + bx + c$
$f'(x) = 2ax + b$
$f''(x) = 2a$
$f''(0) = 2a$ and $f''(0) = 3$, so

$2a = 3$ thus $a = \frac{3}{2}$

$f'(x) = 2\left(\frac{3}{2}\right)x + b = 3x + b$
$f'(0) = 3(0) + b = b$

and $f'(0) = 2$ so $b = 2$.

$f(x) = ax^2 + bx + c = \frac{3}{2}x^2 + 2x + c$

$f(0) = \frac{3}{2}(0)^2 + 2(0) + c = c$

and $f(0) = -2$, so $c = -2$.

Thus $f(x) = \frac{3}{2}x^2 + 2x - 2$.

57. $\frac{d}{dx}(1) = \frac{d}{dx}(x^0) = 0x^{0-1} = 0$

From the graph, we see that the tangent line to $y = f(x) = c$ always has slope 0, so $\frac{d}{dx}(c) = 0$.

59. $d(t)$ represents the national debt. So $d'(t)$ represents the rate of change of the national debt, and $d''(t)$ is the rate at which the rate of change is changing. Although $d''(t)$ has been reduced, the national debt, $d(t)$ is still increasing.

61. $w(b) = cb^{3/2}$

$w'(b) = \frac{3c}{2}b^{1/2}$

$w'(b) \geq 1$ when $\frac{3c}{2}b^{1/2} > 1$

$b^{1/2} > \frac{2}{3c}$

$b > \frac{4}{9c^2}$

63. $f'(x) = 4x^3$
$f(x) = x^4$

65. $f'(x) = x^4$

$f(x) = \frac{x^5}{5}$

67. $f'(x) = 3x - 1$

$f(x) = \dfrac{3}{2}x^2 - x$

69. $f'(x) = \sqrt{x} = x^{1/2}$

$f(x) = \dfrac{2}{3}x^{3/2}$

Section 2.4

5. $f(x) = (x^2 + 3)(x^3 - 3x + 1)$

$f'(x) = \dfrac{d}{dx}(x^2 + 3) \cdot (x^3 - 3x + 1) + (x^2 + 3) \cdot \dfrac{d}{dx}(x^3 - 3x + 1)$

$\qquad = (2x)(x^3 - 3x + 1) + (x^2 + 3)(3x^2 - 3)$

7. $f(x) = (3x + 4)(x^3 - 2x^2 + x)$

$f'(x) = \dfrac{d}{dx}(3x + 4)(x^3 - 2x^2 + x) + (3x + 4)\dfrac{d}{dx}(x^3 - 2x^2 + x)$

$\qquad = 3(x^3 - 2x^2 + x) + (3x + 4)(3x^2 - 4x + 1)$

9. $f(x) = (\sqrt{x} + 3x)\left(5x^2 - \dfrac{3}{x}\right)$

$\qquad = (x^{1/2} + 3x)(5x^2 - 3x^{-1})$

$f'(x) = \dfrac{d}{dx}(x^{1/2} + 3x)(5x^2 - 3x^{-1}) + (x^{1/2} + 3x)\dfrac{d}{dx}(5x^2 - 3x^{-1})$

$\qquad = \left(\dfrac{1}{2}x^{-1/2} + 3\right)(5x^2 - 3x^{-1}) + (x^{1/2} + 3x)(10x + 3x^{-2})$

11. $f(x) = \dfrac{3x - 2}{5x + 1}$

$f'(x) = \dfrac{\frac{d}{dx}(3x - 2)(5x + 1) - (3x - 2)\frac{d}{dx}(5x + 1)}{(5x + 1)^2}$

$\qquad = \dfrac{3(5x + 1) - (3x - 2)5}{(5x + 1)^2}$

$\qquad = \dfrac{15x + 3 - 15x + 10}{(5x + 1)^2} = \dfrac{13}{(5x + 1)^2}$

13. $f(x) = \dfrac{x - 2}{x^2 + x + 1}$

$f'(x) = \dfrac{\frac{d}{dx}(x - 2)(x^2 + x + 1) - (x - 2)\frac{d}{dx}(x^2 + x + 1)}{(x^2 + x + 1)^2}$

$\qquad = \dfrac{1(x^2 + x + 1) - (x - 2)(2x + 1)}{(x^2 + x + 1)^2}$

$\qquad = \dfrac{x^2 + x + 1 - 2x^2 + 3x + 2}{(x^2 + x + 1)^2}$

$\qquad = \dfrac{-x^2 + 4x + 3}{(x^2 + x + 1)^2}$

15. $f(x) = \dfrac{3x - 6\sqrt{x}}{5x^2 - 2} = \dfrac{3x - 6x^{1/2}}{5x^2 - 2}$

$f'(x) = \dfrac{\frac{d}{dx}(3x - 6x^{1/2})(5x^2 - 2) - (3x - 6x^{1/2})\frac{d}{dx}(5x^2 - 2)}{(5x^2 - 2)^2}$

$= \dfrac{(3 - 3x^{-1/2})(5x^2 - 2) - (3x - 6x^{1/2})(10x)}{(5x^2 - 2)^2}$

$= \dfrac{15x^2 - 6 - 15x^{3/2} + 6x^{-1/2} - (30x^2 - 60x^{3/2})}{(5x^2 - 2)^2}$

$= \dfrac{-15x^2 + 45x^{3/2} + 6x^{-1/2} - 6}{(5x^2 - 2)^2}$

17. $f(x) = \dfrac{(x + 1)(x - 2)}{x^2 - 5x + 1} = \dfrac{x^2 - x - 2}{x^2 - 5x + 1}$

$f'(x) = \dfrac{\frac{d}{dx}(x^2 - x - 2)(x^2 - 5x + 1) - (x^2 - x - 2)\frac{d}{dx}(x^2 - 5x + 1)}{(x^2 - 5x + 1)^2}$

$= \dfrac{(2x - 1)(x^2 - 5x + 1) - (x^2 - x - 2)(2x - 5)}{(x^2 - 5x + 1)^2}$

$= \dfrac{2x^3 - 11x^2 + 7x - 1 - (2x^3 - 7x^2 + x + 10)}{(x^2 - 5x + 1)^2}$

$= \dfrac{-4x^2 + 6x - 11}{(x^2 - 5x + 1)^2}$

19. $f(x) = \dfrac{x^2 + 3x - 2}{\sqrt{x}} = \dfrac{x^2}{\sqrt{x}} + \dfrac{3x}{\sqrt{x}} - \dfrac{2}{\sqrt{x}} = x^{3/2} + 3x^{1/2} - 2x^{-1/2}$

$f'(x) = \dfrac{3}{2}x^{1/2} + \dfrac{3}{2}x^{-1/2} + x^{-3/2}$

21. $f(x) = x\left(\sqrt[3]{x} + 3\right) = x^{4/3} + 3x$

$f'(x) = \dfrac{4}{3}x^{1/3} + 3$

23. $f(x) = (x^2 - 1)\dfrac{x^3 + 3x^2}{x^2 + 2} = \dfrac{x^5 + 3x^4 - x^3 - 3x^2}{x^2 + 2}$

$f'(x) = \dfrac{\frac{d}{dx}(x^5 + 3x^4 - x^3 - 3x^2)(x^2 + 2) - (x^5 + 3x^4 - x^3 - 3x^2)\frac{d}{dx}(x^2 + 2)}{(x^2 + 2)^2}$

$= \dfrac{(5x^4 + 12x^3 - 3x^2 - 6x)(x^2 + 2) - (x^5 + 3x^4 - x^3 - 3x^2)(2x)}{(x^2 + 2)^2}$

$= \dfrac{5x^6 + 12x^5 + 7x^4 + 18x^3 - 6x^2 - 12x - 2x^6 - 6x^5 + 2x^4 + 6x^3}{(x^2 + 2)^2}$

$= \dfrac{3x^6 + 6x^5 + 9x^4 + 24x^3 - 6x^2 - 12x}{(x^2 + 2)^2}$

25. $\dfrac{d}{dx}((f(x)g(x)h(x))$

$= \dfrac{d}{dx}f(x)(g(x)h(x)) + f(x)\dfrac{d}{dx}(g(x)h(x))$

$= \dfrac{d}{dx}f(x)\cdot(g(x)h(x)) + f(x)\left[\dfrac{d}{dx}g(x)\cdot h(x) + g(x)\dfrac{d}{dx}h(x)\right]$

$= \dfrac{d}{dx}f(x)(g(x)h(x)) + f(x)\dfrac{d}{dx}g(x)\cdot h(x) + f(x)g(x)\dfrac{d}{dx}h(x)$

$\dfrac{d}{dx}(f_1(x)\cdot f_2(x)\cdot f_3(x)\ldots f_n(x)$

$= \dfrac{d}{dx}f_1(x)\cdot f_2(x)f_3(x)\ldots f_n(x) + f_1(x)\dfrac{d}{dx}f_2(x)\cdot f_3(x)\ldots f_n(x) + \ldots + f_1(x)f_2(x)\ldots f_{n-1}(x)\dfrac{d}{dx}f_n(x)$

There are n terms.

27. One CAS gives $f'(x) = \dfrac{2x}{x^2+1} - \dfrac{2x(x^2-2)}{\left(x^2+1\right)^2}$.

29. One CAS gives $f'(x) = \dfrac{-(4x^2-6x+11)}{\left(x^2-5x+1\right)^2}$.

31. One CAS gives $\dfrac{3x}{2\sqrt{x^2}(3x+1)}$. Factor $\sqrt{x^2}$ out of the denominator as $|x|$; $\dfrac{x}{|x|} = \begin{cases} 1 \text{ if } x > 0 \\ -1 \text{ if } x < 0 \end{cases}$.

33. $f(x) = x^{2/3}(x^2-2)(x^3-x+1)$

$f'(x) = \dfrac{d}{dx}(x^{2/3})(x^2-2)(x^3-x+1) + x^{2/3}\dfrac{d}{dx}(x^2-2)(x^3-x+1) + x^{2/3}(x^2-2)\dfrac{d}{dx}(x^3-x+1)$

$\qquad = \dfrac{2}{3}x^{-1/3}(x^2-2)(x^3-x+1) + x^{2/3}(2x)(x^3-x+1) + x^{2/3}(x^2-2)(3x^2-1)$

$\qquad = \dfrac{2}{3}x^{-1/3}(x^2-2)(x^3-x+1) + 2x^{5/3}(x^3-x+1) + x^{2/3}(x^2-2)(3x^2-1)$

35. $f(x) = (x+1)(x^3+4x)(x^5-3x^2+1)$

$f'(x)$

$= \dfrac{d}{dx}(x+1)(x^3+4x)(x^5-3x^2+1) + (x+1)\dfrac{d}{dx}(x^3+4x)(x^5-3x^2+1) + (x+1)(x^3+4x)\dfrac{d}{dx}(x^5-3x^2+1)$

$= (x^3+4x)(x^5-3x^2+1) + (x+1)(3x^2+4)(x^5-3x^2+1) + (x+1)(x^3+4x)(5x^4-6x)$

37. The rate at which the quantity sold, $Q(t)$, changes is $Q'(t)$. Since the amount sold is decreasing at a rate of 4%, we know the rate of change is –4% of $Q(t)$, or $Q'(t) = -.04Q(t)$.

If the price is increasing at a rate of 3%,
$P'(t) = .03P(t)$.
$R(t) = Q(t)P(t)$ so
$R'(t) = Q'(t)P(t) + Q(t)P'(t)$
$\qquad = -.04Q(t)P(t) + Q(t)(.03)P(t)$
$\qquad = (-.04+.03)(Q(t)P(t))$
$\qquad = -.01(Q(t)P(t))$

So the revenue is decreasing at a rate of 1%.

39. $R'(t) = Q'(t)P(t) + Q(t)P'(t)$

$P(0) = 20$

$Q(0) = 20,000$

$P'(t) = 1.25$

$Q'(t) = 2,000$

$R'(0) = 2,000(20) + (20,000)1.25$

$\qquad = 65,000$

So revenue is increasing by \$65,000/year at present.

41. $u'(m) = \dfrac{7650}{(20m+3)^2}$

43. $u'(m) = \dfrac{-5644}{(20m+1)^2}$; the best club is a very light one that you can swing with a very high velocity.

45. $F(x) = f(x)g(x)$

$F'(x) = f'(x)g(x) + f(x)g'(x)$

$F''(x) = f''(x)g(x) + f'(x)g'(x) + f'(x)g'(x) + f(x)g''(x)$

$\qquad = f''(x)g(x) + 2f'(x)g'(x) + f(x)g''(x)$

$F'''(x) = f'''(x)g(x) + f''(x)g'(x) + 2f''(x)g'(x) + 2f'(x)g''(x) + f'(x)g''(x) + f(x)g'''(x)$

$\qquad = f'''(x)g(x) + 3f''(x)g'(x) + 3f'(x)g''(x) + f(x)g'''(x)$

47. $g(x) = [f(x)]^2 = f(x)f(x)$

$g'(x) = f'(x)f(x) + f(x)f'(x) = 2f(x)f'(x)$

49. $\left(P + \dfrac{n^2 a}{V^2}\right)(V - nb) = nRT$

$$P + \frac{n^2 a}{V^2} = \frac{nRT}{V - nb}$$

$$P = \frac{nRT}{V - nb} - \frac{n^2 a}{V^2}$$

Using a computer algebra system, we find that

$P'(V) = \dfrac{2an^2}{V^3} - \dfrac{nRT}{(V - nb)^2}$ and $P''(V) = -\dfrac{6an^2}{V^4} + \dfrac{2nRT}{(V - nb)^2}$ and the computer is also used to solve the system

of equations $\{P'(V) = 0,\ P''(V) = 0\}$ for T and V.

$T_c = \dfrac{8a}{27Rb}$ and $V_c = 3nb$. Thus, by substitution, $P_c = \dfrac{a}{27b^2}$.

When $R = .08206$, $a = 5.464$ and $b = .03049$, then $T \approx 647^\circ K$.

Section 2.5

3. In Figure 2.26, the graph of $y = \cos x$ has horizontal tangents at $x = -\pi, 0, \pi, 2\pi$, so the derivative must be zero at these x values. The slope is positive for $-\pi < x < 0$, negative for $0 < x < \pi$, and so on.

5. $f(x) = 4\sin x - x$
$f'(x) = 4\cos x - 1$

7. $f(x) = \tan x - \csc x$
$f'(x) = \sec^2 x + \csc x \cot x$

9. $f(x) = x\cos x$
$f'(x) = (1)\cos x + x(-\sin x)$
$\quad\quad = \cos x - x\sin x$

11. $f(x) = 4\sqrt{x} - 2\sin x = 4x^{1/2} - 2\sin x$
$f'(x) = 2x^{-1/2} - 2\cos x$

13. $f(x) = \dfrac{\sin x}{x}$
$f'(x) = \dfrac{\cos x \cdot x - \sin x \cdot 1}{x^2}$
$\quad\quad = \dfrac{x\cos x - \sin x}{x^2}$

15. $f(t) = \sin t \sec t = \tan t$
$f'(t) = \sec^2 t$

17. $f(x) = \dfrac{\cos x - 1}{x^2}$
$f'(x) = \dfrac{-\sin x \cdot x^2 - (\cos x - 1)2x}{x^4}$
$\quad\quad = \dfrac{-x(x\sin x + 2\cos x - 2)}{x^4}$
$\quad\quad = \dfrac{-(x\sin x + 2\cos x - 2)}{x^3}$

19. $f(x) = 2\sin x \cos x$
$f'(x) = 2\cos x \cdot \cos x + 2\sin x(-\sin x)$
$\quad\quad = 2\cos^2 x - 2\sin^2 x$

21. $f(x) = 4x^2 \tan x$
$f'(x) = 8x\tan x + 4x^2 \sec^2 x$

23. $f(x) = 4\sin^2 x + 4\cos^2 x = 4(\cos^2 x + \sin^2 x) = 4$
$f'(x) = 0$

25. $f'(x) = 2\cos^2 x - 2\sin^2 x$

27. $f'(x) = 4\sin x \cos x - 2\sin 2x$; to get the simplified answer 0, expand $\sin 2x$ as $2\sin x \cos x$.

29. $f(x) = \sin x, \; a = \dfrac{\pi}{2}$
$f'(x) = \cos x$
$f'\left(\dfrac{\pi}{2}\right) = \cos\dfrac{\pi}{2} = 0$
$f\left(\dfrac{\pi}{2}\right) = \sin\dfrac{\pi}{2} = 1$

So the tangent line is $y = 0\left(x - \dfrac{\pi}{2}\right) + 1$ or $y = 1$.

31. $f(x) = \cos x, \; a = \dfrac{\pi}{2}$
$f'(x) = -\sin x$
$f'\left(\dfrac{\pi}{2}\right) = -\sin\dfrac{\pi}{2} = -1$
$f\left(\dfrac{\pi}{2}\right) = \cos\dfrac{\pi}{2} = 0$

So the tangent line is $y = -1\left(x - \dfrac{\pi}{2}\right) + 0$ or
$y = -x + \dfrac{\pi}{2}$.

33. $s(t) = t^2 - \sin t, \; a = 0$
$v(t) = s'(t) = 2t - \cos t$
$v(0) = 2\cdot 0 - \cos 0 = 0 - 1 = -1$

35. $s(t) = \dfrac{\cos t}{t} = t^{-1}\cos t, \; a = \pi$
$v(t) = s'(t)$
$\quad\quad = -t^{-2}\cos t + t^{-1}(-\sin t)$
$\quad\quad = -t^{-2}\cos t - t^{-1}\sin t$
$v(\pi) = -\pi^{-2}\cos\pi - \pi^{-1}\sin\pi$
$\quad\quad = -\dfrac{1}{\pi^2}(-1) - \dfrac{1}{\pi}(0)$
$\quad\quad = \dfrac{1}{\pi^2}$

37. Since one revolution takes 2π seconds, t is the angle between the line connecting the radius and the position of the object on the circle and the line from the radius to the object's starting point. So

$$\cos t = \frac{x}{r} = \frac{x}{1}.$$

$y(t) = \sin t$

If the radius is 3, $x(t) = 3 \cos t$, $y(t) = 3 \sin t$.

If a revolution takes π seconds,

$x(t) = \cos 2t$, $y(t) = \sin 2t$.

If a revolution takes 2 seconds, then

$x(t) = \cos(\pi t)$, $y(t) = \sin(\pi t)$.

If the object travels clockwise,

$x(t) = \cos t$, $y(t) = -\sin t$.

If the object is at the top of the circle at $t = 0$,

$$x(t) = \cos\left(t + \frac{\pi}{2}\right), \ y(t) = \sin\left(t + \frac{\pi}{2}\right)$$

39. $f(t) = 4\sin t$

$v(t) = f'(t) = 4\cos t$

41. $Q(t) = 3\sin t + t + 4$

$I(t) = Q'(t) = 3\cos t + 1$

at $t = 0$, $I(0) = 3\cos 0 + 1 = 4$

at $t = 1$, $I(0) = 3\cos 1 + 1 \approx 2.62$

43. $f(x) = \sin 2x = 2\sin x \cos x$

$f'(x) = 2\cos x \cdot \cos x + 2\sin x(-\sin x)$

$\qquad = 2(\cos^2 x - \sin^2 x)$

$\qquad = 2\cos 2x$

45. $f(x) = \sin x$

$f'(x) = \cos x$

$f''(x) = -\sin x$

$f'''(x) = -\cos x$

$f^{(4)}(x) = \sin x$

$f^{(75)}(x) = -\cos x$

$f^{(150)}(x) = -\sin x$

47. For $-\frac{\pi}{2} < \theta < 0$, flipping Figure 2.27 into the fourth quadrant and using opposites to get positive lengths shows that $0 \le -\sin\theta \le -\theta$ and thus $\theta \le \sin\theta \le 0$. The Squeeze Theorem then shows as in Lemma 5.1 that $\lim\limits_{\theta \to 0^-} \sin\theta = 0$.

49. $f'(x)$

$= \lim\limits_{h \to 0} \dfrac{\cos(x+h) - \cos(x)}{h}$

$= \lim\limits_{h \to 0} \dfrac{\cos x \cos h - \sin x \sin h - \cos x}{h}$

$= \lim\limits_{h \to 0}(-\sin x)\left(\dfrac{\sin h}{h}\right) + \lim\limits_{h \to 0}(\cos x)(\cos h - 1)$

$= (-\sin x) \cdot 1 + (\cos x) \cdot 0$

$= -\sin x$

51. a. $\lim\limits_{x \to 0} \dfrac{\sin 3x}{x} = \lim\limits_{x \to 0} 3\dfrac{\sin 3x}{3x}$

$\qquad = 3\lim\limits_{x \to 0}\dfrac{\sin 3x}{3x}$

$\qquad = 3 \cdot 1$

$\qquad = 3$

b. $\lim\limits_{x \to 0}\dfrac{\sin t}{4t} = \dfrac{1}{4}\lim\limits_{t \to 0}\dfrac{\sin t}{t}$

$\qquad = \dfrac{1}{4} \cdot 1$

$\qquad = \dfrac{1}{4}$

c. $\lim\limits_{x \to 0}\dfrac{\cos x - 1}{5x} = \dfrac{1}{5}\lim\limits_{x \to 0}\dfrac{\cos x - 1}{x} = \dfrac{1}{5} \cdot 0 = 0$

d. Let $u = x^2$; then as $x \to 0$, $u \to 0$.

$\lim\limits_{x \to 0}\dfrac{\sin x^2}{x^2} = \lim\limits_{u \to 0}\dfrac{\sin u}{u}$

$\qquad\qquad = 1$

Section 2.6

5. $f(x) = 4e^x - x$

$f'(x) = 4e^x - 1$

7. $f(x) = xe^x$

$f'(x) = 1 \cdot e^x + xe^x = e^x + xe^x$

9. $f(x) = x + 2^x$

$f'(x) = 1 + 2^x \ln 2$

11. $f(x) = 2e^{x+1} = 2e \cdot e^x$

$f'(x) = 2e \cdot e^x = 2e^{x+1}$

13. $f(x) = \left(\dfrac{1}{3}\right)^x$

$f'(x) = \left(\dfrac{1}{3}\right)^x \ln\dfrac{1}{3}$

15. $f(x) = 4^{-x+1} = 4 \cdot 4^{-x} = 4\left(\dfrac{1}{4}\right)^x$

$f'(x) = 4 \cdot \left(\dfrac{1}{4}\right)^x \ln\dfrac{1}{4}$

$\quad\quad = 4^{-x+1} \ln\dfrac{1}{4}$

17. $f(x) = \dfrac{e^x}{x} = x^{-1}e^x$

$f'(x) = -x^{-2}e^x + x^{-1}e^x$

19. $f(x) = \ln 2x = \ln 2 + \ln x$

$f'(x) = \dfrac{1}{x}$

21. $f(x) = \ln x^3 = 3\ln x$

$f'(x) = \dfrac{3}{x}$

23. $f(x) = e^{2x} = e^x e^x$

$f'(x) = e^x e^x + e^x e^x = 2e^x e^x = 2e^{2x}$

25. $f(x) = x^2 e^{-x}$

$f'(x) = 2xe^{-x} + x^2(-e^{-x})$

$\quad\quad = 2xe^{-x} - x^2 e^{-x}$

27. $f(x) = \dfrac{\ln x}{x} = x^{-1}\ln x$

$f'(x) = -x^{-2}\ln x + x^{-1}\cdot\dfrac{1}{x} = x^{-2}(1-\ln x)$

29. $f(x) = 3e^x$

$f'(x) = 3e^x$

$f'(1) = 3e^1 = 3e$

$f(1) = 3e^1 = 3e$

So the tangent line is $y = 3e(x-1) + 3e$
or $y = 3ex$.

31. $f(x) = 3^x$

$f'(x) = 3^x \ln 3$

$f'(1) = 3^1 \ln 3 = 3\ln 3$

$f(1) = 3^1 = 3$

So the tangent line is $y = 3\ln 3(x-1) + 3$
or $y = 3\ln 3 \supseteq x - 3(\ln 3 - 1)$.

33. $f(x) = xe^x$

$f'(x) = 1e^x + xe^x = e^x + xe^x$

$f'(1) = e^1 + 1e^1 = 2e$

$f(1) = 1e^1 = e$

So the tangent line is $y = 2e(x-1) + e$
or $y = 2ex - e$.

35. $f(x) = x^2 \ln x$

$f'(x) = 2x\ln x + x^2 \cdot \dfrac{1}{x} = 2x\ln x + x$

$f'(1) = 2\cdot 1\ln 1 + 1 = 2\cdot 0 + 1 = 1$

$f(1) = 1^2 \ln 1 = 1\cdot 0 = 0$

So the tangent line is $y = 1(x-1) + 0$
or $y = x - 1$.

37. $v(t) = 100 \cdot 3^t$

$v'(t) = 100 \cdot 3^t \ln 3$

$\dfrac{v'(t)}{v(t)} = \dfrac{100 \cdot 3^t \ln 3}{100 \cdot 3^t} = \ln 3 \approx 1.10$

So the percentage change is about 110%.

39. $v(t) = 100e^t$

$v'(t) = 100e^t$

$\dfrac{v'(t)}{v(t)} = \dfrac{100e^t}{100e^t} = 1$

So the percentage change is 100%.

41. $p(t) = 200 \cdot 3^t$

$\ln 3 \approx 1.099$, so the percentage rate of
change of population is about 110%.

43. $f(t) = Ae^{rt}$

a. $APY = \dfrac{f(1)}{A} = \dfrac{Ae^{.05 \cdot 1}}{A} = e^{.05} \approx 1.051$

    APY is 5.1%

b. $APY = \dfrac{f(1)}{A} = \dfrac{Ae^{.1 \cdot 1}}{A} = e^{.1} \approx 1.105$

    APY is 10.5%

c. $APY = \dfrac{f(1)}{A} = \dfrac{Ae^{.2 \cdot 1}}{A} = e^{.2} \approx 1.221$

    APY is 22.1%

d. $APY = \dfrac{f(1)}{A} = \dfrac{Ae^{\ln 2 \cdot 1}}{A} = 2$

    APY is 100%

e. $APY = \dfrac{f(1)}{A} = \dfrac{Ae^{1 \cdot 1}}{A} = e \approx 2.718$

    APY is 171.8%

45. $f(t) = e^{-t} \cos t$

$v(t) = f'(t) = -e^{-t} \cos t - e^{-t} \sin t$

$-e^{-t}(\cos t + \sin t) = 0$

$\cos t + \sin t = 0$

$\cos t = -\sin t$

$t = \dfrac{3\pi}{4}, \dfrac{7\pi}{4}, \ldots$

Position when velocity is zero:

$f\left(\dfrac{3\pi}{4}\right) = e^{-\frac{3\pi}{4}} \cos \dfrac{3\pi}{4} = e^{-\frac{3\pi}{4}}\left(-\dfrac{1}{\sqrt{2}}\right) \approx -.067020$

$f\left(\dfrac{7\pi}{4}\right) = e^{-\frac{7\pi}{4}} \cos \dfrac{7\pi}{4} = e^{-\frac{7\pi}{4}}\left(\dfrac{1}{\sqrt{2}}\right) \approx .002896$

The velocity is zero when the position reaches a minimum or maximum.

47. The maximum velocities occur when the spring passes through the neutral position.

49. $f'(x) = 2x$

51. $f'(x) = \dfrac{3}{2}$

53. $\dfrac{3^{0.001} - 1}{0.001} \approx 1.09922; \quad \ln 3 \approx 1.09861$

55. $\dfrac{\ln(4.001) - \ln 4}{0.001} \approx 0.24997; \quad \dfrac{1}{4} = 0.25$

57. Let $g(x) = e^x$

$g'(x) = e^x$

$g''(x) = e^x$

$g(0) = g'(0) = g''(0) = e^0 = 1$.

$f(x) = \dfrac{a + bx}{1 + cx}$

$f'(x) = \dfrac{b(1 + cx) - (a + bx)(c)}{(1 + cx)^2}$

$= \dfrac{b - ac}{(1 + cx)^2}$

$= \dfrac{b - ac}{1 + 2cx + c^2 x^2}$

$f''(x) = \dfrac{0(1 + 2cx + c^2 x^2) - (b - ac)(2c + 2c^2 x)}{(1 + 2cx + c^2 x^2)^2}$

$= \dfrac{-(b - ac)(2c)(1 + cx)}{(1 + cx)^4}$

$= \dfrac{-2c(b - ac)}{(1 + cx)^3}$

$f(0) = \dfrac{a + b \cdot 0}{1 + c \cdot 0} = a = 1$

$f'(0) = \dfrac{b - ac}{(1 + c \cdot 0)^2} = b - ac = b - c = 1$

$b = 1 + c$

$f''(0) = \dfrac{-2c(b - ac)}{(1 + c \cdot 0)^3} = -2c \cdot 1 = 1$

$-2c = 1, \quad c = -\dfrac{1}{2}$

Since $b = 1 + c, \quad b = 1 - \dfrac{1}{2} = \dfrac{1}{2}$.

So $a = 1, \quad b = \dfrac{1}{2}, \quad c = -\dfrac{1}{2}$.

Section 2.7

5. $f(x) = (x^3 - 1)^2$

Chain rule:

$f'(x) = 2(x^3 - 1)(3x^2) = 6x^2(x^3 - 1)$

Without chain rule:

$f(x) = (x^3 - 1)(x^3 - 1)$

$f'(x) = (3x^2)(x^3 - 1) + (x^3 - 1)(3x^2)$

$= 2(3x^2)(x^3 - 1)$

$= 6x^2(x^3 - 1)$

7. $f(x) = e^{-2x}$

Chain rule:

$f'(x) = -2e^{-2x}$

Without chain rule:

$f(x) = e^{-x}e^{-x}$

$\begin{aligned} f'(x) &= -e^{-x}e^{-x} + e^{-x}(-e^{-x}) \\ &= -2e^{-x}e^{-x} \\ &= -2e^{-2x} \end{aligned}$

9. $f(x) = \sqrt{x^2 + 4} = (x^2 + 4)^{1/2}$

$f'(x) = \dfrac{1}{2}(x^2 + 4)^{-1/2} \cdot 2x = x(x^2 + 4)^{-1/2}$

11. $f(x) = (x^3 + x - 1)^3$

$f'(x) = 3(x^3 + x - 1)^2(3x^2 + 1)$

13. $f(x) = \sin(2x^2 + 3)$

$\begin{aligned} f'(x) &= \cos(2x^2 + 3)(4x) \\ &= 4x\cos(2x^2 + 3) \end{aligned}$

15. $f(x) = \sin^4 x$

$f'(x) = 4\sin^3 x \cos x$

17. $f(x) = \tan^2 x$

$f'(x) = 2\tan x \sec^2 x$

19. $f(x) = e^{x^2}$

$f'(x) = 2xe^{x^2}$

21. $f(x) = e^{1/(3x)} = e^{(3x)^{-1}}$

$f'(x) = e^{(3x)^{-1}} \cdot -3(3x)^{-2} = -\dfrac{1}{3x^2}e^{(1/3x)}$

23. $f(x) = (\ln(x^2 + 1))^8$

$\begin{aligned} f'(x) &= 8(\ln(x^2 + 1))^7 \cdot \dfrac{1}{x^2 + 1} \cdot 2x \\ &= \dfrac{16x}{x^2 + 1}(\ln(x^2 + 1))^7 \end{aligned}$

25. $f(x) = x^2 \sin 4x$

$\begin{aligned} f'(x) &= 2x\sin 4x + x^2(\cos 4x \cdot 4) \\ &= 2x\sin 4x + 4x^2 \cos 4x \end{aligned}$

27. $f(x) = \sec^3 4x$

$\begin{aligned} f'(x) &= 3\sec^2 4x \cdot \dfrac{d}{dx}(\sec 4x) \\ &= 3\sec^2 4x \cdot \sec 4x \tan 4x \cdot 4 \\ &= 12\sec^3 4x \cdot \tan 4x \end{aligned}$

29. $f(x) = \dfrac{\sin x^2}{x^2} = x^{-2}\sin x^2$

$\begin{aligned} f'(x) &= -2x^{-3}\sin x^2 + x^{-2}\cos x^2 \cdot 2x \\ &= -2x^{-3}\sin x^2 + 2x^{-1}\cos x^2 \end{aligned}$

31. $f(x) = \sin(\ln(\cos x^3))$

$\begin{aligned} f'(x) &= \cos(\ln(\cos x^3))\left(\dfrac{1}{\cos x^3}\right)(-\sin x^3)(3x^2) \\ &= -3x^2 \cos(\ln(\cos x^3))\tan x^3 \end{aligned}$

33. $f(x) = \sqrt{\sin x^2} = (\sin x^2)^{1/2}$

$\begin{aligned} f'(x) &= \dfrac{1}{2}(\sin x^2)^{-1/2}\dfrac{d}{dx}(\sin x^2) \\ &= \dfrac{1}{2}(\sin x^2)^{-1/2} \cdot \cos x^2 \cdot 2x \\ &= x\cos x^2(\sin x^2)^{-1/2} \end{aligned}$

35. $f(x) = (\ln x)^8$

$f'(x) = 8(\ln x)^7 \cdot \dfrac{1}{x} = \dfrac{8}{x}(\ln x)^7$

37. $f(x) = \cos\left(\dfrac{4x}{x^2 + 1}\right)$

$\begin{aligned} f'(x) &= -\sin\left(\dfrac{4x}{x^2 + 1}\right)\left(\dfrac{(x^2 + 1)(4) - (4x)(2x)}{(x^2 + 1)^2}\right) \\ &= \dfrac{4x^2 - 4}{(x^2 + 1)^2}\sin\left(\dfrac{4x}{x^2 + 1}\right) \end{aligned}$

39. $f(x) = \ln(\sec x + \tan x)$

$\begin{aligned} f'(x) &= \dfrac{1}{\sec x + \tan x}\left(\sec x \tan x + \sec^2 x\right) \\ &= \dfrac{(\sec x)(\tan x + \sec x)}{\sec x + \tan x} = \sec x \end{aligned}$

41. $f(x) = \sqrt{x^2 + 16} = (x^2 + 16)^{1/2}$, $a = 3$

$f'(x) = \frac{1}{2}(x^2 + 16)^{-1/2}(2x) = \frac{x}{\sqrt{x^2 + 16}}$

$f'(3) = \frac{3}{\sqrt{3^2 + 16}} = \frac{3}{5}$

$f(3) = \sqrt{3^2 + 16} = 5$

So the tangent line is $y = \frac{3}{5}(x - 3) + 5$ or

$y = \frac{3}{5}x + \frac{16}{5}$.

43. $s(t) = e^{-4t}\sin 3t$

$v(t) = s'(t) = e^{-4t}(-4)\cdot\sin 3t + e^{-4t}\cdot\cos 3t \cdot 3$

$\qquad = -4e^{-4t}\sin 3t + 3e^{-4t}\cos 3t$

$v(2) = -4e^{-8}\sin 6 + 3e^{-8}\cos 6 \approx .001341$

45. $s(t) = e^{-2t} - e^{-4t}$

$v(t) = -2e^{-2t} + 4e^{-4t}$

$v(2) = -2e^{-4} + 4e^{-8} \approx -0.0353$

47. $f(x) = \sqrt{2x+1} = (2x+1)^{1/2}$

$f'(x) = \frac{1}{2}(2x+1)^{-1/2}\cdot 2 = (2x+1)^{-1/2}$

$f''(x) = -\frac{1}{2}(2x+1)^{-3/2}(2) = -(2x+1)^{-3/2}$

$f'''(x) = -\left(-\frac{3}{2}\right)(2x+1)^{-5/2}\cdot 2 = 3(2x+1)^{-5/2}$

$f^{(4)}(x) = 3\left(-\frac{5}{2}\right)(2x+1)^{-7/2}\cdot 2$

$\qquad = -15(2x+1)^{-7/2}$

$f^{(n)}(x)$

$= (-1)^{n+1}(1)(3)(5)\ldots(2n-3)(2x+1)^{-(2n-1)/2}$

49. $f(x) = e^{2x}$

$f'(x) = e^{2x}\cdot 2 = 2e^{2x}$

$f''(x) = 2e^{2x}\cdot 2 = 4e^{2x}$

$f'''(x) = 4e^{2x}\cdot 2 = 8e^{2x}$

$f^{(4)}(x) = 8e^{2x}\cdot 2 = 16e^{2x}$

$f^{(n)}(x) = 2^n e^{2x}$

51. $f(x) = (x^2 + 3)^2 \cdot 2x$

$g(x) = \frac{(x^2 + 3)^3}{3}$

53. $f(x) = 3\sin 2x$

$g(x) = -\frac{3}{2}\cos 2x$

55. $D = \sqrt{y^2 + (L-x)^2} + \sqrt{y^2 + (L+x)^2}$

$\quad = \sqrt{(r^2 - x^2) + (L - x^2)} + \sqrt{(r^2 - x^2) + (L + x)^2}$

$\quad = \sqrt{r^2 + L^2 - 2Lx} + \sqrt{r^2 + L^2 + 2Lx}$

Fix $r = 4$. We must have $0 \leq L \leq r$, so draw five graphs of $D(x)$ for $r = 4$ with $L = 0, 1, 2, 3, 4$, on the x-interval $[-4, 4]$. The x-value that maximizes D is $x = 0$.

$D'(x) = \frac{1}{2}(r^2 + L^2 - 2Lx)^{-1/2}(-2L) + \frac{1}{2}(r^2 + L^2 + 2Lx)^{-1/2}(2L)$

$D'(0) = -L(r^2 + L^2)^{-1/2} + L(r^2 + L^2)^{-1/2} = 0$

There are no other points at which the derivative is 0.

57. $a^x = e^{x\ln a}$

$\frac{d}{dx}e^{x\ln a} = e^{x\ln a}\cdot \ln a$

$\qquad = a^x \ln a$

59. $f(x) = e^{-x^2/2}$

$f'(x) = -xe^{-x^2/2}$

$f''(x) = x^2 e^{-x^2/2} - e^{-x^2/2} = (x^2 - 1)e^{-x^2/2}$

The second derivative is 0 at $x = -1$ and at $x = 1$.

61. $f(x) = e^{-(x-m)^2/(2c^2)}$

$f'(x) = -\dfrac{x-m}{c^2} e^{-(x-m)^2/(2c^2)}$

$f''(x)$

$= \dfrac{(x-m)^2}{c^4} e^{-(x-m)^2/(2c^2)} - \dfrac{1}{c^2} e^{-(x-m)^2/(2c^2)}$

$= \dfrac{1}{c^4}((x-m^2-c^2)e^{-(x-m)^2/(2c^2)}$

$= \dfrac{(x-m-c)(x-m+c)}{c^4} e^{-(x-m)^2/(2c^2)}$

The second derivative is 0 at
$x = m+c$ and at $x = m-c$.

Section 2.8

5. $x^2 + 4y^2 = 8$ at (2, 1)

Explicitly:

$4y^2 = 8 - x^2$

$y^2 = \dfrac{8-x^2}{4}$

$y = \pm \dfrac{\sqrt{8-x^2}}{2}$

For $y = \dfrac{\sqrt{8-x^2}}{2} = \dfrac{(8-x)^{1/2}}{2}$

$y'(x) = \dfrac{\frac{1}{2}(8-x^2)^{-1/2}}{2} \cdot (-2x) = \dfrac{-x}{2\sqrt{8-x^2}}$

$y'(2) = \dfrac{-2}{2\sqrt{8-4}} = -\dfrac{1}{2}$

Implicitly: $\dfrac{d}{dx}(x^2+4y^2) = \dfrac{d}{dx}(8)$

$2x + 8y \cdot y'(x) = 0$

$y'(x) = \dfrac{-2x}{8y} = \dfrac{-x}{4y}$

$y'(2) = \dfrac{-2}{4\cdot 1} = -\dfrac{1}{2}$

7. $y - 3x^2 y = \cos x$ at (0, 1)

Explicitly:

$y(1 - 3x^2) = \cos x$

$y = \dfrac{\cos x}{1 - 3x^2}$

$y' = \dfrac{(-\sin x)(1-3x^2) - \cos x(-6x)}{(1-3x^2)^2}$

$= \dfrac{-\sin x + 3x^2 \sin x + 6x\cos x}{(1-3x^2)^2}$

$y'(0) = \dfrac{-\sin(0) + 3(0)^2 \sin 0 + 6\cdot 0\cos 0}{(1 - 3(0)^2)^2}$

$= 0$

Implicitly: $\dfrac{d}{dx}(y - 3x^2 y) = \dfrac{d}{dx}(\cos x)$

$y' - 6xy - 3x^2 y' = (-\sin x)$

$y'(1 - 3x^2) = 6xy - \sin x$

$y' = \dfrac{6xy - \sin x}{1 - 3x^2}$

$y'(0) = \dfrac{6(0)(1) - \sin(0)}{1 - 3(0)^2} = 0$

9. $x^2 y^2 + 3y = 4x$

$\dfrac{d}{dx}(x^2 y^2 + 3y) = \dfrac{d}{dx}(4x)$

$2xy^2 + x^2 2y \cdot y' + 3y' = 4$

$y'(2x^2 y + 3) = 4 - 2xy^2$

$y' = \dfrac{4 - 2xy^2}{2x^2 y + 3}$

11. $\sqrt{xy} - 4y^2 = 12$

$x^{1/2} y^{1/2} - 4y^2 = 12$

$\dfrac{d}{dx}(x^{1/2} y^{1/2} - 4y^2) = \dfrac{d}{dx}(12)$

$\dfrac{1}{2} x^{-1/2} y^{1/2} + x^{1/2} \cdot \dfrac{1}{2} y^{-1/2} y' - 8y \cdot y' = 0$

$y'\left(\dfrac{x^{1/2} y^{-1/2}}{2} - 8y\right) = -\dfrac{1}{2} x^{-1/2} y^{1/2}$

$y'\left(\dfrac{x^{1/2} y^{-1/2} - 16y}{2}\right) = -\dfrac{1}{2} x^{-1/2} y^{1/2}$

$y' = \dfrac{-x^{-1/2} y^{1/2}}{x^{1/2} y^{-1/2} - 16y} = \dfrac{y}{16y\sqrt{xy} - x}$

13. $\dfrac{x+3}{y} = 4x + y^2$

$$\dfrac{d}{dx}\left(\dfrac{x+3}{y}\right) = \dfrac{d}{dx}(4x+y^2)$$

$$\dfrac{1 \cdot y - (x+3)y'}{y^2} = 4 + 2yy'$$

$$y^{-1} - y^{-2}(x+3)y' = 4 + 2yy'$$

$$y'(-2y - y^{-2}(x+3)) = 4 - y^{-1}$$

$$y' = -\dfrac{4 - y^{-1}}{2y + y^{-2}(x+3)}$$

$$= -\dfrac{4y^2 - y}{2y^3 + x + 3}$$

15. $e^{x^2 y} - e^y = x$

$$\dfrac{d}{dx}(e^{x^2 y} - e^y) = \dfrac{d}{dx}(x)$$

$$e^{x^2 y}\dfrac{d}{dx}(x^2 y) - e^y y' = 1$$

$$e^{x^2 y}(2xy + x^2 y') - e^y y' = 1$$

$$y'(x^2 e^{x^2 y} - e^y) = 1 - 2xye^{x^2 y}$$

$$y' = \dfrac{1 - 2xye^{x^2 y}}{x^2 e^{x^2 y} - e^y}$$

17. $\sqrt{x+y} - 4x^2 = y$

$$(x+y)^{1/2} - 4x^2 = y$$

$$\dfrac{d}{dx}((x+y)^{1/2} - 4x^2) = \dfrac{d}{dx}y$$

$$\dfrac{1}{2}(x+y)^{-1/2}(1+y') - 8x = y'$$

$$y'\left(\dfrac{1}{2\sqrt{x+y}} - 1\right) = \dfrac{-1}{2\sqrt{x+y}} + 8x$$

$$y'\left(\dfrac{1 - 2\sqrt{x+y}}{2\sqrt{x+y}}\right) = \dfrac{16x\sqrt{x+y} - 1}{2\sqrt{x+y}}$$

$$y' = \dfrac{16x\sqrt{x+y} - 1}{1 - 2\sqrt{x+y}}$$

19. $e^{4y} - \ln y = 2x$

$$\dfrac{d}{dx}(e^{4y} - \ln y) = \dfrac{d}{dx}(2x)$$

$$e^{4y} \cdot 4y' - \dfrac{1}{y} \cdot y' = 2$$

$$y'\left(4e^{4y} - \dfrac{1}{y}\right) = 2$$

$$y'\left(\dfrac{4ye^{4y} - 1}{y}\right) = 2$$

$$y' = \dfrac{2y}{4ye^{4y} - 1}$$

21. $x^2 - 4y^2 = 0$ at $(2, 1)$

$$\dfrac{d}{dx}(x^2 - 4y^2) = \dfrac{d}{dx}(0)$$

$$2x - 8y \cdot y' = 0$$

$$y' = \dfrac{2x}{8y} = \dfrac{x}{4y}$$

$$y'(2) = \dfrac{2}{4 \cdot 1} = \dfrac{1}{2}$$

So the tangent line is $y = \dfrac{1}{2}(x - 2) + 1$ or

$$y = \dfrac{1}{2}x.$$

23. $x^2 - 4y^3 = 0$ at $(2, 1)$

$$\dfrac{d}{dx}(x^2 - 4y^3) = \dfrac{d}{dx}(0)$$

$$2x - 12y^2 y' = 0$$

$$y' = \dfrac{2x}{12y^2} = \dfrac{x}{6y^2}$$

$$y'(2) = \dfrac{2}{6 \cdot 1^2} = \dfrac{1}{3}$$

So the tangent line is $y = \dfrac{1}{3}(x - 2) + 1$ or

$$y = \dfrac{1}{3}x + \dfrac{1}{3}.$$

25. $x^2 y^2 = 4y$ at $(2, 1)$

$$\frac{d}{dx}(x^2 y^2) = \frac{d}{dx}(4y)$$

$$2xy^2 + x^2 \cdot 2y \cdot y' = 4y'$$

$$y'(2x^2 y - 4) = -2xy^2$$

$$y' = \frac{-2xy^2}{2x^2 y - 4}$$

$$y'(2) = \frac{-2(2)(1)^2}{2(2)^2(1) - 4} = \frac{-4}{4} = -1$$

So the tangent line is $y = (-1)(x - 2) + 1$ or $y = -x + 3$.

27. $x^3 y^3 = 9y$ at $(1, 3)$

$$\frac{d}{dx}(x^3 y^3) = \frac{d}{dx}(9y)$$

$$3x^2 y^3 + x^3 \cdot 3y^2 \cdot y' = 9y'$$

$$y'(3x^3 y^2 - 9) = -3x^2 y^3$$

$$y' = \frac{3x^2 y^3}{9 - 3x^3 y^2}$$

$$y'(1) = \frac{3(1)^2(3)^3}{9 - 3(1)^3(3)^2} = \frac{81}{9 - 27} = -\frac{9}{2}$$

So the tangent line is $y = -\frac{9}{2}(x - 1) + 3$ or

$$y = -\frac{9}{2}x + \frac{15}{2}.$$

29. We have $\tan\theta = \frac{x}{2}$, so

$$\frac{d}{dt}(\tan\theta) = \frac{d}{dt}\left(\frac{x}{2}\right)$$

$$\sec^2\theta \cdot \theta' = \frac{1}{2}x'$$

$$\theta' = \frac{1}{2\sec^2\theta} \cdot x' = \frac{x'\cos^2\theta}{2}$$

at $x = 0$, we have $\tan\theta = \frac{x}{2} = \frac{0}{2}$ so $\theta = 0$, and we have $x' = -130$ ft/s so

$$\theta' = \frac{(-130) \cdot \cos^2 0}{2} = -65 \text{ rad/s}$$

31. t = number of seconds since launch
x = height of rocket in miles after t seconds
θ = camera angle in radians after t seconds

$$\tan\theta = \frac{x}{2}$$

$$\frac{d}{dx}(\tan\theta) = \frac{d}{dx}\left(\frac{x}{2}\right)$$

$$\sec^2\theta \cdot \theta' = \frac{1}{2}x'$$

$$\theta' = \frac{\cos^2\theta \cdot x'}{2}$$

When $x = 3$, $\tan\theta = \frac{3}{2}$, so $\cos\theta = \frac{2}{\sqrt{13}}$

$$\theta' = \frac{\left(\frac{2}{\sqrt{13}}\right)^2 (.2)}{2} \approx .03 \text{ rad/s}$$

33. t = hours elapsed since injury
r = radius of the infected area
A = area of the infection

$$A = \pi r^2$$

$$\frac{d}{dt}(A) = \frac{d}{dt}(\pi r^2)$$

$$A' = 2\pi r \cdot r'$$

When $r = 3$mm, $r' = 1$ mm/hr,

$$A' = 2\pi(3)(1) = 6\pi \text{ mm}^2 \text{/hr}$$

35. t = time
s = distance between plane and airport at time t
x = miles from airport at height h and time t

$$4^2 + x^2 = s^2$$

$$16 + x^2 = s^2$$

$$\frac{d}{dt}(16 + x^2) = \frac{d}{dt}s^2$$

$$2x \cdot x' = 2s \cdot s'$$

$$x' = \frac{s}{x} \cdot s'$$

When $x = 40$, $s = \sqrt{16 + 40^2} = 4\sqrt{101} \approx 40.2$

$$x' \approx \frac{40.2 \cdot (-240)}{40} = -241.2$$

37. If $x'(t) = 0$, $x(t) = \frac{1}{4}$, $y(t) = \frac{1}{2}$, $y'(t) = -50$

$$d'(t) = \frac{x(t) \cdot x'(t) + y(t) \cdot y'(t)}{\sqrt{[x(t)]^2 + [y(t)]^2}}$$

$$= \frac{\frac{1}{4} \cdot 0 + \frac{1}{2}(-50)}{\sqrt{\left(\frac{1}{4}\right)^2 + \left(\frac{1}{2}\right)^2}}$$

$$= \frac{-25}{\sqrt{\frac{1}{16} + \frac{1}{4}}}$$

$$= \frac{-100}{\sqrt{5}}$$

$$\approx -44.7$$

39. For $x = \frac{1}{2}$, $x'(t) = -(\sqrt{2} - 1)50$, $y(t) = \frac{1}{2}$,

$y'(t) = -50$

$$d'(t) = \frac{x(t) \cdot x'(t) + y(t) \cdot y'(t)}{\sqrt{[x(t)]^2 + [y(t)]^2}}$$

$$= \frac{\frac{1}{2}\left(-\sqrt{2} + 1\right)50 + \frac{1}{2}(-50)}{\sqrt{\left(\frac{1}{2}\right)^2 + \left(\frac{1}{2}\right)^2}}$$

$$= -50$$

41. $\bar{C}(x) = 10 + \frac{100}{x} = 10 + 100x^{-1}$

$\bar{C}'(x) = -100x^{-2} \cdot x'$

$\bar{C}'(10) = -100(10)^{-2} \cdot 2 = -2$ dollars/year

43.

Interval	Estimated x'
$(0, 2)$	2000
$(1, 2)$	2000

So $x'(2) \approx 2$ (Thousand)

$x(2) \approx 20$

$s(x) = 60 - 40e^{-.05x}$

$s'(x) = -40e^{-.05x}(-.05x') = 2e^{-.05x} \cdot x'$

$s'(20) = 2e^{-.05(20)} \cdot 2 \approx 1.472$ thousand \$ per year

or \$1,472 per year.

45. Let θ be the angle between the end of the shadow and the top of the lamppost. Then $\tan\theta = \frac{6}{s}$ and

$$\tan\theta = \frac{18}{s + x}$$

So

$$\frac{x + s}{18} = \frac{s}{6}$$

$$\frac{d}{dx}\left(\frac{x + s}{18}\right) = \frac{d}{dx}\left(\frac{s}{6}\right)$$

$$\frac{x' + s'}{18} = \frac{s'}{6}$$

$$x' + s' = 3s'$$

$$s' = \frac{x'}{2}$$

$$x' = 2$$

So $s' = \frac{2}{2} = 1$ ft/s

47. $V = \frac{4}{3}\pi r^3$

$V' = \frac{4}{3}\pi \cdot 3r^2 \cdot r' = 4\pi r^2 r'$

$A = 4\pi r^2$

So $V' = Ar'$

If V' is proportional to the surface area, then $V' = p \cdot A$ for some p, so $V' = Ar' = pA$ so $r' = p$ is constant.

49. $P(t) \cdot V'(t) + P'(t)V(t) = 0$

$$\frac{P'(t)}{V'(t)} = -\frac{P(t)}{V(t)} = -\frac{c}{V(t)^2}$$

51. Let $r(t)$ be the length of the rope at time t and $x(t)$ be the distance (along the water) between the boat and the dock.

$$r(t)^2 = 36 + x(t)^2$$

$$2r(t)r'(t) = 2x(t)x'(t)$$

$$x'(t) = \frac{r(t)r'(t)}{x(t)} = \frac{-2r(t)}{x(t)} = \frac{-2\sqrt{36 + x^2}}{x}$$

When $x = 20$, $x' = -2.088$; when $x = 10$,

$x' = -2.332$.

53. $x^2 + y^3 - 3y = 4$

$$\frac{d}{dx}(x^2 + y^3 - 3y) = \frac{d}{dx}(4)$$
$$2x + 3y^2 y' - 3y' = 0$$
$$y'(3y^2 - 3) = -2x$$
$$y' = \frac{2x}{3 - 3y^2}$$

Horizontal tangents:

$$y' = \frac{2x}{3 - 3y^2} = 0$$
$$x = 0$$

When $x = 0$, we have $0^2 + y^3 - 3y = 4$. Using a computer algebra system to solve this, we find that $y = \left(2 - \sqrt{3}\right)^{1/3} + \left(2 + \sqrt{3}\right)^{1/3} \approx 2.2$ is a horizontal tangent line, tangent to the curve at the (approximate) point (0, 2.2).

Vertical tangents: $3 - 3y^2 = 0$
$$y^2 = 1$$
$$y = \pm 1$$

When $y = 1$, we have
$$x^2 + (1)^3 - 3(1) = 4$$
$$x^2 = 6$$
$$x = \pm\sqrt{6} \approx \pm 2.4$$

Also, when $y = -1$, we have
$$x^2 + (-1)^3 - 3(-1) = 4$$
$$x^2 = 2$$
$$x = \pm\sqrt{2} \approx \pm 1.4$$

Thus, we find 4 vertical tangent lines:
$x = -\sqrt{6}$, $x = -\sqrt{2}$, $x = \sqrt{2}$, $x = \sqrt{6}$, tangent to the curve (respectively) at the points
$\left(-\sqrt{6}, 1\right), \left(-\sqrt{2}, -1\right), \left(\sqrt{2}, -1\right), \left(\sqrt{6}, 1\right)$.

55. $x^2 + y^2 = 4$

$$\frac{d}{dx}(x^2 + y^2) = \frac{d}{dx}(4)$$
$$2x + 2yy' = 0$$
$$y' = \frac{-2x}{2y} = \frac{-x}{y} = -xy^{-1}$$
$$y'' = -1 \cdot y^{-1} - x(-y^{-2})y'$$
$$= -y^{-1} + xy^{-2}y'$$
$$= -\frac{1}{y} + \frac{x}{y^2}\left(\frac{-x}{y}\right)$$
$$= \frac{-y^2 - x^2}{y^3} = \frac{-4}{y^3}$$

57. $x^2 + y^3 - 2y = 3$

$$y'(x) = \frac{-2x}{3y^2 - 2}$$

Tangent line at (2, 1) is $y = -4x + 9$
for $x = 1.9$, $-4(1.9) + 9 = 1.4$ so $y(1.9) \approx 1.4$,
for $x = 2.1$, $-4(2.1) + 9 = 0.6$ so $y(2.1) \approx 0.6$.
Note that for $x = 2.1$ the tangent takes you far away from the curve. The only y corresponding to $x = 2.1$ on the curve is about -1.7.

59. The third point is (3, 7).

61. $x^2 y - 2y = 4$

$$\frac{d}{dx}(x^2 y - 2y) = \frac{d}{dx}(4)$$
$$2xy + x^2 y' - 2y' = 0$$
$$y'(x^2 - 2) = -2xy$$
$$y' = \frac{-2xy}{x^2 - 2} = \frac{2xy}{2 - x^2}$$

Since $y'(\pm\sqrt{2}) = \frac{2xy}{0}$, we would expect vertical tangents here.

When $y = 0$, $y' = \frac{2x \cdot 0}{2 - x^2} = 0$, so we would expect horizontal tangents here.

When $x = \pm\sqrt{2}$, $x^2y - 2y = 4$
$$2y - 2y = 4$$
$$0 = 4$$
so the function is not defined here.
When $y = 0$, $x^2y - 2y = 4$
$$x^2 \cdot 0 - 2 \cdot 0 = 4$$
$$0 = 4$$
so the function is not defined here.
We have vertical asymptotes $x = \pm\sqrt{2}$, and horizontal asymptote $y = 0$.

Section 2.9

5. $f(x) = \dfrac{1}{x}$, $[-1, 1]$

Since $f(x)$ is not defined for $x = 0$, f is not continuous on $[-1, 1]$.

Now, $\dfrac{f(1) - f(-1)}{1 - (-1)} = \dfrac{\frac{1}{1} - \frac{1}{-1}}{1 - (-1)} = 1$ and

$f'(x) = -x^{-2} = \dfrac{-1}{x^2}$

For $f'(c) = \dfrac{-1}{c^2} = 1$ there is no value of c for which this is true.

7. $f(x) = \tan x$, $[0, \pi]$

Since $f(x)$ is not defined for $x = \dfrac{\pi}{2}$, f is not

continuous on $[0, \pi]$.
Now
$$\frac{f(\pi) - f(0)}{\pi - 0} = \frac{\tan \pi - \tan 0}{\pi - 0}$$
$$= \frac{0 - 0}{\pi - 0}$$
$$= 0$$

$f'(x) = \sec^2 x$

For $f'(c) = \sec^2 c = \dfrac{1}{\cos^2 c} = 0$

There is no value of c for which this is true.

9. $f(x) = x^2 + 1$, $[-2, 2]$

$f(x)$ is continuous on $[-2, 2]$, differentiable on $(-2, 2)$, and $f(-2) = (-2)^2 + 1 = 5$,

$f(2) = (2)^2 + 1 = 5$, so $f(-2) = f(2)$.
$$\frac{f(2) - f(-2)}{2 - (-2)} = \frac{5 - 5}{4} = 0$$
$f'(x) = 2x$
For $f'(c) = 2c = 0$, $c = 0$

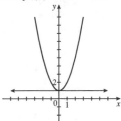

11. $f(x) = x^3 + x^2$, $[0, 1]$

$f(x)$ is continuous on $[0, 1]$, differentiable on $(0, 1)$.
$$f'(x) = 3x^2 + 2x$$
$$\frac{f(1) - f(0)}{1 - 0} = \frac{1^3 + 1^2 - (0^3 + 0^2)}{1 - 0} = 2$$

$f'(c) = 3c^2 + 2c = 2$

for $3c^2 + 2c - 2 = 0$
$$c = \frac{-2 \pm \sqrt{2^2 - 4(3)(-2)}}{2(3)}$$
$$= \frac{-2 \pm \sqrt{28}}{6}$$
$$= \frac{-2 \pm 2\sqrt{7}}{6}$$
$$= \frac{-1 \pm \sqrt{7}}{3}$$
$c \approx -1.22$ or $c \approx .55$
Since $-1.22 \neq (0, 1)$, we have $c \approx .55$.

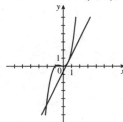

13. $f(x) = \sin x,\ \left[0, \dfrac{\pi}{2}\right]$

$f(x)$ is continuous on $\left[0, \dfrac{\pi}{2}\right]$ and differentiable on

$\left(0, \dfrac{\pi}{2}\right)$

$f'(x) = \cos x$

$\dfrac{f\left(\frac{\pi}{2}\right) - f(0)}{\frac{\pi}{2} - 0} = \dfrac{\sin\frac{\pi}{2} - \sin 0}{\frac{\pi}{2} - 0} = \dfrac{1 - 0}{\frac{\pi}{2}} = \dfrac{2}{\pi}$

$f'(c) = \cos c = \dfrac{2}{\pi}$

So $c = \cos^{-1}\left(\dfrac{2}{\pi}\right) \approx .88$

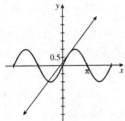

15. $f'(x) > 0$ for all x.

Then for each a, b where $a < b$, we know there exists c (a, b) such that $f'(c) = \dfrac{f(b) - f(a)}{b - a}$ and $f'(c) > 0$. Since $a < b$, $b - a > 0$ so we must have $f(b) - f(a) > 0$, so $f(a) < f(b)$.

17. $f(x) = x^3 + 5x + 1$

$f'(x) = 3x^2 + 5 > 0$ for all x, so $f(x)$ is increasing.

19. $f(x) = -x^3 - 3x + 1$

$f'(x) = -3x^2 - 3 < 0$ for all x, so $f(x)$ is decreasing.

21. $f(x) = e^x$

$f'(x) = e^x > 0$ for all x, so $f(x)$ is increasing.

23. $f(x) = \ln x$

$f'(x) = \dfrac{1}{x}$

$f'(x) > 0$ for $x > 0$, that is, for all x in the domain of f. So $f(x)$ is increasing.

25. $f(x) = x^3 + 5x + 1 = 0$

$f'(x) = 3x^2 + 5 \neq 0$ for all x so it can't be true that $f(x) = 0$ has two (or more) solutions. Thus $f(x) = 0$ has exactly one solution.

27. $f(x) = x^4 + 3x^2 - 2 = 0$

$f'(x) = 4x^3 + 6x$

$f''(x) = 12x^2 + 6 \neq 0$ for all x, so it can't be true that $f'(x) = 0$ has two (or more) solutions. Thus $f'(x) = 0$ has exactly one solution and $f(x) = 0$ has exactly two solutions.

29. $f(x) = x^3 + ax + b = 0$ $a > 0$

$f'(x) = 3x^2 + a \neq 0$ for all x so it can't be true that $f(x) = 0$ has two (or more) solutions. Thus $f(x) = 0$ has exactly one solution.

31. $f(x) = x^5 + ax^3 + bx + c = 0$ $a > 0, b > 0$

$f'(x) = 5x^4 + 3ax^2 + b \neq 0$ for all x, so it can't be true that $f(x) = 0$ has two (or more) solutions. Thus $f(x) = 0$ has exactly one solution.

33. $f(x) = ax^4 + bx^3 + cx^2 + dx + e$

$f'(x) = 4ax^3 + 3bx^2 + 2cx + d$

has at most three zeros, so $f(x)$ has at most four zeros.

35. $f(x) = x^2$

$g(x) = \dfrac{x^3}{3} + c$

37. $f(x) = x^4$

$g(x) = \dfrac{x^5}{5} + c$

39. $f(x) = \dfrac{1}{x^2} = x^{-2}$

$g(x) = -x^{-1} + c$

41. $f(x) = \sin x$

$g(x) = -\cos x + c$

43. $f(0) = f'(0) = 0$ and $f''(0) > 0$

$f''(0) > 0$ implies

$$\lim_{h \to 0^+} \frac{f'(h) - f'(0)}{h} = \lim_{h \to 0^+} \frac{f'(h)}{h} > 0$$

Since h is positive, $f'(h)$ must also be positive for all h in some interval $(0, a)$ with $a > 0$, or this limit could not be positive.

Since $f'(x) > 0$ for all x in the interval $(0, a)$, f is increasing on $(0, a)$. Thus $f(0) < f(x)$ for all x in the interval $(0, a)$. Since $f(0) = 0$, this means $f(x) > 0$ on this interval.

A similar argument shows that $f(x) > 0$ for all x in some interval $(b, 0)$ where b is negative.

45. $f(x) = \begin{cases} 2x & x \le 0 \\ 2x - 4 & x > 0 \end{cases}$

On $(0, 2)$, $f(x) = 2x - 4$, so f is continuous and differentiable on $(0, 2)$. $f(0) = 2 \cdot 0 = 0$ and $f(2) = 2 \cdot 2 - 4 = 0$.

But $f'(x) = 2$ so there is no c such that $f'(c) = 0$. Rolles Theorem requires f continuous on the closed interval, but we have f continuous only on the open interval.

Chapter 2 Review

1. See Example 2.11.

Interval	Average	
$(0,1)$	$\frac{3.0 - 2.0}{1 - 0}$	$= 1$
$(.5,1)$	$\frac{3.0 - 2.6}{.5}$	$= .8$
$(1,1.5)$	$\frac{3.4 - 3.0}{.5}$	$= .8$
$(1,2)$	$\frac{4.0 - 3.0}{1 - 0}$	$= 1$

So $f'(1) = .8$

3. See Example 2.1.

$f'(2)$ for $f(x) = x^2 - 2x$

$$f'(2) = \lim_{h \to 0} \frac{f(2 + h) - f(2)}{h}$$
$$= \lim_{h \to 0} \frac{(2 + h)^2 - 2(2 + h) - (2^2 - 2(2))}{h}$$
$$= \lim_{h \to 0} \frac{4 + 4h + h^2 - 4 - 2h - 0}{h}$$
$$= \lim_{h \to 0} \frac{2h + h^2}{h}$$
$$= \lim_{h \to 0} 2 + h$$
$$= 2$$

5. See Example 2.1.

$f'(1)$ for $f(x) = \sqrt{x}$

$$f'(1) = \lim_{h \to 0} \frac{f(1 + h) - f(1)}{h}$$
$$= \lim_{h \to 0} \frac{\sqrt{1 + h} - \sqrt{1}}{h}$$
$$= \lim_{h \to 0} \frac{\sqrt{1 + h} - 1}{h} \cdot \frac{\sqrt{1 + h} + 1}{\sqrt{1 + h} + 1}$$
$$= \lim_{h \to 0} \frac{1 + h - 1}{h(\sqrt{1 + h} + 1)}$$
$$= \lim_{h \to 0} \frac{1}{\sqrt{1 + h} + 1}$$
$$= \frac{1}{2}$$

7. See Example 2.2.

$f'(x)$ for $f(x) = x^3 + x$

$f'(x)$
$$= \lim_{h \to 0} \frac{f(x + h) - f(x)}{h}$$
$$= \lim_{h \to 0} \frac{(x + h)^3 + (x + h) - (x^3 + x)}{h}$$
$$= \lim_{h \to 0} \frac{x^3 + 3x^2h + 3xh^2 + h^3 + x + h - x^3 - x}{h}$$
$$= \lim_{h \to 0} \frac{3x^2h + 3xh^2 + h^3 + h}{h}$$
$$= \lim_{h \to 0} 3x^2 + 3xh + h^2 + 1$$
$$= 3x^2 + 1$$

9. See Example 1.2.

$y = f(x) = x^4 - 2x + 1$ at $x = 1$

m_{tan}

$= \lim_{h \to 0} \dfrac{f(1+h) - f(1)}{h}$

$= \lim_{h \to 0} \dfrac{(1+h)^4 - 2(1+h) + 1 - (1^4 - 2(1) + 1)}{h}$

$= \lim_{h \to 0} \dfrac{1 + 4h + 6h^2 + 4h^3 + h^4 - 2 - 2h + 1 - 0}{h}$

$= \lim_{h \to 0} \dfrac{2h + 6h^2 + 4h^3 + h^4}{h}$

$= \lim_{h \to 0} 2 + 6h + 4h^2 + h^3$

$= 2$

$f(1) = 1^4 - 2(1) + 1 = 0$

So the tangent line is $y = 2(x - 1) + 0$ or $y = 2x - 2$.

11. See Example 3.5, 7.2.

$y = f(x) = 3e^{2x}$ at $x = 0$

$f'(x) = 6e^{2x}$

$f'(0) = 6e^{2 \cdot 0} = 6$

$f(0) = 3e^{2 \cdot 0} = 3$

So the tangent line is $y = 6(x - 0) + 3$ or $y = 6x + 3$.

17. See Example 3.7.

$s(t) = 10e^{-2t} \sin 4t$

$v(t) = s'(t) = 10(-2e^{-2t} \sin 4t + e^{-2t} 4 \cos 4t)$

$\qquad = 40e^{-2t} \cos 4t - 20e^{-2t} \sin 4t$

$a(t) = v'(t) = 40(-2e^{-2t} \cos 4t - e^{-2t}(4) \sin 4t) - 20(-2e^{-2t} \sin 4t + e^{-2t} \cdot 4 \cos 4t)$

$\qquad = -160e^{-2t} \cos 4t - 120e^{-2t} \sin 4t$

19. See Example 1.4.

$s(t) = -16t^2 + 40t + 10$

$v(t) = -32t + 40$

$v(1) = -32(1) + 40 = 8$. The ball is going up.

$v(2) = -32(2) + 40 = -24$. The ball is going down.

21. See Section 2.1.

$f(x) = \sqrt{x+1}$

13. See Example 8.2.

$y - x^2 y^2 = x - 1$ at $(1, 1)$

$\dfrac{d}{dx}(y - x^2 y^2) = \dfrac{d}{dx}(x - 1)$

$y' - 2xy^2 - x^2 2y \cdot y' = 1$

$y'(1 - x^2 2y) = 1 + 2xy^2$

$y' = \dfrac{1 + 2xy^2}{1 - 2x^2 y}$

$y'(1) = \dfrac{1 + 2(1)(1)^2}{1 - 2(1)^2(1)} = -3$

So the tangent line is $y = -3(x - 1) + 1$ or $y = -3x + 4$.

15. See Example 3.7.

$s(t) = -16t^2 + 40t + 10$

$v(t) = s'(t) = -32t + 40$

$a(t) = v'(t) = -32$

a. $m_{\text{sec}} = \dfrac{f(2) - f(1)}{2 - 1}$

$\qquad = \dfrac{\sqrt{2+1} - \sqrt{2}}{1}$

$\qquad \approx .318$

b. $m_{\text{sec}} = \dfrac{f(1.5) - f(1)}{1.5 - 1} = \dfrac{\sqrt{2.5} - \sqrt{2}}{.5} \approx .334$

c. $m_{\text{sec}} = \dfrac{f(1.1) - f(1)}{1.1 - 1} = \dfrac{\sqrt{2.1} - \sqrt{2}}{.1} \approx .349$

The slope of the tangent line is Approximately .35.

23. See Example 3.3.

$$f(x) = x^4 - 3x^3 + 2x - 1$$
$$f'(x) = 4x^3 - 9x^2 + 2$$

25. See Example 3.3.

$$f(x) = \frac{3}{\sqrt{x}} + \frac{5}{x^2} = 3x^{-1/2} + 5x^{-2}$$
$$f'(x) = -\frac{3}{2}x^{-3/2} - 10x^{-3}$$

27. See Example 7.1.

$$f(t) = t^2(t+2)^3$$
$$f'(t) = 2t(t+2)^3 + t^2 \cdot 3(t+2)^2 \cdot 1$$
$$= 2t(t+2)^3 + 3t^2(t+2)^2$$

29. See Example 4.5.

$$g(x) = \frac{x}{3x^2 - 1}$$
$$g'(x) = \frac{1 \cdot (3x^2 - 1) - x(6x)}{(3x^2 - 1)^2}$$
$$= \frac{3x^2 - 1 - 6x^2}{(3x^2 - 1)^2}$$
$$= -\frac{3x^2 + 1}{(3x^2 - 1)^2}$$

31. See Example 5.1.

$$f(x) = x^2 \sin x$$
$$f'(x) = 2x \sin x + x^2 \cos x$$

33. See Example 7.4.

$$f(x) = \tan \sqrt{x} = \tan x^{1/2}$$
$$f'(x) = \sec^2 x^{1/2} \cdot \frac{1}{2} x^{-1/2} = \frac{\sec^2 \sqrt{x}}{2\sqrt{x}}$$

35. See Example 5.1.

$$f(t) = t \csc t$$
$$f'(t) = 1 \cdot \csc t + t \cdot (-\csc t \cdot \cot t)$$
$$= \csc t - t \csc t \cot t$$

37. See Example 7.2.

$$u(x) = 2e^{-x^2}$$
$$u'(x) = 2e^{-x^2}(-2x) = -4xe^{-x^2}$$

39. See Example 6.3.

$$f(x) = x \ln x^2$$
$$f'(x) = 1 \cdot \ln x^2 + x \cdot \frac{1}{x^2} \cdot 2x$$
$$= \ln x^2 + 2$$

41. See Example 7.4.

$$f(x) = \sqrt{\sin 4x} = (\sin 4x)^{1/2}$$
$$f'(x) = \frac{1}{2}(\sin 4x)^{-1/2} \cdot \cos 4x \cdot 4$$
$$= \frac{2\cos 4x}{\sqrt{\sin 4x}}$$

43. See Example 3.4.

$$f(x) = \left(\frac{x+1}{x-1}\right)^2$$
$$f'(x) = 2\left(\frac{x+1}{x-1}\right)\frac{d}{dx}\left(\frac{x+1}{x-1}\right)$$
$$= 2\left(\frac{x+1}{x-1}\right)\frac{1 \cdot (x-1) - (x+1) \cdot 1}{(x-1)^2}$$
$$= 2\left(\frac{x+1}{x-1}\right)\frac{-2}{(x-1)^2}$$
$$= \frac{-4(x+1)}{(x-1)^3}$$

45. See Example 7.2.

$$f(t) = te^{4t}$$
$$f'(t) = 1e^{4t} + te^{4t} \cdot 4 = e^{4t} + 4te^{4t}$$

47. See Example 2.7.

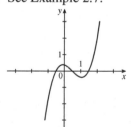

49. See Example 3.6.

$$f(x) = x^4 - 3x^3 + 2x^2 - x - 1$$
$$f'(x) = 4x^3 - 9x^2 + 4x - 1$$
$$f''(x) = 12x^2 - 18x + 4$$

51. See Example 7.2.

$f(x) = xe^{2x}$

$f'(x) = 1 \cdot e^{2x} + xe^{2x} \cdot 2 = e^{2x} + 2xe^{2x}$

$f''(x) = e^{2x} \cdot 2 + 2 \cdot e^{2x} + 2xe^{2x} \cdot 2$
$\quad = 4e^{2x} + 4xe^{2x}$

$f'''(x) = 4e^{2x} \cdot 2 + 4 \cdot e^{2x} + 4x \cdot e^{2x} \cdot 2$
$\quad = 12e^{2x} + 8xe^{2x}$

53. See Example 7.4.

$f(x) = \tan x$

$f'(x) = \sec^2 x$

$f''(x) = 2\sec x \cdot \sec x \tan x$
$\quad = 2\sec^2 x \tan x$

55. See Example 7.4.

$f(x) = \sin 3x$

$f'(x) = \cos 3x \cdot 3 = 3\cos 3x$

$f''(x) = 3(-\sin 3x \cdot 3) = -9\sin 3x$

$f'''(x) = -9\cos 3x \cdot 3 = -27\cos 3x$

$f^{(26)}(x) = -3^{26} \sin 3x$

57. See Example 4.7.

$R'(t) = Q'(t) \cdot P(t) + Q(t) \cdot P'(t)$

$P(0) = 2.4$

$Q(0) = 12$

$Q'(t) = -1.5$

$P'(t) = .1$

$R'(0) = -1.5 \cdot 2.4 + 12 \cdot .1 = -2.4$

Revenue is decreasing at \$2400 per year.

59. See Example 7.3.

$f(t) = 4\cos 2t$

$v(t) = f'(t) = 4(-\sin 2t) \cdot 2 = -8\sin 2t$

a. Velocity is zero when $v(t) = -8\sin 2t = 0$

$2t = 0, \pi, 2\pi, \ldots$ so $t = 0, \dfrac{\pi}{2}, \pi, \dfrac{3\pi}{2}, \ldots$

$f(t) = 4\cos 2t = 4$ for $t = 0, \pi, 2\pi, \ldots$

$f(t) = 4\cos 2t = -4$ for $t = \dfrac{\pi}{2}, \dfrac{3\pi}{2}, \ldots$

b. Velocity is maximum when
$v(t) = -8\sin 2t = 8$

$2t = \dfrac{3\pi}{2}, \dfrac{7\pi}{2}, \ldots$ so $t = \dfrac{3\pi}{4}, \dfrac{7\pi}{4}, \ldots$

$f(t) = 4\cos 2t = 0$ for $t = \dfrac{3\pi}{4}, \dfrac{7\pi}{4}, \ldots$

c. Velocity is at a minimum when
$v(t) = -8\sin 2t = -8$

$2t = \dfrac{\pi}{2}, \dfrac{5\pi}{2}, \ldots$ so $t = \dfrac{\pi}{4}, \dfrac{5\pi}{4}, \ldots$

$f(t) = 4\cos 2t = 0$ for $t = \dfrac{\pi}{4}, \dfrac{5\pi}{4}, \ldots$

61. See Example 8.1.

$x^2 y - 3y^3 = x^2 + 1$

$\dfrac{d}{dx}(x^2 y - 3y^3) = \dfrac{d}{dx}(x^2 + 1)$

$2xy + x^2 y' - 3 \cdot 3y^2 \cdot y' = 2x$

$y'(x^2 - 9y^2) = 2x - 2xy$

$y' = \dfrac{2x - 2xy}{x^2 - 9y^2}$

63. See Example 8.1.

$\dfrac{y}{x+1} - 3y = \tan x$

$y(x+1)^{-1} - 3y = \tan x$

$\dfrac{d}{dx}(y(x+1)^{-1} - 3y) = \dfrac{d}{dx}\tan x$

$y'(x+1)^{-1} + y \cdot (-1)(x+1)^{-2} \cdot 1 - 3y' = \sec^2 x$

$y'((x+1)^{-1} - 3) = \sec^2 x + y(x+1)^{-2}$

$y' = \dfrac{\sec^2 x + y(x+1)^{-2}}{(x+1)^{-1} - 3}$

65. See Example 8.5.

$f'(t) = \dfrac{x(t)x'(t) + y(t)y'(t)}{\sqrt{[x(t)]^2 + [y(t)]^2}}$

$x(t) = \dfrac{1}{2},\ x'(t) = 60,\ y(t) = \dfrac{1}{4},\ y'(t) = 50$

$d'(t) = \dfrac{\frac{1}{2} \cdot 60 + \frac{1}{4} \cdot 50}{\sqrt{\left(\frac{1}{2}\right)^2 + \left(\frac{1}{4}\right)^2}}$

$= \dfrac{30 + \frac{25}{2}}{\sqrt{\frac{1}{4} + \frac{1}{16}}}$

$= \dfrac{170}{\sqrt{5}}$

≈ 76 mph

The radar gun registers 76 mph.

67. See Example 8.1.

$y = x^3 - 6x^2 + 1$

$y' = 3x^2 - 12x = 3x(x - 4)$

$y' = 0$ for $x = 0, 4$ so we have horizontal tangent lines at $(0, 1)$ and $(4, -31)$.

y' is defined for all x, so we have no vertical tangent lines.

69. See Example 8.2.

$x^2y - 4y = x^2$

$\dfrac{d}{dx}(x^2y - 4y) = \dfrac{d}{dx}x^2$

$2xy + x^2y' - 4y' = 2x$

$y'(x^2 - 4) = 2x - 2xy$

$y' = \dfrac{2x - 2xy}{x^2 - 4}$

$y' = 0$ when $2x - 2xy = 0$

$2x(1 - y) = 0$

$x = 0$ or $y = 1$

At $y = 1$ we have $x^2 \cdot 1 - 4 \cdot 1 = x^2$

$x^2 - 4 = x^2$

So there is no x such that $y = 1$.

At $x = 0$, we have $0^2 \cdot y - 4y = 0^2$, so $y = 0$. Thus we have a horizontal tangent line at $(0, 0)$. y' is not defined when $x^2 - 4 = 0$, $x = \pm 2$. At $x = \pm 2$, we have $4y - 4y = 4$ so the function is not defined at $x = \pm 2$. So we have no vertical tangent lines.

71. See Example 9.2.

$f(x) = x^3 + 7x - 1$

$f'(x) = 3x^2 + 7 > 0$ for all x so $f(x) = 0$ has exactly one solution.

73. See Example 9.3.

$f(x) = x^2 - 2x$ on $[0, 2]$

$f'(c) = \dfrac{f(2) - f(0)}{2 - 0} = \dfrac{4 - 4 - (0 - 0)}{2} = 0$

$f'(c) = 2c - 2 = 0$

$c = 1$

75. See Example 9.4.

$f(x) = 3x^2 - \cos x$

$g(x) = x^3 - \sin x + c$

Chapter 3

Section 3.1

5. $f(x_0) = f(1) = \sqrt{1} = 1$

$f'(x) = \dfrac{1}{2}x^{-1/2}$

$f'(x_0) = f'(1) = \dfrac{1}{2}$

So $L(x) = f(x_0) + f'(x_0)(x - x_0)$

$= 1 + \dfrac{1}{2}(x - 1)$

$= \dfrac{1}{2} + \dfrac{1}{2}x$

7. $f(x) = \sqrt{2x + 9}, \; x_0 = 0$

$f(x_0) = f(0) = \sqrt{2 \cdot 0 + 9} = 3$

$f'(x) = \dfrac{1}{2}(2x + 9)^{-1/2} \cdot 2 = (2x + 9)^{-1/2}$

$f'(x_0) = f'(0) = (2 \cdot 0 + 9)^{-1/2} = \dfrac{1}{3}$

So $L(x) = f(x_0) + f'(x_0)(x - x_0)$

$= 3 + \dfrac{1}{3}(x - 0)$

$= 3 + \dfrac{1}{3}x$

9. $f(x) - \sin 3x, \; x_0 - 0$

$f(x_0) = f(0) = \sin 3 \cdot 0 = \sin 0 = 0$

$f'(x) = 3\cos 3x$

$f'(x_0) = f'(0) = 3\cos 3 \cdot 0 = 3$

$L(x) = f(x_0) + f'(x_0)(x - x_0)$

$= 0 + 3(x - 0)$

$= 3x$

11. $f(x) = e^{2x}, \; x_0 = 0$

$f(x_0) = f(0) = e^{2 \cdot 0} = e^0 = 1$

$f'(x) = 2e^{2x}$

$f'(x_0) = f'(0) = 2e^{2 \cdot 0} = 2e^0 = 2$

$L(x) = f(x_0) + f'(x_0)(x - x_0)$

$= 1 + 2(x - 0)$

$= 2x + 1$

13. $f(x) = \tan x$

$f(0) = \tan 0 = 0$

$f'(x) = \sec^2 x$

$f'(0) = \sec^2 0 = 1$

$L(x) = f(0) + f'(0)(x - 0) = 0 + 1(x - 0) = x$

$L(0.01) = .01$

$f(0.01) = \tan 0.01 \approx .0100003$

$L(0.1) = 0.1$

$f(0.1) = \tan(0.1) \approx .1003$

$L(1) = 1$

$f(1) = \tan 1 \approx 1.557$

15. $f(x) = \sqrt{4 + x}$

$f(0) = \sqrt{4 + 0} = 2$

$f'(x) = \dfrac{1}{2}(4 + x)^{-1/2}$

$f'(0) = \dfrac{1}{2}(4 + 0)^{-1/2} = \dfrac{1}{4}$

$L(x) = f(0) + f'(0)(x - 0) = 2 + \dfrac{1}{4}x$

$L(0.01) = 2 + \dfrac{1}{4}(0.01) = 2.0025$

$f(0.01) = \sqrt{4 + 0.01} \approx 2.002498$

$L(0.1) = 2 + \dfrac{1}{4}(0.1) = 2.025$

$f(0.1) = \sqrt{4 + 0.1} \approx 2.0248$

$L(1) = 2 + \dfrac{1}{4}(1) = 2.25$

$f(1) = \sqrt{4 + 1} \approx 2.2361$

17. $f(x) = \sin x, \; x_0 = \dfrac{\pi}{3}$

$f\left(\dfrac{\pi}{3}\right) = \dfrac{\sqrt{3}}{2}$

$f'(x) = \cos x$

$f'\left(\dfrac{\pi}{3}\right) = \cos\dfrac{\pi}{3} = \dfrac{1}{2}$

$L(x) = f\left(\dfrac{\pi}{3}\right) + f'\left(\dfrac{\pi}{3}\right)\left(x - \dfrac{\pi}{3}\right) = \dfrac{\sqrt{3}}{2} + \dfrac{1}{2}\left(x - \dfrac{\pi}{3}\right)$

$L(1) = \dfrac{\sqrt{3}}{2} + \dfrac{1}{2}\left(1 - \dfrac{\pi}{3}\right) \approx .842$

19. $f(x) = \sqrt[4]{16+x}, \ x_0 = 0$

$f(0) = \sqrt[4]{16+0} = 2$

$f'(x) = \dfrac{1}{4}(16+x)^{-3/4}$

$f'(0) = \dfrac{1}{4}(16+0)^{-3/4} = \dfrac{1}{32}$

$L(x) = f(0) + f'(0)(x-0) = 2 + \dfrac{1}{32}x$

$L(0.04) = 2 + \dfrac{1}{32}(0.04) = 2.00125$

21. $f(x) = \sqrt[4]{16+x}, \ x_0 = 0$

$L(x) = 2 + \dfrac{1}{32}x$ (from exercise 19)

$L(.16) = 2 + \dfrac{1}{32}(.16) = 2.005$

23. 19)

$\quad\quad\quad \sqrt[4]{16.04} = 2.0012488$

$\quad\quad\quad L(0.04) = 2.00125$

$\quad\quad\quad |2.0012488 - 2.00125| = .00000117$

   20)

$\quad\quad\quad \sqrt[4]{16.08} = 2.0024953$

$\quad\quad\quad L(.08) = 2.0025$

$\quad\quad\quad |2.0024953 - 2.0025| = .00000467$

   21)

$\quad\quad\quad \sqrt[4]{16.16} = 2.0049814$

$\quad\quad\quad L(.16) = 2.005$

$\quad\quad\quad |2.0049814 - 2.005| = .0000186$

25. The interval is about $(-0.307, 0.307)$.

27. a. $f(24) \approx f(20) + \dfrac{4}{10}(f(30) - f(20))$

$\quad\quad = 18 + 0.4(14 - 18)$

$\quad\quad = 16.4$

   b. $f(36) \approx f(30) + \dfrac{6}{10}(f(40) - f(30))$

$\quad\quad = 14 + 0.6(12 - 14)$

$\quad\quad = 12.8$

29. a. $f(208) \approx f(200) + \dfrac{8}{20}(f(220) - f(200))$

$\quad\quad = 128 + 0.4(142 - 128)$

$\quad\quad = 133.6$

b. $f(232) \approx f(220) + \dfrac{12}{20}(f(240) - f(220))$

$\quad\quad = 142 + 0.6(136 - 142)$

$\quad\quad = 138.4$

31. $\displaystyle\lim_{x\to 2} \dfrac{x-2}{x^2-4} = \lim_{x\to 2}\dfrac{1}{2x} = \dfrac{1}{4}$

33. $\displaystyle\lim_{x\to 0} \dfrac{x^3}{\sin x - x} = \lim_{x\to 0}\dfrac{3x^2}{\cos x - 1}$

$\quad\quad = \displaystyle\lim_{x\to 0}\dfrac{6x}{-\sin x}$

$\quad\quad = \displaystyle\lim_{x\to 0}\dfrac{6}{-\cos x} = -6$

35. $\displaystyle\lim_{x\to 1}\dfrac{x-1}{\ln x} = \lim_{x\to 1}\dfrac{1}{1/x} = 1$

37. $\displaystyle\lim_{x\to 0}\dfrac{e^x-1}{\cos x - 1} = \lim_{x\to 0}\dfrac{e^x}{-\sin x}$; undefined

39. $\displaystyle\lim_{x\to 0}\dfrac{x^2}{\cos x - x} = 0$

41. $\displaystyle\lim_{x\to 1}\dfrac{\ln(\ln x)}{\ln x}$; undefined

43. $\displaystyle\lim_{x\to 0}\dfrac{\sin x^2}{x^2} = \lim_{x\to 0}\dfrac{2x\cos x^2}{2x} = 1$

The result matches Example 1.5.

45. $\displaystyle\lim_{x\to 0}\dfrac{\sin x^3}{x^3} = 1$; $\displaystyle\lim_{x\to 0}\dfrac{1-\cos x^3}{x^6} = \dfrac{1}{2}$

47. The limit of $\cos x$ as x approaches 0 is not 0, so L'Hôpital's rule does not apply and the second limit is not equal to the first (which in fact is ∞).

49. The first step is plausible, since for x close to 0, $\sin x \approx x$. The second step (dividing out the x's) is actually correct.

51. $\displaystyle\lim_{x\to 0}\dfrac{e^{cx}-1}{x} \lim_{x\to 0}\dfrac{ce^x}{1} = c$

53. $\lim\limits_{\omega \to 0} \dfrac{2.5(4\omega t - \sin 4\omega t)}{4\omega^2}$

$= \lim\limits_{\omega \to 0} \dfrac{2.5(4t - 4t\cos 4\omega t)}{8\omega}$

$= \lim\limits_{\omega \to 0} \dfrac{2.5(-16t^2 \sin 4\omega t)}{8} = 0$

55. For small x we approximate e^x by $x+1$.

$\dfrac{Le^{2\pi d/L} - e^{-2\pi d/L}}{e^{2\pi d/L} + e^{-2\pi d/L}}$

$\approx \dfrac{L\left[\left(1 + \dfrac{2\pi d}{L}\right) - \left(1 - \dfrac{2\pi d}{L}\right)\right]}{\left(1 + \dfrac{2\pi d}{L}\right) + \left(1 - \dfrac{2\pi d}{L}\right)}$

$\approx \dfrac{L\left(\dfrac{4\pi d}{L}\right)}{2} = 2\pi d$

$f(d) \approx \dfrac{4.9}{\pi} \cdot 2\pi d = 9.8d$

Section 3.2

5. For $x_0 = 0$ the method fails.

7. $f(x) = x^3 + 3x^2 - 1 = 0,\; x_0 = 1$

$f'(x) = 3x^2 + 6x$

a. $x_1 = x_0 - \dfrac{f(x_0)}{f'(x_0)}$

$= 1 - \dfrac{1^3 + 3\cdot 1^2 - 1}{3\cdot 1^2 + 6\cdot 1}$

$= 1 - \dfrac{3}{9}$

$= \dfrac{2}{3}$

$x_2 = x_1 - \dfrac{f(x_1)}{f'(x_1)}$

$= \dfrac{2}{3} - \dfrac{\left(\dfrac{2}{3}\right)^3 + 3\left(\dfrac{2}{3}\right)^2 - 1}{3\left(\dfrac{2}{3}\right)^2 + 6\left(\dfrac{2}{3}\right)}$

$= \dfrac{79}{144}$

$\approx .5486$

b. .53209

9. $f(x) = x^4 - 3x^2 + 1 = 0,\; x_0 = 1$

$f'(x) = 4x^3 - 6x$

a. $x_1 = x_0 - \dfrac{f(x_0)}{f'(x_0)} = 1 - \left(\dfrac{1^4 - 3\cdot 1^2 + 1}{4\cdot 1^3 - 6\cdot 1}\right) = \dfrac{1}{2}$

$x_2 = x_1 - \dfrac{f(x_1)}{f'(x_1)}$

$= \dfrac{1}{2} - \left(\dfrac{\left(\dfrac{1}{2}\right)^4 - 3\left(\dfrac{1}{2}\right)^2 + 1}{4\left(\dfrac{1}{2}\right)^3 - 6\left(\dfrac{1}{2}\right)}\right)$

$= \dfrac{5}{8}$

b. .61803

11. $f(x) = x^3 + 4x^2 - 3x + 1$

$f'(x) = 3x^2 + 8x - 3$

$\sin 2\theta = \dfrac{32L}{v^2}$

$x_0 = -5$

$x_{n+1} = x_n - \dfrac{f(x_n)}{f'(x_n)}$ for $n = 0, 1, 2, 3, \ldots$

n	x_n
1	−4.718750
2	−4.686202
3	−4.6857796
4	−4.6857795

$x \approx -4.685780$

13. $f(x) = x^5 + 3x^3 + x - 1$

$f'(x) = 5x^4 + 9x^2 + 1$

$x_0 = .5$

$x_{n+1} = x_n - \dfrac{f(x_n)}{f'(x_n)}$ for $n = 0, 1, 2, 3, \ldots$

n	x_n
1	.526316
2	.525262
3	.525261
4	.525261

$x \approx .525260$

15. $f(x) = \cos x - x$
$f'(x) = -\sin x - 1$

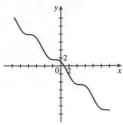

$x_0 = 1$

$x_{n+1} = x_n - \dfrac{f(x_n)}{f'(x_n)}$ for $n = 0, 1, 2, 3, \dots$

n	x_n
1	.750364
2	.739113
3	.739085
4	.739085

$x \approx .739085$

17. $\sin x = x^2 - 1$

$f(x) = \sin x - x^2 + 1$
$f'(x) = \cos x - 2x$

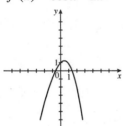

$x_0 = -.5$ or $x = 1.5$

$x_{n+1} = x_n - \dfrac{f(x_n)}{f'(x_n)}$ for $n = 0, 1, 2, 3, \dots$

n	x_n
1	-.644108
2	-.636751
3	-.636733
4	-.636733

$x \approx -.636733$

　　or

n	x_n
1	1.413799
2	1.409634
3	1.409624
4	1.409624

$x \approx 1.409624$

19. $e^x = -x$

$f(x) = e^x + x$
$f'(x) = e^x + 1$
$x_0 = -.5$

$x_{n+1} = x_n - \dfrac{f(x_n)}{f'(x_n)}$ for $n = 0, 1, 2, 3, \dots$

n	x_n
1	-.566311
2	-.567143
3	-.567143
4	-.567143

$x \approx -.567143$

21. $f(x) = 4x^3 - 7x^2 + 1 = 0,\ x_0 = 0$

$f'(x) = 12x^2 - 14x$

$x_1 = x_0 - \dfrac{f(x_0)}{f'(x_0)} = 0 - \dfrac{1}{0}$

The method fails because $f'(x_0) = 0$
Roots are $-.3454,\ .4362,\ 1.659$.

23. $f(x) = x^2 + 1,\ x_0 = 0$

$f'(x) = 2x$

$x_1 = x_0 - \dfrac{f(x_0)}{f'(x_0)} = 0 - \dfrac{1}{0}$

The method fails because $f'(x_0) = 0$.
There are no roots.

25. $f(x) = \dfrac{4x^2 - 8x + 1}{4x^2 - 3x - 7} = 0,\ x_0 = -1$

Note: $f(x_0) = f(-1)$ is undefined, so Newton's Method fails because x_0 is not in the domain of f.

Notice that f(x) = 0 only when $4x^2 - 8x + 1 = 0$. So using Newton's Method on

$g(x) = 4x^2 - 8x + 1$ with $x_0 = -1$ leads to

$x \approx .1339$. The other root is x ≈ 1.8660.

27. $f(x) = x^2 - a = 0$
$f'(x) = 2x$

$x_{n+1} = x_n - \dfrac{f(x_n)}{f'(x_n)}$

$= x_n - \left(\dfrac{x_n^2 - a}{2x_n} \right)$

$= x_n - \dfrac{x_n^2}{2x_n} + \dfrac{a}{2x_n}$

$= \dfrac{x_n}{2} + \dfrac{a}{2x_n}$

$= \dfrac{1}{2}\left(x_n + \dfrac{a}{x_n} \right)$

If $x_0 < \sqrt{a}$, then $\dfrac{a}{x_0} > \sqrt{a}$, so $x_0 < \sqrt{a} < \dfrac{a}{x_0}$.

29. $f(x) = x^2 - 11;\ x_0 = 3;\ \sqrt{11} \approx 3.316625$

31. $f(x) = x^3 - 11;\ x_0 = 2;\ \sqrt[3]{11} \approx 2.22398$

33. $f(x) = x^4 - 24;\ x_0 = 2;\ \sqrt[4]{24} \approx 2.213364$

35. $f(x) = x^{4.4} - 24;\ x_0 = 2;\ \sqrt[4.4]{24} \approx 2.059133$

37. The smallest positive solution of the first equation is 0.132782, and for the second equation the smallest positive solution is 1, so the species modeled by the second equation is certain to go extinct. This is consistent with the models, since the expected number of offspring for the population modeled by the first equation is 2.2, while for the second equation it is only 1.3.

39. The only positive solution is 0.6407

41. $W(x) = \dfrac{PR^2}{(R+x)^2},\ x_0 = 0$

$W'(x) = \dfrac{-2PR^2}{(R+x)^3}$

$L(x) = W(x_0) + W'(x_0)(x - x_0)$

$= \dfrac{PR^2}{(R+0)^2} + \left(\dfrac{-2PR^2}{(R+0)^3} \right)(x-0)$

$= P - \dfrac{2Px}{R}$

$L(x) = 120 - .01(120) = P - \dfrac{2Px}{R} = 120 - \dfrac{2 \cdot 120x}{R}$

$.01 = \dfrac{2x}{R}$

$x = .005R = .005(20,900,000) = 104,500\ \text{ft}$

43. To find the smallest positive solution of $\tan\left(\sqrt{x}\right) = \sqrt{x}$, plot $f(x) = \tan\left(\sqrt{x}\right) - \sqrt{x}$ to see that it crosses the x-axis at approximately x = 20. Newton's method (3 iterations) leads to $L \approx 20.19$.

y
$= \sqrt{L} - \sqrt{L}x - \sqrt{L}\cos\sqrt{L}x + \sin\sqrt{L}x$
$= 4.493 - 4.493x - 4.493\cos 4.493x + \sin 4.493x$

45. $0^3 - 3 \cdot 0^2 + 2 \cdot 0 = 0$
$1^3 - 3 \cdot 1^2 + 2 \cdot 1 = 0$
$2^3 - 3 \cdot 2^2 + 2 \cdot 2 = 0$

47. **a.** 0
b. 2
c. 1

Section 3.3

5. $f(x) = x^2 + 5x - 1$
$f'(x) = 2x + 5$
$2x + 5 = 0$

$x = \dfrac{-5}{2}$ is a critical number,

$x = \dfrac{-5}{2}$ is a local min.

7. $f(x) = x^3 - 3x + 1$

$f'(x) = 3x^2 - 3$

$3x^2 - 3 = 3(x^2 - 1) = 3(x+1)(x-1) = 0$

$x = \pm 1$ are critical numbers

$x = -1$ is a local max, $x = 1$ is a local min.

9. $f(x) = x^3 - 3x^2 + 3x$

$f'(x) = 3x^2 - 6x + 3$

$3x^2 - 6x + 3 = 3(x^2 - 2x + 1) = 3(x-1)^2 = 0$

$x = 1$ is a critical number, but it is neither a local min nor max.

11. $f(x) = x^4 - 3x^3 + 2$

$f'(x) = 4x^3 - 9x^2$

$4x^3 - 9x^2 = x^2(4x - 9) = 0$

$x = 0, \dfrac{9}{4}$ are critical numbers

$x = \dfrac{9}{4}$ is a local min

$x = 0$ is neither a local max nor min.

13. $f(x) = x^{3/4} - 4x^{1/4}$

$f'(x) = \dfrac{3}{4x^{1/4}} - \dfrac{1}{x^{3/4}}$

If $x \neq 0$, $f'(x) = 0$ when $3x^{3/4} = 4x^{1/4}$.

$x = 0, \dfrac{16}{9}$ are critical numbers.

$x = \dfrac{16}{9}$ is a local min,

$x = 0$ is neither a local max nor min.

15. $f(x) = x^3 - 2x^2 - 4x$

$f'(x) = 3x^2 - 4x - 4$

$3x^2 - 4x - 4 = (x-2)(3x+2) = 0$

$x = 2, -\dfrac{2}{3}$ are critical numbers

$x = -\dfrac{2}{3}$ is a local max, $x = 2$ is local min.

17. $f(x) = \sin x \cos x \; [0, 2\pi]$

$f'(x) = \cos x \cos x + \sin x(-\sin x)$

$\quad = \cos^2 x - \sin^2 x$

$\cos^2 x - \sin^2 x = 0$

$\quad \cos^2 x = \sin^2 x$

$\quad \cos x = \pm \sin x$

$x = \dfrac{\pi}{4}, \dfrac{3\pi}{4}, \dfrac{5\pi}{4}, \dfrac{7\pi}{4}$ are critical numbers.

$x = \dfrac{\pi}{4}, \dfrac{5\pi}{4}$ are local max,

$x = \dfrac{3\pi}{4}, \dfrac{7\pi}{4}$ are local min.

19. $f(x) = \dfrac{x+1}{x-1}$

$f'(x) = \dfrac{1(x-1) - (x+1)(1)}{(x-1)^2}$

$\quad = \dfrac{x - 1 - x - 1}{(x-1)^2}$

$\quad = \dfrac{-2}{(x-1)^2}$

$f'(x) \neq 0$ for any x.

$f'(x)$ is not defined for $x = 1$, but $x = 1$ is not in the domain of f, so there are no critical numbers.

21. $f(x) = \dfrac{x}{x^2 + 1}$

$f'(x) = \dfrac{1(x^2 + 1) - x(2x)}{(x^2 + 1)^2}$

$\quad = \dfrac{x^2 + 1 - 2x^2}{(x^2 + 1)^2}$

$\quad = \dfrac{1 - x^2}{(x^2 + 1)^2}$

$f'(x) = 0$ for $1 - x^2 = 0$, $x = 1, -1$

$f'(x)$ is defined for all x,

so $x = 1, -1$ are the critical numbers.

$x = -1$ is local min, $x = 1$ is local max.

23. $f(x) = \dfrac{e^x + e^{-x}}{2}$

$f'(x) = \dfrac{e^x - e^{-x}}{2}$

$\dfrac{e^x - e^{-x}}{2} = 0$ when $e^x = e^{-x}$, that is, x = 0.

$f'(x)$ is defined for all x.

So x = 0 is a critical number.

x = 0 is a local min.

25. $f(x) = x^{4/3} + 4x^{1/3} + 4x^{-2/3}$

$f'(x) = \dfrac{4}{3} \dfrac{(x+2)(x-1)}{x^{5/3}}$

$x = 0, -2, 1$ are critical numbers.

$x = -2$ and $x = 1$ are local minima.

$x = 0$ is neither a max nor a min.

27. $f(x) = 2x\sqrt{x+1} = 2x(x+1)^{1/2}$

$f'(x) = 2(x+1)^{1/2} + 2x\left(\dfrac{1}{2}(x+1)^{-1/2}\right) \cdot 1$

$\quad = \dfrac{2(x+1) + x}{\sqrt{x+1}}$

$\quad = \dfrac{3x+2}{\sqrt{x+1}}$

$f'(x) = 0$ for $3x + 2 = 0$, $x = -\dfrac{2}{3}$

$f'(x)$ is undefined for $\sqrt{x+1} = 0$, $x = -1$

so $x = -\dfrac{2}{3}, -1$ are critical numbers.

$x = -\dfrac{2}{3}$ is a local min.

$x = -1$ is neither a local min nor a local max, though it is a maximum on the interval [−1, 0).

29. $f(x) = e^{-x^2}$

$f'(x) = e^{-x^2}(-2x) = -2xe^{-x^2}$

$f'(x) = 0$ for x = 0

So x = 0 is a critical number.

x = 0 is a local max.

31. $f(x) = \sin x^2$, $[0, \pi]$

$f'(x) = \cos x^2 \cdot 2x = 2x\cos x^2$

$f'(x) = 0$ for $x = 0, \sqrt{\dfrac{\pi}{2}}, \sqrt{\dfrac{3\pi}{2}}, \sqrt{\dfrac{5\pi}{2}}$

$x = \sqrt{\dfrac{3\pi}{2}}$ is a local min.

$f(x)$ is not defined in an open interval containing 0, so 0 is not a local min.

$x = \sqrt{\dfrac{\pi}{2}}, \sqrt{\dfrac{5\pi}{2}}$ are local max.

33. $f(x) = x^3 - 3x + 1$, $[0, 2]$

$f'(x) = 3x^2 - 3 = 3(x^2 - 1)$

$f'(x) = 0$ for x = 1, −1, but only one is in [0, 2]

So 1 is a critical number.

$f(0) = 1$

$f(2) = 3$

$f(1) = -1$

So f(2) = 3 is the absolute maximum,

f(1) = −1 is the absolute minimum.

35. $f(x) = x^4 - 8x^2 + 2$, $[-3, 1]$

$f'(x) = 4x^3 - 16x = 4x(x^2 - 4) = 0$

for x = 0, 2, −2, but only −2, 0 are in $[-3, 1]$.

$f(-3) = 11$

$f(1) = -5$

$f(0) = 2$

$f(-2) = -14$

So $f(-3) = 11$ is the absolute maximum,

$f(-2) = -14$ is the absolute minimum.

37. $f(x) = x^{2/3}$, $[-4, -2]$

$f'(x) = \dfrac{2}{3}x^{-1/3} = \dfrac{2}{3\sqrt[3]{x}}$

$f'(x) \neq 0$ for any x, but $f'(x)$ undefined for x = 0,

so x = 0 is critical number, but $0 \notin [-4, -2]$

$f(-4) = \sqrt[3]{16} \approx 2.52$

$f(-2) = \sqrt[3]{4} \approx 1.59$

So $f(-4) = \sqrt[3]{16}$ is the absolute maximum and

$f(-2) = \sqrt[3]{4}$ is the absolute minimum.

39. $f(x) = \sin x + \cos x, \ [0, 2\pi]$

$f'(x) = \cos x - \sin x$

$f'(x) = 0$ for $x = \dfrac{\pi}{4}, \dfrac{5\pi}{4}$

$f(0) = 1$

$f(2\pi) = 1$

$f\left(\dfrac{\pi}{4}\right) = \sqrt{2}$

$f\left(\dfrac{5\pi}{4}\right) = -\sqrt{2}$

$f\left(\dfrac{\pi}{4}\right) = \sqrt{2}$ is the absolute maximum,

$f\left(\dfrac{5\pi}{4}\right) = -\sqrt{2}$ is the absolute minimum.

41. $f(x) = x \sin x + 3, \ [0, 2\pi]$

$f'(x) = 1 \cdot \sin x + x \cdot \cos x$

$ = \sin x + x \cos x$

$ = 0$

$x = 0, \ 2.0287, \ 4.9131$

$f(0) = 3$

$f(2.0287) = 4.8197$

$f(4.9131) = -1.8144$

$f(2\pi) = 3$

so $f(2.0287) = 4.8197$ is the absolute

maximum, $f(4.9131) = -1.8144$ is the absolute

minimum.

43. a. Absolute min at $(-1, 3)$;

absolute max at $(0.3660, 1.3481)$.

b. Absolute min at $(-1.3660, -3.8481)$;

absolute max at $(-3, 49)$.

45. a. Absolute min at $(0.6371, -1.1305)$;

absolute max at $(-1.2269, 2.7463)$.

b. Absolute min at $(-2.8051, -0.0748)$;

absolute max at $(-5, 29.2549)$.

47. 33) on $(0, 2), f(1) = -1$ is min., no max.

34) on $(-3, 2), f(-1) = 3$ is max, no min.

35) on $(-3, 1), f(-2) = -14$ is min., no max.

36) on $(-1, 3), f(2) = -14$ is min., no max.

37) on $(-4, -2)$, no max or min.

38) on $(-1, 3), f(0) = 0$ is min., no max.

49.

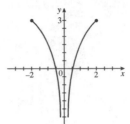

51.

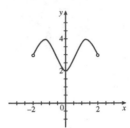

53.

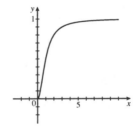

$f(x) = \dfrac{x^2}{x^2 + 1}, \ x > 0$

$f'(x) = \dfrac{2x(x^2 + 1)^2 - 2x \cdot 2(x^2 + 1) \cdot 2x}{(x^2 + 1)^2}$

$ = \dfrac{2x}{(x^2 + 1)^2}$

$f''(x) = \dfrac{2 \cdot (x^2 + 1)^2 - 2x \cdot 2(x^2 + 1) \cdot 2x}{(x^2 + 1)^4}$

$ = \dfrac{2(x^2 + 1)\left[(x^2 + 1) - 4x^2\right]}{(x^2 + 1)^4}$

$ = \dfrac{2\left[1 - 3x^2\right]}{(x^2 + 1)^3}$

$f''(x) = 0$ for $x = \pm\dfrac{1}{\sqrt{3}}, \ x = -\dfrac{1}{\sqrt{3}} \notin (0, \infty)$

$x = \dfrac{1}{\sqrt{3}}$ is steepest point.

55.

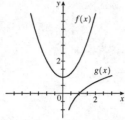

$f(x) = x^2 + 1$

$g(x) = \ln x$

$h(x) = f(x) - g(x) = x^2 + 1 - \ln x$

$h'(x) = 2x - \dfrac{1}{x} = 0$

$2x^2 = 1$

$x = \pm\sqrt{\dfrac{1}{2}}$

$x = \sqrt{\dfrac{1}{2}}$ is min

$f'(x) = 2x$

$g'(x) = \dfrac{1}{x}$

$f'\left(\sqrt{\dfrac{1}{2}}\right) = 2\sqrt{\dfrac{1}{2}} = \sqrt{2}$

$g'\left(\sqrt{\dfrac{1}{2}}\right) = \dfrac{1}{\sqrt{\dfrac{1}{2}}} = \sqrt{2}$

So the tangents are parallel.

57. Let $f(x) = \sec^2 x$

Then $f(0) = 1$ is a min, and $f(\pi) = 1$ is a min, but there is no local max on $(0, \pi)$.

59. $y = x^5 - 4x^3 - x + 10,\ x \in (-2, 2)$

$y' = 5x^4 - 12x^2 - 1$

$x = -1.575,\ 1.575$ are critical numbers.

There is a local max at $x = -1.575$, local min at $x = 1.575$.

$x = -1.575$ represents the top and $x = 1.575$ represents the bottom of the roller coaster.

$y''(x) = 20x^3 - 24x = 4x(5x^2 - 6) = 0$

$x = 0,\ \pm\sqrt{\dfrac{6}{5}}$ are critical numbers

$y'(x)$ has local max at $x = 0$,

local min at $x = \pm\sqrt{\dfrac{6}{5}}$, so $x = \pm\sqrt{\dfrac{6}{5}}$ are the steepest points.

61. $W(t) = a \cdot e^{-be^{-t}}$

as $t \to \infty,\ -be^{-t} \to 0$, so $a \cdot W(t) \to a$. $W'(t) = a \cdot e^{-be^{-t}} \cdot be^{-t}$

as $t \to \infty,\ be^{-t} \to 0$, so $W'(t) \to 0$.

$W''(t) = (a \cdot e^{-be^{-t}} \cdot be^{-t}) \cdot be^{-t} + (a \cdot e^{-be^{-t}}) \cdot (-be^{-t})$

$\quad = a \cdot e^{-be^{-t}} \cdot be^{-t}\left[be^{-t} - 1\right]$

$W''(t) = 0$ when $be^{-t} = 1$

$\qquad\qquad e^{-t} = b^{-1}$

$\qquad\qquad -t = \ln b^{-1}$

$\qquad\qquad t = \ln b$

$W'(\ln b) = a \cdot e^{-be^{-\ln b}} \cdot be^{-\ln b} = a \cdot e^{-b\left(\frac{1}{b}\right)} \cdot b \cdot \dfrac{1}{b} = ae^{-1}$

Maximum growth rate is ae^{-1} when $t = \ln b$.

Section 3.4

5. $y = x^3 - 3x + 2$

   $y' = 3x^2 - 3 = 3(x^2 - 1) = 3(x+1)(x-1)$

   $x = -1, 1$

   $(x+1) > 0$ on $(-1, \infty)$, $(x+1) < 0$ on $(-\infty, -1)$

   $(x-1) > 0$ on $(1, \infty)$, $(x-1) < 0$ on $(-\infty, -1)$

   $3(x+1)(x-1) > 0$ on $(1, \infty) \cup (-\infty, -1)$

   increasing,

   $3(x+1)(x-1) < 0$ on $(-1, 1)$ decreasing.

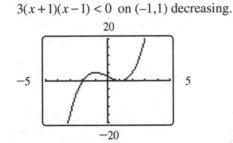

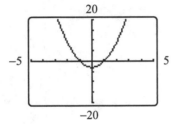

7. $y = x^4 - 8x^2 + 1$

   $y' = 4x^3 - 16x = 4x(x^2 - 4) = 4x(x-2)(x+2)$

   $x = 0, 2, -2$

   $4x > 0$ on $(0, \infty)$, $4x < 0$ on $(-\infty, 0)$

   $(x-2) > 0$ on $(2, \infty)$, $(x-2) < 0$ on $(-\infty, 2)$

   $(x+2) > 0$ on $(-2, \infty)$, $(x+2) < 0$ on $(-\infty, -2)$

   $4(x-2)(x+2) > 0$ on $(-2, 0) \cup (2, \infty)$ increasing

   $4(x-2)(x+2) < 0$ on $(-\infty, -2) \cup (0, 2)$

   decreasing

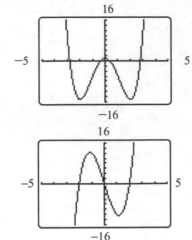

9. $y = (x+1)^{2/3}$

   $y' = \dfrac{2}{3}(x+1)^{-1/3} = \dfrac{2}{3\sqrt[3]{x+1}}$

   y' is not defined for $x = -1$

   $\dfrac{2}{3\sqrt[3]{x+1}} > 0$ on $(-1, \infty)$ increasing

   $\dfrac{2}{3\sqrt[3]{x+1}} < 0$ on $(-\infty, -1)$ decreasing

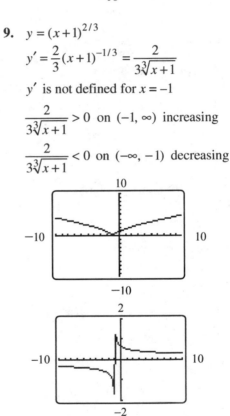

11. $y = \sin 3x$

$y' = \cos 3x \cdot 3 = 3\cos 3x$

$x = \pm\dfrac{\pi}{6},\ \pm\dfrac{3\pi}{6},\ \pm\dfrac{5\pi}{6},\ \pm\dfrac{7\pi}{6},\ \ldots$

$3\cos 3x > 0$ on…

$\ldots \cup \left(-\dfrac{\pi}{6},\ \dfrac{\pi}{6}\right) \cup \left(\dfrac{3\pi}{6},\ \dfrac{5\pi}{6}\right) \cup \ldots$ increasing

$3\cos 3x < 0$ on…

$\ldots \cup \left(-\dfrac{3\pi}{6},\ -\dfrac{\pi}{6}\right) \cup \left(\dfrac{\pi}{6},\ \dfrac{3\pi}{6}\right) \cup \ldots$ decreasing

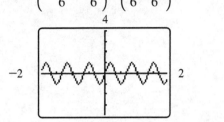

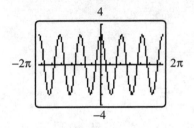

13. $y = e^{x^2-1}$

$y' = e^{x^2-1} \cdot 2x = 2xe^{x^2-1}$

$x = 0$

$2xe^{x^2-1} > 0$ on $(0,\ \infty)$ increasing

$2xe^{x^2-1} < 0$ on $(-\infty, 0)$ decreasing

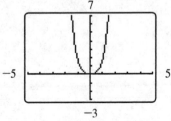

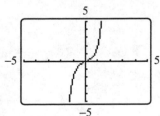

15. $y = x^3 + 2x^2 - x - 1$

$y' = 3x^2 + 4x - 1 = 0$

$x = \dfrac{-4 \pm \sqrt{16+12}}{6} = -\dfrac{2}{3} \pm \dfrac{\sqrt{7}}{3}$

$3x^2 + 4x - 1$

is $\begin{cases} > 0 \text{ on } \left(-\infty,\ -\dfrac{2}{3} - \dfrac{\sqrt{7}}{3}\right) \cup \left(-\dfrac{2}{3} + \dfrac{\sqrt{7}}{3},\ \infty\right) \\[2mm] < 0 \text{ on } \left(-\dfrac{2}{3} - \dfrac{\sqrt{7}}{3},\ -\dfrac{2}{3} + \dfrac{\sqrt{7}}{3}\right) \end{cases}$

So $y = x^3 + 2x^2 - x - 1$ has a local max at

$x = -\dfrac{2}{3} - \dfrac{\sqrt{7}}{3}$ and local min at $-\dfrac{2}{3} + \dfrac{\sqrt{7}}{3}$.

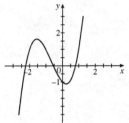

17. $y = x^4 + 2x^2 - x + 2$

$y' = 4x^3 + 4x - 1 = 0$

$x \approx .2367$

$4x^3 + 4x - 1 \begin{cases} > 0 \text{ on } (.2367\ \infty) \\ < 0 \text{ on } (-\infty, .2367) \end{cases}$

So $x = .2367$ is a local minimum.

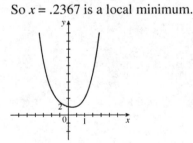

19. $y = x\sqrt{x^2 + 1} = x(x^2 + 1)^{1/2}$

$y' = 1 \cdot (x^2 + 1)^{1/2} + x \cdot \dfrac{1}{2}(x^2 + 1)^{-1/2} \cdot 2x$

$\quad = \sqrt{x^2 + 1} + \dfrac{x^2}{\sqrt{x^2 + 1}}$

$\quad = \dfrac{x^2 + 1 + x^2}{\sqrt{x^2 + 1}}$

$\quad = \dfrac{2x^2 + 1}{\sqrt{x^2 + 1}}$

$y' > 0$ for all x

So $y = x\sqrt{x^2 + 1}$ has no local extrema.

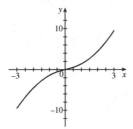

21. $y = xe^{-2x}$

$y' = 1 \cdot e^{-2x} + x \cdot e^{-2x}(-2)$

$\quad = e^{-2x} - 2xe^{-2x}$

$\quad = e^{-2x}(1 - 2x)$

$x = \dfrac{1}{2}$

$e^{-2x}(1 - 2x) > 0$ on $\left(-\infty, \dfrac{1}{2}\right)$

$e^{-2x}(1 - 2x) < 0$ on $\left(\dfrac{1}{2}, \infty\right)$

So $y = xe^{-2x}$ has a local maximum at $x = \dfrac{1}{2}$.

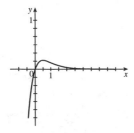

23. $y = \ln x^2$

$y' = \dfrac{1}{x^2} \cdot 2x = \dfrac{2}{x}$

y' is not defined for $x = 0$, but $x = 0$ is not in the domain of y, so there are no critical numbers and thus no extrema.

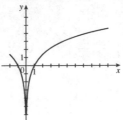

25. $y = \dfrac{x}{x^2 - 1}$

$y' = \dfrac{1(x^2 - 1) - x(2x)}{(x^2 - 1)^2}$

$\quad = \dfrac{x^2 - 1 - 2x^2}{(x^2 - 1)^2}$

$\quad = \dfrac{-x^2 - 1}{(x^2 - 1)^2}$

$y' \neq 0$ for any x, y' not defined for $x = -1, 1$, but $x = -1, 1$ are not in the domain of y. So there are no critical numbers and thus no extrema.

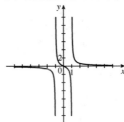

27. $y = \dfrac{x^3}{x^2 - 1}$

$y' = \dfrac{3x^2(x^2-1) - x^3(2x)}{(x^2-1)^2}$

$= \dfrac{3x^4 - 3x^2 - 2x^4}{(x^2-1)^2}$

$= \dfrac{x^4 - 3x^2}{(x^2-1)^2}$

$y' = 0$ for $x^4 - 3x^2$

$= x^2(x^2 - 3)$

$= x^2(x - \sqrt{3})(x + \sqrt{3})$

$= 0$

$x = 0, \sqrt{3}, -\sqrt{3}$

y' undefined for $x = 1, -1$, but $x = 1, -1$ are not in the domain of y.

$x^2 > 0$ on $(-\infty, \infty)$

$\left(x - \sqrt{3}\right) > 0$ on $\left(\sqrt{3}, \infty\right)$

$\left(x + \sqrt{3}\right) > 0$ on $\left(-\sqrt{3}, \infty\right)$

$(x+1) > 0$ on $(-1, \infty)$

$(x-1) > 0$ on $(1, \infty)$

$\dfrac{x^4 - 3x^2}{(x^2 - 1)^2} > 0$ on $\left(-\infty, -\sqrt{3}\right) \cup \left(\sqrt{3}, \infty\right)$

$\dfrac{x^4 - 3x^2}{(x^2 - 1)^2} < 0$ on $\left(-\sqrt{3}, -1\right) \cup (-1, 1) \cup \left(1, \sqrt{3}\right)$

So $y = \dfrac{x^2}{x^2 - 1}$ has local min at $x = \sqrt{3}$, local max

at $x = -\sqrt{3}$.

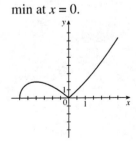

29. $y = \sin x + \cos x$

$y' = \cos x - \sin x = 0$

$\cos x = \sin x$

$x = \dfrac{\pi}{4}, \dfrac{5\pi}{4}, \dfrac{9\pi}{4}$, etc.

$\cos x - \sin x \begin{cases} > 0 \text{ on } \left(-\dfrac{3\pi}{4}, \dfrac{\pi}{4}\right) \cup \left(\dfrac{5\pi}{4}, \dfrac{9\pi}{4}\right) \cup \cdots \\[2mm] < 0 \text{ on } \left(\dfrac{\pi}{4}, \dfrac{5\pi}{4}\right) \cup \left(\dfrac{9\pi}{4}, \dfrac{13\pi}{4}\right) \cup \cdots \end{cases}$

So $y = \sin x + \cos x$ has local max at $x = \dfrac{\pi}{4}, \dfrac{9\pi}{4}$,

etc., local min at $x = \dfrac{5\pi}{4}, \dfrac{13\pi}{4}$, etc.

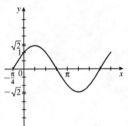

31. $y = \sqrt{x^3 + 3x^2} = (x^3 + 3x^2)^{1/2}$

$y' = \dfrac{1}{2}(x^3 + 3x^2)^{-1/2}(3x^2 + 6x)$

$= \dfrac{3x^2 + 6x}{2\sqrt{x^3 + 3x^2}}$

$= \dfrac{3x(x + 2)}{2\sqrt{x^3 + 3x^2}}$

$= 0$

$x = -2$

y' undefined at $x = 0$

$y' \begin{cases} > 0 \text{ on } (-3, -2) \cup (0, \infty) \\ < 0 \text{ on } (-2, 0) \end{cases}$

So $y = \sqrt{x^3 + 3x^2}$ has local max at $x = -2$, local min at $x = 0$.

33. $y = x^{2/3} - 2x^{-1/3}$

$y' = \dfrac{2}{3}x^{-1/3} + \dfrac{2}{3}x^{-4/3} = \dfrac{2}{3}\dfrac{x+1}{x\sqrt[3]{x}} = 0$

When $x = -1$. y' is undefined when $x = 0$ but 0 is not in the domain of y.

$\dfrac{2}{3}\dfrac{x+1}{x\sqrt[3]{x}} \begin{cases} > 0 \text{ on } (-1, 0) \cup (0, \infty) \\ < 0 \text{ on } (-\infty, -1) \end{cases}$

So $x = -1$ is a local min.

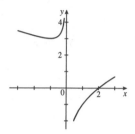

35. Local max at $x = -0.3689$;
local min at $x = 9.0356$

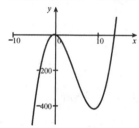

37. Local minima at $x = -0.9474, 11.2599$;
local max at 0.9374

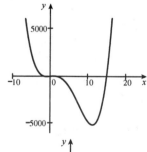

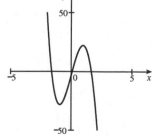

39. Local minima at $x = -1.0084, 10.9079$;
local maxima at $x = -10.9079, 1.0084$

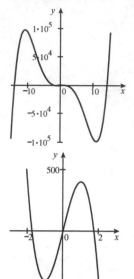

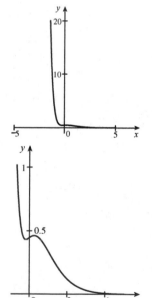

41. Local min at $x = -0.2236$;
local max at $x = 0.2236$

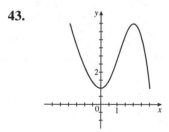

43.

45.

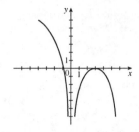

47. Answers will vary.

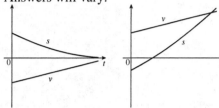

49. f is continuous on $[a, b]$, and $c \in (a, b)$ is a critical number.

 i) If $f'(x) > 0$ for all $x \in (a, c)$ and $f'(x) < 0$ for all $x \in (c, b)$, by Theorem 3.1, f is increasing on (a, c) and decreasing on (c, b), so $f(c) > f(x)$ for all $x \in (a, c)$ and $x \in (c\ b)$. Thus $f(c)$ is a local max.

 ii) If $f'(x) < 0$ for all $x \in (a, c)$ and $f'(x) > 0$ for all $x \in (c, b)$, by Theorem 3.1, f is decreasing on (a, c) and increasing on (c, b). So $f(c) < f(x)$ for all $x \in (a, c)$ and $x \in (c, b)$. Thus $f(c)$ is a local min.

 iii) If $f'(x) > 0$ on (a, c) and (c, b), then $f(c) > f(x)$ for all $x \in (a, c)$ and $f(c) < f(x)$ for all $x \in (c, b)$, so c is not a local extremum. If $f'(x) < 0$ on (a, c) and (c, b), then $f(c) < f(x)$ for all $x \in (a, c)$ and $f(c) > f(x)$ for all $x \in (c, b)$, so c is not a local extremum.

51. Let $f(x) = 2\sqrt{x}$, $g(x) = 3 - \dfrac{1}{x}$

Then $f(1) = 2\sqrt{1} = 2$, and $g(1) = 3 - \dfrac{1}{1} = 2$,

so $f(1) = g(1)$

$$f'(x) = \frac{1}{\sqrt{x}}$$

$$g'(x) = \frac{1}{x^2}$$

So $f'(x) > g'(x)$ for all $x > 1$, and

$$f(x) = 2\sqrt{x} > 3 - \frac{1}{x} = g(x) \text{ for all } x > 1.$$

53. Let $f(x) = e^x$, $g(x) = x + 1$.

Then $f(0) = e^0 = 1$, $g(0) = 0 + 1 = 0$,

so $f(0) = g(0)$.

$f'(x) = e^x$, $g'(x) = 1$

So $f'(x) > g'(x)$ for $x > 0$.

Thus $f(x) = e^x > x + 1 = g(x)$ for $x > 0$.

55. Let $f(x) = 3 + e^{-x}$; then $f(0) = 4$,

$f'(x) = -e^{-x} < 0$, so f is decreasing. But

$f(x) = 3 + e^{-x} = 0$ has no solution.

57. $s(t) = \sqrt{t + 4} = (t + 4)^{1/2}$

$$s'(t) = \frac{1}{2}(t + 4)^{-1/2} = \frac{1}{2\sqrt{t + 4}} > 0$$

So total sales are always increasing at the rate of $\dfrac{1}{2\sqrt{t + 4}}$ thousand dollars per month.

Section 3.4

5. The function is concave up for $x < -\dfrac{1}{2}$, $x > \dfrac{1}{2}$, and concave down for $-\dfrac{1}{2} < x < \dfrac{1}{2}$.

7. The function is concave up for $x > 1$, and concave down for $x < 1$.

9. $f(x) = x^3 - 3x^2 + 4$

$f'(x) = 3x^2 - 6x = 3x(x-2)$

$f''(x) = 6x - 6 = 6(x-1)$

$f'(x) > 0$ on $(-\infty, 0) \cup (2, \infty)$

$f'(x) < 0$ on $(0, 2)$

$f''(x) > 0$ on $(1, \infty)$

$f''(x) < 0$ on $(-\infty, 1)$

$f''(0) = -6$

$f''(2) = 6$

So f is increasing on $(-\infty, 0) \cup (2, \infty)$, decreasing on $(0, 2)$. $x = 0$ is a local maximum, and $x = 2$ is a local minimum. f is concave up on $(1, \infty)$, concave down on $(-\infty, 1)$, and $x = 1$ is an inflection point.

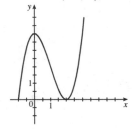

$f''(0) = -4$

$f''(-1) = 8$

$f''(1) = 8$

f is increasing on $(-1, 0) \cup (1, \infty)$, decreasing on $(-\infty, -1) \cup (0, 1)$, concave up on

$\left(-\infty, -\dfrac{1}{\sqrt{3}}\right) \cup \left(\dfrac{1}{\sqrt{3}}, \infty\right)$, concave down on

$\left(-\dfrac{1}{\sqrt{3}}, \dfrac{1}{\sqrt{3}}\right)$, $x = 0$ is a local max, $x = -1, 1$ are

local min, $x = -\dfrac{1}{\sqrt{3}}, \dfrac{1}{\sqrt{3}}$ are inflection points.

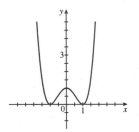

11. $f(x) = x^4 - 2x^2 + 1$

$f'(x) = 4x^3 - 4x$

$\qquad = 4x(x^2 - 1)$

$\qquad = 4x(x+1)(x-1)$

$f''(x) = 12x^2 - 4$

$\qquad = 4(3x^2 - 1)$

$\qquad = 4\left(\sqrt{3}x + 1\right)\left(\sqrt{3}x - 1\right)$

$f'(x) \begin{cases} > 0 \text{ on } (-1, 0) \cup (1, \infty) \\ < 0 \text{ on } (-\infty, -1) \cup (0, 1) \end{cases}$

$f''(x) \begin{cases} > 0 \text{ on } \left(-\infty, -\dfrac{1}{\sqrt{3}}\right) \cup \left(\dfrac{1}{\sqrt{3}}, \infty\right) \\[4mm] < 0 \text{ on } \left(-\dfrac{1}{\sqrt{3}}, \dfrac{1}{\sqrt{3}}\right) \end{cases}$

13. $f(x) = x + \dfrac{1}{x} = x + x^{-1}$

$f'(x) = 1 - x^{-2}$

$f''(x) = 2x^{-3}$

$f'(x) \begin{cases} > 0 \text{ on } (-\infty, -1) \cup (1, \infty) \\ < 0 \text{ on } (-1, 0) \cup (0, 1) \end{cases}$

$f''(x) \begin{cases} > 0 \text{ on } (0, \infty) \\ < 0 \text{ on } (-\infty, 0) \end{cases}$

$f''(-1) = -2$

$f''(1) = 2$

$f''(x)$ is not defined for $x = 0$.

f is increasing on $(-\infty, -1) \cup (1, \infty)$, decreasing on $(-1, 0) \cup (0, 1)$, concave up on $(0, \infty)$, concave down on $(-\infty, 0)$, $x = -1$ is a local maximum and $x = 1$ is a local minimum, and there are no inflection points.

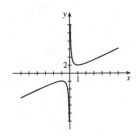

15. $f(x) = x^3 - 6x + 1$

$f'(x) = 3x^2 - 6 = 3(x^2 - 2)$

$\quad = 3\left(x + \sqrt{2}\right)\left(x - \sqrt{2}\right)$

$f''(x) = 6x$

$f'(x) \begin{cases} > 0 \text{ on } \left(-\infty, -\sqrt{2}\right) \cup \left(\sqrt{2}, \infty\right) \\ > 0 \text{ on } \left(-\sqrt{2}, \sqrt{2}\right) \end{cases}$

$f''(x) \begin{cases} > 0 \text{ on } (0, \infty) \\ < 0 \text{ on } (-\infty, 0) \end{cases}$

$f''\left(\sqrt{2}\right) = 6\sqrt{2}$

$f''\left(-\sqrt{2}\right) = -6\sqrt{2}$

f is increasing on $\left(-\infty, -\sqrt{2}\right) \cup \left(\sqrt{2}, \infty\right)$,

decreasing on $\left(-\sqrt{2}, \sqrt{2}\right)$, concave up on $(0, \infty)$,

concave down on $(-\infty, 0)$. $x = \sqrt{2}$ is a local min,

$x = -\sqrt{2}$ is a local max, $x = 0$ is an inflection
point.

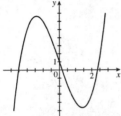

17. $f(x) = x^4 + 4x^3 - 1$

$f'(x) = 4x^3 + 12x^2 = 4x^2(x + 3)$

$f''(x) = 12x^2 + 24x = 12x(x + 2)$

$f'(x) \begin{cases} > 0 \text{ on } (-3, \infty) \\ < 0 \text{ on } (-\infty, -3) \end{cases}$

$f''(x) \begin{cases} > 0 \text{ on } (-\infty, -2) \cup (0, \infty) \\ < 0 \text{ on } (-2, 0) \end{cases}$

$f''(0) = 0$

$f''(-3) = 36$

f is increasing on $(-3, \infty)$, decreasing on
$(-\infty, -3)$, concave up on $(-\infty, -2) \cup (0, \infty)$,

concave down on $(-2, 0)$. $x = -3$ is a local min,
$x = -2, 0$ are inflection points.

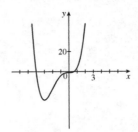

19. $f(x) = xe^{-x}$

$f'(x) = 1 \cdot e^{-x} + x \cdot (-e^{-x})$

$\quad = e^{-x} - xe^{-x}$

$\quad = e^{-x}(1 - x)$

$f''(x) = -e^{-x} - 1 \cdot e^{-x} - x(-e^{-x}) = e^{-x}(x - 2)$

$f'(x) \begin{cases} > 0 \text{ on } (-\infty, 1) \\ < 0 \text{ on } (1, \infty) \end{cases}$

$f''(x) \begin{cases} > 0 \text{ on } (2, \infty) \\ < 0 \text{ on } (-\infty, 2) \end{cases}$

$f''(1) = e^{-1}(1 - 2) = -e^{-1}$

f is increasing on $(-\infty, 1)$, decreasing on $(1, \infty)$,
concave up on $(2, \infty)$, concave down on
$(-\infty, 2)$, $x = 1$ is a local max, $x = 2$ is an inflection
point.

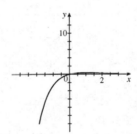

21. $f(x) = x^2\sqrt{x^2-9}$ f is undefined on $(-3, 3)$

$$f'(x) = 2x\sqrt{x^2-9} + x^2\left(\frac{1}{2}(x^2-9)^{-1/2}\cdot 2x\right)$$

$$= 2x\sqrt{x^2-9} + \frac{x^3}{\sqrt{x^2-9}}$$

$$= \frac{2x(x^2-9)+x^3}{\sqrt{x^2-9}}$$

$$= \frac{3x^3-18x}{\sqrt{x^2-9}}$$

$$= \frac{3x(x^2-6)}{\sqrt{x^2-9}}$$

$$= \frac{3x\left(x+\sqrt{6}\right)\left(x-\sqrt{6}\right)}{\sqrt{x^2-9}}$$

Critical points are $x = 0$, $\pm\sqrt{6}$, ± 3. f is undefined at $x = 0$, $\pm\sqrt{6}$.

$$f''(x) = \frac{(9x^2-18)\sqrt{x^2-9}-(3x^3-18x)\cdot\frac{1}{2}(x^2-9)^{-1/2}\cdot 2x}{x^2-9}$$

$$= \frac{(9x^2-18)(x^2-9)-x(3x^3-18x)}{(x^2-9)^{3/2}}$$

$$= \frac{(6x^4-81x^2+162)}{(x^2-9)^{3/2}}$$

$f''(x) = 0$ when $x^2 = \dfrac{81\pm\sqrt{81^2-4(6)(162)}}{2(6)} = \dfrac{81\pm\sqrt{2673}}{12} = \dfrac{3}{12}(27\pm\sqrt{297})$

So $x \approx \pm 3.325$ or $x \approx \pm 1.562$, but these latter values are not in the same domain. So only ± 3.325 are potential inflection points

$$f'(x)\begin{cases} > 0 \text{ on } (3, \infty) \\ < 0 \text{ on } (-\infty, -3) \end{cases}$$

$$f''(x)\begin{cases} > 0 \ (-\infty, -3.3)\cup(3.3, \infty) \\ < 0 \ (-3.3, 3)\cup(3, 3.3) \end{cases}$$

f is increasing on $(3, \infty)$, decreasing on $(-\infty, -3)$, concave up on $(-\infty, -3.3)\cup(3.3, \infty)$, concave down on $(-3.3, -3)\cup(3, 3.3)$. $x = \pm 3.3$ are inflection points.

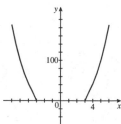

23. $f(x) = (x^2 + 1)^{2/3}$

$f'(x) = \frac{2}{3}(x^2 + 1)^{-1/3} \cdot 2x = \frac{4}{3}\frac{x}{\sqrt[3]{x^2 + 1}}$

$f''(x)$

$= \frac{4}{3}(x^2 + 1)^{-1/3} + \frac{4}{3}x \cdot \left(-\frac{1}{3}(x^2 + 1)^{-4/3} \cdot 2x\right)$

$= \frac{4}{9}\left(\frac{x^2 + 3}{(x^2 + 1)^{4/3}}\right)$

$f'(x)\begin{cases} > 0 \text{ on } (0, \infty) \\ < 0 \text{ on } (-\infty, 0) \end{cases}$

$f''(x) > 0$ for all x

f is increasing on $(0, \infty)$, decreasing on $(-\infty, 0)$, concave up on $(-\infty, \infty)$. $x = 0$ is a local min.

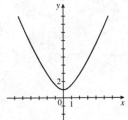

25. $f(x) = \frac{x^2}{x^2 - 9}$

$f'(x) = \frac{2x(x^2 - 9) - x^2(2x)}{(x^2 - 9)^2}$

$= \frac{-18x}{(x^2 - 9)^2}$

$= \frac{-18x}{\{(x+3)(x-3)\}^2}$

$f''(x) = \frac{-18(x^2 - 9)^2 + 18x \cdot 2(x^2 - 9) \cdot 2x}{(x^2 - 9)^4}$

$= \frac{54x^2 + 162}{(x^2 - 9)^3}$

$= \frac{54(x^2 + 3)}{(x^2 - 9)^3}$

$f'(x)\begin{cases} > 0 \text{ on } (-\infty, -3) \cup (-3, 0) \\ < 0 \text{ on } (0, 3) \cup (3, \infty) \end{cases}$

$f''(x)\begin{cases} > 0 \text{ on } (-\infty, -3) \cup (3, \infty) \\ < 0 \text{ on } (-3, 3) \end{cases}$

$f''(0) = \frac{162}{(-9)^3}$

f is increasing on $(-\infty, -3) \cup (-3, 0)$, decreasing on $(0, 3) \cup (3, \infty)$, concave up on $(-\infty, -3) \cup (3, \infty)$, concave down on $(-3, 3)$, $x = 0$ is a local max.

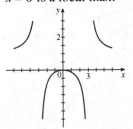

27. $f(x) = \sin x + \cos x$

$f'(x) = \cos x - \sin x$

$f''(x) = -\sin x - \cos x$

$f'(x)\begin{cases} > 0 \text{ on } \left(-\frac{3\pi}{4}, \frac{\pi}{4}\right) \cup \left(\frac{5\pi}{4}, \frac{9\pi}{4}\right) \text{ etc.} \\ < 0 \text{ on } \left(\frac{\pi}{4}, \frac{5\pi}{4}\right) \text{ etc.} \end{cases}$

$f''(x)\begin{cases} > 0 \text{ on } \left(\frac{3\pi}{4}, \frac{7\pi}{4}\right) \text{ etc.} \\ < 0 \text{ on } \left(-\frac{\pi}{4}, \frac{3\pi}{4}\right) \cup \left(\frac{7\pi}{4}, \frac{11\pi}{4}\right) \text{ etc.} \end{cases}$

f is increasing on $\left(-\dfrac{3\pi}{4},\dfrac{\pi}{4}\right)\cup\left(\dfrac{5\pi}{4},\dfrac{9\pi}{4}\right)$, etc.,

decreasing on $\left(\dfrac{\pi}{4},\dfrac{5\pi}{4}\right)$ etc., concave up on

$\left(\dfrac{3\pi}{4},\dfrac{7\pi}{4}\right)$ etc., concave down on

$\left(-\dfrac{\pi}{4},\dfrac{3\pi}{4}\right)\cup\left(\dfrac{7\pi}{4},\dfrac{11\pi}{4}\right)$, etc.

$x=\dfrac{\pi}{4},\dfrac{9\pi}{4}$ are local max,

$x=\dfrac{5\pi}{4}$ local min,

$x=-\dfrac{\pi}{4},\dfrac{3\pi}{4},\dfrac{7\pi}{4}$ are inflection points, etc.

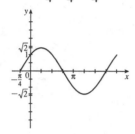

29. $f(x)=e^{-x}\sin x$

$f'(x)=-e^{-x}(\sin x-\cos x)$

$f''(x)=-2e^{-x}\cos x$

f is increasing on $\left(-\dfrac{3\pi}{4},\dfrac{\pi}{4}\right)\cup\left(\dfrac{5\pi}{4},\dfrac{9\pi}{4}\right)\dots,$

decreasing on $\left(\dfrac{\pi}{4},\dfrac{5\pi}{4}\right)\cup\left(\dfrac{9\pi}{4},\dfrac{13\pi}{4}\right)\dots,$

concave up on $\left(-\dfrac{3\pi}{2},-\dfrac{\pi}{2}\right)\cup\left(\dfrac{\pi}{2},\dfrac{3\pi}{2}\right)\dots,$

concave down on $\left(-\dfrac{\pi}{2},\dfrac{\pi}{2}\right)\cup\left(\dfrac{3\pi}{2},\dfrac{5\pi}{2}\right)\dots,$

local maxima at $\dfrac{\pi}{4},\dfrac{9\pi}{4},\dots,$

local minima at $-\dfrac{3\pi}{4},\dfrac{5\pi}{4},\dots,$

inflection points at $\dfrac{\pi}{2},\dfrac{3\pi}{2},\dots.$

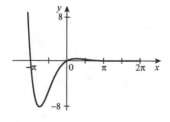

31. $f(x)=x^{3/4}-4x^{1/4}$

$f'(x)=\dfrac{3\sqrt{x}-4}{4x^{3/4}}$

$f''(x)=\dfrac{-3(\sqrt{x}-4)}{16x^{7/4}}$

f is decreasing on $\left(0,\dfrac{16}{9}\right)$

and increasing on $\left(\dfrac{16}{9},\infty\right)$.

f is concave up on $(0,16)$

and concave down on $(16,\infty)$.

There is a local minimum at $x=\dfrac{16}{9}$

and an inflection point at $x=16$.

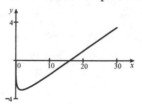

33. $f(x)=\sqrt[3]{2x^2-1}$

$f'(x)=\dfrac{4x}{3(2x^2-1)^{2/3}}$

$f''(x)=\dfrac{-4(2x^2+3)}{9(2x^2-1)^{5/3}}$

f is decreasing on $(-\infty,0)$

and increasing on $(0,\infty)$.

f is concave down on $(-\infty,-1)\cup(1,\infty)$

and concave up on $(-1,1)$

$x=0$ is a minimum and $x=\pm1$ are inflection points.

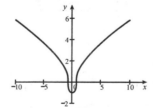

35. $f(x) = x^4 - 26x^3 + x$

$f'(x) = 4x^3 - 78x^2 + 1$

$f''(x) = 12x^2 - 156x$

$f'(x) \begin{cases} > 0 \text{ on } (-.1129, .1135) \cup (19.4993, \infty) \\ < 0 \text{ on } (-\infty, -.1129) \cup (.1135, 19.4993) \end{cases}$

$f''(x) \begin{cases} > 0 \text{ on } (-\infty, 0) \cup (13, \infty) \\ < 0 \text{ on } (0, 13) \end{cases}$

f is increasing on $(-.1129, .1135) \cup (19.4993, \infty)$, decreasing on $(-\infty, -.1129) \cup (.1135, 19.4993)$, concave up on $(-\infty, 0) \cup (13, \infty)$, concave down on $(0, 13)$, $x = -.1129$, $x = 19.4993$ are local min, $x = .1135$ is local max, $x = 0, 13$ are inflection points.

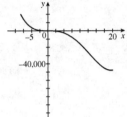

37. $f(x) = \dfrac{x^2 - 5x + 4}{x} = x - 5 + 4x^{-1}$

$f'(x) = 1 - 4x^{-2} = \dfrac{x^2 - 4}{x^2} = \dfrac{(x+2)(x-2)}{x^2}$

$f''(x) = 8x^{-3} = \dfrac{8}{x^3}$

$f'(x) \begin{cases} > 0 \text{ on } (-\infty, -2) \cup (2, \infty) \\ < 0 \text{ on } (-2, 0) \cup (0, 2) \end{cases}$

$f''(x) \begin{cases} > 0 \text{ on } (0, \infty) \\ < 0 \text{ on } (-\infty, 0) \end{cases}$

f is increasing on $(-\infty, -2) \cup (2, \infty)$, decreasing on $(-2, 0) \cup (0, 2)$, concave up on $(0, \infty)$, concave down on $(-\infty, 0)$, $x = -2$ is local max, $x = 2$ is local min.

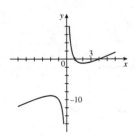

39. $f(x) = x^4 - 16x^3 + 42x^2 - 39.6x + 14$

$f'(x) = 4x^3 - 48x^2 + 84x - 39.6$

$f''(x) = 12x^2 - 96x + 84$

$ = 12(x^2 - 8x + 7)$

$ = 12(x - 7)(x - 1)$

$f'(x) \begin{cases} > 0 \text{ on } (.8952, 1.106) \cup (9.9987, \infty) \\ < 0 \text{ on } (-\infty, .8952) \cup (1.106, 9.9987) \end{cases}$

$f''(x) \begin{cases} > 0 \text{ on } (-\infty, 1) \cup (7, \infty) \\ < 0 \text{ on } (1, 7) \end{cases}$

f is increasing on $(.8952, 1.106) \cup (9.9987, \infty)$, decreasing on $(-\infty, .8952) \cup (1.106, 9.9987)$, concave up on $(-\infty, 1) \cup (7, \infty)$, concave down on $(1, 7)$, $x = .8952, 9.9987$ are local min, $x = 1.106$ is local max, $x = 1, 7$ are inflection points.

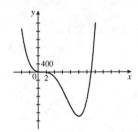

41.

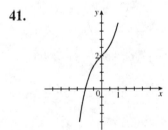

43.

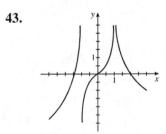

45.

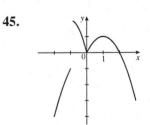

47. f is increasing on $(0, \infty)$, decreasing on $(-\infty, 0)$, concave up on $(-\infty, \infty)$, local min at $x = 0$.

49. f is increasing on $(-\infty, -1) \cup (1, \infty)$, decreasing on $(-1, 1)$, concave down on $(-\infty, 0)$, concave up on $(0, \infty)$, local min at $x = 1$, local max at $x = -1$, inflection point at $x = 0$.

51. We need to know $w'(0)$ to know if the depth is increasing.

53. $s(x) = -3x^3 + 270x^2 - 3600x + 18000$

$s'(x) = -9x^2 + 540x - 3600$

$s''(x) = -18x + 540 = 0$

$x = 30$

Spend $30,000 on advertising to maximize the rate of change of sales.

55. $C(x) = .01x^2 + 40x + 3600$

$\overline{C}(x) = \dfrac{C(x)}{x} = .01x + 40 + 3600x^{-1}$

$\overline{C}'(x) = .01 - 3600x^{-2} = 0$

$x = 600$

Manufacture 600 units to minimize average cost.

57. $f(x) = x^4 + cx^3$

$f'(x) = 4x^3 + 3cx^2 = x^2(4x + 3c)$

$f''(x) = 12x^2 + 6cx = 6x(2x + c)$

So $x = 0$ is inflection point, and $-\dfrac{3}{4}c$ is local

min, $-\dfrac{1}{2}c$ is inflection point.

59. Let $f(x) = -1 - x^2$. Then $f'(x) = -2x$
$\qquad\qquad\qquad\qquad\qquad\qquad f''(x) = -2$

so f is concave down for all x.

But $-1 - x^2 = 0$ has no solution.

61. Since the tangent line points above the sun, the sun appears higher in the sky than it really is.

Section 3.6

5. $f(x) = x^3 - 3x^2 + 3x$

$f'(x) = 3x^2 - 6x + 3 = 3(x^2 - 2x + 1) = 3(x - 1)^2$

$f''(x) = 6x - 6 = 6(x - 1)$

$f'(x) > 0$ on $(-\infty, 1) \cup (1, \infty)$

$f''(x) \begin{cases} > 0 \text{ on } (1, \infty) \\ < 0 \text{ on } (-\infty, 1) \end{cases}$

f is increasing on $(-\infty, \infty)$, and concave up on $(1, \infty)$, concave down on $(-\infty, 1)$, $x = 1$ is an inflection point.

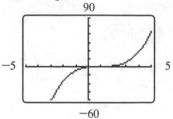

7. $f(x) = x^4 - 3x^2 + 2x$

$f'(x) = 4x^3 - 6x + 2 = 2(2x^3 - 3x + 1)$

$f''(x) = 12x^2 - 6 = 6(2x^2 - 1)$

$f'(x) \begin{cases} > 0 \text{ on } \left(-\dfrac{1}{2} - \dfrac{\sqrt{3}}{2}, -\dfrac{1}{2} + \dfrac{\sqrt{3}}{2}\right) \\ < 0 \text{ on } \left(-\infty, -\dfrac{1}{2} - \dfrac{\sqrt{3}}{2}\right) \cup \left(-\dfrac{1}{2} + \dfrac{\sqrt{3}}{2}, 1\right) \end{cases}$

$f''(x) \begin{cases} > 0 \text{ on } \left(-\infty, -\sqrt{\dfrac{1}{2}}\right) \cup \left(\sqrt{\dfrac{1}{2}}, \infty\right) \\ < 0 \text{ on } \left(-\sqrt{\dfrac{1}{2}}, \sqrt{\dfrac{1}{2}}\right) \end{cases}$

f is increasing on $\left(-\dfrac{1}{2} - \dfrac{\sqrt{3}}{2}, -\dfrac{1}{2} + \dfrac{\sqrt{3}}{2}\right) \cup (1, \infty)$,

decreasing on $\left(-\infty, -\dfrac{1}{2} - \dfrac{\sqrt{3}}{2}\right) \cup \left(-\dfrac{1}{2} + \dfrac{\sqrt{3}}{2}, 1\right)$,

concave up on $\left(-\infty, -\sqrt{\dfrac{1}{2}}\right) \cup \left(\sqrt{\dfrac{1}{2}}, \infty\right)$, concave

down on $\left(-\sqrt{\dfrac{1}{2}}, \sqrt{\dfrac{1}{2}}\right)$, local min at

$x = -\dfrac{1}{2} - \dfrac{\sqrt{3}}{2}, 1$, local max at $x = -\dfrac{1}{2} + \dfrac{\sqrt{3}}{2}$,

inflection points at $x = \pm\sqrt{\dfrac{1}{2}}$.

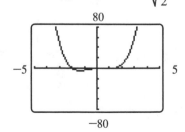

9. $f(x) = x^5 - 2x^3 + 1$

$f'(x) = 5x^4 - 6x^2 = x^2(5x^2 - 6)$

$f''(x) = 20x^3 - 12x = x(20x^2 - 12)$

$f'(x)\begin{cases} > 0 \text{ on } \left(-\infty, -\sqrt{\dfrac{6}{5}}\right) \cup \left(\sqrt{\dfrac{6}{5}}, \infty\right) \\ < 0 \text{ on } \left(-\sqrt{\dfrac{6}{5}}, 0\right) \cup \left(0, \sqrt{\dfrac{6}{5}}\right) \end{cases}$

$f''(x)\begin{cases} > 0 \text{ on } \left(-\sqrt{\dfrac{3}{5}}, 0\right) \cup \left(\sqrt{\dfrac{3}{5}}, \infty\right) \\ < 0 \text{ on } \left(-\infty, -\sqrt{\dfrac{3}{5}}\right) \cup \left(0, \sqrt{\dfrac{3}{5}}\right) \end{cases}$

f is increasing on $\left(-\infty, -\sqrt{\dfrac{6}{5}}\right) \cup \left(\sqrt{\dfrac{6}{5}}, \infty\right)$,

decreasing on $\left(-\sqrt{\dfrac{6}{5}}, \sqrt{\dfrac{6}{5}}\right)$, concave up on

$\left(-\sqrt{\dfrac{3}{5}}, 0\right) \cup \left(\sqrt{\dfrac{3}{5}}, \infty\right)$, concave down on

$\left(-\infty, -\sqrt{\dfrac{3}{5}}\right) \cup \left(0, \sqrt{\dfrac{3}{5}}\right)$, $x = -\sqrt{\dfrac{6}{5}}$ is local max,

local min at $x = \sqrt{\dfrac{6}{5}}$ $x = -\sqrt{\dfrac{3}{5}}, 0 \ \sqrt{\dfrac{3}{5}}$ are

inflection points.

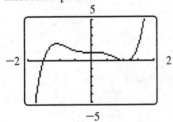

11. $f(x) = x\sqrt{x^2 - 4}$ f undefined on $(-2, 2)$

$f'(x) = \sqrt{x^2 - 4} + x\left(\dfrac{1}{2}\right)(x^2 - 4)^{-1/2}(2x)$

$= \sqrt{x^2 - 4} + \dfrac{x^2}{\sqrt{x^2 - 4}}$

$= \dfrac{2x^2 - 4}{\sqrt{x^2 - 4}}$

$f''(x) = \dfrac{4x\sqrt{x^2 - 4} - (2x^2 - 4)\frac{1}{2}(x^2 - 4)^{-1/2}(2x)}{x^2 - 4}$

$= \dfrac{4x(x^2 - 4) - x(2x^2 - 4)}{(x^2 - 4)^{3/2}} = \dfrac{2x^3 - 12x}{(x^2 - 4)^{3/2}}$

$= \dfrac{2x(x^2 - 6)}{(x^2 - 4)^{3/2}}$

$f'(x) > 0(-\infty, -2) \cup (2, \infty)$

$f''(x)\begin{cases} > 0 \text{ on } \left(-\sqrt{6}, -2\right) \cup \left(\sqrt{6}, \infty\right) \\ < 0 \text{ on } \left(-\infty, -\sqrt{6}\right) \cup \left(2, \sqrt{6}\right) \end{cases}$

f is increasing on $(-\infty, -2) \cup (2, \infty)$, concave up

on $\left(-\sqrt{6}, -2\right) \cup \left(\sqrt{6}, \infty\right)$, concave down on

$\left(-\infty, -\sqrt{6}\right) \cup \left(2, \sqrt{6}\right)$, $x = \pm\sqrt{6}$ are inflection

points.

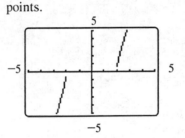

13. $f(x) = x + \dfrac{4}{x}$

$f'(x) = 1 - \dfrac{4}{x^2}$

$f''(x) = \dfrac{8}{x^3}$

$f'(x)\begin{cases} > 0 \text{ on } (-\infty, -2) \cup (2, \infty) \\ < 0 \text{ on } (-2, 0) \cup (0, 2) \end{cases}$

$f''(x)\begin{cases} > 0 \text{ on } (0, \infty) \\ < 0 \text{ on } (-\infty, 0) \end{cases}$

f is decreasing on $(-2, 0) \cup (0, 2)$, increasing on

$(-\infty, -2) \cup (2, \infty)$, concave up on $(0, \infty)$,

concave down on $(-\infty, 0)$, $x = -2$ local max,

$x = 2$ local min.

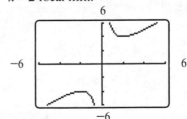

15. $f(x) = \sin x - \cos x$
$f'(x) = \cos x + \sin x$
$f''(x) = -\sin x + \cos x$

$f'(x) \begin{cases} > 0 ... \left(-\dfrac{9\pi}{4}, -\dfrac{5\pi}{4}\right) \cup \left(-\dfrac{\pi}{4}, \dfrac{3\pi}{4}\right)... \\[4mm] < 0 ... \left(-\dfrac{5\pi}{4}, -\dfrac{\pi}{4}\right) \cup \left(\dfrac{3\pi}{4}, \dfrac{7\pi}{4}\right)... \end{cases}$

$f''(x) \begin{cases} < 0 ... \left(\dfrac{\pi}{4}, \dfrac{5\pi}{4}\right) \cup \left(\dfrac{9\pi}{4}, \dfrac{13\pi}{4}\right)... \\[4mm] > 0 ... \left(-\dfrac{3\pi}{4}, -\dfrac{\pi}{4}\right) \cup \left(\dfrac{5\pi}{4}, \dfrac{9\pi}{4}\right)... \end{cases}$

f is increasing on $\left(-\dfrac{9\pi}{4}, -\dfrac{5\pi}{4}\right) \cup \left(-\dfrac{\pi}{4}, \dfrac{3\pi}{4}\right)$,

decreasing on $\left(-\dfrac{5\pi}{4}, -\dfrac{\pi}{4}\right) \cup \left(\dfrac{3\pi}{4}, \dfrac{7\pi}{4}\right)$,

concave down on $\left(\dfrac{\pi}{4}, \dfrac{5\pi}{4}\right) \cup \left(\dfrac{9\pi}{4}, \dfrac{13\pi}{4}\right)$,

concave up on $\left(-\dfrac{3\pi}{4}, \dfrac{\pi}{4}\right) \cup \left(\dfrac{5\pi}{4}, \dfrac{9\pi}{4}\right)$, local

max at $-\dfrac{5\pi}{4}, \dfrac{3\pi}{4}$, local min at $-\dfrac{\pi}{4}, \dfrac{7\pi}{4}$,

inflection points at $\dfrac{\pi}{4}, \dfrac{5\pi}{4}$, etc.

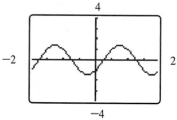

17. $f(x) = e^{-x^2/4}$

$f'(x) = e^{-x^2/4}\left(-\dfrac{1}{2}x\right) = -\dfrac{1}{2}xe^{-x^2/4}$

$f''(x) = -\dfrac{1}{2}e^{-x^2/4} + \dfrac{1}{4}x^2 e^{-x^2/4}$

$f'(x) \begin{cases} < 0 \text{ on } (0, \infty) \\ > 0 \text{ on } (-\infty, 0) \end{cases}$

$f''(x) \begin{cases} > 0 \text{ on } \left(-\infty, -\sqrt{2}\right) \cup \left(\sqrt{2}, \infty\right) \\ < 0 \text{ on } \left(-\sqrt{2}, \sqrt{2}\right) \end{cases}$

f is decreasing on $(0, \infty)$, increasing on $(-\infty, 0)$,

concave up on $\left(-\infty, -\sqrt{2}\right) \cup \left(\sqrt{2}, \infty\right)$, concave

down on $\left(-\sqrt{2}, \sqrt{2}\right)$, local max at

$x = 0$, $x = \pm\sqrt{2}$ inflection points.

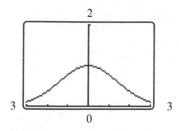

19. $f(x) = x \ln x$

$f'(x) = \ln x + x \cdot \dfrac{1}{x} = \ln x + 1$

$f''(x) = 1/x$

$f'(x) \begin{cases} > 0 \text{ on } \left(\dfrac{1}{e}, \infty\right) \\[3mm] < 0 \text{ on } \left(0, \dfrac{1}{e}\right) \end{cases}$

$f''(x) > 0$ on $(0, \infty)$

f is increasing on $\left(\dfrac{1}{e}, \infty\right)$, decreasing on $\left(0, \dfrac{1}{e}\right)$,

concave up on $(0, \infty)$, local min at $x = \dfrac{1}{e}$.

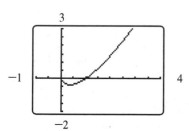

21. $f(x) = \sqrt{x^2 + 1}$

$f'(x) = \dfrac{2x}{2\sqrt{x^2+1}} = \dfrac{x}{\sqrt{x^2+1}}$

$f''(x) = \dfrac{\sqrt{x^2+1} - \dfrac{x^2}{\sqrt{x^2+1}}}{x^2+1} = \dfrac{1}{(x^2+1)^{3/2}}$

$f'(x) \begin{cases} > 0 \text{ on } (0, \infty) \\ < 0 \text{ on } (-\infty, 0) \end{cases}$

$f''(x) > 0$ on $(-\infty, \infty)$

f is increasing on $(0, \infty)$, decreases on $(-\infty, 0)$,

concave up on $(-\infty, \infty)$, $x = 0$ is local min.

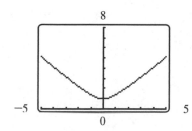

23. $f(x) = \dfrac{x^2+1}{3x^2-1}$

$f'(x) = \dfrac{2x(3x^2-1)-(x^2+1)(6x)}{(3x^2-1)^2} = \dfrac{-8x}{(3x^2-1)^2}$

$f''(x) = \dfrac{-8(3x^2-1)^2+8x\cdot 2(3x^2-1)\cdot 6x}{(3x^2-1)^4}$

$\qquad = \dfrac{72x^2+8}{(3x^2-1)^4}$

$f'(x) = \begin{cases} >0 \text{ on } \left(-\infty,\, -\sqrt{\tfrac{1}{3}}\right) \cup \left(-\sqrt{\tfrac{1}{3}},\, 0\right) \\[2ex] <0 \text{ on } \left(0,\, \sqrt{\tfrac{1}{3}}\right) \cup \left(\sqrt{\tfrac{1}{3}},\, \infty\right) \end{cases}$

$f''(x) = \begin{cases} >0 \text{ on } \left(-\infty,\, -\sqrt{\tfrac{1}{3}}\right) \cup \left(\sqrt{\tfrac{1}{3}},\, \infty\right) \\[2ex] <0 \text{ on } \left(-\sqrt{\tfrac{1}{3}},\, \sqrt{\tfrac{1}{3}}\right) \end{cases}$

f increasing on $\left(-\infty,\, -\sqrt{\tfrac{1}{3}}\right) \cup \left(-\sqrt{\tfrac{1}{3}},\, 0\right)$,

decreasing on $\left(0,\, \sqrt{\tfrac{1}{3}}\right) \cup \left(\sqrt{\tfrac{1}{3}},\, \infty\right)$, concave up

on $\left(-\infty,\, -\sqrt{\tfrac{1}{3}}\right) \cup \left(\sqrt{\tfrac{1}{3}},\, \infty\right)$, concave down on

$\left(-\sqrt{\tfrac{1}{3}},\, \sqrt{\tfrac{1}{3}}\right)$, local max at $x=0$.

f is undefined at $x = \pm\sqrt{\tfrac{1}{3}}$

$\displaystyle \lim_{x\to\sqrt{\frac{1}{3}}^+} \frac{x^2+1}{3x^2-1} = \infty, \text{ and } \lim_{x\to\sqrt{\frac{1}{3}}^-} \frac{x^2+1}{3x^2-1} = -\infty$

Since f is even, $\displaystyle \lim_{x\to-\sqrt{\frac{1}{3}}^-} f(x) = \infty$

and $\displaystyle \lim_{x\to-\sqrt{\frac{1}{3}}^+} f(x) = -\infty$.

So f has vertical asymptotes at $x = \pm\sqrt{\tfrac{1}{3}}$.

$\displaystyle \lim_{x\to\infty} \frac{x^2+1}{3x^2-1} = \lim_{x\to-\infty} \frac{x^2+1}{3x^2-1} = \frac{1}{3}$

So f has a horizontal asymptote at $y = \dfrac{1}{3}$.

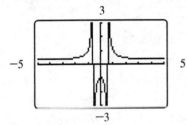

25. $f(x) = \dfrac{2x^2}{x^3+1}$

$f'(x) = \dfrac{4x(x^3+1)-2x^2(3x^2)}{(x^3+1)^2} = \dfrac{4x-2x^4}{(x^3+1)^2}$

$f''(x) = \dfrac{(4-8x^3)(x^3+1)^2-(4x-2x^4)\cdot 2(x^3+1)\cdot 3x^2}{(x^3+1)^4} = \dfrac{4x^6-28x^3+4}{(x^3+1)^3}$

$f'(x) = \begin{cases} >0 \text{ on } \left(0,\, \sqrt[3]{2}\right) \\[2ex] <0 \text{ on } (-\infty,\, -1) \cup (-1,\, 0) \cup \left(\sqrt[3]{2},\, \infty\right) \end{cases}$

$f''(x) = \begin{cases} >0 \text{ on } (-1,\, .5264) \cup (1.8995,\, \infty) \\[1ex] <0 \text{ on } (-\infty,\, -1) \cup (.5264,\, 1.8995) \end{cases}$

f increasing on $\left(0,\, \sqrt[3]{2}\right)$, decreasing on $(-\infty,\, -1) \cup (-1,\, 0) \cup \left(\sqrt[3]{2},\, \infty\right)$, concave up on

$(-1,\, .5264) \cup (1.8995,\, \infty)$, concave down on $(-\infty,\, -1) \cup (.5264, 1.8995)$, local min at $x=0$, max at $x = \sqrt[3]{2}$,

inflection points at $x = .5264,\ 1.8995$.
f is undefined at $x = -1$

$$\lim_{x\to -1^+}\frac{\overset{+}{2x^2}}{\underset{+}{x^3+1}}=\infty, \text{ and } \lim_{x\to -1^-}\frac{\overset{+}{2x^2}}{\underset{-}{x^3+1}}=-\infty \text{ So } f \text{ has a vertical asymptote at } x=-1.$$

$$\lim_{x\to \infty}\frac{2x^2}{x^3+1}=\lim_{x\to -\infty}\frac{2x^2}{x^3+1}=0 \text{ So } f \text{ has a horizontal asymptote at } y=0.$$

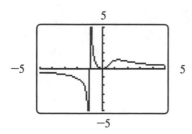

27. $f(x)=\dfrac{4x}{x^2-x+1}$

$$f'(x)=\frac{4(x^2-x+1)-4x(2x-1)}{(x^2-x+1)^2}=\frac{4-4x^2}{(x^2-x+1)^2}$$

$$f''(x)=\frac{-8x(x^2-x+1)^2-(4-4x^2)\cdot 2(x^2-x+1)(2x-1)}{(x^2-x+1)^4}=\frac{8x^3-24x+8}{(x^2-x+1)^3}$$

$$f'(x)=\begin{cases}>0 & (-1,1)\\ <0 & (-\infty,-1)\cup(1,\infty)\end{cases}$$

$$f''(x)=\begin{cases}>0 & (-1.879,0.347)\cup(1.532,\infty)\\ <0 & (-\infty,-1.879)\cup(0.347,1.532)\end{cases}$$

f increasing on $(-1, 1)$, decreasing on $(-\infty,-1)\cup(1,\infty)$, concave up on $(-1.879, 0.347)\cup(1.532,\infty)$, concave down on $(-\infty,-1.879)\cup(0.347,1.532)$, local min at $x=-1$, local max at $x=1$ inflection points at $x=-1.879$, 0.347, 1.532.

$$\lim_{x\to\infty}\frac{4x}{x^2-x+1}=\lim_{x\to -\infty}\frac{4x}{x^2-x+1}=0$$

So f has a horizontal asymptote at $y=0$.

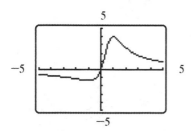

29. $f(x)=\dfrac{x-1}{x^2-x-2}=\dfrac{x-1}{(x+1)(x-2)}$

$$f'(x)=\frac{-(x^2-2x+3)}{(x+1)^2(x-2)^2}$$

$$f''(x)=\frac{2(x^3-3x^2+9x-5)}{(x+1)^3(x-2)^3}$$

The numerator of f' is always negative, so sign changes are determined by the denominator. The numerator of f'' has one real root at $x\approx 0.672520$.
f is decreasing on $(-\infty,-1)\cup(-1,2)\cup(2,\infty)$.
f is concave up on $(-1, 0.67252)\cup(2,\infty)$ and concave down on $(-\infty,-1)\cup(0.67252,2)$.

There is a horizontal asymptote at $y = 0$
and there are vertical asymptotes at $x = -1$
and $x = 2$.

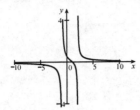

31. $f(x) = x + \sin x$
$f'(x) = 1 + \cos x$
$f''(x) = -\sin x$
f is increasing on $(-\infty, \infty)$.
There are inflection points at $n\pi$.
f is concave up between each odd multiple of π
and the following even multiple of π.
f is concave down between each even multiple of π
and the following odd multiple of π.

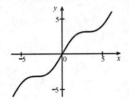

33. $f(x) = x^5 - 5x$
$f'(x) = 5x^4 - 5$
$f''(x) = 20x^3$
$f'(x) \begin{cases} > 0 \text{ on } (-\infty, -1) \cup (1, \infty) \\ < 0 \text{ on } (-1, 1) \end{cases}$
$f''(x) \begin{cases} > 0 \text{ on } (0, \infty) \\ < 0 \text{ on } (-\infty, 0) \end{cases}$

f increasing on $(-\infty, -1) \cup (1, \infty)$, decreasing on
$(-1, 1)$, concave up on $(0, \infty)$, concave down on
$(-\infty, 0)$, local max at $x = -1$, local min at
$x = 1$, inflection point at $x = 0$.

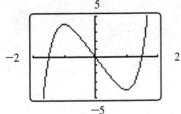

35. $f(x) = \dfrac{2x}{\sqrt{x^2 + 2}}$

$f'(x) = \dfrac{4}{(x^2 + 2)^{3/2}}$

$f''(x) = \dfrac{-12x}{(x^2 + 2)^{5/2}}$

$f'(x) > 0$ on $(-\infty, \infty)$

$f''(x) \begin{cases} > 0 \text{ on } (-\infty, 0) \\ < 0 \text{ on } (0, \infty) \end{cases}$

f increasing on $(-\infty, \infty)$, concave up on $(-\infty, 0)$,
concave down on $(0, \infty)$, $x = 0$ inflection point.'

$$\lim_{x \to \infty} \frac{2x}{\sqrt{x^2 + 2}} = \lim_{x \to \infty} \frac{x \cdot 2}{\sqrt{x^2} \cdot \sqrt{1 + \frac{2}{x^2}}}$$

$$= \lim_{x \to \infty} \frac{x \cdot 2}{x \cdot \sqrt{1 + \frac{2}{x^2}}}$$

$$= \lim_{x \to \infty} \frac{2}{x \cdot \sqrt{1 + \frac{2}{x^2}}}$$

$$= 2$$

$$\lim_{x \to -\infty} \frac{2x}{\sqrt{x^2 + 2}} = \lim_{x \to -\infty} \frac{x \cdot 2}{\sqrt{x^2} \cdot \sqrt{1 + \frac{2}{x^2}}}$$

$$= \lim_{x \to -\infty} \frac{x \cdot 2}{(-x) \cdot \sqrt{1 + \frac{2}{x^2}}}$$

$$= \lim_{x \to -\infty} \frac{-2}{x \cdot \sqrt{1 + \frac{2}{x^2}}}$$

$$= -2$$

So f has horizontal asymptotes at $y = 2$ and
$y = -2$.

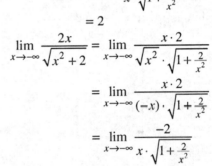
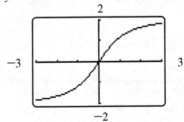

37. $f(x) = x^{1/5}(x+1) = x^{6/5} + x^{1/5}$

$f'(x) = \dfrac{6}{5}x^{1/5} + \dfrac{1}{5}x^{-4/5}$

$f''(x) = \dfrac{6}{25}x^{-4/5} - \dfrac{4}{25}x^{-9/5}$

$f'(x) \begin{cases} >0 \text{ on } \left(-\dfrac{1}{6}, 0\right) \cup (0, \infty) \\ <0 \text{ on } \left(-\infty, -\dfrac{1}{6}\right) \end{cases}$

$f''(x) \begin{cases} >0 \text{ on } (-\infty, 0) \cup \left(\dfrac{2}{3}, \infty\right) \\ <0 \text{ on } \left(\dfrac{2}{3}, \infty\right) \cup \left(0, \dfrac{2}{3}\right) \end{cases}$

f increasing on $\left(-\dfrac{1}{6}, \infty\right)$, decreasing on

$\left(-\infty, -\dfrac{1}{6}\right)$, concave up on $(-\infty, 0) \cup \left(\dfrac{2}{3}, \infty\right)$,

concave down on $\left(0, \dfrac{2}{3}\right)$, local min at $x = -\dfrac{1}{6}$,

inflection points at $x = \dfrac{2}{3}$ and $x = 0$.

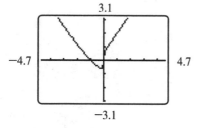

39. $f(x) = \sin x - \dfrac{1}{2}\sin 2x$

$f'(x) = \cos x - \cos 2x$

$f''(x) = -\sin x + 2\sin 2x$

$f'(x) \begin{cases} >0 \text{ on } \left(0, \dfrac{2\pi}{3}\right) \cup \left(\dfrac{4\pi}{3}, 2\pi\right) \\ <0 \text{ on } \left(\dfrac{2\pi}{3}, \dfrac{4\pi}{3}\right) \end{cases}$

$f''(x) \begin{cases} >0 \text{ on } (0, 1.318) \cup (\pi, 2\pi - 1.318) \\ <0 \text{ on } (1.318, \pi) \cup (2\pi - 1.318, 2\pi) \end{cases}$

f increasing on $\left(0, \dfrac{2\pi}{3}\right) \cup \left(\dfrac{4\pi}{3}, 2\pi\right)$, etc.,

decreasing on $\left(\dfrac{2\pi}{3}, \dfrac{4\pi}{3}\right)$, etc., concave up on

$(0, 1.318) \cup (\pi, 2\pi - 1.318)$, concave down on

$(1.318, \pi) \cup (2\pi - 1.318, 2\pi)$ etc., local max at

$x = \dfrac{2\pi}{3}$, local min at $\dfrac{4\pi}{3}$, etc., inflection points at

$x = 1.318$, π, $2\pi - 1.318$, etc.

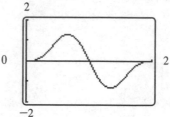

41. $f(x) = e^{-2/x}$

$f'(x) = e^{-2/x}\left(\dfrac{2}{x^2}\right) = \dfrac{2}{x^2}e^{-2/x}$

$f''(x) = \dfrac{-4}{x^3}e^{-2/x} + \dfrac{2}{x^2}e^{-2/x}\left(\dfrac{2}{x^2}\right)$

$\qquad = \dfrac{4}{x^4}e^{-2/x} - \dfrac{4}{x^3}e^{-2/x}$

$f'(x) > 0$ on $(-\infty, 0) \cup (0, \infty)$

$f''(x) \begin{cases} >0 \text{ on } (-\infty, 0) \cup (0, 1) \\ <0 \text{ on } (1, \infty) \end{cases}$

f increasing on $(-\infty, 0) \cup (0, \infty)$, concave up on

$(-\infty, 0) \cup (0, 1)$, concave down on $(1, \infty)$,

inflection point at $x = 1$.

f is undefined at $x = 0$.

$\displaystyle\lim_{x \to 0^+} e^{-2/x} = \lim_{x \to 0^+} \dfrac{1}{e^{2/x}} = 0$, and

$\displaystyle\lim_{x \to 0^-} e^{-2/x} = \infty$

So f has a vertical asymptote at $x = 0$.

$\displaystyle\lim_{x \to \infty} e^{-2/x} = \lim_{x \to -\infty} e^{-2/x} = 1$

So f has a horizontal asymptote at $y = 1$.

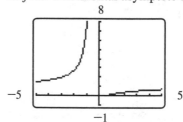

43. $f(x) = xe^{-x^2/2}$

$f'(x) = e^{-x^2/2} + xe^{-x^2/2}(-x) = e^{-x^2/2}(1 - x^2)$

$f''(x) = -xe^{-x^2/2}(1 - x^2) + e^{-x^2/2}(-2x)$

$\qquad = xe^{-x^2/2}(x^2 - 3)$

$f'(x)\begin{cases} > 0 \text{ on } (-1, 1) \\ < 0 \text{ on } (-\infty, -1) \cup (1, \infty) \end{cases}$

$f''(x)\begin{cases} > 0 \text{ on } \left(-\sqrt{3}, 0\right) \cup \left(\sqrt{3}, \infty\right) \\ < 0 \text{ on } \left(-\infty, -\sqrt{3}\right) \cup \left(0, \sqrt{3}\right) \end{cases}$

f increasing on $(-1, 1)$, decreasing on $(-\infty, -1) \cup (1, \infty)$, concave up on

$\left(-\sqrt{3}, 0\right) \cup \left(\sqrt{3}, \infty\right)$, concave down on

$\left(-\infty, -\sqrt{3}\right) \cup \left(0, \sqrt{3}\right)$, local min at $x = -1$, local

max at $x = 1$, inflection points at $x = 0, \pm\sqrt{3}$.

Note: for all x, $\dfrac{x^2}{2} + 1 \le e^{x^2/2}$. So

$0 < e^{-x^2/2} \le \dfrac{2}{x^2 + 2}$. Hence for $x > 0$,

$0 < xe^{-x^2/2} \le \dfrac{2x}{x^2 + 2}$ Since $\displaystyle\lim_{x \to \infty} \dfrac{2x}{x^2 + 2} = 0$ it

must be that $\displaystyle\lim_{x \to \infty} xe^{-x^2/2} = 0$ too.

When $x < 0$, we have $\dfrac{2x}{x^2 + 2} \le xe^{-x^2/2} < 0$ and

hence $\displaystyle\lim_{x \to -\infty} xe^{-x^2/2} = 0$. So f has a horizontal

asymptote at $y = 0$.

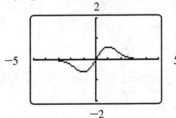

45. $f(x) = x^4 + cx^2$

$f'(x) = 4x^3 + 2cx$

$f''(x) = 12x^2 + 2c$

$c = 0$: 1 extremum, 0 inflection points

$c < 0$: 3 extrema, 2 inflection points

$c > 0$: 1 extremum, 0 inflection points

$c \to -\infty$: the graph widens and lowers

$c \to +\infty$: the graph narrows

47. $f(x) = \dfrac{x^2}{x^2 + c^2}$

$f'(x) = \dfrac{2c^2 x}{(x^2 + c^2)^2}$

$f''(x) = \dfrac{2c^4 - 6c^2 x^2}{(x^2 + c^2)^3}$

$c = 0$: $f(x) = 1$

$c < 0, c > 0$: horizontal asymptote at $y = 1$, local min at $x = 0$, 2 inflection points

$c \to -\infty, c \to +\infty$: the graph widens.

For $c = 0$, f is undefined at $x = 0$.

49. $f(x) = \sin(cx)$

$f'(x) = c \cos cx$

$f''(x) = -c^2 \sin cx$

$c = 0$: $f(x) = 0$

$|c|$ is the frequency of oscillation.

The graph for negative c is a reflection in the y-axis of the graph for the positive c with the same absolute value.

51. $f(x) = xe^{-bx}$

$f'(x) = e^{-bx} - bxe^{-bx}$

f has local max at $x = \dfrac{1}{b}$. For time since

conception, $\dfrac{1}{b}$ represents the most common

gestation time. For survival time, $\dfrac{1}{b}$ represents

the most common life span.

53. Let $p(x) = 2x$, $q(x) = x^2 + 1$. Then

$\dfrac{p(x)}{q(x)} = \dfrac{2x}{x^2 + 1}$ has no vertical asymptotes.

Let $p(x) = x^2 + 1$, $q(x) = 2x$. Then

$\dfrac{p(x)}{q(x)} = \dfrac{x^2 + 1}{2x}$ has no horizontal asymptotes.

55. $f(x) = \dfrac{3x^2 - 1}{x} = 3x - \dfrac{1}{x}$

$y = 3x$ is slant asymptote.

57. $f(x) = \dfrac{x^3 - 2x^2 + 1}{x^2} = x - 2 + \dfrac{1}{x^2}$

$y = x - 2$ is slant asymptote.

59. $f(x) = \dfrac{x^4}{x^3 + 1} = x - \dfrac{x}{x^3 + 1}$

$y = x$ is slant asymptote.

61. asymptotes $x = 1,\ x = 2,\ y = 3$

$f(x) = \dfrac{3x^2}{(x-1)(x-2)}$

63. asymptotes $x = -1,\ x = 1,\ y = -2,\ y = 2$

$f(x) = \dfrac{2x}{\sqrt{(x-1)(x+1)}}$

Section 3.7

5. $f(x) = x^2 + 1$ has a minimum at $x = 0$, while $\sin(x^2 + 1)$ has minima where

$x^2 + 1 = \dfrac{3\pi}{2} + 2np\pi.$

7. $A = xy = 1800$

$y = \dfrac{1800}{x}$

$P = 2x + y = 2x + \dfrac{1800}{x}$

$P' = 2 - \dfrac{1800}{x^2} = 0$

$2x^2 = 1800$

$x = 30$

$P'(x) \begin{cases} > 0 \text{ for } x > 30 \\ < 0 \text{ for } 0 < x < 30 \end{cases}$

So $x = 30$ is min.

$y = \dfrac{1800}{x} = \dfrac{1800}{30} = 60$

So the dimensions are $30' \times 60'$ and the minimum perimeter is 120 ft.

9. $P = 2x + 3y = 120$

$3y = 120 - 2x$

$y = 40 - \dfrac{2}{3}x$

$A = x \cdot y$

$A(x) = x\left(40 - \dfrac{2}{3}x\right)$

$A'(x) = 1\left(40 - \dfrac{2}{3}x\right) + x\left(-\dfrac{2}{3}\right)$

$= 40 - \dfrac{4}{3}x = 0$

$40 = \dfrac{4}{3}x$

$x = 30$

$A'(x) > 0$ for $0 < x < 30$

$A'(x) < 0$ for $x > 30$

So $x = 30$ is max, $y = 40 - \dfrac{2}{3} \cdot 30 = 20$

So the dimensions are $20' \times 30'$.

11. $A = xy$

$P = 2x + 2y$

$2y = P - 2x$

$y = \dfrac{P}{2} - x$

$A(x) = x\left(\dfrac{P}{2} - x\right)$

$A'(x) = 1 \cdot \left(\dfrac{P}{2} - x\right) + x(-1) = \dfrac{P}{2} - 2x = 0$

$P = 4x$

$x = \dfrac{P}{4}$

$A'(x) \begin{cases} > 0 \text{ for } 0 < x < \dfrac{P}{4} \\ < 0 \text{ for } x > \dfrac{P}{4} \end{cases}$

So $x = \dfrac{P}{4}$ is max, $y = \dfrac{P}{2} - x = \dfrac{P}{2} - \dfrac{P}{4} = \dfrac{P}{4}$

So the dimensions are $\dfrac{P}{4} \times \dfrac{P}{4}$. Thus we have a square.

13. $d = \sqrt{(x-0)^2 + (y-1)^2}$

$y = x^2$

$d = \sqrt{x^2 + (x^2 - 1)^2}$

$f(x) = [d(x)]^2 = x^2 + (x^2 - 1)^2, -1 \le x \le 1$

$f'(x) = 2x + 2(x^2 - 1) \cdot 2x$

$\qquad = 2x(1 + 2x^2 - 2)$

$\qquad = 2x(2x^2 - 1) = 0$

$x = 0, \pm\sqrt{\dfrac{1}{2}}$

so $f(0) = 1$

$f(1) = 1 \qquad f(-1) = 1$

$f\left(\sqrt{\dfrac{1}{2}}\right) = \dfrac{3}{4} \qquad f\left(-\sqrt{\dfrac{1}{2}}\right) = \dfrac{3}{4}$

Thus $x = \pm\sqrt{\dfrac{1}{2}}$ are min, and the points on $y = x^2$

closest to (0, 1) are $\left(\sqrt{\dfrac{1}{2}}, \dfrac{1}{2}\right)$ and $\left(-\sqrt{\dfrac{1}{2}}, \dfrac{1}{2}\right)$.

15. $d = \sqrt{(x-0)^2 + (y-0)^2}$

$y = \cos x$

$d = \sqrt{x^2 + \cos^2 x}$

$f(x) = [d(x)]^2 = x^2 + \cos^2 x, \ 0 \le x \le \dfrac{\pi}{2}$

$f'(x) = 2x + 2\cos x(-\sin x)$

$\qquad = 2x - 2\cos x \sin x = 0$

$x = \cos x \sin x$

$x = 0$

$f(0) = 0^2 + \cos^2 0 = 1$

$f\left(\dfrac{\pi}{2}\right) = \left(\dfrac{\pi}{2}\right)^2 + \cos^2 \dfrac{\pi}{2} = \dfrac{\pi^2}{4}$

So $x = 0$ is min and the point on $y = \cos x$ closest to (0, 0) is (0, 1).

17. For (0, 1), $\left(\sqrt{\dfrac{1}{2}}, \dfrac{1}{2}\right)$ on $y = x^2$, we have $y' = 2x$, $y'\left(\sqrt{\dfrac{1}{2}}\right) = 2 \cdot \sqrt{\dfrac{1}{2}} = \sqrt{2}$

and $m = \dfrac{\dfrac{1}{2} - 1}{\sqrt{\dfrac{1}{2}} - 0} = \dfrac{-1}{\sqrt{2}}$

For (0, 1), $\left(-\sqrt{\dfrac{1}{2}}, \dfrac{1}{2}\right)$ on $y = x^2$, we have $y'\left(-\sqrt{\dfrac{1}{2}}\right) = 2\left(-\sqrt{\dfrac{1}{2}}\right) = -\sqrt{2}$

and $m = \dfrac{\dfrac{1}{2} - 1}{-\sqrt{\dfrac{1}{2}} - 0} = \dfrac{1}{\sqrt{2}}$.

For (3, 4), (2.06, 4.2436) on $y = x^2$, we have

$y'(2.06) = 2(2.06) = 4.12$ and $m = \dfrac{4.2436 - 4}{2.06 - 3} = -0.2591 \approx -\dfrac{1}{4.12}$

19. $V = l \cdot w \cdot h$

$V(x) = (10 - 2x)(6 - 2x) \cdot x, \ 0 \le x \le 3$

$V'(x) = -2(6 - 2x) \cdot x + (10 - 2x)(-2) \cdot x + (10 - 2x)(6 - 2x)$

$\qquad = 60 - 64x + 12x^2 = 4(3x^2 - 16x + 15)$

$\qquad = 0$

$x = \dfrac{16 \pm \sqrt{(-16)^2 - 4 \cdot 3 \cdot 15}}{6} = \dfrac{8}{3} \pm \dfrac{\sqrt{19}}{3}$

$x = \dfrac{8}{3} + \dfrac{\sqrt{19}}{3} > 3.$

$V'(x) > 0$ for $x < \dfrac{8}{3} - \dfrac{\sqrt{19}}{3}$

$V'(x) < 0$ for $x > \dfrac{8}{3} - \dfrac{\sqrt{19}}{3}$

So $x = \dfrac{8}{3} - \dfrac{\sqrt{19}}{3}$ is a max

$x = \dfrac{8}{3} - \dfrac{\sqrt{19}}{3} \approx 1.2137$

21. $f(x) = \sqrt{3^2 + (5-x)^2} + \sqrt{4^2 + x^2},\ 0 \le x \le 5$

$f'(x) = \dfrac{1}{2}(9 + (5-x)^2)^{-1/2}(2)(5-x)(-1) + \dfrac{1}{2}(16 + x^2)^{-1/2}(2x)$

$\qquad = \dfrac{x-5}{\sqrt{9 + (5-x)^2}} + \dfrac{x}{\sqrt{16 + x^2}}$

$\qquad = 0$

$x = \dfrac{20}{7} \approx 2.857$

$f(0) = 4 + \sqrt{34} \approx 9.831$

$f\left(\dfrac{20}{7}\right) = \sqrt{74} \approx 8.602$

$f(5) = 3 + \sqrt{41} \approx 9.403$

So $x = \dfrac{20}{7}$ is min.

The water line should be $\dfrac{20}{7}$ miles west of the second development, $5 - \dfrac{20}{7} = \dfrac{15}{7}$ miles east of the first development.

23. $C(x) = 5\sqrt{16 + x^2} + 2\sqrt{36 + (8-x)^2},\ 0 \le x \le 8$

$C(x) = 5\sqrt{16 + x^2} + 2\sqrt{100 - 16x + x^2}$

$C'(x) = 5\left(\dfrac{1}{2}\right)(16 + x^2)^{-1/2} \cdot 2x + 2\left(\dfrac{1}{2}\right)(100 - 16x + x^2)^{-1/2}(2x - 16)$

$\qquad = \dfrac{5x}{\sqrt{16 + x^2}} + \dfrac{2x - 16}{\sqrt{100 - 16x + x^2}}$

$\qquad = 0$

$x \approx 1.2529$

$C(0) = 40$

$C(1.2529) \approx 39.0162$

$C(8) \approx 56.7214$

The highway should emerge from the marsh 1.2529 miles east of the bridge. If we build a straight line to the interchange, we have $x = (3.2)$. Since $C(3.2) - C(1.2529) \approx 1.963$, we save \$1.963 million.

25. $C(x) = 5\sqrt{16 + x^2} + 3\sqrt{36 + (8 - x)^2}$, $0 \le x \le 8$

$C'(x) = \dfrac{5x}{\sqrt{16 + x^2}} + \dfrac{3x - 24}{\sqrt{100 - 16x + x^2}} = 0$

$x \approx 1.8941$

$C(0) = 50$

$C(1.8941) \approx 47.8104$

$C(8) \approx 62.7214$

The highway should emerge from the marsh 1.8941 miles east of the bridge.
So if we must use the path from exercise 21, the extra cost is

$C(1.2529) - C(1.8941) = 48.0452 - 47.8104$

$\qquad\qquad\qquad\qquad\qquad = 0.2348$

or about \$234.8 thousand.

27. $T(x) = \dfrac{\sqrt{1 + x^2}}{v_1} + \dfrac{\sqrt{1 + (2 - x)^2}}{v_2}$

$T'(x) = \dfrac{1}{v_1} \cdot \dfrac{1}{2}(1 + x^2)^{-1/2} \cdot 2x + \dfrac{1}{v_2}(1 + (2 - x)^2)^{-1/2} \cdot (2 - x)(-1)$

$\qquad = \dfrac{x}{v_1\sqrt{1 + x^2}} + \dfrac{x - 2}{v_2\sqrt{1 + (2 - x)^2}}$

Note that $T'(x) = \dfrac{1}{v_1} \cdot \dfrac{x}{\sqrt{1 + x^2}} - \dfrac{1}{v_2} \cdot \dfrac{(2 - x)}{\sqrt{1 + (2 - x)^2}}$

$\qquad\qquad\qquad = \dfrac{1}{v_1}\sin\theta_1 - \dfrac{1}{v_2}\sin\theta_2$

When $T'(x) = 0$, we have $\dfrac{1}{v_1}\sin\theta_1 = \dfrac{1}{v_2}\sin\theta_2$

$\qquad\qquad\qquad\qquad\qquad \dfrac{\sin\theta_1}{\sin\theta_2} = \dfrac{v_1}{v_2}$

29. Cost $= 2(2\pi r^2) + 2\pi rh$

$12 \text{ fl oz} = 12 \text{ fl oz} \cdot \dfrac{1.80469 \text{ in}^3}{\text{fl oz}} = 21.65628 \text{ in}^3$

$\text{Vol} = \pi r^2 h$, $h = \dfrac{\text{Vol}}{\pi r^2} = \dfrac{21.65628}{\pi r^2}$, so

$\text{Cost} = 4\pi r^2 + 2\pi r\left(\dfrac{21.65628}{\pi r^2}\right)$ or $C(r) = 4\pi r^2 + 43.31256r^{-1}$

so $C'(r) = 8\pi r - 43.31256r^{-2} = \dfrac{8\pi r^3 - 43.31256}{r^2}$

$r = \sqrt[3]{\dfrac{43.31256}{8\pi}} = 1.1989''$ when $C'(r) = 0$.

$C'(r)\begin{cases} < 0 \text{ on } (0, 1.1989) \\ > 0 \text{ on } (1.1989, \infty) \end{cases}$

Thus $r = 1.1989$ minimizes the cost.

$h = \dfrac{21.65628}{\pi(1.1989)^2} = 4.7957''$

31. $V(r) = cr^2(r_0 - r)$

$V'(r) = 2cr(r_0 - r) + cr^2(-1)$

$\qquad = 2crr_0 - 3cr^2$

$\qquad = cr(2r_0 - 3r)$

$\qquad = 0$

$r = \dfrac{2r_0}{3}$ when $V'(r) = 0$

$V'(r) \begin{cases} > 0 \text{ on } \left(0, \dfrac{2r_0}{3}\right) \\ < 0 \text{ on } \left(\dfrac{2r_0}{3}, \infty\right) \end{cases}$

Thus $r = \dfrac{2r_0}{3}$ maximizes the velocity.

$r = \dfrac{2r_0}{3} < r_0$, so the windpipe contracts.

33. $p(x) = \dfrac{V^2 x}{(R+x)^2}$

$p'(x) = \dfrac{V^2(R+x)^2 - V^2 x \cdot 2(R+x)}{(R+x)^4}$

$\qquad = \dfrac{V^2 R^2 - V^2 x^2}{(R+x)^4}$

$\qquad = 0$

$x = R$ when $p'(x) = 0$

$p'(x) \begin{cases} > 0 \text{ on } (0, R) \\ < 0 \text{ on } (R, \infty) \end{cases}$

Thus $x = R$ maximizes the power absorbed.

35. $\pi r + 4r + 2w = 8 + \pi$

$\qquad w = \dfrac{8 + \pi - r(\pi + 4)}{2}$

$A(r) = \dfrac{\pi r^2}{2} + 2rw$

$\qquad = \dfrac{\pi r^2}{2} + r(8 + \pi - r(\pi + 4))$

$\qquad = r^2\left(-4 - \dfrac{\pi}{2}\right) + r(8 + \pi)$

$A'(r) = -2r\left(4 + \dfrac{\pi}{2}\right) + (8 + \pi) = 0$

$r = 1, 2r = 2$ when $A'(r) = 0$

$A'(r) \begin{cases} > 0 \text{ on } (0, 1) \\ < 0 \text{ on } (1, \infty) \end{cases}$

Thus $r = 1$ maximizes the area.

so $w = \dfrac{8 + \pi - (\pi + 4)}{2} = 2$

So the dimensions of the rectangle are 2×2.

37. $l \times w = 92, \quad w = \dfrac{92}{l}$

$A(l) = (l + 4)(w + 2)$

$\qquad = (l + 4)\left(\dfrac{92}{l} + 2\right)$

$\qquad = 92 + \dfrac{368}{l} + 2l + 8$

$\qquad = 100 + 368l^{-1} + 2l$

$A'(l) = -368l^{-2} + 2$

$\qquad = \dfrac{2l^2 - 368}{l^2}$

$l = \sqrt{184} = 2\sqrt{46}$ when $A'(l) = 0$

$A'(l) \begin{cases} < 0 \text{ on } (0, 2\sqrt{46}) \\ > 0 \text{ on } (2\sqrt{46}, \infty) \end{cases}$

So $l = 2\sqrt{46}$ minimizes the total area.

When $l = 2\sqrt{46}$, $w = \dfrac{92}{2\sqrt{46}} = \sqrt{46}$.

For the minimum total area, the printed area has width $\sqrt{46}$ in. and length $2\sqrt{46}$ in., and the advertisement has overall width $\sqrt{46} + 2$ in. and overall length $2\sqrt{46} + 4$ in.

39. $I(A) = 2A^3 - 33A^2 + 108A - 310$

$I'(A) = 6A^2 - 66A + 108 = 6(A^2 - 11A + 18)$

$\qquad = 6(A - 9)(A - 2)$

$I''(A) = 12A - 66$

$I''(9) = 42$, 9 is a min

$I''(2) = -42$, 2 is a max.

The farmer should plant 2 acres.

41. $R = \dfrac{2v^2 \cos^2 \theta}{g}(\tan \theta - \tan \beta)$

$R'(\theta) = \dfrac{2v^2}{g}\left[2\cos\theta(-\sin\theta)(\tan\theta - \tan\beta) + \cos^2\theta \cdot \sec^2\theta\right]$

$\qquad = \dfrac{2v^2}{g}\left[-2\cos\theta\sin\theta \cdot \dfrac{\sin\theta}{\cos\theta} + 2\cos\theta\sin\theta\tan\beta + \cos^2\theta \cdot \dfrac{1}{\cos^2\theta}\right]$

$\qquad = \dfrac{2v^2}{g}\left[-2\sin^2\theta + \sin(2\theta)\tan\beta + 1\right]$

$\qquad = \dfrac{2v^2}{g}\left[-2\sin^2\theta + \sin(2\theta)\tan\beta + (\sin^2\theta + \cos^2\theta)\right]$

$\qquad = \dfrac{2v^2}{g}\left[\sin(2\theta)\tan\beta + (\cos^2\theta - \sin^2\theta)\right]$

$\qquad = \dfrac{2v^2}{g}\left[\sin(2\theta)\tan\beta + \cos(2\theta)\right]$

$R'(\theta) = 0$ when $\tan\beta = \dfrac{-\cos(2\theta)}{\sin(2\theta)} = -\cot(2\theta) = -\tan\left(\dfrac{\pi}{2} - 2\theta\right) = \tan\left(2\theta - \dfrac{\pi}{2}\right)$

Hence $\beta = 2\theta - \dfrac{\pi}{2}$, so $\theta = \dfrac{1}{2}\left(\beta + \dfrac{\pi}{2}\right) = \dfrac{\beta}{2} + \dfrac{\pi}{4} = \dfrac{\beta°}{2} + 45°$

a. $\beta = 10°,\ \theta = 50°$

b. $\beta = 0°,\ \theta = 45°$

c. $\beta = -10°,\ \theta = 40°$

43. $T = \dfrac{-1}{c}\ln\left(1 - c \cdot \dfrac{b-a}{v_0}\right)$

$b = 300,\ a = 0,\ v_0 = 125,\ c = 0.1$

$T = \dfrac{-1}{0.1}\ln\left(1 - 0.1 \cdot \dfrac{300-0}{125}\right) = 2.744$ sec

$T(x) = -10\ln(1 - 0.0008(300 - x)) - 10\ln(1 - 0.0008x) + 0.1$

$T'(x) = -10\left(\dfrac{0.0008}{0.76 + 0.0008x} - \dfrac{0.0008}{1 - 0.0008x}\right) = 0$

$0.0008(1 - 0.0008x) = 0.0008(0.76 + 0.0008x)$

$x = \dfrac{1 - 0.76}{0.0016} = 150$ ft when $T'(x) = 0$.

$T'(x)\begin{cases} < 0 \text{ on } (0, 150) \\ > 0 \text{ on } (150, 300) \end{cases}$

Hence $x = 150$ minimizes the total time.

$T(150) = -10\ln(1 - 0.0008(300 - 150)) - 10\ln(1 - 0.0008(150)) + 0.1$

$\qquad\qquad = 2.656$ sec.

So the relay is faster.

If the delay is 0.2 sec, the relay takes longer.

Section 3.8

5. $Q(t) = e^{-2t}(\cos 3t - 2\sin 3t)$ Coulombs

$Q'(t) = e^{-2t} \cdot (-2)(\cos 3t - 2\sin 3t) + e^{-2t}((-\sin 3t \cdot 3) - 2\cos 3t \cdot 3)$

$\qquad = e^{-2t}(-8\cos 3t + \sin 3t)$ amps

7. $Q(t) = e^{-3t}\cos 2t + 4\sin 3t$

as $t \to \infty$, $Q(t) \to 4\sin 3t$, so $e^{-3t}\cos 2t$ is called the transient term and $4\sin 3t$ is called the steady-state value.

$Q'(t) = e^{-3t} \cdot (-3)\cos 2t + e^{-3t}(-\sin 2t \cdot 2) + 4\cos 3t \cdot 3$

$\qquad = e^{-3t}(-3\cos 2t - 2\sin 2t) + 12\cos 3t$

The transient term is $e^{-3t}(-3\cos 2t - 2\sin 2t)$.

The steady-state value is $12\cos 3t$.

9. $x'(t) = 2x(t)[4 - x(t)] = 0$

$f(x) = 2x(4 - x)$

$f'(x) = 2(4 - x) + 2x(-1) = 8 - 4x = 4(2 - x) = 0$

$x = 2$, so $x(t) = 2$.

This is a maximum since $f(x)$ is a downward opening parabola.

11. $x'(t) = 2x(t)[4 - x(t)] = 0$

$x(t) = 0, x(t) = 4$ are critical numbers.

$x'(t)\begin{cases} > 0 \text{ for } 0 < x(t) < 4 \\ < 0 \text{ for } x(t) > 4 \end{cases}$

So $x(t) = 4$ is the maximum concentration.

$x'(t) = 0.5x(t)[5 - x(t)]$

$x(t) = 0,\ x(t) = 5$ are critical numbers.

$x'(t)\begin{cases} > 0 \text{ for } 0 < x(t) < 5 \\ < 0 \text{ for } x(t) > 5 \end{cases}$

So $x(t) = 5$ is the maximum concentration.

13. $y'(t) = c \cdot y(t)[K - y(t)]$

$y(t) = Kx(t)$

$y'(t) = Kx'(t)$

$Kx'(t) = c \cdot Kx(t)[K - Kx(t)]$

$\quad x'(t) = c \cdot Kx(t)[1 - x(t)]$

$\qquad\quad = rx(t)[1 - x(t)]$

$\quad r = cK$

15. $x'(t) = [a - x(t)][b - x(t)]$

for $x(t) = a$,

$x'(t) = [a - a][b - a] = 0$

So the concentration of product is staying the same.

If $a < b$ and $x(0) = 0$ then

$x'(t)\begin{cases} > 0 \text{ for } 0 < x < a \text{ and } x > b \\ < 0 \text{ for } a < x < b \end{cases}$

Thus $x(t) = a$ is a maximum.

17. $x(t) = \dfrac{a[1 - e^{-(b-a)t}]}{1 - \left(\frac{a}{b}\right)e^{-(b-a)t}},\ a < b$

$x(0) = \dfrac{a[1 - e^{-(b-a)\cdot 0}]}{1 - \left(\frac{a}{b}\right)e^{-(b-a)\cdot 0}} = \dfrac{a[1-1]}{1 - \left(\frac{a}{b}\right)} = 0$

$\displaystyle\lim_{t \to \infty} x(t) = \dfrac{a[1-0]}{1-0} = a$

19. $m(x) = 4x - \sin x$ dynes for $0 \le x \le 6$

$m'(x) = 4 - \cos x$, so the rod is less dense at the ends.

21. $m(x) = 4x$ dynes for $0 \le x \le 2$

$m'(x) = 4$, so the rod is homogeneous.

23. healthy

25. in danger

27. $C(x) = x^3 + 20x^2 + 90x + 15$

$C'(x) = 3x^2 + 40x + 90$

$C'(50) = 3(50)^2 + 40 \cdot 50 + 90 = 9590$

$C(50) = 50^3 + 20(50)^2 + 90 \cdot 50 + 15 = 179515$

$C(49) = 49^3 + 20(49)^2 + 90 \cdot 49 + 15 = 170094$

$C(50) - C(49) = 9421$

29. $C(x) = x^3 + 21x^2 + 110x + 20$

$C'(x) = 3x^2 + 42x + 110$

$C'(100) = 3(100)^2 + 42 \cdot 100 + 110 = 34310$

$C(100) = 100^3 + 21(100)^2 + 110 \cdot 100 + 20$
$\quad = 1221020$

$C(99) = 99^3 + 21(99)^2 + 110 \cdot 99 + 20 = 1187030$
$C(100) - C(99) = 33990$

31. $C(x) = x^3 - 30x^2 + 300x + 100$

$C'(x) = 3x^2 - 60x + 300$
$C''(x) = 6x - 60 = 0$

$x = 10$ is the inflection point because $C''(x)$ changes from negative to positive at this value. After this point, cost rises more sharply.

33. $f(t) = \dfrac{80}{1 + 3e^{-0.4t}} = 80(1 + 3e^{-0.4t})^{-1}$

$\lim_{t \to \infty} f(t) = \dfrac{80}{1+0} = 80$

$f'(t) = 80(-1(1 + 3e^{-0.4t})^{-2}[3e^{-0.4t}(-0.4)])$

$\quad = \dfrac{96e^{-0.4t}}{(1 + 3e^{-0.4t})^2} > 0 \text{ for } t > 0$

It is reasonable for a learning curve to increase and approach a limit.

35. $f(t) = \dfrac{180}{2 + 4e^{-0.2t}} = 180(2 + 4e^{-0.2t})^{-1}$

$f(3) = \dfrac{180}{2 + 4e^{-0.2 \cdot 3}} \approx 43$ is the score after

studying 3 hours.

$f'(t) = 180(-1)(2 + 4e^{-0.2t})^{-2}(4e^{-0.2t})(-0.2)$

$\quad = \dfrac{144e^{-0.2t}}{(2 + 4e^{-0.2t})^2}$

$f'(3) = \dfrac{144e^{-0.2 \cdot 3}}{(2 + 4e^{-0.2 \cdot 3})^2} \approx 4.49$

So the person would earn about $4\dfrac{1}{2}$ points by studying a fourth hour.

$f'(10) = \dfrac{144e^{-0.2 \cdot 10}}{(2 + 4e^{-0.2 \cdot 10})^2} \approx 3.017$

So the person would earn about 3 points by studying an eleventh hour.

37. $f(t) = \dfrac{a}{1 + 3e^{-bt}}$

for $a = 70$, $b = 0.2$,

$f(t) = \dfrac{70}{1 + 3e^{-0.2t}} = 70(1 + 3e^{-0.2t})^{-1}$

$f(2) = \dfrac{70}{1 + 3e^{-0.2 \cdot 2}} \approx 23$

$f'(t) = 70(-1)(1 + 3e^{-0.2t})^{-2}(3e^{-0.2t})(-0.2)$

$\quad = \dfrac{42e^{-0.2t}}{(1 + 3e^{-0.2t})^2}$

$f'(2) = \dfrac{42e^{0.2 \cdot 2}}{(1 + 3e^{-0.2 \cdot 2})^2} \approx 3.105$

So about 3% of the population will hear the rumor in the 3rd hour.

$\lim_{t \to \infty} f(t) = \dfrac{70}{1+0} = 70$ so 70% of the population

will eventually hear the rumor.

39. $f(x) = \dfrac{160x^{-0.4} + 90}{4x^{-0.4} + 15}$

$f'(x) = \dfrac{-64x^{-1.4}(4x^{-0.4} + 15) - (160x^{-0.4} + 90)(-1.6x^{-1.4})}{(4x^{-0.4} + 15)^2}$

$\quad = \dfrac{-816x^{-1.4}}{(4x^{-0.4} + 15)^2} < 0$

So $f(x)$ is decreasing. This shows that pupils shrink as light increases.

41. $C(x) = 0.01x^2 + 40x + 3600$

$C'(x) = 0.02x + 40$

$\overline{C}(x) = \dfrac{C(x)}{x} = 0.01x + 40 + \dfrac{3600}{x}$

$C'(100) = 42$

$\overline{C}(100) = 77$

so $C'(100) < \overline{C}(100)$

$\overline{C}(101) = 76.65 < \overline{C}(100)$

43. $\overline{C}'(x) = 0.01 - \dfrac{3600}{x^2} = 0$

so $x = 600$ is min and

$\overline{C}'(600) = 52$

$\overline{C}(600) = 52$

45. $P(x) = R(x) - C(x)$

$P'(x) = R'(x) - C'(x) = 0$

$R'(x) = C'(x)$

47. $p(f) = c + \ln\dfrac{f}{1-f},\ 0 < f < 1$

$p'(f) = \dfrac{1}{\frac{f}{1-f}}\left(\dfrac{1-f-f(-1)}{(1-f)^2}\right) = \dfrac{1}{f(1-f)}$

$\qquad = (f - f^2)^{-1}$

$p''(f) = -(f - f^2)^{-2}(1 - 2f) = \dfrac{2f - 1}{(f - f^2)^2} = 0$

$f = \dfrac{1}{2}.\ p''(f)\begin{cases} > 0 \text{ when } \dfrac{1}{2} < f < 1 \\ < 0 \text{ when } 0 < f < \dfrac{1}{2} \end{cases}$

so $f = \dfrac{1}{2}$ is a min.

As $f \to 1$, $p(f) = c + \ln\dfrac{f}{1-f} \to \infty$

Chapter 3 Review

1. See Example 1.1.

$f(x) = e^{3x},\ x_0 = 0$

$f'(x) = 3e^{3x}$

$L(x) = f(x_0) + f'(x_0)(x - x_0)$

$\qquad = f(0) + f'(0)(x - 0)$

$\qquad = e^{3 \cdot 0} + 3e^{3 \cdot 0}x$

$\qquad = 1 + 3x$

3. See Example 1.2.

$f(x) = \sqrt[3]{x} = x^{1/3},\ x_0 = 8$

$f'(x) = \dfrac{1}{3}x^{-2/3}$

$L(x) = f(x_0) + f'(x_0)(x - x_0)$

$\qquad = f(8) + f'(8)(x - 8)$

$\qquad = \sqrt[3]{8} + \dfrac{1}{3}(8)^{-2/3}(x - 8)$

$\qquad = 2 + \dfrac{1}{12}(x - 8)$

$L(7.96) = 2 + \dfrac{1}{12}(7.96 - 8) \approx 1.99666$

5. See Example 2.1.

$f(x) = x^3 + 5x - 1 = 0$

$f'(x) = 3x^2 + 5$

$x_0 = 0$

$x_{n+1} = x_n - \dfrac{f(x_n)}{f'(x_n)}$ for $n = 0, 1, 2, 3, \ldots$

n	x_n
1	.2
2	.198437
3	.198437

$x \approx .198437$

7. See Example 2.4.

$f(x) = x^3 + 2 = 0,\ x_0 = 1$

$f'(x) = 3x^2 - 3x$

$x_1 = x_0 - \dfrac{f(x_0)}{f'(x_0)} = 1 - \dfrac{1 - 3 + 2}{3 - 3} = 1 - \dfrac{0}{0}$

Newton's method fails because $f'(1) = 0$.

For Exercises 9-17, see the Examples in Sections 3.3, 3.4, and 3.5.

9. $f(x) = x^3 + 3x^2 - 9x$

$f'(x) = 3x^2 + 6x - 9$

$\qquad = 3(x^2 + 2 - 3)$

$\qquad = 3(x + 3)(x - 1)$

$f''(x) = 6x + 6 = 6(x + 1)$

$f'(x)\begin{cases} > 0 \text{ on } (-\infty, -3) \cup (1, \infty) \\ < 0 \text{ on } (-3, 1) \end{cases}$

$f''(x)\begin{cases} > 0 \text{ on } (-1, \infty) \\ < 0 \text{ on } (-\infty, -1) \end{cases}$

$f''(-3) < 0$

$f''(1) > 0$

a. $x = -3$, $x = 1$ are critical numbers

b. $f(x)$ increasing on $(-\infty, -3) \cup (1, \infty)$, decreasing on $(-3, 1)$

c. $x = -3$ is a max, $x = 1$ is a min

d. f is concave up on $(-1, \infty)$, concave down on $(-\infty, -1)$

e. $x = -1$ is an inflection point.

11. $f(x) = x^4 - 4x^3 + 2$

$f'(x) = 4x^3 - 12x^2 = 4x^2(x - 3)$

$f''(x) = 12x^2 - 24x = 12x(x - 2)$

$f'(x) \begin{cases} > 0 \text{ on } (3, \infty) \\ < 0 \text{ on } (-\infty, 0) \cup (0, 3) \end{cases}$

$f''(x) \begin{cases} > 0 \text{ on } (-\infty, 0) \cup (2, \infty) \\ < 0 \text{ on } (0, 2) \end{cases}$

$f''(3) > 0$

a. $x = 0, 3$ are critical numbers

b. f increasing on $(3, \infty)$, decreasing on $(-\infty, 3)$

c. $x = 3$ is a local min

d. f is concave up on $(-\infty, 0) \cup (2, \infty)$, concave down on $(0, 2)$

e. $x = 0, 2$ are inflection points.

13. $f(x) = xe^{-4x}$

$f'(x) = 1 \cdot e^{-4x} + xe^{-4x}(-4) = e^{-4x}(1 - 4x)$

$f''(x) = e^{-4x}(-4)(1 - 4x) + e^{-4x}(-4)$

$\quad = -4e^{-4x}(2 - 4x)$

$f'(x) \begin{cases} > 0 \text{ on } \left(-\infty, \dfrac{1}{4}\right) \\ < 0 \text{ on } \left(\dfrac{1}{4}, \infty\right) \end{cases}$

$f''(x) \begin{cases} > 0 \text{ on } \left(\dfrac{1}{2}, \infty\right) \\ < 0 \text{ on } \left(-\infty, \dfrac{1}{2}\right) \end{cases}$

$f''\left(\dfrac{1}{4}\right) < 0$

a. $x = \dfrac{1}{4}$ is a critical number

b. f increasing on $\left(-\infty, \dfrac{1}{4}\right)$, decreasing on $\left(-\dfrac{1}{4}, \infty\right)$

c. $x = \dfrac{1}{4}$ is a local max

d. f is concave up on $\left(\dfrac{1}{2}, \infty\right)$, concave down on $\left(-\infty, \dfrac{1}{2}\right)$

e. $x = \dfrac{1}{2}$ is inflection point

15. $f(x) = x\sqrt{x^2 - 4}$ f undefined on $(-2, 2)$

$f'(x) = \sqrt{x^2 - 4} + x \cdot \dfrac{1}{2}(x^2 - 4)^{-1/2} \cdot 2x$

$\quad = \dfrac{2x^2 - 4}{\sqrt{x^2 - 4}}$

$f'(x) = 0$ when $x = \pm\sqrt{2}$ but $\pm\sqrt{2}$ are not in the domain.

$f''(x) = \dfrac{4x\sqrt{x^2 - 4} - (2x^2 - 4)\frac{1}{2}(x^2 - 4)^{-1/2} \cdot 2x}{x^2 - 4}$

$\quad = \dfrac{2x^3 - 12x}{(x^2 - 4)^{3/2}}$

$f'(x) > 0$ for all x in the domain.

$f''(x) \begin{cases} > 0 \text{ on } (-\sqrt{6}, -2) \cup (\sqrt{6}, \infty) \\ < 0 \text{ on } (-\infty, -\sqrt{6}) \cup (2, \sqrt{6}) \end{cases}$

a. none

b. f increasing on $(-\infty, -2) \cup (2, \infty)$

c. no local min/max

d. f is concave up on $\left(-\sqrt{6}, -2\right) \cup \left(\sqrt{6}, \infty\right)$, concave down on $\left(-\infty, -\sqrt{6}\right) \cup \left(2, \sqrt{6}\right)$

e. $x = \pm\sqrt{6}$ are inflection points.

17. $f(x) = \dfrac{x}{x^2 + 4}$

$f'(x) = \dfrac{x^2 + 4 - x(2x)}{(x^2 + 4)^2} = \dfrac{4 - x^2}{(x^2 + 4)^2}$

$f''(x) = \dfrac{-2x(x^2 + 4)^2 - (4 - x^2)[2(x^2 + 4) \cdot 2x]}{(x^2 + 4)^4}$

$ = \dfrac{2x^3 - 24x}{(x^2 + 4)^3}$

$f'(x) \begin{cases} > 0 \text{ on } (-2, 2) \\ < 0 \text{ on } (-\infty, -2) \cup (2, \infty) \end{cases}$

$f''(x) \begin{cases} > 0 \text{ on } \left(-\sqrt{12}, 0\right) \cup \left(\sqrt{12}, \infty\right) \\ < 0 \text{ on } \left(-\infty, -\sqrt{12}\right) \cup \left(0, \sqrt{12}\right) \end{cases}$

a. $x = \pm 2$ are critical numbers

b. f increasing on $(-2, 2)$, decreasing on $(-\infty, -2) \cup (2, \infty)$

c.   local min at $x = -2$, local max at $x = 2$

d. f is concave up on $\left(-\sqrt{12}, 0\right) \cup \left(\sqrt{12}, \infty\right)$, concave down on $\left(-\infty, -\sqrt{12}\right) \cup \left(0, \sqrt{12}\right)$

e. $x = \pm\sqrt{12}, 0$ are inflection points.

19.   See Example 3.11.

$f(x) = x^3 + 3x^2 - 9x$ on $[0, 4]$

$f'(x) = 3x^2 + 6x - 9$

$ = 3(x^2 + 2x - 3)$

$ = 3(x + 3)(x - 1)$

$x = -3, x = 1$ are critical numbers.
$x = -3 \notin [0, 4]$

$f(0) = 0^3 + 3 \cdot 0^2 - 9 \cdot 0 = 0$

$f(4) = 4^3 + 3 \cdot 4^2 - 9 \cdot 4 = 76$

$f(1) = 1^3 + 3 \cdot 1^2 - 9 \cdot 1 = -5$

So $x = 4$ is absolute max on $[0, 4]$,
$x = 1$ is absolute min.

21.   See Example 3.12.

$f(x) = x^{4/5}$ on $[-2, 3]$

$f'(x) = \dfrac{4}{5}x^{-1/5}$

$x = 0$ is critical number.

$f(-2) = (-2)^{4/5} \approx 1.74$

$f(3) = (3)^{4/5} \approx 2.41$

$f(0) = (0)^{4/5} = 0$

$x = 0$ is absolute min, $x = 3$ is absolute max.

23.   See Example 3.6.

$f(x) = x^3 + 4x^2 + 2x$

$f'(x) = 3x^2 + 8x + 2 = 0$

$x = \dfrac{-8 \pm \sqrt{64 - 24}}{6} = -\dfrac{4}{3} \pm \dfrac{\sqrt{10}}{3}$

$x = -\dfrac{4}{3} - \dfrac{\sqrt{10}}{3}$ is local max, $x = -\dfrac{4}{3} + \dfrac{\sqrt{10}}{3}$ is local min.

25.   See Example 3.6.

$f(x) = x^5 - 2x^2 + x$

$f'(x) = 5x^4 - 4x + 1 = 0$

$x \approx 0.2553, \ 0.8227$

local min at $x \approx 0.8227$
local max at $x \approx 0.2553$

27.   See Theorem 4.1.

29.   See Example 6.1.

$f(x) = x^4 + 4x^3$

$f'(x) = 4x^3 + 12x^2 = 4x^2(4x + 3)$

$f''(x) = 12x^2 + 12x = 12x(x + 2)$

$f'(x) \begin{cases} > 0 \text{ on } (-3, 0) \cup (0, \infty) \\ < 0 \text{ on } (-\infty, -3) \end{cases}$

$f''(x) \begin{cases} > 0 \text{ on } (-\infty, -2) \cup (0, \infty) \\ < 0 \text{ on } (-2, 0) \end{cases}$

f increasing on $(-3, \infty)$, decreasing on $(-\infty, -3)$, concave up on $(-\infty, -2) \cup (0, \infty)$, concave down on $(-2, 0)$, local min at $x = -3$, inflection points at $x = -2, 0$.

$x = -\dfrac{1}{2} - \dfrac{\sqrt{3}}{2}$

31. See Example 6.1.

$$f(x) = x^4 + 4x$$

$$f'(x) = 4x^3 + 4 = 4(x^3 + 1)$$

$$f''(x) = 12x^2$$

$$f'(x) \begin{cases} > 0 \text{ on } (-1, \infty) \\ < 0 \text{ on } (-\infty, -1) \end{cases}$$

$$f''(x) > 0 \text{ on } (-\infty, 0) \cup (0, \infty)$$

f increasing on $(-1, \infty)$, decreasing on $(-\infty, -1)$, concave up on $(-\infty, \infty)$, local min at $x = -1$.

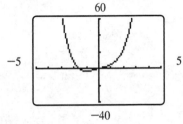

33. See Example 6.2.

$$f(x) = \frac{x}{x^2 + 1}$$

$$f'(x) = \frac{x^2 + 1 - x(2x)}{(x^2 + 1)^2} = \frac{1 - x^2}{(x^2 + 1)^2}$$

$$f''(x) = \frac{-2x(x^2 + 1)^2 - (1 - x^2)2(x^2 + 1)2x}{(x^2 + 1)^4}$$

$$= \frac{2x(x^2 - 3)}{(x^2 + 1)^4}$$

$$f'(x) \begin{cases} > 0 \text{ on } (-1, 1) \\ < 0 \text{ on } (-\infty, -1) \cup (1, \infty) \end{cases}$$

$$f''(x) \begin{cases} > 0 \text{ on } \left(-\sqrt{3}, 0\right) \cup \left(\sqrt{3}, \infty\right) \\ < 0 \text{ on } \left(-\infty, -\sqrt{3}\right) \cup \left(0, \sqrt{3}\right) \end{cases}$$

f increasing on $(-1, 1)$, decreasing on $(-\infty, -1) \cup (1, \infty)$, concave up on $\left(-\sqrt{3}, 0\right) \cup \left(\sqrt{3}, \infty\right)$, concave down on $\left(-\infty, -\sqrt{3}\right) \cup \left(0, \sqrt{3}\right)$, local min at $x = -1$, local max at $x = 1$, inflection points at $0, \pm\sqrt{3}$

$$\lim_{x \to \infty} \frac{x}{x^2 + 1} = \lim_{x \to -\infty} \frac{x}{x^2 + 1} = 0$$

So f has a horizontal asymptote at $y = 0$.

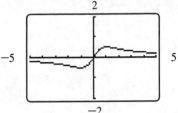

35. See Example 6.2.

$$f(x) = \frac{x^2}{x^2 + 1}$$

$$f'(x) = \frac{(2x)(x^2 + 1) - x^2(2x)}{(x^2 + 1)^2} = \frac{2x}{(x^2 - 1)^2}$$

$$f''(x) = \frac{2(x^2 + 1)^2 - 2x \cdot 2(x^2 + 1)2x}{(x^2 + 1)^4}$$

$$= \frac{2 - 6x^2}{(x^2 + 1)^3}$$

$$f'(x) \begin{cases} > 0 \text{ on } (0, \infty) \\ < 0 \text{ on } (-\infty, 0) \end{cases}$$

$$f''(x) \begin{cases} > 0 \text{ on } \left(-\sqrt{\frac{1}{3}}, \sqrt{\frac{1}{3}}\right) \\ < 0 \text{ on } \left(-\infty, -\sqrt{\frac{1}{3}}\right) \cup \left(\sqrt{\frac{1}{3}}, \infty\right) \end{cases}$$

f increasing on $(0, \infty)$ decreasing on $(-\infty, 0)$, concave up on $\left(-\sqrt{\frac{1}{3}}, \sqrt{\frac{1}{3}}\right)$, concave down on $\left(-\infty, -\sqrt{\frac{1}{3}}\right) \cup \left(\sqrt{\frac{1}{3}}, \infty\right)$, local min at $x = 0$, inflection points at $x = \pm\sqrt{\frac{1}{3}}$.

$$\lim_{x \to \infty} \frac{x^2}{x^2 + 1} = \lim_{x \to -\infty} \frac{x^2}{x^2 + 1} = 1$$

So f has a horizontal asymptote at $y = 1$.

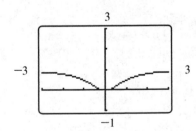

37. See Example 6.2.

$$f(x) = \frac{x^3}{x^2 - 1}$$

$$f'(x) = \frac{3x^2(x^2 - 1) - x^3(2x)}{(x^2 - 1)^2} = \frac{x^4 - 3x^2}{(x^2 - 1)^2}$$

$$f''(x) = \frac{(4x^3 - 6x)(x^2 - 1)^2 - (x^4 - 3x^2)2(x^2 + 1)2x}{(x^2 - 1)^4} = \frac{2x^3 + 6x}{(x^2 - 1)^4}$$

$$f'(x) \begin{cases} > 0 \text{ on } \left(-\infty, -\sqrt{3}\right) \cup \left(\sqrt{3}, \infty\right) \\ < 0 \text{ on } \left(-\sqrt{3}, -1\right) \cup (-1, 0) \cup (0, 1) \cup \left(1, \sqrt{3}\right) \end{cases}$$

$$f''(x) \begin{cases} > 0 \text{ on } (-1, 0) \cup (1, \infty) \\ < 0 \text{ on } (-\infty, -1) \cup (0, 1) \end{cases}$$

f increasing on $\left(-\infty, -\sqrt{3}\right) \cup \left(\sqrt{3}, \infty\right)$ decreasing on $\left(-\sqrt{3}, -1\right) \cup (-1, 1) \cup \left(1, \sqrt{3}\right)$, concave up on $(-1, 0) \cup (1, \infty)$, concave down on $(-\infty, -1) \cup (0, 1)$, $x = -\sqrt{3}$ local max, $x = \sqrt{3}$ local min, $x = 0$ inflection point. f is undefined at $x = -1$ and $x = 1$.

$$\lim_{x \to -1^+} \frac{\overset{-}{x^3}}{\underset{-}{x^2} - 1} = \infty, \text{ and } \lim_{x \to -1^-} \frac{\overset{-}{x^3}}{-\underset{+}{x^2} - 1} = -\infty$$

$$\lim_{x \to 1^+} \frac{\overset{+}{x^3}}{\underset{+}{x^2} - 1} = \infty, \text{ and } \lim_{x \to 1^-} \frac{\overset{+}{x^3}}{\underset{-}{x^2} - 1} = -\infty$$

So f has vertical asymptotes at $x = 1$ and $x = -1$.

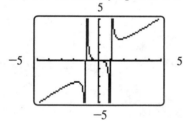

39. See Example 7.4.

$$y = 2x^2, (2, 1)$$

$$d = \sqrt{(x - 2)^2 + (y - 1)^2} = \sqrt{(x - 2)^2 + (2x^2 - 1)^2}$$

$$f(x) = (x - 2)^2 + (2x^2 - 1)^2$$

$$f'(x) = 2(x - 2) + 2(2x^2 - 1)4x$$

$$= 16x^3 - 6x - 4$$

$x \approx 0.8237$ when $f'(x) = 0$

$$f'(x) \begin{cases} < 0 \text{ on } (-\infty, 0.8237) \\ > 0 \text{ on } (0.8237, \infty) \end{cases}$$

So $x \approx 0.8237$ corresponds to the closest point.

$$y = 2x^2 = 2(0.8237)^2 = 1.3570$$

$(0.8237, 1.3570)$ is closest to $(2, 1)$.

41. See Example 7.6.

$$C(x) = 6\sqrt{4^2 + (4-x)^2} + 2\sqrt{2^2 + x^2}$$

$$C'(x) = 6 \cdot \frac{1}{2}[16 + (4-x)^2]^{-1/2} \cdot 2(4-x)(-1) + 2\frac{1}{2}(4+x^2)^{-1/2} \cdot 2x$$

$$= \frac{6(x-4)}{\sqrt{16 + (4-x)^2}} + \frac{2x}{\sqrt{4+x^2}}$$

$x \approx 2.864$ when $C'(x) = 0$

$$C'(x)\begin{cases} < 0 \text{ on } (0, 2.864) \\ > 0 \text{ on } (2.864) \end{cases}$$

So $x \approx 2.864$ gives the minimum cost.
Locate highway corner $4 - 2.864 = 1.136$ miles east of point A.

43. See Example 7.5.

Area $= 2\pi r^2 + 2\pi rh$

16 fl oz = 16 fl oz $\cdot$ 1.80469 $\dfrac{in^3}{fl\ oz}$

$\qquad$ = 28.87504 in^3

Vol $= \pi r^2 h$

$$h = \frac{\text{Vol}}{\pi r^2} = \frac{28.87504}{\pi r^2}$$

$$A(r) = 2\pi\left(r^2 + \frac{28.87504}{\pi r} \right)$$

$$A'(r) = 2\pi\left(2r - \frac{28.87504}{\pi r^2} \right)$$

$2\pi r^3 = 28.87504$

$$r = \sqrt[3]{\frac{28.87504}{2\pi}} \approx 1.663$$

$$A'(r)\begin{cases} < 0 \text{ on } (0, 1.663) \\ > 0 \text{ on } (1.663, \infty) \end{cases}$$

So $r \approx 1.663$ gives the minimum surface area.

$$h = \frac{28.87504}{\pi(1.663)^2} \approx 3.325$$

45. See Example 8.1.

$$Q(t) = e^{-3t}\sin 2t$$

$$Q'(t) = -3e^{-3t}\sin 2t + e^{-3t}\cos 2t \cdot 2$$

$$= e^{-3t}(2\cos 2t - 3\sin 2t) \text{ amps}$$

47. See Example 8.3.

$$m(x) = 20 + x^2,\ 0 \le x \le 4$$

49. See Example 8.5.

$$C(x) = 0.02x^2 + 20x + 1800$$
$$C'(x) = 0.04x + 20$$
$$C'(20) = 0.04(20) + 20 = 20.8$$
$$C(20) - C(19) = 0.02(20)^2 + 20(20) + 1800$$
$$\qquad\qquad -[0.02(19)^2 + 20(19) + 1800]$$
$$= 20.78$$

Chapter 4

Section 4.1

5. $\int 3x^4 dx = \dfrac{3x^5}{5} + c$

7. $\int (3x^4 - 3x) dx = \int 3x^4 dx - \int 3x dx$

$\qquad = \dfrac{3x^5}{5} - \dfrac{3x^2}{2} + c$

9. $\int 3\sqrt{x} dx = 2x^{3/2} + c$

11. $\int \left(3 - \dfrac{1}{x^4}\right) dx = \int 3 dx - \int x^{-4} dx = 3x + \dfrac{x^{-3}}{3} + c$

13. $\int \dfrac{x^{1/3} - 3}{x^{2/3}} dx = \int x^{-1/3} dx - \int 3x^{-2/3} dx$

$\qquad = \dfrac{3}{2} x^{2/3} - 9x^{1/3} + c$

15. $\int (2 \sin x + \cos x) dx = \int 2 \sin x \, dx + \int \cos x \, dx$

$\qquad = -2 \cos x + \sin x + c$

17. $\int 2 \sec x \tan x \, dx = 2 \sec x + c$

19. $\int 5 \sec^2 x \, dx = 5 \tan x + c$

21. $\int (3e^x - 2) dx = \int 3e^x dx - \int 2 dx = 3e^x - 2x + c$

23. $\int \left(3 \cos x - \dfrac{1}{x}\right) dx = \int 3 \cos x \, dx - \int \dfrac{1}{x} dx$

$\qquad = 3 \sin x - \ln|x| + c$

25. $\int \dfrac{4x}{x^2 + 4} dx = 2 \ln\left|x^2 + 4\right| + c$

27. $\int \left(5x - \dfrac{3}{e^x}\right) dx = \int 5x dx - \int 3e^{-x} dx$

$\qquad = \dfrac{5x^2}{2} + 3e^{-x} + c$

29. $\int 5 \sin 2x \, dx = -\dfrac{5}{2} \cos 2x + c$

31. $\int (e^{3x} - x) dx = \dfrac{e^{3x}}{3} - \dfrac{x^2}{2} + c$

33. $\int 3 \sec 2x \tan 2x \, dx = \dfrac{3}{2} \sec 2x + c$

35. $\int \dfrac{e^x}{e^x + 3} dx = \ln\left|e^x + 3\right| + c$

37. $\int \dfrac{e^x + 3}{e^x} dx = \int dx + \int 3e^{-x} dx$

$\qquad = x - 3e^{-x} + c$

39. $\int x^{1/4} (x^{5/4} - 4) dx = \int x^{3/2} dx - \int 4x^{1/4} dx$

$\qquad = \dfrac{2}{5} x^{5/2} - \dfrac{16}{5} x^{5/4} + c$

41. $\int \sqrt{x^3 + 4} dx$: N/A

43. $\int \dfrac{3x^2 - 4}{x^2} dx = \int 3 dx - \int 4x^{-2} dx$

$\qquad = 3x + 4x^{-1} + c$

45. $\int 2 \sec x \, dx$: N/A

47. $\int 2 \sin 4x \, dx = -\dfrac{1}{2} \cos 4x + c$

49. $\int e^{x^2} dx$: N/A

51. $\int \left(\dfrac{1}{x^2} - 1\right) dx = \int x^{-2} dx - \int dx = -x^{-1} - x + c$

53. Example 1.12b. $\ln|\sec x + \tan x| + c$

Example 1.12f. $\dfrac{1}{4} \sin 2x - \dfrac{1}{2} x \cos 2x + c$

(The CAS does not show the absolute value bars or the constant c.

55. $f'(x) = 4x^2 - 1, f(0) = 2$

$f(x) = \int (4x^2 - 1) dx = \dfrac{4}{3} x^3 - x + c$

$f(0) = \dfrac{4}{3} \cdot 0^3 - 0 + c = 2$

$c = 2$

$f(x) = \dfrac{4}{3} x^3 - x + 2$

57. $f'(x) = 3e^x + x, f(0) = 4$

$$f(x) = \int \left(3e^x + x\right)dx = 3e^x + \frac{x^2}{2} + c$$

$$f(0) = 3e^0 + \frac{0^2}{2} + c = 4$$

$$c = 1$$

$$f(x) = 3e^x + \frac{x^2}{2} + 1$$

59. $f''(x) = 12, f'(0) = 2, f(0) = 3$

$$f'(x) = \int 12\,dx = 12x + c_1$$

$$f'(0) = 12 \cdot 0 + c_1 = 2, \text{ so } c_1 = 2$$

and $f'(x) = 12x + 2$

$$f(x) = \int (12x + 2)dx = 6x^2 + 2x + c_2$$

$$f(0) = 6 \cdot 0^2 + 2 \cdot 0 + c_2 = 3, \text{ so } c_2 = 3$$

$$f(x) = 6x^2 + 2x + 3$$

61. $f''(x) = 3\sin x + 4x^2$

$$f'(x) = \int \left(3\sin x + 4x^2\right)dx = -3\cos x + \frac{4}{3}x^3 + c_1$$

$$f(x) = \int \left(-3\cos x + \frac{4}{3}x^3 + c_1\right)dx$$

$$= -3\sin x + \frac{1}{3}x^4 + c_1 x + c_2$$

63. $f'''(x) = 4 - \dfrac{2}{x^3} = 4 - 2x^{-3}$

$$f''(x) = \int \left(4 - 2x^{-3}\right)dx = 4x + x^{-2} + c_1$$

$$f'(x) = \int \left(4x + x^{-2} + c_1\right)dx$$

$$= 2x^2 - x^{-1} + c_1 x + c_2$$

$$f(x) = \int \left(2x^2 - \frac{1}{x} + c_1 x + c_2\right)dx$$

$$= \frac{2}{3}x^3 - \ln|x| + \frac{c_1}{2}x^2 + c_2 x + c_3$$

65. $v(t) = 3 - 12t, s(0) = 3$

$$s(t) = \int v(t)dt = \int (3 - 12t)dt = 3t - 6t^2 + c$$

$$s(0) = 3 \cdot 0 - 6 \cdot 0^2 + c = 3$$

$$c = 3$$

$$s(t) = 3t - 6t^2 + 3$$

67. $a(t) = 3\sin t + 1, v(0) = 0, s(0) = 4$

$$v(t) = \int a(t)dt$$

$$= \int (3\sin t + 1)dt$$

$$= -3\cos t + t + c_1$$

$$v(0) = -3\cos 0 + 0 + c = 0$$

$$c_1 = 3$$

$$v(t) = -3\cos t + t + 3$$

$$s(t) = \int v(t)dt$$

$$= \int (-3\cos t + t + 3)dt$$

$$= -3\sin t + \frac{1}{2}t^2 + 3t + c_2$$

$$s(0) = -3\sin 0 + \frac{1}{2}0^2 + 3 \cdot 0 + c_2 = 4$$

$$c_2 = 4$$

$$s(t) = -3\sin t + \frac{1}{2}t^2 + 3t + 4$$

69. $v(0) = 30 \text{ mph} = 30 \cdot \dfrac{1 \text{ hr}}{3600 \text{ sec}} = \dfrac{1}{120}\text{miles/sec}$

$$v(4) = 50 \text{ mph} = 50 \cdot \frac{1 \text{ hr}}{3600 \text{ sec}} = \frac{1}{72}\text{miles/sec}$$

$$a = \frac{v(4) - v(0)}{4 - 0} = \frac{\frac{1}{72} - \frac{1}{120}}{4} = \frac{1}{720}\text{miles/sec}^2$$

$$v(t) = \int a(t)dt = \int \frac{1}{720}dt = \frac{t}{720} + c_1$$

$$v(0) = \frac{1}{120} = \frac{0}{720} + c_1$$

$$c_1 = \frac{1}{120}$$

$$v(t) = \frac{t}{720} + \frac{1}{120}$$

$$s(t) = \int v(t)dt$$

$$= \int \left(\frac{t}{720} + \frac{1}{120}\right)dt$$

$$= \frac{t^2}{1440} + \frac{t}{120} + c_2$$

$$s(0) = 0, \text{ so } c_2 = 0$$

$$s(t) = \frac{t^2}{1440} + \frac{t}{120}$$

$$s(4) = \frac{4^2}{1440} + \frac{4}{120} = \frac{1}{90} + \frac{4}{120} = \frac{4 + 12}{360}$$

$$= \frac{2}{45} \text{ miles}$$

71.

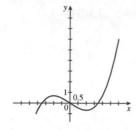

75. $y(t) = -16t^2 - 100t + 100,000 = 0$
  $t \approx 75.99 \approx 76$ sec
  $v(76) = -32(76) - 100 = -2532$ ft/s

77. Use derivative formulas that you know.

73. All functions that have the derivative shown in Exercise 71 are vertical translations of the graph given as the answer for Exercise 71.

Section 4.2

3. $\displaystyle\sum_{i=1}^{50} i^2 = \frac{50(50+1)(2 \cdot 50 + 1)}{6} = 42,925$

5. $\displaystyle\sum_{i=1}^{10} \sqrt{i} = \sqrt{1} + \sqrt{2} + \sqrt{3} + \sqrt{4} + \sqrt{5} + \sqrt{6} + \sqrt{7} + \sqrt{8} + \sqrt{9} + \sqrt{10}$
  ≈ 22.47

7. $\displaystyle\sum_{i=1}^{6} 3i^2 = 3 \cdot 1^2 + 3 \cdot 2^2 + 3 \cdot 3^2 + 3 \cdot 4^2 + 3 \cdot 5^2 + 3 \cdot 6^2$
  $= 3 + 12 + 27 + 48 + 75 + 108$
  $= 273$

9. $\displaystyle\sum_{i=3}^{7} (i^2 + i) = (3^2 + 3) + (4^2 + 4) + (5^2 + 5) + (6^2 + 6) + (7^2 + 7)$
  $= 12 + 20 + 30 + 42 + 56$
  $= 160$

11. $\displaystyle\sum_{i=0}^{7} (3i - 1) = (3 \cdot 0 - 1) + (3 \cdot 1 - 1) + (3 \cdot 2 - 1) + (3 \cdot 3 - 1) + (3 \cdot 4 - 1) + (3 \cdot 5 - 1) + (3 \cdot 6 - 1) + (3 \cdot 7 - 1)$
  $= -1 + 2 + 5 + 8 + 11 + 14 + 17 + 20$
  $= 76$

13. $\displaystyle\sum_{i=2}^{5} (\sqrt{i} + i) = (\sqrt{2} + 2) + (\sqrt{3} + 3) + (\sqrt{4} + 4) + (\sqrt{5} + 5)$
  ≈ 21.38

15. $\displaystyle\sum_{i=1}^{70} (3i - 1) = 3\sum_{i=1}^{70} i - \sum_{i=1}^{70} 1$
  $= 3 \cdot \dfrac{70(71)}{2} - 70 \cdot 1$
  $= 7385$

17. $\displaystyle\sum_{i=1}^{40} (4 - i^2) = \sum_{i=1}^{40} 4 - \sum_{i=1}^{40} i^2$
  $= 4 \cdot 40 - \dfrac{40(41)(81)}{6}$
  $= -21,980$

19. $\displaystyle\sum_{i=1}^{100}(i^2-3i+2)$

$\displaystyle=\sum_{i=1}^{100}i^2-3\sum_{i=1}^{100}i+\sum_{i=1}^{100}2$

$\displaystyle=\frac{100(101)(201)}{6}-\frac{3\cdot100(101)}{2}+2\cdot100$

$=323,400$

21. $\displaystyle\sum_{i=1}^{200}(4-3i-i^2)$

$\displaystyle=\sum_{i=1}^{200}4-3\sum_{i=1}^{200}i-\sum_{i=1}^{200}i^2$

$\displaystyle=4\cdot200-\frac{3(200)(201)}{2}-\frac{200(201)(401)}{6}$

$=-2,746,200$

23. $\displaystyle\sum_{i=1}^{n}(i^2-3)=\sum_{i=1}^{n}i^2-\sum_{i=1}^{n}3$

$\displaystyle=\frac{n(n+1)(2n+1)}{6}-3n$

25. $\displaystyle\sum_{i=1}^{n}(4i^2-i)=4\sum_{i=1}^{n}i^2-\sum_{i=1}^{n}i$

$\displaystyle=\frac{4n(n+1)(2n+1)}{6}-\frac{n(n+1)}{2}$

27. $\displaystyle\sum_{i=1}^{n}\frac{1}{n}\left[\left(\frac{i}{n}\right)^2+2\left(\frac{i}{n}\right)\right]$

$\displaystyle=\frac{1}{n^3}\sum_{i=1}^{n}i^2+\frac{2}{n^2}\sum_{i=1}^{n}i$

$\displaystyle=\frac{1}{n^3}\cdot\frac{n(n+1)(2n+1)}{6}+\frac{2}{n^2}\cdot\frac{n(n+1)}{2}$

$\displaystyle=\frac{(n+1)(2n+1)}{6n^2}+\frac{n+1}{n}\to\frac{2}{6}+1$

$\displaystyle=\frac{4}{3}$ as $n\to\infty$

29. $\displaystyle\sum_{i=1}^{n}\frac{1}{n}\left[4\left(\frac{2i}{n}\right)^2-\left(\frac{2i}{n}\right)\right]$

$\displaystyle=\frac{16}{n^3}\sum_{i=1}^{n}i^2-\frac{2}{n^2}\sum_{i=1}^{n}i$

$\displaystyle=\frac{16}{n^3}\cdot\frac{n(n+1)(2n+1)}{6}-\frac{2}{n^2}\cdot\frac{n(n+1)}{2}$

$\displaystyle=\frac{8}{3}\frac{(n+1)(2n+1)}{n^2}-\frac{(n+1)}{n}\to\frac{16}{3}-1$

$\displaystyle=\frac{13}{3}$ as $n\to\infty$

31. $f(x)=x^2+4x$

$\displaystyle\sum_{k=1}^{5}f\left(\frac{k}{5}\right)=\frac{71}{5}$

33. $f(x)=4x^2-2$

$\displaystyle\sum_{k=1}^{10}f\left(2+\frac{k}{10}\right)=\frac{1217}{5}=243.4$

35. $d=50(2)+60(1)+70(.5)+60(3)$

$=100+60+35+180$

$=375$ miles

37. $\displaystyle d=15\left(\frac{1}{3}\right)+18\left(\frac{1}{2}\right)+16\left(\frac{1}{6}\right)+12\left(\frac{2}{3}\right)$

$\displaystyle=5+9+\frac{8}{3}+8$

$\displaystyle=\frac{74}{3}$ miles

39. For $n=k+1,$

$\displaystyle\sum_{i=1}^{k+1}i^3=\sum_{i=1}^{k}i^3+(k+1)^3=\frac{k^2(k+1)^2}{4}+(k+1)^3$

$\displaystyle=\frac{k^4+6k^3+13k^2+12k+4}{4}$

$\displaystyle=\frac{(k+1)^2(k+2)^2}{4}$

41. $\displaystyle\sum_{i=1}^{10}\left(i^3-3i+1\right)$

$\displaystyle=\sum_{i=1}^{10}i^3-3\sum_{i=1}^{10}i+\sum_{i=1}^{10}1$

$\displaystyle=\frac{10^2(10+1)^2}{4}-\frac{3\cdot10(10+1)}{2}+1\cdot10$

$=2870$

43. $\displaystyle\sum_{i=1}^{100}\left(i^5-2i^2\right)$

$\displaystyle=\sum_{i=1}^{100}i^5-2\sum_{i=1}^{100}i^2$

$\displaystyle=\frac{100^2(100+1)^2(2\cdot100^2+2\cdot100-1)}{12}$

$\displaystyle\qquad-\frac{2\cdot100(101)(201)}{6}$

$=171,707,655,800$

45. $\displaystyle\sum_{i=1}^{n}(ca_i+db_i)$

$=(ca_1+db_1)+(ca_2+db_2)+\cdots+(ca_n+db_n)$

$=ca_1+ca_2+\cdots+ca_n+db_1+db_2+\cdots+db_n$

$=c(a_1+a_2+\cdots+a_n)+d(b_1+b_2+\cdots+b_n)$

$\displaystyle=c\sum_{i=1}^{n}a_i+d\sum_{i=1}^{n}b_i$

Section 4.3

3. $f(x)=x^2+1,\ [0,1],\ n=4$

The evaluation points are
.125, .375, .625, .875

$\displaystyle A\approx A_4=\sum_{i=1}^{4}f(x_i)\Delta x$

$\qquad=(f(.125)+f(.375)+f(.625)+f(.875))\cdot(.25)$

$\qquad=1.328125$

5. $f(x)=x^3-1,\ [1,\,2],\ n=4$

The evaluation points are
1.125, 1.375, 1.625, 1.875

$\displaystyle A\approx A_4=\sum_{i=1}^{4}f(x_i)\Delta x$

$\qquad=(f(1.25)+f(1.375)+f(1.625)+f(1.875))\cdot(.25)$

$\qquad=2.7265625$

7. $f(x)=\sin x,\ [0,\,\pi],\ n=4$

The evaluation points are
$\dfrac{\pi}{8},\dfrac{3\pi}{8},\dfrac{5\pi}{8},\dfrac{7\pi}{8}$

$\displaystyle A\approx A_4=\sum_{i=1}^{4}f(x_i)\Delta x$

$\qquad=\left(f\left(\dfrac{\pi}{8}\right)+f\left(\dfrac{3\pi}{8}\right)+f\left(\dfrac{5\pi}{8}\right)+f\left(\dfrac{7\pi}{8}\right)\right)\cdot\left(\dfrac{\pi}{4}\right)$

$\qquad=2.05234$

9. $f(x)=4-x^2,\ [-1,\,1],\ n=4$

The evaluation points are
$-.75,-.25,.25,.75$

$\displaystyle A\approx A_4=\sum_{i=1}^{4}f(x_i)\Delta x$

$\qquad=\big(f(-.75)+f(-.25)+f(.25)+f(.75)\big)\cdot(.5)$

$\qquad=7.375$

11. $y = x^2$ on $[0, 1]$, $n = 8$, midpoint evaluation

$\Delta x = \dfrac{1-0}{8} = .125$

$x_i = .0625 + .125i$ for $i = 0, 1, 2, \ldots, 7$

$A \approx A_8 \approx \displaystyle\sum_{i=0}^{7} f(x_i)(.125)$

$\qquad = \displaystyle\sum_{i=0}^{7} x_i^2 (.125)$

$\qquad = .125 \displaystyle\sum_{i=0}^{7} x_i^2$

$\qquad = .125\left(.0625^2 + .1875^2 + .3125^2 + .4375^2 + .5625^2 + .6875^2 + .8125^2 + .9375^2\right)$

$\qquad \approx .3320$

13. $y = x^2$ on $[-1, 1]$, $n = 8$, left-endpoint evaluation

$\Delta x = \dfrac{1-(-1)}{8} = .25$

$x_i = -1 + .25i$, $i = 0, 1, \ldots, 7$

$A \approx A_8 \approx \displaystyle\sum_{i=0}^{7} f(x_i)(.25)$

$\qquad = .25 \displaystyle\sum_{i=0}^{7} x_i^2$

$\qquad = .25((-1)^2 + (-.75)^2 + (-.5)^2 + (-.25)^2 + 0^2 + .25^2 + .5^2 + .75^2)$

$\qquad = .6875$

15. $y = \sqrt{x+2}$ on $[1, 4]$, $n = 8$, midpoint evaluation

$\Delta x = \dfrac{4-1}{8} = .375$

$x_i = 1.1875 + .375i$, $i = 0, 1, \ldots, 7$

$A \approx A_8 \approx \displaystyle\sum_{i=0}^{7} f(x_i)(.375)$

$\qquad = .375 \displaystyle\sum_{i=0}^{7} \sqrt{x_i + 2}$

$\qquad = .375\left(\sqrt{3.1875} + \sqrt{3.5625} + \sqrt{3.9375} + \sqrt{4.3125} + \sqrt{4.6875} + \sqrt{5.0625} + \sqrt{5.4375} + \sqrt{5.8125}\right)$

$\qquad = 6.3343$

17. $y = e^{-2x}$ on $[-1, 1]$, $n = 16$, left-endpoint evaluation

$\Delta x = \dfrac{1-(-1)}{16} = .125$

$x_i = -1 + .125i,\ i = 0, 1, \ldots, 15$

$A \approx A_{16} \approx \displaystyle\sum_{i=0}^{15} f(x_i)(.125)$

$\quad = .125 \displaystyle\sum_{i=0}^{15} e^{-2x_i}$

$\quad = .125\left(e^{-2(-1)} + e^{-2(-.875)} + e^{-2(-.75)} + e^{-2(-.625)} + e^{-2(-.5)} + e^{-2(-.375)} + e^{-2(-.25)} + e^{-2(-.125)} + e^{-2(0)}\right.$

$\qquad \left. + e^{-2(.125)} + e^{-2(.25)} + e^{-2(.375)} + e^{-2(.5)} + e^{-2(.625)} + e^{-2(.75)} + e^{-2(.875)}\right)$

$\quad \approx 4.0991$

19. $y = \cos x$ on $\left[0, \dfrac{\pi}{2}\right]$, $n = 50$, midpoint evaluation

$\Delta x = \dfrac{\frac{\pi}{2} - 0}{50} = \dfrac{\pi}{100}$

$x_i = \dfrac{\pi}{200} + \dfrac{\pi}{100}i,\ i = 0, 1, \ldots, 49$

$A \approx A_{50} \approx \displaystyle\sum_{i=0}^{49} f(x_i)\left(\dfrac{\pi}{100}\right)$

$\quad = \dfrac{\pi}{100} \displaystyle\sum_{i=0}^{49} \cos x_i$

$\quad = \dfrac{\pi}{100} \displaystyle\sum_{i=0}^{49} \cos\left(\dfrac{\pi}{200} + \dfrac{\pi i}{100}\right)$

$\quad = \dfrac{\pi}{100}\left(\cos\dfrac{\pi}{200} + \cos\dfrac{3\pi}{200} + \cos\dfrac{5\pi}{200} + \ldots + \cos\dfrac{99\pi}{200}\right)$

$\quad \approx 1.00004$

21. $y = 3x - 2$ on $[1, 4]$, $n = 4$, midpoint evaluation

$\Delta x = \dfrac{4-1}{4} = .75$

$x_i = 1.375 + .75i,\ i = 0, 1, 2, 3$

$A \approx A_4 \approx \displaystyle\sum_{i=0}^{3} f(x_i)(.75)$

$\quad = .75 \displaystyle\sum_{i=0}^{3} (3x_i - 2)$

$\quad = 2.25 \displaystyle\sum_{i=0}^{3} x_i + .75 \displaystyle\sum_{i=0}^{3} -2$

$\quad = 2.25(1.375 + 2.125 + 2.875 + 3.625) - 2(.75)(4)$

$\quad = 16.5$

23. $y = x^3 - 1$ on $[1, 3]$, $n = 100$, midpoint evaluation

$\Delta x = \dfrac{3-1}{100} = .02$

$x_i = 1.01 + .02i.\ i = 0, 1, \ldots, 99$

$A \approx A_{100} \approx \displaystyle\sum_{i=0}^{99} f(x_i)(.02)$

$\phantom{A \approx A_{100}} = \displaystyle\sum_{i=0}^{99} \left(x_i^3 - 1\right)(.02)$

$\phantom{A \approx A_{100}} = .02\displaystyle\sum_{i=0}^{99} x_i^3 - .02(100)$

$\phantom{A \approx A_{100}} = .02\displaystyle\sum_{i=0}^{99} (1.01 + .02i)^3 - 2$

$\phantom{A \approx A_{100}} = .02\displaystyle\sum_{i=0}^{99} \left(1.030301 + .061206i + .001212i^2 + .000008i^3\right) - 2$

$\phantom{A \approx A_{100}} = .02(100)(1.030301) + .02(.061206)\displaystyle\sum_{i=1}^{99} i + .02(.001212)\displaystyle\sum_{i=1}^{99} i^2 + (.02)(.000008)\displaystyle\sum_{i=1}^{99} i^3 - 2$

$\phantom{A \approx A_{100}} = 2.060602 + .00122412 \cdot \dfrac{99(100)}{2} + .00002424 \cdot \dfrac{99(100)(199)}{6} + .00000016 \cdot \dfrac{99^2(100)^2}{4} - 2$

$\phantom{A \approx A_{100}} = 17.9996$

25. $y = x^3 - 1$ on $[-1, 1]$, $n = 100$, left-endpoint evaluation

$\Delta x = \dfrac{1 - (-1)}{100} = .02$

$x_i = -1 + .02i,\ i = 0, 1, \ldots, 99$

$A \approx A_{100} \approx \displaystyle\sum_{i=0}^{99} f(x_i)(.02)$

$\phantom{A \approx A_{100}} = .02\displaystyle\sum_{i=0}^{99} \left((-1 + .02i)^3 - 1\right)$

$\phantom{A \approx A_{100}} = .02\displaystyle\sum_{i=0}^{99} \left(-2 + .06i - .0012i^2 + .000008i^3\right)$

$\phantom{A \approx A_{100}} = .02\left(-2\displaystyle\sum_{i=1}^{100} 1 + .06\displaystyle\sum_{i=1}^{99} i - .0012\displaystyle\sum_{i=1}^{99} i^2 + .000008\displaystyle\sum_{i=1}^{99} i^3\right)$

$\phantom{A \approx A_{100}} = .02\left(-2(100) + .06\dfrac{(99)(100)}{2} - .0012\dfrac{(99)(100)(199)}{6} + .000008\dfrac{(99)^2(100)^2}{4}\right)$

$\phantom{A \approx A_{100}} = -2.02$

27.

n	left endpoint	midpoint	right endpoint
10	10.56	10.72	10.56
100	10.6656	10.6672	10.6656
1000	10.6667	10.6667	10.6667

29.

n	left endpoint	midpoint	right endpoint
10	15.48	17.96	20.68
100	17.7408	17.9996	18.2608
1000	17.974	18.0000	18.026

31. $f(x) = x^2 + 2;\ a = 0,\ b = 1.$

The Riemann sum is

$$\frac{b-a}{n}\sum_{k=0}^{n-1} f\left(\frac{b-a}{n}\cdot k + a\right) = \frac{1}{n}\sum_{k=0}^{n-1}\left(\left(\frac{k}{n}\right)^2 + 2\right)$$

$$\lim_{n\to\infty}\frac{1}{n}\sum_{k=0}^{n-1}\left(\frac{k^2}{n^2} + 2\right)$$

$$= \lim_{n\to\infty}\frac{2n^2 - 3n + 1}{6n^2} + 2$$

$$= \lim_{n\to\infty}\frac{14n^2 - 3n + 1}{6n^2}$$

$$= \frac{7}{3}$$

33. $f(x) = 2x^2 + 1;\ a = 1,\ b = 3.$

The Riemann sum is

$$\frac{b-a}{n}\sum_{k=0}^{n-1} f\left(\frac{b-a}{n}\cdot k + a\right) = \frac{2}{n}\sum_{k=0}^{n-1}\left(2\left(\frac{2k}{n} + 1\right)^2 + 1\right)$$

$$\lim_{n\to\infty}\frac{2}{n}\sum_{k=0}^{n-1}\left(\frac{8k^2}{n^2} + \frac{8k}{n} + 3\right)$$

$$= \lim_{n\to\infty}\frac{2}{n}\left(\frac{8(2n^2 - 3n + 1)}{6n} + \frac{8(n-1)}{2} + 3n\right)$$

$$= \lim_{n\to\infty}\frac{2(29n^2 - 24n + 4)}{3n^2}$$

$$= \frac{58}{3}$$

35. left endpoint evaluation

$A \approx .1(2.0 + 2.4 + 2.6 + 2.7 + 2.6 + 2.4 + 2.0 + 1.4) = 1.81$

right endpoint evaluation

$A \approx .1(2.4 + 2.6 + 2.7 + 2.6 + 2.4 + 2.0 + 1.4 + 0.6) = 1.67$

37. left endpoint evaluation
$$A \approx .2(1 + 1.4 + 2.1 + 2.7 + 2.6 + 2.8 + 3 + 3.4) = 3.8$$
right endpoint evaluation
$$A \approx .2(1.4 + 2.1 + 2.7 + 2.6 + 2.8 + 3 + 3.4 + 3.6) = 4.32$$

39. left endpoint evaluation
$$A \approx .1(1.8 + 1.4 + 1.1 + .7 + 1.2 + 1.4 + 1.8 + 2.4) = 1.18$$
right endpoint evaluation
$$A \approx .1(1.4 + 1.1 + .7 + 1.2 + 1.4 + 1.8 + 2.4 + 2.6) = 1.26$$

41. L and M are lower, R is higher.

43. L is higher, M and R are lower.

45. For example, use $x = \dfrac{1}{\sqrt{6}}$ on $[0, 0.5]$ and use $x = \dfrac{1}{\sqrt{2}}$ on $[0.5, 1]$.

47. $\Delta x = \dfrac{(b - a)}{n}$, $[a, b]$ interval for right endpoint evaluation, the evaluation points are
$$c_1 = a + \Delta x$$
$$c_2 = c_1 + \Delta x = (a + \Delta x) + \Delta x = a + 2\Delta x$$
$$c_3 = c_2 + \Delta x = (a + 2\Delta x) + \Delta x = a + 3\Delta x$$
$$\vdots$$
$$c_i = a + i \cdot \Delta x, \ i = 1, \ldots, n$$

49. $\Delta x = \dfrac{(b - a)}{n}$, $[a, b]$ interval for midpoint evaluation, the evaluation points are
$$c_1 = a + \dfrac{\Delta x}{2}$$
$$c_2 = c_1 + \Delta x = \left(a + \dfrac{\Delta x}{2}\right) + \Delta x = a + \dfrac{3}{2}\Delta x$$
$$c_3 = c_2 + \Delta x = \left(a + \dfrac{3}{2}\Delta x\right) + \Delta x = a + \dfrac{5}{2}\Delta x$$
$$\vdots$$
$$c_i = a + \left(i - \dfrac{1}{2}\right)\Delta x, \ i = 1, \ldots, n$$

Section 4.4

5. $\displaystyle\int_0^3 \left(x^3 + x\right)dx \approx \sum_{i=1}^n \left(c_i^3 + c_i\right)\Delta x$
$$= \sum_{i=1}^n \left(c_i^3 + c_i\right)\left(\frac{3 - 0}{n}\right)$$
$$= \sum_{i=1}^n \frac{3}{n}\left(c_i^3 + c_i\right),$$
$$c_i = \frac{x_i + x_{i-1}}{2}$$
$$n \geq 50 \Rightarrow \text{Riemann sum} \approx 24.75$$

7. $\displaystyle\int_2^4 \frac{1}{x^2}dx \approx \sum_{i=1}^n \left(\frac{1}{c_i^2}\right)\Delta x$
$$= \sum_{i=1}^n \frac{1}{c_i^2}\left(\frac{4 - 2}{n}\right)$$
$$= \sum_{i=1}^n \frac{2}{n}\frac{1}{c_i^2},$$
$$c_i = \frac{x_i + x_{i-1}}{2}$$
$$n \geq 10 \Rightarrow \text{Riemann sum} \approx .25$$

9. $\int_0^\pi \sin x^2 dx = \sum_{i=1}^n \sin c_i^2 \Delta x$

$$= \sum_{i=1}^n \sin c_i^2 \left(\frac{\pi - 0}{n} \right)$$

$$= \sum_{i=1}^n \frac{\pi}{n} \sin c_i^2,$$

$c_i = \dfrac{x_i + x_{i-1}}{2}$

$n \geq 40 \Rightarrow$ Riemann sum $\approx .77$

11. $\int_0^1 2x\, dx$

$\Delta x = \dfrac{1-0}{n} = \dfrac{1}{n}$

$x_0 = 0,\ x_1 = x_0 + \Delta x = \dfrac{1}{n},$

$x_2 = x_1 + \Delta x = \dfrac{1}{n} + \dfrac{1}{n} = \dfrac{2}{n},\ x_i = \dfrac{i}{n}$

$R_n = \sum_{i=1}^n f(x_i)\Delta x$

$\quad = \sum_{i=1}^n (2x_i)\Delta x$

$\quad = \sum_{i=1}^n (2x_i)\dfrac{1}{n}$

$\quad = \sum_{i=1}^n \dfrac{2i}{n} \cdot \dfrac{1}{n}$

$\quad = \sum_{i=1}^n \dfrac{2i}{n^2}$

$\quad = \dfrac{2}{n^2} \sum_{i=1}^n i$

$\quad = \dfrac{2}{n^2} \dfrac{(n)(n+1)}{2}$

$\quad = \dfrac{n^2 + n}{n^2}$

$\int_0^1 2x\, dx = \lim_{n \to \infty} \left(\dfrac{n^2 + n}{n^2} \right) = 1$

13. $\int_0^2 x^2 dx$

$\Delta x = \dfrac{2-0}{n} = \dfrac{2}{n}$

$x_0 = 0,\ x_1 = \dfrac{2}{n},\ x_2 = \dfrac{4}{n},\ x_i = \dfrac{2i}{n}$

$R_n = \sum_{i=1}^n f(x_i)\Delta x$

$\quad = \sum_{i=1}^n x_i^2 \Delta x$

$\quad = \sum_{i=1}^n \left(\dfrac{2i}{n} \right)^2 \left(\dfrac{2}{n} \right)$

$\quad = \dfrac{8}{n^3} \sum_{i=1}^n i^2$

$\quad = \dfrac{8}{n^3} \dfrac{n(n+1)(2n+1)}{6}$

$\int_0^2 x^2 dx = \lim_{n \to \infty} \left(\dfrac{8}{6n^2}(n+1)(2n+1) \right)$

$\quad\quad = \dfrac{8}{3}$

15. $\int_1^3 (2x-1)dx$

$\Delta x = \dfrac{3-1}{n} = \dfrac{2}{n}$

$x_0 = 1,\ x_1 = 1 + \dfrac{2}{n},\ x_2 = 1 + \dfrac{4}{n},$

$x_i = 1 + \dfrac{2i}{n} = \dfrac{n+2i}{n}$

$R_n = \sum_{i=1}^n f(x_i)\Delta x$

$\quad = \sum_{i=1}^n (2x_i - 1)\Delta x$

$\quad = \sum_{i=1}^n \left[2\left(\dfrac{n+2i}{n} \right) - 1 \right] \cdot \dfrac{2}{n}$

$\quad = \dfrac{2}{n} \sum_{i=1}^n 1 + \dfrac{8}{n^2} \sum_{i=1}^n i$

$\quad = \dfrac{2}{n} \cdot n + \dfrac{8}{n^2} \cdot \dfrac{n(n+1)}{2}$

$\quad = 2 + 4 + \dfrac{4}{n}$

$\quad = 6 + \dfrac{4}{n}$

$\int_1^3 (2x-1)dx = \lim_{n \to \infty} \left(6 + \dfrac{4}{n} \right) = 6$

17. $\int_{-2}^{2}\left(4-x^2\right)dx$

19. $-\int_{-2}^{2}\left(x^2-4\right)dx$

21. $\int_{0}^{2}x^2dx$

23. $\int_{0}^{\pi}\sin x\,dx$

25. $\int_{0}^{1}\left(x^3-3x^2+2x\right)dx-\int_{1}^{2}\left(x^3-3x^2+2x\right)dx$

27. $v(t)=60-16t,\ [0,2]$

distance traveled over $[0,2]$

$=\int_{0}^{2}(60-16t)dt$

$\approx\sum_{i=1}^{n}(60-16c_i)\Delta t$

$=\sum_{i=1}^{n}(60-16c_i)\dfrac{2-0}{n}$

$=\sum_{i=1}^{n}\dfrac{2}{n}(60-16c_i),$

$c_i=\dfrac{t_i+t_{i-1}}{2}$

Any value of n gives a Riemann sum = 88.
Since the starting position is 2, the final position is $2+88=90$.

29. $v(t)=40\left(1-e^{-2t}\right),\ [0,4]$

distance traveled over $[0,4]$

$=\int_{0}^{4}40\left(1-e^{-2t}\right)dt$

$\approx 40\sum_{i=1}^{n}\left(1-e^{-2c_i}\right)\Delta t$

$=40\sum_{i=1}^{n}\left(1-e^{-2c_i}\right)\dfrac{4}{n},\ c_i=\dfrac{t_i+t_{i-1}}{2}$

$n\geq 40\ \Rightarrow$ Riemann sum ≈ 140.0. Since the starting position is 0, the final position is 140.0.

31. $\int_{0}^{2}f(x)dx+\int_{2}^{3}f(x)dx=\int_{0}^{3}f(x)dx$

33. $\int_{0}^{2}f(x)dx+\int_{2}^{1}f(x)dx=\int_{0}^{1}f(x)dx$

35.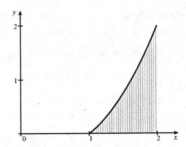

37.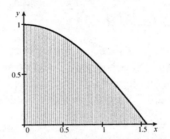

39. Max of the integrand is ≈ 1.36981;
min of the integrand is ≈ -2.343636.
The interval of integration has length

$\dfrac{\pi}{2}-\dfrac{\pi}{3}=\dfrac{\pi}{6}.$

$-2.343636\cdot\dfrac{\pi}{6}\leq I\leq 1.36981\cdot\dfrac{\pi}{6}$

$-1.227\leq I\leq 0.718$

41. Max of the integrand is 3;
min of the integrand is 0.
The interval of integration has length 2.

$0\cdot 2\leq I\leq 3\cdot 2$

$0\leq I\leq 6$

43. We need $3x^2=4$, so $x=\dfrac{2}{\sqrt{3}}$.

45. Substituting g' for f in the Integral Mean Value Theorem, the theorem says that for g' continuous on $[a, b]$, there is a c in (a, b) such that

$$g'(c) = \frac{1}{b-a}\int_a^b g'(x)\,dx.$$

The Mean Value Theorem says that if g is continuous $[a, b]$ and differentiable on (a, b), there is some c in (a, b) (perhaps a different c) such that

$$g'(c) = \frac{g(b) - g(a)}{b-a}.$$

Note that since g' satisfies the hypotheses of the Integral Mean Value theorem, g' is continuous on $[a, b]$, so certainly g will be continuous on $[a, b]$, and thus the conditions of the Mean Value Theorem are fulfilled for the function g.

47. Since f is integrable on $[a, b]$, we have

$$\int_a^b f(x)\,dx = \lim_{n\to\infty}\sum_{i=1}^n f(c_i)\Delta x \quad \text{where}$$

$c_i \in [a + (i-1)\Delta x,\, a + i\Delta x]$ and $\Delta x = \dfrac{b-a}{n}$.

So for n sufficiently large, we have

$$\int_a^b f(x)\,dx \approx \sum_{i=1}^n f(c_i)\Delta x,\text{ and we might as well}$$

take n to be even.

When n is even, then $c = \dfrac{a+b}{2} = a + \dfrac{n}{2}\Delta x$, so we can rewrite the Riemann sum

$$\sum_{i=1}^n f(c_i)\Delta x = \sum_{i=1}^{n/2} f(c_i)\Delta x + \sum_{i=\frac{n}{2}+1}^{n} f(c_i)\Delta x$$

Note:

$$\Delta x = \frac{b-a}{n} = \frac{(2c-a)-a}{n} = \frac{2(c-a)}{n} = \frac{c-a}{\left(\frac{n}{2}\right)},\text{ and}$$

$$\Delta x = \frac{b-a}{n} = \frac{b-(2c-b)}{n} = \frac{2(b-a)}{n} = \frac{b-c}{\left(\frac{n}{2}\right)}$$

Thus,

$$\sum_{i=1}^{n/2} f(c_i)\Delta x = \sum_{i=1}^{n/2} f(c_i)\cdot\frac{c-a}{\left(\frac{n}{2}\right)} \approx \int_a^c f(x)\,dx,\text{ and}$$

$$\sum_{i=\frac{n}{2}+1}^{n} f(c_i)\Delta x = \sum_{i=\frac{n}{2}+1}^{n} f(c_i)\cdot\frac{b-c}{\left(\frac{n}{2}\right)} \approx \int_c^b f(x)\,dx$$

So as $n \to \infty$, we have

$$\int_a^b f(x)\,dx = \int_a^c f(x)\,dx + \int_c^b f(x)\,dx.$$

49. positive

51. negative

53. If n is even and we use midpoint evaluation, we never have an evaluation at $x = 1$, which is the only x for which functions have different values. Since the approximate Riemann sums with midpoint evaluation are equal for all even values of n, $\displaystyle\lim_{n\to\infty}\sum_{i=1}^n f(c_i)\Delta x = \lim_{n\to\infty}\sum_{i=1}^n g(c_i)\Delta x$, so the integrals are equal.

55. $\displaystyle\int_0^4 f(x)\,dx = \int_0^1 2x\,dx + \int_1^4\left(x^2 + 2\right)dx$

$\displaystyle\int_0^1 2x\,dx = 1$ (by calculating area of triangle formed by $y = 2x$, $x = 1$, and the x-axis)

$\displaystyle\int_1^4 (x^2 + 2)\,dx \approx 27$

(using midpoint evaluation with $n = 10$)

So $\displaystyle\int_0^4 f(x)\,dx \approx 1 + 27 = 28$

57. $\displaystyle\int_0^4 f(x)\,dx = \int_0^3 2x^2\,dx + \int_3^4 (x+1)\,dx$

$\displaystyle\int_0^3 2x^2\,dx \approx 18$

(using midpoint evaluation with $n = 10$)

$\displaystyle\int_3^4 (x+1)\,dx = 4.5$

(by calculating area of trapezoid formed by $y = x + 1$, $x = 3$, $x = 4$ and the x-axis)

So $\displaystyle\int_0^4 f(x)\,dx \approx 18 + 4.5 = 22.5$

59. $\displaystyle\int_0^{12}[b(t) - a(t)]\,dt$ is the births minus deaths in the first year, which is the net change in population.

$b(t) = 410 - .3t$

$a(t) = 390 + .2t$

$b(t) > a(t)$

$410 - .3t > 390 + .2t$

$20 > .5t$

For $40 > t, b(t) > a(t)$

The population is increasing for $t < 40$, decreasing for $t > 40$, max at $t = 40$.

61. $PV = 10$, $P = \dfrac{10}{v}$

$\int_2^4 P(V) dV = \int_2^4 \dfrac{10}{V} dV \approx 6.93$

(using midpoint evaluation with $n = 10$).

63. $f(x) = 2x + 1, [0, 4]$

$f_{ave} = \dfrac{1}{4-0} \int_0^4 (2x+1) dx = \dfrac{1}{4} \int_0^4 (2x+1) dx = 5$

(by calculating area of trapezoid formed by $y = 2x + 1$, $x = 0$, $x = 4$ and x-axis)

65. $f(x) = x^2 - 1, [1, 3]$

$f_{ave} = \dfrac{1}{3-1} \int_1^3 (x^2 - 1) dx = \dfrac{1}{2} \int_1^3 (x^2 - 1) dx \approx 3.33$

(using midpoint evaluation with $n = 10$)

67. $f(x) = \cos x, \left[0, \dfrac{\pi}{2}\right]$

$f_{ave} = \dfrac{1}{\frac{\pi}{2} - 0} \int_0^{\frac{\pi}{2}} \cos x \, dx$

$= \dfrac{1}{\frac{\pi}{2}} \int_0^{\frac{\pi}{2}} \cos x \, dx$

$= \dfrac{2}{\pi} \int_0^{\frac{\pi}{2}} \cos x \, dx$

$\approx .64$

(using midpoint evaluation with n = 10)

69. Since $F(t) = 9 - 10^8 (t - 0.0003)^2$ is a quadratic function, with a graph that is a parabola opening downward, $F(t)$ has a maximum value and this occurs at its vertex; that is, when $t = 0.0003$. So $F(0.0003) = 9$ thousand pounds is the maximum force on the ball.

$m\Delta v = \int_a^b F(t) dt$

$m = 0.01$, $a = 0$, $b = 0.0006$

$v(0.0006) - v(0)$

$= \dfrac{1}{0.01} \int_0^{0.0006} (9 - 10^8 (t - 0.0003)^2) \, dt$

≈ 0.36 thousand ft/s

$= 360$ ft/s

(using midpoint evaluation with $n = 20$)

71. If f is continuous on $[a, b]$, there is a $c \in [a, b]$ such that $f(c) = \dfrac{1}{b-a} \int_a^b f(x) dx$.

So let $n = 1$, and $x_i = c$, $\Delta x = b - a$.

Then $\sum_{i=1}^n f(x_i) \Delta x = f(c)(b-a) = \int_a^b f(x) dx$.

Section 4.5

5. $\int_0^2 (2x - 3) dx = \left(x^2 - 3x\right)\Big|_0^2 = (4 - 6) - (0) = -2$

7. $\int_{-1}^1 \left(x^3 + 2x\right) dx = \left(\dfrac{x^4}{4} + x^2\right)\Big|_{-1}^1$

$= \left(\dfrac{1}{4} + 1\right) - \left(\dfrac{1}{4} + 1\right)$

$= 0$

9. $\int_0^4 \left(\sqrt{x} + 3x\right) dx = \left(\dfrac{2}{3} x^{3/2} + \dfrac{3}{2} x^2\right)\Big|_0^4$

$= \left(\dfrac{16}{3} + 24\right) - (0)$

$= \dfrac{88}{3}$

11. $\int_0^1 \left(x\sqrt{x} + x^{-1/2}\right) dx = \left(\dfrac{2}{5} x^{5/2} + 2x^{1/2}\right)\Big|_0^1$

$= \left(\dfrac{2}{5} + 2\right) - (0)$

$= \dfrac{12}{5}$

13. $\int_0^{\pi/2} 2\sin x \, dx = -2 \cos x \Big|_0^{\pi/2}$

$= -2(0) - (-2)(1)$

$= 2$

15. $\int_0^{\pi/4} \sec x \tan x \, dx = \sec x \Big|_0^{\pi/4} = \sqrt{2} - 1$

17. $\int_{\pi/2}^{\pi} (2 \sin x - \cos x) dx = (-2 \cos x - \sin x)\Big|_{\pi/2}^{\pi}$

$= (2 - 0) - (0 - 1)$

$= 3$

19. $\int_0^1 \left(e^x - e^{-x} \right) dx = \left(e^x + e^{-x} \right)\Big|_0^1$

$\qquad\qquad = \left(e^1 + e^{-1} \right) - \left(e^0 + e^0 \right)$

$\qquad\qquad = e + \dfrac{1}{e} - 2$

21. $\int_0^3 \left(3e^{2x} - x^2 \right) dx = \left(\dfrac{3}{2} e^{2x} - \dfrac{x^3}{3} \right)\Big|_0^3$

$\qquad\qquad = \left(\dfrac{3}{2} e^6 - 9 \right) - \left(\dfrac{3}{2} - 0 \right)$

$\qquad\qquad = \dfrac{3}{2} e^6 - \dfrac{21}{2}$

23. $\int_{-1}^1 \left(e^x + e^{-x} \right)^2 dx$

$\qquad = \int_{-1}^1 \left(e^{2x} + 2 + e^{-2x} \right) dx$

$\qquad = \left(\dfrac{e^{2x}}{2} + 2x - \dfrac{e^{-2x}}{2} \right)\Big|_{-1}^1$

$\qquad = \left(\dfrac{e^2}{2} + 2 - \dfrac{e^{-2}}{2} \right) - \left(\dfrac{e^{-2}}{2} - 2 - \dfrac{e^2}{2} \right)$

$\qquad = e^2 + 4 - e^{-2}$

25. $\int_1^4 \dfrac{x-3}{x} dx = \int_1^4 \left(1 - 3x^{-1} \right) dx$

$\qquad = \left(x - 3 \ln |x| \right)\Big|_1^4$

$\qquad = (4 - 3 \ln 4) - (1 - 3 \ln 1)$

$\qquad = 3 - 3 \ln 4$

27. $\int_0^4 x(x-2)dx = \int_0^4 \left(x^2 - 2x \right) dx \Big|_0^4$

$\qquad = \left(\dfrac{x^3}{3} - x^2 \right)$

$\qquad = \left(\dfrac{64}{3} - 16 \right) - (0)$

$\qquad = \dfrac{16}{3}$

29. $\int_0^3 \left(x^3 - \sin x \right) dx = \left(\dfrac{x^4}{4} + \cos x \right)\Big|_0^3$

$\qquad = \left(\dfrac{81}{4} + \cos 3 \right) - (0+1)$

$\qquad = \dfrac{77}{4} + \cos 3$

31. $\int_0^2 \sqrt{x^2 + 1}\, dx \approx \sum_{i=1}^n \sqrt{c_i^2 + 1} \left(\dfrac{2}{n} \right),\ c_i = \dfrac{x_i + x_{i-1}}{2}$

$\qquad n \geq 10 \Rightarrow$ Riemann sum ≈ 2.96

33. $\int_0^1 \left(e^x + 1 \right)^2 dx = \int_0^1 \left(e^{2x} + 2e^x + 1 \right) dx$

$\qquad = \left(\dfrac{e^{2x}}{2} + 2e^x + x \right)\Big|_0^1$

$\qquad = \left(\dfrac{e^2}{2} + 2e + 1 \right) - \left(\dfrac{1}{2} + 2 + 0 \right)$

$\qquad = \dfrac{e^2}{2} + 2e - \dfrac{3}{2}$

35. $\int_1^4 \dfrac{x^2}{x^2 + 4} dx \approx \sum_{i=1}^n \dfrac{c_i^2}{c_i^2 + 4} \left(\dfrac{3}{n} \right),\ c_i = \dfrac{x_i + x_{i-1}}{2}$

$\qquad n \geq 10 \Rightarrow$ Riemann sum ≈ 1.71

37. $\int_0^\pi \sin x^2\, dx \approx \sum_{i=1}^n \sin c_i^2 \left(\dfrac{\pi}{n} \right),\ c_i = \dfrac{x_i + x_{i-1}}{2}$

$\qquad n \geq 40 \Rightarrow$ Riemann sum $\approx .77$

39. $\int_0^{\pi/4} \dfrac{\sin x}{\cos^2 x} dx = \int_0^{\pi/4} \sec x \tan x\, dx$

$\qquad = \sec x \Big|_0^{\pi/4}$

$\qquad = \sqrt{2} - 1$

41. $f(x) = \int_0^x \left(t^2 - 3t + 2 \right) dt = x^2 - 3x + 2$

43. $f(x) = \int_0^{x^2} \left(e^{-t^2} + 1 \right) dt$

$\qquad$ Let $u = x^2$, then

$\qquad f(x) = \int_0^u \left(e^{-t^2} + 1 \right) dt$

$\qquad f'(x) = \left(e^{-u^2} + 1 \right) \dfrac{du}{dx} = \left(e^{-x^4} + 1 \right) 2x$

45. $f(x) = \int_x^{-1} \ln \left(t^2 + 1 \right) dt = -\int_{-1}^x \ln \left(t^2 + 1 \right) dt$

$\qquad f'(x) = -\ln \left(x^2 + 1 \right)$

47. $\int_{-2}^2 4 - x^2\, dx = \dfrac{32}{3}$

49. $-\int_{-2}^2 x^2 - 4\, dx = \dfrac{32}{3}$

51. $\int_0^2 x^2 \, dx = \frac{8}{3}$

53. $\int_0^\pi \sin x \, dx = 2$

55. $y = \int_0^x \sin \sqrt{t^2 + \pi^2} \, dt, \ x = 0$

$y'(x) = \sin \sqrt{x^2 + \pi^2}$

$y'(0) = \sin \sqrt{0 + \pi^2} = 0$

$y(0) = \int_0^0 \sin \sqrt{t^2 + \pi^2} \, dt = 0$

The tangent line is $y = 0$.

57. $y = \int_2^x \cos\left(\pi t^3\right) dt, \ x = 2$

$y'(x) = \cos\left(\pi x^3\right)$

$y'(2) = \cos 8\pi = 1$

$y(2) = \int_2^2 \cos\left(\pi t^3\right) dt = 0$

The tangent line is $y = x - 2$.

59.

$f(x) = \int_0^x \left(t^2 - 3t + 2\right) dt$

$f'(x) = x^2 - 3x + 2 = (x - 2)(x - 1)$

$x = 2, 1$ are critical points

$f''(x) = 2x - 3$

$f''(2) = 1$

$f''(1) = -1$

so $x = 2$ is local min, $x = 1$ is local max

61. a. $\frac{1}{x^2} > 0$, so the area can't be negative.

  b. The integral is incorrect because $\frac{1}{x^2}$ is not defined at $x = 0$.

63. $v(t) = 40 - \sin t, \ s(0) = 2$

$s(t) = \int v(t) \, dt = \int (40 - \sin t) \, dt$

$\quad = 40t + \cos t + c$

$s(0) = 40 \cdot 0 + \cos 0 + c = 2$

$c = 1$

$s(t) = 40t + \cos t + 1$

65. $v(t) = 25\left(1 - e^{-2t}\right), \ s(0) = 0$

$s(t) = \int v(t) \, dt$

$\quad = \int 25\left(1 - e^{-2t}\right) dt$

$\quad = 25t + \frac{25}{2} e^{-2t} + c$

$s(0) = 0 + \frac{25}{2} + c$

$c = -\frac{25}{2}$

$s(t) = 25t + \frac{25}{2} e^{-2t} - \frac{25}{2}$

67. $a(t) = 4 - t, \ v(0) = 8, \ s(0) = 0$

$v(t) = \int a(t) \, dt = \int (4 - t) \, dt = 4t - \frac{t^2}{2} + c_1$

$v(0) = 0 - 0 + c_1 = 8, \ c_1 = 8$

$v(t) = 4t - \frac{t^2}{2} + 8$

$s(t) = \int v(t) \, dt$

$\quad = \int \left(4t - \frac{t^2}{2} + 8\right) dt$

$\quad = 2t^2 - \frac{t^3}{6} + 8t + c_2$

$s(0) = 0 - 0 + 0 + c_2 = 0, \ c_2 = 0$

$s(t) = 2t^2 - \frac{t^3}{6} + 8t$

69. $a(t) = 24 + e^{-t}, \ v(0) = 0, \ s(0) = 0$

$v(t) = \int a(t) \, dt$

$\quad = \int \left(24 + e^{-t}\right) dt$

$\quad = 24t - e^{-t} + c_1$

$v(0) = 0 - 1 + c_1 = 0, \ c_1 = 1$

$v(t) = 24t - e^{-t} + 1$

$s(t) = \int v(t) \, dt$

$\quad = \int \left(24t - e^{-t} + 1\right) dt$

$\quad = 12t^2 + e^{-t} + t + c_2$

$s(0) = 0 + 1 + 0 + c_2 = 0, \ c_2 = -1$

$s(t) = 12t^2 + e^{-t} + t - 1$

71. $\alpha(t) = 10 \text{ rad/} s^2$

$\omega(t) = \int a(t)dt = \int 10 dt = 10t + c_1$

$\omega(0) = 10 \cdot 0 + c_1 = 0, \ \omega(t) = 10t$

$\omega(.8) = 10(.8) = 8 \text{ rad/s}$

$v = 3\omega = 3 \cdot 8 = 24 \text{ ft/s}$

$\theta(t) = \int \omega(t)dt = \int 10t \, dt = 5t^2 + c_2$

$\theta(0) = 5 \cdot 0^2 + c_2 = 0$

$\theta(t) = 5t^2$

$\theta(.8) = 5(.8)^2 = 3.2 \text{ rad}$

73. $t = 4$ corresponds to 1974.

$16.1e^{.07(4)} \approx 21.3024$

$21.3e^{.04(4-4)} \approx 21.3$

$\int_4^{10} 16.1e^{.07t} dt = \frac{16.1}{.07}e^{.07t}\Big|_4^{10} \approx 158.84$

$\int_4^{10} 21.3e^{.04(t-4)} dt = \frac{21.3}{.04}e^{.04(t-4)}\Big|_4^{10} \approx 144.44$

$158.84 - 144.44 = 14.4$ million barrels saved.

75. $F(x) = 1000 \text{ lb}, \ b'(t) = 130 \text{ ft/s}$

$F\big(b(t)\big)b'(t) = 1000 \text{ lb} \cdot 130 \text{ ft/s}$

$= 130,000 \text{ ft-lb/s}$

$= \frac{130,000}{550} \text{hp}$

$\approx 236.36 \text{ hp}$

77. $f(x) = x^2 - 1, \ [1, 3]$

$f_{ave} = \frac{1}{3-1}\int_1^3 \left(x^2 - 1\right)dx$

$= \frac{1}{2}\left(\frac{x^3}{3} - x\right)\Big|_1^3$

$= \frac{1}{2}\left[(9-3) - \left(\frac{1}{3} - 1\right)\right]$

$= \frac{10}{3}$

79. $f(x) = 2x - 2x^2, \ [0, 1]$

$f_{ave} = \frac{1}{1-0}\int_0^1 \left(2x - 2x^2\right)dx$

$= \left(x^2 - \frac{2}{3}x^3\right)\Big|_0^1$

$= \left(1 - \frac{2}{3}\right) - (0)$

$= \frac{1}{3}$

81. $f(x) = \cos x, \ \left[0, \frac{\pi}{2}\right]$

$f_{ave} = \frac{1}{\frac{\pi}{2} - 0}\int_0^{\pi/2} \cos x \, dx$

$= \frac{2}{\pi}(\sin x)\Big|_0^{\pi/2}$

$= \frac{2}{\pi}(1 - 0)$

$= \frac{2}{\pi}$

83. smallest to largest:

$\int_0^3 f(x)dx, \ \int_0^2 f(x)dx, \ \int_0^1 f(x)dx$

85. Let F be an antiderivative of f. Then by the Fundamental Theorem,

$\int_{a(x)}^{b(x)} f(t)\,dt = F(b(x)) - F(a(x)).$

Now by the chain rule,

$\frac{d}{dx}\int_{a(x)}^{b(x)} f(t)\,dt$

$= \frac{d}{dx}\big(F(b(x)) - F(a(x))\big)$

$= f(b(x))b'(x) - f(a(x))a'(x).$

Section 4.6

5. $u = x^3 + 2, \ du = 3x^2 dx$

$\int x^2 \sqrt{x^3 + 2}\,dx = \frac{1}{3}\int \sqrt{u}\,du$

$= \frac{1}{3} \cdot \frac{2}{3}u^{3/2} + c$

$= \frac{2}{9}\left(x^3 + 2\right)^{3/2} + c$

7. $u = \sqrt{x} + 2, \ du = \frac{1}{2\sqrt{x}}dx$

$\int \frac{\left(\sqrt{x} + 2\right)^3}{\sqrt{x}}\,dx = 2\int u^3 du$

$= \frac{1}{2}u^4 + c$

$= \frac{1}{2}\left(\sqrt{x} + 2\right)^4 + c$

9. $u = x^2 - 3, du = 2x\, dx$

$$\int x\left(x^2 - 3\right)^4 dx = \int \frac{1}{2}u^4 du$$

$$= \frac{1}{10}u^5 + c$$

$$= \frac{\left(x^2 - 3\right)^5}{10} + c$$

11. $u = x^2 + x, du = (2x+1)dx$

$$\int (2x+1)\left(x^2 + x\right)^3 dx = \int u^3 du$$

$$= \frac{u^4}{4} + c$$

$$= \frac{1}{4}\left(x^2 + x\right)^4 + c$$

13. $u = \sin x + 1, du = \cos x\, dx$

$$\int \cos x \sqrt{\sin x + 1}\, dx = \int \sqrt{u}\, du$$

$$= \frac{2}{3}u^{3/2} + c$$

$$= \frac{2}{3}(\sin x + 1)^{3/2} + c$$

15. $u = \cos x, du = -\sin x\, dx$

$$\int \frac{\sin x}{\sqrt{\cos x}}\, dx = -\int \frac{du}{\sqrt{u}}$$

$$= -2\sqrt{u} + c$$

$$= -2\sqrt{\cos x} + c$$

17. $u = x^3, du = 3x^2 dx$

$$\int x^2 \cos x^3 dx = \frac{1}{3}\int \cos u\, du$$

$$= \frac{1}{3}\sin u + c$$

$$= \frac{1}{3}\sin x^3 + c$$

19. $u = \cos x + 3, du = -\sin x\, dx$

$$\int \sin x (\cos x + 3)^{3/4} dx = -\int u^{3/4} du$$

$$= -\frac{4}{7}u^{7/4} + c$$

$$= -\frac{4}{7}(\cos x + 3)^{7/4} + c$$

21. $u = x^2 + 1, du = 2x\, dx$

$$\int xe^{x^2+1} dx = \frac{1}{2}\int e^u du = \frac{1}{2}e^u + c = \frac{1}{2}e^{x^2+1} + c$$

23. $u = \sqrt{x}, du = \frac{1}{2\sqrt{x}}dx$

$$\int \frac{e^{\sqrt{x}}}{\sqrt{x}}dx = 2\int e^u du = 2e^u + c = 2e^{\sqrt{x}} + c$$

25. $u = x^3 - 2, du = 3x^2 dx$

$$\int \frac{x^2}{\sqrt{x^3 - 2}}dx = \frac{1}{3}\int \frac{1}{\sqrt{u}}du$$

$$= \frac{2}{3}\sqrt{u} + c$$

$$= \frac{2}{3}\sqrt{x^3 - 2} + c$$

27. $u = \ln x + 2, du = \frac{1}{x}dx$

$$\int \frac{(\ln x + 2)^2}{x}dx = \int u^2 du$$

$$= \frac{1}{3}u^3 + c$$

$$= \frac{1}{3}(\ln x + 2)^3 + c$$

29. $u = x^2 + x - 1, du = (2x+1)dx$

$$\int \frac{2x+1}{x^2 + x - 1}dx = \int \frac{1}{u}du$$

$$= \ln|u| + c$$

$$= \ln\left|x^2 + x - 1\right| + c$$

31. $u = \sqrt{x} + 1, du = \frac{1}{2\sqrt{x}}dx$

$$\int \frac{1}{\sqrt{x}\left(\sqrt{x} + 1\right)^2}dx = 2\int u^{-2} du$$

$$= -2u^{-1} + c$$

$$= -2\left(\sqrt{x} + 1\right)^{-1} + c$$

33. $u = \sin x, du = \cos x\, dx$

$$\int \cos x \cdot e^{\sin x} dx = \int e^u du = e^u + c = e^{\sin x} + c$$

35. $u = \cos x - 1, du = -\sin x\, dx$

$$\int \sin x (\cos x - 1)^3 dx = -\int u^3 du$$

$$= -\frac{1}{4}u^4 + c$$

$$= -\frac{1}{4}(\cos x - 1)^4 + c$$

37. $u = \ln x + 1, \; du = \dfrac{1}{x} du$

$$\int \frac{4}{x(\ln x + 1)^2} dx = 4 \int u^{-2} du$$

$$= -4u^{-1} + c$$

$$= -4(\ln x + 1)^{-1} + c$$

39. $u = e^x + e^{-x}, \; du = \left(e^x - e^{-x}\right) dx$

$$\int \frac{e^x - e^{-x}}{e^x + e^{-x}} dx = \int \frac{1}{u} du$$

$$= \ln|u| + c$$

$$= \ln\left(e^x + e^{-x}\right) + c$$

41. $u = \ln \sqrt{x}, \; du = \dfrac{1}{2x} dx$

$$\int \frac{1}{x \ln \sqrt{x}} dx = 2 \int \frac{1}{u} du$$

$$= 2 \ln|u| + c$$

$$= 2 \ln\left|\ln \sqrt{x}\right| + c$$

43. $u = x + 7, \; du = dx, \; x = u - 7$

$$\int \frac{2x + 3}{x + 7} dx = \int \frac{2(u - 7) + 3}{u} du$$

$$= \int \left(2 - \frac{11}{u}\right) du$$

$$= 2u - 11 \ln|u| + c$$

$$= 2(x + 7) - 11 \ln|x + 7| + c$$

45. $\displaystyle\int \frac{x^2}{\sqrt[3]{x + 3}} dx$

$$= \int \left[\frac{(x + 3)^2 - 6(x + 3) + 9}{\sqrt[3]{x + 3}} \right] dx$$

$$= \int \left[(x + 3)^{5/3} - 6(x + 3)^{2/3} + 9(x + 3)^{-1/3} \right] dx$$

$$= I$$

$u = x + 3, \; du = dx$

$$I = \int \left[u^{5/3} - 6u^{2/3} + 9u^{-1/3} \right] du$$

$$= \frac{3}{8} u^{8/3} - \frac{18}{5} u^{5/3} + \frac{27}{2} u^{2/3} + c$$

$$= \frac{3}{8}(x + 3)^{8/3} - \frac{18}{5}(x + 3)^{5/3} + \frac{27}{2}(x + 3)^{2/3} + c$$

47. $u = x^2 + 1, \; du = 2x \, dx, \; u(0) = 1, \; u(2) = 5$

$$\int_0^2 x\sqrt{x^2 + 1}\, dx = \frac{1}{2} \int_1^5 \sqrt{u}\, du$$

$$= \frac{1}{2} \cdot \frac{2}{3} u^{3/2} \Big|_1^5$$

$$= \frac{1}{3}\left(\sqrt{125} - 1\right)$$

$$= \frac{5}{3}\sqrt{5} - \frac{1}{3}$$

49. $u = x^2 + 1, \; du = 2x \, dx, \; u(-1) = 2, \; u(1) = 2$

$$\int_{-1}^1 \frac{x}{\left(x^2 + 1\right)^2} dx = \frac{1}{2} \int_2^2 \frac{1}{u^2} du = 0$$

51. $u = \sin x + 1, \; du = \cos x \, dx, \; u\left(\dfrac{\pi}{2}\right) = 2, \; u(\pi) = 1$

$$\int_{\pi/2}^{\pi} \frac{4\cos x}{(\sin x + 1)^2} dx = 4 \int_2^1 \frac{1}{u^2} du = -4u^{-1} \Big|_2^1 = -2$$

53. $u = \sin x, \; du = \cos x \, dx, \; u\left(\dfrac{\pi}{4}\right) = \dfrac{1}{\sqrt{2}}, \; u\left(\dfrac{\pi}{2}\right) = 1$

$$\int_{\pi/4}^{\pi/2} \cot x \, dx = \int_{\pi/4}^{\pi/2} \frac{\cos x}{\sin x} dx$$

$$= \int_{1/\sqrt{2}}^1 \frac{1}{u} du$$

$$= \left(\ln|u|\right)\Big|_{1/\sqrt{2}}^1$$

$$= \ln|1| - \ln\left|\frac{1}{\sqrt{2}}\right|$$

$$= -\ln \frac{1}{\sqrt{2}}$$

$$= \frac{1}{2} \ln 2$$

55. $\displaystyle\int_1^4 \frac{x - 1}{\sqrt{x}} dx = \int_1^4 \left(x^{1/2} - x^{-1/2}\right) dx$

$$= \left(\frac{2}{3} x^{3/2} - 2x^{1/2}\right)\Big|_1^4$$

$$= \left(\frac{16}{3} - 4\right) - \left(\frac{2}{3} - 2\right)$$

$$= \frac{8}{3}$$

57. $\displaystyle\int_0^{\pi} \sin x^2 \, dx \approx .77$

using midpoint evaluation with $n \geq 40$

59. $u = x^2$, $du = 2x\,dx$, $u(-1) = 1$, $u(1) = 1$

$$\int_{-1}^{1} xe^{-x^2}\,dx = \frac{1}{2}\int_{1}^{1} e^{-u}\,du = 0$$

61. $\int_{0}^{2} \dfrac{4}{\left(x^2+1\right)^2}\,dx \approx 3.01$

using midpoint evaluation with $n \geq 10$

63. $\int_{0}^{2} \dfrac{4x^2}{\left(x^2+1\right)^2}\,dx \approx 1.4$

using midpoint evaluation with $n \geq 10$

65. $\int_{0}^{\pi/4} \sec x\,dx \approx .88$

using midpoint evaluation with $n \geq 10$

67. $\dfrac{1}{2}\int_{0}^{4} f(u)\,du$

69. $\int_{0}^{1} f(u)\,du$

71. Assuming we already know the trigonometric identity $\cos(a-b) = \cos(a)\cos(b) - \sin(a)\sin(b)$

we have

$\cos(-x) = \cos(0-x)$

$\qquad = \cos(0)\cos(x) - \sin(0)\sin(x)$

$\qquad = \cos x$

so $\cos x$ is an even function.

Thus, if $f(x) = x\cos x$, then

$f(-x) = (-x)\cos(-x) = -x\cos(x) = -f(x)$

so $f(x)$ is an odd function.

Assuming we already know the trigonometric identity $\sin(a-b) = \sin(a)\cos(b) - \sin(b)\cos(a)$

we have $\sin(-x) = \sin(0-x)$

$\qquad\qquad = \sin(0)\cos(x) - \sin(x)\cos(0)$

$\qquad\qquad = -\sin(x)$

so $\sin x$ is an odd function.

Thus, if $g(x) = x\sin x$, then

$g(-x) = (-x)\sin(-x)$

$\qquad = (-x)(-\sin(x))$

$\qquad = x\sin x$

$\qquad = g(x)$

so $g(x)$ is an even function.

73. $\int_{-1}^{1} x\cos x\,dx = I$, $x\cos x$ is odd, so $I = 0$.

75. $\int_{-1}^{1}(x^4 - 2x^2 + 1)\,dx = I$,

$x^4 - 2x^2 + 1$ is even, so

$I = 2\int_{0}^{1} x^4 - 2x^2 + 1$

$\quad = 2\left(\dfrac{x^5}{5} - \dfrac{2}{3}x^3 + x\right)\Big|_{0}^{1}$

$\quad = 2\left[\left(\dfrac{1}{5} - \dfrac{2}{3} + 1\right) - (0 - 0 + 0)\right]$

$\quad = 2\left(\dfrac{8}{15}\right)$

$\quad = \dfrac{16}{15}$

77. $\int_{-1}^{1}\left(x^2 + \sin x\right)dx = I$

$x^2 + \sin x$ is neither even nor odd.

$I = \left(\dfrac{x^3}{3} - \cos x\right)\Big|_{-1}^{1}$

$\quad = \left(\dfrac{1}{3} - \cos 1\right) - \left(-\dfrac{1}{3} - \cos(-1)\right)$

$\quad = \dfrac{1}{3} - \cos 1 - \left(-\dfrac{1}{3} - \cos 1\right)$

$\quad = \dfrac{2}{3}$

79. $y = f(x) = \sqrt{4 - x^2}$

$\bar{x} = \dfrac{\int_{a}^{b} xf(x)\,dx}{\int_{a}^{b} f(x)\,dx}$

$\int_{a}^{b} xf(x)\,dx = \int_{-2}^{2} x\sqrt{4-x^2} = 0$ since

$x\sqrt{4-x^2}$ odd.

$\int_{a}^{b} f(x)\,dx = \int_{-2}^{2}\sqrt{4-x^2} = 2\pi$

since 2π is the area of a semi-circle of radius 2.

So $\bar{x} = 0$.

$$\bar{y} = \frac{\int_a^b [f(x)]^2 dx}{2\int_a^b f(x) dx}$$

$$\int_a^b [f(x)]^2 dx = \int_{-2}^{2}(4-x^2)dx = 2\int_0^2 (4-x^2)dx$$

$$\bar{y} = \frac{2\int_0^2 (4-x^2)dx}{2(2\pi)}$$

$$= \frac{1}{2\pi}\int_0^2 (4-x^2)dx$$

$$= \frac{1}{2\pi}\left(4x - \frac{x^3}{3}\right)\Big|_0^2$$

$$= \frac{1}{2\pi}\left[\left(8 - \frac{8}{3}\right) - (0-0)\right]$$

$$= \frac{8}{3\pi}$$

81. $rms = \sqrt{f\int_0^{1/f} V^2(t)dt}$

$V(t) = V_p \sin(2\pi ft)$

$V^2(t) = V_p^2 \sin^2(2\pi ft)$

$$= V_p^2\left(\frac{1}{2} - \frac{1}{2}\cos(4\pi ft)\right)$$

$$= \frac{V_p^2}{2}(1 - \cos(4\pi ft))$$

$$\int_0^{1/f} V^2(t)dt$$

$$= \int_0^{1/f} \frac{V_p^2}{2}(1-\cos(4\pi ft))dt$$

$$= \frac{V_p^2}{2}\left(t - \frac{\sin(4\pi ft)}{4\pi f}\right)\Bigg|_0^{1/f}$$

$$= \frac{V_p^2}{2}\left[\left(\frac{1}{f} - \frac{\sin\left(4\pi f \cdot \frac{1}{f}\right)}{4\pi f}\right) - \left(0 - \frac{\sin 0}{4\pi f}\right)\right]$$

$$= \frac{V_p^2}{2}\left(\frac{1}{f}\right)$$

$$rms = \sqrt{f \cdot \frac{V_p^2}{2} \cdot \frac{1}{f}} = \sqrt{\frac{V_p^2}{2}} = \frac{V_p}{\sqrt{2}}$$

Section 4.7

5. $\int_0^1 (x^2 + 1)dx$

Midpoint:

$$\frac{1}{4}\left[f\left(\frac{1}{8}\right) + f\left(\frac{3}{8}\right) + f\left(\frac{5}{8}\right) + f\left(\frac{7}{8}\right)\right]$$

$$= \frac{85}{64}$$

Trapezoidal:

$T_4(f)$

$$= \frac{1-0}{2(4)}\left[f(0) + 2f\left(\frac{1}{4}\right) + 2f\left(\frac{1}{2}\right) + 2f\left(\frac{3}{4}\right) + f(1)\right]$$

$$= \frac{43}{32}$$

Simpson:

$S_4(f)$

$$= \frac{1-0}{3(4)}\left[f(0) + 4f\left(\frac{1}{4}\right) + 2f\left(\frac{1}{2}\right) + 4f\left(\frac{3}{4}\right) + f(1)\right]$$

$$= \frac{4}{3}$$

7. $\int_1^3 \frac{1}{x}dx$

Midpoint:

$$\frac{3-1}{4}\left[f\left(\frac{5}{4}\right) + f\left(\frac{7}{4}\right) + f\left(\frac{9}{4}\right) + f\left(\frac{11}{4}\right)\right]$$

$$= \frac{1}{2}\left(\frac{4}{5} + \frac{4}{7} + \frac{4}{9} + \frac{4}{11}\right) = \frac{3776}{3465}$$

Trapezoidal:

$T_4(f)$

$$= \frac{3-1}{2(4)}\left[f(1) + 2f\left(\frac{3}{2}\right) + 2f(2) + 2f\left(\frac{5}{2}\right) + f(3)\right]$$

$$= \frac{1}{4}\left(1 + \frac{4}{3} + 1 + \frac{4}{5} + \frac{1}{3}\right) = \frac{67}{60}$$

Simpson:

$S_4(f)$

$$= \frac{3-1}{3(4)}\left[f(1) + 4f\left(\frac{3}{2}\right) + 2f(2) + 4f\left(\frac{5}{2}\right) + f(3)\right]$$

$$= \frac{1}{6}\left(1 + \frac{8}{3} + 1 + \frac{8}{5} + \frac{1}{3}\right) = \frac{11}{10}$$

9. $\int_0^2 f(x)dx$

    a. $\dfrac{2-0}{4}\left[f(0)+f(.5)+f(1)+f(1.5)\right]$

        $=\dfrac{1}{2}(1+.25+0+.25)$

        $=.75$

    b. $\dfrac{2-0}{4}\left[f(.25)+f(.75)+f(1.25)+f(1.75)\right]$

        $=\dfrac{1}{2}(.65+.15+.15+.65)$

        $=.8$

    c. $\dfrac{2-0}{2(4)}\left[f(0)+2f(.5)+2f(1)+2f(1.5)+f(2)\right]$

        $=\dfrac{1}{4}(1+.5+0+.5+1)$

        $=.75$

11. $\int_0^\pi \cos x^2\, dx \approx .565694$

n	midpoint	trapezoidal	Simpson
10	.5538	.5889	.5660
20	.5629	.5713	.5655
50	.5652	.5666	.5657

13. $\int_0^2 e^{-x^2}\, dx \approx .882081$

n	midpoint	trapezoidal	Simpson
10	.88220	.88184	.88207
20	.88211	.88202	.88208
50	.88209	.88207	.88208

15. $\int_0^\pi e^{\cos x}\, dx \approx 3.97746$

n	midpoint	trapezoidal	Simpson
10	3.9775	3.9775	3.9775
20	3.9775	3.9775	3.9775
50	3.9775	3.9775	3.9775

17. $\int_0^1 5x^4\,dx = x^5\Big|_0^1 = 1 - 0 = 1$

n	midpoint error	trapezoidal error	Simpson error
10	.00832	−.01665	−.00007
20	.00208	−.00417	-4.2×10^{-6}
40	.00052	−.00104	-2.6×10^{-7}
80	.00013	−.00026	-1.6×10^{-8}

19. $\int_0^\pi \cos x\,dx = \sin x\Big|_0^\pi = 0$

n	midpoint error	trapezoidal error	Simpson error
10	-5.5×10^{-17}	0	0
20	-2.7×10^{-17}	1.6×10^{-16}	1.1×10^{-16}
40	-2.9×10^{-16}	-6.9×10^{-17}	-1.3×10^{-16}
80	-1.7×10^{-16}	-3.1×10^{-16}	1.5×10^{-16}

21. 4, 4 and 16

23. **a.** 1.366162
    b. 1.428091
    c. 1.391621

25. **a.** 0.843666
    b. 0.837084
    c. 0.841489

27. Using Theorems 7.1 and 7.2, we find the following lower bounds for the number of steps needed to guarantee accuracy of 10^{-7} in Exercise 23:

Midpoint: $\sqrt{\dfrac{2\cdot3^3}{24\cdot10^{-7}}} \approx 4744$

Trapezoidal: $\sqrt{\dfrac{2\cdot3^3}{14\cdot10^{-7}}} \approx 6709$

Simpson's: $\sqrt[4]{\dfrac{24\cdot3^5}{180\cdot10^{-7}}} \approx 135$

29. Error bounds and actual errors in Exercise 17:

Number of steps	Midpoint		Trapezoidal		Simpson's	
	Error bound	Error	Error bound	Error	Error bound	Error
10	0.025	0.008319	0.05	0.01665	6.6667×10^{-5}	6.6667×10^{-5}
20	0.00625	0.002082	0.0125	0.004166	4.1667×10^{-6}	4.1667×10^{-6}
40	0.001563	0.000521	0.003125	0.001042	2.6042×10^{-7}	2.6042×10^{-7}
80	0.000391	0.00013	0.000781	0.00026	1.6276×10^{-8}	1.6276×10^{-8}

31. $\int_0^2 f(x)\,dx$

    a. Trapezoidal:

$$T_8(f) = \frac{2-0}{2(8)}\left[f(0) + 2f(.25) + 2f(.5) + 2f(.75) + 2f(1) + 2f(1.25) + 2f(1.5) + 2f(1.75) + f(2)\right]$$

$$= \frac{1}{8}\left[4.0 + 9.2 + 10.4 + 9.6 + 10 + 9.2 + 8.8 + 7.6 + 4.0\right]$$

$$= 9.1$$

    b. Simpson:

$$S_8(f) = \frac{2-0}{3(8)}\left[f(0) + 4f(.25) + 2f(.5) + 4f(.75) + 2f(1) + 4f(1.25) + 2f(1.5) + 4f(1.75) + f(2)\right]$$

$$= \frac{1}{12}(4.0 + 18.4 + 10.4 + 19.2 + 10 + 18.4 + 8.8 + 15.2 + 4.0)$$

$$\approx 9.033$$

33. $\int_0^2 f(x)\,dx$

    a. Trapezoidal:

$$T_{10}(f) = \frac{2-0}{2(10)}[f(0) + 2f(.2) + 2f(.4) + 2f(.6) + 2f(.8) + 2f(1) + 2f(1.2) + 2f(1.4)$$
$$+ 2f(1.6) + 2f(1.8) + f(2)]$$

$$= \frac{1}{10}(2.4 + 5.2 + 5.8 + 6.4 + 6.8 + 7.2 + 7.6 + 7.8 + 8 + 8.2 + 4.2)$$

$$= 6.96$$

    b. Simpson:

$$S_{10}(f) = \frac{2-0}{3(10)}[f(0) + 2f(.2) + 2f(.4) + 2f(.6) + 2f(.8) + 2f(1) + 2f(1.2)$$
$$+ 2f(1.4) + f(1.6) + 2f(1.8) + f(2)]$$

$$= \frac{1}{15}(2.4 + 10.4 + 5.8 + 12.8 + 6.8 + 14.4 + 7.6 + 15.6 + 8 + 16.4 + 4.2)$$

$$= 6.96$$

35. $S_{12}(f) = \dfrac{120-0}{3(12)}\big[f(0) + 4f(10) + 2f(20) + 4f(30) + 2f(40) + 4f(50) + 2f(60) + 4f(70) + 2f(80) + 4f(90)$

$$+\, 2f(100) + 4f(110) + f(120)\big]$$

$$= \dfrac{10}{3}(56 + 216 + 116 + 248 + 116 + 232 + 124 + 224 + 104 + 192 + 80 + 128 + 22)$$

$$\approx 6193 \ \text{ft}^2$$

37. $S_{12}(f) = \dfrac{12-0}{3(12)}\big[f(0) + 4f(1) + 2f(2) + 4f(3) + 2f(4) + 4f(5) + 2f(6) + 4f(7) + 2f(8) + 4f(9) + 2f(10)$

$$+\, 4f(11) + f(12)\big]$$

$$= \dfrac{1}{3}(40 + 168 + 80 + 176 + 96 + 200 + 92 + 184 + 84 + 176 + 80 + 168 + 42)$$

$$\approx 529 \ \text{ft}$$

39. $S_{12}(f) = \dfrac{2.4-0}{3(12)}\big[f(0) + 4f(.2) + 2f(.4) + 4f(.6) + 2f(.8) + 4f(1) + 2f(1.2) + 4f(1.4) + 2f(1.6)$

$$+\, 4f(1.8) + 2f(2) + 4f(2.2) + f(2.4)\big]$$

$$= \dfrac{1}{15}(0 + .8 + .8 + 4 + 3.2 + 8 + 4.4 + 8 + 3.2 + 4.8 + 1.2 + .8 + 0)$$

$$\approx 2.6 \ \text{liters}$$

41. $f''(x) > 0,\ f'(x) > 0$

So f is increasing and concave up.

    a. Midpoint rule would underestimate.

    b. Trapezoidal rule would overestimate.

    c. Simpson's rule—not enough information

43. $f''(x) < 0,\ f'(x) > 0$

So f is increasing and concave down.

    a. Midpoint rule would overestimate.

    b. Trapezoidal rule would underestimate.

    c. Simpson's rule—not enough information

45. $f''(x) = 4,\ f'(x) > 0$

So f is quadratic, concave up and increasing.

    a. Midpoint rule would underestimate.

    b. Trapezoidal rule would overestimate.

    c. Simpson's rule would be exact.

47. $A = Lh_2 + \dfrac{L(h_1 - h_2)}{2}$

$$= Lh_2 + \dfrac{Lh_1}{2} - \dfrac{Lh_2}{2}$$

$$= \dfrac{Lh_1 + Lh_2}{2}$$

$$= \dfrac{L(h_1 + h_2)}{2}$$

49.

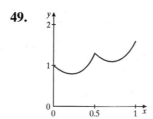

51. Let I be the exact integral. Then we have

$$T_n - I \approx -2(M_n - I)$$

$$T_n - I = 2I - 2M_n$$

$$T_n + 2M_n \approx 3I$$

$$\dfrac{T_n}{3} + \dfrac{2}{3}M_n = I$$

Chapter 4 Review

1. See Example 1.8.

$$\int\left(4x^2-3\right)dx=\frac{4}{3}x^3-3x+c$$

3. See page 328.

$$\int\frac{4}{x}dx=4\ln|x|+c$$

5. See Example 1.9.

$$\int 2\sin 4x\,dx=-\frac{1}{2}\cos 4x+c$$

7. See Example 1.9.

$$\int\left(x-e^{4x}\right)dx=\frac{x^2}{2}-\frac{e^{4x}}{4}+c$$

9. See page 328.

$$\int\frac{x^2+4}{x}dx=\int\left(x+\frac{4}{x}\right)dx=\frac{x^2}{2}+4\ln|x|+c$$

11. See page 324.

$$\int e^x\left(1-e^{-x}\right)dx=\int\left(e^x-1\right)dx=e^x-x+c$$

13. See Example 6.2.

$$u=x^2+4,\ du=2x\,dx$$
$$\int x\sqrt{x^2+4}=\frac{1}{2}\int\sqrt{u}\,du$$
$$=\frac{1}{3}u^{3/2}+c$$
$$=\frac{1}{3}\left(x^2+4\right)^{3/2}+c$$

15. See Example 6.3.

$$u=x^3,\ du=3x^2dx$$
$$\int 6x^2\cos x^3dx=2\int\cos u\,du$$
$$=2\sin u+c$$
$$=2\sin x^3+c$$

17. See Example 6.2.

$$u=\frac{1}{x},\ du=-\frac{1}{x^2}dx$$
$$\int\frac{e^{1/x}}{x^2}dx=-\int e^u\,du$$
$$=-e^u+c$$
$$=-e^{1/x}+c$$

19. See Example 6.7.

$$u=\cos x,\ du=-\sin x\,dx$$
$$\int\tan x\,dx=\int\frac{\sin x}{\cos x}dx$$
$$=-\int\frac{1}{u}du$$
$$=-\ln|u|+c$$
$$=-\ln|\cos x|+c$$

21. See Example 1.13.

$$f'(x)=3x^2+1$$
$$\int f'(x)=\int\left(3x^2+1\right)dx=x^3+x+c$$
$$f(0)=0^3+0+c=2$$
$$f(x)=x^3+x+2$$

23. See Example 1.13.

$$v(t)=-32t+10$$
$$\int(-32t+10)dt=-16t^2+10t+c$$
$$s(0)=-16\cdot 0^2+10\cdot 0+c=2$$
$$s(t)=-16t^2+10t+2$$

25. See Example 2.3.

$$\sum_{i=1}^{6}\left(i^2+3i\right)=\left(1^2+3(1)\right)+\left(2^2+3(2)\right)+\left(3^2+3(3)\right)+\left(4^2+3(4)\right)+\left(5^2+3(5)\right)+\left(6^2+3(6)\right)$$
$$=4+10+18+28+40+54$$
$$=154$$

27. See Example 2.5.

$$\sum_{i=1}^{100}\left(i^2-1\right)=\sum_{i=1}^{100}i^2-\sum_{i=1}^{100}1=\frac{100(101)(201)}{6}-100$$
$$=338,250$$

29. See Example 3.3.

$$\frac{1}{n^3}\sum_{i=1}^{n}\left(i^2-i\right)=\frac{1}{n^3}\left[\sum_{i=1}^{n}i^2-\sum_{i=1}^{n}i\right]=\frac{1}{n^3}\left[\frac{n(n+1)(2n+1)}{6}-\frac{n(n+1)}{2}\right]$$

$$\lim_{n\to\infty}\left(\frac{1}{n^3}\right)\left[\frac{n(n+1)(2n+1)}{6}-\frac{n(n+1)}{2}\right]=\lim_{n\to\infty}\left(\frac{1}{n^3}\right)\left[\frac{(n+1)(2n^2+n-3n)}{6}\right]$$

$$=\lim_{n\to\infty}\frac{(n+1)(2n^2-2n)}{6n^3}=\frac{1}{3}$$

31. See Example 3.1.

$f(x)=x^2$, $[0, 2]$, $n=8$, midpoint evaluation

$$R_8=\frac{2-0}{8}\left[f\left(\frac{1}{8}\right)+f\left(\frac{3}{8}\right)+f\left(\frac{5}{8}\right)+f\left(\frac{7}{8}\right)+f\left(\frac{9}{8}\right)+f\left(\frac{11}{8}\right)+f\left(\frac{13}{8}\right)+f\left(\frac{15}{8}\right)\right]$$

$$=\frac{1}{4}(.015625+.140625+.390625+.765625+1.265625+1.890625+2.640625+3.515625)$$

$$=2.65625$$

33. See Example 3.1.

$f(x)=\sqrt{x+1}$ on $[0, 3]$, $n=8$, midpoint evaluation

$$R_8=\frac{3-0}{8}\left[f\left(\frac{3}{16}\right)+f\left(\frac{9}{16}\right)+f\left(\frac{15}{16}\right)+f\left(\frac{21}{16}\right)+f\left(\frac{27}{16}\right)+f\left(\frac{33}{16}\right)+f\left(\frac{39}{16}\right)+f\left(\frac{45}{16}\right)\right]$$

$$\approx\frac{3}{8}(12.4483)\approx 4.668$$

35. See Example 7.9.

a. $A_8=\dfrac{1.6-0}{8}\left[f(0)+f(.2)+f(.4)+f(.6)+f(.8)+f(1)+f(1.2)+f(1.4)\right]$

$$=\frac{1}{5}(1+1.4+1.6+2+2.2+2.4+2+1.6)$$

$$=2.84$$

b. $A_8=\dfrac{1.6-0}{8}\left[f(.2)+f(.4)+f(.6)+f(.8)+f(1)+f(1.2)+f(1.4)+f(1.6)\right]$

$$=\frac{1}{5}(1.4+1.6+2+2.2+2.4+2+1.6+1.4)$$

$$=2.92$$

c. $T_8=\dfrac{1.6-0}{2(8)}\left[f(0)+2f(.2)+2f(.4)+2f(.6)+2f(.8)+2f(1)+2f(1.2)+2f(1.4)+f(1.6)\right]$

$$=2.88$$

d. $S_8=\dfrac{1.6-0}{3(8)}\left[f(0)+4f(.2)+2f(.4)+4f(.6)+2f(.8)+4f(1)+2f(1.2)+4f(1.4)+f(1.6)\right]$

$$\approx 2.907$$

37. See Example 7.8.

Simpson's Rule is expected to be most accurate.

39. See Example 4.3.

$$\int_0^1 2x^2 dx$$

$$\Delta x = \frac{1-0}{n} = \frac{1}{n}, \ x_i = \frac{i}{n}$$

$$R_n = \sum_{i=1}^{n} f(x_i) \Delta x$$

$$= \sum_{i=1}^{n} \left(2x_i^2\right) \Delta x$$

$$= \sum_{i=1}^{n} \left(2\left(\frac{i}{n}\right)^2\right) \frac{1}{n}$$

$$= \frac{2}{n^3} \sum_{i=1}^{n} i^2$$

$$= \frac{2}{n^3} \cdot \frac{n(n+1)(2n+1)}{6}$$

$$\lim_{n\to\infty} (R_n) = \lim_{n\to\infty} \left(\frac{1}{3} \cdot \frac{(n+1)(2n+1)}{n^2}\right) = \frac{2}{3}$$

41. See Example 4.4 and Example 5.3.

$$\int_0^3 \left(3x - x^2\right) dx = \left(\frac{3x^2}{2} - \frac{x^3}{3}\right)\bigg|_0^3$$

$$= \left(\frac{27}{2} - \frac{27}{3}\right) - (0 - 0)$$

$$= \frac{27}{6}$$

$$= \frac{9}{2}$$

43. See Example 5.10.

$$v(t) = 40 - 10t, \ [1, 2]$$

$$d = \int_1^2 (40 - 10t) dt$$

$$= \left(40t - 5t^2\right)\bigg|_1^2$$

$$= (80 - 20) - (40 - 5)$$

$$= 25$$

45. See Example 4.7.

$$f_{ave} = \frac{1}{2} \int_0^2 e^x dx$$

$$= \frac{1}{2} e^x \bigg|_0^2$$

$$= \frac{1}{2} \left(e^2 - e^0\right)$$

$$= \frac{e^2}{2} - \frac{1}{2}$$

47. See Example 5.1.

$$\int_0^2 \left(x^2 - 2\right) dx = \left(\frac{x^3}{3} - 2x\right)\bigg|_0^2$$

$$= \left(\frac{8}{3} - 4\right) - (0)$$

$$= -\frac{4}{3}$$

49. See Example 5.1.

$$\int_0^{\pi/2} \sin 2x \, dx = -\frac{1}{2} \cos 2x \bigg|_0^{\pi/2}$$

$$= -\frac{1}{2}(-1) - \left(-\frac{1}{2}\right)(1)$$

$$= 1$$

51. See Example 5.4.

$$\int_0^{10} \left(1 - e^{-t/4}\right) dt = \left(t + 4e^{-t/4}\right)\bigg|_0^{10}$$

$$= \left(10 + 4e^{-10/4}\right) - (0 + 4)$$

$$= 6 + 4e^{-5/2}$$

53. See Example 6.9.

$$\int_0^2 \frac{x}{x^2 + 1} dx = I$$

$$u = x^2 + 1, \ du = 2x \, dx, \ u(0) = 1, \ u(2) = 5$$

$$I = \frac{1}{2} \int_1^5 \frac{1}{u} du = \frac{1}{2} \ln|u| \bigg|_1^5 = \frac{1}{2} \ln 5 - 0 = \frac{1}{2} \ln 5$$

55. See Example 6.9.

$$\int_0^2 x\sqrt{x^2 + 4} dx = I$$

$$u = x^2 + 4, \ du = 2x \, dx, \ u(0) = 4, \ u(2) = 8$$

$$I = \frac{1}{2} \int_4^8 \sqrt{u} du = \frac{1}{3} u^{3/2} \bigg|_4^8 = \frac{1}{3}\left(8^{3/2} - 8\right)$$

57. See Example 5.4.

$$\int_0^1 \left(e^x - 2\right)^2 dx = \int_0^1 \left(e^{2x} - 4e^x + 4\right) dx$$

$$= \left(\frac{e^{2x}}{2} - 4e^x + 4x\right)\Bigg|_0^1$$

$$= \left(\frac{e^2}{2} - 4e + 4\right) - \left(\frac{1}{2} - 4 + 0\right)$$

$$= \frac{15}{2} + \frac{1}{2}e^2 - 4e$$

59. See Example 5.7.

$$f(x) = \int_2^x \left(\sin t^2 - 2\right) dt$$

$$f'(x) = \sin x^2 - 2$$

61. See Examples 7.1, 7.5, 7.6.

$$\int_0^1 \sqrt{x^2 + 4}\, dx$$

a. Midpoint:

$$\frac{1-0}{4}\left[f\left(\frac{1}{8}\right) + f\left(\frac{3}{8}\right) + f\left(\frac{5}{8}\right) + f\left(\frac{7}{8}\right)\right]$$

$$\approx 2.079$$

b. Trapezoidal:

$$\frac{1-0}{2(4)}\left[f(0) + 2f\left(\frac{1}{4}\right) + 2f\left(\frac{1}{2}\right) + 2f\left(\frac{3}{4}\right) + f(1)\right]$$

$$\approx 2.083$$

c. Simpson:

$$\frac{1-0}{3(4)}\left[f(0) + 4f\left(\frac{1}{4}\right) + 2f\left(\frac{1}{2}\right) + 4f\left(\frac{3}{4}\right) + f(1)\right]$$

$$\approx 2.080$$

63. (Example 7.8)

n	midpoint	trapezoidal	Simpson
20	2.08041	2.08055	2.08046
40	2.08045	2.08048	2.08046

Chapter 5

Section 5.1

5. $y = x^3$, $y = x^2 - 1$, $1 \le x \le 3$

$$\int_1^3 \left[x^3 - \left(x^2 - 1 \right) \right] dx$$

$$= \left(\frac{x^4}{4} - \frac{x^3}{3} + x \right) \Bigg|_1^3$$

$$= \left(\frac{81}{4} - \frac{27}{3} + 3 \right) - \left(\frac{1}{4} - \frac{1}{3} + 1 \right)$$

$$= \frac{160}{12}$$

$$= \frac{40}{3}$$

7. $y = e^x$, $y = x - 1$, $-2 \le x \le 0$

$$\int_{-2}^0 \left[e^x - (x - 1) \right] dx$$

$$= \left(e^x - \frac{x^2}{2} + x \right) \Bigg|_{-2}^0$$

$$= (1 - 0 + 0) - \left(e^{-2} - \frac{4}{2} - 2 \right)$$

$$= 5 - e^{-2}$$

9. $y = x^2 - 1$, $y = 1 - x$, $0 \le x \le 2$

$$\int_0^1 \left[1 - x - \left(x^2 - 1 \right) \right] dx + \int_1^2 \left[x^2 - 1 - (1 - x) \right] dx$$

$$= \int_0^1 \left(2 - x - x^2 \right) dx + \int_1^2 \left(x^2 + x - 2 \right) dx$$

$$= \left(2x - \frac{x^2}{2} - \frac{x^3}{3} \right) \Bigg|_0^1 + \left(\frac{x^3}{3} + \frac{x^2}{2} - 2x \right) \Bigg|_1^2$$

$$= \left(2 - \frac{1}{2} - \frac{1}{3} \right) - (0 - 0 - 0) + \left(\frac{8}{3} + \frac{4}{2} - 4 \right) - \left(\frac{1}{3} + \frac{1}{2} - 2 \right)$$

$$= 3$$

11. $y = x^3 - 1$, $y = 1 - x$, $-2 \le x \le 2$

$$\int_{-2}^1 \left[1 - x - \left(x^3 - 1 \right) \right] dx + \int_1^2 \left[x^3 - 1 - (1 - x) \right] dx$$

$$= \left(2x - \frac{x^2}{2} - \frac{x^4}{4} \right) \Bigg|_{-2}^1 + \left(\frac{x^4}{4} - 2x + \frac{x^2}{2} \right) \Bigg|_1^2$$

$$= \left(2 - \frac{1}{2} - \frac{1}{4} \right) - (-4 - 2 - 4) + (4 - 4 + 2) - \left(\frac{1}{4} - 2 + \frac{1}{2} \right)$$

$$= \frac{29}{2}$$

13. $y = x^2 - 1,\ y = 7 - x^2$

$$\int_{-2}^{2}\left[7 - x^2 - \left(x^2 - 1\right)\right]dx$$

$$= \left(8x - \frac{2x^3}{3}\right)\Big|_{-2}^{2}$$

$$= \left(16 - \frac{16}{3}\right) - \left(-16 + \frac{16}{3}\right)$$

$$= \frac{64}{3}$$

15. $y = x^2 + 1,\ y = 3x - 1$

$$\int_{1}^{2}\left[3x - 1 - \left(x^2 + 1\right)\right]dx$$

$$= \left(\frac{3x^2}{2} - 2x - \frac{x^3}{3}\right)\Big|_{1}^{2}$$

$$= \left(\frac{12}{2} - 4 - \frac{8}{3}\right) - \left(\frac{3}{2} - 2 - \frac{1}{3}\right)$$

$$= \frac{1}{6}$$

17. $y = x^3,\ y = 3x + 2$

$$\int_{-1}^{2}\left(3x + 2 - x^3\right)dx$$

$$= \left(\frac{3x^2}{2} + 2x - \frac{x^4}{4}\right)\Big|_{-1}^{2}$$

$$= (6 + 4 - 4) - \left(\frac{3}{2} - 2 - \frac{1}{4}\right)$$

$$= \frac{27}{4}$$

19. $y = x^3,\ y = x^2$

$$\int_{0}^{1}\left(x^2 - x^3\right)dx = \left(\frac{x^3}{3} - \frac{x^4}{4}\right)\Big|_{0}^{1}$$

$$= \left(\frac{1}{3} - \frac{1}{4}\right) - (0 - 0)$$

$$= \frac{1}{12}$$

21. $y = e^x,\ y = 1 - x^2$

$$\int_{-.7145}^{0}(1 - x^2) - e^x\,dx$$

$$= \left(-e^x + x - \frac{x^3}{3}\right)\Big|_{-.7145}^{0}$$

$$= (-1 + 0 - 0) - (-1.08235)$$

$$= .08235$$

23. $y = \sin x,\ y = x^2$

$$\int_{0}^{.8767}\left(\sin x - x^2\right)dx = \left(-\cos x - \frac{x^3}{3}\right)\Big|_{0}^{.8767}$$

$$= .135697$$

25. $y = x^4,\ y = 2 + x$

$$\int_{-1}^{1.3532}\left(2 + x - x^4\right)dx = \left(2x + \frac{x^2}{2} - \frac{x^5}{5}\right)\Big|_{-1}^{1.3532}$$

$$= 4.01449$$

27. $y = x,\ y = 2 - x,\ y = 0$
$x = 2 - y$

$$\int_{1}^{2}y - (2 - y)\,dy = \int_{1}^{2}2y - 2\,dy$$

$$= \left(y^2 - 2y\right)\Big|_{1}^{2}$$

$$= (4 - 4) - (1 - 2)$$

$$= 1$$

29. $x = 3y,\ x = 2 + y^2$

$$\int_{1}^{2}\left[3y - \left(2 + y^2\right)\right]dy$$

$$= \left(\frac{3}{2}y^2 - 2y - \frac{y^3}{3}\right)\Big|_{1}^{2}$$

$$= \left(6 - 4 - \frac{8}{3}\right) - \left(\frac{3}{2} - 2 - \frac{1}{3}\right)$$

$$= \frac{1}{6}$$

31. $x = y,\ x = -y,\ x = 1$

$$2\int_{0}^{1}x\,dx = x^2\Big|_{0}^{1} = 1 - 0 = 1$$

33. $y = x,\ y = 2,\ y = 6 - x,\ y = 0$

$$\int_{0}^{2}(6 - y - y)\,dy = \int_{0}^{2}(6 - 2y)\,dy$$

$$= \left(6y - y^2\right)\Big|_{0}^{2}$$

$$= (12 - 4) - (0 - 0)$$

$$= 8$$

35. $A = \frac{1}{b - a}\int_{a}^{b}f(x)\,dx,\ f(x) = x^2$ on $[0, 3]$

$$A = \frac{1}{3 - 0}\int_{0}^{3}x^2\,dx = \left(\frac{1}{3}\cdot\frac{x^3}{3}\right)\Big|_{0}^{3} = 3 - 0 = 3$$

$f(x) = x^2 = 3,\ x = \pm\sqrt{3},$ on $[0, 3],\ x = \sqrt{3}$

$$\int_0^{\sqrt{3}}\left(3-x^2\right)dx = \left(3x - \frac{x^3}{3}\right)\Bigg|_0^{\sqrt{3}}$$

$$= \left(3\sqrt{3} - \sqrt{3}\right) - (0 - 0)$$

$$= 2\sqrt{3}$$

$$\int_{\sqrt{3}}^{3}\left(x^2-3\right)dx = \left(\frac{x^3}{3} - 3x\right)\Bigg|_{\sqrt{3}}^{3}$$

$$= (9-9) - \left(\sqrt{3} - 3\sqrt{3}\right)$$

$$= 2\sqrt{3}$$

37. $f(4) = 16.1e^{.07(4)} = 21.3$

$g(4) = 21.3e^{.04(4-4)} = 21.3$

21.3 million barrels per year represents the consumption for 1974.

$$\int_4^{10}\left(16.1e^{.07t} - 21.3e^{.04(t-4)}\right)dt$$

$$= \left(230e^{.07t} - 532.5e^{.04(t-4)}\right)\Bigg|_4^{10}$$

$$= 14.4$$

14.4 million barrels saved

39. For $t \geq 0$,

$$b(t) = 2e^{.04t} \geq 2e^{.02t} = d(t)$$

$$\int_0^{10}\left(2e^{.04t} - 2e^{.02t}\right)dt$$

$$= \left(50e^{.04t} - 100e^{.02t}\right)\Bigg|_0^{10}$$

$$= 2.45 \text{ million people per year}$$

41. $\int_0^{.4} f_c(x) = \frac{.4}{3(4)}\left(f_c(0) + 4f_c(.1) + 2f_c(.2) + 4f_c(.3) + f_c(.4)\right) \approx 291.67$

$\int_0^{.4} f_e(x) \approx \frac{.4}{3(4)}\left(f_e(0) + 4f_e(.1) + 2f_e(.2) + 4f_e(.3) + f_e(.4)\right) \approx 102.33$

$$\frac{\int_0^{.4} f_c(x) - \int_0^{.4} f_e(x)}{\int_0^{.4} f_c(x)} \approx \frac{291.67 - 102.33}{291.67} = .6491$$

$1 - .6491 = .3508$, so the proportion of energy retained is about 35.08%.

43. $\int_0^{3} f_s(x) \approx \frac{3}{3(4)}\left[f_s(0) + 4f_s(.75) + 2f_s(1.5) + 4f_s(2.25) + f_s(3)\right] = 860$

$\int_0^{3} f_r(x) \approx \frac{3}{3(4)}\left[f_r(0) + 4f_r(.75) + 2f_r(1.5) + 4f_r(2.25) + f_r(3)\right] = 800$

$1 - \left(\frac{860 - 800}{860}\right) = .9302$. Energy returned by the tendon is 93.02%.

45. $f(t) = -40 + 32t$; $g(t) = -30 - 32t$

$\int_0^{10}[g(t) - f(t)]dt = \int_0^{10} 10\,dt = 100$

The area between the curves is 100, representing the 100 feet that separate the objects after 10 seconds. This is just the difference in their initial velocities multiplied by 10 second.

47. $f(t) = 40\left(1 - e^{-t}\right)$; $g(t) = 20t$

$f(t) = g(t)$ at $t = 0$ and $t \approx 1.593624$.

Therefore the largest lead is

$\int_0^{1.594}[f(t) - g(t)]dt \approx 6.476$ miles

By trial and error we find that

$\int_0^{T}[f(t) - g(t)]dt = 0$ when $T \approx 2.557$, so the

second car catches up after 2.577 hours.

Section 5.2

5. $A(x) = x + 2, -1 \le x \le 3$

$V = \int_{-1}^{3} A(x)dx = \int_{-1}^{3}(x + 2)dx$

$= \left(\dfrac{x^2}{2} + 2x\right)\Big|_{-1}^{3}$

$= \left(\dfrac{9}{2} + 6\right) - \left(\dfrac{1}{2} - 2\right)$

$= 12$

7. $A(x) = \pi(4 - x)^2, 0 \le x \le 2$

$V = \int_0^2 \left(16\pi - 8\pi x + \pi x^2\right)dx$

$= \left(16\pi x - 4\pi x^2 - \pi \dfrac{x^3}{3}\right)\Big|_0^2$

$= \left(32\pi - 16\pi + \dfrac{8}{3}\pi\right) - (0 - 0 + 0)$

$= \dfrac{56\pi}{3}$

9. $V = \int_{-3}^{3} 2\left(\dfrac{x^2}{\sqrt{x^4 + 1}} + 1\right)\left(6 + 3\sin\dfrac{\pi}{6}x\right)dx$

≈ 123.8 ft^3

11. $V = \int_0^{500}\left(2 \cdot \dfrac{375}{500}y\right)^2 dy$

$= \dfrac{750^2}{500^2}\int_0^{500} y^2 dy$

$= \dfrac{750^2}{500^2} \dfrac{y^3}{3}\Big|_0^{500}$

$= 93{,}750{,}000$ ft^3

13. $V = \int_0^{30}\left(3 - \dfrac{1}{12}y\right)^2 dy$

$= \int_0^{30}\left(9 - \dfrac{1}{2}y + \dfrac{1}{144}y^2\right)dy$

$= \left(9y - \dfrac{1}{4}y^2 + \dfrac{1}{432}y^3\right)\Big|_0^{30}$

$= \dfrac{215}{2}$ ft^3

15. $V = \int_0^{2\pi}\left(4 + \sin\dfrac{x}{2}\right)^2 \pi dx$

$= \pi\int_0^{2\pi}\left(16 + 8\sin\dfrac{x}{2} + \sin^2\dfrac{x}{2}\right)dx$

$= \pi\left(16x - 16\cos\dfrac{x}{2} + \dfrac{1}{2}x - \dfrac{1}{2}\sin x\right)\Big|_0^{2\pi}$

$= 33\pi^2 + 32\pi$ in.^{3}

17. $V = \int_0^1 A(x)\,dx$

$\approx \dfrac{1}{3(10)}\big[A(0) + 4A(.1) + 2A(.2) + 4A(.3) + 2A(.4) + 4A(.5) + 2A(.6) + 4A(.7) + 2A(.8) + 4A(.9) + A(1.0)\big]$

$\approx 0.5133 \text{ cm}^3$

19. $V = \int_0^2 A(x)\,dx \approx \dfrac{2}{3(4)}\big[A(0) + 4A(.5) + 2A(1) + 4A(1.5) + A(2)\big] = 2.5 \text{ ft}^3$

21. **a.** $y = 2 - x$, $y = 0$, $x = 0$, about the x-axis

$V = \int_0^2 \pi(2 - x)^2\,dx$

$= \int_0^2 \left(4\pi - 4\pi x + \pi x^2\right)dx$

$= \left(4\pi x - 2\pi x^2 + \pi\dfrac{x^3}{3}\right)\Bigg|_0^2$

$= \left(8\pi - 8\pi + \dfrac{8\pi}{3}\right) - (0 - 0 + 0)$

$= \dfrac{8\pi}{3}$

b. $y = 2 - x$, $y = 0$, $x = 0$, about $y = 3$

$V = \int_0^2 \pi(3)^2\,dx - \int_0^2 \pi\big(3 - (2 - x)\big)^2\,dx$

$= \pi\int_0^2 \left(8 - 2x - 2x^2\right)dx$

$= \pi\left(8x - x^2 \dfrac{x^3}{3}\right)\Bigg|_0^2$

$= \pi\left[\left(16 - 4 - \dfrac{8}{3}\right) - (0 - 0 - 0)\right]$

$= \dfrac{28\pi}{3}$

23. a. $y = \sqrt{x}$, $y = 2$, $x = 0$, about y-axis

$V = \int_0^2 \pi (y^2)^2 \, dy$

$\quad = \pi \int_0^2 y^4 \, dy$

$\quad = \pi \left(\dfrac{y^5}{5} \right) \Big|_0^2$

$\quad = \dfrac{32\pi}{5}$

b. $y = \sqrt{x}$, $y = 2$, $x = 0$, about $x = 4$

$V = \int_0^2 \pi (4)^2 \, dy - \int_0^2 \pi (4 - y^2)^2 \, dy$

$\quad = \int_0^2 (-y^4 + 8y^2) \, dy$

$\quad = \pi \left(-\dfrac{x^5}{5} + \dfrac{8x^3}{3} \right) \Big|_0^2$

$\quad = \pi \left[\left(-\dfrac{32}{5} + \dfrac{64}{3} \right) - (0 + 0) \right]$

$\quad = \dfrac{224\pi}{15}$

25. a. $y = e^x$, $x = 0$, $x = 2$, $y = 0$, about y-axis

$V = \int_0^{e^2} \pi (2)^2 \, dy - \int_1^{e^2} \pi (\ln y)^2 \, dy$

$\quad = \pi (4y) \big|_0^{e^2} - \pi \left(y(\ln y)^2 - 2y \ln y + 2y \right) \Big|_1^{e^2}$

$\quad = 4\pi e^2 - \pi \left[\left(4e^2 - 4e^2 + 2e^2 \right) - (0 - 0 + 2) \right]$

$\quad = 2\pi e^2 + 2\pi$

b. $y = e^x$, $x = 0$, $x = 2$, $y = 0$, about $y = -2$

$V = \int_0^2 \pi \left(e^x + 2 \right)^2 \, dx - \int_0^2 \pi (2)^2 \, dx$

$\quad = \pi \int_0^2 \left(e^{2x} + 4e^x \right) dx$

$\quad = \pi \left(\dfrac{e^{2x}}{2} + 4e^x \right) \Big|_0^2$

$\quad = \pi \left[\left(\dfrac{e^4}{2} + 4e^2 \right) - \left(\dfrac{1}{2} + 4 \right) \right]$

$\quad = \pi \left(\dfrac{e^4}{2} + 4e^2 - \dfrac{9}{2} \right)$

27. a. $y = x^3$, $y = 0$, $x = 1$, about y-axis

$V = \int_0^1 \pi (1)^2 \, dy - \int_0^1 \pi \left(\sqrt[3]{y} \right)^2 \, dy$

$\quad = \pi \int_0^1 \left(1 - y^{2/3} \right) dy$

$\quad = \pi \left(y - \dfrac{3}{5} y^{5/3} \right) \Big|_0^1$

$\quad = \pi \left(1 - \dfrac{3}{5} \right)$

$\quad = \dfrac{2\pi}{5}$

b. $y = x^3$, $y = 0$, $x = 1$, about x-axis

$V = \int_0^1 \pi \left(x^3 \right)^2 \, dx$

$\quad = \pi \int_0^1 x^6 \, dx$

$\quad = \pi \dfrac{x^7}{7} \Big|_0^1$

$\quad = \dfrac{\pi}{7}$

29. $y = 3 - x$, x-axis, y-axis

a. about y-axis

$V = \int_0^3 \pi (3 - y)^2 \, dy$

$\quad = \pi \int_0^3 \left(9 - 6y + y^2 \right) dy$

$\quad = \pi \left(9y - 3y^2 + \dfrac{y^3}{3} \right) \Big|_0^3$

$\quad = 9\pi$

b. about x-axis

$V = \int_0^3 \pi (3 - x)^2 \, dx$

$\quad = \pi \int_0^3 \left(9 - 6x + x^2 \right) dx$

$\quad = \pi \left(9x - 3x^2 + \dfrac{x^3}{3} \right) \Big|_0^3$

$\quad = 9\pi$

c. about $y = 3$

$$V = \int_0^3 \pi(3)^2\, dx - \int_0^3 \pi\big(3-(3-x)\big)^2\, dx$$

$$= \pi\int_0^3 \big(9 - x^2\big)dx$$

$$= \pi\left(9x - \frac{x^3}{3}\right)\Bigg|_0^3$$

$$= 18\pi$$

d. about $y = -3$

$$V = \int_0^3 \pi\big((3-x)+3\big)^2\, dx - \int_0^3 \pi(3)^2\, dx$$

$$= \pi\int_0^3 \big(27 - 12x + x^2\big)dx$$

$$= \pi\left(27x - 6x^2 + \frac{x^3}{3}\right)\Bigg|_0^3$$

$$= 36\pi$$

e. about $x = 3$

$$V = \int_0^3 \pi(3)^2\, dy - \int_0^3 \pi\big(3-(3-y)\big)^2\, dy$$

$$= \pi\int_0^3 \big(9 - y^2\big)dy$$

$$= \pi\left(9y - \frac{y^3}{3}\right)\Bigg|_0^3$$

$$= 18\pi$$

f. about $x = -3$

$$V = \pi\int_0^3 \big((3-y)+3\big)^2\, dy - \pi\int_0^3 (3)^2\, dy$$

$$= \pi\int_0^3 \big(27 - 12y + y^2\big)dy$$

$$= \pi\left(27y - 6y^2 + \frac{y^3}{3}\right)\Bigg|_0^3$$

$$= 36\pi$$

31. $y = x^2$, $y = 0$, $x = 1$

a. about y–axis

$$V = \int_0^1 \pi(1)^2\, dy - \int_0^1 \pi\big(\sqrt{y}\big)^2\, dy$$

$$= \pi\int_0^1 (1 - y)dy$$

$$= \pi\left(y - \frac{y^2}{2}\right)\Bigg|_0^1$$

$$= \frac{\pi}{2}$$

b. about x–axis

$$V = \int_0^1 \pi\big(x^2\big)^2\, dx = \pi\frac{x^5}{5}\Bigg|_0^1 = \frac{\pi}{5}$$

c. about $x = 1$

$$V = \int_0^1 \pi\big(1 - \sqrt{y}\big)^2\, dy$$

$$= \pi\int_0^1 \big(1 - 2y^{1/2} + y\big)dy$$

$$= \pi\left(y - \frac{4}{3}y^{3/2} + \frac{y^2}{2}\right)\Bigg|_0^1$$

$$= \frac{\pi}{6}$$

d. about $y = 1$

$$V = \int_0^1 \pi(1)^2\, dx - \int_0^1 \pi\big(1 - x^2\big)^2\, dx$$

$$= \pi\int_0^1 \big(2x^2 - x^4\big)dx$$

$$= \pi\left(\frac{2}{3}x^3 - \frac{x^5}{5}\right)\Bigg|_0^1$$

$$= \frac{7\pi}{15}$$

e. about $x = -1$

$$V = \int_0^1 \pi(2)^2\, dy - \int \pi\big(1 + \sqrt{y}\big)^2\, dy$$

$$= \pi\int_0^1 \big(3 - 2y^{1/2} - y\big)dy$$

$$= \pi\left(3y - \frac{4}{3}y^{3/2} - \frac{y^2}{2}\right)\Bigg|_0^1$$

$$= \frac{7\pi}{6}$$

f. about $y = -1$

$$V = \int_0^1 \pi\big(x^2 + 1\big)^2\, dx - \int_0^1 \pi(1)^2\, dx$$

$$= \pi\int_0^1 \big(x^4 + 2x^2\big)dx$$

$$= \pi\left(\frac{x^5}{5} + \frac{2}{3}x^3\right)\Bigg|_0^1$$

$$= \frac{13\pi}{15}$$

33. $y = ax^2$, $y = h$, y-axis, revolved about y-axis

$$V = \int_0^h \pi \left(\sqrt{\frac{y}{a}} \right)^2 dy$$

$$= \pi \int_0^h \frac{y}{a} dy$$

$$= \frac{\pi y^2}{2a} \Big|_0^h$$

$$= \frac{\pi h^2}{2a}$$

For cylinder of height h, radius $\sqrt{\frac{h}{a}}$,

$$V_c = \pi \left(\sqrt{\frac{h}{a}} \right)^2 \cdot h = \frac{\pi h^2}{a}$$

$$\frac{1}{2} V_c = \frac{\pi h^2}{2a} = V$$

35. $V = \int_{-1}^1 \pi (1)^2 dy = \pi y \Big|_{-1}^1 = 2\pi$

37. $V = \int_{-1}^1 \pi \left(\frac{1-y}{2} \right)^2 dy$

$$= \pi \int_{-1}^1 \left(\frac{1}{4} - \frac{y}{2} + \frac{y^2}{4} \right) dy$$

$$= \pi \left(\frac{y}{4} - \frac{y^2}{4} + \frac{y^3}{12} \right) \Big|_{-1}^1$$

$$= \pi \left[\left(\frac{1}{4} - \frac{1}{4} + \frac{1}{12} \right) - \left(-\frac{1}{4} - \frac{1}{4} - \frac{1}{12} \right) \right]$$

$$= \frac{2\pi}{3}$$

39. $V = \int_{-r}^r \pi \left(\sqrt{r^2 - y^2} \right)^2 dy$

$$= \pi \int_{-r}^r \left(r^2 - y^2 \right) dy$$

$$= \pi \left(r^2 y - \frac{y^3}{3} \right) \Big|_{-r}^r$$

$$= \frac{4}{3} \pi r^3$$

41. If we compute the two volumes using disks parallel to the base, we'll get identical integrals, so the volumes are the same.

Section 5.3

5. $y = x^2$, x-axis, $-1 \le x \le 1$, about $x = 2$

$r = 2 - x$, $h = x^2$

$$V = \int_{-1}^1 2\pi (2 - x) x^2 dx$$

$$= 2\pi \left(\frac{2x^3}{3} - \frac{x^4}{4} \right) \Big|_{-1}^1$$

$$= \frac{8\pi}{3}$$

7. $y = x$, $y = -x$, $x = 1$, about y–axis

$r = x$, $h = 2x$

$$V = \int_0^1 2\pi x (2x) dx = \frac{4\pi}{3} x^3 \Big|_0^1 = \frac{4\pi}{3}$$

9. $y = x$, $y = -x$, $y = 1$, about $y = 2$

$r = 2 - y$, $h = 2y$

$$V = \int_0^1 2\pi (2 - y)(2y) dy$$

$$= 2\pi \left(2y^2 - \frac{2y^3}{3} \right) \Big|_0^1$$

$$= \frac{8\pi}{3}$$

11. right half of $x^2 + (y-1)^2 = 1$, about x-axis

$r = y$, $h = \sqrt{1 - (y-1)^2}$

$$V = \int_0^2 2\pi y \sqrt{1 - (y-1)^2} \, dy = \pi^2 \approx 9.8696$$

(using CAS or calculator)

13. $y = x^2$, $y = 2 - x^2$, about $x = -2$

$$V = \int_{-1}^1 2\pi (x + 2)\left(2 - x^2 - x^2 \right) dx$$

$$= 2\pi \int_{-1}^1 \left(4 + 2x - 4x^2 - 2x^3 \right) dx$$

$$= 2\pi \left(4x + x^2 - \frac{4x^3}{3} - \frac{x^4}{2} \right) \Big|_{-1}^1$$

$$= \frac{32\pi}{3}$$

15. $x = y^2$, $x = 1$, about $y = -2$

$$V = \int_{-1}^{1} 2\pi(y+2)\left(1-y^2\right)dy$$

$$= 2\pi\int_{-1}^{1}\left(2+y-2y^2-y^3\right)dy$$

$$= 2\pi\left(2y+\frac{y^2}{2}-\frac{2y^3}{3}-\frac{y^4}{4}\right)\Bigg|_{-1}^{1}$$

$$= \frac{16\pi}{3}$$

19. $x = (y-1)^2$, $x = 1$, about x – axis

$$V = \int_{0}^{2} 2\pi(y)\left(1-(y-1)^2\right)dy$$

$$= 2\pi\int_{0}^{2}(-y^3+2y^2)dy$$

$$= 2\pi\left(-\frac{y^4}{4}+\frac{2y^3}{3}\right)\Bigg|_{0}^{2}$$

$$= \frac{8\pi}{3}$$

17. $y = x$, $y = x^2-2$, about $x = 2$

$$V = \int_{-1}^{2} 2\pi(2-x)\left(x-\left(x^2-2\right)\right)dx$$

$$= 2\pi\int_{-1}^{2}\left(4-3x^2+x^3\right)dx$$

$$= 2\pi\left(4x-x^3+\frac{x^4}{4}\right)\Bigg|_{-1}^{2}$$

$$= \frac{27\pi}{2}$$

21. $y = 4-x$, $y = 4$, $y = x$

   a. x-axis

$$V = \int_{2}^{4} 2\pi(y)\left(y-(4-y)\right)dy$$

$$= 2\pi\int_{2}^{4}\left(2y^2-4y\right)dy$$

$$= 2\pi\left(\frac{2y^3}{3}-2y^2\right)\Bigg|_{2}^{4}$$

$$= \frac{80\pi}{3}$$

   b. y-axis

$$V = \int_{0}^{2} 2\pi(x)\left(4-(4-x)\right)dx + \int_{2}^{4} 2\pi(x)(4-x)dx$$

$$= 2\pi\left(\frac{x^3}{3}\right)\Bigg|_{0}^{2} + 2\pi\left(2x^2-\frac{x^3}{3}\right)\Bigg|_{2}^{4}$$

$$= 2\pi\left(\frac{8}{3}+\frac{16}{3}\right)$$

$$= 16\pi$$

c. $x = 4$

$$V = \int_2^4 \pi\left(4 - (4 - y)\right)^2 dy - \int_2^4 \pi(4 - y)^2 dy$$

$$= \pi\int_2^4 y^2 dy - \pi\int_2^4 (16 - 8y + y^2)dy$$

$$= \pi\int_2^4 (-16 + 8y)dy$$

$$= \pi(-16y + 4y^2)\Big|_2^4$$

$$= 16\pi$$

d. $y = 4$

$$V = \int_2^4 2\pi(4 - y)\left(y - (4 - y)\right)dy$$

$$= 2\pi\int_2^4 \left(-2y^2 + 12y - 16\right)dy$$

$$= 2\pi\left(-\frac{2y^3}{3} + 6y^2 - 16y\right)\Bigg|_2^4$$

$$= \frac{16\pi}{3}$$

23. $y = x,\ y = x^2 - 6$

a. about $x = 3$

$$V = \int_{-2}^3 2\pi(3 - x)\left(x - \left(x^2 - 6\right)\right)dx$$

$$= 2\pi\int_{-2}^3 \left(x^3 - 4x^2 - 3x + 18\right)dx$$

$$= 2\pi\left(\frac{x^4}{4} - \frac{4x^3}{3} - \frac{3x^2}{2} + 18x\right)\Bigg|_{-2}^3$$

$$= \frac{625\pi}{6}$$

b. about $y = 3$

$$V = \int_{-2}^3 \pi\left(3 - (x^2 - 6)\right)^2 dx - \int_{-2}^3 \pi(3 - x)^2 dx$$

$$= \pi\int_{-2}^3 (x^4 - 18x^2 + 81)dx - \pi\int_{-2}^3 (9 - 6x + x^2)dx$$

$$= \pi\int_{-2}^3 (x^4 - 19x^2 + 6x + 72)dx$$

$$= \pi\left(\frac{x^5}{5} - \frac{19x^3}{3} + 3x^2 + 72x\right)\Bigg|_{-2}^3$$

$$= \frac{625\pi}{3}$$

c.   about $x = -3$

$$V = \int_{-2}^{3} 2\pi(x+3)\left(x - \left(x^2 - 6\right)\right)dx$$

$$= 2\pi\int_{-2}^{3}\left(-x^3 - 2x^2 + 9x + 18\right)dx$$

$$= 2\pi\left(-\frac{x^4}{4} - \frac{2x^3}{3} + \frac{9x^2}{2} + 18x\right)\Bigg|_{-2}^{3}$$

$$= \frac{875\pi}{6}$$

d.   about $y = -6$

$$V = \int_{-2}^{3}\pi(x+6)^2 dx - \int_{-2}^{3}\pi\left((x^2 - 6) + 6\right)^2 dx$$

$$= \pi\int_{-2}^{3}(x^2 + 12x + 36)dx - \pi\int_{-2}^{3}x^4 dx$$

$$= \pi\int_{-2}^{3}(-x^4 + x^2 + 12x + 36)dx$$

$$= \pi\left(-\frac{x^5}{5} + \frac{x^3}{3} + 6x^2 + 36x\right)\Bigg|_{-2}^{3}$$

$$= \frac{500\pi}{3}$$

25. $y = \cos x, \ y = x^4$

a.   about $x = 2$

$$V = \int_{-.8906}^{.8906} 2\pi(2 - x)\left(\cos x - x^4\right)dx \approx 16.723$$

b.   about $y = 2$

$$V = 2\int_{0}^{.8906}\pi(2 - x^4)^2 dx - 2\int_{0}^{.8906}\pi(2 - \cos x)^2 dx \approx 12.635$$

c.   about x-axis

$$V = 2\int_{0}^{.8906}\pi(\cos x)^2 dx - 2\int_{0}^{.8906}\pi(x^4)^2 dx \approx 4.088$$

d.   about y-axis

$$V = \int_{0}^{.8906} 2\pi(x)\left(\cos x - x^4\right)dx \approx 1.496$$

27. $y = x^2$, $y = 2 - x$, $x = 0$

  a. about x-axis

$$V = \int_0^1 \pi(2-x)^2\,dx - \int_0^1 \pi\left(x^2\right)^2 dx$$

$$= \pi\int_0^1 (x^2 - 4x + 4)dx - \pi\int_0^1 x^4 dx$$

$$= \pi\int_0^1 (-x^4 + x^2 - 4x + 4)dx$$

$$= \pi\left(-\frac{x^5}{5} + \frac{x^3}{3} - 2x^2 + 4x \right)\Big|_0^1$$

$$= \frac{32\pi}{15}$$

  b. about y-axis

$$V = \int_0^1 2\pi x\left(2 - x - x^2\right)dx$$

$$= 2\pi\int_0^1 \left(2x - x^2 - x^3\right)dx$$

$$= 2\pi\left(x^2 - \frac{x^3}{3} - \frac{x^4}{4} \right)\Big|_0^1$$

$$= \frac{5\pi}{6}$$

  c. about $x = 1$

$$V = \int_0^1 2\pi(1-x)(2 - x - x^2)dx$$

$$= 2\pi\int_0^1 \left(x^3 - 3x + 2\right)dx$$

$$= 2\pi\left(\frac{x^4}{4} - \frac{3x^2}{2} + 2x \right)\Big|_0^1$$

$$= \frac{3\pi}{2}$$

  d. about $y = 2$

$$V = \int_0^1 \pi\left(2 - 2x^2\right)^2 dx - \int_0^1 \pi(2 - (2-x))^2\,dx$$

$$= \pi\int_0^1 (x^4 - 4x^2 + 4)dx - \pi\int_0^1 x^2 dx$$

$$= \pi\int_0^1 (x^4 - 5x^2 + 4)dx$$

$$= \pi\left(\frac{x^5}{5} - \frac{5x^3}{3} + 4x \right)\Big|_0^1$$

$$= \frac{38\pi}{15}$$

29. $y = 2 - x$, $y = x - 2$, $x = y^2$

  a. about x-axis

$$V = \int_0^1 2\pi(y)\left(2 - y - y^2\right)dy$$

$$= 2\pi\int_0^1 (2y - y^2 - y^3)dy$$

$$= 2\pi\left(y^2 - \frac{y^3}{3} - \frac{y^4}{4} \right)\Big|_0^1$$

$$= \frac{5\pi}{6}$$

  b. about y-axis

$$V = 2\int_0^1 \pi(2 - y)^2\,dy - 2\int_0^1 \pi(y^2)^2 dy$$

$$= 2\pi\int_0^1 (y^2 - 4y + 4)dy - 2\pi\int_0^1 y^4 dy$$

$$= 2\pi\int_0^1 (-y^4 + y^2 - 4y + 4)dy$$

$$= 2\pi\left(-\frac{y^5}{5} + \frac{y^3}{3} - 2y^2 + 4y \right)\Big|_0^1$$

$$= \frac{64\pi}{15}$$

31. Rotation around the x-axis.

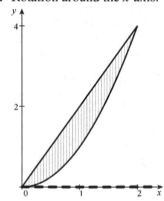

33. Rotation around the y-axis.

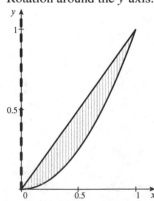

35. Rotation around the y-axis.

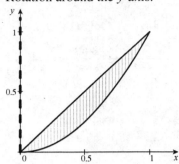

37. $A = \lim\limits_{n \to \infty} \sum\limits_{i=1}^{n} 2\pi \dfrac{i}{n} R \, \Delta r = \int_0^R 2\pi r \, dr = \pi R^2$

39. $\int_{1/2}^1 4\pi x \sqrt{1-x^2} \, dx = I$

$u = 1-x^2, \; du = -2x \, dx, \; u\!\left(\dfrac{1}{2}\right) = \dfrac{3}{4}, \; u(1) = 0$

$I = -2\pi \int_{3/4}^0 \sqrt{u} \, du = -\dfrac{4\pi}{3} u^{3/2} \Big|_{3/4}^0 = \dfrac{\sqrt{3}}{2}\pi$

41. $V_{\text{hill}} = \int_0^1 2\pi x (1-x^2) \, dx$

$= 2\pi \int_0^1 (x - x^3) \, dx$

$= 2\pi \left(\dfrac{x^2}{2} - \dfrac{x^4}{4} \right) \Big|_0^1$

$= \dfrac{\pi}{2}$

$V_{\text{core}} = \int_0^r 2\pi x (1-x^2) \, dx = 2\pi \left(\dfrac{r^2}{2} - \dfrac{r^4}{4} \right) = \dfrac{1}{10} \cdot \dfrac{\pi}{2}$

$r^2 - \dfrac{r^4}{2} = \dfrac{1}{20}$

$10 r^4 - 20 r^2 + 1 = 0$

$r^2 = \dfrac{20 \pm \sqrt{400 - 40}}{20}$

$r^2 = \dfrac{20 \pm 6\sqrt{10}}{20} = \dfrac{100 \pm 30\sqrt{10}}{100}$

Since $r < 1$, we must have $r = \dfrac{\sqrt{100 - 30\sqrt{10}}}{10}$

$r = \dfrac{\sqrt{100 - 30\sqrt{10}}}{10} = .2265$

Section 5.4

5. $y = x^2, \; 0 \le x \le 1$

$n = 2 : \Delta x = \dfrac{1-0}{2} = \dfrac{1}{2}$

$s_1 \approx \sqrt{\left(\dfrac{1}{2} - 0\right)^2 + \left(\left(\dfrac{1}{2}\right)^2 - (0)^2\right)^2} \approx .5590$

$s_2 \approx \sqrt{\left(1 - \dfrac{1}{2}\right)^2 + \left((1)^2 - \left(\dfrac{1}{2}\right)^2\right)^2} \approx .9014$

$s \approx s_1 + s_2 \approx 1.4604$

$n = 4 : \Delta x = \dfrac{1-0}{4} = \dfrac{1}{4}$

$s_1 \approx \sqrt{\left(\dfrac{1}{4} - 0\right)^2 + \left(\left(\dfrac{1}{4}\right)^2 - (0)^2\right)^2} \approx .2577$

$s_2 \approx \sqrt{\left(\dfrac{1}{2} - \dfrac{1}{4}\right)^2 + \left(\left(\dfrac{1}{2}\right)^2 - \left(\dfrac{1}{4}\right)^2\right)^2} \approx .3125$

$s_3 \approx \sqrt{\left(\dfrac{3}{4} - \dfrac{1}{2}\right)^2 + \left(\left(\dfrac{3}{4}\right)^2 - \left(\dfrac{1}{2}\right)^2\right)^2} \approx .4002$

$s_4 \approx \sqrt{\left(1 - \dfrac{3}{4}\right)^2 + \left((1)^2 - \left(\dfrac{3}{4}\right)^2\right)^2} \approx .5039$

$s \approx s_1 + s_2 + s_3 + s_4 \approx 1.4743$

7. $y = \cos x, \; 0 \le x \le \pi$

$n = 2 : \Delta x = \dfrac{\pi - 0}{2} = \dfrac{\pi}{2}$

$s_1 \approx \sqrt{\left(\dfrac{\pi}{2} - 0\right)^2 + \left(\cos\dfrac{\pi}{2} - \cos 0\right)^2} \approx 1.8621$

$s_2 \approx \sqrt{\left(1 - \dfrac{\pi}{2}\right)^2 + \left(\cos\pi - \cos\dfrac{\pi}{2}\right)^2} \approx 1.8621$

$s \approx s_1 + s_2 \approx 3.7242$

$n = 4 : \Delta x = \dfrac{\pi - 0}{4} = \dfrac{1}{4}$

$s_1 \approx \sqrt{\left(\dfrac{\pi}{4} - 0\right)^2 + \left(\cos\dfrac{\pi}{4} - \cos 0\right)^2} \approx .8382$

$s_2 \approx \sqrt{\left(\dfrac{\pi}{2} - \dfrac{\pi}{4}\right)^2 + \left(\cos\dfrac{\pi}{2} - \cos\dfrac{\pi}{4}\right)^2} \approx 1.0568$

$s_3 \approx \sqrt{\left(\dfrac{3\pi}{4} - \dfrac{\pi}{2}\right)^2 + \left(\cos\dfrac{3\pi}{4} - \cos\dfrac{3\pi}{2}\right)^2} \approx 1.0568$

$s_4 \approx \sqrt{\left(1 - \dfrac{3\pi}{4}\right)^2 + \left(\cos\pi - \cos\dfrac{3\pi}{4}\right)^2} \approx .8382$

$s \approx s_1 + s_2 + s_3 + s_4 \approx 3.7900$

9. $s = \int_{-1}^{1}\sqrt{1 + \left(3x^2\right)^2}\,dx = \int_{-1}^{1}\sqrt{1 + 9x^4}\,dx \approx 3.0957$

11. $s = \int_{0}^{2}\sqrt{1 + (2 - 2x)^2}\,dx \approx 2.9578$

13. $\int_{0}^{3}\sqrt{1 + \left(3x^2 + 1\right)^2}\,dx \approx 30.3665$

15. $\int_{0}^{\pi}\sqrt{1 + (-\sin x)^2}\,dx = \int_{0}^{\pi}\sqrt{1 + \sin^2 x}\,dx \approx 3.8201$

17. $\int_{0}^{\pi}\sqrt{1 + (x\sin x)^2}\,dx = 4.6984$

19. $s = \int_{-10}^{10}\sqrt{1 + \left(\dfrac{2}{5}\left(e^{x/10} - e^{-x/10}\right)\right)^2}\,dx$

≈ 22.346 ft

21. $y(x) = 5\left(e^{x/10} + e^{-x/10}\right)$

$y(0) = 5\left(e^0 + e^0\right) = 10$

$y(-10) = y(10) = 5\left(e^1 + e^{-1}\right) = 15.43$

$sag = 15.43 - 10 = 5.43$ ft

23. $y = \dfrac{1}{15}x(60 - x) = 0$

$x = 0, 60$, so the punt traveled 60 yds horizontally.

$y'(x) = 4 - \dfrac{2}{15}x = 0$

$x = 30,\ y(30) = \dfrac{1}{15}\cdot 30(60 - 30) = 60$

The punt was 60 yds high.

$s = \int_{0}^{60}\sqrt{1 + \left(4 - \dfrac{2}{15}x\right)^2}\,dx \approx 139.4$ yards

$v = \dfrac{s}{4\ \text{sec}} = \dfrac{139.4\ \text{yards}}{4\ \text{sec}}\cdot\dfrac{3\ \text{feet}}{1\ \text{yard}} = 104.55$ ft/s

25. The arc-length integrand simplifies to
$\dfrac{1}{2}e^{x/10} + \dfrac{1}{2}e^{-x/10}$. An antiderivative is

$5e^{x/10} - 5e^{-x/10}$.

$5e^{x/10} - 5e^{-x/10}\Big|_{-10}^{10} = 10\left(e^1 - e^{-1}\right)$

27. $\dfrac{d}{dx}\sqrt{2}\displaystyle\int_{0}^{x}\sqrt{1 - \dfrac{\sin^2 u}{2}}\,du = \dfrac{1}{2}\sqrt{2}\cdot\sqrt{4 - 2\sin^2 x}$

$= \sqrt{1 + \cos^2 x}$

29. The antiderivatives returned by some CAS still include an integral, indicating that the CAS can't find an antiderivative in closed form (see Exercise 30). In this case numerical integration is the method of choice.

31. $y = x^2,\ 0 \le x \le 1$ revolved about x-axis

$S = \int_{0}^{1}2\pi x^2\sqrt{1 + (2x)^2}\,dx \approx 3.8097$

33. $y = 2x - x^2,\ 0 \le x \le 2$, about x-axis

$S = \int_{0}^{2}2\pi\left(2x - x^2\right)\sqrt{1 + (2 - 2x)^2}\,dx \approx 10.9654$

35. $y = e^x,\ 0 \le x \le 1$, about x-axis

$S = \int_{0}^{1}2\pi e^x\sqrt{1 + e^{2x}}\,dx \approx 22.9430$

37. $y = \cos x,\ 0 \le x \le \dfrac{\pi}{2}$, about x-axis

$S = \int_{0}^{\pi/2}2\pi\cos x\sqrt{1 + \sin^2 x}\,dx \approx 7.2117$

39. $L_1 = \int_{-\pi/6}^{\pi/6}\sqrt{1 + \cos^2 x}\,dx \approx 1.44829$

$L_2 = \sqrt{\left(\sin\dfrac{\pi}{6} - \sin\left(-\dfrac{\pi}{6}\right)\right)^2 + \left(\dfrac{\pi}{6} - \left(-\dfrac{\pi}{6}\right)\right)^2}$

≈ 1.44797

$\dfrac{L_2}{L_1} = \dfrac{1.44797}{1.44829} \approx .9998$

41. $L_1 = \int_{\pi/6}^{\pi/2}\sqrt{1+\cos^2 x}\,dx \approx 1.18595$

$L_2 = \sqrt{\left(\sin\dfrac{\pi}{2}-\sin\dfrac{\pi}{6}\right)^2 + \left(\dfrac{\pi}{2}-\dfrac{\pi}{6}\right)^2} \approx 1.16044$

$\dfrac{L_2}{L_1} = \dfrac{1.16044}{1.18595} \approx .9785$

43. $L_1 = \int_3^5 \sqrt{1+e^{2x}}\,dx \approx 128.34914$

$L_2 = \sqrt{\left(e^5-e^3\right)^2 + (5-3)^2} \approx 128.34320$

$\dfrac{L_2}{L_1} = \dfrac{128.34320}{128.34914} \approx .99995$

45. $\int_0^1 \sqrt{1+\left(6x^5\right)^2}\,dx = \int_0^1 \sqrt{1+36x^{10}}\,dx \approx 1.672$

$\int_0^1 \sqrt{1+\left(8x^7\right)^2}\,dx = \int_0^1 \sqrt{1+64x^{14}}\,dx \approx 1.720$

$\int_0^1 \sqrt{1+\left(10x^9\right)^2}\,dx = \int_0^1 \sqrt{1+100x^{18}}\,dx \approx 1.754$

As $n \to \infty$, the length $\to 2$.

47. $y = x^4,\ y' = 4x^3$

$y = x^2,\ y' = 2x$

$4x^3 > 2x$

$x^2 > \dfrac{1}{2}$

$x > \sqrt{\dfrac{1}{2}}$

49. $S = \int_{-1}^{1} 2\pi(1)\sqrt{1+0^2}\,dy$

$= 2\pi y\Big|_{-1}^{1}$

$= 4\pi$

The solid of revolution is a cylinder with radius 1, so its top and bottom each have area $\pi(1)^2 = \pi$. Hence the total surface area is $4\pi + \pi + \pi = 6\pi$.

51. $\int_{-1}^{1} 2\pi\left(\dfrac{1-y}{2}\right)\sqrt{1+\left(-\dfrac{1}{2}\right)^2}\,dy$

$= \int_{-1}^{1} 2\pi\left(\dfrac{1-y}{2}\right)\sqrt{\dfrac{5}{4}}$

$= \dfrac{\pi\sqrt{5}}{2}\left(y-\dfrac{y^2}{2}\right)\Bigg|_{-1}^{1}$

$= \sqrt{5}\pi$

The solid of revolution is a cone with base of radius 1, so its base has area $\pi(1)^2 = \pi$. Hence the total surface area is $\sqrt{5}\pi + \pi$.

53.

$x^{2/3} + y^{2/3} = 1$

$y = \pm\left(1-x^{2/3}\right)^{3/2}$

Taking the positive function, we have

$y' = \dfrac{3}{2}\left(1-x^{2/3}\right)^{1/2}\left(-\dfrac{2}{3}x^{-1/3}\right)$

$= -x^{-1/3}\left(1-x^{2/3}\right)^{1/2}$

$s = 4\int_0^1 \sqrt{1+\left(-x^{-1/3}\left(1-x^{2/3}\right)^{1/2}\right)^2}\,dx$

$= 4\int_0^1 \sqrt{1+x^{-2/3}\left(1-x^{2/3}\right)}\,dx$

$= 4\int_0^1 \sqrt{1+x^{-2/3}-1}\,dx$

$= 4\int_0^1 x^{-1/3}\,dx$

Note that $x^{-1/3}$ is not continuous on $[0, 1]$. In fact $x^{-1/3}$ has a vertical asymptote at $x = 0$. On the other hand, $x^{-1/3}$ is continuous on $[a, 1]$ for any $a > 0$. So to compute the above integral, we can integrate over $[a, 1]$, which allows us to use the Fundamental Theorem of Calculus, and then take the limit as a approaches 0 from the right.

$4\int_0^1 x^{-1/3}\,dx = 4\cdot\lim_{a\to 0^+}\int_a^1 x^{-1/3}\,dx$

$= 4\cdot\lim_{a\to 0^+}\left[\dfrac{3}{2}x^{2/3}\Big|_a^1\right]$

$= 4\cdot\lim_{a\to 0^+}\left[\dfrac{3}{2}-\dfrac{3}{2}a^{2/3}\right]$

$= 4\cdot\dfrac{3}{2}$

$= 6$

Section 5.5

5. $y(0) = 80$, $y'(0) = 0$

7. $y(0) = 60$, $y'(0) = 10$

9. $h''(t) = -32$
$h(0) = 30$
$h'(0) = 0$
$\int h''(t)dt = \int -32dt = -32t + c_1$
$h'(t) = -32t + c_1$
$h'(0) = -32(0) + c_1 = 0$, $c_1 = 0$
$\int h'(t)dt = \int -32t\,dt = -16t^2 + c_2$
$h(0) = -16(0)^2 + c_2 = 30$, $c_2 = 30$
$h(t) = -16t^2 + 30 = 0$
$t = \dfrac{\sqrt{30}}{4}$
$h'\left(\dfrac{\sqrt{30}}{4}\right) = -32\left(\dfrac{\sqrt{30}}{4}\right) = -8\sqrt{30}$ ft/s

11. $v(t) = -32 - 32t$
$h(t) = \int_0^t -32 - 32u\,du = -32t - 16t^2$
Set $h(t) = -6$ and solve for t :
$t = -1 + \dfrac{\sqrt{22}}{4}$
$v\left(-1 + \dfrac{\sqrt{22}}{4}\right) \approx -37.52$
The impact velocity is about 37.52 ft/s.

13. object dropped from 30 ft, impact velocity
$= -8\sqrt{30}$
object dropped from 120 ft, impact velocity is
$-16\sqrt{30} = -8\sqrt{120}$
object dropped from 3000 ft, impact velocity is
$-8\sqrt{3000}$
So velocity increases by a factor of $\sqrt{h}$ when height increases by a factor of h.

15. $h''(t) = 32$
$h'(0) = 0$
$h(4) = 0$
$\int h''(t)dt = \int -32dt = -32t + c_1$
$h'(0) = -32(0) + c_1$, $c_1 = 0$
$\int h'(t)dt = \int -32t\,dt = -16t^2 + c_2$
$h(4) = -16(4)^2 + c_2 = 0$, $c_2 = 256$
$h(t) = -16t^2 + 256$
$h(0) = 16(0)^2 + 256 = 256$ ft

17. $h'(0) = 64$
$h''(t) = -32$
$h(0) = 0$
$h'(t) = -32t + c_1$
$h'(0) = -32(0) + c_1 = 64$, $c_1 = 64$,
$h'(t) = -32t + 64$
$h(t) = \int h'(t)dt = \int(-32t + 64)dt = -16t^2 + 64t + c_2$
$h(0) = -16(0)^2 + 64(0) + c_2 = 0$, $c_2 = 0$
$h(t) = -16t^2 + 64t$
$h'(t) = -32t + 64 = 0$, $t = 2$
$h(2) = -16(2)^2 + 64(2) = 64$ feet

$h(t) = 0 = -16t^2 + 64t$
$t = 0, 4\,\text{sec}$
$h'(4) = -32(4) + 64 = -64$ ft/s

19. $h''(t) = -32$, $h'(0) = v_0$, $h(0) = 0$
$h'(t) = -32t + c_1$
$v_0 = h'(0) = -32(0) + c_1 = c_1$
$h'(t) = v_0 - 32t$
$h(t) = v_0 t - 16t^2 + c_2$
$h(0) = v_0(0) - 16(0)^2 + c_2$, $c_2 = 0$
$h(t) = v_0 t - 16t^2$
$h'(t) = 0 = v_0 - 32t$, $t = \dfrac{v_0}{32}$
$h\left(\dfrac{v_0}{32}\right) = v_0\left(\dfrac{v_0}{32}\right) - 16\left(\dfrac{v_0}{32}\right)^2 = \dfrac{v_0^2}{64}$ feet
$\dfrac{v_0^2}{64}$ feet $= 20\,\text{in} \cdot \dfrac{1\,\text{ft}}{12\,\text{in}}$
$v_0 = 8\sqrt{\dfrac{5}{3}}$ ft/s ≈ 10.3 ft/s

21. $h''(t) = -32$

$h'(t) = -32t + 5 = 0$

$t = \dfrac{5}{32}$

$h\left(\dfrac{5}{32}\right) = -16\left(\dfrac{5}{32}\right)^2 + 5\left(\dfrac{5}{32}\right) = \dfrac{25}{64}$ ft $\approx$ 4.7 in.

23. $h(0) = H$

$h'(0) = 0$

$h''(t) = -32$

$h'(t) = -32t$

$h(t) = -16t^2 + H = 0$

$t = \dfrac{\sqrt{H}}{4}$ sec

$h'\left(\dfrac{\sqrt{H}}{4}\right) = -32\dfrac{\sqrt{H}}{4} = -8\sqrt{H}$ ft/sec

25. $\theta = \dfrac{\pi}{3}$, $v_0 = 98$ m/s

$y''(t) = -9.8$

$y'(0) = 98\sin\dfrac{\pi}{3} = 49\sqrt{3}$

$y(0) = 0$

$y'(t) = \int y''(t)dt$

$= \int -9.8\,dt$

$= -9.8t + c_1$

$= -9.8t + 49\sqrt{3}$

$y(t) = \int y'(t)dt$

$= -4.9t^2 + 49\sqrt{3}t + c_2$

$= -.49t^2 + 49\sqrt{3}t$

$= 0$

$49t\left(\sqrt{3} - .1t\right) = 0$

$t = 0, 10\sqrt{3}$

The flight time is $10\sqrt{3} \approx 17$ seconds.

$x''(t) = 0$

$x'(0) = 98\cos\dfrac{\pi}{3} = 49$

$x(0) = 0$

$x'(t) = 49$

$x(t) = 49t$

$x\left(10\sqrt{3}\right) = 49\left(10\sqrt{3}\right) = 490\sqrt{3} \approx 849$ m

27. $\theta = \dfrac{\pi}{6}$, $v_0 = 120$ ft/s

$y''(t) = -32$

$y'(0) = 120\sin\dfrac{\pi}{6} = 60$

$y(0) = 0$

$y'(t) = \int y''(t)dt = -32t + c_1 = -32t + 60$

$y(t) = \int y'(t)dt$

$= -16t^2 + 60t + c_2$

$= -16t^2 + 60t$

$= 0$

$4t(15 - 4t) = 0$

$t = 0, \dfrac{15}{4}$

flight time is $\dfrac{15}{4}$ sec

$x''(t) = 0$

$x'(0) = 120\cos\dfrac{\pi}{6} = 60\sqrt{3}$

$x(0) = 0$

$x'(t) = 60\sqrt{3}$

$x(t) = 60\sqrt{3}t$

$x\left(\dfrac{15}{4}\right) = 60\sqrt{3}\cdot\dfrac{15}{4} = 225\sqrt{3}$ ft

29. $y(t) = -16t^2 - 18.397t + 8$

$x(t) = 175.036t$

$x(t) = 39$ when $t = 0.223$.

$y(0.223) \approx 3.11$ so the serve clears the net.

$y(t) = 0$ when $t = 0.336$.

$x(0.336) \approx 58.89$ so the serve goes in.

31. $y(t) = -16t^2 - 14.816t + 8$

$x(t) = 169.353t$

$x(t) = 39$ when $t = 0.230$.

$y(0.230) \approx 3.74$, so the serve clears the net.

$y(t) = 0$ when $t = 0.382$.

$x(0.382) \approx 64.73$, so the serve is out.

33. $\theta = 0$, $v_0 = 130$

$y(0) = 6, x(0) = 0$

$y'(0) = 130\sin 0 = 0$

$x'(0) = 130\cos 0 = 130$

$x''(t) = 0$

$x'(t) = 130$

$x(t) = 130t = 60$

$t = \dfrac{6}{13}$

$y''(t) = -32$

$y'(t) = -32t$

$y(t) = -16t^2 + 6$

$y\left(\dfrac{6}{13}\right) = -16\left(\dfrac{6}{13}\right)^2 + 6 = \dfrac{438}{169} \approx 2.59$ ft

35. $y(0) = 5$, $x(0) = 0$

$v_0 = 120$, $\theta = \dfrac{\pi}{36}$

$y'(0) = 120\sin\dfrac{\pi}{36} \approx 10.46$

$x'(0) = 120\cos\dfrac{\pi}{36} \approx 119.54$

$x''(0) = 0$

$x'(t) = 119.54$

$x(t) = 119.54t = 120$

$t = \dfrac{120}{119.54}$

$y''(t) = -32$

$y'(t) = -32t + 10.46$

$y(t) = -16t^2 + 10.46t + 5$

$y\left(\dfrac{120}{119.54}\right) = -16\left(\dfrac{120}{119.54}\right)^2 + 10.46\left(\dfrac{120}{119.54}\right) + 5$

$\approx -.62$ ft

37. We need $t = \dfrac{1}{\cos\theta}$ so that

$y(t) = -16t^2 + 120\sin(\theta)t + 5 = 5$.

So we need $-16\left(\dfrac{1}{\cos\theta}\right)^2 + 120\sin\theta\left(\dfrac{1}{\cos\theta}\right) = 0$

$-16\left(\dfrac{1}{\cos\theta}\right)\cdot\left[\dfrac{1}{\cos\theta} - 7.5\sin\theta\right] = 0$

Plotting $\dfrac{1}{\cos\theta} - 7.5\sin\theta$ on $\left[\dfrac{\pi}{36}, \dfrac{\pi}{18}\right]$ and

tracing reveals that this quantity equals 0 when $\theta \approx .135$ radians ≈ 7.7 degrees.

To find the aim, we need the length of the vertical leg of a right triangle with opposite angle 7.7 degrees, and adjacent leg 120 ft. Thus the player should aim $120\tan(7.7°) \approx 120\tan(.135) \approx 16.2$ ft above the first baseman's head.

39. $\theta = \dfrac{\pi}{6}, x(0) = 0$, $y(0) = 0$

$y'(0) = v_0\sin\dfrac{\pi}{6} = \dfrac{v_0}{2}$

$x'(0) = v_0\cos\dfrac{\pi}{6} = \dfrac{\sqrt{3}}{2}v_0$

$y''(t) = -32, x''(t) = 0$

$x'(t) = \dfrac{\sqrt{3}}{2}v_0$

$x(t) = \dfrac{\sqrt{3}}{2}v_0 t = 5(25)$

$t = \dfrac{250}{\sqrt{3}v_0}$

$y'(t) = -32t + \dfrac{v_0}{2}$

$y(t) = -16t^2 + \dfrac{v_0}{2}t$

$y\left(\dfrac{250}{\sqrt{3}v_0}\right) = -16\left(\dfrac{250}{\sqrt{3}v_0}\right)^2 + \dfrac{v_0}{2}\left(\dfrac{250}{\sqrt{3}v_0}\right) = 0$

$v_0 = \sqrt{\dfrac{8000}{\sqrt{3}}} \approx 68$ ft/s

41. $y(0) = 256$, $x(0) = 0$

$y'(0) = 0$, $x'(0) = 100$

$y''(t) = -32$, $x''(t) = 0$

$y'(t) = -32t$

$y(t) = -16t^2 + 256 = 0$

$t = 4$

$x'(t) = 100$

$x(t) = 100t$

$x(4) = 100(4) = 400$ ft

43. $\theta_0 = 0$, $\omega = 1$

$x''(t) = -25\sin(4t)$

$x'(0) = x(0) = 0$

$x'(t) = \dfrac{25}{4}\cos 4t - \dfrac{25}{4}$

$x(t) = \dfrac{25}{16}\sin 4t - \dfrac{25}{4}t$

45. $\theta_0 = \dfrac{\pi}{4},\ \omega = 2$

$x''(t) = -25\sin\left(8t + \dfrac{\pi}{4}\right)$

$x'(0) = 0 = x(0)$

$x'(t) = \dfrac{25}{8}\cos\left(8t + \dfrac{\pi}{4}\right) - \dfrac{25\sqrt{2}}{16}$

$x(t) = \dfrac{25}{64}\sin\left(8t + \dfrac{\pi}{4}\right) - \dfrac{25\sqrt{2}}{16}t - \dfrac{25\sqrt{2}}{128}$

47. From Exercise 9, time of impact is $t = \dfrac{\sqrt{30}}{4}$

$2\dfrac{1}{2}$ somersaults $= 5\pi$

$f''(t) = 0,\ f'(t) = v_0,$

$f(t) = v_0 t$

$f\left(\dfrac{\sqrt{30}}{4}\right) = v_0\dfrac{\sqrt{30}}{4} = 5\pi$

$v_0 = \dfrac{20\pi}{\sqrt{30}}$

49. $y(0) = 6,\ x(0) = 0$

$\theta = 52° = \dfrac{13\pi}{45}, v_0 = 25$

$y'(0) = 25\sin\dfrac{13\pi}{45} \approx 19.70$

$x'(0) = 25\cos\dfrac{13\pi}{45} \approx 15.39$

$y''(t) = -32$

$x''(t) = 0$

$x'(t) = 15.39$

$x(t) = 15.39t = 15$

$t \approx \dfrac{15}{15.39}$

$y'(t) = -32t + 19.70$

$y(t) = -16t^2 + 19.70t + 6$

$y\left(\dfrac{15}{15.39}\right) \approx 10.001$

$y(0) = 6\ x(0) = 0$

θ unknown, $v_0 = 25$

$y'(0) = 25\sin\theta$

$x'(0) = 25\cos\theta$

$x''(t) = 0$

$x'(t) = 25\cos\theta$

$x(t) = (25\cos\theta)t$

$y''(t) = -32$

$y'(t) = -32t + 25\sin\theta$

$y(t) = -16t^2 + (25\sin\theta)t + 6$

Let's start with $x = 14.65$.

$x(t) = (25\cos\theta)t = 14.65,\ t = \dfrac{14.65}{25\cos\theta}$

$y\left(\dfrac{14.65}{25\cos\theta}\right)$

$= -16\left(\dfrac{14.65}{25\cos\theta}\right)^2 + (25\sin\theta)\left(\dfrac{14.65}{25\cos\theta}\right) + 6$

$= 10$

Solving graphically, we find $\theta \approx .8379 \approx 48°$ or $\theta \approx .9994 \approx 57°$ and from the graph, it's clear that $y \geq 10$ when $48° \leq \theta \leq 57°$. By trying a few other values of x in the range [14.65, 15.35], we find that x-values up to 15 give progressively smaller θ-intervals on which $y \geq 10$. At $x = 15$, the only θ (to the nearest degree) to give $y = 10$ is $\theta = 52°$. For $x > 15$, $y < 10$, so these free throws are no good.

51. $y(t) = 99.62t$

$x(t) = -10t^2 + 8.72t$

$y(t) = 90$ when $t = 0.903$.

$x(0.903) \approx -0.29$

The ball just gets into the goal.

53. $y''(t) = -g$

$y'(t) = -gt + y'(0)$

$y(t) = \dfrac{-gt^2}{2} + y'(0)t + y(0)$

$x'(t) = c$

$x(t) = ct + x(0)$

Solving for t, we have $\dfrac{1}{c}(x - x(0)) = t$.

Substituting this expression for t in $y(t)$, we have

$$y = -\frac{g}{2}\left[\frac{1}{c}(x - x(0))\right]^2 + y'(0)\left[\frac{1}{c}(x - x(0))\right] + y(0)$$

$$y = -\frac{g}{2c^2}(x - x(0))^2 + \frac{y'(0)}{c}(x - x(0)) + y(0)$$

Hence the path is a parabola.

Let $x(0) = 0$, $y(0) = 0$. then $y(t) = 2500$ when $y'(t) = 0$.

$$y'(t) = -gt + y'(0) = 0 \text{ when } t = \frac{y'(0)}{g}$$

$$y\left(\frac{y'(0)}{g}\right) = \frac{-g}{2}\left(\frac{y'(0)}{g}\right)^2 + y'(0)\left(\frac{y'(0)}{g}\right) = 2500$$

$$\frac{(y'(0))^2}{2g} = 2500$$

$$y'(0) = \sqrt{5000g}$$

So $y(t) = \dfrac{-g}{2}t^2 + \sqrt{5000g}\,t$

The time to complete the path can be found by finding when $y(t) = 0$.

$$y(t) = \frac{-g}{2}t^2 + \sqrt{5000g}\,t = t\left(\frac{-g}{2}t + \sqrt{5000g}\,\right) = 0$$

$$t = \frac{2\sqrt{5000g}}{g} = 25\sec$$

55. $y(0) = 0$, $x(0) = 0$,

$y''(t) = -9.8$

$y'(t) = -9.8t + v_0 \sin\theta = 0$

$t = \dfrac{v_0 \sin\theta}{9.8}$

$y(t) = -4.9t^2 + v_0 \sin\theta\, t$

$x''(t) = 0$

$x'(t) = v_0 \cos\theta$

$x(t) = v_0 \cos\theta\, t$

When $t = \dfrac{v_0 \sin\theta}{9.8}$ we have

$x\left(\dfrac{v_0 \sin\theta}{9.8}\right) = v_0 \cos\theta\left(\dfrac{v_0 \sin\theta}{9.8}\right) = \dfrac{v_0^2}{9.8}\sin\theta\cos\theta$

and

$y\left(\dfrac{v_o \sin\theta}{9.8}\right) = -4.9\left(\dfrac{v_0 \sin\theta}{9.8}\right)^2 + v_0 \sin\theta\left(\dfrac{v_0 \sin\theta}{9.8}\right)$

$= \dfrac{v_0^2}{9.8}\sin^2\theta\left(\dfrac{-4.9}{9.8}+1\right)$

So

$\dfrac{x\left(\frac{v_0 \sin\theta}{9.8}\right)}{y\left(\frac{v_0 \sin\theta}{9.8}\right)} = \dfrac{\left(\frac{v_0^2}{9.8}\sin\theta\cos\theta\right)}{\left(\frac{v_0^2}{9.8}\sin^2\theta\cdot\frac{1}{2}\right)} = 2\dfrac{\cos\theta}{\sin\theta} = 2\cot\theta$

On the other hand, since $t = \dfrac{v_0 \sin\theta}{9.8}$ corresponds to the arrow landing in the cauldron, we know that $x\left(\dfrac{v_0 \sin\theta}{9.8}\right) = 70$ and $y\left(\dfrac{v_0 \sin\theta}{9.8}\right) = 30$.

Thus $2\cot\theta = \dfrac{70}{30} = \dfrac{7}{3}$.

Solving for θ, we have

$\cot\theta = \dfrac{7}{6}$

$\tan\theta = \dfrac{6}{7}$

$\theta = \tan^{-1}\left(\dfrac{6}{7}\right) \approx .7086 \approx 40.6°$

Note: if $\theta = \tan^{-1}\left(\dfrac{6}{7}\right)$, then

$\sin\theta = \sin\left(\tan^{-1}\left(\dfrac{6}{7}\right)\right) = \dfrac{6}{\sqrt{85}}$ and

$\cos\left(\tan^{-1}\left(\dfrac{6}{7}\right)\right) = \dfrac{7}{\sqrt{85}}$

Since $x\left(\dfrac{v_0 \sin\theta}{9.8}\right) = v_0 \cos\theta \cdot \left(\dfrac{v_0 \sin\theta}{9.8}\right) = 70$

$v_0 \cdot \dfrac{7}{\sqrt{85}} \cdot \dfrac{v_0}{9.8} \cdot \dfrac{6}{\sqrt{85}} = 70$

$v_0^2 = \dfrac{(70)(85)(9.8)}{42} = \dfrac{4165}{3}$

$v_0 = \sqrt{\dfrac{4165}{3}} \approx 37.26 \text{ m/s}$

Section 5.6

5. $F(x) = Kx$

$5 = K \cdot \dfrac{4}{12}$

$K = 15$

$F(x) = 15x$

$W = \displaystyle\int_0^{.5} 15x\,dx = \dfrac{15x^2}{2}\Big|_0^{.5} = \dfrac{15}{8} \text{ ft-lb}$

7. $F(x) = Kx$

$20 = K \cdot \dfrac{1}{2}$

$K = 40$

$F(x) = 40x$

$W = \displaystyle\int_0^1 40x\,dx = 20x^2\Big|_0^1 = 20 \text{ ft-lb}$

9. $W = 250 \cdot \dfrac{20}{12} = \dfrac{1250}{3} \text{ ft-lb}$

11. $W = 100 \cdot 3 = 300 \text{ ft-lb}$

13. If x is between 0 and 30,000 ft, then the weight of the rocket at altitude x is $10000 - \dfrac{1}{15}x$.

$\displaystyle\int_0^{30,000}\left(10,000 - \dfrac{x}{15}\right)dx = \left(10,000x - \dfrac{x^2}{30}\right)\Big|_0^{30,000}$

$= 270,000,000 \text{ ft-lb}$

15. $W = \displaystyle\int_0^1 800x(1-x)dx = \left(400x^2 - \dfrac{800}{3}x^3\right)\Big|_0^1$

$= \dfrac{400}{3} \text{ mile-lb}$

$\dfrac{400}{3} \text{ mile-lb} \cdot 5280 \text{ ft/mile} = 704,000 \text{ ft-lb}$

17. $W = \int_0^{100} 62.4\pi(100x - x^2)(200 + x)dx$

$= 62.4\pi \int_0^{100} \left(20,000x - 100x^2 - x^3\right)dx$

$= 62.4\pi \left(10,000x^2 - \dfrac{100x^3}{3} - \dfrac{x^4}{4}\right)\Big|_0^{100}$

$= 8,168,140,899$ ft-lb

19. $W = \int_{10}^{20} 62.4\pi x(20 - x)^2 dx$

$= 62.4\pi \int_{10}^{20} \left(400x - 40x^2 + x^3\right)dx$

$= 62.4\pi \left(200x^2 - 40\dfrac{x^3}{3} + \dfrac{x^4}{4}\right)\Big|_{10}^{20}$

$= 816,814$ ft-lb

21. $J \approx \dfrac{.0008}{3(8)}\left[0 + 4(1000) + 2(2100) + 4(4000) + 2(5000) + 4(5200) + 2(2500) + 4(1000) + 0\right]$

≈ 2.133

$2.13 = J = m\Delta v = .01\Delta v$

$\Delta v = 213$

$213 - 100 = 113$ ft/sec after impact

23. $J \approx \dfrac{.6}{3(6)}\left[0 + 4(8000) + 2(16,000) + 4(24,000) + 2(15,000) + 4(9000) + 0\right]$

≈ 7533.3

$7533.3 = J = m\Delta v = 200\Delta v$

$\Delta v = 37.7$ ft/sec

25. $\rho(x) = \dfrac{x}{6} + 2$ kg/m, $0 \le x \le 6$

$M = \int_0^6 x\left(\dfrac{x}{6} + 2\right)dx = \left(\dfrac{x^3}{18} + x^2\right)\Big|_0^6 = 48$

$m = \int_0^6 \left(\dfrac{x}{6} + 2\right)dx = \left(\dfrac{x^2}{12} + 2x\right)\Big|_0^6 = 15$ kg

$\bar{x} = \dfrac{M}{m} = \dfrac{48}{15} = \dfrac{16}{5}$ m

27. $\rho(x) = 4 + \dfrac{x^2}{4}$ kg/m, $-2 \le x \le 2$

$M = \int_{-2}^2 x\left(4 + \dfrac{x^2}{4}\right)dx = \left(2x^2 + \dfrac{x^4}{16}\right)\Big|_{-2}^2 = 0$

$m = \int_{-2}^2 \left(4 + \dfrac{x^2}{4}\right)dx = \left(4x + \dfrac{x^3}{12}\right)\Big|_{-2}^2 = \dfrac{52}{3}$ kg

$\bar{x} = \dfrac{M}{m} = \dfrac{0}{\frac{52}{3}} = 0$ m

29. $m = \int_{-3}^{27} \left(\dfrac{1}{46} + \dfrac{x+3}{690}\right)^2 dx$

$= \dfrac{690}{3}\left(\dfrac{1}{46} + \dfrac{x+3}{690}\right)^3 \Big|_{-3}^{27}$

$= .0614$ slugs

$(.0614)(32)(16) \approx 31.5$ oz

31. $M = \int_{-3}^{27} x\left(\dfrac{1}{46} + \dfrac{x+3}{690}\right)^2 dx$

$= \int_{-3}^{27} x \cdot \dfrac{(x+18)^2}{690^2}dx$

$= \dfrac{1}{476,100}\int_{-3}^{27}(x^3 + 36x^2 + 324x)dx$

$= \dfrac{1}{476,100}\left(\dfrac{x^4}{4} + 12x^3 + 162x^2\right)\Big|_{-3}^{27}$

≈ 1.0208

$\bar{x} = \dfrac{M}{m} = \dfrac{1.0208}{.0614} \approx 16.6$ in.

33. $m = \int_0^{30} .00468\left(\frac{3}{16} + \frac{x}{60}\right)dx$

$= .00468\left(\frac{3}{16}x + \frac{x^2}{120}\right)\Big|_0^{30}$

$\approx .0614$ slugs

$M = \int_0^{30} .00468x\left(\frac{3}{16} + \frac{x}{60}\right)dx$

$= .00468\left(\frac{3x^2}{32} + \frac{x^3}{180}\right)\Big|_0^{30}$

≈ 1.0969

weight $= m(32)(16) = 31.4$ oz

$\bar{x} = \frac{M}{m} = \frac{1.0969}{.0614} \approx 17.8$ in.

35. Area of the base is $\frac{1}{2}(3+1) = 2$.

Area of the body is $1 \times 4 = 4$.

Area of the tip is $\frac{1}{2}(1 \times 1) = \frac{1}{2}$.

Base:

$m = \int_0^1 \rho(3 - 2x)dx = \rho(3x - x^2)\Big|_0^1 = 2\rho$

$M = \int_0^1 \rho x(3 - 2x)dx = \rho\left(\frac{3x^2}{2} - \frac{2x^3}{3}\right)\Big|_0^1 = \frac{5}{6}\rho$

$\bar{x} = \frac{M}{m} \approx .4167$

Body:

$m = \int_1^5 \rho\, dx = \rho x\Big|_1^5 = 4\rho$

$M = \int_1^5 \rho x\, dx = \rho\frac{x^2}{2}\Big|_1^5 = 12\rho$

$\bar{x} = \frac{M}{m} = 3$

Tip:

$m = \int_5^6 \rho(6 - x)dx = \rho\left(6x - \frac{x^2}{2}\right)\Big|_5^6 = .5\rho$

$M = \int_5^6 \rho x(6 - x)dx = \rho\left(3x^2 - \frac{x^3}{3}\right)\Big|_5^6 \approx 2.67\rho$

$\bar{x} = \frac{M}{m} \approx 5.33$

37. $F = \int_0^{60} 62.4x(x + 40)dx$

$= 62.4\left(\frac{x^3}{3} + 20x^2\right)\Big|_0^{60}$

$= 8,985,600$ lb

39. $F = \int_0^{10} 62.4(35 + x) \cdot 2\sqrt{5^2 - (5 - x)^2}\, dx$

$= \int_0^{10} 124.8(35 + x)\sqrt{10x - x^2}\, dx$

$\approx 196,035$ lb

(approximated numerically)

41. Assuming that the *center* of the circular window descends to 1000 feet, and measuring the radius of the window in feet, we have

$F = \int_0^{.5} 62.4(999.75 + x) \cdot 2\sqrt{(.25)^2 - (.25 - x)^2}\, dx$

$= \int_0^{.5} 124.8(999.75 + x)\sqrt{.5x - x^2}\, dx$

$\approx 12,252$ lb

(approximated numerically)

43. $h(0) = 20, h'(0) = 0, h''(0) = -32$

$h'(t) = -32t, h(t) = -16t^2 + 20$

$h(t) = 0$ for $t = \sqrt{\frac{20}{16}} = \frac{\sqrt{5}}{2}$

$v = h'\left(\frac{\sqrt{5}}{2}\right) = -16\sqrt{5} \approx -35.7$ ft/s

$\frac{1}{2}mv^2 = \frac{1}{2}\left(\frac{200}{32}\right)(16^2 \cdot 5) = 4000$ ft-lb

45. (100 tons)(20 miles/hr)

$= \frac{(100 \cdot 2000 \text{ lbs})(20 \cdot 5280 \text{ ft})}{3600 \text{ sec}}$

$\approx 5,866,667$ ft-lb/s

$= \frac{5,866,667}{550}$ hp

$\approx 10,667$ hp

47. Recall that the bat in Exercise 29 models the bat of Example 6.5 choked up 3 in.
From Example 6.5:

$$f(x) = \left(\frac{1}{46} + \frac{x}{690}\right)^2 ; \int_0^{30} f(x) \cdot x^2 \, dx \approx 27.22$$

From Exercise 29:

$$f(x) = \left(\frac{1}{46} + \frac{x+3}{690}\right)^2 ; \int_{-3}^{27} f(x) \cdot x^2 \, dx \approx 20.54$$

Reduction in moment:

$$\frac{27.22 - 20.54}{27.22} \approx 24.5\%$$

49. A CAS gives

$$\int_{-a}^{a} 2\rho x^2 b \sqrt{1 - \frac{x^2}{a^2}} \, dx = \frac{1}{4}\rho \pi a^3 b .$$

51. Using the formula in Exercise 50, we find that the moments are 1323.8 for the wooden racket, 1792.9 for the midsized racket, and 2361.0 for the oversized racket.

Section 5.7

5. $f(x) = 4x^3$, $[0, 1]$

 i) $f(x) = 4x^3 \geq 0$ for $0 \leq x \leq 1$

 ii) $\int_0^1 4x^3 \, dx = x^4 \Big|_0^1 = 1 - 0 = 1$

7. $f(x) = x + 2x^3$, $[0, 1]$

 i) $x + 2x^3 \geq 0$ for $0 \leq x \leq 1$

 ii) $\int_0^1 \left(x + 2x^3\right) dx = \frac{x^2}{2} + \frac{x^4}{2} \Big|_0^1 = 1$

9. $f(x) = \frac{1}{2}\sin x$, $[0, \pi]$

 i) $\frac{1}{2}\sin x \geq 0$ for $0 \leq x \leq \pi$

 ii) $\int_0^\pi \frac{1}{2}\sin x \, dx = -\frac{1}{2}\cos x \Big|_0^\pi$

$$= -\frac{1}{2}(-1) + \frac{1}{2}(1)$$
$$= 1$$

11. $f(x) = e^{-x/2}$, $[0, \ln 4]$

 i) $e^{-x/2} \geq 0$ for $0 \leq x \leq \ln 4$

 ii) $\int_0^{\ln 4} e^{-x/2} dx = -2e^{-x/2} \Big|_0^{\ln 4} = -1 + 2 = 1$

13. $f(x) = cx^3$, $[0, 1]$

$$1 = \int_0^1 cx^3 dx = \frac{cx^4}{4} \Big|_0^1 = \frac{c}{4} - 0$$

$$c = 4$$

15. $f(x) = ce^{-4x}$, $[0, 1]$

$$1 = \int_0^1 ce^{-4x} dx = -\frac{1}{4}ce^{-4x} \Big|_0^1 = -\frac{1}{4}ce^{-4} + \frac{1}{4}c$$

$$c = \frac{4}{1 - e^{-4}}$$

17. $f(x) = 2ce^{-cx}$, $[0, 2]$

$$1 = \int_0^2 2ce^{-cx} dx = -2e^{-cx} \Big|_0^2 = -2e^{-2c} + 2$$

$$c = \frac{\ln 2}{2} \approx 0.346$$

19. $P(70 \leq x \leq 72) = \int_{70}^{72} \frac{.4}{\sqrt{2\pi}} e^{-.08(x-68)^2} \, dx \approx .157$

21. $P(84 \leq x \leq 120) = \int_{84}^{120} \frac{.4}{\sqrt{2\pi}} e^{-.08(x-68)^2} \, dx$

$$\approx 7.76 \times 10^{-11}$$

23. $P\left(0 \leq x \leq \frac{1}{4}\right) = \int_0^{1/4} 6e^{-6x} dx$

$$= -e^{-6x} \Big|_0^{1/4}$$
$$= (-e^{-3/2} + 1)$$
$$\approx .77687$$

25. $P(1 \le x \le 2) = \int_1^2 6e^{-6x} dx$

$$= -e^{-6x}\Big|_1^2$$

$$= (-e^{-12} + e^{-6})$$

$$\approx .00247$$

27. $P\left(\dfrac{1}{12} \le x \le \dfrac{1}{6}\right) = \int_{1/12}^{1/6} 8e^{-8x} dx$

$$= -e^{-8x}\Big|_{1/12}^{1/6}$$

$$= (-e^{-4/3} + e^{-2/3})$$

$$\approx .24982$$

29. $P\left(0 \le x \le \dfrac{1}{2}\right) = \int_0^{1/2} 8e^{-8x} dx$

$$= -e^{-8x}\Big|_0^{1/2}$$

$$= (-e^{-4} + 1)$$

$$\approx .9817$$

31. $P(0 \le x \le 1) = \int_0^1 4xe^{-2x} dx$

$$\approx .594$$

(using CAS or calculator)

33. $\mu = \int_0^{10} x\left(4xe^{-2x}\right) dx \sim .9999995$

35. $f(x) = 3x^2,\ 0 \le x \le 1$

$$\mu = \int_0^1 x \cdot 3x^2 dx = \dfrac{3}{4}x^4\Big|_0^1 = \dfrac{3}{4}$$

$$.5 = \int_0^c 3x^2 dx = x^3\Big|_0^c = c^3$$

$$c = \sqrt[3]{.5}$$

37. $f(x) = \dfrac{1}{2}\sin x,\ 0 \le x \le \pi$

$$\mu = \int_0^\pi x \cdot \dfrac{1}{2}\sin x\, dx = \dfrac{\pi}{2} \approx 1.57$$

(using CAS or calculator)

$$.5 = \int_0^c \dfrac{1}{2}\sin x\, dx$$

$$= -\dfrac{1}{2}\cos x\Big|_0^c$$

$$= -\dfrac{1}{2}\cos c + \dfrac{1}{2},\ c = \dfrac{\pi}{2}$$

39. $f(x) = \dfrac{1}{2}(\ln 3)e^{-kx},\ k = \dfrac{1}{3}\ln 3,\ 0 \le x \le 3$

$$\int_0^3 x \cdot \dfrac{1}{2}(\ln 3)e^{-kx} dx \approx 1.23$$

(using CAS or calculator)

$$.5 = \int_0^c \dfrac{1}{2}(\ln 3)e^{-kx} dx = \dfrac{1}{2}\ln 3\left(-\dfrac{1}{k}\right)e^{-kx}\Big|_0^c$$

$$= \dfrac{1}{2k}\ln 3\left(1 - e^{-ck}\right)$$

$$= \dfrac{3}{2}(1 - e^{-c/3\ln 3})$$

So

$$\dfrac{1}{3} = 1 - e^{-c/3\ln 3}$$

$$\dfrac{2}{3} = e^{-c/3\ln 3}$$

$$\ln\dfrac{2}{3} \doteq -\dfrac{c}{3}\ln 3$$

$$-\dfrac{3\ln\frac{2}{3}}{\ln 3} = c$$

$$c \approx 1.11$$

41. $f(x) = \dfrac{4}{1 - e^{-4}}e^{-4x},\ 0 \le x \le 1$

$$\mu = \int_0^1 x\dfrac{4}{1 - e^{-4}}e^{-4x} dx \approx .2313$$

(using CAS or calculator)

$$.5 = \int_0^c \dfrac{4}{1 - e^{-4}}e^{-4x} dx$$

$$= \dfrac{-1}{1 - e^{-4}}e^{-4x}\Big|_0^c$$

$$= \dfrac{1}{e^{-4} - 1}(e^{-4c} - 1)$$

So

$$\dfrac{1}{2}(e^{-4} - 1) = e^{-4c} - 1$$

$$\dfrac{1}{2}(e^{-4} + 1) = e^{-4c}$$

$$\ln\left(\dfrac{1}{2}(e^{-4} + 1)\right) = -4c$$

$$-\dfrac{1}{4}\ln\left(\dfrac{e^{-4} + 1}{2}\right) = c$$

$$c \approx .1687$$

43. $f(x) = ce^{-4x}, [0, b], b > 0$

$$1 = \int_0^b ce^{-4x}dx = -\frac{c}{4}e^{-4x}\Big]_0^b = -\frac{c}{4}\left(e^{-4b} - 1\right)$$

$$c = \frac{4}{1 - e^{-4b}}$$

As $b \to \infty$, $c \to 4$

45. $f(x) = ce^{-6x}, [0, b], b > 0$

$$1 = \int_0^b ce^{-6x}dx = \frac{-c}{6}e^{-6x}\Big|_0^b = -\frac{c}{6}\left(e^{-6b} - 1\right)$$

$$c = \frac{6}{1 - e^{-6b}}$$

As $b \to \infty$, $c \to 6$

$$\mu = \int_0^b xce^{-6x}dx = \frac{ce^{-6c}}{36}(-6x - 1)\Big|_0^b$$

$$= \frac{ce^{-6b}}{36}(-6b - 1) + \frac{c}{36}$$

As $b \to \infty$, $\mu \to \frac{1}{6}$

47. a. $P(h \leq 3) = P(0) + P(1) + P(2) + P(3)$

$$= \frac{1}{256} + \frac{8}{256} + \frac{28}{256} + \frac{56}{256}$$

$$= \frac{93}{256}$$

b. $P(h > 4) = P(5) + P(6) + P(7) + P(8)$

$$= \frac{56}{256} + \frac{28}{256} + \frac{8}{256} + \frac{1}{256}$$

$$= \frac{93}{256}$$

c. $P(0) + P(8) = \frac{1}{256} + \frac{1}{256} = \frac{2}{256}$

d. $P(1) + P(3) + P(5) + P(7)$

$$= \frac{8}{256} + \frac{56}{256} + \frac{56}{256} + \frac{8}{256}$$

$$= \frac{1}{2}$$

49. a. $P(4/3) + P(4/2) + P(4/1) + P(4/0)$

$$= .1659 + .2073 + .2073 + .1296$$

$$= .7101$$

b. $P(0/4) + P(1/4) + P(2/4) + P(3/4)$

$$= .0256 + .0615 + .0922 + .1106$$

$$= .2899$$

c. $P(0/4) + P(4/0) = .0256 + .1296$

$$= .1552$$

d. $P(2/4) + P(3/4) + P(4/3) + P(4/2)$

$$= .0922 + .1106 + .1659 + .2073$$

$$= .576$$

51. $e(p) = cp^{-2}, 1 < p < 100$

$$1 = \int_1^{100} cp^{-2}dp = -cp^{-1}\Big|_1^{100} = -c\left(\frac{1}{100} - 1\right)$$

$$c = \frac{100}{99} \quad e(p) = \frac{100}{99}p^{-2}$$

$$\int_{60}^{70}\frac{100}{99}p^{-2}dp = -\frac{100}{99}p^{-1}\Big|_{60}^{70}$$

$$= -\frac{100}{99}\left(\frac{1}{70} - \frac{1}{60}\right)$$

$$\approx .0024$$

53. $f(x) = \frac{.4}{\sqrt{2\pi}}e^{-.08(x-68)^2}$

$$f'(x) = \frac{-.064}{\sqrt{2\pi}}(x - 68)e^{-.08(x-68)^2}$$

$$f''(x) = \frac{-.064}{\sqrt{2\pi}}e^{-.08(x-68)^2}\left(1 - .16(x-68)^2\right) = 0$$

$$x = 68 \pm 2.5$$

standard deviation is $\frac{5}{2}$

55. If the entire class does well, the middle score of C will be close to the other scores which will not be A's.

57. $f(t) = t^{-3/2}e^{0.38t - 100/t}$

$$\int_0^{40} k \cdot f(t)\,dt = 1 \text{ for } k = 0.000318.$$

$$\int_{20}^{30} 0.000318 \cdot f(t)\,dt \approx 0.0134$$

Chapter 5 Review

1. See Example 1.1.

$y = x^2 + 2, \; y = \sin x, \; \text{for } 0 \le x \le \pi$

$\int_0^\pi \left(x^2 + 2 - \sin x \right) dx$

$= \left(\dfrac{x^3}{3} + 2x + \cos x \right) \Big|_0^\pi$

$= \dfrac{\pi^3}{3} + 2\pi - 1 - (0 + 0 + 1)$

$= \dfrac{\pi^3}{3} + 2\pi - 2$

3. See Example 1.2.

$y = x^3, \; y = 2x^2 - x$

$\int_0^1 x^3 - \left(2x^2 - x \right)$

$= \left(\dfrac{x^4}{4} - \dfrac{2}{3}x^3 + \dfrac{x^2}{2} \right) \Big|_0^1$

$= \left(\dfrac{1}{4} - \dfrac{2}{3} + \dfrac{1}{2} \right) - (0 - 0 + 0)$

$= \dfrac{1}{12}$

5. See Example 1.3.

$y = e^{-x}, \; y = 2 - x^2$

$\int_{-.537}^{1.316} \left(2 - x^2 - e^{-x} \right) dx = \left(2x - \dfrac{x^3}{3} + e^{-x} \right) \Big|_{-.537}^{1.316}$

≈ 1.452

7. See Example 1.4.

$y = x^2, \; y = 2 - x, \; y = 0$

$\int_0^1 x^2 dx + \int_1^2 (2 - x)dx$

$= \dfrac{x^3}{3} \Big|_0^1 + \left(2x - \dfrac{x^2}{2} \right) \Big|_1^2$

$= \dfrac{1}{3} + (4 - 2) - \left(2 - \dfrac{1}{2} \right)$

$= \dfrac{5}{6}$

9. See Example 1.1, Exercises 1.39 and 1.40.

$A = \int_0^6 (10 + 2t) - (4 + t)dt$

$= \int_0^6 (6 + t)dt$

$= \left(6t + \dfrac{t^2}{2} \right) \Big|_0^6$

$= 54$

population $= 10{,}000 + 54 = 10{,}054$

11. See Example 2.1.

$V = \int_0^2 \pi(3 + x)^2 dx$

$= \pi \int_0^2 \left(9 + 6x + x^2 \right) dx$

$= \pi \left(9x + 3x^2 + \dfrac{x^3}{3} \right) \Big|_0^2$

$= \dfrac{98\pi}{3}$

13. See Example 2.2.

$V = 0.4 \left(\dfrac{0.4}{2} + 1.4 + 1.8 + 2.0 + 2.1 + 1.8 + 1.1 + \dfrac{0.4}{2} \right)$

≈ 4.2

15. See Examples 2.6 and 2.7.

a. $V = \int_{-2}^2 \pi(4)^2 dx - \int_{-2}^2 \pi(x^2)^2 dx$

$= \pi \int_{-2}^2 (16 - x^4) dx$

$= \pi \left(16x - \dfrac{x^5}{5} \right) \Big|_{-2}^2$

$= \dfrac{256\pi}{5}$

b. $V = \int_0^4 \pi(\sqrt{y})^2 dy = \pi\int_0^4 y\, dy = \dfrac{\pi y^2}{2}\bigg|_0^4 = 8\pi$

c. $V = \int_0^4 \pi(2+\sqrt{y})^2 dy - \int_0^4 \pi(2-\sqrt{y})^2 dy$

$\quad = \pi\int_0^4 (4+4y^{1/2}+y)dy - \pi\int_0^4 (4-4y^{1/2}+y)dy$

$\quad = \pi\int_0^4 (8y^{1/2})dy$

$\quad = 8\pi\cdot\dfrac{2}{3}y^{3/2}\bigg|_0^4$

$\quad = \dfrac{128\pi}{3}$

d. $V = \int_{-2}^2 \pi(6)^2 dx - \int_{-2}^2 \pi(x^2+2)^2 dx$

$\quad = \pi\int_{-2}^2 (-x^4 - 4x^2 + 32)dx$

$\quad = \pi\left(-\dfrac{x^5}{5} - \dfrac{4x^3}{3} + 32x\right)\bigg|_{-2}^2$

$\quad = \dfrac{1408\pi}{15}$

17. See Examples 3.3 and 3.4.

a.

$V = \int_0^1 2\pi y((2-y)-y)dy$

$\quad = 2\pi\int_0^1 (2y - 2y^2)dy$

$\quad = 2\pi\left(y^2 - \dfrac{2y^3}{3}\right)\bigg|_0^1$

$\quad = \dfrac{2\pi}{3}$

b. $V = \int_0^1 \pi(2-y)^2 dy - \int_0^1 \pi(y)^2 dy$

$\quad = \pi\int_0^1 (4-4y)dy$

$\quad = \pi(4y - 2y^2)\bigg|_0^1$

$\quad = 2\pi$

c. $V = \int_0^1 \pi((2-y)+1)^2 dy - \int_0^1 \pi(y+1)^2 dy$

$= \pi\int_0^1 (9-6y+y^2)dy - \pi\int_0^1 (y^2+2y+1)dy$

$= \pi\int_0^1 (8-8y)dy$

$= \pi(8y-4y^2)\Big|_0^1$

$= 4\pi$

d. $V = \int_0^1 2\pi(4-y)((2-y)-y)dy$

$= 2\pi\int_0^1 (8-10y+2y^2)dy$

$= 2\pi\left(8y-5y^2+\frac{2y^3}{3}\right)\Big|_0^1$

$= \frac{22\pi}{3}$

19. See Example 4.2.

$y = x^4, -1 \le x \le 1$

$s = \int_{-1}^1 \sqrt{1+\left(4x^3\right)^2}\, dx \approx 3.2$

21. See Example 4.2.

$y = e^{x/2}, -2 \le x \le 2$

$\int_{-2}^2 \sqrt{1+\left(\frac{e^{x/2}}{2}\right)^2}\, dx \approx 4.767$

23. See Example 4.5.

$s = \int_0^1 2\pi\left(1-x^2\right)\sqrt{1+4x^2}\, dx \approx 5.48$

25. See Example 5.1.

$h''(t) = -32$

$h(0) = 64, h'(0) = 0$

$h'(t) = -32t$

$h(t) = -16t^2 + 64 = 0$

$t = 2$

$h'(2) = -32(2) = -64$ ft/s

27. See Example 5.4.

$y''(t) = -32, x''(t) = 0, y(0) = 0, x(0) = 0$

$y'(0) = 48\sin\left(\frac{\pi}{9}\right), x'(0) = 48\cos\left(\frac{\pi}{9}\right)$

$y'(0) \approx 16.42, x'(0) \approx 45.11$

$y'(t) = -32t + 16.42$

$y(t) = -16t^2 + 16.42t = 0$

$t \approx 1.026$ sec

$x'(t) = 45.11$

$x(t) = 45.11t$

$x(1.026) = 45.11(1.026) \approx 46.3$ ft

29. See Example 5.4.

$y(0) = 6, x(0) = 0$

$y'(0) = 80\sin\left(\frac{2\pi}{45}\right) \approx 11.13,$

$x'(0) = 80\cos\left(\frac{2\pi}{45}\right) \approx 79.22$

$y''(t) = -32, x''(t) = 0$

$y'(t) = -32t + 11.13$

$y(t) = -16t^2 + 11.13t + 6$

$x'(t) = 79.22$

$x(t) = 79.22t = 120$

$t \approx 1.51$

$y(1.51) = -16(1.51)^2 + 11.13(1.51) + 6$

$\qquad = -13.6753,$

So the ball bounces and is not catchable.

31. See Example 5.3.

$h''(t) = -32$

$h'(0) = v_0$

$h(0) = 0$

$h'(t) = -32t + v_0 = 0, \ t = \dfrac{v_0}{32}$

$h(t) = -16t^2 + v_0 t$

$h\left(\dfrac{v_0}{32}\right) = -16\left(\dfrac{v_0^2}{32^2}\right) + \dfrac{v_0^2}{32}$

$= \dfrac{-16v_0^2 + 32v_0^2}{32^2}$

$= \dfrac{16v_0^2}{32^2}$

$= 128$

$v_0 = 64\sqrt{2}$ ft/s

$h(t) = -16t^2 + 64\sqrt{2}\,t = 0$

$t = 4\sqrt{2}$

$h'\left(4\sqrt{2}\right) = -32 \cdot 4\sqrt{2} + 64\sqrt{2} = -64\sqrt{2}$ ft/s

33. See Example 6.1.

$60 = k \cdot 1, \ k = 60$

$W = \displaystyle\int_0^{2/3} 60x\,dx = 30x^2\Big|_0^{2/3} = \dfrac{30\cdot 4}{9} = \dfrac{40}{3}$ ft-lb

35. See Example 6.6.

$m = \displaystyle\int_0^4 \left(x^2 - 2x + 8\right)dx = \left(\dfrac{x^3}{3} - x^2 + 8x\right)\Big|_0^4 = \dfrac{112}{3}$

$M = \displaystyle\int_0^4 x\left(x^2 - 2x + 8\right)dx$

$= \displaystyle\int_0^4 \left(x^3 - 2x^2 + 8x\right)dx$

$= \left(\dfrac{x^4}{4} - \dfrac{2x^3}{3} + 4x^2\right)\Big|_0^4$

$= \dfrac{256}{3}$

$\overline{x} = \dfrac{M}{m} = \dfrac{\frac{256}{3}}{\frac{112}{3}} = \dfrac{256}{112} = \dfrac{16}{7}$

Center of mass is greater than 2 because the object has greater density on the right side of the interval [0, 4].

37. See Example 6.7.

$F = \displaystyle\int_0^{80} 62.4x(60 + x)dx$

$= 62.4\displaystyle\int_0^{80} (60x + x^2)dx$

$= 62.4\left(30x^2 + \dfrac{x^3}{3}\right)\Big|_0^{80}$

$= 22,630,400$ lb

39. See Example 6.4.

$J \approx \dfrac{.0008}{3(8)}(0 + 4(800) + 2(1600) + 4(2400) + 2(3000) + 4(3600) + 2(2200) + 4(1200) + 0)$

$= 1.52$

$J = m\Delta v$

$1.52 = .01\Delta v$

$\Delta v = 152$ ft/s

$152 - 120 = 32$ ft/s

41. See Example 7.1.

$f(x) = x + 2x^3$ on [0, 1]

$f(x) \geq 0$ for $0 \leq x \leq 1$

and $\displaystyle\int_0^1 \left(x + 2x^3\right)dx = \left(\dfrac{x^2}{2} + \dfrac{x^4}{2}\right)\Big|_0^1 = 1$

43. See Example 7.4.

$1 = \displaystyle\int_1^2 \dfrac{c}{x^2}dx = \dfrac{-c}{x}\Big|_1^2 = \dfrac{-c}{2} + c$

$c = 2$

45. See Example 7.3.

a.
$$P(x < .5) = \int_0^{.5} 4e^{-4x}\,dx$$
$$= -e^{-4x}\Big|_0^{.5}$$
$$= 1 - e^{-2}$$
$$\approx .864$$

b.
$$P(.5 \le x \le 1) = \int_{.5}^1 4e^{-4x}\,dx$$
$$= -e^{-4x}\Big|_{.5}^1$$
$$= -e^{-4} + e^{-2}$$
$$\approx .117$$

47. See Example 7.5.

a.
$$\mu = \int_0^1 x\left(x + 2x^3\right)dx = \frac{x^3}{3} + \frac{2x^5}{5}\Big|_0^1 = \frac{11}{15}$$

b.
$$.5 = \int_0^c \left(x + 2x^3\right)dx = \frac{x^2}{2} + \frac{x^4}{2}\Big|_0^c = \frac{c^2}{2} + \frac{c^4}{2}$$
$$c^2 + c^4 = 1$$
$$c = \sqrt{\frac{-1 + \sqrt{5}}{2}} \approx .786$$

Chapter 6

Section 6.1

5. $\ln 4 = \int_1^4 \frac{1}{t}\,dt$

7. $\ln 8.2 = \int_1^{8.2} \frac{1}{t}\,dt$

9. $(\ln 4x)' = \frac{1}{4x}\cdot 4 = \frac{1}{x}$

11. $\left[\ln(\cos x)\right]' = \frac{1}{\cos x}(-\sin x) = -\tan x$

13. $\left[x\ln x\right]' = x\cdot\frac{1}{x} + \ln x = 1 + \ln x$

15. $\left[\ln(x^2+2)\right]' = \frac{1}{x^2+2}\cdot 2x = \frac{2x}{x^2+2}$

17. $\ln\sqrt{2} + 3\ln 2 = \ln 2^{1/2} + 3\ln 2$
$= \frac{1}{2}\ln 2 + 3\ln 2$
$= \frac{7}{2}\ln 2$

19. $2\ln 3 - \ln 9 + \ln\sqrt{3} = 2\ln 3 - \ln 3^2 + \ln 3^{1/2}$
$= 2\ln 3 - 2\ln 3 + \frac{1}{2}\ln 3$
$= \frac{1}{2}\ln 3$

21. Use $u = x^2+1$, then $du = 2x\,dx$.
$\int \frac{2x}{x^2+1}\,dx = \int\frac{1}{u}\,du$
$= \ln|u| + c$
$= \ln\left|x^2+1\right| + c$

23. Use $u = \cos x$, then $du = -\sin x\,dx$.
$\int\frac{\sin x}{\cos x}\,dx = -\int\frac{-\sin x}{\cos x}\,dx$
$= -\int\frac{1}{u}\,du$
$= -\ln|u| + c$
$= -\ln|\cos x| + c$

25. Use $u = x^2+2x-1$, then
$du = (2x+2)dx = 2(x+1)dx$.
$\int\frac{x+1}{x^2+2x-1}\,dx = \frac{1}{2}\int\frac{2(x+1)}{x^2+2x-1}\,dx$
$= \frac{1}{2}\int\frac{1}{u}\,du$
$= \frac{1}{2}\ln|u| + c$
$= \frac{1}{2}\ln\left|x^2+2x-1\right| + c$

27. Use $u = \ln x$, then $du = \frac{1}{x}\,dx$.
$\int\frac{1}{x\ln x}\,dx = \int\frac{1}{\ln x}\frac{1}{x}\,dx$
$= \int\frac{1}{u}\,du$
$= \ln|u| + c$
$= \ln|\ln x| + c$

29. Use $u = \ln x + 1$, then $du = \frac{1}{x}\,dx$.
$\int\frac{(\ln x+1)^2}{x}\,dx = \int(\ln x+1)^2\frac{1}{x}\,dx$
$= \int u^2\,du$
$= \frac{1}{3}u^3 + c$
$= \frac{1}{3}(\ln x+1)^3 + c$

31. $\int_0^2\frac{1}{x+1}\,dx = \left[\ln|x+1|\right]_0^2 = \ln 3$

33. $\int_0^1\frac{x^2}{x^3-4}\,dx = \frac{1}{3}\left[\ln\left|x^3-4\right|\right]_0^1$
$= \frac{1}{3}(\ln 3 - \ln 4)$
$= \frac{1}{3}\ln\frac{3}{4}$

35. $\frac{d}{dx}\left[\ln\sqrt{x^2+1}\right] = \frac{d}{dx}\left[\ln(x^2+1)^{1/2}\right]$
$= \frac{d}{dx}\left[\frac{1}{2}\ln(x^2+1)\right]$
$= \frac{1}{2}\frac{1}{x^2+1}2x$
$= \frac{x}{x^2+1}$

37. $\dfrac{d}{dx}\left[\ln\dfrac{x^4}{x^5+1}\right] = \dfrac{d}{dx}\left[\ln x^4 - \ln(x^5+1)\right]$

$$= \dfrac{d}{dx}\left[\ln x^4\right] - \dfrac{d}{dx}\left[\ln(x^5+1)\right]$$

$$= \dfrac{4}{x} - \dfrac{5x^4}{x^5+1}$$

39. $\displaystyle\int \dfrac{1}{x\ln x^2}\,dx = \int\dfrac{1}{x\cdot 2\ln|x|}\,dx = \dfrac{1}{2}\int\dfrac{1}{\ln|x|}\dfrac{1}{x}\,dx$

Use $u = \ln|x|$, then $du = \dfrac{1}{x}dx$.

$$\dfrac{1}{2}\int\dfrac{1}{\ln|x|}\dfrac{1}{x}\,dx = \dfrac{1}{2}\int\dfrac{1}{u}\,du$$

$$= \dfrac{1}{2}\ln|u| + c$$

$$= \dfrac{1}{2}\ln\left|\ln|x|\right| + c$$

41.

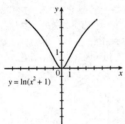

Increasing and concave down on $(2, \infty)$.

43.

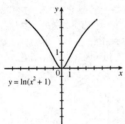

Increasing on $(0, \infty)$, decreasing on $(-\infty, 0)$.
Concave up on $(-1, 1)$;
concave down on $(-\infty, -1)\cup(1, \infty)$.

45.

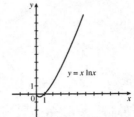

Decreasing on $\left(0, \dfrac{1}{e}\right)$, increasing on $\left(\dfrac{1}{e}, \infty\right)$.
Concave up on $(0, \infty)$.

47. We will make use of the Reciprocal Property,

$$\ln\dfrac{1}{b} = -\ln b,$$ in our proof. To prove this property,

notice that

$$\ln\dfrac{b}{b} = \ln b\cdot\dfrac{1}{b} = \ln b + \ln\dfrac{1}{b}$$

And since

$$\ln\dfrac{b}{b} = \ln 1 = 0$$

we must have that $\ln\dfrac{1}{b} = -\ln b$ (since the

additive inverse of a real number is unique and
since $y = \ln x$ is one-to-one).
Therefore,

$$\ln\dfrac{a}{b} = \ln a\cdot\dfrac{1}{b}$$

$$= \ln a + \ln\dfrac{1}{b}$$

$$= \ln a - \ln b$$

as desired.

49. $s(x) = x^2\ln\dfrac{1}{x} = -x^2\ln x$

$$s'(x) = -x^2\dfrac{1}{x} - \ln x\cdot 2x$$

$$= -x - 2x\ln x$$

$$= -x(1 + 2\ln x)$$

The critical values are
$x = 0$ and $1 + 2\ln x = 0$

$$\ln x = -\dfrac{1}{2}$$

$$x = e^{-1/2}$$

The function is undefined for $x = 0$ and the
derivative changes from $+$ to $-$ at $x = e^{-1/2}$.
Therefore, there is a maximum at $x = e^{-1/2}$.

51. $f(x) = \dfrac{x}{\ln x}$

$f'(x) = \dfrac{\ln x - 1}{(\ln x)^2}$, positive for $x > e$.

$f''(x) = \dfrac{2 - \ln x}{x(\ln x)^3}$, negative for $x > e^2$.

As x increases, the number of primes less than x increases, but at a decreasing rate.

Section 6.2

5. $f'(x) = 3x^2 \geq 0$ for all x, so f is always increasing, hence one-to-one.

7. $f'(x) = 3x^2 - 2$, which is positive for $|x| > \sqrt{\dfrac{2}{3}}$

and negative for $|x| < \sqrt{\dfrac{2}{3}}$. Hence f has a local

maximum at $x = -\sqrt{\dfrac{2}{3}}$ and a local minimum at

$x = \sqrt{\dfrac{2}{3}}$, so is not one-to-one.

9. The function is not one-to-one since it's periodic.

11. $f'(x) = \dfrac{1}{x} > 0$ for all $x > 0$, that is, for all x in the

domain of f. So f is always increasing, hence one-to-one.

13. $f'(x) = 5x^4 + 6x^2 \geq 0$ for all x, so f is always increasing, hence one-to-one.

15. $f'(x) = \dfrac{1}{2}(x^3 + 2x)^{-1/2}(3x^2 + 2)$

$= \dfrac{3x^2 + 2}{2\sqrt{x^3 + 2x}} > 0$

for all $x > 0$, that is, for all x in the domain of f, so f is always increasing, hence one-to-one.

17. $f'(x) = 3x^2 \geq 0$ for all x, so f is always increasing, hence one-to-one.

$y = x^3 - 2$

$y + 2 = x^3$

$\sqrt[3]{y + 2} = x$

Therefore, $f^{-1}(x) = \sqrt[3]{x + 2}$.

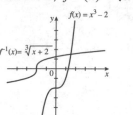

19. $f'(x) = 5x^4 \geq 0$ for all x, so f is always increasing, hence one-to-one.

$y = x^5 - 1$

$y + 1 = x^5$

$\sqrt[5]{y + 1} = x$

Therefore, $f^{-1}(x) = \sqrt[5]{x + 1}$.

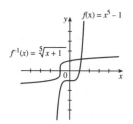

21. The function is not one-to-one since it's an even function.

23. $f'(x) = \dfrac{1}{2}(x^3 + 1)^{-1/2}(3x^2) = \dfrac{3x^2}{2\sqrt{x^3 + 1}} \geq 0$ for

all $x > -1$, that is, for all x in the domain of f, so f is always increasing, hence one-to-one.

$y = \sqrt{x^3 + 1}$

$y^2 = x^3 + 1$

$y^2 - 1 = x^3$

$\sqrt[3]{y^2 - 1} = x$

Therefore, $f^{-1}(x) = \sqrt[3]{x^2 - 1}$ and $x \geq 0$.

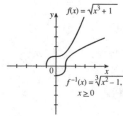

25. $f(x) = x^3 + 4x - 1$, $a = -1$

$f'(x) = 3x^2 + 4$

If $f(x) = x^3 + 4x - 1 = -1$ then $x = 0$ and

$f'(0) = 3(0)^2 + 4 = 4$. Therefore,

$(f^{-1})'(-1) = \dfrac{1}{f'(0)} = \dfrac{1}{4}$

27. $f(x) = x^5 + 3x^3 + x$, $a = 5$

$f'(x) = 5x^4 + 9x^2 + 1$

If $f(x) = x^5 + 3x^3 + x = 5$ then $x = 1$ and

$f'(1) = 15$. Therefore,

$(f^{-1})'(5) = \dfrac{1}{f'(1)} = \dfrac{1}{15}$

29. $f(x) = \sqrt{x^3 + 2x + 1}$, $a = 2$

$f'(x) = \dfrac{3x^2 + 2}{2\sqrt{x^3 + 2x + 1}}$

If $f(x) = \sqrt{x^3 + 2x + 1} = 2$ then $x = 1$ and

$f'(1) = \dfrac{5}{4}$. Therefore, $(f^{-1})'(2) = \dfrac{1}{f'(1)} = \dfrac{4}{5}$

31. $(x_1, y_1) = (-1, 0)$ and $m = \dfrac{1}{4}$

$y - y_1 = m(x - x_1)$

$y - 0 = \dfrac{1}{4}[x - (-1)]$

$y = \dfrac{1}{4}(x + 1)$

33. Increasing, concave down.

35. Increasing, concave up.

37.

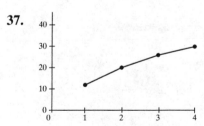

Since 23 is halfway between 20 and 26, the x-value should be halfway between 2 and 3: $f^{-1}(23) = 2.5$. Since the curve is concave down and increasing, the estimate is too high.

39.

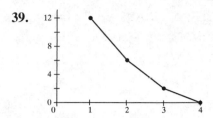

Since 5 is three-quarters of the way from 2 to 6, the x-value should be three-quarters of the way from 3 to 2: $f^{-1}(5) = 2.25$. Since the curve is concave up and decreasing, the estimate is too high.

41. A company's income will rise and fall over time, so the function is probably not one-to-one. Hence, the function doesn't have an inverse.

43. Until the ball bounces, the height is always decreasing (during which time, an inverse exists), and than the height rises and falls with each bounce, so overall the function isn't one-to-one. Hence the function doesn't have an inverse.

45. Two three-dimensional shapes with congruent faces, but different depths, will cast identical shadows if their congruent faces are towards the light source. So, the function is not one-to-one. Hence, the function doesn't have an inverse.

47. $f(x) = x^3 + kx + 1$ has an inverse if it is one-to-one or if the function is either increasing or decreasing on its domain. This is equivalent to the derivative being either positive or negative on its domain.

$f'(x) = 3x^2 + k$

The graph of f' is a parabola, opening up, with vertex at $(0, k)$. Thus $f'(x) \geq 0$ for all x, when $k \geq 0$. Thus f is always increasing, hence one-to-one, when $k \geq 0$.

49. Let your salary before the raise be x. After a 10% raise, your salary is $x + .10x = 1.1x$. After a 10% pay cut, your salary is $1.1x - .10(1.1x) = .9(1.1x) = .99x$, not x. The inverse of adding 10% would be $\dfrac{1}{1.1}x \approx .91x$.

51. Exponential fit via a CAS gives the function
$f(x) = 4.8197e^{0.6832x} + 2.0057$. This is equal to
30 when $x \approx 2.5751$.

Section 6.3

5.

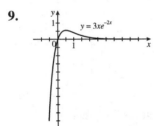

7.

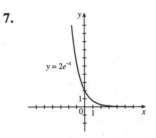

9.

$y = 3xe^{-2x}$

11.

$y = 2e^{-x^2}$

13. $\dfrac{d}{dx}\left[4e^{3x}\right] = 4e^{3x}(3) = 12e^{3x}$

15. $\dfrac{d}{dx}\left[3xe^{-2x}\right] = 3\left[xe^{-2x}(-2) + e^{-2x}\right]$
$= 3e^{-2x} - 6xe^{-2x}$

17. $\dfrac{d}{dx}\left[2e^{-x^2}\right] = 2e^{-x^2}(-2x) = -4xe^{-x^2}$

19. $\dfrac{d}{dx}\left[x^2 e^{-x}\right] = x^2 e^{-x}(-1) + e^{-x}(2x)$
$= 2xe^{-x} - x^2 e^{-x}$

21. $\dfrac{d}{dx}\left[2e^{x^3 - 3x}\right] = 2e^{x^3 - 3x}(3x^2 - 3)$
$= 2(3x^2 - 3)e^{x^3 - 3x}$

23. $\dfrac{d}{dx}\left[\sqrt{e^{2x}}\right] = \dfrac{1}{2\sqrt{e^{2x}}}e^{2x}(2) = \sqrt{e^{2x}} = e^x$

25. $\dfrac{d}{dx}\left[e^{\cos x}\right] = e^{\cos x}(-\sin x) = -\sin x e^{\cos x}$

27. Let $u = 3x$, then $du = 3\, dx$.
$$\int e^{3x}\,dx = \frac{1}{3}\int e^{3x}\, 3dx$$
$$= \frac{1}{3}\int e^u\, du$$
$$= \frac{1}{3}e^u + c$$
$$= \frac{1}{3}e^{3x} + c$$

29. Let $u = x^2$, then $du = 2x\, dx$.
$$\int xe^{x^2}\,dx = \frac{1}{2}\int e^{x^2}\, 2x\,dx$$
$$= \frac{1}{2}\int e^u\, du$$
$$= \frac{1}{2}e^u + c$$
$$= \frac{1}{2}e^{x^2} + c$$

31. Let $u = \cos x$, then $du = -\sin x\, dx$.
$$\int \sin x e^{\cos x}\,dx = -\int e^{\cos x}(-\sin x)\,dx$$
$$= -\int e^u\, du$$
$$= -e^u + c$$
$$= -e^{\cos x} + c$$

33. Let $u = \dfrac{1}{x}$, then $du = -\dfrac{1}{x^2}dx$.
$$\int \frac{e^{1/x}}{x^2}\,dx = -\int e^{1/x}\left(-\frac{1}{x^2}\right)dx$$
$$= -\int e^u\, du$$
$$= -e^u + c$$
$$= -e^{1/x} + c$$

35. $\int (1+e^x)^2\, dx = \int (1 + 2e^x + e^{2x})dx$

$$= x + 2e^x + \frac{1}{2}e^{2x} + c$$

37. $\int \frac{4}{e^{2x}}\, dx = \int 4e^{-2x}\, dx$

$$= -\frac{1}{2}(4)e^{-2x} + c$$

$$= -2e^{-2x} + c$$

39. $\int \ln e^{x^2}\, dx = \int x^2\, dx = \frac{1}{3}x^3 + c$

41. $\int_0^1 e^{3x}\, dx = \frac{1}{3}\Big[e^{3x}\Big]_0^1 = \frac{1}{3}(e^3 - 1) = \frac{1}{3}e^3 - \frac{1}{3}$

43. $\int_{-2}^2 xe^{-x^2}\, dx = -\frac{1}{2}\Big[e^{-x^2}\Big]_{-2}^2 = -\frac{1}{2}(e^{-4} - e^{-4}) = 0$

45. $f(x) = xe^{-2x}$

$f'(x) = xe^{-2x}(-2) + e^{-2x} = e^{-2x}(1-2x)$

critical numbers: $e^{-2x}(1-2x) = 0$

$e^{-2x} = 0$ or $1 - 2x = 0$

    none $x = \frac{1}{2}$

$f'(x)$ $+$ $\frac{1}{2}$

Therefore, $f\left(\frac{1}{2}\right) = \frac{1}{2}e^{-1}$ is a maximum

$f''(x) = e^{-2x}(-2) + (1-2x)e^{-2x}(-2)$

$$= -2e^{-2x}(2 - 2x)$$

$$= -4e^{-2x}(1-x)$$

inflection points: $-4e^{-2x}(1-x) = 0$

$e^{-2x} = 0$ or $1 - x = 0$

    none $x = 1$

$f''(x)$ $-$ 1 $+$

Therefore, $f(1) = e^{-2}$ is an inflection point.

47. $f(x) = x^2 e^{-2x}$

$f'(x) = x^2 e^{-2x}(-2) + e^{-2x}(2x)$

$$= 2xe^{-2x}(1-x)$$

critical numbers: $2xe^{-2x}(1-x) = 0$

$x = 0$ or $e^{-2x} = 0$ or $1 - x = 0$

             none $x = 1$

$f'(x)$ $-$ 0 $+$ 1 $-$

Therefore, $f(0) = 0$ is a minimum and

$f(1) = e^{-2}$ is a maximum.

$f''(x) = 2e^{-2x}(1 - 2x) + 2(x - x^2)e^{-2x}(-2)$

$$= 2e^{-2x}(2x^2 - 4x + 1)$$

inflection points: $2e^{-2x}(2x^2 - 4x + 1) = 0$

$e^{-2x} = 0$ or $2x^2 - 4x + 1 = 0$

   none $x = 1 \pm \frac{\sqrt{2}}{2}$

$f''(x)$ $+$ $1 - \frac{\sqrt{2}}{2}$ $-$ $1 + \frac{\sqrt{2}}{2}$ $+$

Therefore,

$$f\left(1 - \frac{\sqrt{2}}{2}\right) \approx .096 \text{ and } f\left(1 + \frac{\sqrt{2}}{2}\right) \approx .048 \text{ are}$$

inflection points.

49. As in the proof of part (i) of Theorem 1.1, we make use of the rules of logarithms.

(ii) To show $\dfrac{e^r}{e^s} = e^{r-s}$, note that

$$\ln \frac{e^r}{e^s} = \ln e^r - \ln e^s = r - s$$

Therefore, $\dfrac{e^r}{e^s} = e^{r-s}$

(iii) To show $(e^r)^t = e^{r \cdot t}$, note that

$$\ln(e^r)^t = t \ln(e^r) = t \cdot r$$

Therefore, $(e^r)^t = e^{r \cdot t}$.

51.

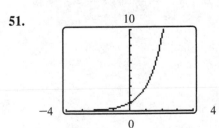

53.

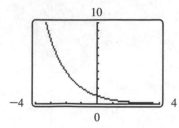

55.

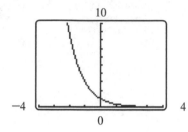

57.

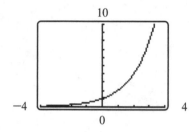

59. For $b > 1$, the graph of $y = b^x$ is increasing and concave up. For $0 < b < 1$, the graph of $y = b^x$ is decreasing and concave up.

61. $\dfrac{d}{dx}\left[2^x\right] = (\ln 2)2^x$

63. $\dfrac{d}{dx}\left[3^{2x}\right] = (\ln 3)3^{2x}(2) = 2(\ln 3)3^{2x}$

65. $\dfrac{d}{dx}\left[3^{x^2}\right] = (\ln 3)3^{x^2}(2x) = 2x(\ln 3)3^{x^2}$

67. $\dfrac{d}{dx}\left[\log_3 x\right] = \dfrac{1}{\ln 3}\dfrac{1}{x} = \dfrac{1}{x\ln 3}$

69. $\dfrac{d}{dx}\left[\log_4 x^2\right] = \dfrac{1}{\ln 4}\dfrac{1}{x^2}(2x) = \dfrac{2}{x\ln 4}$

71. $\displaystyle\int 2^x\,dx = \dfrac{1}{\ln 2}2^x + c$

73. Let $u = x^2$, then $du = 2x\,dx$.

$$\int x\,2^{x^2}\,dx = \frac{1}{2}\int 2^{x^2}\cdot 2x\,dx$$
$$= \frac{1}{2}\int 2^u\,du$$
$$= \frac{1}{2}\frac{1}{\ln 2}2^u + c$$
$$= \frac{1}{2\ln 2}2^{x^2} + c$$

75. $\displaystyle\int 4^{3x}\,dx = \dfrac{1}{3}\dfrac{1}{\ln 4}4^{3x} + c$

$$= \frac{1}{3\ln 4}4^{3x} + c$$

77.

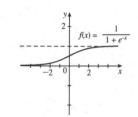

$$\lim_{x\to\infty} f(x) = \lim_{x\to\infty}\frac{1}{1+e^{-x}} = \frac{1}{1+0} = 1$$

$$\lim_{x\to-\infty} f(x) = \lim_{x\to-\infty}\frac{1}{1+e^{-x}} = \frac{1}{1+\infty} = 0$$

If $g(x)$ is $f(x)$ rounded to the nearest integer, then $g(x) = \begin{cases} 0, & x < 0 \\ 1, & x \geq 0 \end{cases}$ and the threshold for switching from "off" to "on" is $x = 0$. To move the threshold to $x = 4$, redefine the function as

$$f(x) = \frac{1}{1+e^{-(x-4)}}.$$

79. We solve the equation $\displaystyle\int_0^3 2ke^{-kx}\,dx = 1$ for k.

$$\int_0^3 2ke^{-kx}\,dx = -2\left[e^{-kx}\right]_0^3 = 2(1 - e^{-3k})$$
$$2(1 - e^{-3k}) = 1$$
$$1 - e^{-3k} = \frac{1}{2}$$
$$e^{-3k} = \frac{1}{2}$$
$$-3k = \ln\frac{1}{2}$$
$$k = -\frac{1}{3}\ln\frac{1}{2} = \frac{1}{3}\ln 2$$

The pdf $f(x)$ can be written in simpler form:

$$f(x) = 2ke^{-kx} = 2\left(\frac{1}{3}\ln 2\right)e^{-\left(\frac{1}{3}\ln 2\right)x} = \left(\frac{2}{3}\ln 2\right)2^{-x/3}$$

The probability a given cell is between 1 and 2 days old is given by

$$\int_1^2 \left(\frac{2}{3}\ln 2\right)2^{-x/3}\,dx = \left(\frac{2}{3}\ln 2\right)\cdot\frac{1}{\ln 2}(-3)\left[2^{-x/3}\right]_1^2$$
$$= -2(2^{-2/3} - 2^{-1/3})$$
$$= 2^{2/3} - 2^{1/3}$$
$$\approx .327,\ \text{or about } 32.7\%$$

81. For $h > 0$, $h = \ln e^h = \int_1^{e^h} \frac{1}{x}\,dx$.

Since $\frac{1}{x}$ is continuous on $[1, e^h]$, the Integral Mean Value Theorem (Theorem 4.4.3) implies there is a number $\overline{x} \in (1, e^h)$ for which

$\frac{1}{\overline{x}} = \frac{1}{e^h - 1}\int_1^{e^h}\frac{1}{x}\,dx$. Hence,

$\frac{e^h - 1}{\overline{x}} = \int_1^{e^h}\frac{1}{x}\,dx = h$, and this implies

$\frac{e^h - 1}{h} = \overline{x}$. So $\lim_{h\to 0^+}\frac{e^h - 1}{h} = \lim_{h\to 0^+}\overline{x}$. As

$h \to 0^+$, $e^h \to 1^+$ and since $\overline{x} \in (1, e^h)$ it must

be that $\overline{x} \to 1$. Hence $\lim_{h\to 0^+}\frac{e^h - 1}{h} = \lim_{h\to 0^+}\overline{x} = 1$.

A similar argument works for $h < 0$, but with

$h < 0$, $e^h < 1$, so $h = \ln e^h = \int_1^{e^h}\frac{1}{x}\,dx = -\int_{e^h}^1\frac{1}{x}\,dx$.

Now, the Integral Mean Value Theorem implies there is a number $\overline{x} \in (e^h, 1)$ for which

$-\frac{1}{\overline{x}} = \frac{1}{1 - e^h}\int_{e^h}^1 -\frac{1}{x}\,dx$. Hence,

$-\frac{1 - e^h}{\overline{x}} = \int_{e^h}^1 -\frac{1}{x}\,dx = h$, and we have $\frac{e^h - 1}{h} = \overline{x}$.

As $h \to 0^-$, $e^h \to 1^-$ and since $\overline{x} \in (e^h, 1)$ it must be that $\overline{x} \to 1$. Hence

$\lim_{h\to 0^-}\frac{e^h - 1}{h} = \lim_{h\to 0^-}\overline{x} = 1$. We conclude that

$\lim_{h\to 0}\frac{e^h - 1}{h}$ exists and equals 1.

83. $\lim_{n\to\infty}\dfrac{\ln\left(1 + \dfrac{1}{n}\right)}{1/n} = \lim_{n\to\infty}\dfrac{\dfrac{d}{dn}\ln\left(1 + \dfrac{1}{n}\right)}{\dfrac{d}{dn}(1/n)}$

$= \lim_{n\to\infty}\dfrac{\dfrac{-1}{\left(1 + \dfrac{1}{n}\right)}(n^2)}{\dfrac{-1}{n^2}} = \lim_{n\to\infty}\dfrac{1}{1 + \dfrac{1}{n}} = 1$

Section 6.4

5.
$$y' = 4y$$
$$\frac{1}{y}y' = 4$$
$$\int\frac{1}{y}y'\,dx = \int 4\,dx$$
$$\ln|y| = 4x + c$$
$$y = e^{4x+c} = ce^{4x}$$

If $y(0) = 2$, then $2 = ce^0 = c$. Therefore,

$$y = 2e^{4x}.$$

7.
$$y' = -3y$$
$$\frac{1}{y}y' = -3$$
$$\int\frac{1}{y}y'\,dx = \int(-3)\,dx$$
$$\ln|y| = -3x + c$$
$$y = e^{-3x+c} = ce^{-3x}$$

If $y(0) = 5$, then $5 = ce^0 = c$. Therefore,

$$y = 5e^{-3x}.$$

9.
$$y' = 2y$$
$$\frac{1}{y}y' = 2$$
$$\int\frac{1}{y}y'\,dx = \int 2\,dx$$
$$\ln|y| = 2x + c$$
$$y = e^{2x+c} = ce^{2x}$$

If $y(1) = 2$, then $2 = ce^2$ so $c = 2e^{-2}$. Therefore,

$$y = 2e^{-2}e^{2x} = 2e^{2x-2} = 2e^{2(x-1)}.$$

11.
$$y' = -3$$
$$\int y' dx = \int (-3) dx$$
$$y = -3x + c$$
If $y(0) = 3$, then $3 = -3(0) + c = c$. Therefore, $y = -3x + 3$.

13. We first solve for k using the doubling time.
$$y(t) = Ae^{kt}$$
$$2A = Ae^{k \cdot 4}$$
$$e^{4k} = 2$$
$$4k = \ln 2$$
$$k = \frac{\ln 2}{4}$$
And, if the initial population is 100, $A = 100$.

We have $y = 100e^{\left(\frac{\ln 2}{4}\right)t}$

The population will reach 6000 after t hours, where
$$100e^{\left(\frac{\ln 2}{4}\right)t} = 6000$$
$$e^{\left(\frac{\ln 2}{4}\right)t} = 60$$
$$\left(\frac{\ln 2}{4}\right)t = \ln 60$$
$$t = \frac{4 \ln 60}{\ln 2} \approx 23.628 \text{ hours}$$

15. We first solve for k using the doubling time.
$$y(t) = Ae^{kt}$$
$$800 = 400e^{k \cdot 1}$$
$$e^k = 2$$
$$k = \ln 2$$
And, if the initial population is 400, then $y(t) = 400e^{(\ln 2)t}$. After 10 hours, the population will be $y(10) = 400e^{(\ln 2) \cdot 10} = 409,600$ bacteria

17. If the growth constant is .44, then $y(t) = Ae^{.44t}$
The doubling time is t, where
$$2A = Ae^{.44t}$$
$$e^{.44t} = 2$$
$$.44t = \ln 2$$
$$t = \frac{\ln 2}{.44} \approx 1.575 \text{ hours}$$

19. Given $y(t) = Ae^{-1.3863t}$, the half-life is t, where
$$\frac{1}{2}A = Ae^{-1.3863t}$$
$$e^{-1.3863t} = \frac{1}{2}$$
$$-1.3863t = \ln\frac{1}{2} = -\ln 2$$
$$t = \frac{\ln 2}{1.3863} \approx .500 \text{ day}$$

21. Given $y(t) = Ae^{rt}$, the doubling time is t, where
$$2A = Ae^{rt}$$
$$2 = e^{rt}$$
$$rt = \ln 2$$
$$t = \frac{\ln 2}{r}$$
as desired.

23. From exercise 22
$$-\frac{\ln 2}{r} = 3$$
$$r = -\frac{\ln 2}{3}$$
And, if the initial amount is .4, then
$y(t) = .4e^{\left(-\frac{\ln 2}{3}\right)t}$. To determine when the amount drops below .01, solve the equation
$$.4e^{\left(-\frac{\ln 2}{3}\right)t} = .01$$
$$e^{\left(-\frac{\ln 2}{3}\right)t} = .025$$
$$-\frac{\ln 2}{3}t = \ln .025$$
$$t = -\frac{3 \ln .025}{\ln 2} \approx 15.96 \text{ hours}$$

25. If the doubling time is 20 minutes, from exercise 21, we have
$$\frac{\ln 2}{r} = 20$$
$$r = \frac{\ln 2}{20}$$
For an initial population of 10^8, the population at any time t is given by $y(t) = 10^8 e^{\left(\frac{\ln 2}{20}\right)t}$. After T minutes, the population is $y(T) = 10^8 e^{\left(\frac{\ln 2}{20}\right)T}$.

The treatment then reduces the population by 90% so that only 10% remains and, if this is 10^8, we have

$$.1\left[10^8 e^{\left(\frac{\ln 2}{20}\right)T}\right] = 10^8$$

$$e^{\frac{\ln 2}{20}T} = 10$$

$$\frac{\ln 2}{20}T = \ln 10$$

$$T = \frac{20\ln 10}{\ln 2} \approx 66.4$$

or, after about 66.4 minutes.

27. From exercise 22,

$$-\frac{\ln 2}{r} = 28$$

$$r = -\frac{\ln 2}{28}$$

Therefore, $y(t) = Ae^{\left(-\frac{\ln 2}{28}\right)t}$ gives the amount of strontium-90 remaining after t years. After 50 years, $y(50) = Ae^{\left(-\frac{\ln 2}{28}\right)50} \approx .290A$ Or, about 29% will remain.

29. If the half life is 5730 years, then

$$-\frac{\ln 2}{r} = 5730$$

$$r = -\frac{\ln 2}{5730}$$

If 20% of the amount is left, then

$$.2A = Ae^{\left(-\frac{\ln 2}{5730}\right)t}$$

$$e^{\left(-\frac{\ln 2}{5730}\right)t} = .2$$

$$-\frac{\ln 2}{5730}t = \ln.2$$

$$t = -\frac{5730\ln.2}{\ln 2} \approx 13,304.648$$

The fossil is about 13,305 years old.

31. We use Newton's Law of Cooling,

$$y(t) = Ae^{kt} + T_a$$

First,

$$200 = Ae^0 + 70$$

$$A = 130$$

so that $y(t) = 130e^{kt} + 70$. After one minute,

$$180 = 130e^{k\cdot 1} + 70$$

$$e^k = \frac{110}{130} = \frac{11}{13}$$

$$k = \ln\frac{11}{13}$$

Therefore, $y(t) = 130e^{\left(\ln\frac{11}{13}\right)t} + 70$. The temperature will be $120°$ after t minutes where

$$120 = 130e^{\left(\ln\frac{11}{13}\right)t} + 70$$

$$e^{\left(\ln\frac{11}{13}\right)t} = \frac{50}{130} = \frac{5}{13}$$

$$\left(\ln\frac{11}{13}\right)t = \ln\frac{5}{13}$$

$$t = \frac{\ln\frac{5}{13}}{\ln\frac{11}{13}} \approx 5.720 \text{ minutes}$$

33. Using Newton's Law of Cooling with the ambient temperature of $70°$ and initial temperature is $50°$,

$$50 = Ae^0 + 70$$

$$A = -20$$

so that $y(t) = -20e^{kt} + 70$. If, after two minutes, the temperature is $56°$,

$$56 = -20e^{k\cdot 2} + 70$$

$$e^{2k} = \frac{14}{20} = .7$$

$$2k = \ln.7$$

$$k = \frac{1}{2}\ln.7$$

Therefore, $y(t) = -20e^{\frac{1}{2}(\ln.7)t} + 70$.

35. Using Newton's Law of Cooling with ambient temperature $70°$ and initial temperature is $60°$

$$60 = Ae^{kt} + 70$$
$$A = -10$$

so that $y(t) = -10e^{kt} + 70$. In two minutes, the temperature is $61°$.

$$61 = -10e^{k \cdot 2} + 70$$
$$e^{2k} = \frac{9}{10}$$
$$2k = \ln\frac{9}{10}$$
$$k = \frac{1}{2}\ln\frac{9}{10} = \frac{1}{2}\ln .9$$

Therefore, $y(t) = -10e^{\left(\frac{1}{2}\ln .9\right)t} + 70$. The temperature is $40°$ after t minutes where

$$40 = -10e^{\left(\frac{1}{2}\ln .9\right)t} + 70$$
$$e^{\left(\frac{1}{2}\ln .9\right)t} = \frac{30}{10} = 3$$
$$\left(\frac{1}{2}\ln .9\right)t = \ln 3$$
$$t = \frac{2\ln 3}{\ln .9} \approx -20.854$$

or about 21 minutes. The time was 9:46.

37. With the ambient temperature $68°$ and, after 20 minutes, if the temperature is $160°$, then

$$y(t) = Ae^{kt} + T_a$$
$$160 = Ae^{k \cdot 20} + 68$$
$$Ae^{20k} = 92$$

After 22 minutes the temperature is $158°$.

$$158 = Ae^{k \cdot 22} + 68$$
$$Ae^{22k} = 90$$
$$\frac{Ae^{22k}}{Ae^{20k}} = \frac{90}{92}$$
$$e^{2k} = \frac{90}{92}$$
$$k = \frac{1}{2}\ln\frac{90}{92}$$

Therefore, $y(t) = Ae^{\frac{1}{2}\left(\ln\frac{90}{92}\right)t} + 68$. Using the first set of numbers,

$$Ae^{20 \cdot \frac{1}{2}\ln\frac{90}{92}} = 92$$
$$A = \frac{92}{e^{10\ln\frac{90}{92}}} \approx 114.615$$

So, $y(t) = 114.615e^{\frac{1}{2}\left(\ln\frac{90}{92}\right)t} + 68$. At time $t = 0$, the serving temperature is

$$y(0) = 114.615e^0 + 68 = 182.615,$$ or about $182.6°$.

39. Annual: $A = 100(1 + .08)^1 = \$1080.00$

Monthly: $A = 1000\left(1 + \frac{.08}{12}\right)^{12} = \$1083.00$

Daily: $A = 1000\left(1 + \frac{.08}{365}\right)^{365} = \$1083.28$

Continuous: $A = 1000e^{.08 \cdot 1} = \$1083.29$

41. Person A: $A = 10,000e^{.12 \cdot 20} = \$110,231.76$

Person B: $B = 20,000e^{.12 \cdot 10} = \$66,402.34$

43. Let t be the number of years after 1985. Then, assuming continuous compounding,

$$9800 = 34e^{r \cdot 10}$$
$$e^{10r} = \frac{9800}{34}$$
$$10r = \ln\left(\frac{9800}{34}\right)$$
$$r = \frac{1}{10}\ln\left(\frac{9800}{34}\right)$$

Therefore, $A = 34e^{\frac{1}{10}\ln\left(\frac{9800}{34}\right)t}$. In 2005, $t = 20$.

$$A = 34e^{\frac{1}{10}\ln\left(\frac{9800}{34}\right) \cdot 20} = \$2,824,705.88$$

45. Assuming a 3% annual inflation rate, \$20,000 in 1975 dollars was worth only about \$9,800 in 1975 dollars by 1988. Thus people in the \$16,000 to \$20,000 bracket in 1975 would be in a higher bracket in 1975 if their earnings stayed constant in purchasing power.

47. $T_1 = 30,000 \cdot .15 + (40,000 - 30,000) \cdot .28$

       $= \$7300$

   $T_2 = 30,000 \cdot .15 + (42,000 - 30,000) \cdot .28$

       $= \$7860$

   $T_1 + .05 T_1 = \$7665$

The tax on the new salary is greater than that on the adjusted old salary.

49. In the interval of time t, a drop from 1000 to 500 would be represented by a slope of $\dfrac{500}{t}$ while a drop from 10 to 5 would be represented by a slope of $\dfrac{5}{t}$. A drop from ln 1000 to ln 500 would result in a slope of

$$\frac{\ln 1000 - \ln 500}{t} = \frac{\ln \frac{1000}{500}}{t} = \frac{\ln 2}{t}$$

while a drop from ln 10 to ln 5 would result in a slope of

$$\frac{\ln 10 - \ln 5}{t} = \frac{\ln \frac{10}{5}}{t} = \frac{\ln 2}{t}.$$

So the slopes of the drops in the logarithms are the same. If a population were changing at a constant percentage rate, the graph of population versus time would appear exponential while the graph of the logarithm of population versus time would appear linear.

51. Fitting a line to the first two data points on the plot of time vs. the natural log of the population produces the line $y = 1.468x + 0.182$, which is equivalent to fitting the original date with the exponential function $P(x) = e^{1.468x + 0.182}$ or

$P(x) = 1.200 e^{1.468}$.

53. Using the first and last data points on the time vs. log(population) graph gives the exponential fit

$P(x) = 10.004 e^{0.397x}$.

Section 6.5

5. $y' = (3x + 1) \cos y$

Yes, it is separable.

7. $y' = (3x + y) \cos y$

No, it isn't separable.

9. $y' = x^2 y + y \cos x$

   $y' = y(x^2 + \cos x)$

Yes, it is separable.

11. $y' = x^2 y - x \cos y$

No, it is not separable.

13.
$$y' = (x^2 + 1)y$$
$$\frac{1}{y} y' = x^2 + 1$$
$$\int \frac{1}{y} y' dx = \int (x^2 + 1) dx$$
$$\ln |y| = \frac{1}{3} x^3 + x + c$$
$$y = e^{\frac{1}{3}x^3 + x + c}$$
$$y = c e^{\frac{1}{3}x^3 + x}$$

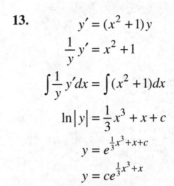

15.
$$y' = 2x^2 y^2$$
$$\frac{1}{y^2} y' = 2x^2$$
$$\int \frac{1}{y^2} y' dx = \int 2x^2 dx$$
$$-\frac{1}{y} = \frac{2}{3} x^3 + c$$
$$y = -\frac{1}{\frac{2}{3} x^3 + c}$$

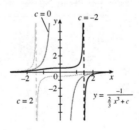

17.

$$y' = \frac{3}{y(1+x^2)}$$

$$yy' = \frac{3}{1+x^2}$$

$$\int yy'\,dx = \int \frac{3}{1+x^2}\,dx$$

$$\frac{1}{2}y^2 = 3\tan^{-1}x + c$$

$$y = \pm\sqrt{6\tan^{-1}x + c}$$

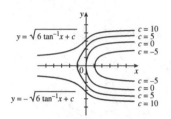

19.

$$y' = \frac{4x^2}{y^2}$$

$$y^2 y' = 4x^2$$

$$\int y^2 y'\,dx = \int 4x^2\,dx$$

$$\frac{1}{3}y^3 = \frac{4}{3}x^3 + c$$

$$y^3 = 4x^3 + c$$

$$y = \sqrt[3]{4x^3 + c}$$

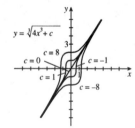

21.

$$y' = y^2 - y$$

$$\frac{1}{y^2 - y}y' = 1$$

$$\int \frac{1}{y^2 - y}y'\,dx = \int dx$$

$$\int\left(\frac{1}{y-1} - \frac{1}{y}\right)dx = \int dx$$

$$\ln|y-1| - \ln|y| = x + c$$

$$\ln\left|\frac{y-1}{y}\right| = x + c$$

$$\frac{y-1}{y} = ce^x$$

$$y - 1 = yce^x$$

$$y - yce^x = 1$$

$$y(1 - ce^x) = 1$$

$$y = \frac{1}{1 - ce^x}$$

Since c is an arbitrary constant, $y = \dfrac{1}{1 + ce^x}$.

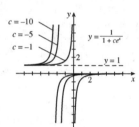
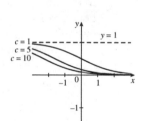

23.

$$y' = \frac{xy}{1+x^2}$$

$$\frac{1}{y}y' = \frac{x}{1+x^2}$$

$$\int \frac{1}{y}y'\,dx = \int \frac{x}{1+x^2}\,dx$$

$$\ln|y| = \frac{1}{2}\ln(1+x^2) + c$$

$$\ln|y| = \ln c\sqrt{1+x^2}, \, c > 0$$

$$y = c\sqrt{1+x^2}, \, c \text{ any real number}$$

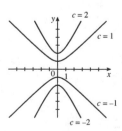

25.
$$y' = \frac{4x\sin x}{y^3}$$
$$y^3 y' = 4x\sin x$$
$$\int y^3 y' dx = \int 4x\sin x\, dx$$
$$\frac{1}{4}y^4 = 4(\sin x - x\cos x) + c$$
$$y = \pm\sqrt[4]{16\sin x - 16x\cos x + c}$$

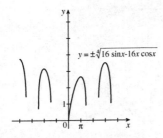

$y = \pm\sqrt[4]{16\sin x - 16x\cos x}$

27.
$$y' = 3(x+1)^2 y$$
$$\frac{1}{y}y' = 3(x+1)^2$$
$$\int \frac{1}{y}y' dx = \int 3(x+1)^2 dx$$
$$\ln|y| = (x+1)^3 + c$$
$$y = ce^{(x+1)^3}$$

If $y(0) = 1$, then $y(0) = ce^1 = 1$, so $c = \frac{1}{e}$.

Therefore, $y(t) = \frac{1}{e}e^{(x+1)^3}$.

29.
$$y' = \frac{4x^2}{y}$$
$$yy' = 4x^2$$
$$\int yy' dx = \int 4x^2 dx$$
$$\frac{1}{2}y^2 = \frac{4}{3}x^3 + c$$
$$y^2 = \frac{8}{3}x^3 + c$$
$$y = \pm\sqrt{\frac{8}{3}x^3 + c}$$

If $y(0) = 2$, then $y(0) = \sqrt{c} = 2$, so $c = 4$.

Therefore, $y(x) = \sqrt{\frac{8}{3}x^3 + 4}$.

31.
$$y' = \frac{4y}{x+3}$$
$$\frac{1}{y}y' = \frac{4}{x+3}$$
$$\int \frac{1}{y}y' dx = \int \frac{4}{x+3} dx$$
$$\ln|y| = 4\ln|x+3| + c$$
$$y = e^{4\ln|x+3|+c}$$
$$y = c(x+3)^4$$

If $y(-2) = 1$, then $y(-2) = c = 1$, so $c = 1$.
Therefore, $y(x) = (x+3)^4$.

33.
$$y' = \frac{4x}{\cos y}$$
$$\cos y y' = 4x$$
$$\int \cos y y' dx = \int 4x\, dx$$
$$\sin y = 2x^2 + c$$

If $y(0) = 0$, then $\sin 0 = c = 0$. Therefore,
$\sin y = 2x^2$.

35.
$$y' = 3y(2-y)$$
$$\frac{1}{y(2-y)}y' = 3$$
$$\int \frac{1}{y(2-y)}y' dt = \int 3\, dt$$
$$\int \left(\frac{1}{2y} + \frac{1}{2(2-y)}\right) dy = \int 3\, dt$$
$$\frac{1}{2}\ln|y| - \frac{1}{2}\ln|2-y| = 3t + c$$
$$\frac{1}{2}\ln\left|\frac{y}{2-y}\right| = 3t + c$$
$$\ln\left|\frac{y}{2-y}\right| = 6t + c$$
$$\frac{y}{2-y} = ce^{6t}$$

If $y(0) = 1$, then $1 = c$. Therefore
$$\frac{y}{2-y} = e^{6t}$$
$$y = 2e^{6t} - ye^{6t}$$
$$y(1 + e^{6t}) = 2e^{6t}$$
$$y = \frac{2e^{6t}}{1 + e^{6t}}$$

37.

$$y' = 2y(5-y)$$

$$\frac{1}{y(5-y)}y' = 2$$

$$\int \frac{1}{y(5-y)}y'\,dt = \int 2\,dt$$

$$\int \left(\frac{1}{5y} + \frac{1}{5(5-y)}\right)dy = \int 2\,dt$$

$$\frac{1}{5}\ln|y| - \frac{1}{5}\ln|5-y| = 2t + c$$

$$\frac{1}{5}\ln\left|\frac{y}{5-y}\right| = 2t + c$$

$$\ln\left|\frac{y}{5-y}\right| = 10t + c$$

$$\frac{y}{5-y} = ce^{10t}$$

If $y(0) = 4$, then $\dfrac{4}{5-4} = ce^0$, or $c = 4$. Therefore,

$$\frac{y}{5-y} = 4e^{10t} \text{ or } y = \frac{20e^{10t}}{1+4e^{10t}}.$$

39.

$$y' = y(1-y)$$

$$\frac{1}{y(1-y)}y' = 1$$

$$\int \frac{1}{y(1-y)}y'\,dt = \int dt$$

$$\int \left(\frac{1}{y} + \frac{1}{1-y}\right)dy = \int dt$$

$$\ln|y| - \ln|1-y| = t + c$$

$$\ln\left|\frac{y}{1-y}\right| = t + c$$

$$\frac{y}{1-y} = ce^t$$

If $y(0) = \dfrac{3}{4}$, then $\dfrac{\frac{3}{4}}{1-\frac{3}{4}} = ce^0$, or $c = 3$.

Therefore, $\dfrac{y}{1-y} = 3e^t$ or $y = \dfrac{3e^t}{1+3e^t}$.

41.

$$y'(t) = r \cdot y \cdot \left(1 - \frac{y}{M}\right)$$

$$y'(t) = ry \cdot \frac{1}{M}(M-y)$$

$$y'(t) = \frac{r}{M} \cdot y(M-y)$$

$$\frac{1}{y(M-y)}y'(t) = \frac{r}{M}$$

Letting $\dfrac{r}{M} = k$, we have equation 5.6:

$$\frac{1}{y(M-y)}y'(t) = k$$

For the halibut data, for convenience, the values will be substituted later. We have

$$\int \frac{1}{y(M-y)}y'(t)\,dt = \int \frac{r}{M}\,dt$$

$$\int \left(\frac{1}{My} + \frac{1}{M(M-y)}\right)dy = \int \frac{r}{M}\,dt$$

$$\frac{1}{M}\ln|y| - \frac{1}{M}\ln|M-y| = \frac{r}{M}t + c$$

$$\frac{1}{M}\ln\left|\frac{y}{M-y}\right| = \frac{r}{M}t + c$$

$$\ln\left|\frac{y}{M-y}\right| = rt + c$$

$$\frac{y}{M-y} = ce^{rt}$$

Letting $r = .71$ and $M = 8 \times 10^7$ and using $y(0) = 2 \times 10^7$, we solve for c:

$$\frac{2 \times 10^7}{8 \times 10^7 - 2 \times 10^7} = ce^{.71 \cdot 0} = c$$

$$c = \frac{1}{3}$$

Therefore

$$\frac{y}{M-y} = \frac{1}{3}e^{rt}$$

$$y = \frac{1}{3}Me^{rt} - \frac{1}{3}ye^{rt}$$

$$y\left(1 + \frac{1}{3}e^{rt}\right) = \frac{1}{3}Me^{rt}$$

$$y = \frac{\frac{1}{3}Me^{rt}}{1 + \frac{1}{3}e^{rt}} = \frac{Me^{rt}}{3 + e^{rt}}$$

Using the given values,

$$y = \frac{(8 \times 10^7)e^{.71t}}{3 + e^{.71t}}$$

43.
$$\frac{1}{y(M-y)}y' = k$$

$$\int \frac{1}{y(M-y)}y'\,dt = \int k\,dt$$

$$\int \left(\frac{1}{My} + \frac{1}{M(M-y)}\right)dy = \int k\,dt$$

$$\frac{1}{M}\ln|y| - \frac{1}{M}\ln|M-y| = kt + c$$

$$\frac{1}{M}\ln\frac{y}{y-M} = kt + c$$

$$\ln\frac{y}{y-M} = Mkt + Mc$$

$$\frac{y}{y-M} = Ae^{Mkt}$$

$$y = (y-M)Ae^{Mkt}$$

$$y(1 - Ae^{Mkt}) = -MAe^{Mkt}$$

$$y = \frac{MAe^{Mkt}}{Ae^{Mkt} - 1}$$

45. $y = 0, 2$

47. $y = 0, -2, 2$

49. $y = 1$

51. $y' = 3y(2-y)$

    a. If $y = 0$: $y' = 3(0)(2) = 0$

$$\int y'\,dt = \int 0\,dt$$
$$y = 0$$

    b. If $y = 2$: $y' = 3(2)(0) = 0$

$$\int y'\,dt = \int 0\,dt$$
$$y = 2$$

53. The equilibrium solution is found by setting $v' = 0$ and solving for v.

$$32 - .4v^2 = 0$$
$$.4v^2 = 32$$
$$v^2 = \frac{32}{.4} = 80$$
$$v = \sqrt{80} \approx 8.94 \text{ ft/s}$$

55. Since the rate of formation of X approaches 0 as x approaches 10, the amount $x(t)$ of X will always be less than 10. Note that the expression $\ln\left(\frac{12-x}{10-x}\right)$ is defined only when x is less than 10.

$$x'(t) = r(10-x)(12-x)$$

$$\frac{1}{(10-x)(12-x)}x'(t) = r$$

$$\int \frac{1}{(10-x)(12-x)}x'(t)\,dt = \int r\,dt$$

$$\int \left[\frac{1}{2(10-x)} - \frac{1}{2(12-x)}\right]dx = \int r\,dt$$

$$-\frac{1}{2}\ln|10-x| + \frac{1}{2}\ln|12-x| = rt + c$$

$$\frac{1}{2}\ln\frac{12-x}{10-x} = rt + c$$

$$\ln\frac{12-x}{10-x} = 2rt + c$$

$$\frac{12-x}{10-x} = ce^{2rt}$$

$$x = \frac{10ce^{2rt} - 12}{ce^{2rt} - 1}$$

If $x(0) = 0$, then

$$0 = \frac{10ce^0 - 12}{ce^0 - 1}$$
$$10c = 12$$
$$c = \frac{6}{5}$$

Therefore, $x = \dfrac{12e^{2rt} - 12}{\frac{6}{5}e^{2rt} - 1}$ or, $x = \dfrac{10e^{2rt} - 10}{e^{2rt} - \frac{5}{6}}$

57. $y' = ky(M-y)$

$$y'' = k[y'(M-y) + y(-y')]$$
$$= ky'(M - 2y)$$

So $y'' = 0$ if and only if $y' = 0$ or

$$M - 2y = 0$$
$$2y = M$$
$$y = \frac{1}{2}M$$

When $y' = 0$, either $y = 0$ or $y = M$ and at neither value is there an inflection point. (See Figure 6.24.) Thus, $y = \frac{1}{2}M$ gives us the only inflection point.

59. $\dfrac{dv}{dt} = -32 + 0.003v^2 = \dfrac{3}{1000}\left(v^2 - \dfrac{32{,}000}{3}\right)$

$\dfrac{3}{1000}dt = \dfrac{1}{v^2 - \dfrac{32{,}000}{3}}dv = \left(\dfrac{1/(2a)}{v-a} - \dfrac{1/(2a)}{v+a}\right)dv$

where $a = \sqrt{\dfrac{32{,}000}{3}} = \dfrac{80\sqrt{15}}{3}$

Now integrate on both sides:

$\dfrac{3}{1000}t + c = \dfrac{1}{2a}\left(\ln|v-a| - \ln|v-a|\right)$

$= \dfrac{1}{2a}\ln\left|\dfrac{v-a}{v+a}\right|$

Since $v(0) = 0$, we must have $c = 0$.
Since v is negative and less than a in absolute

value, $\left|\dfrac{v-a}{v+a}\right| = \dfrac{a-v}{a+v}$.

Multiply through by $2a$ and exponentiate:

$e^{kt} = \dfrac{a-v}{a+v}$ where $k = \dfrac{4\sqrt{15}}{25}$

Solve for v:

$v(t) = a\dfrac{1 - e^{kt}}{1 + e^{kt}}$

Section 6.6

5.

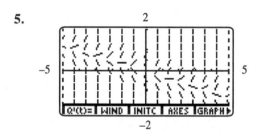

7.

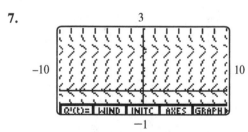

9.

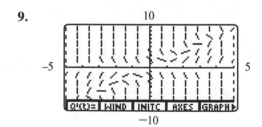

11.

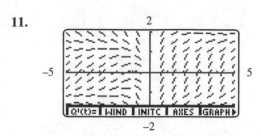

13.

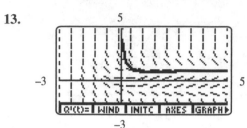

15.

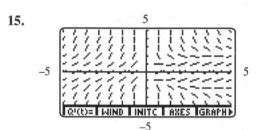

17.

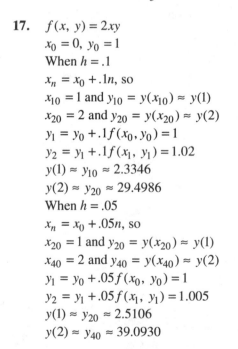

$f(x, y) = 2xy$
$x_0 = 0,\ y_0 = 1$
When $h = .1$
$x_n = x_0 + .1n$, so
$x_{10} = 1$ and $y_{10} = y(x_{10}) \approx y(1)$
$x_{20} = 2$ and $y_{20} = y(x_{20}) \approx y(2)$
$y_1 = y_0 + .1f(x_0, y_0) = 1$
$y_2 = y_1 + .1f(x_1, y_1) = 1.02$
$y(1) \approx y_{10} \approx 2.3346$
$y(2) \approx y_{20} \approx 29.4986$
When $h = .05$
$x_n = x_0 + .05n$, so
$x_{20} = 1$ and $y_{20} = y(x_{20}) \approx y(1)$
$x_{40} = 2$ and $y_{40} = y(x_{40}) \approx y(2)$
$y_1 = y_0 + .05f(x_0, y_0) = 1$
$y_2 = y_1 + .05f(x_1, y_1) = 1.005$
$y(1) \approx y_{20} \approx 2.5106$
$y(2) \approx y_{40} \approx 39.0930$

19. $f(x, y) = 4y - y^2$

$x_0 = 0, \ y_0 = 1$

When $h = .1$

$x_n = x_0 + .1n$, so

$x_{10} = 1$ and $y_{10} = y(x_{10}) \approx y(1)$

$x_{20} = 2$ and $y_{20} = y(x_{20}) \approx y(2)$

$y_1 = y_0 + .1f(x_0, y_0) = 1.3$

$y_2 = y_1 + .1f(x_1, y_1) = 1.651$

$y(1) \approx y_{10} \approx 3.8478$

$y(2) \approx y_{20} \approx 3.9990$

When $h = .05$

$x_n = x_0 + .05n$, so

$x_{20} = 1$ and $y_{20} = y(x_{20}) \approx y(1)$

$x_{40} = 2$ and $y_{40} = y(x_{40}) \approx y(2)$

$y_1 = y_0 + .05f(x_0, y_0) = 1.15$

$y_2 = y_1 + .05f(x_1, y_1) \approx 1.3139$

$y(1) \approx y_{20} \approx 3.8188$

$y(2) \approx y_{40} \approx 3.9978$

21. $f(x, y) = 1 - y + e^{-x}$

$x_0 = 0, \ y_0 = 3$

When $h = .1$

$x_n = x_0 + .1n$, so

$x_{10} = 1$ and $y_{10} = y(x_{10}) \approx y(1)$

$x_{20} = 2$ and $y_{20} = y(x_{20}) \approx y(2)$

$y_1 = y_0 + .1f(x_0, y_0) = 2.9$

$y_2 = y_1 + .1f(x_1, y_1) \approx 2.8005$

$y(1) \approx y_{10} \approx 2.0943$

$y(2) \approx y_{20} \approx 1.5275$

When $h = .05$

$x_n = x_0 + .05n$, so

$x_{20} = 1$ and $y_{20} = y(x_{20}) \approx y(1)$

$x_{40} = 2$ and $y_{40} = y(x_{40}) \approx y(2)$

$y_1 = y_0 + .05f(x_0, y_0) = 2.95$

$y_2 = y_1 + .05f(x_1, y_1) \approx 2.9001$

$y(1) \approx y_{20} \approx 2.0990$

$y(2) \approx y_{40} \approx 1.5345$

23. $f(x, y) = \sqrt{x + y}$

$x_0 = 0, \ y_0 = 1$

When $h = .1$

$x_n = x_0 + .1n$, so

$x_{10} = 1$ and $y_{10} = y(x_{10}) \approx y(1)$

$x_{20} = 2$ and $y_{20} = y(x_{20}) \approx y(2)$

$y_1 = y_0 + .1f(x_0, y_0) = 1.1$

$y_2 = y_1 + .1f(x_1, y_1) \approx 1.2095$

$y(1) \approx y_{10} \approx 2.3960$

$y(2) \approx y_{20} \approx 4.5688$

When $h = .05$

$x_n = x_0 + .05n$, so

$x_{20} = 1$ and $y_{20} = y(x_{20}) \approx y(1)$

$x_{40} = 2$ and $y_{40} = y(x_{40}) \approx y(2)$

$y_1 = y_0 + .05f(x_0, y_0) = 1.05$

$y_2 = y_1 + .05f(x_1, y_1) \approx 1.1024$

$y(1) \approx y_{20} \approx 2.4210$

$y(2) \approx y_{40} \approx 4.6203$

25. Exercise 17:

$$y' = 2xy$$

$$\frac{1}{y}y' = 2x$$

$$\int \frac{1}{y}y' \, dx = \int 2x \, dx$$

$$\ln|y| = x^2 + c$$

$$y = ce^{x^2}$$

If $y(0) = 1$, then $1 = ce^0 = c$; therefore

$$y(x) = e^{x^2} \ .$$

$y(1) = e^1 \approx 2.7183$

$y(2) = e^4 \approx 54.5982$

Exercise 18:

$$y' = x / y$$

$$yy' = x$$

$$\int yy' \, dx = \int dx$$

$$y^2 = x^2 + c$$

If $y(0) = 2$, then $4 = c$ and so

$$y(x) = \sqrt{x^2 + 4}$$

$y(1) = \sqrt{5} \approx 2.236068$

$y(2) = \sqrt{8} \approx 2.828427$

27.

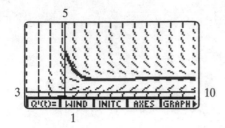

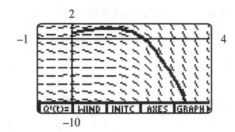

29. $y' = 2y - y^2$

$y' = y(2 - y)$

Equilibrium solutions are $y = 0$ and $y = 2$.

If $y < 0$, then $y' < 0$ so solution decreases.

If $0 < y < 2$, then $y' > 0$ so solution increases.

If $y > 2$, then $y' < 0$ so solution decreases. Thus $y = 0$ is unstable and $y = 2$ is stable.

31. $y' = y^2 - y^4$

$y' = y^2(1 + y)(1 - y)$

Equilibrium solutions are $y = -1$, $y = 0$, and $y = 1$.

When $y < -1$ or $y > 1$, $y' < 0$ and solution decreases. When $-1 < y < 0$ or $0 < y < 1$, $y' > 0$ and solution increases. Thus, $y = -1$ and $y = 0$ are unstable, while $y = 1$ is stable.

33. $y' = (1 - y)\sqrt{1 + y^2}$

Equilibrium solution is $y = 1$.

If $y < 1$, then $y' > 0$ and solution increases. If $y > 1$, then $y' < 0$ and solution decreases. Thus $y = 1$ is stable.

35.

$$g' = -g + \frac{3g^2}{1 + g^2}$$

$$g' = \frac{-g(1 + g^2) + 3g^2}{1 + g^2}$$

$$\frac{-g^3 + 3g^2 - g}{1 + g^2} = 0$$

$$-g(g^2 - 3g + 1) = 0$$

Equilibrium states: $g = 0$, $g = \dfrac{3 - \sqrt{5}}{2} = a$, and

$g = \dfrac{3 + \sqrt{5}}{2} = b$.

Therefore $g' > 0$ for $a < g < b$ and $g' < 0$ for $0 < g < a$ or $g > b$.

If $0 < g(0) < a$, then g will decrease towards 0; hence $\lim\limits_{t \to \infty} g(t) = 0$. If $a < g(0) < b$ then g will increase towards b, and if $g(0) > b$, then g will decrease towards b; hence $\lim\limits_{t \to \infty} g(t) = b$.

Where $g(0) > a = \dfrac{1}{2}\left(3 - \sqrt{5}\right)$, there will be an

eventual activated-gene level of $b = \dfrac{1}{2}\left(3 + \sqrt{5}\right)$,

hence white coloring. Where

$0 < g(0) < a = \dfrac{1}{2}\left(3 - \sqrt{5}\right)$, there will be an

eventual activated gene level of 0, hence black coloring.

$$g(0) = \frac{3}{2} + \frac{3}{2}\sin x = \frac{1}{2}\left(3 - \sqrt{5}\right)$$

$$\sin x = -\frac{\sqrt{5}}{3}$$

$$x \approx 3.98,\ 5.44,\ 10.27,\ 11.73$$

For $x \in (3.98,\ 5.44) \cup (10.27,\ 11.73)$,

$g(0) = \dfrac{3}{2} + \dfrac{3}{2}\sin x < a$, so these intervals

correspond to black stripes. The other intervals $(0,\ 3.98) \cup (5.44,\ 10.27) \cup (11.73,\ 4\pi)$

correspond to white stripes.

37. For $h = 0.1$, $y(0.5) \approx 31.64$;

for $h = 0.05$, $f(0.5) \approx 218.12$;

for $h = 0.01$, evaluating $f(0.5)$ produces overflow. We conjecture that the solution has a vertical asymptote somewhere in [0, 0.5]. The asymptote is actually located at $\dfrac{\ln 2}{2} \approx 0.345$.

39. The vertical asymptote is at $x = \dfrac{\ln 2}{2}$. If Euler's method takes large steps, it will step right over this asymptote and land on another branch of the solution.

Section 6.7

5. On the unit circle, the angle $\theta \in \left[-\dfrac{\pi}{2}, \dfrac{\pi}{2} \right]$ for which $\sin \theta = 0$ is $\theta = 0$. Hence, $\sin^{-1} 0 = 0$.

7. On the unit circle, the angle $\theta \in \left[-\dfrac{\pi}{2}, \dfrac{\pi}{2} \right]$ for which $\tan \theta = 1$ is $\theta = \dfrac{\pi}{4}$. Hence, $\tan^{-1} 1 = \dfrac{\pi}{4}$.

9. On the unit circle, the angle $\theta \in \left[-\dfrac{\pi}{2}, \dfrac{\pi}{2} \right]$ for which $\sin \theta = -1$ is $\theta = -\dfrac{\pi}{2}$. Hence,

$$\sin^{-1}(-1) = -\dfrac{\pi}{2}.$$

11. On the unit circle, the angle $\theta \in \left[0, \dfrac{\pi}{2} \right) \cup \left(\dfrac{\pi}{2}, \pi \right]$ for which $\sec \theta = 1$ is $\theta = 0$. Hence, $\sec^{-1} 1 = 0$.

13. On the unit circle, the angle $\theta \in \left[-\dfrac{\pi}{2}, \dfrac{\pi}{2} \right]$ for which $\tan \theta = -1$ is $\theta = -\dfrac{\pi}{4}$. Hence,

$$\tan^{-1}(-1) = -\dfrac{\pi}{4}.$$

15. On the unit circle, the angle

$\theta \in 0, \left[0, \dfrac{\pi}{2} \right) \cup \left(\dfrac{\pi}{2}, \pi \right]$ for which

$\sec \theta = 2$ is $\theta = \dfrac{\pi}{3}$. Hence, $\sec^{-1} 2 = \dfrac{\pi}{3}$.

17. On the unit circle, the angle $\theta \in \left[-\dfrac{\pi}{2}, \dfrac{\pi}{2} \right]$ for which $\sin \theta = -\dfrac{1}{2}$ is $\theta = -\dfrac{\pi}{6}$. Hence,

$$\sin^{-1} \left(-\dfrac{1}{2} \right) = -\dfrac{\pi}{6}.$$

19. Let $\sin^{-1} x = \theta$, then $\cos(\sin^{-1} x) = \cos \theta$

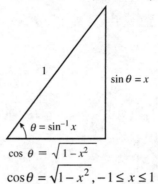

$$\cos \theta = \sqrt{1 - x^2}$$
$$\cos \theta = \sqrt{1 - x^2}, -1 \le x \le 1$$

21. $|\cos x| \le 1$ and $|\sec \theta| \ge 1$, so $\sec^{-1}(\cos x)$ is only defined when $\cos x = \pm 1$; that is, when $x = n\pi$ for n an integer.

$$\sec^{-1}(\cos x) = \begin{cases} 0 \text{ if } x = n\pi \ (n \text{ even}) \\ \pi \text{ if } x = n\pi \ (n \text{ odd}) \end{cases}$$

23. Let $\sec^{-1} x = \theta$, then $\tan(\sec^{-1} x) = \tan \theta$

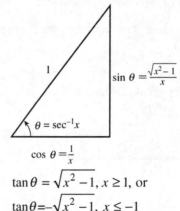

$$\tan \theta = \sqrt{x^2 - 1}, x \ge 1, \text{ or}$$
$$\tan \theta = -\sqrt{x^2 - 1}, x \le -1$$

25. No triangle is needed here. For $-\dfrac{\pi}{2} \le x \le \dfrac{\pi}{2}$,

$$\sin^{-1}(\sin x) = x.$$

27. Let $\cos^{-1}\dfrac{1}{2} = \theta$, then $\sin\left(\cos^{-1}\dfrac{1}{2}\right) = \sin\theta$

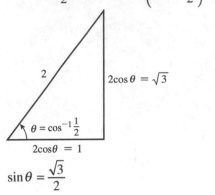

$2\cos\theta = \sqrt{3}$

$\theta = \cos^{-1}\dfrac{1}{2}$

$2\cos\theta = 1$

$\sin\theta = \dfrac{\sqrt{3}}{2}$

29. Let $\cos^{-1}\dfrac{3}{5} = \theta$, then $\tan\left(\cos^{-1}\dfrac{3}{5}\right) = \tan\theta$

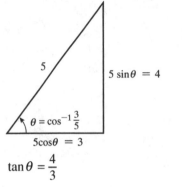

$5\sin\theta = 4$

$\theta = \cos^{-1}\dfrac{3}{5}$

$5\cos\theta = 3$

$\tan\theta = \dfrac{4}{3}$

31. The domain of the inverse cosine is

$0 \le x \le \pi$, and $\dfrac{\pi}{8}$ has the same cosine as $-\dfrac{\pi}{8}$, so

$\cos^{-1}(\cos(-\pi/8)) = \pi/8$.

33. Expanding by the double angle formula,

$\sin\left(2\sin^{-1}(x/3)\right)$

$= 2\sin\left(\sin^{-1}(x/3)\right)\cos\left(\sin^{-1}(x/3)\right)$

$= 2\cdot\dfrac{x}{3}\cdot\sqrt{1-\left(\dfrac{x}{3}\right)^2}$

$= \dfrac{2x}{9}\sqrt{9-x^2}$ for $|x| \le 3$

35.

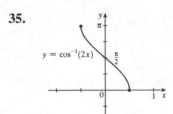

37.

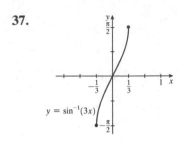

39.

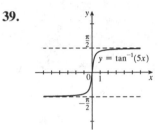

41. $\cos(2x) = 0$

$\qquad 2x = \dfrac{\pi}{2} + \pi n$

$\qquad x = \dfrac{\pi}{4}n$, n any odd integer

43. $\sin(3x) = 0$

$\qquad 3x = 0 + \pi n$

$\qquad x = 0 + \dfrac{\pi}{3}n = \dfrac{\pi}{3}n$, n any integer

45. $\tan(5x) = 1$

$\qquad 5x = \dfrac{\pi}{4} + \pi n$

$\qquad x = \dfrac{\pi}{20} + \dfrac{\pi}{5}n$, n any integer

47. $\sin(3x) = \dfrac{1}{2}$

$\qquad 3x = \dfrac{\pi}{6} + 2\pi n, \dfrac{5\pi}{6} + 2\pi n$

$\qquad x = \dfrac{\pi}{18} + \dfrac{2\pi}{3}n, \dfrac{5\pi}{18} + \dfrac{2\pi}{3}n$, n any integer

49. $\sec(2x) = 2$

$$2x = \frac{\pi}{3} + 2\pi n, -\frac{\pi}{3} + 2\pi n$$

$$x = \frac{\pi}{6} + \pi n, -\frac{\pi}{6} + \pi n, n \text{ any integer}$$

51. $y = \csc^{-1} x$ if and only if $\csc y = x$ and

$$y \in \left[-\frac{\pi}{2}, 0\right) \cup \left(0, \frac{\pi}{2}\right]$$

$y = \cot^{-1} x$ if and only if $\cot y = x$ and

$y \in (0, \pi)$.

53. $\tan A = \dfrac{20}{12x}$

$$A = \tan^{-1}\left(\frac{20}{12x}\right)$$

Section 6.8

3. $\dfrac{d}{dx}\left[\sin^{-1}(3x^2)\right] = \dfrac{1}{\sqrt{1-9x^4}}\dfrac{d}{dx}(3x^2)$

$$= \frac{6x}{\sqrt{1-9x^4}}$$

5. $\dfrac{d}{dx}\left[\cos^{-1} x^3\right] = \dfrac{-1}{\sqrt{1-(x^3)^2}}\dfrac{d}{dx}(x^3)$

$$= \frac{-3x^2}{\sqrt{1-x^6}}$$

7. $\dfrac{d}{dx}\left[\tan^{-1} x^2\right] = \dfrac{1}{1+(x^2)^2}\dfrac{d}{dx}(x^2)$

$$= \frac{2x}{1+x^4}$$

9. $\dfrac{d}{dx}\left[\sec^{-1} x^2\right] = \dfrac{1}{\left|x^2\right|\sqrt{(x^2)^2-1}}\dfrac{d}{dx}(x^2)$

$$= \frac{2}{x\sqrt{x^4-1}}$$

11. $\dfrac{d}{dx}\left[x\cos^{-1}(2x)\right] = x\dfrac{-1}{\sqrt{1-(2x)^2}}2 + \cos^{-1}(2x)$

$$= \frac{-2x}{\sqrt{1-4x^2}} + \cos^{-1}(2x)$$

13. $\dfrac{d}{dx}\left[\cos^{-1}(\sin x)\right] = \dfrac{-1}{\sqrt{1-(\sin x)^2}}\cos x$

$$= -\frac{\cos x}{\sqrt{\cos^2 x}}$$

$$= -\frac{\cos x}{|\cos x|} = \pm 1$$

15. $\dfrac{d}{dx}\left[\tan^{-1}(\sec x)\right] = \dfrac{1}{1+(\sec x)^2}\cdot(\sec x \tan x)$

$$= \frac{\sec x \tan x}{1+\sec^2 x}$$

17. $\displaystyle\int \frac{6}{1+x^2}dx = 6\int\frac{1}{1+x^2}dx$

$$= 6\tan^{-1} x + c$$

19. $\displaystyle\int \frac{4}{\sqrt{1-x^2}}dx = 4\int\frac{1}{\sqrt{1-x^2}}dx$

$$= 4\sin^{-1} x + c$$

21. $\displaystyle\int \frac{2x}{1+x^4}dx = \int\frac{2x}{1+(x^2)^2}dx$

$$= \tan^{-1}(x^2) + c$$

23. $\displaystyle\int \frac{4x}{\sqrt{1-x^4}}dx = \int\frac{4x}{\sqrt{1-(x^2)^2}}dx$

Let $u = x^2$. Then $du = 2x\,dx$.

$$\int \frac{4x}{\sqrt{1-(x^2)^2}}dx = 2\int\frac{2x}{\sqrt{1-(x^2)^2}}dx$$

$$= 2\int\frac{1}{\sqrt{1-u^2}}du$$

$$= 2\sin^{-1}(u) + c$$

$$= 2\sin^{-1}(x^2) + c$$

25. $\int \dfrac{2x}{x^2\sqrt{x^4-1}}\,dx = \int \dfrac{2x}{x^2\sqrt{(x^2)^2-1}}\,dx$

Let $u = x^2$. Then $du = 2x\,dx$.

$\int \dfrac{1}{u\sqrt{u^2-1}}\,du = \sec^{-1}(u) + c = \sec^{-1}(x^2) + c$

27. $\int \dfrac{2}{4+x^2}\,dx = \int \dfrac{\frac{1}{2}}{1+\left(\frac{x}{2}\right)^2}\,dx$

Let $u = \dfrac{x}{2}$. Then $du = \dfrac{1}{2}dx$.

$\int \dfrac{\frac{1}{2}}{1+\left(\frac{x}{2}\right)^2}\,dx = \int \dfrac{1}{1+u^2}\,du$

$= \tan^{-1}u + c$

$= \tan^{-1}\left(\dfrac{x}{2}\right) + c$

29. $\int \dfrac{e^x}{\sqrt{1-e^{2x}}}\,dx = \int \dfrac{e^x}{\sqrt{1-(e^x)^2}}\,dx$

Let $u = e^x$. Then $du = e^x dx$.

$\int \dfrac{e^x}{\sqrt{1-(e^x)^2}}\,dx = \int \dfrac{1}{\sqrt{1-u^2}}\,du$

$= \sin^{-1}(u) + c$

$= \sin^{-1}(e^x) + c$

31. $\int \dfrac{\cos x}{4+\sin^2 x}\,dx = \dfrac{1}{4}\int \dfrac{\cos x}{1+\left(\frac{\sin x}{2}\right)^2}\,dx$

Let $u = \dfrac{\sin x}{2}$. Then $du = \dfrac{1}{2}\cos x\,dx$.

$\dfrac{1}{2}\int \dfrac{\frac{1}{2}\cos x}{1+\left(\frac{\sin x}{2}\right)^2}\,dx = \dfrac{1}{2}\int \dfrac{1}{1+u^2}\,du$

$= \dfrac{1}{2}\tan^{-1}(u) + c$

$= \dfrac{1}{2}\tan^{-1}\left(\dfrac{\sin x}{2}\right) + c$

33. $\displaystyle\int_0^2 \dfrac{6}{4+x^2}\,dx = \dfrac{6}{4}\int_0^2 \dfrac{1}{1+\left(\frac{x}{2}\right)^2}\,dx$

Let $u = \dfrac{x}{2}$. Then $du = \dfrac{1}{2}dx$,

and the limits are $\dfrac{0}{2} = 0$ and $\dfrac{2}{2} = 1$.

The integral is

$\dfrac{6}{4} \cdot 2 \cdot \displaystyle\int_0^1 \dfrac{1}{1+u^2}\,du$

$= 3\left(\left.\tan^{-1}u\right|_0^1\right)$

$= 3\left(\dfrac{\pi}{4} - 0\right) = \dfrac{3\pi}{4}$

35. $\displaystyle\int_{\sqrt{2}}^2 \dfrac{4}{x\sqrt{x^2-1}}\,dx = 4\int_{\sqrt{2}}^2 \dfrac{1}{x\sqrt{x^2-1}}\,dx$

This is an inverse secant integral. Note that $x > 1$ over the interval of integration.

$4\displaystyle\int_{\sqrt{2}}^2 \dfrac{1}{x\sqrt{x^2-1}}\,dx$

$= 4\left(\left.\sec^{-1}x\right|_{\sqrt{2}}^2\right)$

$= 4\left(\dfrac{\pi}{3} - \dfrac{\pi}{4}\right) = \dfrac{\pi}{3}$

37. $\displaystyle\int_0^{\frac{1}{4}} \dfrac{3}{\sqrt{1-4x^2}}\,dx = 3\int_0^{\frac{1}{4}} \dfrac{1}{\sqrt{1-(2x)^2}}\,dx$

Let $u = 2x$. Then $du = 2dx$ and the

limits are $2 \cdot 0 = 0$ and $2 \cdot \dfrac{1}{4} = \dfrac{1}{2}$.

$\dfrac{3}{2}\displaystyle\int_0^{\frac{1}{2}} \dfrac{1}{\sqrt{1-u^2}}\,du$

$= \dfrac{3}{2}\left(\left.\sin^{-1}u\right|_0^{1/2}\right) = \dfrac{3}{2}\left(\dfrac{\pi}{6} - 0\right) = \dfrac{\pi}{4}$

39. Let $\cos^{-1} x = y$. We want to find

$\dfrac{d}{dx}\cos^{-1} x = \dfrac{dy}{dx}$. If $y = \cos^{-1} x$ then $\cos y = x$

and we have the following:

$$\frac{d}{dx}\cos y = \frac{d}{dx}x$$

$$-\sin y \frac{dy}{dx} = 1$$

$$\frac{dy}{dx} = -\frac{1}{\sin y}$$

And, if $x = \cos y$, using a Pythagorean identity,

$\sin y = \sqrt{1-x^2}$. Therefore,

$$\frac{dy}{dx} = -\frac{1}{\sin y} = -\frac{1}{\sqrt{1-x^2}} \text{ as desired.}$$

41. Let $x = a\tan u$. Then $dx = a\sec^2 u\, du$ and

$$a^2 + x^2 = a^2 + (a\tan u)^2$$
$$= a^2(1 + \tan^2 u)$$
$$= a^2 \sec^2 u$$

And

$$\int \frac{1}{a^2 + x^2}\,dx = \int \frac{1}{a^2 \sec^2 u}\cdot a\sec^2 u\, du$$

$$= \frac{1}{a}\int du$$

$$= \frac{1}{a}u + c$$

If $x = a\tan u$, then $\tan u = \dfrac{x}{a}$ or $u = \tan^{-1}\left(\dfrac{x}{a}\right)$ so

that

$$\int \frac{1}{a^2 + x^2}\,dx = \frac{1}{a}u + c$$

$$= \frac{1}{a}\tan^{-1}\left(\frac{x}{a}\right) + c$$

as desired.

43. From Example 8.2, we have

$$\theta(t) = \tan^{-1}\left[\frac{d(t)}{2}\right]$$

$$\theta'(t) = \frac{2d'(t)}{4 + \left[d(t)\right]^2}$$

$$-3 = \frac{2(-130)}{4 + \left[d(t)\right]^2}$$

$$4 + \left[d(t)\right]^2 = \frac{2\cdot 130}{3} = \frac{260}{3}$$

$$\left[d(t)\right]^2 = \frac{260}{3} - 4 = \frac{248}{3}$$

$$d(t) = \sqrt{\frac{248}{3}} \approx 9.09 \text{ feet}$$

45. Label the triangles as illustrated.

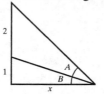

$$\tan(A + B) = \frac{3}{x}$$

$$A + B = \tan^{-1}\left(\frac{3}{x}\right)$$

$$\tan B = \frac{1}{x}$$

$$B = \tan^{-1}\frac{1}{x}$$

Therefore,

$$A = (A + B) - B$$

$$A = \tan^{-1}\left(\frac{3}{x}\right) - \tan^{-1}\left(\frac{1}{x}\right)$$

$$\frac{dA}{dx} = \frac{-\frac{3}{x^2}}{1 + \left(\frac{3}{x}\right)^2} - \frac{-\frac{1}{x^2}}{1 + \left(\frac{1}{x}\right)^2} = \frac{1}{x^2 + 1} - \frac{3}{x^2 + 9}$$

The maximum viewing angle will occur at a critical value.

$$\frac{dA}{dx} = 0$$

$$\frac{1}{x^2 + 1} = \frac{3}{x^2 + 9}$$

$$x^2 + 9 = 3x^2 + 3$$

$$2x^2 = 6$$

$$x^2 = 3$$

$$x = \sqrt{3} \text{ ft} \approx 1.73 \text{ ft}$$

Section 6.9

5.

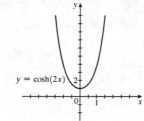

$y = \cosh(2x)$

207

7.

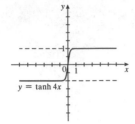

$y = \tanh 4x$

9.

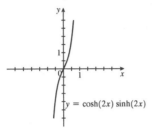

$y = \cosh(2x)\sinh(2x)$

11.

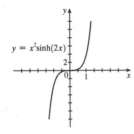

$y = x^2\sinh(2x)$

13. $\dfrac{d}{dx}\left[\cosh 4x\right] = \sinh 4x \dfrac{d}{dx}(4x) = 4\sinh 4x$

15. $\dfrac{d}{dx}\left[\sinh 2x\right] = \cosh 2x \dfrac{d}{dx}(2x) = 2\cosh 2x$

17. $\dfrac{d}{dx}\left[\tanh 4x\right] = \operatorname{sech}^2 4x \dfrac{d}{dx}(4x) = 4\operatorname{sech}^2 4x$

19. $\dfrac{d}{dx}\left[\cosh^{-1} 2x\right] = \dfrac{1}{\sqrt{(2x)^2 - 1}}\dfrac{d}{dx}(2x) = \dfrac{2}{\sqrt{4x^2 - 1}}$

21. $\dfrac{d}{dx}\left[x^2 \sinh 2x\right] = x^2 \cosh 2x \cdot 2 + \sinh 2x \cdot 2x$

$\qquad\qquad = 2x^2 \cosh 2x + 2x \sinh 2x$

23. $\dfrac{d}{dx}\left[\tanh^{-1} 4x\right] = \dfrac{1}{1-(4x)^2}\dfrac{d}{dx}(4x) = \dfrac{4}{1-16x^2}$

25. $\displaystyle\int \cosh 6x\,dx$

Let $u = 6x$. Then $du = 6\,dx$.

$\dfrac{1}{6}\displaystyle\int \cosh u \cdot 6\,dx = \dfrac{1}{6}\sinh u + c = \dfrac{1}{6}\sinh 6x + c$

27. $\displaystyle\int \tanh 3x\,dx = \int \dfrac{\sinh 3x}{\cosh 3x}\,dx$

Let $u = \cosh 3x$. Then $du = \sinh 3x \cdot 3\,dx$.

$\dfrac{1}{3}\displaystyle\int \dfrac{1}{u}\,du = \dfrac{1}{3}\ln|u| + c = \dfrac{1}{3}\ln(\cosh 3x) + c$

29. $\displaystyle\int_0^1 \dfrac{e^{4x} - e^{-4x}}{2}\,dx = \int_0^1 \sinh 4x\,dx$

$\qquad\qquad = \dfrac{1}{4}\left[\cosh 4x\right]_0^1$

$\qquad\qquad = \dfrac{1}{4}(\cosh 4 - 1)$

31. $\displaystyle\int \dfrac{2}{\sqrt{1+x^2}}\,dx = 2\sinh^{-1} x + c$

33. $\displaystyle\int \cos x \sinh(\sin x)\,dx$

Let $u = \sin x$. Then $du = \cos x\,dx$.

$\displaystyle\int \sinh u\,du = \cosh u + c = \cosh(\sin x) + c$

35. $\displaystyle\int_0^1 \cosh x\, e^{\sinh x}\,dx$

Let $u = \sinh x$. Then $du = \cosh x\,dx$.

$\displaystyle\int_a^b e^u\,du = \left[e^u\right]_a^b = \left[e^{\sinh x}\right]_0^1 = e^{\sinh(1)} - 1$

37. $\dfrac{d}{dx}\cosh x = \dfrac{d}{dx}\dfrac{e^x + e^{-x}}{2}$

$\qquad\qquad = \dfrac{e^x - e^{-x}}{2} = \sinh x$

$\dfrac{d}{dx}\tanh x = \dfrac{d}{dx}\dfrac{\sinh x}{\cosh x}$

$\qquad\qquad = \dfrac{\cosh x \frac{d}{dx}\sinh x - \sinh x \frac{d}{dx}\cosh x}{\cosh^2 x}$

$\qquad\qquad = \dfrac{\cosh^2 x - \sinh^2 x}{\cosh^2 x}$

$\qquad\qquad = \dfrac{1}{\cosh^2 x}$

$\qquad\qquad = \operatorname{sech}^2 x$

39. First, $e^x > e^{-x}$ if $x > 0$ and $e^x < e^{-x}$ if $x < 0$.

Since $\sinh x = \dfrac{e^x - e^{-x}}{2}$, we have that

$e^x - e^{-x} > 0$ if $x > 0$

and

$e^x - e^{-x} < 0$ if $x < 0$.

Therefore, $\sinh x > 0$ if $x > 0$

and $\sinh x < 0$ if $x < 0$.

41. $\dfrac{d}{dx}\cosh x = \sinh x$

critical values: $\sinh x = 0$

$x = 0$

$\dfrac{dy}{dx}:$

There is a minimum at $\cosh(0) = 1$. $y = \cosh x$ is decreasing on $(-\infty, 0)$ and increasing on $(0, \infty)$.

$\dfrac{d^2}{dx^2}\cosh x = \dfrac{d}{dx}\sinh x = \cosh x$

points of inflection:

$\cosh x = 0$ no solution

There are no points of inflection. The graph is concave up on $(-\infty, \infty)$.

43. If $y = \cosh^{-1} x$ then $x = \cosh y$ and

$x = \dfrac{e^y + e^{-y}}{2}$

Also $\sinh y = \dfrac{e^y - e^{-y}}{2}$.

$e^y = \cosh y + \sinh y$

$e^y = \cosh y + \sqrt{\sinh^2 y}$

$e^y = \cosh y + \sqrt{\cosh^2 y - 1}$

$e^y = x + \sqrt{x^2 - 1}$

$y = \cosh^{-1} x = \ln\left(x + \sqrt{x^2 - 1}\right)$

45. For the function

$f(x) = 10\cosh\left(\dfrac{x}{10}\right)$, $-20 \le x \le 20$, the amount of

sag will occur at the minimum value of the function. Find the critical values.

$f'(x) = \sinh\left(\dfrac{x}{10}\right) = 0$

$x = 0$

The lowest point of the cable is at

$f(0) = 10\cosh(0) = 10$

The highest point of the cable is at

$f(20) = 10\cosh(2) \approx 37.62$.

Therefore, the amount of sag is

$37.622 - 10 = 27.62$ feet

47. $f(x) = 15\cosh\left(\dfrac{x}{15}\right)$, $-25 \le x \le 25$

$f'(x) = \sinh\left(\dfrac{x}{15}\right)$

The critical value is $x = 0$.

The lowest point of the cable is at

$f(0) = 15\cosh(0) = 15$.

The highest point of the cable is at

$f(25) = 15\cosh\left(\dfrac{25}{15}\right) \approx 41.13$.

Therefore, the amount of sag is

$41.13 - 15 = 26.13$

49. $f(x) = a\cosh\left(\dfrac{x}{a}\right)$, $-b \le x \le b$

$f'(x) = \sinh\left(\dfrac{x}{a}\right)$

The critical value is $x = 0$.

The lowest point of the cable is at

$f(0) = a\cosh(0) = a$.

The highest point of the cable is at

$f(b) = a\cosh\left(\dfrac{b}{a}\right)$.

Therefore, the amount of sag is

$f(b) - f(0) = a\cosh\left(\dfrac{b}{a}\right) - a$

$L = \displaystyle\int_{-b}^{b}\sqrt{1 + \left[\sinh\left(\dfrac{x}{a}\right)\right]^2}\, dx$

$= \displaystyle\int_{-b}^{b}\cosh\left(\dfrac{x}{a}\right)dx$

$= a\left[\sinh\left(\dfrac{x}{a}\right)\right]_{-b}^{b}$

$= a\left[\sinh\left(\dfrac{b}{a}\right) - \sinh\left(-\dfrac{b}{a}\right)\right]$

Since $y = \sinh x$ is an odd function,

$\sinh\left(-\dfrac{b}{a}\right) = -\sinh\left(\dfrac{b}{a}\right)$. So,

$L = a\left[\sinh\left(\dfrac{b}{a}\right) + \sinh\left(\dfrac{b}{a}\right)\right] = 2a\sinh\left(\dfrac{b}{a}\right)$.

51. $\cosh x + \sinh x = \dfrac{e^x + e^{-x}}{2} + \dfrac{e^x - e^{-x}}{2} = e^x$

53. $\displaystyle\lim_{x\to\infty}\dfrac{e^x - e^{-x}}{e^x + e^{-x}} = 1$ since both e^{-x} terms tend to 0.

$\displaystyle\lim_{x\to-\infty}\dfrac{e^x - e^{-x}}{e^x + e^{-x}} = \lim_{x\to\infty}\dfrac{e^{-x} - e^x}{e^{-x} + e^x} = -1$.

Chapter 6   Exponentials, Logarithms and other Transcendental Functions

55. b. $\dfrac{1}{v^2-a^2}\dfrac{dv}{dt}=\dfrac{k}{m}$, where $a=\sqrt{\dfrac{mg}{k}}$

$\dfrac{dv}{v^2-a^2}=\dfrac{dv}{2a}\left(\dfrac{1}{v-1}-\dfrac{1}{v+a}\right)=\dfrac{k}{m}dt$

Integrating,

$\dfrac{dv}{2a}\left(\dfrac{1}{v-a}-\dfrac{1}{v+a}\right)=\dfrac{k}{m}dt$

Integrating,

$\dfrac{1}{2a}\ln\left|\dfrac{v-a}{v+a}\right|=\dfrac{k}{m}t+C.$

v is negative and less than in absolute value, so

$\ln\left(\dfrac{a-v}{a+v}\right)=\dfrac{2ak}{m}t+C'.$

Exponentiating,

$\dfrac{a-v}{a+v}=ce^{\frac{2ak}{m}}$, where $c=e^{C'}$ is a postive constant.

Solving for v and using $\dfrac{2ak}{m}=2\sqrt{\dfrac{kg}{m}}$, we get

$v(t)=-a\dfrac{ce^{2\sqrt{kg/m}\,t}-1}{ce^{2\sqrt{kg/m}\,t}+1}$, which is the answer

given in part c.

d. $v(0)=0$ implies that

$ce^{2\sqrt{kg/m}\,(0)}-1=0$ or $c=1.$

e. The fraction tends to 1, so the limiting velocity is $-a$.

57. For the first skydiver:
terminal velocity, -250.52 m/s;
distance in 2 seconds, 19.59 m;
distance in 4 seconds, 78.13 m..
For the second skydiver:
terminal velocity, -125.26 m/s;
distance in 2 seconds, 19.53 m;
distance in 4 seconds, 77.20 m.

59. For an initial velocity $v_0=2000$, we set the derivative of the velocity equal to 0 and solve the resulting equation in a CAS. The maximum acceleration of -9.797 occurs at about 206 seconds.

Chapter 6 Review

1. See Example 1.1.

$\dfrac{d}{dx}\left[\ln(x^3+5)\right]=\dfrac{1}{x^3+5}\dfrac{d}{dx}(x^3+5)=\dfrac{3x^2}{x^3+5}$

3. See Example 1.1.

$\dfrac{d}{dx}\left[\ln\sqrt{x^4+x}\right]=\dfrac{d}{dx}\left[\dfrac{1}{2}\ln(x^4+x)\right]$

$=\dfrac{1}{2}\dfrac{1}{x^4+x}\dfrac{d}{dx}(x^4+x)$

$=\dfrac{4x^3+1}{2(x^4+x)}$

5. See Example 3.1.

$\dfrac{d}{dx}\left[e^{-x^2}\right]=e^{-x^2}\dfrac{d}{dx}[-x^2]=-2xe^{-x^2}$

7. See Exercises 3.61–3.66.

$\dfrac{d}{dx}\left[4^{x^3}\right]=(\ln 4)4^{x^3}\dfrac{d}{dx}[x^3]=3x^2(\ln 4)4^{x^3}$

9. See Example 8.1.

$\dfrac{d}{dx}\left[\sin^{-1}2x\right]=\dfrac{1}{\sqrt{1-(2x)^2}}\dfrac{d}{dx}[2x]$

$=\dfrac{2}{\sqrt{1-4x^2}}$

11. See Example 8.1.

$\dfrac{d}{dx}\left[\tan^{-1}(\cos 2x)\right]=\dfrac{1}{1+(\cos 2x)^2}\dfrac{d}{dx}[\cos 2x]$

$=-\dfrac{2\sin 2x}{1+\cos^2 2x}$

13. See Example 9.1.

$\dfrac{d}{dx}\left[\cosh\sqrt{x}\right]=\sinh\sqrt{x}\dfrac{d}{dx}\left[\sqrt{x}\right]=\dfrac{\sinh\sqrt{x}}{2\sqrt{x}}$

15. See Exercises 9.19, 9.20, 9.23, 9.24.

$\dfrac{d}{dx}\left[\sinh^{-1}3x\right]=\dfrac{1}{\sqrt{1+(3x)^2}}\dfrac{d}{dx}[3x]=\dfrac{3}{\sqrt{1+9x^2}}$

17. See Example 1.2.

$\displaystyle\int\dfrac{x^2}{x^3+4}dx$

Let $u=x^3+4$. Then $du=3x^2dx$.

$\dfrac{1}{3}\displaystyle\int\dfrac{1}{u}du=\dfrac{1}{3}\ln|u|+c=\dfrac{1}{3}\ln\left|x^3+4\right|+c$

210

19. See Example 1.2.

$$\int_0^1 \frac{x}{x^2+1}\,dx$$

Let $u = x^2 +1$. Then $du = 2x\,dx$. If $x = 1$, then $u = 2$. If $x = 0$, then $u = 1$.

$$\frac{1}{2}\int_1^2 \frac{1}{u}\,du = \frac{1}{2}\Big[\ln|u|\Big]_1^2 = \frac{1}{2}\ln 2$$

21. See Exercises 1.29 and 1.30.

$$\int \frac{\sin(\ln x)}{x}\,dx$$

Let $u = \ln x$. Then $du = \frac{1}{x}\,dx$.

$$\int \sin(u)\,du = -\cos(u) + c = -\cos(\ln x) + c$$

23. See Example 3.2.

$$\int e^{-4x}\,dx$$

Let $u = -4x$. Then $du = -4\,dx$.

$$-\frac{1}{4}\int e^u\,du = -\frac{1}{4}e^u + c = -\frac{1}{4}e^{-4x} + c$$

25. See Example 3.2.

$$\int \frac{e^{\sqrt{x}}}{\sqrt{x}}\,dx$$

Let $u = \sqrt{x}$. Then $du = \frac{1}{2\sqrt{x}}\,dx$.

$$2\int e^u\,du = 2e^u + c = 2e^{\sqrt{x}} + c$$

27. See Example 3.2.

$$\int_0^2 e^{3x}\,dx$$

Let $u = 3x$. Then $du = 3\,dx$. If $x = 2$, then $u = 6$. If $x = 0$, then $u = 0$.

$$\frac{1}{3}\int_0^6 e^u\,du = \frac{1}{3}\Big[e^u\Big]_0^6 = \frac{1}{3}(e^6 - 1)$$

29. See Exercises 3.71–3.76.

$$\int 3^{4x}\,dx$$

Let $u = 4x$. Then $du = 4\,dx$.

$$\frac{1}{4}\int 3^u\,du = \frac{1}{4}\frac{1}{\ln 3}3^u + c = \frac{1}{4\ln 3}3^{4x} + c$$

31. See Example 8.3.

$$\int \frac{3}{x^2+4}\,dx = \frac{3}{4}\int \frac{1}{\left(\frac{x}{2}\right)^2 + 1}\,dx$$

Let $u = \frac{x}{2}$. Then $du = \frac{1}{2}\,dx$.

$$2\cdot\frac{3}{4}\int \frac{1}{u^2+1}\,du = \frac{3}{2}\tan^{-1}\left(\frac{x}{2}\right) + c$$

33. See Example 8.5.

$$\int \frac{x^2}{\sqrt{1-x^6}}\,dx = \int \frac{x^2}{\sqrt{1-(x^3)^2}}\,dx$$

Let $u = x^3$. Then $du = 3x^2\,dx$.

$$\frac{1}{3}\int \frac{1}{\sqrt{1-u^2}}\,du = \frac{1}{3}\sin^{-1}(u) + c = \frac{1}{3}\sin^{-1}(x^3) + c$$

35. See Exercise 8.25 and Example 8.5.

$$\int \frac{9x}{x^2\sqrt{x^4-1}}\,dx = \int \frac{9x}{x^2\sqrt{(x^2)^2-1}}\,dx$$

Let $u = x^2$. Then $du = 2x\,dx$.

$$\frac{9}{2}\int \frac{1}{u\sqrt{u^2-1}}\,du = \frac{9}{2}\sec^{-1}(u) + c$$
$$= \frac{9}{2}\sec^{-1}(x^2) + c$$

37. See Exercises 9.31 and 9.32.

$$\int \frac{4}{\sqrt{1+x^2}}\,dx = 4\sinh^{-1}x + c$$

39. See Example 9.2.

$$\int \cosh 4x\,dx = \frac{1}{4}\sinh 4x + c$$

41. See Examples 2.3, 2.4 and 2.5.

$$\frac{d}{dx}[x^3 - 1] = 3x^2$$

Since $3x^2 \geq 0$ for all x, the function $f(x) = x^3 - 1$ is increasing on $(-\infty, \infty)$. The function is one-to-one.

$$y = x^3 - 1$$
$$y + 1 = x^3$$
$$(y+1)^{1/3} = x$$

Therefore, $f^{-1}(x) = (x+1)^{1/3}$.

43. See Examples 2.3, 2.4 and 2.5.

$$\frac{d}{dx}\left[e^{2x^2}\right] = 4xe^{2x^2}$$

Since $4xe^{2x^2} < 0$ on $(-\infty, 0)$ and $4xe^{2x^2} > 0$ on $(0, \infty)$, the function is not one-to-one. There is no inverse.

45. See Examples 2.7 and 2.8.

a. $f(x) = x^5 + 2x^3 - 1$

$f'(x) = 5x^4 + 6x^2$

If $a = f(x) = 2$ then $x = 1$.

$$(f^{-1})'(2) = \frac{1}{f'(1)} = \frac{1}{11}$$

b.

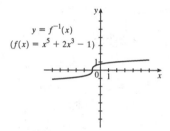

47. See Examples 2.7 and 2.8.

a. $f(x) = \sqrt{x^3 + 4x}$

$$f'(x) = \frac{3x^2 + 4}{2\sqrt{x^3 + 4x}}$$

If $a = f(x) = 4$ then $x = 2$.

$$(f^{-1})'(4) = \frac{1}{f'(2)} = \frac{1}{\frac{3(2)^2 + 4}{2\sqrt{(2)^3 + 4(2)}}} = \frac{8}{16} = \frac{1}{2}$$

b.

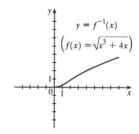

49. See Examples 5.4 and 5.5.

$$y' = 2y$$

$$\frac{1}{y}y' = 2$$

$$\int \frac{1}{y}y'\,dx = \int 2\,dx$$

$$\ln|y| = 2x + c$$

$$y = ce^{2x}$$

If $y(0) = 3$, then $3 = ce^0 = c$. Therefore,

$$y = 3e^{2x}.$$

51. See Examples 5.4 and 5.5.

$$y' = \frac{2x}{y}$$

$$yy' = 2x$$

$$\int yy'\,dx = \int 2x\,dx$$

$$\frac{1}{2}y^2 = x^2 + c$$

$$y^2 = 2x^2 + c$$

$$y = \pm\sqrt{2x^2 + c}$$

If $y(0) = 2$, then $2 = \sqrt{0 + c}$ or $c = 4$. Therefore,

$$y = \sqrt{2x^2 + 4}.$$

53. See Examples 5.4 and 5.5.

$$y' = \sqrt{xy}$$

$$\frac{1}{\sqrt{y}}y' = \sqrt{x}$$

$$\int \frac{1}{\sqrt{y}}y'\,dx = \int \sqrt{x}\,dx$$

$$2\sqrt{y} = \frac{2}{3}x^{3/2} + c$$

$$\sqrt{y} = \frac{1}{3}x^{3/2} + c$$

$$y = \left(\frac{1}{3}x^{3/2} + c\right)^2$$

If $y(1) = 4$, then

$$4 = \left(\frac{1}{3} + c\right)^2$$

$$2 = \frac{1}{3} + c$$

$$c = \frac{5}{3}$$

Therefore, $y = \left(\frac{1}{3}x^{3/2} + \frac{5}{3}\right)^2.$

55. See Example 4.1.

$$y(t) = Ae^{kt}$$

If the population doubles every two hours, then $y(2) = 2A$.

$$2A = Ae^{k \cdot 2}$$
$$e^{2k} = 2$$
$$2k = \ln 2$$
$$k = \frac{\ln 2}{2}$$

Therefore, $y(t) = (10^4)e^{\left(\frac{\ln 2}{2}\right)t}$.

The population reaches 10^6 after t hours, where

$$10^6 = 10^4 e^{\left(\frac{\ln 2}{2}\right)t}$$
$$e^{\left(\frac{\ln 2}{2}\right)t} = 100$$
$$\left(\frac{\ln 2}{2}\right)t = \ln 100$$
$$t = \frac{2\ln 100}{\ln 2} \approx 13.29 \text{ hours}$$

57. See Example 4.2.

If the half-life is 2 hours, then

$$y(2) = \frac{1}{2}A = Ae^{k \cdot 2}$$
$$e^{2k} = \frac{1}{2}$$
$$2k = \ln\left(\frac{1}{2}\right) = -\ln 2$$
$$k = -\frac{\ln 2}{2}$$

Therefore, $y(t) = Ae^{-\left(\frac{\ln 2}{2}\right)t}$

If there is 2 mg initially, then $y(t) = 2e^{-\left(\frac{\ln 2}{2}\right)t}$.

The nicotine level reaches 0.1 mg after t hours, where

$$2e^{-\left(\frac{\ln 2}{2}\right)t} = 0.1$$
$$e^{-\left(\frac{\ln 2}{2}\right)t} = 0.05$$
$$-\frac{\ln 2}{2}t = \ln 0.05$$
$$t = -\frac{2\ln 0.05}{\ln 2} \approx 8.64 \text{ hours}$$

59. See Example 4.4.

$$y(t) = Pe^{rt}$$
$$4000 = 2000e^{.08t}$$
$$e^{.08t} = 2$$
$$.08t = \ln 2$$
$$t = \frac{\ln 2}{.08} \approx 8.66 \text{ years}$$

61. See Example 4.3.

$$y(t) = Ae^{kt} + T_a$$

At time $t = 0$, $y(0) = 180$.

$$180 = Ae^0 + 68$$
$$A = 112$$

Therefore, $y(t) = 112e^{kt} + 68$.

At time $t = 1$, $y(1) = 176$.

$$176 = 112e^k + 68$$
$$112e^k = 108$$
$$e^k = \frac{108}{112}$$
$$k = \ln\frac{108}{112}$$
$$y(t) = 112e^{\left(\ln\frac{108}{112}\right)t} + 68$$

The temperature will reach $120°$ after t minutes, where

$$120 = 112e^{\left(\ln\frac{108}{112}\right)t} + 68$$
$$e^{\left(\ln\frac{108}{112}\right)t} = \frac{52}{112}$$
$$\left(\ln\frac{108}{112}\right)t = \ln\left(\frac{52}{112}\right)$$
$$t = \frac{\ln\left(\frac{52}{112}\right)}{\ln\left(\frac{108}{112}\right)} \approx 21.10 \text{ minutes}$$

63. See Example 5.3.

$$y' = 2x^3 y$$
$$\frac{1}{y}y' = 2x^3$$
$$\int \frac{1}{y}y'dx = \int 2x^3 dx$$
$$\ln|y| = \frac{1}{2}x^4 + c$$
$$y = ce^{\frac{1}{2}x^4}$$

65. See Example 5.3.

$$y' = \frac{4}{(y^2 + y)(1 + x^2)}$$

$$(y^2 + y)y' = \frac{4}{1 + x^2}$$

$$\int (y^2 + y)y' dx = \int \frac{4}{1 + x^2} dx$$

$$\frac{1}{3}y^3 + \frac{1}{2}y^2 = 4\tan^{-1} x + c$$

67. See Example 6.3.

$y' = 3y(2 - y)$

Equilibrium solutions are $y = 0$ and $y = 2$.
When $y < 0$ or $y > 3$, $y' < 0$, so the solution decreases. When $0 < y < 3$, $y' > 0$, so the solution increases. Hence $y = 0$ is unstable and $y = 2$ is stable.

69. See Example 6.3.

$y' = -y\sqrt{1 + y^2}$

Equilibrium solution is $y = 0$.
When $y < 0$, $y' > 0$, so the solution increases.
When $y > 0$, $y' < 0$, so the solution decreases.
Hence $y = 0$ is stable.

71. See Examples 6.1 and 6.2.

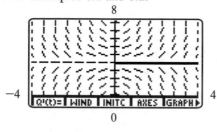

73. See Examples 6.1 and 6.2.

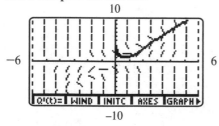

75. See Example 6.4.

a. $f(x, y) = 2x - y$
$x_0 = 0$, $y_0 = 3$
When $h = .1$
$x_n = x_0 + .1n$, so
$x_{10} = 1$ and $y_{10} = y(x_{10}) \approx y(1)$
$x_{20} = 2$ and $y_{20} = y(x_{20}) \approx y(2)$
$y_1 = y_0 + .1f(x_0, y_0) = 2.7$
$y_2 = y_1 + .1f(x_1, y_1) = 2.45$
$y(1) \approx y_{10} \approx 1.7434$
$y(2) \approx y_{20} \approx 2.6079$

b. When $h = .05$
$x_n = x_0 + .05n$, so
$x_{20} = 1$ and $y_{20} = y(x_{20}) \approx y(1)$
$x_{40} = 2$ and $y_{40} = y(x_{40}) \approx y(2)$
$y_1 = y_0 + .05f(x_0, y_0) = 2.85$
$y_2 = y_1 + .05f(x_1, y_1) = 2.7125$
$y(1) \approx y_{20} \approx 1.7924$
$y(2) \approx y_{40} \approx 2.6426$

77. See Examples 7.1–7.5.

On the unit circle, the angle $\theta \in \left[-\frac{\pi}{2}, \frac{\pi}{2}\right]$ for which $\sin\theta = 1$ is $\theta = \frac{\pi}{2}$. Hence, $\sin^{-1}1 = \frac{\pi}{2}$.

79. See Examples 7.1–7.5.

On the unit circle, the angle $\theta \in \left[-\frac{\pi}{2}, \frac{\pi}{2}\right]$ for which $\tan\theta = -1$ is $\theta = -\frac{\pi}{4}$. Hence,

$$\tan^{-1}(-1) = -\frac{\pi}{4}.$$

81. See Example 7.8.

$$\sin(\sec^{-1}2) = \sin\frac{\pi}{3} = \frac{\sqrt{3}}{2}$$

83. See Example 7.8.

$$\sin^{-1}\left(\sin\left(\frac{3\pi}{4}\right)\right) = \sin^{-1}\left(\frac{\sqrt{2}}{2}\right) = \frac{\pi}{4}$$

85. See Exercises 7.41–7.50.

$\sin 2x = 1$

$$2x = \frac{\pi}{2} + 2\pi n$$

$$x = \frac{\pi}{4} + \pi n, \text{ for any integer } n$$

87. See Section 6.9.

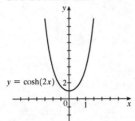

$y = \cosh(2x)$

89. See Section 6.7.

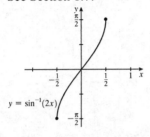

$y = \sin^{-1}(2x)$

91. See Section 6.3.

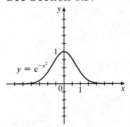

$y = e^{-x^2}$

93. See Example 9.3.

$$f(x) = 20\cosh\left(\frac{x}{20}\right), -25 \leq x \leq 25$$

$$f'(x) = \sinh\frac{x}{20}$$

The critical value is $x = 0$.
The lowest point of the cable is at
$f(0) = 20\cosh(0) = 20$.
The highest point of the cable is at

$$f(25) = 20\cosh\left(\frac{25}{20}\right) \approx 37.77 .$$

Therefore, the amount of sag is
$37.77 - 20 = 17.77$ feet.

Chapter 7

Section 7.1

3. $\int \sin 6x\, dx = -\dfrac{1}{6}\cos 6x + c$

5. $\int \sec 2x \tan 2x\, dx$

Let $u = 2x,\ du = 2dx$

$\int \sec u \tan u \cdot \dfrac{1}{2}\, du = \dfrac{1}{2}\int \sec u \tan u\, du$

$\qquad = \dfrac{1}{2}\sec u + c$

$\qquad = \dfrac{1}{2}\sec 2x + c$

7. $\int e^{3-2x}\, dx$

Let $u = 3 - 2x,\ du = -2dx$

$\int e^u \left(-\dfrac{1}{2}\, du\right) = -\dfrac{1}{2}\int e^u\, du$

$\qquad = -\dfrac{1}{2}e^u + c$

$\qquad = -\dfrac{1}{2}e^{3-2x} + c$

9. $\int \dfrac{\sin \sqrt{x}}{\sqrt{x}}\, dx$

Let $u = \sqrt{x},\ du = \dfrac{1}{2\sqrt{x}}\, dx$

$\int \sin u\, (2du) = 2\int \sin u\, du$

$\qquad = -2\cos u + c$

$\qquad = -2\cos \sqrt{x} + c$

11. $\int_0^{\pi} \cos x\, e^{\sin x}\, dx$

Let $u = \sin x,\ du = \cos x\, dx$

The new limits are $\sin 0 = 0$ and $\sin \pi = 0$.

$\int_0^0 e^u\, du = 0$

13. $\displaystyle\int_{-\pi/4}^{0} \dfrac{\sin x}{\cos^2 x}\, dx$

Let $u = \cos x,\ du = -\sin x\, dx$

The new limits are

$\cos\left(-\dfrac{\pi}{4}\right) = \dfrac{1}{\sqrt{2}}$ and $\cos 0 = 0$.

$\displaystyle\int_{1/\sqrt{2}}^{1} -\dfrac{du}{u^2} = \dfrac{1}{u}\bigg|_{1/\sqrt{2}}^{1} = 1 - \sqrt{2}$

15. $\displaystyle\int \dfrac{3}{16 + x^2}\, dx = \int \dfrac{3}{16\left(1 + \left(\frac{x}{4}\right)^2\right)}\, dx$

$\qquad = \dfrac{3}{16}\int \dfrac{1}{1 + \left(\frac{x}{4}\right)^2}\, dx$

Let $u = \dfrac{x}{4},\ du = \dfrac{1}{4}\, dx$

$\dfrac{3}{16}\int \dfrac{1}{1 + u^2}\, (4\, du) = \dfrac{3}{4}\int \dfrac{1}{1 + u^2}\, du$

$\qquad = \dfrac{3}{4}\tan^{-1} u + c$

$\qquad = \dfrac{3}{4}\tan^{-1}\left(\dfrac{x}{4}\right) + c$

17. $\displaystyle\int \dfrac{x^2}{1 + x^6}\, dx = \int \dfrac{x^2}{1 + (x^3)^2}\, dx$

Let $u = x^3,\ du = 3x^2\, dx$

$\displaystyle\int \dfrac{1}{1 + u^2}\cdot\dfrac{1}{3}\, du = \dfrac{1}{3}\int \dfrac{1}{1 + u^2}\, du$

$\qquad = \dfrac{1}{3}\tan^{-1} u + c$

$\qquad = \dfrac{1}{3}\tan^{-1} x^3 + c$

19. $\displaystyle\int \dfrac{1}{\sqrt{4 - x^2}}\, dx = \int \dfrac{1}{2\sqrt{1 - \left(\frac{x}{2}\right)^2}}\, dx$

Let $u = \dfrac{x}{2},\ du = \dfrac{1}{2}\, dx$

$\displaystyle\int \dfrac{1}{\sqrt{1 - u^2}}\, du = \sin^{-1} u + c$

$\qquad = \sin^{-1}\left(\dfrac{x}{2}\right) + c$

21. $\displaystyle\int \frac{x}{\sqrt{1-x^4}}\,dx = \int \frac{x}{\sqrt{1-(x^2)^2}}\,dx$

Let $u = x^2$, $du = 2x\,dx$

$\displaystyle\int \frac{1}{\sqrt{1-u^2}}\cdot\frac{1}{2}\,du = \frac{1}{2}\int \frac{1}{\sqrt{1-u^2}}\,du$

$\displaystyle = \frac{1}{2}\sin^{-1}u + c$

$\displaystyle = \frac{1}{2}\sin^{-1}x^2 + c$

23. $\displaystyle\int \frac{4}{5+2x+x^2}\,dx = \int \frac{4}{5+(x^2+2x+1)-1}\,dx$

$\displaystyle = \int \frac{4}{4+(x+1)^2}\,dx$

$\displaystyle = \int \frac{4}{4\left(1+\left(\frac{x+1}{2}\right)^2\right)}\,dx$

$\displaystyle = \int \frac{1}{1+\left(\frac{x+1}{2}\right)^2}\,dx$

Let $u = \dfrac{x+1}{2}$, $du = \dfrac{1}{2}\,dx$

$\displaystyle\int \frac{1}{1+u^2}\cdot 2\,du = 2\int \frac{1}{1+u^2}\,du$

$\displaystyle = 2\tan^{-1}u + c$

$\displaystyle = 2\tan^{-1}\left(\frac{x+1}{2}\right) + c$

25. $\displaystyle\int \frac{4x}{5+2x+x^2}\,dx = \int \frac{4x}{5+(x^2+2x+1)-1}\,dx$

$\displaystyle = \int \frac{4x}{4+(x+1)^2}\,dx$

$\displaystyle = \int \frac{4x}{4\left(1+\left(\frac{x+1}{2}\right)^2\right)}\,dx$

$\displaystyle = \int \frac{x}{1+\left(\frac{x+1}{2}\right)^2}\,dx$

Let $u = \dfrac{x+1}{2}$, $du = \dfrac{1}{2}\,dx$, $x = 2u-1$

$\displaystyle\int \frac{2u-1}{1+u^2}\cdot 2\,du$

$\displaystyle = 2\int \frac{2u-1}{1+u^2}\,du$

$\displaystyle = 2\int \frac{2u}{1+u^2}\,du - 2\int \frac{1}{1+u^2}\,du$

$\displaystyle = 2\ln(1+u^2) - 2\tan^{-1}u + c$

$\displaystyle = 2\ln\left(1+\left(\frac{x+1}{2}\right)^2\right)$

$\displaystyle\quad -2\tan^{-1}\left(\frac{x+1}{2}\right) + c$

$\displaystyle = 2\ln\left(\frac{x^2+2x+5}{4}\right) - 2\tan^{-1}\left(\frac{x+1}{2}\right) + c$

$\displaystyle = 2\ln(x^2+2x+5) - 2\tan^{-1}\left(\frac{x+1}{2}\right) + c$

27. $\displaystyle\int (x^2+4)^2\,dx = \int (x^4+8x^2+16)\,dx$

$\displaystyle = \frac{1}{5}x^5 + \frac{8}{3}x^3 + 16x + c$

29. $\displaystyle\int \frac{1}{\sqrt{3-2x-x^2}}\,dx = \int \frac{1}{\sqrt{3-(x^2+2x+1)+1}}\,dx$

$\displaystyle = \int \frac{1}{\sqrt{4-(x+1)^2}}\,dx$

$\displaystyle = \int \frac{1}{2\sqrt{1-\left(\frac{x+1}{2}\right)^2}}\,dx$

Let $u = \dfrac{x+1}{2}$, $du = \dfrac{1}{2}\,dx$

$\displaystyle\int \frac{1}{\sqrt{1-u^2}}\,du = \sin^{-1}u + c$

$\displaystyle = \sin^{-1}\left(\frac{x+1}{2}\right) + c$

31. $\displaystyle\int \frac{2}{\sqrt{4x-x^2}}\,dx = \int \frac{2}{\sqrt{-(x^2-4x+4)+4}}\,dx$

$$= \int \frac{2}{\sqrt{4-(x-2)^2}}\,dx$$

$$= \int \frac{2}{2\sqrt{1-\left(\frac{x-2}{2}\right)^2}}\,dx$$

Let $u = \dfrac{x-2}{2}$, $du = \dfrac{1}{2}dx$

$$\int \frac{2}{\sqrt{1-u^2}}\,du = 2\sin^{-1}u + c$$

$$= 2\sin^{-1}\left(\frac{x-2}{2}\right) + c$$

33. $\displaystyle\int_{-2}^{-1} e^{\ln(x^2+1)}\,dx = \int_{-2}^{-1} x^2 + 1\,dx$

$$= \left(\frac{1}{3}x^3 + x\right)\Big|_{-2}^{-1}$$

$$= -\frac{1}{3} - 1 - \left(-\frac{8}{3} - 2\right)$$

$$= \frac{10}{3}$$

35. $\displaystyle\int_3^4 x\sqrt{x-3}\,dx$

Let $u = x-3$, $du = dx$, $x = u+3$
The new limits are $3-3=0$ and $4-3=1$.

$$\int_0^1 (u+3)\sqrt{u}\,du = \int_0^1 \left(u^{3/2} + 3u^{1/2}\right)du$$

$$= \frac{2}{5}u^{5/2} + 2u^{3/2}\Big|_0^1$$

$$= \frac{2}{5} + 2 = \frac{12}{5}$$

37. $\displaystyle\int_0^2 \frac{e^x}{1+e^{2x}}\,dx$

Let $u = e^x$, $du = e^x dx$

The new limits are $e^0 = 1$ and e^2.

$$\int_1^{e^2} \frac{1}{1+u^2} = \tan^{-1}u\Big|_1^{e^2}$$

$$= \tan^{-1}e^2 - \tan^{-1}1$$

$$= \tan^{-1}e^2 - \frac{\pi}{4}$$

39. $\displaystyle\int_1^4 \frac{x^2+1}{\sqrt{x}}\,dx = \int_1^4 (x^{3/2} + x^{-1/2})\,dx$

$$= \frac{2}{5}x^{5/2} + 2x^{1/2}\Big|_1^4$$

$$= \frac{2}{5}\cdot 32 + 2\cdot 2 - \left(\frac{2}{5} + 2\right)$$

$$= \frac{72}{5}$$

41. $\displaystyle\int \sinh 2x\,dx = \frac{1}{2}\cosh 2x + c$

43. $\displaystyle\int \tanh 3x\,dx = \int \frac{\sinh 3x}{\cosh 3x}\,dx$

Let $u = \cosh 3x$, $du = 3\sinh 3x\,dx$

$$\int \frac{1}{u}\cdot\frac{1}{3}\,du = \frac{1}{3}\int \frac{1}{u}\,du$$

$$= \frac{1}{3}\ln|u| + c$$

$$= \frac{1}{3}\ln|\cosh 3x| + c$$

45. $\displaystyle\int x\cosh x^2\,dx$

Let $u = x^2$, $du = 2x\,dx$

$$\int \cosh u\cdot\frac{1}{2}\,du = \frac{1}{2}\int \cosh u\,du$$

$$= \frac{1}{2}\sinh u + c$$

$$= \frac{1}{2}\sinh x^2 + c$$

47. $\displaystyle\int \frac{3}{\sqrt{x^2-1}}\,dx = 3\int \frac{1}{\sqrt{x^2-1}}\,dx$

$$= 3\cosh^{-1}x + c$$

49. $\displaystyle\int \frac{3x}{x^2\sqrt{x^4-1}}\,dx$

Let $u = x^2$, $du = 2x\,dx$

$$\int \frac{3}{u\sqrt{u^2-1}}\cdot\frac{1}{2}\,du = \frac{3}{2}\int \frac{1}{u\sqrt{u^2-1}}\,du$$

$$= \frac{3}{2}\sec^{-1}u + c$$

$$= \frac{3}{2}\sec^{-1}x^2 + c$$

51. $\displaystyle\int \frac{5}{3+x^2}\,dx = \int \frac{5}{3\left(1+\left(\frac{x}{\sqrt{3}}\right)^2\right)}\,dx$

Let $u = \dfrac{x}{\sqrt{3}},\ du = \dfrac{1}{\sqrt{3}}\,dx$

$\displaystyle\int \frac{5}{3(1+u^2)}\cdot\sqrt{3}\,du = \frac{5}{\sqrt{3}}\int \frac{1}{1+u^2}\,du$

$\qquad = \dfrac{5}{\sqrt{3}}\tan^{-1} u + c$

$\qquad = \dfrac{5}{\sqrt{3}}\tan^{-1}\left(\dfrac{x}{\sqrt{3}}\right) + c$

53. $\displaystyle\int \frac{\ln x}{2x}\,dx$

Let $u = \ln x,\ du = \dfrac{1}{x}\,dx$

$\displaystyle\int \frac{u}{2}\,du = \frac{1}{2}\int u\,du$

$\qquad = \dfrac{1}{4}u^2 + c$

$\qquad = \dfrac{1}{4}(\ln x)^2 + c$

55. $\displaystyle\int xe^{-x^2}\,dx$

Let $u = -x^2,\ du = -2x\,dx$

$\displaystyle\int e^u\left(-\frac{1}{2}\,du\right) = -\frac{1}{2}\int e^u\,du$

$\qquad = -\dfrac{1}{2}e^u + c$

$\qquad = -\dfrac{1}{2}e^{-x^2} + c$

57. $\displaystyle\int_0^2 f(x)\,dx = \int_0^1 \frac{x}{x^2+1}\,dx + \int_1^2 \frac{x^2}{x^2+1}\,dx$

For the first integral, use $u = x^2+1,\ du = 2x\,dx$.

Rewrite the second integrand as $1 - \dfrac{1}{1+x^2}$. Then

$\displaystyle\int_0^2 f(x)\,dx = \frac{1}{2}\int_1^2 \frac{1}{u}\,du + \int_1^2 1 - \frac{1}{1+x^2}\,dx$

$\qquad = \dfrac{1}{2}\left(\ln|u|\,\Big|_1^2\right) + \left(x - \tan^{-1} x\,\Big|_1^2\right)$

$\qquad \dfrac{\ln 2}{2} + 1 - \tan^{-1} 2 + \dfrac{\pi}{4}$

Section 7.2

3. $\displaystyle\int x\cos\,dx \qquad u = x \qquad dv = \cos x\,dx$
$\qquad\qquad\qquad\qquad du = dx \qquad v = \sin x$

$\qquad = x\sin x - \displaystyle\int \sin x\,dx$

$\qquad = x\sin x + \cos x + c$

5. $\displaystyle\int xe^{2x}\,dx \qquad u = x \qquad dv = e^{2x}\,dx$
$\qquad\qquad\qquad\qquad du = dx \qquad v = \dfrac{1}{2}e^{2x}$

$\qquad = \dfrac{1}{2}xe^{2x} - \displaystyle\int \frac{1}{2}e^{2x}\,dx$

$\qquad = \dfrac{1}{2}xe^{2x} - \dfrac{1}{4}e^{2x} + c$

7. $\displaystyle\int x^2 \ln x\,dx \qquad u = \ln x \qquad dv = x^2\,dx$
$\qquad\qquad\qquad\qquad du = \dfrac{1}{x}\,dx \qquad v = \dfrac{1}{3}x^3$

$\qquad = \dfrac{1}{3}x^3 \ln x - \displaystyle\int \frac{1}{3}x^3 \cdot \frac{1}{x}\,dx$

$\qquad = \dfrac{1}{3}x^3 \ln x - \dfrac{1}{3}\displaystyle\int x^2\,dx$

$\qquad = \dfrac{1}{3}x^3 \ln x - \dfrac{1}{9}x^3 + c$

9. $\displaystyle\int x^2 e^{-3x}\,dx \qquad u = x^2 \qquad dv = e^{-3x}\,dx$
$\qquad\qquad\qquad\qquad du = 2x\,dx \qquad v = -\dfrac{1}{3}e^{-3x}$

$\qquad = -\dfrac{1}{3}x^2 e^{-3x} - \displaystyle\int\left(-\frac{1}{3}e^{-3x}\right)\cdot 2x\,dx$

$\qquad = -\dfrac{1}{3}x^2 e^{-3x} + \dfrac{2}{3}\displaystyle\int xe^{-3x}\,dx \qquad u = x \qquad dv = e^{-3x}\,dx$
$\qquad\qquad\qquad\qquad\qquad\qquad\qquad\qquad du = dx \qquad v = -\dfrac{1}{3}e^{-3x}$

$\qquad = -\dfrac{1}{3}x^2 e^{-3x} + \dfrac{2}{3}\left[-\frac{1}{3}xe^{-3x} - \displaystyle\int\left(-\frac{1}{3}e^{-3x}\right)dx\right]$

$\qquad = -\dfrac{1}{3}x^2 e^{-3x} - \dfrac{2}{9}xe^{-3x} + \dfrac{2}{9}\displaystyle\int e^{-3x}\,dx$

$\qquad = -\dfrac{1}{3}x^2 e^{-3x} - \dfrac{2}{9}xe^{-3x} - \dfrac{2}{27}e^{-3x} + c$

11. $\int x^2 \cos 2x\,dx \qquad u = x^2 \qquad dv = \cos 2x\,dx$

$$du = 2x\,dx \qquad v = \frac{1}{2}\sin 2x$$

$$= \frac{1}{2}x^2 \sin 2x - \int \frac{1}{2}(\sin 2x)2x\,dx$$

$$= \frac{1}{2}x^2 \sin 2x - \int x \sin 2x\,dx \qquad u = x \qquad dv = \sin 2x\,dx$$

$$du = dx \qquad v = -\frac{1}{2}\cos 2x$$

$$= \frac{1}{2}x^2 \sin 2x - \left[-\frac{1}{2}x\cos 2x - \int\left(-\frac{1}{2}\cos 2x\right)dx \right]$$

$$= \frac{1}{2}x^2 \sin 2x + \frac{1}{2}x\cos 2x - \frac{1}{4}\sin 2x + c$$

13. $\int e^x \sin 4x\,dx \qquad u = e^x \qquad dv = \sin 4x\,dx$

$$du = e^x dx \qquad v = -\frac{1}{4}\cos 4x$$

$$= -\frac{1}{4}e^x \cos 4x - \int\left(-\frac{1}{4}\cos 4x\right)e^x dx$$

$$= -\frac{1}{4}e^x \cos 4x + \frac{1}{4}\int e^x \cos 4x\,dx \qquad u = e^x \qquad dv = \cos 4x\,dx$$

$$du = e^x\,dx \qquad v = \frac{1}{4}\sin 4x$$

$$= -\frac{1}{4}e^x \cos 4x + \frac{1}{4}\left(\frac{1}{4}e^x \sin 4x - \int \frac{1}{4}(\sin 4x)e^x\,dx \right)$$

$$\int e^x \sin 4x\,dx = -\frac{1}{4}e^x \cos 4x + \frac{1}{16}e^x \sin 4x - \frac{1}{16}\int e^x \sin 4x\,dx$$

$$\frac{17}{16}\int e^x \sin 4x\,dx = -\frac{1}{4}e^x \cos 4x + \frac{1}{16}e^x \sin 4x + K$$

$$\int e^x \sin 4x\,dx = -\frac{4}{17}e^x \cos 4x + \frac{1}{17}e^x \sin 4x + c$$

15. $\int \cos x \cos 2x\,dx \qquad u = \cos x \qquad dv = \cos 2x\,dx$

$$du = -\sin x\,dx \qquad v = \frac{1}{2}\sin 2x$$

$$= \frac{1}{2}\cos x \sin 2x - \int \frac{1}{2}\sin 2x(-\sin x)dx$$

$$= \frac{1}{2}\cos x \sin 2x + \frac{1}{2}\int \sin x \sin 2x\,dx \qquad u = \sin x \qquad dv = \sin 2x\,dx$$

$$du = \cos x\,dx \qquad v = -\frac{1}{2}\cos 2x$$

$$= \frac{1}{2}\cos x \sin 2x + \frac{1}{2}\left[-\frac{1}{2}\cos 2x \sin x - \int\left(-\frac{1}{2}\cos 2x\right)\cos x\,dx \right]$$

$$\int \cos x \cos 2x\,dx = \frac{1}{2}\cos x \sin 2x - \frac{1}{4}\cos 2x \sin x + \frac{1}{4}\int \cos x \cos 2x\,dx$$

$$\frac{3}{4}\int \cos x \cos 2x\,dx = \frac{1}{2}\cos x \sin 2x - \frac{1}{4}\cos 2x \sin x + K$$

$$\int \cos x \cos 2x\,dx = \frac{2}{3}\cos x \sin 2x - \frac{1}{3}\cos 2x \sin x + c$$

17. $\int x\sec^2 x\, dx$ $\qquad u = x \qquad dv = \sec^2 x\, dx$
$\qquad\qquad\qquad\qquad\quad du = dx \qquad v = \tan x$

$= x\tan x - \int \tan x\, dx$

$= x\tan x - \int \dfrac{\sin x}{\cos x}\, dx$

Let $u = \cos x,\ du = -\sin x\, dx$

$= x\tan x + \int \dfrac{1}{u}\, du$

$= x\tan x + \ln|u| + c$

$= x\tan x + \ln|\cos x| + c$

19. $\int (\ln x)^2\, dx \qquad u = (\ln x)^2 \qquad dv = dx$
$\qquad\qquad\qquad\qquad du = 2\dfrac{\ln x}{x}\, dx \qquad v = x$

$= x(\ln x)^2 - \int x\cdot 2\dfrac{\ln x}{x}\, dx$

$= x(\ln x)^2 - 2\int \ln x\, dx \qquad u = \ln x \qquad dv = dx$
$\qquad\qquad\qquad\qquad\qquad\quad du = \dfrac{1}{x}\, dx \qquad v = x$

$= x(\ln x)^2 - 2\left[x\ln x - \int x\cdot\dfrac{1}{x}\, dx \right]$

$= x(\ln x)^2 - 2x\ln x + 2\int dx$

$= x(\ln x)^2 - 2x\ln x + 2x + c$

21. $\int \cos x\ln(\sin x)\, dx$

$u = \ln(\sin x) \qquad\qquad dv = \cos x\, dx$

$du = \dfrac{1}{\sin x}\cdot \cos x\, dx \qquad v = \sin x$

$= \sin x\ln(\sin x) - \int \sin x\cdot\dfrac{1}{\sin x}\cdot\cos x\, dx$

$= \sin x\ln(\sin x) - \int \cos x\, dx$

$= \sin x\ln(\sin x) - \sin x + c$

23. $\int x\cosh 2x\, dx \qquad u = x \qquad dv = \cosh 2x\, dx$
$\qquad\qquad\qquad\qquad\qquad du = dx \qquad v = \dfrac{1}{2}\sinh 2x$

$= \dfrac{1}{2}x\sinh 2x - \int \dfrac{1}{2}\sinh 2x\, dx$

$= \dfrac{1}{2}x\sinh 2x - \dfrac{1}{4}\cosh 2x + c$

25. $\int_0^1 x\sin 2x\, dx \qquad u = x \qquad dv = \sin 2x\, dx$
$\qquad\qquad\qquad\qquad\qquad\quad du = dx \qquad v = -\dfrac{1}{2}\cos 2x$

$= -\dfrac{1}{2}x\cos 2x \Big|_0^1 - \int_0^1\left(-\dfrac{1}{2}\cos 2x \right)dx$

$= -\dfrac{1}{2}(1\cos 2 - 0\cos 0) + \dfrac{1}{2}\int_0^1 \cos 2x\, dx$

$= -\dfrac{1}{2}\cos 2 + \dfrac{1}{2}\left(\dfrac{1}{2}\sin 2x \right)\Big|_0^1$

$= -\dfrac{1}{2}\cos 2 + \dfrac{1}{4}(\sin 2 - \sin 0)$

$= -\dfrac{1}{2}\cos 2 + \dfrac{1}{4}\sin 2$

27. $\int_0^1 x\cos \pi x\, dx \qquad u = x \qquad dv = \cos \pi x\, dx$
$\qquad\qquad\qquad\qquad\qquad\quad du = dx \qquad v = \dfrac{1}{\pi}\sin \pi x$

$= \dfrac{1}{\pi}x\sin \pi x \Big|_0^1 - \int_0^1 \dfrac{1}{\pi}\sin \pi x\, dx$

$= \dfrac{1}{\pi}\sin \pi - 0 + \dfrac{1}{\pi^2}\cos \pi x \Big|_0^1$

$= 0 + \dfrac{1}{\pi^2}(\cos \pi - \cos 0)$

$= \dfrac{1}{\pi^2}(-1-1)$

$= -\dfrac{2}{\pi^2}$

29. $\int_0^1 x\sin \pi x\, dx \qquad u = x \qquad dv = \sin \pi x\, dx$
$\qquad\qquad\qquad\qquad\qquad\quad du = dx \qquad v = -\dfrac{1}{\pi}\cos \pi x$

$= -\dfrac{1}{\pi}x\cos \pi x \Big|_0^1 - \int_0^1\left(-\dfrac{1}{\pi}\cos \pi x \right)dx$

$= -\dfrac{1}{\pi}\cos \pi - 0 + \dfrac{1}{\pi^2}\sin \pi x \Big|_0^1$

$= -\dfrac{1}{\pi}(-1) + \dfrac{1}{\pi^2}(\sin \pi - \sin 0)$

$= \dfrac{1}{\pi}$

31. $\displaystyle\int_1^{10} \ln x \, dx$ $u = \ln x$ $dv = dx$

$\qquad\qquad\qquad du = \dfrac{1}{x} dx$ $v = x$

$= x \ln x \Big|_1^{10} - \displaystyle\int_1^{10} x \cdot \dfrac{1}{x} dx$

$= 10 \ln 10 - 1 \ln 1 - \displaystyle\int_1^{10} dx$

$= 10 \ln 10 - 0 - x \Big|_1^{10}$

$= 10 \ln 10 - (10 - 1)$

$= 10 \ln 10 - 9$

33. $\displaystyle\int \cos^{-1} x \, dx$ $u = \cos^{-1} x$ $dv = dx$

$\qquad\qquad\qquad du = -\dfrac{1}{\sqrt{1-x^2}} dx$ $v = x$

$= x \cos^{-1} x - \displaystyle\int x \left(-\dfrac{1}{\sqrt{1-x^2}} \right) dx$

$= x \cos^{-1} x + \displaystyle\int \dfrac{x}{\sqrt{1-x^2}} dx$

Let $u = 1 - x^2$, $du = -2x \, dx$

$= x \cos^{-1} x + \displaystyle\int \dfrac{1}{\sqrt{u}} \left(-\dfrac{1}{2} du \right)$

$= x \cos^{1} x - \dfrac{1}{2} \displaystyle\int u^{-1/2} du$

$= x \cos^{-1} x - \dfrac{1}{2} \cdot 2u^{1/2} + c$

$= x \cos^{-1} x - \sqrt{1-x^2} + c$

35. $\displaystyle\int \sin \sqrt{x} \, dx$ $u = \sin \sqrt{x}$ $dv = dx$

$\qquad\qquad\qquad du = (\cos \sqrt{x}) \left(\dfrac{1}{2\sqrt{x}} \right) dx$ $v = x$

$= x \sin \sqrt{x} - \displaystyle\int x (\cos \sqrt{x}) \left(\dfrac{1}{2\sqrt{x}} \right) dx$

$= x \sin \sqrt{x} - \dfrac{1}{2} \displaystyle\int \sqrt{x} \cos \sqrt{x} \, dx$

Let $p = \sqrt{x}$, $dp = \dfrac{1}{2\sqrt{x}} dx$

$= x \sin \sqrt{x} - \dfrac{1}{2} \displaystyle\int 2p^2 \cos p \, dp$

$= x \sin \sqrt{x} - \displaystyle\int p^2 \cos p \, dp$ $u = p^2$ $dv = \cos p \, dp$

$\qquad\qquad\qquad\qquad\qquad du = 2p \, dp$ $v = \sin p$

$= x \sin \sqrt{x} - (p^2 \sin p - \displaystyle\int 2p \sin p \, dp)$

$= x \sin \sqrt{x} - p^2 \sin p + 2 \displaystyle\int p \sin p \, dp$ $u = p$ $dv = \sin p \, dp$

$\qquad\qquad\qquad\qquad\qquad du = dp$ $v = -\cos p$

$= x \sin \sqrt{x} - p^2 \sin p + 2(-p \cos p + \displaystyle\int \cos p \, dp)$

$= x \sin \sqrt{x} - p^2 \sin p - 2p \cos p + 2 \sin p + c$

$= x \sin \sqrt{x} - x \sin \sqrt{x} - 2\sqrt{x} \cos \sqrt{x} + 2 \sin \sqrt{x} + c$

$= -2\sqrt{x} \cos \sqrt{x} + 2 \sin \sqrt{x} + c$

37. n times

39. $\int \cos^n x\,dx \qquad u = \cos^{n-1} x \qquad\qquad dv = \cos x\,dx$

$\qquad\qquad\qquad du = (n-1)(\cos^{n-2} x)(-\sin x)\,dx \qquad v = \sin x$

$= \sin x\cos^{n-1} x - \int (\sin x)(n-1)(\cos^{n-2} x)(-\sin x)\,dx$

$= \sin x\cos^{n-1} x + \int (n-1)(\cos^{n-2} x)(\sin^2 x)\,dx$

$= \sin x\cos^{n-1} x + \int (n-1)(\cos^{n-2} x)(1-\cos^2 x)\,dx$

$= \sin x\cos^{n-1} x + \int (n-1)(\cos^{n-2} x - \cos^n x)\,dx$

$\int \cos^n x\,dx = \sin x\cos^{n-1} x + \int (n-1)\cos^{n-2} x\,dx - (n-1)\int \cos^n x\,dx$

$n\int \cos^n x\,dx = \sin x\cos^{n-1} x + (n-1)\int \cos^{n-2} x\,dx$

$\int \cos^n x\,dx = \dfrac{1}{n}\sin x\cos^{n-1} x + \dfrac{n-1}{n}\int \cos^{n-2} x\,dx$

41. $\int x^3 e^x\,dx$

Three applications of the formula in Equation (2.4) produce, after factoring out e^x, $e^x(x^3 - 3x^2 + 6x - 6) + c$.

43. $\int \cos^3 x\,dx = \dfrac{1}{3}\cos^2 x \sin x + \dfrac{2}{3}\int \cos x\,dx$

$\qquad\qquad = \dfrac{1}{3}\cos^2 x\sin x + \dfrac{2}{3}\sin x + c$

45. Four applications of the formula in Equation (2.4) produce, after factoring out e^x, $e^x(x^4 - 4x^3 + 12x^2 - 24x + 24) + c$.

$e^x(x^4 - 4x^3 + 12x^2 - 24x + 24)\Big|_0^1 = 9e - 24$

47. $\int_0^{\pi/2} \sin^5 x\,dx$

$= -\dfrac{1}{5}\sin^4 x \cos x \Big|_0^{\pi/2} + \dfrac{4}{5}\int_0^{\pi/2} \sin^3 x\,dx$

$= -\dfrac{1}{5}\sin^4 x\cos x \Big|_0^{\pi/2} + \dfrac{4}{5}\left(-\dfrac{1}{3}\sin^2 x\cos x - \dfrac{2}{3}\cos x\right)\Big|_0^{\pi/2}$ (Using Exercise 43)

$= -\dfrac{1}{5}\left(\sin^4\left(\dfrac{\pi}{2}\right)\cos\dfrac{\pi}{2} - \sin^4 0\cos 0\right) + \dfrac{4}{5}\left[\left(-\dfrac{1}{3}\sin^2\left(\dfrac{\pi}{2}\right)\cos\dfrac{\pi}{2} - \dfrac{2}{3}\cos\dfrac{\pi}{2}\right) - \left(-\dfrac{1}{3}\sin^2 0\cos 0 - \dfrac{2}{3}\cos 0\right)\right]$

$= -\dfrac{1}{5}(1\cdot 0 - 0\cdot 1) + \dfrac{4}{5}\left[\left(-\dfrac{1}{3}\cdot 1\cdot 0 - \dfrac{2}{3}\cdot 0\right) - \left(-\dfrac{1}{3}\cdot 0\cdot 1 - \dfrac{2}{3}\cdot 1\right)\right]$

$= \dfrac{4}{5}\left(\dfrac{2}{3}\right) = \dfrac{8}{15}$

49. m even: $\dfrac{(m-1)(m-3)\cdots 1}{m(m-2)\cdots 2}\cdot\dfrac{\pi}{2}$; m odd: $\dfrac{(m-1)(m-3)\cdots 2}{m(m-2)\cdots 3}$

51. First column: derivatives; second column: antiderivatives

53.

	$\cos x$	
x^4	$\sin x$	$+$
$4x^3$	$-\cos x$	$-$
$12x^2$	$-\sin x$	$+$
$24x$	$\cos x$	$-$
24	$\sin x$	$+$

$$\int x^4 \cos x\,dx = x^4 \sin x + 4x^3 \cos x - 12x^2 \sin x - 24x \cos x + 24\sin x + c$$

55.

	e^{2x}	
x^4	$\dfrac{1}{2}e^{2x}$	$+$
$4x^3$	$\dfrac{1}{4}e^{2x}$	$-$
$12x^2$	$\dfrac{1}{8}e^{2x}$	$+$
$24x$	$\dfrac{1}{16}e^{2x}$	$-$
24	$\dfrac{1}{32}e^{2x}$	$+$

$$\int x^4 e^{2x}dx = \left(\frac{1}{2}x^4 - x^3 + \frac{3}{2}x^2 - \frac{3}{2}x + \frac{3}{4}\right)e^{2x} + c$$

57.

	e^{-3x}	
x^3	$-\dfrac{1}{3}e^{-3x}$	$+$
$3x^2$	$\dfrac{1}{9}e^{-3x}$	$-$
$6x$	$-\dfrac{1}{27}e^{-3x}$	$+$
6	$\dfrac{1}{81}e^{-3x}$	$-$

$$\int x^3 e^{-3x}dx = \left(-\frac{1}{3}x^3 - \frac{1}{3}x^2 - \frac{2}{9}x - \frac{2}{27}\right)e^{-3x} + c$$

59. $\displaystyle\int_{-\pi}^{\pi} \cos(mx)\cos(nx)dx \qquad u = \cos(mx) \qquad dv = \cos(nx)dx$

$$du = -m\sin(mx) \qquad v = \frac{1}{n}\sin(nx)$$

$$= \frac{1}{n}\cos(mx)\sin(nx)\Big|_{-\pi}^{\pi} + \frac{m}{n}\int_{-\pi}^{\pi} \sin(mx)\sin(nx)dx \qquad u = \sin(mx) \qquad dv = \sin(nx)dx$$

$$du = m\cos(mx)dx \qquad v = -\frac{1}{n}\cos(nx)$$

$$= \frac{1}{n}\left[\cos(m\pi)\sin(m\pi) - \cos(-m\pi)\sin(-m\pi)\right] + \frac{m}{n}\left[-\frac{1}{n}\cos(nx)\sin(mx)\Big|_{-\pi}^{\pi} + \frac{m}{n}\int_{-\pi}^{\pi}\cos(mx)\cos(nx)dx\right]$$

$$\left(1 - \frac{m^2}{n^2}\right)\int_{-\pi}^{\pi}\cos(mx)\cos(nx)\,dx = 0 - \frac{m}{n^2}[\cos(n\pi)\sin(m\pi) - \cos(-n\pi)\sin(-m\pi)]$$

$$= 0 \quad (\sin(n\pi) = 0 \text{ when } n \text{ is an integer})$$

61. $\int_{-\pi}^{\pi} \cos{(mx)}\sin{(nx)}\,dx$ $\qquad u = \cos{(mx)}$ $\qquad dv = \sin{(nx)}\,dx$

$\qquad\qquad\qquad\qquad\qquad du = -m\sin{(mx)}dx \qquad v = -\dfrac{1}{n}\cos{(nx)}$

$= -\dfrac{1}{n}\cos{(mx)}\cos{(nx)}\Big|_{-\pi}^{\pi} - \dfrac{m}{n}\int_{-\pi}^{\pi}\sin{(mx)}\cos{(nx)}\,dx \qquad u = \sin{(mx)} \qquad dv = \cos{(nx)}\,dx$

$\qquad\qquad\qquad\qquad\qquad\qquad\qquad\qquad\qquad\qquad\qquad\qquad\qquad\qquad du = m\cos{(mx)}\,dx \qquad v = \dfrac{1}{n}\sin{(nx)}$

$= -\dfrac{1}{n}\cos{(mx)}\cos{(nx)}\Big|_{-\pi}^{\pi} - \dfrac{m}{n}\left[\dfrac{1}{n}\sin{(mx)}\sin{(nx)}\Big|_{-\pi}^{\pi} - \dfrac{m}{n}\int_{-\pi}^{\pi}\cos{(mx)}\sin{(nx)}\,dx\right]$

$\left(1 - \dfrac{m^2}{n^2}\right)\int_{-\pi}^{\pi}\cos{(mx)}\sin{(nx)}dx$

$= -\dfrac{1}{n}[\cos{(m\pi)}\cos{(n\pi)} - \cos{(-m\pi)}\cos{(-n\pi)}] - \dfrac{m}{n^2}[\sin{(m\pi)}\sin{(n\pi)} - \sin{(-m\pi)}\sin{(-n\pi)}]$

$= 0$ ($\cos{x} = \cos{(-x)}$ and $\sin{(n\pi)} = 0$ when n is an integer)

Section 7.3

3. $\int \cos{x}\sin^4{x}\,dx$

Let $u = \sin{x}$, $du = \cos{x}\,dx$

$\int u^4\,du = \dfrac{1}{5}u^5 + c$

$= \dfrac{1}{5}\sin^5{x} + c$

5. $\int_0^{\pi/4}\cos{x}\sin^3{x}\,dx$

Let $u = \sin{x}$, $du = \cos{x}\,dx$

The new limits are $\sin{0} = 0$ and $\sin{(\pi/4)} = 1/\sqrt{2}$.

$\int_0^{1/\sqrt{2}} u^3\,du = \dfrac{1}{4}u^4\Big|_0^{1/\sqrt{2}}$

$= \dfrac{1}{4\cdot\left(\sqrt{2}\right)^4}$

$= \dfrac{1}{16}$

7. $\int \cos^2{x}\sin{x}\,dx$

Let $u = \cos{x}$, $du = -\sin{x}\,dx$

$\int u^2(-du) = -\dfrac{1}{3}u^3 + c$

$= -\dfrac{1}{3}\cos^3{x} + c$

9. $\int_0^{\pi/2}\cos{x}\sin{x}\,dx$

Let $u = \sin{x}$, $du = \cos{x}\,dx$

The new limits are $\sin{0} = 0$ and $\sin{(\pi/2)} = 1$.

$\int_0^1 u\,du = \dfrac{1}{2}u^2\Big|_0^1 = \dfrac{1}{2}$

11. $\int \cos^2{x}\,dx = \dfrac{1}{2}\int(1 + \cos{2x})\,dx$

$= \dfrac{1}{2}x + \dfrac{1}{4}\sin{2x} + c$

13. $\int \tan{x}\sec^3{x}\,dx = \int \tan{x}\sec{x}\sec^2{x}\,dx$

Let $u = \sec{x}$, $du = \sec{x}\tan{x}\,dx$

$\int u^2\,du = \dfrac{1}{3}u^3 + c$

$= \dfrac{1}{3}\sec^3{x} + c$

15. $\int_0^{\pi/4}\tan^4{x}\sec^4{x}\,dx$

$= \int_0^{\pi/4}\tan^4{x}\sec^2{x}\sec^2{x}\,dx$

$= \int_0^{\pi/4}\tan^4{x}(1 + \tan^2{x})\sec^2{x}\,dx$

Let $u = \tan{x}$, $du = \sec^2{x}\,dx$

The new limits are $\tan{0} = 0$ and $\tan{(\pi/4)} = 1$.

$\int_0^1 u^4(1 + u^2)\,du = \int_0^1(u^4 + u^6)\,du$

$= \dfrac{1}{5}u^5 + \dfrac{1}{7}u^7\Big|_0^1$

$\dfrac{12}{35}$

17. $\int \cos^2 x \sin^2 x\, dx$

$= \int \frac{1}{2}(1+\cos 2x)\cdot\frac{1}{2}(1-\cos 2x)\,dx$

$= \frac{1}{4}\int(1-\cos^2 2x)\,dx$

$= \frac{1}{4}\int\left[1-\frac{1}{2}(1+\cos 4x)\right]dx$

$= \frac{1}{4}\left(\frac{1}{2}x-\frac{1}{8}\sin 4x\right)+c$

$= \frac{1}{8}x-\frac{1}{32}\sin 4x+c$

19. $\int \cot^3 x \csc^3 x\, dx$

$= \int \cot^2 x\csc^2 x\cot x\csc x\,dx$

$= \int(\csc^2 x-1)\csc^2 x\cot x\csc x\,dx$

Let $u=\csc x$, $du=-\csc x\cot x\,dx$

$\int(u^2-1)u^2(-du)=\int(u^2-u^4)du$

$= \frac{1}{3}u^3-\frac{1}{5}u^5+c$

$= \frac{1}{3}\csc^3 x-\frac{1}{5}\csc^5 x+c$

21. $\int_{-\pi/3}^{0}\sqrt{\cos x}\sin^3 x\,dx$

$= \int_{\pi/3}^{0}\sqrt{\cos x}(1-\cos^2 x)\sin x\,dx$

Let $u=\cos x$, $du=-\sin x\,dx$
The new limits are $\cos(-\pi/3)=1/2$ and $\cos 0=1$.

$\int_{1/2}^{1}\sqrt{u}(1-u^2)(-du)=\int_{1/2}^{1}(u^{5/2}-u^{1/2})du$

$= \frac{2}{7}u^{7/2}-\frac{2}{3}u^{3/2}\Big|_{1/2}^{1}$

$= \frac{25}{168}\sqrt{2}-\frac{8}{21}$

23. $\int \cot^2 x\csc^4 x\,dx=\int \cot^2 x\csc^2 x\csc^2 x\,dx$

$= \int \cot^2 x(1+\cot^2 x)\csc^2 x\,dx$

Let $u=\cot x$, $du=-\csc^2 x\,dx$

$\int u^2(1+u^2)(-du)=-\int(u^2+u^4)du$

$= -\frac{1}{3}u^3-\frac{1}{5}u^5+c$

$= -\frac{1}{3}\cot^3 x-\frac{1}{5}\cot^5 x+c$

25. $\int \frac{\sin x}{\tan x}dx=\int \cos x\,dx=\sin x+c$

27. $\int \frac{1}{x^2\sqrt{9-x^2}}dx$

Let $x=3\sin\theta$, $-\frac{\pi}{2}<\theta<\frac{\pi}{2}$

$dx=3\cos\theta\,d\theta$

$\int \frac{3\cos\theta}{9\sin^2\theta\sqrt{9-9\sin^2\theta}}d\theta$

$= \int \frac{\cos\theta}{(3\sin^2\theta)3\sqrt{1-\sin^2\theta}}d\theta$

$= \int \frac{\cos\theta}{9\sin^2\theta\cos\theta}d\theta$

$= \frac{1}{9}\int \csc^2\theta\,d\theta$

$= -\frac{1}{9}\cot\theta+c$

$\cot\theta=\frac{\cos\theta}{\sin\theta}=\frac{\sqrt{9-x^2}}{x}$

$\int \frac{1}{x^2\sqrt{9-x^2}}dx=-\frac{1}{9}\cot\theta+c$

$= -\frac{1}{9}\frac{\sqrt{9-x^2}}{x}+c$

29. $\int_0^1\sqrt{4-x^2}\,dx$ represents the area of one fourth
of a circle of radius 4, so it has the value
$\frac{1}{4}(\pi\cdot 2^2)=\pi$.

31. $\int \frac{x^2}{\sqrt{x^2-9}}dx$ For $\theta\in\left[0,\frac{\pi}{2}\right)\cup\left(\frac{\pi}{2},\pi\right]$,

let $x=3\sec\theta$, $dx=3\sec\theta\tan\theta$, so $x^2=9\sec^2\theta$.

$\int \frac{x^2}{\sqrt{x^2-9}}dx=\int \frac{(9\sec^2\theta)(3\sec\theta\tan\theta)}{\sqrt{9\sec^2\theta-9}}d\theta$

$= \int \frac{(9\sec^2\theta)(3\sec\theta\tan\theta)}{3\sqrt{\tan^2\theta}}d\theta$

$= 9\int \sec^3\theta\,d\theta$

Use the reduction formula in Exercise 43 and the result in Example 3.8 to get

$$9\int \sec^3\theta\,d\theta$$

$$= \frac{9}{2}\sec\theta\tan\theta + \frac{9}{2}\ln\left|\sec\theta + \tan\theta\right| + c.$$

$$\tan\theta = \pm\sqrt{\sec^2\theta - 1} = \pm\sqrt{\left(\frac{x}{3}\right)^2 - 1},$$

but for $\theta \in \left[0, \frac{\pi}{2}\right) \cup \left(\frac{\pi}{2}, \pi\right]$, tan and sec have the same sign, so our integral is equal to

$$\frac{9}{2}\cdot\frac{x}{3}\sqrt{\left(\frac{x}{3}\right)^2 - 1} + \frac{9}{2}\ln\left|\frac{x}{3} + \sqrt{\left(\frac{x}{3}\right)^2 - 1}\right| + c$$

$$= \frac{3x}{2}\sqrt{\left(\frac{x}{3}\right)^2 - 1} + \frac{9}{2}\ln\left|\frac{x}{3} + \sqrt{\left(\frac{x}{3}\right)^2 - 1}\right| + c$$

33. $\displaystyle\int\frac{2}{\sqrt{x^2 - 4}}\,dx = \int\frac{2}{2\sqrt{\left(\frac{x}{2}\right)^2 - 1}}\,dx$

$$= \int\frac{1}{\sqrt{\left(\frac{x}{2}\right)^2 - 1}}\,dx$$

Let $u = \frac{x}{2}$, $du = \frac{1}{2}dx$

$$2\int\frac{1}{\sqrt{u^2 - 1}}\,du = 2\cosh^{-1}u + c$$

$$= 2\cosh^{-1}\left(\frac{x}{2}\right) + c$$

35. $\displaystyle\int\frac{x^2}{\sqrt{x^2 + 9}}\,dx$

Let $x = 3\tan\theta$ for $-\frac{\pi}{2} < \theta < \frac{\pi}{2}$, $dx = 3\sec^2\theta\,d\theta$.

$$\int\frac{x^2}{\sqrt{x^2 + 9}}\,dx = \int\frac{9\tan^2\theta}{\sqrt{9\tan^2\theta + 9}}\,3\sec^2\theta\,d\theta$$

$$= \int\frac{27\tan^2\theta\sec^2\theta}{\sqrt{9\sec^2\theta}}\,d\theta$$

Since sec is positive for $\frac{\pi}{2} < \theta < \frac{\pi}{2}$,

$$\sqrt{\sec^2\theta} = \sec\theta, \text{ so the last integral is}$$

$$\int 9\tan^2\theta\sec\theta\,d\theta = 9\int(\sec^2\theta - 1)\sec\theta\,d\theta$$

$$= 9\int\sec^3\theta\,d\theta - 3\int\sec\theta\,d\theta$$

Applying Example3.8 and the reduction formula in Exercise 43 to the last expression yields

$$9\left(\frac{1}{2}\sec\theta\tan\theta + \frac{1}{2}\ln\left|\sec\theta + \tan\theta\right|\right)$$

$$-9\ln\left|\sec\theta + \tan\theta\right| + c$$

$$= \frac{9}{2}\sec\theta\tan\theta - \frac{9}{2}\ln\left|\sec\theta + \tan\theta\right| + c$$

Using $\tan\theta = \frac{x}{3}$ and $\sec\theta = \sqrt{\left(\frac{x}{3}\right)^2 + 1}$ we get

$$\frac{3x}{2}\sqrt{\left(\frac{x}{3}\right)^2 + 1} - \frac{9}{2}\ln\left(\frac{x}{3} + \sqrt{\left(\frac{x}{3}\right)^2 + 1}\right) + c.$$

Note that the absolute value is not needed in the log because the argument of ln is always positive.

37. $\displaystyle\int\sqrt{x^2 + 16}\,dx$

Let $x = 4\tan\theta$, $-\frac{\pi}{2} < \theta < \frac{\pi}{2}$

$$dx = 4\sec^2\theta\,d\theta$$

$$\int\sqrt{16\tan^2\theta + 16}\cdot 4\sec^2\theta\,d\theta$$

$$= 16\int\sqrt{\tan^2\theta + 1}\sec^2\theta\,d\theta$$

$$= 16\int\sec^3\theta\,d\theta$$

$$- 16\left(\frac{1}{2}\sec\theta\tan\theta + \frac{1}{2}\int\sec\theta\,d\theta\right)$$

(Using the reduction formula in Exercise 43)

$$= 8\sec\theta\tan\theta + 8\int\sec\theta\,d\theta$$

$$= 8\sec\theta\tan\theta + 8\ln\left|\sec\theta + \tan\theta\right| + c$$

(Example 3.8)

$$\tan\theta = \frac{x}{4}, \sec\theta = \sqrt{1 + \left(\frac{x}{4}\right)^2} = \frac{1}{4}\sqrt{16 + x^2}$$

$$\int\sqrt{x^2 + 16}\,dx$$

$$= 8\sec\theta\tan\theta + 8\ln\left|\sec\theta + \tan\theta\right| + c$$

$$= \frac{1}{2}x\sqrt{16 + x^2} + 8\ln\left|\frac{1}{4}\sqrt{16 + x^2} + \frac{x}{4}\right| + c$$

39. $\int_0^1 x\sqrt{x^2+8}\,dx$

Let $u = x^2 + 8$, $du = 2x\,dx$

The new limits are $0^2 + 8 = 8$ and $1^2 + 8 = 9$.

$\int_8^9 \sqrt{u} \cdot \frac{1}{2}\,du = \frac{1}{2}\int_8^9 \sqrt{u}\,dx$

$\quad = \frac{1}{2}\cdot\frac{2}{3}u^{3/2}\Big|_8^9$

$\quad = \frac{1}{3}u^{3/2}\Big|_8^9$

$\quad = 9 - \frac{16\sqrt{2}}{3}$

41. $\int \tan x \sec^4 x\,dx$

Let $u = \tan x$, $du = \sec^2 x\,dx$

$\int \tan x(1 + \tan^2 x)\sec^2 x\,dx$

$\quad = \int u(1 + u^2)\,du$

$\quad = \int (u + u^3)\,du$

$\quad = \frac{1}{2}u^2 + \frac{1}{4}u^4 + c$

$\quad = \frac{1}{2}\tan^2 x + \frac{1}{4}\tan^4 x + c$

Let $u = \sec x$, $du = \sec x \tan x\,dx$

$\int \tan x \sec x \sec^3 x\,dx = \int u^3\,du$

$\quad\quad = \frac{1}{4}u^4 + c = \frac{1}{4}\sec^4 x + c$

43. $\int \sec^n x\,dx \qquad u = \sec^{n-2} x \qquad\qquad dv = \sec^2 x\,dx$

$\qquad\qquad du = (n-2)\sec^{n-3} x \tan x \sec x\,dx \qquad v = \tan x$

$\qquad\qquad\quad = (n-2)\sec^{n-2} x \tan x\,dx$

$= \sec^{n-2} x \tan x - (n-2)\int \sec^{n-2} x \tan^2 x\,dx$

$= \sec^{n-2} x \tan x - (n-2)\int \sec^{n-2} x(\sec^2 x - 1)\,dx$

$\int \sec^n x\,dx = \sec^{n-2} x \tan x - (n-2)\int \sec^n x\,dx + (n-2)\int \sec^{n-2} x\,dx$

$(n-1)\int \sec^n x\,dx = \sec^{n-2} x \tan x + (n-2)\int \sec^{n-2} x\,dx$

$\int \sec^n x\,dx = \frac{1}{n-1}\sec^{n-2} x \tan x + \frac{n-2}{n-1}\int \sec^{n-2} x\,dx$

45. $\frac{1}{\frac{2\pi}{\omega}}\int_0^{2\pi/\omega} RI^2 \cos^2(\omega t)\,dt = \frac{\omega RI^2}{2\pi}\int_0^{2\pi/\omega} \frac{1}{2}[1 + \cos(2\omega t)]\,dt$

$\qquad\qquad = \frac{\omega RI^2}{4\pi}\left[t + \frac{1}{2\omega}\sin(2\omega t)\right]\Big|_0^{2\pi/\omega}$

$\qquad\qquad = \frac{\omega RI^2}{4\pi}\left[\frac{2\pi}{\omega} + \frac{1}{2\omega}\sin\left(\frac{4\omega\pi}{\omega}\right) - 0\right]$

$\qquad\qquad = \frac{1}{2}RI^2$

47. Answers will depend on the CAS used.

49. Write $\frac{1}{7}\cos^7 x$ as $\frac{1}{7}\cos^5 x(1 - \sin^2 x)$. Then our antiderivative is

$-\frac{1}{5}\cos^5 x + \frac{1}{7}\cos^5 x(1 - \sin^2 x) + c$

$\quad = \left(-\frac{1}{5} + \frac{1}{7}\right)\cos^5 x - \frac{1}{7}\cos^5 x \sin^2 x + c.$

This is equal to the CAS answer when $c = 0$.

Section 7.4

3. $\int \dfrac{x-5}{x^2-1} = \dfrac{x-5}{(x+1)(x-1)} = \dfrac{A}{x+1} + \dfrac{B}{x-1}$

$\qquad\qquad x-5 = A(x-1) + B(x+1)$

$x = -1 : -6 = -2A;\ A = 3$

$x = 1 : -4 = 2B;\ B = -2$

$\qquad \dfrac{x-5}{x^2-1} = \dfrac{3}{x+1} - \dfrac{2}{x-1}$

$\qquad \int \dfrac{x-5}{x^2-1}\,dx = \int\left(\dfrac{3}{x+1} - \dfrac{2}{x-1}\right)dx$

$\qquad\qquad\qquad = 3\ln|x+1| - 2\ln|x-1| + c$

5. $\dfrac{6x}{x^2-x-2} = \dfrac{6x}{(x-2)(x+1)} = \dfrac{A}{x-2} + \dfrac{B}{x+1}$

$\qquad\qquad 6x = A(x+1) + B(x-2)$

$x = 2 : 12 = 3A;\ A = 4$

$x = -1 : -6 = -3B;\ B = 2$

$\qquad \dfrac{6x}{x^2-x-2} = \dfrac{4}{x-2} + \dfrac{2}{x+1}$

$\qquad \int \dfrac{6x}{x^2-x-2}\,dx = \int\left(\dfrac{4}{x-2} + \dfrac{2}{x+1}\right)dx$

$\qquad\qquad\qquad = 4\ln|x-2| + 2\ln|x+1| + c$

9. $\dfrac{-x+5}{x^3-x^2-2x} = \dfrac{-x+5}{x(x^2-x-2)} = \dfrac{-x+5}{x(x-2)(x+1)} = \dfrac{A}{x} + \dfrac{B}{x-2} + \dfrac{C}{x+1}$

$\qquad\qquad -x+5 = A(x-2)(x+1) + Bx(x+1) + Cx(x-2)$

$x = 0 : 5 = -2A : A = -\dfrac{5}{2}$

$x = 2 : 3 = 6B : B = \dfrac{1}{2}$

$x = -1 : 6 = 3C : C = 2$

$\qquad \dfrac{-x+5}{x^3-x^2-2x} = -\dfrac{\frac{5}{2}}{x} + \dfrac{\frac{1}{2}}{x-2} + \dfrac{2}{x+1}$

$\qquad \int \dfrac{-x+5}{x^3-x^2-2x}\,dx = \int\left(-\dfrac{\frac{5}{2}}{x} + \dfrac{\frac{1}{2}}{x-2} + \dfrac{2}{x+1}\right)dx$

$\qquad\qquad\qquad = -\dfrac{5}{2}\ln|x| + \dfrac{1}{2}\ln|x-2| + 2\ln|x+1| + c$

7. $\dfrac{x+1}{x^2-x-6} = \dfrac{x+1}{(x-3)(x+2)} = \dfrac{A}{x-3} + \dfrac{B}{x+2}$

$\qquad\qquad x+1 = A(x+2) + B(x-3)$

$x = 3 : 4 = 5A;\ A = \dfrac{4}{5}$

$x = -2 : -1 = -5B;\ B = \dfrac{1}{5}$

$\qquad \dfrac{x+1}{x^2-x-6} = \dfrac{\frac{4}{5}}{x-3} + \dfrac{\frac{1}{5}}{x+2}$

$\qquad \int \dfrac{x+1}{x^2-x-6}\,dx = \int\left(\dfrac{\frac{4}{5}}{x-3} + \dfrac{\frac{1}{5}}{x+2}\right)dx$

$\qquad\qquad\qquad = \dfrac{4}{5}\ln|x-3| + \dfrac{1}{5}\ln|x+2| + c$

11.

$$x^2+2x-8 \overline{\smash{\big)}\ x^3+x+2}$$
$$\underline{-(x^3+2x^2-8x)}$$
$$-2x^2+9x+2$$
$$\underline{-(-2x^2-4x+16)}$$
$$13x-14$$

$$\frac{x^3+x+2}{x^2+2x-8}=x-2+\frac{13x-14}{x^2+2x-8}=x-2+\frac{13x-14}{(x+4)(x-2)}=x-2+\frac{A}{x+4}+\frac{B}{x-2}$$
$$13x-14=A(x-2)+B(x+4)$$
$$x=-4:-66=-6A;\ A=11$$
$$x=2:12=6B;\ B=2$$

$$\frac{x^3+x+2}{x^2+2x-8}=x-2+\frac{11}{x+4}+\frac{2}{x-2}$$
$$\int\frac{x^3+x+2}{x^2+2x-8}dx=\int\left(x-2+\frac{11}{x+4}+\frac{2}{x-2}\right)dx$$
$$=\frac{1}{2}x^2-2x+11\ln|x+4|+2\ln|x-2|+c$$

13. $\dfrac{-3x-1}{x^3-x}=\dfrac{-3x-1}{x(x^2-1)}=\dfrac{-3x-1}{x(x+1)(x-1)}=\dfrac{A}{x}+\dfrac{B}{x+1}+\dfrac{C}{x-1}$

$$-3x-1=A(x+1)(x-1)+Bx(x-1)+Cx(x+1)$$
$$x=0:-1=-A;\ A=1$$
$$x=-1:2=2B;\ B=1$$
$$x=1:-4=2C;\ C=-2$$

$$\frac{-3x-1}{x^3-x}=\frac{1}{x}+\frac{1}{x+1}-\frac{2}{x-1}$$
$$\int\frac{-3x-1}{x^3-x}dx=\int\left(\frac{1}{x}+\frac{1}{x+1}-\frac{2}{x-1}\right)dx$$
$$=\ln|x|+\ln|x+1|-2\ln|x-1|+c$$

15. $\dfrac{2x+3}{(x+2)^2}=\dfrac{A}{x+2}+\dfrac{B}{(x+2)^2}$

$$2x+3=A(x+2)+B=Ax+2A+B$$
$$A=2;B=-1$$

$$\frac{2x+3}{(x+2)^2}=\frac{2}{x+2}-\frac{1}{(x+2)^2}$$
$$\int\frac{2x+3}{(x+2)^2}dx=\int\left(\frac{2}{x+2}-\frac{1}{(x+2)^2}\right)dx$$
$$=2\ln|x+2|+\frac{1}{x+2}+c$$

17. $\dfrac{x-1}{x^3+4x^2+4x} = \dfrac{x-1}{x(x^2+4x+4)} = \dfrac{x-1}{x(x+2)^2} = \dfrac{A}{x} + \dfrac{B}{x+2} + \dfrac{C}{(x+2)^2}$

$\qquad x-1 = A(x+2)^2 + Bx(x+2) + Cx$

$x = 0: -1 = 4A;\ A = -\dfrac{1}{4}$

$x = -2: -3 = -2C;\ C = \dfrac{3}{2}$

$x = 1: 0 = 9A + 3B + C = -\dfrac{9}{4} + 3B + \dfrac{3}{2};\ B = \dfrac{1}{4}$

$\dfrac{x-1}{x^3+4x^2+4x} = -\dfrac{\frac{1}{4}}{x} + \dfrac{\frac{1}{4}}{x+2} + \dfrac{\frac{3}{2}}{(x+2)^2}$

$\displaystyle\int \dfrac{x-1}{x^3+4x^2+4x}\,dx = \int\left(-\dfrac{\frac{1}{4}}{x} + \dfrac{\frac{1}{4}}{x+2} + \dfrac{\frac{3}{2}}{(x+2)^2}\right)dx$

$\qquad\qquad = -\dfrac{1}{4}\ln|x| + \dfrac{1}{4}\ln|x+2| - \dfrac{3}{2(x+2)} + c$

19. $\dfrac{x+4}{x^3+3x^2+2x} = \dfrac{x+4}{x(x^2+3x+2)} = \dfrac{x+4}{x(x+2)(x+1)} = \dfrac{A}{x} + \dfrac{B}{x+2} + \dfrac{C}{x+1}$

$\qquad x+4 = A(x+2)(x+1) + Bx(x+1) + Cx(x+2)$

$x = 0: 4 = 2A;\ A = 2$

$x = -2: 2 = 2B;\ B = 1$

$x = -1: 3 = -C;\ C = -3$

$\dfrac{x+4}{x^3+3x^2+2x} = \dfrac{2}{x} + \dfrac{1}{x+2} - \dfrac{3}{x+1}$

$\displaystyle\int \dfrac{x+4}{x^3+3x^2+2x}\,dx = \int\left(\dfrac{2}{x} + \dfrac{1}{x+2} - \dfrac{3}{x+1}\right)dx$

$\qquad\qquad = 2\ln|x| + \ln|x+2| - 3\ln|x+1| + c$

21. $\dfrac{x+2}{x^3+x} = \dfrac{x+2}{x(x^2+1)} = \dfrac{A}{x} + \dfrac{Bx+C}{x^2+1}$

$\qquad x+2 = A(x^2+1) + (Bx+C)x = Ax^2 + A + Bx^2 + Cx = (A+B)x^2 + Cx + A$

$\qquad A = 2;\ C = 1;\ B = -2$

$\qquad\qquad\qquad \dfrac{x+2}{x^3+x} = \dfrac{2}{x} + \dfrac{-2x+1}{x^2+1}$

$\displaystyle\int \dfrac{x+2}{x^3+x}\,dx = \int\left(\dfrac{2}{x} + \dfrac{-2x+1}{x^2+1}\right)dx = \int\left(\dfrac{2}{x} - \dfrac{2x}{x^2+1} + \dfrac{1}{x^2+1}\right)dx$

$\qquad\qquad = 2\ln|x| - \ln(x^2+1) + \tan^{-1}x + c$

23. $\dfrac{4x-2}{x^4-1} = \dfrac{4x-2}{(x^2+1)(x^2-1)} = \dfrac{4x-2}{(x^2+1)(x+1)(x-1)} = \dfrac{Ax+B}{x^2+1} + \dfrac{C}{x+1} + \dfrac{D}{x-1}$

$4x-2 = (Ax+B)(x+1)(x-1) + C(x^2+1)(x-1) + D(x^2+1)(x+1)$

$4x-2 = Ax^3 - Ax + Bx^2 - B + Cx^3 - Cx^2 + Cx - C + Dx^3 + Dx^2 + Dx + D$

$x=1: 2 = 4D; \; D = \dfrac{1}{2}$

$x=-1: -6 = -4C; \; C = \dfrac{3}{2}$

$A+C+D = 0: A = -2$

$B-C+D = 0: B = 1$

$\dfrac{4x-2}{x^4-1} = \dfrac{-2x+1}{x^2+1} + \dfrac{\frac{3}{2}}{x+1} + \dfrac{\frac{1}{2}}{x-1}$

$\displaystyle \int \dfrac{4x-2}{x^4-1}\,dx = \int \left(\dfrac{-2x+1}{x^2+1} + \dfrac{\frac{3}{2}}{x+1} + \dfrac{\frac{1}{2}}{x-1} \right) dx$

$\displaystyle = \int \left(-\dfrac{2x}{x^2+1} + \dfrac{1}{x^2+1} + \dfrac{\frac{3}{2}}{x+1} + \dfrac{\frac{1}{2}}{x-1} \right) dx$

$= -\ln(x^2+1) + \tan^{-1}x + \dfrac{3}{2}\ln|x+1| + \dfrac{1}{2}\ln|x-1| + c$

25. $\dfrac{3x^2-6}{x^2-x-2}$

$\begin{array}{r} 3 \\ x^2-x-2 \overline{\smash{\big)}\, 3x^2 -6} \\ \underline{-(-3x^2-3x-6)} \\ 3x \end{array}$

$\dfrac{3x^2-6}{x^2-x-2} = 3 + \dfrac{3x}{x^2-x-2}$

$= 3 + \dfrac{3x}{(x-2)(x+1)}$

$= 3 + \dfrac{A}{x-2} + \dfrac{B}{x+1}$

$3x = A(x+1) + B(x-2)$

$x=-1: -3 = -3B; \; B = 1$

$x=2: 6 = 3A; \; A = 2$

$\dfrac{3x^2-6}{x^2-x-2} = 3 + \dfrac{2}{x-2} + \dfrac{1}{x+1}$

$\displaystyle \int \dfrac{3x^2-6}{x^2-x-2}\,dx = \int \left(3 + \dfrac{2}{x-2} + \dfrac{1}{x+1} \right) dx$

$= 3x + 2\ln|x-2| + \ln|x+1| + c$

27. $\dfrac{2x+3}{x^2+2x+1} = \dfrac{2x+3}{(x+1)^2} = \dfrac{A}{x+1} + \dfrac{B}{(x+1)^2}$

$2x+3 = A(x+1) + B$

$x=-1: B=1; \; A=2$

$\dfrac{2x+3}{x^2+2x+1} = \dfrac{2}{x+1} + \dfrac{1}{(x+1)^2}$

$\displaystyle \int \dfrac{2x+3}{x^2+2x+1}\,dx = \int \left(\dfrac{2}{x+1} + \dfrac{1}{(x+1)^2} \right) dx$

$= 2\ln|x+1| - \dfrac{1}{x+1} + c$

29. $\dfrac{x^2+2x+1}{x^3+x} = \dfrac{x^2+2x+1}{x(x^2+1)} = \dfrac{A}{x} + \dfrac{Bx+C}{x^2+1}$

$x^2+2x+1 = A(x^2+1) + (Bx+C)x$

$x^2+2x+1 = Ax^2 + A + Bx^2 + Cx$

$A=1; \; C=2; \; B=0$

$\displaystyle \int \dfrac{x^2+2x+1}{x^3+x} = \dfrac{1}{x} + \dfrac{2}{x^2+1}$

$\displaystyle \int \dfrac{x^2+2x+1}{x^3+x}\,dx = \int \left(\dfrac{1}{x} + \dfrac{2}{x^2+1} \right) dx$

$= \ln|x| + 2\tan^{-1}x + c$

31. $\dfrac{4x^2+3}{x^3+x^2+x} = \dfrac{4x^2+3}{x(x^2+x+1)} = \dfrac{A}{x} + \dfrac{Bx+C}{x^2+x+1}$

$\quad\quad 4x^2+3 = A(x^2+x+1) + (Bx+C)x = Ax^2 + Ax + A + Bx^2 + Cx$

$A = 3; \; C = -3; \; B = 1$

$$\dfrac{4x^2+3}{x^3+x^2+x} = \dfrac{3}{x} + \dfrac{x-3}{x^2+x+1}$$

$$\int \dfrac{4x^2+3}{x^3+x^2+x}\,dx = \int \left(\dfrac{3}{x} + \dfrac{x-3}{x^2+x+1} \right) dx$$

$$= \int \left(\dfrac{3}{x} + \dfrac{x+\frac{1}{2}}{x^2+x+1} - \dfrac{\frac{7}{2}}{x^2+x+1} \right) dx$$

$$= 3\ln|x| + \dfrac{1}{2}\ln\left|x^2+x+1\right| - \dfrac{7}{\sqrt{3}}\tan^{-1}\left(\dfrac{2x+1}{\sqrt{3}}\right) + c$$

33.
$$\begin{array}{r} 3 \\ x^3 - x^2 + x - 1 \overline{\smash{\big)}\, 3x^3 \quad\quad\quad +1 } \\ \underline{-\,(3x^3 - 3x^2 + 3x - 3)} \\ 3x^2 - 3x + 4 \end{array}$$

$$\dfrac{3x^3+1}{x^3-x^2+x-1} = 3 + \dfrac{3x^2-3x+4}{x^3-x^2+x-1}$$

$$= 3 + \dfrac{3x^2-3x+4}{x^2(x-1)+x-1}$$

$$- 3 + \dfrac{3x^2-3x+4}{(x^2+1)(x-1)}$$

$$= 3 + \dfrac{Ax+B}{x^2+1} + \dfrac{C}{x-1}$$

$3x^2 - 3x + 4 = (Ax+B)(x-1) + C(x^2+1) = Ax^2 - Ax + Bx - B + Cx^2 + C$

$x = 1 : 4 = 2C; \; C = 2$

$A + C = 3 : A = 1$

$-A + B = -3 : B = -2$

$$\dfrac{3x^3+1}{x^3-x^2+x-1} = 3 + \dfrac{x-2}{x^2+1} + \dfrac{2}{x-1}$$

$$\int \dfrac{3x^3+1}{x^3-x^2+x-1}\,dx = \int \left(3 + \dfrac{x-2}{x^2+1} + \dfrac{2}{x-1} \right) dx$$

$$= \int \left(3 + \dfrac{x}{x^2+1} - \dfrac{2}{x^2+1} + \dfrac{2}{x-1} \right) dx$$

$$= 3x + \dfrac{1}{2}\ln(x^2+1) - 2\tan^{-1}x + 2\ln|x-1| + c$$

35. $A + C = 0 : C = 0$

$B + D = 2 : D = -2$

$$\dfrac{4x^2+2}{(x^2+1)^2} = \dfrac{4}{x^2+1} - \dfrac{2}{(x^2+1)^2}$$

37. $\dfrac{2x^2+4}{(x^2+4)^2} = \dfrac{Ax+B}{x^2+4} + \dfrac{Cx+D}{(x^2+4)^2}$

$\quad\quad 2x^2+4 = (Ax+B)(x^2+4)+Cx+D$

$\quad\quad\quad\quad = Ax^3+4Ax+Bx^2+4B+Cx+D$

$A=0; \; B=2$

$4A+C=0 : C=0$

$4B+D=4 : D=-4$

$\dfrac{2x^2+4}{(x^2+4)^2} = \dfrac{2}{x^2+4} - \dfrac{4}{(x^2+4)^2}$

39. $\dfrac{4x^2+3}{(x^2+x+1)^2} = \dfrac{Ax+B}{x^2+x+1} + \dfrac{Cx+D}{(x^2+x+1)^2}$

$\quad\quad 4x^2+3 = (Ax+B)(x^2+x+1)+Cx+D$

$\quad\quad\quad\quad = Ax^3+Ax^2+Ax+Bx^2+Bx+B+Cx+D$

$A=0$

$A+B=4 : B=4$

$A+B+C=0 : C=-4$

$B+D=3 : D=-1$

$\dfrac{4x^2+3}{(x^2+x+1)^2} = \dfrac{4}{x^2+x+1} - \dfrac{4x+1}{(x^2+x+1)^2}$

41. $\dfrac{1}{y(1-y)} = \dfrac{A}{y} + \dfrac{B}{1-y}$

$1 = A(1-y)+By$

$A=1; \; B=1$

$\dfrac{1}{y(1-y)} = \dfrac{1}{y} + \dfrac{1}{1-y}$

$\displaystyle\int \dfrac{1}{y(1-y)}\,dy = \int\left(\dfrac{1}{y}+\dfrac{1}{1-y}\right)dy$

$\quad\quad\quad\quad = \ln|y| - \ln|1-y| + c$

$\quad\quad\quad\quad = \ln\dfrac{y}{1-y} + c$

$\quad at-c = \ln\dfrac{y}{1-y}$

$\quad e^{at-c} = \dfrac{y}{1-y}$

$\quad e^{c-at} = \dfrac{1-y}{y} = \dfrac{1}{y} - 1$

$\quad 1+e^{c-at} = \dfrac{1}{y}$

$\quad \dfrac{1}{1+e^{c-at}} = y$

$\quad\quad y = \dfrac{ce^{at}}{1+ce^{at}}$

Section 7.5

3. $\displaystyle\int \frac{x}{(2+4x)^2}\,dx$

$$\int \frac{u\,du}{(a+bu)^2} = \frac{a}{b^2(a+bu)} + \frac{1}{b^2}\ln|a+bu| + c$$

$$\int \frac{x}{(2+4x)^2}\,dx = \frac{2}{16(2+4x)} + \frac{1}{16}\ln|2+4x| + c = \frac{1}{8(2+4x)} + \frac{1}{16}\ln|2+4x| + c$$

5. $\displaystyle\int e^{2x}\sqrt{1+e^x}\,dx$

Let $u = 1+e^x$, $du = e^x dx$, $e^x = u-1$

$$\int (u-1)\sqrt{u}\,du = \int (u^{3/2} - u^{1/2})\,du$$

$$= \frac{2}{5}u^{5/2} - \frac{2}{3}u^{3/2} + c$$

$$= \frac{2}{5}(1+e^x)^{5/2} - \frac{2}{3}(1+e^x)^{3/2} + c$$

7. $\displaystyle\int \frac{x^2}{\sqrt{1+4x^2}}\,dx$

Let $u = 2x$, $du = 2\,dx$, $x = \dfrac{u}{2}$

$$\int \frac{\left(\frac{u}{2}\right)^2}{\sqrt{1+u^2}} \cdot \frac{1}{2}\,du = \frac{1}{8}\int \frac{u^2}{\sqrt{1+u^2}}\,du$$

From the table,

$$\int \frac{u^2\,du}{\sqrt{a^2+u^2}} = \frac{u}{2}\sqrt{a^2+u^2} - \frac{a^2}{2}\ln\left(u+\sqrt{a^2+u^2}\right) + c$$

$$\frac{1}{8}\int \frac{u^2}{\sqrt{1+u^2}}\,du = \frac{1}{8}\left[\frac{u}{2}\sqrt{1+u^2} - \frac{1}{2}\ln\left(u+\sqrt{1+u^2}\right)\right] + c$$

$$= \frac{1}{8}x\sqrt{1+4x^2} - \frac{1}{16}\ln\left(2x+\sqrt{1+4x^2}\right) + c$$

9. $\int x^8 \sqrt{4 - x^6}\, dx$

Let $u = x^3$, $du = 3x^2 dx$

$\int u^2 \sqrt{4 - u^2} \cdot \frac{1}{3} du = \frac{1}{3} \int u^2 \sqrt{4 - u^2}\, du$

From the table,

$\int u^2 \sqrt{a^2 - u^2}\, du = \frac{u}{8}(2u^2 - a^2)\sqrt{a^2 - u^2} + \frac{a^4}{8}\sin^{-1}\frac{u}{a} + c$

$\frac{1}{3}\int u^2 \sqrt{4 - u^2}\, du = \frac{1}{3}\left[\frac{u}{8}(2u^2 - 4)\sqrt{4 - u^2} + \frac{16}{8}\sin^{-1}\frac{u}{2}\right] + c$

$= \frac{1}{24} x^3 (2x^6 - 4)\sqrt{4 - x^6} + \frac{2}{3}\sin^{-1}\frac{x^3}{2} + c$

Using a CAS to evaluate the definite integral we get

$\int_0^1 x^8 \sqrt{4 - x^6}\, dx = \frac{\pi}{9} - \frac{\sqrt{3}}{12}$.

11. $\int \frac{e^x}{\sqrt{e^{2x} + 4}}\, dx$

Let $u = e^x$, $du = e^x dx$

$\int \frac{1}{\sqrt{u^2 + 4}}\, du$

From the table,

$\int \frac{du}{\sqrt{a^2 + u^2}} = \ln\left(u + \sqrt{a^2 + u^2}\right) + c$

$\int \frac{1}{\sqrt{u^2 + 4}}\, du = \ln\left(u + \sqrt{4 + u^2}\right) + c$

$= \ln\left(e^x + \sqrt{4 + e^{2x}}\right) + c$

Using a CAS to evaluate the definite integral we get

$\int_0^{\ln 2} \frac{e^x}{\sqrt{e^{2x} + 4}}\, dx = \ln\left(\frac{2\sqrt{2} + 2}{1 + \sqrt{5}}\right)$.

13. $\displaystyle\int \frac{\sqrt{6x-x^2}}{(x-3)^2}\,dx = \int \frac{\sqrt{x(6-x)}}{(x-3)^2}\,dx$

Let $u = x - 3$, $du = dx$, $x = u + 3$

$$\int \frac{\sqrt{(u+3)(6-(u+3))}}{u^2}\,du = \int \frac{\sqrt{(u+3)(3-u)}}{u^2}\,du$$

$$= \int \frac{\sqrt{9-u^2}}{u^2}\,du$$

From the table,

$$\int \frac{\sqrt{a^2-u^2}}{u^2}\,du = -\frac{1}{u}\sqrt{a^2-u^2} - \sin^{-1}\frac{u}{a} + c$$

$$\int \frac{\sqrt{9-u^2}}{u^2}\,du = -\frac{1}{u}\sqrt{9-u^2} - \sin^{-1}\frac{u}{3} + c$$

$$= -\frac{1}{x-3}\sqrt{9-(x-3)^2} - \sin^{-1}\left(\frac{x-3}{3}\right) + c$$

15. $\displaystyle\int \tan^6 x\,dx$

$$\int \tan^n u\,du = \frac{1}{n-1}\tan^{n-1}u - \int \tan^{n-2}u\,du$$

$$\int \tan^6 x\,dx = \frac{1}{5}\tan^5 x - \int \tan^4 x\,dx$$

$$= \frac{1}{5}\tan^5 x - \left[\frac{1}{3}\tan^3 x - \int \tan^2 x\,dx\right]$$

$$= \frac{1}{5}\tan^5 x - \frac{1}{3}\tan^3 x + \tan x - x + c$$

$$\left(\text{Using the formula for } \int \tan^2 x\,dx.\right)$$

17. $\displaystyle\int \frac{\cos x}{\sin x\sqrt{4+\sin x}}\,dx$

Let $u = \sin x$, $du = \cos x\,dx$

$$\int \frac{1}{u\sqrt{4+u}}\,du$$

From the table,

$$\int \frac{du}{u\sqrt{a+bu}} = \frac{1}{\sqrt{a}}\ln\left|\frac{\sqrt{a+bu}-\sqrt{a}}{\sqrt{a+bu}+\sqrt{a}}\right| + c$$

$$\int \frac{1}{u\sqrt{4+u}}\,du = \frac{1}{\sqrt{4}}\ln\left|\frac{\sqrt{4+u}-2}{\sqrt{4+u}+2}\right| + c$$

$$= \frac{1}{2}\ln\left|\frac{\sqrt{4+\sin x}-2}{\sqrt{4+\sin x}+2}\right| + c$$

19. $\int x^3 \cos x^2 \, dx$

Let $u = x^2$, $du = 2x \, dx$

$\int u \cos u \cdot \dfrac{1}{2} du = \dfrac{1}{2} \int u \cos u \, du$

From the table,

$\int u \cos u \, du = \cos u + u \sin u + c$

$\dfrac{1}{2} \int u \cos u \, du = \dfrac{1}{2}(\cos u + u \sin u) + c$

$\qquad\qquad = \dfrac{1}{2}\cos x^2 + \dfrac{1}{2}x^2 \sin x^2 + c$

21. $\int \dfrac{\sin x \cos x}{\sqrt{1 + \cos x}} \, dx$

Let $u = \cos x$, $du = -\sin x \, dx$

$-\int \dfrac{u}{\sqrt{1 + u}} \, du$

From the table,

$\int \dfrac{u \, du}{\sqrt{a + bu}} = \dfrac{2}{3b^2}(bu - 2a)\sqrt{a + bu} + c$

$-\int \dfrac{u}{\sqrt{1 + u}} \, du = -\dfrac{2}{3}(u - 2)\sqrt{1 + u} + c$

$\qquad\qquad = -\dfrac{2}{3}(\cos x - 2)\sqrt{1 + \cos x} + c$

23. $\int \dfrac{\sin^2 x \cos x}{\sqrt{\sin^2 x + 4}} \, dx$

Let $u = \sin x$, $du = \cos x \, dx$

$\int \dfrac{u^2}{\sqrt{u^2 + 4}} \, du$

From the table,

$\int \dfrac{u^2 \, du}{\sqrt{a^2 + u^2}} = \dfrac{u}{2}\sqrt{a^2 + u^2} - \dfrac{a^2}{2}\ln(u + \sqrt{a^2 + u^2}) + c$

$\int \dfrac{u^2}{\sqrt{u^2 + 4}} \, du = \dfrac{u}{2}\sqrt{4 + u^2} - \dfrac{4}{2}\ln(u + \sqrt{4 + u^2}) + c$

$\qquad\qquad = \dfrac{1}{2}\sin x \sqrt{4 + \sin^2 x} - 2\ln(\sin x + \sqrt{4 + \sin^2 x}) + c$

25. $\int \dfrac{e^{-2/x^2}}{x^3}\,dx$

Let $u = -\dfrac{2}{x^2}$, $du = \dfrac{4}{x^3}\,dx$

$\dfrac{1}{4}\int e^u\,du = \dfrac{1}{4}e^u + c$

$= \dfrac{1}{4}e^{-2/x^2} + c$

27. $\int \dfrac{x}{\sqrt{4x-x^2}}\,dx$

From the table,

$\int \dfrac{u\,du}{\sqrt{2au-u^2}} = -\sqrt{2au-u^2} + a\cos^{-1}\left(\dfrac{a-u}{a}\right) + c$

$\int \dfrac{x}{\sqrt{4x-x^2}}\,dx = -\sqrt{4x-x^2} + 2\cos^{-1}\left(\dfrac{2-x}{2}\right) + c$

29. $\int e^x \tan^{-1}(e^x)\,dx$

Let $u = e^x$, $du = e^x\,dx$

$\int \tan^{-1} u\,du = u\tan^{-1} u - \dfrac{1}{2}\ln(1+u^2) + c$

$= e^x \tan^{-1} e^x - \dfrac{1}{2}\ln(1+e^{2x}) + c$

31. Answer depends on CAS used.

33. Any answer is wrong because the integrand is undefined for all $x \neq 1$.

35. Answer depends on CAS used.

37. Answer depends on CAS used.

Section 7.6

3. $\displaystyle\lim_{x\to-2}\dfrac{x+2}{x^2-4} = \lim_{x\to-2}\dfrac{x+2}{(x+2)(x-2)}$

$= \lim_{x\to-2}\dfrac{1}{x-2}$

$= \dfrac{1}{-4} = -\dfrac{1}{4}$

5. $\displaystyle\lim_{x\to 2}\dfrac{x+1}{x^2+4x+3} = \dfrac{-2+1}{4-8+3} = \dfrac{-1}{-1} = 1$

7. $\displaystyle\lim_{x\to\infty}\dfrac{3x^2+2}{x^2-4} = \lim_{x\to\infty}\dfrac{x^2\left(3+\frac{2}{x^2}\right)}{x^2\left(1-\frac{4}{x^2}\right)}$

$= \lim_{x\to\infty}\dfrac{3+\frac{2}{x^2}}{1-\frac{4}{x^2}} = 3$

9. $\displaystyle\lim_{x\to-\infty}\dfrac{x+1}{x^2+4x+3} = \lim_{x\to-\infty}\dfrac{x\left(1+\frac{1}{x}\right)}{x^2\left(1+\frac{4}{x}+\frac{3}{x^2}\right)}$

$= \lim_{x\to-\infty}\dfrac{1+\frac{1}{x}}{x\left(1+\frac{4}{x}+\frac{3}{x^2}\right)}$

$= 0$

11. $\displaystyle\lim_{x\to 0}\dfrac{\sin x}{x}\left(\dfrac{0}{0}\right) = \lim_{x\to 0}\dfrac{\cos x}{1}$

$= \dfrac{1}{1} = 1$

13. $\displaystyle\lim_{x\to 0}\dfrac{\sin x - x}{x^3}\left(\dfrac{0}{0}\right) = \lim_{x\to 0}\dfrac{\cos x - 1}{3x^2}\left(\dfrac{0}{0}\right)$

$= \lim_{x\to 0}\dfrac{-\sin x}{6x}\left(\dfrac{0}{0}\right)$

$= \lim_{x\to 0}\dfrac{-\cos x}{6}$

$= \dfrac{-1}{6} = -\dfrac{1}{6}$

15. $\displaystyle\lim_{x\to 1}\dfrac{\sqrt{x}-1}{x-1}\left(\dfrac{0}{0}\right) = \lim_{x\to 1}\dfrac{\frac{1}{2}x^{-1/2}}{1}$

$= \dfrac{1}{2}$

17. $\displaystyle\lim_{x\to\infty}\dfrac{x^3}{e^x}\left(\dfrac{\infty}{\infty}\right) = \lim_{x\to\infty}\dfrac{3x^2}{e^x}\left(\dfrac{\infty}{\infty}\right)$

$= \lim_{x\to\infty}\dfrac{6x}{e^x}\left(\dfrac{\infty}{\infty}\right)$

$= \lim_{x\to\infty}\dfrac{6}{e^x}$

$= 0$

19. $\displaystyle\lim_{x\to 0}\dfrac{e^x-1}{x}\left(\dfrac{0}{0}\right) = \lim_{x\to 0}\dfrac{e^x}{1}$

$= \dfrac{1}{1}$

$= 1$

21. $\lim\limits_{x\to 1} \dfrac{\sin \pi x}{x-1}\left(\dfrac{0}{0}\right) = \lim\limits_{x\to 1}\dfrac{\pi\cos \pi x}{1}$

$\qquad\qquad = \dfrac{-\pi}{1}$

$\qquad\qquad = -\pi$

23. $\lim\limits_{x\to\infty}\dfrac{\ln x}{x^2}\left(\dfrac{\infty}{\infty}\right) = \lim\limits_{x\to\infty}\dfrac{\frac{1}{x}}{2x}$

$\qquad\qquad = 0$

27. $\lim\limits_{x\to 0}\dfrac{x\sin x}{\cos x - 1}\left(\dfrac{0}{0}\right)$

$\qquad = \lim\limits_{x\to 0}\dfrac{x\cos x + \sin x}{-\sin x}\left(\dfrac{0}{0}\right)$

$\qquad = \lim\limits_{x\to 0}\dfrac{-x\sin x + \cos x + \cos x}{-\cos x}$

$\qquad = \dfrac{0+1+1}{-1}$

$\qquad = -2$

29. $\lim\limits_{x\to 0^+} x\ln x(0\cdot(-\infty)) = \lim\limits_{x\to 0^+}\dfrac{\ln x}{\frac{1}{x}}\left(-\dfrac{\infty}{\infty}\right)$

$\qquad\qquad\qquad = \lim\limits_{x\to 0^+}\dfrac{\frac{1}{x}}{-\frac{1}{x^2}}$

$\qquad\qquad\qquad = \lim\limits_{x\to 0^+}(-x)$

$\qquad\qquad\qquad = 0$

31. $\lim\limits_{x\to 0^+}\dfrac{\ln x}{\cot x}\left(\dfrac{-\infty}{\infty}\right) = \lim\limits_{x\to 0^+}\dfrac{\frac{1}{x}}{-\csc^2 x}$

$\qquad\qquad\qquad = \lim\limits_{x\to 0^+}\dfrac{-\sin^2 x}{x}\left(\dfrac{0}{0}\right)$

$\qquad\qquad\qquad = \lim\limits_{x\to 0^+}\dfrac{-2\sin x\cos x}{1}$

$\qquad\qquad\qquad = 0$

33. $\lim\limits_{x\to\infty}(\sqrt{x^2+1} - x)$

$\qquad = \lim\limits_{x\to\infty}\left[(\sqrt{x^2+1} - x)\cdot\dfrac{\sqrt{x^2+1}+x}{\sqrt{x^2+1}+x}\right]$

$\qquad = \lim\limits_{x\to\infty}\left(\dfrac{x^2+1-x^2}{\sqrt{x^2+1}+x}\right)$

$\qquad = \lim\limits_{x\to\infty}\dfrac{1}{\sqrt{x^2+1}+x}$

$\qquad = 0$

25. $\lim\limits_{x\to\infty} xe^{-x} = \lim\limits_{x\to\infty}\dfrac{x}{e^x}\left(\dfrac{\infty}{\infty}\right)$

$\qquad\qquad = \lim\limits_{x\to\infty}\dfrac{1}{e^x}$

$\qquad\qquad = 0$

35. $\lim\limits_{x\to\infty}\left(1+\dfrac{1}{x}\right)^x (1^\infty)$

$y = \left(1+\dfrac{1}{x}\right)^x$

$\ln y = x\ln\left(1+\dfrac{1}{x}\right)$

$\lim\limits_{x\to\infty}\ln y = \lim\limits_{x\to\infty} x\ln\left(1+\dfrac{1}{x}\right)(\infty\cdot 0)$

$\qquad = \lim\limits_{x\to\infty}\dfrac{\ln\left(1+\frac{1}{x}\right)}{\frac{1}{x}}\left(\dfrac{0}{0}\right)$

$\qquad = \lim\limits_{x\to\infty}\dfrac{\frac{1}{1+\frac{1}{x}}\left(-\frac{1}{x^2}\right)}{-\frac{1}{x^2}}$

$\qquad = \lim\limits_{x\to\infty}\dfrac{1}{1+\frac{1}{x}}$

$\qquad = 1$

$\lim\limits_{x\to\infty} y = \lim\limits_{x\to\infty} e^{\ln y} = e$

37. $\lim\limits_{x\to 0^+}\left(\ln x + \dfrac{1}{x}\right)(-\infty + \infty)$

$y = \ln x + \dfrac{1}{x}$

$e^y = e^{\ln x + 1/x} = e^{\ln x}e^{1/x} = xe^{1/x}$

$\lim\limits_{x\to 0^+} e^y = \lim\limits_{x\to 0^+} xe^{1/x} = \lim\limits_{x\to 0^+}\dfrac{e^{1/x}}{\frac{1}{x}}\left(\dfrac{\infty}{\infty}\right)$

$\qquad = \lim\limits_{x\to 0^+}\dfrac{e^{1/x}\left(-\frac{1}{x^2}\right)}{-\frac{1}{x^2}}$

$\qquad = \lim\limits_{x\to 0^+} e^{1/x}$

$\qquad = \infty$

$\lim\limits_{x\to 0^+} y = \lim\limits_{x\to 0^+}(\ln e^y) = \infty$

39. $\lim_{x\to0^+}\left(\frac{1}{x}\right)^x (\infty^0)$

$y = \left(\frac{1}{x}\right)^x;\ \ln y = x\ln\left(\frac{1}{x}\right)$

$\lim_{x\to0^+}\ln y = \lim_{x\to0^+}\left[x\ln\left(\frac{1}{x}\right)\right]$

$\quad = \lim_{x\to0^+}\frac{\ln\left(\frac{1}{x}\right)}{\frac{1}{x}}\left(-\frac{\infty}{\infty}\right)$

$\quad = \lim_{x\to0^+}\frac{\frac{1}{\left(\frac{1}{x}\right)}\cdot\left(-\frac{1}{x^2}\right)}{-\frac{1}{x^2}}$

$\quad = \lim_{x\to0^+} x$

$\quad = 0$

$\lim_{x\to0^+}\ln y = \lim_{x\to0^+} e^{\ln y} = 1$

41. $\lim_{x\to\infty}\frac{e^x}{x^n} = \infty$ because after n applications of

L'Hôpital's Rule we have $\lim_{x\to\infty}\frac{e^x}{n!} = \infty$. e^x

dominates x^n.

43. $\frac{\cos 0}{0^2} = \frac{1}{0}$ is not indeterminate

45. $\lim_{x\to0}\frac{\sin kx^2}{x^2} = k\lim_{x\to0}\frac{\sin kx^2}{kx^2}$

$\quad = k\lim_{kx^2\to0}\frac{\sin kx^2}{kx^2}$

$\quad = k\cdot 1$

$\quad = k$

47. a. $\frac{x(2+\sin x)}{x(2+\cos x)}$

b. $\frac{x}{e^x}$

c. $\frac{3x+1}{x-7}$

d. $\frac{3-8x}{1+2x}$

Section 7.7

3. improper

5. not improper

7. improper

9. improper

11. $\int_0^1 x^{-1/3}dx = \lim_{R\to0^+}\int_R^1 x^{-1/3}dx$

$\quad = \lim_{R\to0^+}\frac{3}{2}x^{2/3}\Big|_R^1$

$\quad = \lim_{R\to0^+}\frac{3}{2}(1-R^{2/3})$

$\quad = \frac{3}{2}$

13. $\int_0^1 x^{-4/3}dx = \lim_{R\to0^+}\int_R^1 x^{-4/3}dx$

$\quad = \lim_{R\to0^+}(-3x^{-1/3})\Big|_R^1$

$\quad = \lim_{R\to0^+}(-3)(1-R^{-1/3})$

$\quad = \infty$

diverges

15. $\int_1^\infty x^{-1/3}dx = \lim_{R\to\infty}\int_1^R x^{-1/3}dx$

$\quad = \lim_{R\to\infty}\frac{3}{2}x^{2/3}\Big|_1^R$

$\quad = \lim_{R\to\infty}\frac{3}{2}(R^{2/3}-1)$

$\quad = \infty$

diverges

17. $\int_1^\infty x^{-4/3}dx = \lim_{R\to\infty}\int_1^R x^{-4/3}dx$

$\quad = \lim_{R\to\infty}(-3x^{-1/3})\Big|_1^R$

$\quad = \lim_{R\to\infty}(-3)(R^{-1/3}-1)$

$\quad = 3$

19. $\displaystyle\int_1^3 \frac{1}{\sqrt{x-1}}\,dx = \lim_{R\to 1^+}\int_R^3 \frac{1}{\sqrt{x-1}}\,dx$

$$= \lim_{R\to 1^+} 2\sqrt{x-1}\,\Big|_R^3$$

$$= \lim_{R\to 1^+} 2(\sqrt{2}-\sqrt{R-1})$$

$$= 2\sqrt{2}$$

21. $\displaystyle\int_0^1 \ln x\,dx = \lim_{R\to 0^+}\int_R^1 \ln x\,dx$

$\quad u = \ln x \qquad dv = dv$

$\quad du = \dfrac{1}{x}dx \qquad v = x$

$$= \lim_{R\to 0^+}\left[x\ln x\,\Big|_R^1 - \int_R^1 dx\right]$$

$$= \lim_{R\to 0^+}(x\ln x - x)\,\Big|_R^1$$

$$= \lim_{R\to 0^+}[(0-1)-(R\ln R - R)]$$

$$= -1 - \lim_{R\to 0^+}(R\ln R - R)$$

$$= -1 - \lim_{R\to 0^+}\frac{\ln R}{\frac{1}{R}} + \lim_{R\to 0} R$$

$$= -1 - \lim_{R\to 0^+}\frac{\frac{1}{R}}{-\frac{1}{R^2}}$$

$$= -1 - \lim_{R\to 0^+}(-R)$$

$$= -1$$

23. $\displaystyle\int_0^3 \frac{1}{x^2-1}\,dx = \int_0^3 \frac{1}{(x-1)(x+1)}\,dx$

$$= \lim_{R\to 1^-}\int_0^R \frac{1}{(x-1)(x+1)}\,dx$$

$$+ \lim_{R\to 1^+}\int_R^3 \frac{1}{(x-1)(x+1)}\,dx$$

Both of these integrals behave like

$$\lim_{R\to 0^+}\int_R^1 \frac{1}{x}\,dx = \lim_{R\to 0^+}(\ln 1 - \ln R)$$

$$= \lim_{R\to 0^+}\ln\left(\frac{1}{R}\right) = \infty,$$

so the original integral diverges.

25. $\displaystyle\int_0^\infty xe^x\,dx = \lim_{R\to\infty}\int_0^R xe^x\,dx \qquad u=x \qquad dv=e^x dx$

$\qquad\qquad\qquad\qquad\qquad\qquad\qquad\quad du=dx \qquad v=e^x$

$$= \lim_{R\to\infty}\left(xe^x\,\Big|_0^R - \int_0^R e^x dx\right)$$

$$= \lim_{R\to\infty}\left(Re^R - 0 - e^x\,\Big|_0^R\right)$$

$$= \lim_{R\to\infty}[(Re^R - (e^R - 1)]$$

$$= \lim_{R\to\infty}[e^R(R-1)+1]$$

$$= \infty$$

diverges

27. $\displaystyle\int_{-\infty}^{0} xe^x\,dx = \lim_{R\to-\infty}\int_R^0 xe^x\,dx \qquad u = x \qquad dv = e^x\,dx$

$$\qquad\qquad\qquad\qquad\qquad\qquad du = dx \qquad v = e^x$$

$$= \lim_{R\to-\infty}\left(xe^x\,\Big|_R^0 - \int_R^0 e^x\,dx \right)$$

$$= \lim_{R\to-\infty}(xe^x - e^x)\,\Big|_R^0$$

$$= \lim_{R\to-\infty}[(0-1) - (Re^R - e^R)]$$

$$= \lim_{R\to-\infty}[-1 - e^R(R-1)]$$

$$= -1 - \lim_{R\to-\infty}\frac{R-1}{e^{-R}}$$

$$= -1 - \lim_{R\to-\infty}\frac{1}{-e^{-R}}$$

$$= -1$$

29. $\displaystyle\int_{-\infty}^{\infty}\frac{1}{x^2}\,dx = \int_{-\infty}^{-1}\frac{1}{x^2}\,dx + \int_{-1}^{0}\frac{1}{x^2}\,dx + \int_{0}^{1}\frac{1}{x^2}\,dx + \int_{1}^{\infty}\frac{1}{x^2}\,dx$

$$= \lim_{R\to-\infty}\int_R^{-1}\frac{1}{x^2}\,dx + \lim_{R\to0^-}\int_{-1}^{R}\frac{1}{x^2}\,dx + \lim_{R\to0^+}\int_R^{1}\frac{1}{x^2}\,dx + \lim_{R\to\infty}\int_1^{R}\frac{1}{x^2}\,dx$$

$$= \lim_{R\to-\infty}\left(-\frac{1}{x}\right)\Big|_R^{-1} + \lim_{R\to0^-}\left(-\frac{1}{x}\right)\Big|_{-1}^{R} + \lim_{R\to0^+}\left(-\frac{1}{x}\right)\Big|_R^{1} + \lim_{R\to\infty}\left(-\frac{1}{x}\right)\Big|_1^{R}$$

Since $\displaystyle\lim_{R\to0^-}\left(-\frac{1}{x}\right)\Big|_{-1}^{R} = \lim_{R\to0^-}\left(-\frac{1}{R}+\frac{1}{-1}\right) = \infty$, the integral diverges

31. $\displaystyle\int_{-\infty}^{\infty} e^{-2x}\,dx = \int_{-\infty}^{0} e^{-2x}\,dx + \int_{0}^{\infty} e^{-2x}\,dx$

$$= \lim_{R\to-\infty}\int_R^0 e^{-2x}\,dx + \lim_{R\to-\infty}\int_0^R e^{-2x}\,dx$$

$$= \lim_{R\to-\infty}\left(-\frac{1}{2}e^{-2x}\right)\Big|_R^0 + \lim_{R\to\infty}\left(-\frac{1}{2}e^{-2x}\right)\Big|_0^R$$

$$= \lim_{R\to-\infty}\left[-\frac{1}{2}(1-e^{-2R})\right] + \lim_{R\to\infty}\left[-\frac{1}{2}(e^{-2R}-1)\right]$$

$$= \infty + \lim_{R\to\infty}\left[-\frac{1}{2}(e^{-2R}-1)\right]$$

$$= \infty$$

diverges

33. $\displaystyle\int_{-\infty}^{\infty}\frac{1}{1+x^2}\,dx = \int_{-\infty}^{0}\frac{1}{1+x^2}\,dx + \int_{0}^{\infty}\frac{1}{1+x^2}\,dx$

$$= \lim_{R\to-\infty}\int_R^0\frac{1}{1+x^2}\,dx + \lim_{R\to\infty}\int_0^R\frac{1}{1+x^2}\,dx$$

$$= \lim_{R\to\infty}\tan^{-1}x\,\Big|_R^0 + \lim_{R\to\infty}\tan^{-1}x\,\Big|_0^R$$

$$= \lim_{R\to-\infty}(\tan^{-1}0 - \tan^{-1}R) + \lim_{R\to\infty}(\tan^{-1}R - \tan^{-1}0)$$

$$= 0 - \left(-\frac{\pi}{2}\right) + \frac{\pi}{2} - 0$$

$$= \pi$$

35. $\displaystyle\int_0^{\pi}\tan x\,dx=\int_0^{\pi/2}\tan x\,dx+\int_{\pi/2}^{\pi}\tan x\,dx$

$\displaystyle=\lim_{R\to\frac{\pi}{2}^{-}}\int_0^R\tan x\,dx+\lim_{R\to\frac{\pi}{2}^{+}}\int_R^{\pi}\tan x\,dx$

$\displaystyle=\lim_{R\to\frac{\pi}{2}^{-}}\ln\left|\sec x\right|\Big|_0^R+\lim_{R\to\frac{\pi}{2}^{+}}\ln\left|\sec x\right|\Big|_R^{\pi}$

$\displaystyle=\lim_{R\to\frac{\pi}{2}^{-}}\left(\ln\left|\sec R\right|-\ln\left|\sec 0\right|\right)+\lim_{R\to\frac{\pi}{2}^{+}}\left(\ln\left|\sec \pi\right|-\ln\left|\sec R\right|\right)$

$=\infty+\infty$

diverges

37. $\displaystyle\int_0^2\frac{x}{x^2-1}dx=\int_0^1\frac{x}{x^2-1}dx+\int_1^2\frac{x}{x^2-1}dx$

$\displaystyle=\lim_{R\to1^{-}}\int_0^R\frac{x}{x^2-1}dx+\lim_{R\to1^{+}}\int_R^2\frac{x}{x^2-1}dx$

$\displaystyle=\lim_{R\to1^{-}}\frac{1}{2}\ln\left|x^2-1\right|\Big|_0^R+\lim_{R\to1^{+}}\frac{1}{2}\ln\left|x^2-1\right|\Big|_R^2$

$\displaystyle\lim_{R\to1^{-}}\frac{1}{2}\ln\left|x^2-1\right|\Big|_0^R=\lim_{R\to1^{-}}\left(\frac{1}{2}\ln\left|R^2-1\right|-\frac{1}{2}\ln\left|-1\right|\right)=-\infty$

diverges

39. $\displaystyle\int_0^1\frac{2}{\sqrt{1-x^2}}dx=\lim_{R\to1^{-}}\int_0^R\frac{2}{\sqrt{1-x^2}}dx$

$\displaystyle=\lim_{R\to1^{-}}2\sin^{-1}x\Big|_0^R$

$\displaystyle=\lim_{R\to1^{-}}2(\sin^{-1}R-\sin^{-1}0)$

$\displaystyle=2\left(\frac{\pi}{2}-0\right)$

$=\pi$

41. $\displaystyle\int_0^{\infty}\frac{1}{(x-2)^2}dx=\int_0^2\frac{1}{(x-2)^2}dx+\int_2^3\frac{1}{(x-2)^2}dx+\int_3^{\infty}\frac{1}{(x-2)^2}dx$

$\displaystyle\int_0^2\frac{1}{(x-2)^2}dx=\lim_{R\to2^{-}}\int_0^R\frac{1}{(x-2)^2}dx=\lim_{R\to2^{-}}\left(-\frac{1}{x-2}\right)\Big|_0^R=\lim_{R\to2^{-}}\left(-\frac{1}{R-2}+\frac{1}{-2}\right)=\infty$

$\displaystyle\int_0^{\infty}\frac{1}{(x-2)^2}dx$ diverges

43. $\int_0^\infty \dfrac{1}{\sqrt{x}e^{\sqrt{x}}}\,dx$

$\quad = \lim\limits_{R\to 0^+}\int_R^1 \dfrac{1}{\sqrt{x}e^{\sqrt{x}}}\,dx + \lim\limits_{R\to\infty}\int_1^R \dfrac{1}{\sqrt{x}e^{\sqrt{x}}}\,dx$

Let $u = \sqrt{x}$, $du = \dfrac{1}{2\sqrt{x}}\,dx$.

$\int \dfrac{1}{\sqrt{x}e^{\sqrt{x}}}\,dx = 2\int e^{-u}\,du = -2e^{-u}$

So the original integral is equal to

$\lim\limits_{R\to 0^+}\left(\dfrac{-2}{e^u}\Big|_R^1\right) + \lim\limits_{R\to\infty}\left(\dfrac{-2}{e^u}\Big|_1^R\right)$

$\quad = \lim\limits_{R\to 0^+}\left(\dfrac{-2}{e} + \dfrac{2}{e^R}\right) + \lim\limits_{R\to\infty}\left(\dfrac{-2}{e} + \dfrac{2}{e^R}\right)$

$\quad = 1 + 1 = 2$

45. $\int_0^\infty \cos x\,dx = \lim\limits_{R\to\infty}\int_0^R \cos x\,dx$

$\quad = \lim\limits_{R\to\infty} \sin x\Big|_0^R$

$\quad = \lim\limits_{R\to\infty}(\sin R - \sin 0)$

diverges

47. $r = 1$

49. $\int_0^\infty xe^{cx}\,dx$ converges for $c < 0$

$\int_{-\infty}^0 xe^{cx}\,dx$ converges for $c > 0$

51. $\int_1^\infty \dfrac{x}{1+x^3}\,dx$

$0 < \dfrac{x}{1+x^3} < \dfrac{x}{x^3} = \dfrac{1}{x^2}$

$\int_1^\infty \dfrac{1}{x^2}\,dx = \lim\limits_{R\to\infty}\int_1^R \dfrac{1}{x^2}\,dx$

$\quad = \lim\limits_{R\to\infty}\left(-\dfrac{1}{x}\right)\Big|_1^R$

$\quad = \lim\limits_{R\to\infty}\left(-\dfrac{1}{R} + 1\right)$

$\quad = 1$

So $\int_1^\infty \dfrac{x}{1+x^3}\,dx$ converges

53. $\int_2^\infty \dfrac{x}{x^{3/2}-1}\,dx$

$\dfrac{x}{x^{3/2}-1} > \dfrac{x}{x^{3/2}} = \dfrac{1}{x^{1/2}} > 0$

$\int_2^\infty x^{-1/2}\,dx = \lim\limits_{R\to\infty}\int_2^R x^{-1/2}\,dx$

$\quad = \lim\limits_{R\to\infty} 2\sqrt{x}\Big|_2^R$

$\quad = \lim\limits_{R\to\infty}(2\sqrt{R} - 2\sqrt{2})$

$\quad = \infty$

So $\int_2^\infty \dfrac{x}{x^{3/2}-1}\,dx$ diverges

55. $\int_0^\infty \dfrac{3}{x+e^x}\,dx$

$0 < \dfrac{3}{x+e^x} < \dfrac{3}{e^x}$

$\int_0^\infty \dfrac{3}{e^x}\,dx = \lim\limits_{R\to\infty}\int_0^R \dfrac{3}{e^x}\,dx$

$\quad = \lim\limits_{R\to\infty}\left(-\dfrac{3}{e^x}\right)\Big|_0^R$

$\quad = \lim\limits_{R\to\infty}\left(-\dfrac{3}{e^R} + 3\right)$

$\quad = 3$

So $\int_0^\infty \dfrac{3}{x+e^x}\,dx$ converges

57. $\int_0^\infty \dfrac{\sin^2 x}{1+e^x}\,dx$

$\dfrac{\sin^2 x}{1+e^x} \le \dfrac{1}{1+e^x} < \dfrac{1}{e^x}$

$\int_0^\infty \dfrac{1}{e^x}\,dx = \lim\limits_{R\to\infty}\int_0^R \dfrac{1}{e^x}\,dx$

$\quad = \lim\limits_{R\to\infty}\int(-e^{-x})\Big|_0^R$

$\quad = \lim\limits_{R\to\infty}(-e^{-R} + 1)$

$\quad = 1$

So $\int_0^\infty \dfrac{\sin^2 x}{1+e^x}\,dx$ converges

59. $\displaystyle\int_2^\infty \frac{x^2 e^x}{\ln x}\,dx$

$\dfrac{x^2 e^x}{\ln x} > e^x$

$\displaystyle\int_2^\infty e^x\,dx = \lim_{R\to\infty}\int_2^R e^x\,dx$

$\displaystyle\qquad\quad = \lim_{R\to\infty} e^x\Big|_2^R$

$\displaystyle\qquad\quad = \lim_{R\to\infty}(e^R - e^2)$

$\displaystyle\qquad\quad = \infty$

So $\displaystyle\int_2^\infty \frac{x^2 e^x}{\ln x}\,dx$ diverges

61. $\displaystyle\int_0^1 x\ln 4x\,dx = \lim_{R\to 0^+}\int_R^1 x\ln 4x\,dx \qquad u = \ln 4x \qquad dv = x\,dx$

$$\qquad\qquad\qquad\qquad\qquad du = \frac{4}{4x}\,dx \qquad v = \frac{x^2}{2}$$

$\displaystyle = \lim_{R\to 0^+}\left[\frac{1}{2}x^2\ln 4x\Big|_R^1 - \int_R^1 \frac{1}{2}x\,dx\right]$

$\displaystyle = \lim_{R\to 0^+}\left[\left(\frac{1}{2}\ln 4 - \frac{1}{2}R^2\ln 4R\right) - \frac{1}{4}x^2\Big|_R^1\right]$

$\displaystyle = \lim_{R\to 0^+}\left[\frac{1}{2}\ln 4 - \frac{\ln 4R}{\frac{2}{R^2}} - \left(\frac{1}{4} - \frac{R^2}{4}\right)\right]$

$\displaystyle = \frac{1}{2}\ln 4 - \frac{1}{4} + \lim_{R\to 0^+}\frac{\frac{4}{4R}}{-\frac{4}{R^3}}$

$\displaystyle = \frac{1}{2}\ln 4 - \frac{1}{4} + \lim_{R\to 0^+}\left(-\frac{R^2}{4}\right)$

$\displaystyle = \frac{1}{2}\ln 4 - \frac{1}{4}$

63. $V = \int_1^\infty \pi\left(\dfrac{1}{x}\right)^2 dx = \lim_{R\to\infty} \pi\int_1^R \dfrac{1}{x^2}dx$

$$= \lim_{R\to\infty} \pi\left(-\dfrac{1}{x}\right)\Big|_1^R$$

$$= \lim_{R\to\infty} \pi\left(-\dfrac{1}{R}+1\right)$$

$$= \pi$$

$SA = \int_1^\infty 2\pi\cdot\dfrac{1}{x}\sqrt{1+\left(-\dfrac{1}{x^2}\right)^2}\,dx$

$$= \int_1^\infty 2\pi\cdot\dfrac{1}{x}\sqrt{1+\dfrac{1}{x^4}}\,dx$$

$2\pi\dfrac{1}{x}\sqrt{1+\dfrac{1}{x^4}} > \dfrac{1}{x}$

$\int_1^\infty \dfrac{1}{x}dx = \lim_{R\to\infty}\int_1^R \dfrac{1}{x}dx$

$$= \lim_{R\to\infty} \ln x\Big|_1^R$$

$$= \lim_{R\to\infty} (\ln R - 0)$$

$$= \infty$$

So $\int_1^\infty 2\pi\cdot\dfrac{1}{x}\sqrt{1+\dfrac{1}{x^4}}dx$ diverges

65. The integral of an odd function over an interval symmetric around 0 is always 0.

67. a. $\int_0^\infty ke^{-2x}dx = \lim_{R\to\infty}\int_0^R ke^{-2x}dx$

$$= \lim_{R\to\infty}\left(-\dfrac{k}{2}e^{-2x}\Big|_0^R\right)$$

$$= \lim_{R\to\infty}\left(-\dfrac{k}{2}e^{-2R}+\dfrac{k}{2}\right)$$

$$= \dfrac{k}{2}$$

$$= 1$$

$$k = 2$$

b. $\int_0^\infty ke^{-4x}dx = \lim_{R\to\infty}\int_0^R ke^{-4x}dx$

$$= \lim_{R\to\infty}\left(-\dfrac{k}{4}e^{-4x}\Big|_0^R\right)$$

$$= \lim_{R\to\infty}\left(-\dfrac{k}{4}e^{-4R}+\dfrac{k}{4}\right)$$

$$= \dfrac{k}{4}$$

$$= 1$$

$$k = 4$$

c. $\int_0^\infty ke^{-rx}dx = \lim_{R\to\infty}\int_0^R ke^{-rx}dx$

$$= \lim_{R\to\infty}\left(-\dfrac{k}{r}e^{-rx}\Big|_0^R\right)$$

$$= \lim_{R\to\infty}\left(-\dfrac{k}{r}e^{-rR}+\dfrac{k}{r}\right)$$

$$= \dfrac{k}{r}$$

$$= 1$$

$$k = r$$

69. $\mu = \int_0^\infty rxe^{-rx}dx = r\lim_{R\to\infty}\int_0^R xe^{-rx}dx$

$\quad u = x \qquad dv = e^{-rx}dx$

$\quad du = dx \qquad v = -\dfrac{1}{r}e^{-rx}dx$

$$= r\lim_{R\to\infty}\left[-\dfrac{1}{r}xe^{-rx}\Big|_0^R + \dfrac{1}{r}\int_0^R e^{-rx}dx\right]$$

$$= r\lim_{R\to\infty}\left[-\dfrac{1}{r}Re^{-rR} - \dfrac{1}{r^2}e^{-rx}\Big|_0^R\right]$$

$$= r\lim_{R\to\infty}\left(-\dfrac{1}{r}Re^{-rR} - \dfrac{1}{r^2}e^{-rR} + \dfrac{1}{r^2}\right)$$

$$= \dfrac{r}{r^2}$$

$$= \dfrac{1}{r}$$

71. $\int_{1/r}^\infty re^{-rx}dx = \lim_{R\to\infty}\int_{1/r}^R re^{-rx}dx$

$$= \lim_{R\to\infty}\left(-e^{-rx}\Big|_{1/r}^R\right)$$

$$= \lim_{R\to\infty}(-e^{-rR}+e^{-r/r})$$

$$= e^{-1}$$

Chapter 7 Review

1. See Section 7.1.

$$\int \frac{e^{\sqrt{x}}}{\sqrt{x}}\,dx$$

Let $u = \sqrt{x}$, $du = \left(\dfrac{1}{2\sqrt{x}}\right)dx$

$$\int e^u \cdot 2\,du = 2e^u + c$$
$$= 2e^{\sqrt{x}} + c$$

3. See Section 7.3.

$$\int \frac{x^2}{\sqrt{1-x^2}}\,dx$$

Let $x = \sin\theta$
$\quad dx = \cos\theta\,d\theta$

$$\int \frac{\sin^2\theta}{\sqrt{1-\sin^2\theta}}\cos\theta\,d\theta = \int \sin^2\theta\,d\theta$$
$$= \frac{1}{2}\int (1-\cos 2\theta)\,d\theta$$
$$= \frac{1}{2}\left(\theta - \frac{1}{2}\sin 2\theta\right) + c$$

$0 = \sin^{-1} x$, $\sin 2\theta = 2\sin\theta\cos\theta = 2x\sqrt{1-x^2}$

$$\int \frac{x^2}{\sqrt{1-x^2}}\,dx = \frac{1}{2}\theta - \frac{1}{4}\sin 2\theta + c$$
$$= \frac{1}{2}\sin^{-1} x - \frac{1}{2}x\sqrt{1-x^2} + c$$

5. See Section 7.2.

$$\int x^2 e^{-3x}\,dx \qquad u = x^2 \qquad dv = e^{-3x}\,dx$$
$$\qquad\qquad du = 2x\,dx \qquad v = -\frac{1}{3}e^{-3x}$$

$$= -\frac{1}{3}x^2 e^{-3x} + \frac{2}{3}\int xe^{-3x}\,dx \qquad u = x \qquad dv = e^{-3x}\,dx$$
$$\qquad\qquad\qquad\qquad du = dx \qquad v = -\frac{1}{3}e^{-3x}$$

$$= -\frac{1}{3}x^2 e^{-3x} + \frac{2}{3}\left(-\frac{1}{3}xe^{-3x} + \frac{1}{3}\int e^{-3x}\,dx\right)$$
$$= -\frac{1}{3}x^2 e^{-3x} - \frac{2}{9}xe^{-3x} - \frac{2}{27}e^{-3x} + c$$

7. See Section 7.1.

$$\int \frac{x}{1+x^4} dx$$

Let $u = x^2$, $du = 2x\,dx$

$$\int \frac{1}{1+u^2} \cdot \frac{1}{2} du = \frac{1}{2} \tan^{-1} u + c$$

$$= \frac{1}{2} \tan^{-1} x^2 + c$$

9. See Section 7.1.

$$\int \frac{x^3}{4+x^4} dx$$

Let $u = 4 + x^4$, $du = 4x^3 dx$

$$\int \frac{1}{u} \cdot \frac{1}{4} du = \frac{1}{4} \ln|u| + c$$

$$= \frac{1}{4} \ln(4 + x^4) + c$$

11. See Section 7.1.

$$\int e^{2\ln x} dx = \int e^{\ln x^2} dx$$

$$= \int x^2 dx$$

$$= \frac{1}{3} x^3 + c$$

13. See Section 7.2.

$$\int_0^1 x \sin 3x\,dx \qquad u = x \qquad dv = \sin 3x\,dx$$

$$du = dx \qquad v = -\frac{1}{3}\cos 3x$$

$$= -\frac{1}{3} x \cos 3x \Big|_0^1 + \frac{1}{3} \int_0^1 \cos 3x\,dx$$

$$= -\frac{1}{3}\cos 3 + \frac{1}{3} \cdot \frac{1}{3}\sin 3x \Big|_0^1$$

$$= -\frac{1}{3}\cos 3 + \frac{1}{9}\sin 3$$

15. See Section 7.3.

$$\int_0^{\pi/2} \sin^4 x\,dx$$

$$= \int_0^{\pi/2} \frac{1}{2}(1 - \cos 2x) \cdot \frac{1}{2}(1 - \cos 2x)\,dx$$

$$= \frac{1}{4} \int_0^{\pi/2} (1 - 2\cos 2x + \cos^2 2x)\,dx$$

$$= \frac{1}{4} \int_0^{\pi/2} \left[1 - 2\cos 2x + \frac{1}{2}(1 + \cos 4x)\right] dx$$

$$= \frac{1}{4} \int_0^{\pi/2} \left(\frac{3}{2} - 2\cos 2x + \frac{1}{2}\cos 4x\right) dx$$

$$= \frac{1}{4}\left(\frac{3}{2}x - \sin 2x + \frac{1}{8}\sin 4x\right)\Big|_0^{\pi/2}$$

$$= \frac{1}{4}\left(\frac{3\pi}{4} - \sin \pi + \frac{1}{8}\sin 2\pi\right) - \frac{1}{4}(0 - 0 + 0)$$

$$= \frac{3\pi}{16}$$

17. See Section 7.2.

$$\int_{-1}^1 x \sin \pi x\,dx \qquad u = x \qquad dv = \sin \pi x\,dx$$

$$du = dx \qquad v = -\frac{1}{\pi}\cos \pi x$$

$$= -\frac{1}{\pi} x \cos \pi x \Big|_{-1}^1 + \frac{1}{\pi} \int_{-1}^1 \cos \pi x\,dx$$

$$= -\frac{1}{\pi}\cos \pi + \frac{1}{\pi}(-1)\cos(-\pi) + \frac{1}{\pi^2}\sin \pi x \Big|_{-1}^1$$

$$= \frac{1}{\pi} + \frac{1}{\pi} + \frac{1}{\pi^2}(\sin \pi - \sin(-\pi))$$

$$= \frac{2}{\pi}$$

19. See Section 7.2.

$$\int_1^2 x^3 \ln x\,dx \qquad u = \ln x \qquad dv = x^3 dx$$

$$du = \frac{1}{x} dx \qquad v = \frac{1}{4} x^4$$

$$= \frac{1}{4} x^4 \ln x \Big|_1^2 - \frac{1}{4} \int_1^2 x^3 dx$$

$$= \frac{1}{4} \cdot 16\ln 2 - \frac{1}{4}\ln 1 - \frac{1}{16} x^4 \Big|_1^2$$

$$= 4\ln 2 - \frac{1}{16}(16 - 1)$$

$$= 4\ln 2 - \frac{15}{16}$$

21. See Section 7.3.

$$\int \cos x \sin^2 x \, dx$$

Let $u = \sin x$, $du = \cos x \, dx$

$$\int u^2 \, du = \frac{1}{3} u^3 + c$$

$$= \frac{1}{3} \sin^3 x + c$$

23. See Section 7.3.

$$\int \cos^3 x \sin^3 x \, dx$$

$$= \int \cos x (1 - \sin^2 x)(\sin^3 x) \, dx$$

Let $u = \sin x$, $du = \cos x \, dx$

$$\int (1 - u^2) u^3 \, du = \int (u^3 - u^5) \, du$$

$$= \frac{1}{4} u^4 - \frac{1}{6} u^6 + c$$

$$= \frac{1}{4} \sin^4 x - \frac{1}{6} \sin^6 x + c$$

25. See Section 7.3.

$$\int \tan^2 x \sec^4 x \, dx$$

$$= \int \tan^2 x (1 + \tan^2 x) \sec^2 x \, dx$$

Let $u = \tan x$, $du = \sec^2 x \, dx$

$$\int u^2 (1 + u^2) \, du = \int (u^2 + u^4) \, du$$

$$= \frac{1}{3} u^3 + \frac{1}{5} u^5 + c$$

$$= \frac{1}{3} \tan^3 x + \frac{1}{5} \tan^5 x + c$$

27. See Section 7.3.

$$\int \sqrt{\sin x} \cos^3 x \, dx$$

$$= \int \sqrt{\sin x} (1 - \sin^2 x) \cos x \, dx$$

Let $u = \sin x$, $du = \cos x \, dx$

$$\int \sqrt{u} (1 - u^2) \, du$$

$$= \int (u^{1/2} - u^{5/2}) \, du$$

$$= \frac{2}{3} u^{3/2} - \frac{2}{7} u^{7/2} + c$$

$$= \frac{2}{3} (\sin x)^{3/2} - \frac{2}{7} (\sin x)^{7/2} + c$$

29. See Section 7.1, Example 1.4.

$$\int \frac{2}{8 + 4x + x^2} \, dx = \int \frac{2}{x^2 + 4x + 4 + 8 - 4} \, dx$$

$$= \int \frac{2}{(x + 2)^2 + 4} \, dx$$

$$= \int \frac{2}{4\left[\left(\frac{x+2}{2}\right)^2 + 1\right]} \, dx$$

Let $u = \frac{x + 2}{2}$, $du = \frac{1}{2} dx$

$$\int \frac{1}{u^2 + 1} \, du = \tan^{-1} u + c$$

$$= \tan^{-1}\left(\frac{x + 2}{2}\right) + c$$

31. See Section 7.3, Example 3.9.

$$\int \frac{2}{x^2 \sqrt{4 - x^2}} \, dx$$

Let $x = 2\sin\theta$, $-\frac{\pi}{2} < \theta < \frac{\pi}{2}$

$$dx = 2\cos\theta \, d\theta$$

$$\int \frac{2}{4\sin^2\theta \sqrt{4 - 4\sin^2\theta}} \cdot 2\cos\theta \, d\theta$$

$$= \int \frac{\cos\theta}{2\sin^2\theta \sqrt{1 - \sin^2\theta}} \, d\theta$$

$$= \int \frac{\cos\theta}{2\sin^2\theta \cos\theta} \, d\theta$$

$$= \int \frac{1}{2\sin^2\theta} \, d$$

$$= \frac{1}{2} \int \csc^2\theta \, d\theta$$

$$= -\frac{1}{2} \cot\theta + c$$

$$\sin\theta = \frac{x}{2}, \quad \cos\theta = \frac{1}{2}\sqrt{4 - x^2},$$

$$\cot\theta = \frac{\cos\theta}{\sin\theta} = \frac{\sqrt{4 - x^2}}{x}$$

$$\int \frac{2}{x^2 \sqrt{4 - x^2}} \, dx = -\frac{1}{2} \cot\theta + c$$

$$= -\frac{\sqrt{4 - x^2}}{2x} + c$$

33. See Section 7.1.

$$\int \frac{x^3}{\sqrt{9-x^2}}\,dx = \int \frac{x^2}{\sqrt{9-x^2}}\,x\,dx$$

Let $u = 9 - x^2$, $du = -2x\,dx$, $x^2 = 9 - u$

$$\int \frac{9-u}{\sqrt{u}} \cdot \left(-\frac{1}{2}\right) du$$

$$= -\frac{1}{2}\int (9u^{-1/2} - u^{1/2})\,du$$

$$= -\frac{1}{2}\left(18u^{1/2} - \frac{2}{3}u^{3/2}\right)du$$

$$= -9\sqrt{9-x^2} + \frac{1}{3}(9-x^2)^{3/2} + c$$

35. See Section 7.1.

$$\int \frac{x^3}{\sqrt{x^2+9}}\,dx = \int \frac{x^2}{\sqrt{x^2+9}}\,x\,dx$$

Let $u = x^2 + 9$, $du = 2x\,dx$, $x^2 = u - 9$

$$\int \frac{u-9}{\sqrt{u}} \cdot \frac{1}{2}\,du = \frac{1}{2}\int (u^{1/2} - 9u^{-1/2})\,du$$

$$= \frac{1}{2}\left(\frac{2}{3}u^{3/2} - 18u^{1/2}\right) + c$$

$$= \frac{1}{3}(x^2+9)^{3/2} - 9\sqrt{x^2+9} + c$$

39. See Section 7.4.

$$\int \frac{4x^2+6x-12}{x^3-4x}\,dx = \int \frac{4x^2+6x-12}{x(x+2)(x-2)}\,dx$$

$$\frac{4x^2+6x-12}{x(x+2)(x-2)} = \frac{A}{x} + \frac{B}{x+2} + \frac{C}{x-2}$$

$$4x^2+6x-12 = A(x+2)(x-2) + Bx(x-2) + Cx(x+2)$$

$x = 0 : -12 = -4A;\ A = 3$

$x = -2 : -8 = 8B;\ B = -1$

$x = 2 : 16 = 8C;\ C = 2$

$$\frac{4x^2+6x-12}{x(x+2)(x-2)} = \frac{3}{x} - \frac{1}{x+2} + \frac{2}{x-2}$$

$$\int \frac{4x^2+6x-12}{x(x+2)(x-2)}\,dx = \int \left(\frac{3}{x} - \frac{1}{x+2} + \frac{2}{x-2}\right)dx$$

$$= 3\ln|x| - \ln|x+2| + 2\ln|x-2| + c$$

37. See Section 7.4.

$$\int \frac{x+4}{x^2+3x+2}\,dx = \int \frac{x+4}{(x+2)(x+1)}\,dx$$

$$\frac{x+4}{(x+2)(x+1)} = \frac{A}{x+2} + \frac{B}{x+1}$$

$$x+4 = A(x+1) + B(x+2)$$

$x = -1 : 3 = B$

$x = -2 : 2 = -A;\ A = -2$

$$\frac{x+4}{(x+2)(x+1)} = \frac{-2}{x+2} + \frac{3}{x+1}$$

$$\int \frac{x+4}{x^2+3x+2}\,dx = \int \left(-\frac{2}{x+2} + \frac{3}{x+1}\right)dx$$

$$= -2\ln|x+2| + 3\ln|x+1| + c$$

41. See Section 7.2, Example 2.5.

$$\int e^x \cos 2x\, dx \qquad u = e^x \qquad dv = \cos 2x\, dx$$

$$du = e^x dx \qquad v = \frac{1}{2}\sin 2x$$

$$= \frac{1}{2}e^x \sin 2x - \frac{1}{2}\int e^x \sin 2x\, dx \qquad u = e^x \qquad dv = \sin 2x\, dx$$

$$du = e^x dx \qquad v = -\frac{1}{2}\cos 2x$$

$$= \frac{1}{2}e^x \sin 2x - \frac{1}{2}\left(-\frac{1}{2}e^x \cos 2x + \frac{1}{2}\int e^x \cos 2x\, dx\right)$$

$$\int e^x \cos 2x\, dx = \frac{1}{2}e^x \sin 2x + \frac{1}{4}e^x \cos 2x - \frac{1}{4}\int e^x \cos 2x\, dx$$

$$\frac{5}{4}\int e^x \cos 2x\, dx = \frac{1}{2}e^x \sin 2x + \frac{1}{4}e^x \cos 2x + K$$

$$\int e^x \cos 2x\, dx = \frac{2}{5}e^x \sin 2x + \frac{1}{5}e^x \cos 2x + c$$

43. See Section 7.1.

$$\int x\sqrt{x^2 + 1}\, dx$$

Let $u = x^2 + 1$, $du = 2x\, dx$

$$\frac{1}{2}\int \sqrt{u}\, du = \frac{1}{2}\cdot\frac{2}{3}u^{3/2} + c$$

$$= \frac{1}{3}(x^2 + 1)^{3/2} + c$$

45. See Example 4.1.

$$\frac{4}{x^2 - 3x - 4} = \frac{4}{(x-4)(x+1)}$$

$$= \frac{A}{x-4} + \frac{B}{x+1}$$

$$4 = A(x+1) + B(x-4)$$

$$x = 4: A = \frac{4}{5}$$

$$x = -1: B = -\frac{4}{5}$$

$$\frac{4}{x^2 - 3x - 4} = \frac{\frac{4}{5}}{x-4} - \frac{\frac{4}{5}}{x+1}$$

47. See Example 4.2.

$$\frac{-6}{x^3 + x^2 - 2x} = \frac{-6}{x(x^2 + x - 2)}$$

$$= \frac{-6}{x(x+2)(x-1)}$$

$$= \frac{A}{x} + \frac{B}{x+2} + \frac{C}{x-1}$$

$$-6 = A(x+2)(x-1) + Bx(x-1) + Cx(x+2)$$

$$x = 0: -6 = -2A; \ A = 3$$

$$x = -2: -6 = 6B; \ B = -1$$

$$x = 1: -6 = 3C; \ C = -2$$

$$\frac{-6}{x^3 + x^2 - 2x} = \frac{3}{x} - \frac{1}{x+2} - \frac{2}{x-1}$$

49. See Example 4.4.

$$\frac{x-2}{x^2+4x+4}=\frac{x-2}{(x+2)^2}=\frac{A}{x+2}+\frac{B}{(x+2)^2}$$
$$x-2=A(x+2)+B=Ax+2A+B$$
$$A=1;\ B=-2-2A=-4$$
$$\frac{x-2}{x^2+4x+4}=\frac{1}{x+2}-\frac{4}{(x+2)^2}$$

51. See Example 5.1.

$$\int e^{3x}\sqrt{4+e^{2x}}\,dx=\int e^x e^{2x}\sqrt{4+e^{2x}}\,dx$$

Let $u=e^x$, $du=e^x dx$

$$\int u^2\sqrt{4+u^2}\,du$$

From the table,

$$\int u^2\sqrt{a^2+u^2}\,du=\frac{u}{8}(a^2+2u^2)\sqrt{a^2+u^2}-\frac{a^4}{8}\ln(u+\sqrt{a^2+u^2})+c$$
$$\int u^2\sqrt{4+u^2}\,du=\frac{u}{8}(4+2u^2)\sqrt{4+u^2}-2\ln(u+\sqrt{4+u^2})+c$$
$$=\frac{1}{8}e^x(4+2e^{2x})\sqrt{4+e^{2x}}-2\ln(e^x+\sqrt{4+e^{2x}})+c$$

53. See Example 5.2.

$$\int\sec^4 x\,dx$$

From the table,

$$\int\sec^n x\,dx=\frac{1}{n-1}\tan u\sec^{n-2}u+\frac{n-2}{n-1}\int\sec^{n-2}u\,du$$
$$\int\sec^4 x\,dx=\frac{1}{3}\tan x\sec^2 x+\frac{2}{3}\int\sec^2 x\,dx$$
$$=\frac{1}{3}\tan x\sec^2 x+\frac{2}{3}\tan x+c$$

55. See Example 5.1.

$$\int\frac{4}{x(3-x)^2}\,dx=4\int\frac{1}{x(3-x)^2}\,dx$$

From the table,

$$\int\frac{du}{u(a+bu)^2}=\frac{1}{a(a+bu)}-\frac{1}{a^2}\ln\left|\frac{a+bu}{u}\right|+c$$
$$4\int\frac{1}{x(3-x)^2}\,dx=4\left[\frac{1}{3(3-x)}-\frac{1}{9}\ln\left|\frac{3-x}{x}\right|\right]+c$$
$$=\frac{4}{3(3-x)}-\frac{4}{9}\ln\left|\frac{3-x}{x}\right|+c$$

57. See Example 5.1.

$$\int \frac{\sqrt{9+4x^2}}{x^2}dx = \int \frac{2\sqrt{\frac{9}{4}+x^2}}{x^2}dx$$

From the table,

$$\int \frac{\sqrt{a^2+u^2}}{u^2}du = -\frac{\sqrt{a^2+u^2}}{u} + \ln(u+\sqrt{a^2+u^2})+c$$

$$2\int \frac{\sqrt{\frac{9}{4}+x^2}}{x^2}dx = 2\left[-\frac{\sqrt{\frac{9}{4}+x^2}}{x} + \ln\left(x+\sqrt{\frac{9}{4}+x^2}\right)\right]+c$$

$$= -\frac{\sqrt{9+4x^2}}{x} + 2\ln\left(x+\sqrt{\frac{9}{4}+x^2}\right)+c$$

59. See Example 5.1.

$$\int \frac{\sqrt{4-x^2}}{x}dx$$

From the table,

$$\int \frac{\sqrt{a^2-u^2}}{u}du = \sqrt{a^2-u^2} - a\ln\left|\frac{a+\sqrt{a^2-u^2}}{u}\right|+c$$

$$\int \frac{\sqrt{4-x^2}}{x}dx = \sqrt{4-x^2} - 2\ln\left|\frac{2+\sqrt{4-x^2}}{x}\right|+c$$

61. See Example 6.1.

$$\lim_{x\to1}\frac{x^3-1}{x^2-1}\left(\frac{0}{0}\right) = \lim_{x\to1}\frac{3x^2}{2x}$$
$$= \frac{3}{2}$$

63. See Example 6.2.

$$\lim_{x\to\infty}\frac{e^{2x}}{x^4+2}\left(\frac{\infty}{\infty}\right) = \lim_{x\to\infty}\frac{2e^{2x}}{4x^3}\left(\frac{\infty}{\infty}\right)$$
$$= \lim_{x\to\infty}\frac{4e^{2x}}{12x^2}\left(\frac{\infty}{\infty}\right)$$
$$= \lim_{x\to\infty}\frac{8e^{2x}}{24x}\left(\frac{\infty}{\infty}\right)$$
$$= \lim_{x\to\infty}\frac{16e^{2x}}{24}$$
$$= \infty$$

65. See Example 6.8.

$$\lim_{x\to\infty}\left(1+\frac{3}{x}\right)^x (1^\infty)$$

$$y = \left(1+\frac{3}{x}\right)^x$$

$$\ln y = x\ln\left(1+\frac{3}{x}\right)$$

$$\lim_{x\to\infty}\ln y = \lim_{x\to\infty} x\ln\left(1+\frac{3}{x}\right)(\infty\cdot0)$$

$$= \lim_{x\to\infty}\frac{\ln\left(1+\frac{3}{x}\right)}{\frac{1}{x}}\left(\frac{0}{0}\right)$$

$$= \lim_{x\to\infty}\frac{\frac{1}{1+\frac{3}{x}}\cdot\left(-\frac{3}{x^2}\right)}{-\frac{1}{x^2}}$$

$$= \lim_{x\to\infty}\frac{3}{1+\frac{3}{x}}$$

$$= 3$$

$$\lim_{x\to\infty}y = \lim_{x\to\infty}e^{\ln y} = e^3$$

254

67. See Example 6.1.

$$\lim_{x\to 0}\frac{\tan x}{x}\left(\frac{0}{0}\right)=\lim_{x\to 0}\frac{\sec^2 x}{1}$$
$$=1$$

69. See Example 7.1.

$$\int_0^1\frac{x}{x^2-1}\,dx=\lim_{R\to 1^-}\int_0^R\frac{x}{x^2-1}\,dx$$
$$=\lim_{R\to 1^-}\frac{1}{2}\ln\left|x^2-1\right|\bigg|_0^R$$
$$=\lim_{R\to 1^-}\left(\frac{1}{2}\ln\left|R^2-1\right|-\frac{1}{2}\ln\left|-1\right|\right)$$
$$=-\infty$$

diverges

71. See Example 7.6.

$$\int_1^\infty\frac{3}{x^2}\,dx=\lim_{R\to\infty}\int_1^R\frac{3}{x^2}\,dx$$
$$=\lim_{R\to\infty}\left(-\frac{3}{x}\bigg|_1^R\right)$$
$$=\lim_{R\to\infty}\left(-\frac{3}{R}+3\right)$$
$$=3$$

73. See Example 7.6.

$$\int_0^\infty\frac{4}{4+x^2}\,dx=\lim_{R\to\infty}\int_0^R\frac{4}{4+x^2}\,dx$$
$$=\lim_{R\to\infty}\int_0^R\frac{1}{1+\left(\frac{x}{2}\right)^2}\,dx$$
$$=\lim_{R\to\infty}2\tan^{-1}\left(\frac{x}{2}\right)\bigg|_0^R$$
$$=\lim_{R\to\infty}\left[2\tan^{-1}\left(\frac{R}{2}\right)-2\tan^{-1}(0)\right]$$
$$=2\left(\frac{\pi}{2}\right)$$
$$=\pi$$

75. See Example 7.5.

$$\int_{-2}^2\frac{3}{x^2}\,dx=\int_{-2}^0\frac{3}{x^2}\,dx+\int_0^2\frac{3}{x^2}\,dx$$
$$=\lim_{R\to 0^-}\int_{-2}^R\frac{3}{x^2}\,dx+\lim_{R\to 0^+}\int_R^2\frac{3}{x^2}\,dx$$
$$=\lim_{R\to 0^-}\left(-\frac{3}{x}\right)\bigg|_{-2}^R+\lim_{R\to 0^+}\left(-\frac{3}{x}\right)\bigg|_R^2$$
$$=\lim_{R\to 0^-}\left(-\frac{3}{R}+\frac{3}{2}\right)+\lim_{R\to 0^+}\left(-\frac{3}{2}+\frac{3}{R}\right)$$
$$=\infty+\infty$$

diverges

Chapter 8

Section 8.1

5. $1, \dfrac{3}{4}, \dfrac{5}{9}, \dfrac{7}{16}, \dfrac{9}{25}, \dfrac{11}{36}$

7. $4, 2, \dfrac{2}{3}, \dfrac{1}{6}, \dfrac{1}{30}, \dfrac{1}{180}$

9. a. $\lim\limits_{n\to\infty} \dfrac{1}{n^3} = 0$

b. Let $\varepsilon > 0$ be given. We must find N sufficiently large so that, for every

$n > N$, $\left| \dfrac{1}{n^3} - 0 \right| < \varepsilon$. This is equivalent to

$\dfrac{1}{n^3} < \varepsilon$ which implies $\dfrac{1}{\varepsilon} < n^3$, so $\sqrt[3]{\dfrac{1}{\varepsilon}} < n$.

Thus choosing $N > \sqrt[3]{\dfrac{1}{\varepsilon}}$, we have, for all

$n > N$, $\dfrac{1}{n^3} < \varepsilon$. Hence, $\left\{ \dfrac{1}{n^3} \right\}_{n=1}^{\infty}$ converges to zero.

11. a. $\lim\limits_{n\to\infty} \dfrac{n}{n+1} = \lim\limits_{n\to\infty} \dfrac{1}{1+\frac{1}{n}} = 1$

b. Let $\varepsilon > 0$ be given. We must find N sufficiently large so that for every $n > N$,

$\left| \dfrac{n}{n+1} - 1 \right| < \varepsilon$. This is equivalent to

$\left| \dfrac{-1}{n+1} \right| < \varepsilon$ which implies that $\dfrac{1}{n+1} < \varepsilon$ so

that $\dfrac{1}{\varepsilon} - 1 < n$. Thus choosing $N > \dfrac{1}{\varepsilon} - 1$ we

have for every $n > N$, $\left| \dfrac{n}{n+1} - 1 \right| < \varepsilon$. Hence,

$\left\{ \dfrac{n}{n+1} \right\}_{n=1}^{\infty}$ converges to 1.

13. a. $\lim\limits_{n\to\infty} \dfrac{2}{\sqrt{n}} = 0$

b. Let $\varepsilon > 0$ be given. We must find N sufficiently large so that for every $n > N$,

$\left| \dfrac{2}{\sqrt{n}} - 0 \right| < \varepsilon$. This is equivalent to $\dfrac{2}{\sqrt{n}} < \varepsilon$

which implies that $\dfrac{2}{\varepsilon} < \sqrt{n}$ so that $\dfrac{4}{\varepsilon^2} < n$.

Thus choosing $N > \dfrac{4}{\varepsilon^2}$ we have for every

$n > N$, $\dfrac{2}{\sqrt{n}} < \varepsilon$. Hence, $\left\{ \dfrac{2}{\sqrt{n}} \right\}_{n=1}^{\infty}$

converges to zero.

15. $\lim\limits_{n\to\infty} \dfrac{3n^2+1}{2n^2-1} = \lim\limits_{n\to\infty} \dfrac{3+\frac{1}{n^2}}{2-\frac{1}{n^2}} = \dfrac{3}{2}$; converges to $\dfrac{3}{2}$

17. $\lim\limits_{n\to\infty} \dfrac{n^2+1}{n+1} = \lim\limits_{n\to\infty} \dfrac{n+\frac{1}{n}}{1+\frac{1}{n}} = \infty$; diverges

19. $\lim\limits_{n\to\infty} \dfrac{n+2}{3n-1} = \lim\limits_{n\to\infty} \dfrac{1+\frac{2}{n}}{3-\frac{1}{n}} = \dfrac{1}{3}$; converges to $\dfrac{1}{3}$

21. $\lim\limits_{n\to\infty} (-1)^n \dfrac{n+2}{3n-1} = \lim\limits_{n\to\infty} (-1)^n \dfrac{1+\frac{2}{n}}{3-\frac{1}{n}} = \pm\dfrac{1}{3}$,

so $\lim\limits_{n\to\infty} (-1)^n \dfrac{n+2}{3n-1}$ does not exist; diverges

23. $\lim\limits_{n\to\infty} \left| (-1)^n \dfrac{n+2}{n^2+4} \right| = \lim\limits_{n\to\infty} \dfrac{n+2}{n^2+4}$

$= \lim\limits_{n\to\infty} \dfrac{\frac{1}{n}+\frac{2}{n^2}}{1+\frac{4}{n^2}} = 0$

so by Corollary 1.1, $\lim\limits_{n\to\infty} \dfrac{(-1)^2(n+2)}{n^2+4} = 0$;

converges to 0

25. $\{\cos \pi n\}_{n=1}^{\infty} = \{-1, 1, -1, 1, \ldots\}$; diverges

27. $\lim\limits_{x\to\infty} \dfrac{x}{e^x} = \lim\limits_{x\to\infty} \dfrac{1}{e^x}$ by l'Hopital's Rule and

$\lim\limits_{x\to\infty} \dfrac{1}{e^x} = 0$, so by Theorem 1.2 $\lim\limits_{n\to\infty} \dfrac{n}{e^n} = 0$;

converges to 0

29. $\displaystyle\lim_{n\to\infty}\frac{e^n+2}{e^{2n}-1}=\lim_{n\to\infty}\frac{\frac{1}{e^n}+\frac{2}{e^{2n}}}{1-\frac{1}{e^{2n}}}=0$; converges to 0

31. For $n\ge 1$, $e^n+1<2e^n$,

so $\dfrac{3^n}{e^n+1}\ge\dfrac{1}{2}\cdot\dfrac{3^n}{e^n}=\dfrac{1}{2}\cdot\left(\dfrac{3}{e}\right)^n$.

Since $\dfrac{3}{e}>1$, $\displaystyle\lim_{n\to\infty}\frac{1}{2}\left(\frac{3}{e}\right)^n=\infty$; so

$\displaystyle\lim_{n\to\infty}\frac{3^n}{e^n+1}=\infty$; diverges

33. $-1\le\cos n\le 1\Rightarrow\dfrac{-1}{n!}\le\dfrac{\cos n}{n!}\le\dfrac{1}{n!}$ for all n, and

$\displaystyle\lim_{n\to\infty}\frac{-1}{n!}=\lim_{n\to\infty}\frac{1}{n!}=0$ so by the Squeeze

Theorem, $\displaystyle\lim_{n\to\infty}\frac{\cos n}{n!}=0$; converges to 0

35. $-1\le\cos n\le 1\Rightarrow\dfrac{-1}{n^2}\le\dfrac{\cos n}{n^2}\le\dfrac{1}{n^2}$ for all n, and

$\displaystyle\lim_{n\to\infty}\frac{-1}{n^2}=\lim_{n\to\infty}\frac{1}{n^2}=0$ so by the Squeeze

Theorem $\displaystyle\lim_{n\to\infty}\frac{\cos n}{n^2}=0$; converges to 0

37. $0\le|a_n|=\dfrac{1}{ne^n}\le\dfrac{1}{n}$ and $\displaystyle\lim_{n\to\infty}\frac{1}{n}=0$ so by the

Squeeze Theorem and Corollary 1.1,

$\displaystyle\lim_{n\to\infty}\frac{(-1)^n}{ne^n}=0$; converges to 0

39. $\dfrac{a_{n+1}}{a_n}=\left(\dfrac{n+4}{n+3}\right)\cdot\left(\dfrac{n+2}{n+3}\right)=\dfrac{n^2+6n+8}{n^2+6n+9}<1$ for

all n, so $a_{n+1}<a_n$ for all n, so $\{a_n\}_{n=1}^{\infty}$ is decreasing.

41. $\dfrac{a_{n+1}}{a_n}=\left(\dfrac{e^{n+1}}{n+1}\right)\cdot\left(\dfrac{n}{e^n}\right)=\dfrac{e\cdot n}{n+1}>1$ for all n, so

$a_{n+1}>a_n$ for all n, so $\{a_n\}_{n=1}^{\infty}$ is increasing.

43. $\dfrac{a_{n+1}}{a_n}=\left(\dfrac{2^{n+1}}{(n+2)!}\right)\cdot\left(\dfrac{(n+1)!}{2^n}\right)=\dfrac{2}{n+2}<1$ for all

n, so $a_{n+1}<a_n$ for all n, so $\{a_n\}_{n=1}^{\infty}$ is

decreasing.

45. $\left(\dfrac{a_{n+1}}{a_n}\right)=\left(\dfrac{10^{n+1}}{(n+1)!}\right)\cdot\left(\dfrac{n!}{10^n}\right)=\dfrac{10}{n+1}<1$ for

$n>9$, so $a_{n+1}<a_n$ for $n>9$ so as

$n\to\infty$, $\{a_n\}_{n=1}^{\infty}$ is decreasing.

47. $|a_n|=\left|\dfrac{3n^2-2}{n^2+1}\right|=\dfrac{3n^2-2}{n^2+1}<\dfrac{3n^2}{n^2+1}<\dfrac{3n^2}{n^2}=3$ so

$\{a_n\}_{n=1}^{\infty}$ is bounded by 3; $|a_n|<3$

49. $|a_n|=\left|\dfrac{\sin(n^2)}{n+1}\right|<\dfrac{1}{n+1}<\dfrac{1}{2}$ for $n>1$, so $\{a_n\}_{n=1}^{\infty}$

is bounded by $\dfrac{1}{2}$; $|a_n|<\dfrac{1}{2}$.

51. $a_n=\left(1+\dfrac{1}{n}\right)^n$; $a_{1000}\approx 2.716924$; $e\approx 2.718282$

$b_n=\left(1-\dfrac{1}{n}\right)^n$; $b_{1000}\approx 0.367695$; $e^{-1}\approx 0.367879$

53. A plausible argument goes like this:

$$\lim_{n\to\infty}\left(1+\frac{r}{n}\right)^n=\lim_{m\to\infty}\left(1+\frac{r}{r\cdot m}\right)^{r\cdot m}$$

$$=\lim_{m\to\infty}\left[\left(1+\frac{1}{m}\right)^m\right]^r$$

Note that $m=rn$ will generally not be an

integer, so we need to proceed as follows:

Let $y=x/r$. Then

$$\lim_{x\to\infty}\left(1+\frac{r}{x}\right)^x=\lim_{y\to\infty}\left(1+\frac{1}{y}\right)^{ry}$$

$$=\left[\lim_{y\to\infty}\left(1+\frac{1}{y}\right)^y\right]^r$$

$$=e^r$$

Now apply Theorem 1.2 to show that

$$\lim_{n\to\infty}\left(1+\frac{r}{n}\right)^n=e^r.$$

55. If side $s = 12$" and the diameter $D = \dfrac{12}{n}$ then the number of disks that fit along one side is $\dfrac{12}{\left(\frac{12}{n}\right)} = n$. Thus, the total number of disks is

$$\frac{12}{\left(\frac{12}{n}\right)} \cdot \frac{12}{\left(\frac{12}{n}\right)} = n \cdot n = n^2.$$

a_n = wasted area in box with n^2 disks, so

a_n = Area of box–Area of disks

$$a_n = 12 \cdot 12 - n^2 \left(\frac{6}{n}\right)^2 \pi = 144 - 36\pi$$

$$a_n \approx 30.9$$

57. $a_0 = 3.049$

$a_1 = 3.049 + .005(3.049)^{2.01} = 3.096$

$a_2 = 3.096 + .005(3.096)^{2.01} = 3.144$

$a_3 = 3.144 + .005(3.144)^{2.01} = 3.194$

$a_{10} = 3.594 < 3.721 =$ population in 1970

$a_{20} = 4.376 < 4.473 =$ population in 1980

$a_{30} = 5.589 > 5.333 =$ population 1990

Estimated population in 2035 $= a_{75}$

$$= 4.131 \times 10^{53}$$

59. In the 3rd month, only the adult rabbits have newborns, so $a_3 = 2 + 1 = 3$.

In the 4th month, only the 2 pairs of adult rabbits from a_2 can have newborns, so $a_4 = 3 + 2 = 5$.

In general, $a_n = a_{n-1} + a_{n-2}$

61. $\displaystyle\lim_{n \to \infty} \frac{\ln n}{n} = \lim_{n \to \infty} \frac{\frac{1}{n}}{1} = \lim_{n \to \infty} \frac{1}{n} = 0$ by l'Hopital's

Rule so $\displaystyle\lim_{n \to \infty} \sqrt[n]{n} = \lim_{n \to \infty} e^{(1/n)\ln n} = e^0 = 1$ so the sequence converges to 1.

63. Letting $L = \displaystyle\lim_{n \to \infty} a_{n+1} = \lim_{n \to \infty} a_n$, we have

$$L = \frac{1}{2}\left(L + \frac{c}{L}\right) \Rightarrow 2L^2 = L^2 + c \Rightarrow L^2 = c$$

$$\Rightarrow L = \sqrt{c}.$$

Section 8.2

5. $\displaystyle\sum_{k=0}^{\infty} 3\left(\frac{1}{5}\right)^k$ is a geometric series with $a = 3$ and $|r| = \frac{1}{5} < 1$, so it converges to $\dfrac{3}{1 - \frac{1}{5}} = \dfrac{15}{4}$.

7. $\displaystyle\sum_{k=0}^{\infty} \frac{1}{2}\left(-\frac{1}{3}\right)^k$ is a geometric series with $a = \frac{1}{2}$ and $|r| = \frac{1}{3} < 1$, so it converges to $\dfrac{\frac{1}{2}}{1 - \left(-\frac{1}{3}\right)} = \dfrac{3}{8}$.

9. $\displaystyle\sum_{k=0}^{\infty} \frac{1}{2}(3)^k$ is a geometric series with $|r| = 3 > 1$, so it diverges.

11. Using partial fractions,

$$S_n = \sum_{k=1}^{n} \frac{4}{k(k+2)} = \sum_{k=1}^{n} \left(\frac{2}{k} - \frac{2}{k+2}\right)$$

$$= \left(2 - \frac{2}{3}\right) + \left(1 - \frac{2}{4}\right) + \left(\frac{2}{3} - \frac{2}{5}\right) + \cdots + \left(\frac{2}{n-1} - \frac{2}{n+1}\right) + \left(\frac{2}{n} - \frac{2}{n+2}\right)$$

$$= 2 + 1 - \frac{2}{n+1} - \frac{2}{n+2}$$

$$= 3 - \frac{4n+6}{n^2+3n+2}$$

and $\displaystyle\lim_{n \to \infty} S_n = \lim_{n \to \infty}\left(3 - \frac{4n+6}{n^2+3n+2}\right) = 3$. Thus, the series converges to 3.

13. $\lim\limits_{k\to\infty}\dfrac{3k}{k+4}=3\neq 0$, so by the kth-Term Test for Divergence, the series diverges.

15. $\sum\limits_{k=1}^{\infty}\dfrac{2}{k}=2\sum\limits_{k=1}^{\infty}\dfrac{1}{k}$ and from Example 2.7, $\sum\limits_{k=1}^{\infty}\dfrac{1}{k}$ diverges, so $2\sum\limits_{k=1}^{\infty}\dfrac{1}{k}$ diverges.

17. Using partial fractions

$$S_n=\sum_{k=1}^{n}\frac{2k+1}{k^2(k+1)^2}$$

$$=\sum_{k=1}^{n}\left[\frac{1}{k^2}-\frac{1}{(k+1)^2}\right]$$

$$=\left(1-\frac{1}{4}\right)+\left(\frac{1}{4}-\frac{1}{9}\right)+\left(\frac{1}{9}-\frac{1}{16}\right)+\ldots+\left(\frac{1}{(n-1)^2}-\frac{1}{n^2}\right)+\left(\frac{1}{n^2}-\frac{1}{(n+1)^2}\right)$$

$$=1-\frac{1}{(n+1)^2}\text{ and }\lim_{n\to\infty}S_n$$

$$=\lim_{n\to\infty}\left[1-\frac{1}{(n+1)^2}\right]$$

$$=1$$

Thus the series converges to 1.

19. $\sum\limits_{k=0}^{\infty}\dfrac{1}{3^k}$ is a geometric series with $a=1$, and $|r|=\dfrac{1}{3}<1$ so it converges to $\dfrac{1}{1-\frac{1}{3}}=\dfrac{3}{2}$.

21. $\sum\limits_{k=0}^{\infty}\left(\dfrac{1}{2^k}-\dfrac{1}{k+1}\right)=\sum\limits_{k=0}^{\infty}\dfrac{1}{2^k}-\sum\limits_{k=0}^{\infty}\dfrac{1}{k+1}$. The first series is a convergent geometric series but the second series is the divergent harmonic series, so the original series diverges.

23. $\sum\limits_{k=0}^{\infty}\left(\dfrac{2}{3^k}+\dfrac{1}{2^k}\right)=\sum\limits_{k=0}^{\infty}\dfrac{2}{3^k}+\sum\limits_{k=0}^{\infty}\dfrac{1}{2^k}$. The first series is a geometric series with $a=2$ and $|r|=\dfrac{1}{3}<1$ so it converges to $\dfrac{2}{1-\frac{1}{3}}=3$. The second series is a geometric series with $a=1$ and $|r|=\dfrac{1}{2}<1$, so it converges to $\dfrac{1}{1-\frac{1}{2}}=2$. Thus $\sum\limits_{k=0}^{\infty}\left(\dfrac{2}{3^k}+\dfrac{1}{2^k}\right)$ converges to $3+2=5$.

25. $\sum\limits_{k=0}^{\infty}3\left(-\dfrac{1}{2}\right)^k$ is a geometric series with $a=3$ and $|r|=\dfrac{1}{2}<1$ so it converges to $\dfrac{3}{1-\left(-\frac{1}{2}\right)}=2$.

27. $\lim\limits_{k\to\infty}|a_k|=\lim\limits_{k\to\infty}\dfrac{3k}{k+1}=3\neq 0$ so by the kth-Term Test for Divergence, the series diverges.

29.

n	$S_n = \sum\limits_{k=1}^{n} \dfrac{1}{k^2}$
1	1
2	1.25
3	1.3611
4	1.4236
5	1.4636
6	1.4914
7	1.5118
8	1.5274
9	1.5398
10	1.5498
11	1.5580
12	1.5650
13	1.5709
14	1.5760
15	1.5804
16	1.5843
17	1.5878
18	1.5909
19	1.5937
20	1.5962

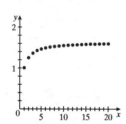

converges

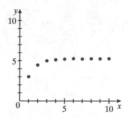

converges

33.

n	$S_n = \sum\limits_{k=1}^{n} \dfrac{4^k}{k^2}$
1	4
2	8
3	15.1111
4	31.1111
5	72.0711
6	185.8489
7	520.2162
8	1544.2162
9	4780.5619
10	15,266.3219

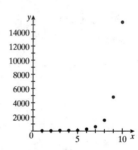

diverges

31.

n	$S_n = \sum\limits_{k=1}^{n} \dfrac{3}{k!}$
1	3
2	4.5
3	5.0
4	5.125
5	5.150
6	5.15417
7	5.15476
8	5.15484
9	5.154844
10	5.154845

35. Assume $\sum\limits_{k=1}^{\infty} a_k$ converges to L. Then for any m,

$$L = \sum_{k=1}^{\infty} a_k = \sum_{k=1}^{m-1} a_k + \sum_{k=m}^{\infty} a_k = S_{m-1} + \sum_{k=m}^{\infty} a_k .$$

So $\sum\limits_{k=m}^{\infty} a_k = L - S_{m-1}$, and thus converges.

37. Assume $\sum_{k=1}^{\infty} a_k$ converges to A and $\sum_{k=1}^{\infty} b_k$ converges to B. Then the sequences of partial sums converge, and letting

$$S_n = \sum_{k=1}^{n} a_k \text{ and } T_n = \sum_{k=1}^{n} b_k,$$

we have $\lim_{n \to \infty} S_n = A$ and $\lim_{n \to \infty} T_n = B$.

Let $Q_n = \sum_{k=1}^{n} (a_k + b_k)$, the sequence of partial

sums for $\sum_{k=1}^{\infty} (a_k + b_k)$. Since S, T, and Q are all

finite sums, $Q_n = S_n + T_n$.

Then by Theorem 1.1(i),

$$A + B = \lim_{n \to \infty} S_n + \lim_{n \to \infty} T_n$$
$$= \lim_{n \to \infty} (S_n + T_n)$$
$$= \lim_{n \to \infty} Q_n$$
$$= \sum_{k=1}^{\infty} (a_k + b_k)$$

The proofs for $\sum_{k=1}^{\infty} (a_k - b_k)$ and $\sum_{k=1}^{\infty} ca_k$ are

similar.

39. Let $S_n = \sum_{k=1}^{\infty} \frac{1}{k}$. Then $S_1 = \frac{1}{1} = 1$ and

$S_2 = 1 + \frac{1}{2} = \frac{3}{2}$. Since $S_8 > \frac{5}{2}$, we have

$$S_{16} = S_8 + \frac{1}{9} + \frac{1}{10} + \dots + \frac{1}{16} > S_8 + 8\left(\frac{1}{16}\right)$$

$$= S_8 + \frac{1}{2} > \frac{5}{2} + \frac{1}{2} = 3. \text{ So } S_{16} > 3.$$

$$S_{32} = S_{16} + \frac{1}{17} + \frac{1}{18} + \dots + \frac{1}{32} > S_{16} + 16\left(\frac{1}{32}\right)$$

$$= S_{16} + \frac{1}{2} > 3 + \frac{1}{2} = \frac{7}{2}. \text{ So } S_{32} > \frac{7}{2}$$

If $n = 64$, then $S_{64} > 4$. If $n = 256$, then

$S_{256} > 5$. If $n = 4^{m-1}$, then $S_n > m$.

41. $.9 + .09 + .009 + \dots = \sum_{k=0}^{\infty} .9(.1)^k$ which is a

geometric series with $a = .9$ and $|r| = .1 < 1$ so it

converges to $\frac{.9}{1 - .1} = 1$.

43. The amount of overhang is $\sum_{k=0}^{n-1} \frac{L}{2(n-k)}$. So if

$n = 8$, then $\sum_{k=0}^{7} \frac{L}{2(8-k)} = 1.3589L$.

When $n = 4$,

$$L\sum_{k=0}^{n-1} \frac{1}{2(n-k)} = L\sum_{k=0}^{3} \frac{1}{2(4-k)} = 1.0417L > L.$$

$$\lim_{n \to \infty} \sum_{k=0}^{n-1} \frac{L}{2(n-k)} = \lim_{n \to \infty} \sum_{k=1}^{n} \frac{L}{2k}$$

$$= \frac{L}{2} \lim_{n \to \infty} \sum_{k=1}^{n} \frac{1}{k}$$

$$= \infty$$

Diverges

45. If $0 < 2r < 1$ then $1 + 2r + (2r)^2 + \dots = \sum_{k=0}^{\infty} (2r)^k$

which is a geometric series with $a = 1$ and

$|2r| = 2r < 1$, so it converges to $\frac{1}{1 - 2r}$.

If $r = \frac{1}{1000}$, then

$$1 + .002 + .000004 + \dots = \frac{1}{1 - 2\left(\frac{1}{1000}\right)}$$

$$= \frac{500}{499}$$

$$= 1.002004008.$$

47. $p^2 + 2p(1-p)p^2 + [2p(1-p)]^2 p^2 + \dots$

$$= \sum_{k=0}^{\infty} p^2 [2p(1-p)]^k$$

is a geometric series with $a = p^2$ and

$|r| = 2p(1-p) < 1$ because $2p(1-p)$ is a

probability and therefore must be between 0 and

1. So the series converges to

$$\frac{p^2}{1 - [2p(1-p)]} = \frac{p^2}{1 - 2p + 2p^2}.$$

If $p = .6$, $\dfrac{.6^2}{1 - 2(.6) + 2(.6)^2} = .692 > .6$.

If $p > \frac{1}{2}$, $\dfrac{p^2}{1 - 2p(1-p)} > p$.

49. $30 + 15 + \dfrac{15}{2} + \ldots = \displaystyle\sum_{k=0}^{\infty} 30\left(\dfrac{1}{2}\right)^k$

$= \dfrac{30}{1 - \frac{1}{2}}$

$= 60$ miles

The bikes meet after 1 hour. In that time, a fly flying 60 mph will have traveled 60 miles.

51. $\displaystyle\sum_{k=0}^{\infty} 100{,}000\left(\dfrac{3}{4}\right)^k = \dfrac{100{,}000}{1 - \frac{3}{4}} = 400{,}000;$

$\$400{,}000$

53. $d + de^{-r} + de^{-2r} + \ldots = \displaystyle\sum_{k=0}^{\infty} d\left(e^{-r}\right)^k$ which is a

geometric series with $a = d$ and $\left|e^{-r}\right| = e^{-r} < 1$ if

$r > 0$. So $\displaystyle\sum_{k=0}^{\infty} d(e^{-r})^k = \dfrac{d}{1 - e^{-r}}.$

If $r = .1$, $\displaystyle\sum_{k=0}^{\infty} d(e^{-.1})^k = \dfrac{d}{1 - e^{-.1}} = 2$

so $d = 2(1 - .905) \approx .19$

55. $\displaystyle\sum_{k=0}^{\infty} ce^{-r}\left(\dfrac{1}{e^{rk}}\right) = \dfrac{ce^{-r}}{1 - e^{-r}} = \dfrac{c}{e^r - 1}$

Section 8.3

5. $\displaystyle\sum_{k=1}^{\infty} \dfrac{4}{\sqrt[3]{k}}$ is a divergent p–series $\left(p = \dfrac{1}{3} < 1\right);$

diverges

7. $\displaystyle\sum_{k=1}^{\infty} \dfrac{1}{k^{11/10}}$ is a convergent p–series

$\left(p = \dfrac{11}{10} > 1\right);$ converges

9. Using the Limit Comparison Test, let

$a_k = \dfrac{k+1}{k^2 + 2k + 3}$ and $b_k = \dfrac{1}{k}$, so

$\displaystyle\lim_{k\to\infty} \dfrac{a_k}{b_k} = \lim_{k\to\infty}\left(\dfrac{k+1}{k^2 + 2k + 3}\right)\left(\dfrac{k}{1}\right)$

$= \displaystyle\lim_{k\to\infty} \dfrac{k^2 + k}{k^2 + 2k + 3}$

$= 1 > 0,$

and since $\displaystyle\sum_{k=1}^{\infty} \dfrac{1}{k}$ is the divergent harmonic series,

$\displaystyle\sum_{k=0}^{\infty} \dfrac{k+1}{k^2 + 2k + 3}$ diverges.

11. Using the Limit Comparison Test, let

$a_k = \dfrac{4}{2 + 4k}$ and $b_k = \dfrac{1}{k}$, so

$\displaystyle\lim_{k\to\infty} \dfrac{a_k}{b_k} = \lim_{k\to\infty}\left(\dfrac{4}{2 + 4k}\right)\left(\dfrac{k}{1}\right)$

$= \displaystyle\lim_{k\to\infty} \dfrac{4k}{2 + 4k}$

$= 1 > 0$

and since $\displaystyle\sum_{k=1}^{\infty} \dfrac{1}{k}$ is the divergent harmonic series,

$\displaystyle\sum_{k=1}^{\infty} \dfrac{4}{2 + 4k}$ diverges.

13. Let $f(x) = \dfrac{2}{x \ln x}$. Then f is continuous and

positive on $[2, \infty]$ and $f'(x) = \dfrac{-2(1 + \ln x)}{x^2 (\ln x)^2} < 0$

for $x \in [2, \infty)$, so f is decreasing. So we can use

the Integral Test,

$\displaystyle\int_2^{\infty} \dfrac{2}{x \ln x}\,dx = \lim_{R\to\infty} \int_2^R \dfrac{2}{x \ln x}\,dx$

$= 2 \displaystyle\lim_{R\to\infty} [\ln(\ln x)]_2^R$

$= 2 \displaystyle\lim_{R\to\infty} [\ln(\ln R) - \ln(\ln 2)]$

$= \infty$, so $\displaystyle\sum_{k=2}^{\infty} \dfrac{2}{k \ln k}$ diverges.

15. Using the Limit Comparison Test, let $a_k = \dfrac{2k}{k^3+1}$

and $b_k = \dfrac{1}{k^2}$. Then

$$\lim_{k\to\infty}\frac{a_k}{b_k} = \lim_{k\to\infty}\left(\frac{2k}{k^3+1}\right)\cdot\left(\frac{k^2}{1}\right)$$
$$= \lim_{k\to\infty}\frac{2k^3}{k^3+1}$$
$$= 2 > 0$$

and $\displaystyle\sum_{k=1}^{\infty}\frac{1}{k^2}$ is a convergent p-series $(p=2>1)$,

so $\displaystyle\sum_{k=1}^{\infty}\frac{2k}{k^3+1}$ converges.

17. Let $f(x) = \dfrac{e^{1/x}}{x^2}$. Then f is continuous and

positive on $[1,\infty)$ and $f'(x) = \dfrac{-e^{1/x}(1+2x)}{x^4} < 0$

for all $x\in[1,\infty)$, so f is decreasing. Therefore, we
can use the Integral Test.

$$\int_1^{\infty}\frac{e^{1/x}}{x^2}dx = \lim_{R\to\infty}\int_1^{R}\frac{e^{1/x}}{x^2}dx$$
$$= \lim_{R\to\infty}\left[-e^{1/x}\right]_1^R$$
$$= \lim_{R\to\infty}\left[e - e^{1/R}\right]$$
$$= e - 1$$

So the series $\displaystyle\sum_{k=1}^{\infty}\frac{e^{1/k}}{k^2}$ converges.

19. Let $f(x) = \dfrac{e^{-\sqrt{x}}}{\sqrt{x}}$. Then f is continuous and

positive on $[1,\infty)$ and $f'(x) = \dfrac{-(\sqrt{x}-1)}{2x^{3/2}e^{\sqrt{x}}} < 0$ for

$x\in[1,\infty)$. So f is decreasing. Therefore, we can
use the Integral Test.

$$\int_1^{\infty}\frac{e^{-\sqrt{x}}}{\sqrt{x}}dx = \lim_{R\to\infty}\int_1^{R}\frac{e^{-\sqrt{x}}}{\sqrt{x}}dx$$
$$= \lim_{R\to\infty}\left[-2e^{-\sqrt{x}}\right]_1^R$$
$$= \lim_{R\to\infty}\left[\frac{2}{e} - \frac{2}{e^{\sqrt{R}}}\right] = \frac{2}{e}$$

So $\displaystyle\sum_{k=1}^{\infty}\frac{e^{\sqrt{k}}}{\sqrt{k}}$ converges.

21. Using the Limit Comparison Test, let
$a_k = \dfrac{3k}{k^{3/2}+2}$ and $b_k = \dfrac{1}{k^{1/2}}$. Then

$$\lim_{k\to\infty}\frac{a_k}{b_k} = \lim_{k\to\infty}\left(\frac{3k}{k^{3/2}+2}\right)\left(\frac{k^{1/2}}{1}\right)$$
$$= \lim_{k\to\infty}\frac{3k^{3/2}}{k^{3/2}+2}$$
$$= 3 > 0$$

and $\displaystyle\sum_{k=1}^{\infty}\frac{1}{k^{1/2}}$ is a divergent p-series

$\left(p=\dfrac{1}{2}<1\right)$, so $\displaystyle\sum_{k=1}^{\infty}\frac{3k}{k^{3/2}+2}$ diverges.

23. Using the Limit Comparison Test, let
$a_k = \dfrac{2}{\sqrt{k^2+4}}$ and $b_k = \dfrac{1}{k}$. Then

$$\lim_{k\to\infty}\frac{a_k}{b_k} = \lim_{k\to\infty}\left(\frac{2}{\sqrt{k^2+4}}\right)\left(\frac{k}{1}\right)$$
$$= \lim_{k\to\infty}\frac{2k}{\sqrt{k^2+4}}$$
$$= 2 > 0$$

and $\displaystyle\sum_{k=1}^{\infty}\frac{1}{k}$ is the divergent harmonic series, so

$\displaystyle\sum_{k=0}^{\infty}\frac{2}{\sqrt{k^2+4}}$ diverges.

25. Using the Limit Comparison Test, let
$a_k = \dfrac{k^2+1}{\sqrt{k^5+1}}$ and $b_k = \dfrac{1}{k^{1/2}}$. Then

$$\lim_{k\to\infty}\frac{a_k}{b_k} = \lim_{k\to\infty}\left(\frac{k^2+1}{\sqrt{k^5+1}}\right)\left(\frac{k^{1/2}}{1}\right)$$
$$= \lim_{k\to\infty}\frac{k^{5/2}+k^{1/2}}{\sqrt{k^5+1}}$$
$$= 1 > 0$$

and $\displaystyle\sum_{k=1}^{\infty}\frac{1}{k^{1/2}}$ is a divergent p-series

$\left(p=\dfrac{1}{2}<1\right)$, so $\displaystyle\sum_{k=0}^{\infty}\frac{k^2+1}{\sqrt{k^5+1}}$ diverges.

27. Let $f(x) = \dfrac{\tan^{-1} x}{1 + x^2}$ which is continuous and

positive on $[1, \infty)$ and $f'(x) = \dfrac{1 - 2x \tan^{-1} x}{(1 + x^2)^2} < 0$

for $x \in [1, \infty)$, so f is decreasing. So we can use the Integral Test.

$$\int_1^\infty \frac{\tan^{-1} x}{1 + x^2} dx = \lim_{R \to \infty} \int_1^R \frac{\tan^{-1} x}{1 + x^2} dx$$

$$= \lim_{R \to \infty} \left[\frac{1}{2} (\tan^{-1} x)^2 \right]_1^R$$

$$= \lim_{R \to \infty} \left[\frac{1}{2} (\tan^{-1} R)^2 - \frac{1}{2} (\tan^{-1} 1)^2 \right]$$

$$= \frac{1}{2} \left(\frac{\pi}{2} \right)^2 - \frac{1}{2} \left(\frac{\pi}{4} \right)^2$$

$$= \frac{3\pi^2}{32},$$

so $\displaystyle\sum_{k=1}^\infty \frac{\tan^{-1} k}{1 + k^2}$ converges.

29. $\left| \cos^2 k \right| < k$ for all

$k \Rightarrow \cos^2 k < k \Rightarrow \dfrac{1}{\cos^2 k} > \dfrac{1}{k}$ and $\displaystyle\sum_{k=1}^\infty \frac{1}{k}$ is the

divergent harmonic series so by the Comparison

Test, $\displaystyle\sum_{k=1}^\infty \frac{1}{\cos^2 k}$ diverges.

31. $-1 \le \sin k \le 1 \Rightarrow 0 < \dfrac{\sin k + 2}{k^2} \le \dfrac{3}{k^2}$ and $\displaystyle\sum_{k=1}^\infty \frac{1}{k^2}$

is convergent p–series $(p = 2 > 1)$, so by the

Comparison Test, $\displaystyle\sum_{k=1}^\infty \frac{\sin k + 2}{k^2}$ converges.

33. Let $f(x) = \dfrac{\ln x}{x}$ which is continuous and positive

on $[2, \infty)$ and $f'(x) = \dfrac{1 - \ln x}{x^2} < 0$ for $x \in [2, \infty)$,

so f is decreasing. Therefore, we can use the Integral Test.

$$\int_2^\infty \frac{\ln x}{x} dx = \lim_{R \to \infty} \int_2^R \frac{\ln x}{x} dx$$

$$= \lim_{R \to \infty} \left[\frac{(\ln x)^2}{2} \right]_2^R$$

$$= \lim_{R \to \infty} \left[\frac{(\ln R)^2}{2} - \frac{(\ln 2)^2}{2} \right]$$

$$= \infty,$$

so $\displaystyle\sum_{k=2}^\infty \frac{\ln k}{k}$ diverges.

35. Using the Limit Comparison Test, let

$a_k = \dfrac{k^4 + 2k - 1}{k^5 + 3k^2 + 1}$ and $b_k = \dfrac{1}{k}$. Then

$$\lim_{k \to \infty} \frac{a_k}{b_k} = \lim_{k \to \infty} \left(\frac{k^4 + 2k - 1}{k^5 + 3k^2 + 1} \right) \left(\frac{k}{1} \right)$$

$$= \lim_{k \to \infty} \frac{k^5 + 2k^2 - k}{k^5 + 3k^2 + 1}$$

$$= 1 > 0.$$

Since $\displaystyle\sum_{k=1}^\infty \frac{1}{k}$ is the divergent harmonic series,

$$\sum_{k=1}^\infty \frac{k^4 + 2k - 1}{k^5 + 3k^2 + 1} \text{ diverges.}$$

37. $\displaystyle\lim_{k \to \infty} \frac{k+1}{k+2} = 1 \ne 0$, so by the kth-Term Test for

Divergence, $\displaystyle\sum_{k=1}^\infty \frac{k+1}{k+2}$ diverges.

39. Using the Limit Comparison Test, let

$a_k = \dfrac{k+1}{k^3+2}$ and $b_k = \dfrac{1}{k^2}$. Then

$$\lim_{k\to\infty}\frac{a_k}{b_k} = \lim_{k\to\infty}\left(\frac{k+1}{k^3+2}\right)\left(\frac{k^2}{1}\right)$$

$$= \lim_{k\to\infty}\frac{k^3+k^2}{k^3+2}$$

$$= 1 > 0$$

and $\displaystyle\sum_{k=1}^{\infty}\frac{1}{k^2}$ is a convergent p-series ($p = 2 > 1$),

so $\displaystyle\sum_{k=1}^{\infty}\frac{k+1}{k^3+2}$ converges.

41. We are only concerned with what happens to the terms as $k \to \infty$, so the first few terms don't matter.

43. Since $\displaystyle\lim_{k\to\infty}\frac{a_k}{b_k} = 0$, there exists N, so that for all

$k > N$, $\left|\dfrac{a_k}{b_k} - 0\right| < 1$. Since a_k and b_k are

positive, $\dfrac{a_k}{b_k} < 1$, so $a_k < b_k$. Thus, since $\displaystyle\sum_{k=1}^{\infty}b_k$

converges, $\displaystyle\sum_{k=1}^{\infty}a_k$ converges by the Comparison

Test.

45. If $p < 0$, let $f(x) = \dfrac{1}{x(\ln x)^p} = \dfrac{(\ln x)^{-p}}{x}$. f is

continuous and positive on $[2, \infty)$ and

$f'(x) = \dfrac{-(\ln x)^{-p-1}(p+\ln x)}{x^2} < 0$ for all

$x \in [2, \infty)$, so f is decreasing. Thus we can use the Integral Test.

$$\int_2^\infty \frac{(\ln x)^{-p}}{x}dx = \lim_{R\to\infty}\int_2^R \frac{(\ln x)^{-p}}{x}dx$$

$$= \lim_{R\to\infty}\left[\frac{(\ln x)^{-p+1}}{-p+1}\right]_2^R$$

$$= \lim_{R\to\infty}\left[\frac{(\ln R)^{-p+1}}{-p+1} - \frac{(\ln 2)^{-p+1}}{-p+1}\right]$$

$$= \infty$$

since $p < 0$.

So when $p < 0$, $\displaystyle\sum_{k=2}^{\infty}\frac{1}{k(\ln k)^p}$ diverges. If

$0 \le p \le 1$, then the series diverges because

$\dfrac{1}{k(\ln k)^p} \ge \dfrac{1}{k\ln k}$ and $\displaystyle\sum_{k=2}^{\infty}\frac{1}{k\ln k}$ diverges (from

exercise 13). If $p > 1$, then let $f(x) = \dfrac{1}{x(\ln x)^p}$. f

is continuous and positive on $[2, \infty)$, and

$f'(x) = \dfrac{-(\ln(x))^{-p-1}(p+\ln x)}{x^2} < 0$ so f is

decreasing. Thus, we can use the Integral Test.

$$\int_2^\infty \frac{1}{x(\ln x)^p}dx$$

$$= \lim_{R\to\infty}\int_2^R \frac{1}{x(\ln x)^p}dx$$

$$= \lim_{R\to\infty}\left[\frac{1}{(1-p)(\ln x)^{p-1}}\right]_2^R$$

$$= \lim_{R\to\infty}\left[\frac{1}{(1-p)(\ln R)^{p-1}} - \frac{1}{(1-p)(\ln 2)^{p-1}}\right]$$

$$= \frac{1}{(1-p)(\ln 2)^{p-1}},$$

so $\displaystyle\sum_{k=2}^{\infty}\frac{1}{k(\ln k)^p}$ converges when $p > 1$.

47. If $p < 0$, then the series diverges by the kth-Term Test of Divergence because . If

$0 \le p \le 1$, then $k^p \le k$ for all $k \ge 2$, so $\dfrac{1}{k^p} \ge \dfrac{1}{k}$ for all $k \ge 2$, and $\dfrac{\ln k}{k^p} \ge \dfrac{\ln k}{k}$ for all $k \ge 2$, and $\displaystyle\sum_{k=2}^{\infty} \dfrac{\ln k}{k}$ diverges.

So by the Comparison Test, $\displaystyle\sum_{k=2}^{\infty} \dfrac{\ln k}{k^p}$ diverges. If $p > 1$, then let $f(x) = \dfrac{\ln x}{x^p}$. Then f is continuous and positive

on $[2, \infty)$ and $f'(x) = \dfrac{k^{p-1}(1 - p \ln k)}{k^2 p} < 0$ for $k > 2$, so f is decreasing. Thus, we can use the Integral Test.

$\displaystyle\int_2^{\infty} \dfrac{\ln x}{x^p} dx = \lim_{R \to \infty} \int_2^{R} \dfrac{\ln x}{x^p} dx$

$= \displaystyle\lim_{R \to \infty} \left[\dfrac{\ln x}{(1-p)x^{p-1}} - \dfrac{1}{(1-p)^2 x^{p-1}} \right]_2^R$

$= \displaystyle\lim_{R \to \infty} \left[\dfrac{\ln R}{(1-p)R^{p-1}} - \dfrac{1}{(1-p)^2 R^{p-1}} - \dfrac{\ln 2}{(1-p)2^{p-1}} + \dfrac{1}{(1-p)2^{p-1}} \right]$

$= \dfrac{1}{(1-p)2^{p-1}} - \dfrac{\ln 2}{(1-p)2^{p-1}}$

because $\displaystyle\lim_{R \to \infty} \dfrac{\ln R}{(1-p)R^{p-1}} = \lim_{R \to \infty} \dfrac{\frac{1}{R}}{-(1-p)^2 R^{p-2}} = \lim_{R \to \infty} \dfrac{-1}{(1-p)^2 R^{p-1}} = 0$ by l'Hopital's Rule and because

$p > 1$, and $\displaystyle\lim_{R \to \infty} \dfrac{1}{(1-p)^2 R^{p-1}} = 0$ because $p > 1$. Thus by the Integral Test, $\displaystyle\sum_{k=R}^{\infty} \dfrac{\ln k}{k^p}$ converges when $p > 1$.

49. $\dfrac{1}{3 \cdot 100^3} = 3.33 \times 10^{-7}$

51. $\dfrac{6}{7 \cdot 50^7} = 1.097 \times 10^{-12}$

53. To estimate $\left(\dfrac{1}{2} e^{-40^2} \right)$, take the logarithm:

$\log\left(\dfrac{1}{2} e^{-40^2} \right) \approx -695.2$, so the error is less than 10^{-695}.

55. We need $\dfrac{3}{3n^2} < 10^{-6}$, so $n > 100$ will do.

57. We need $\dfrac{1}{2} e^{-n^2} < 10^{-6}$. Taking the natural logarithm,
we need $n > \sqrt{\ln 500000}$ or $n > 4$.

59. a. can't tell

b. converge

c. converges

d. can't tell

61. $1 = \dfrac{1}{3} + \dfrac{1}{5} + \dfrac{1}{7} + \ldots = \displaystyle\sum_{k=0}^{\infty} \dfrac{1}{2k+1}$. Using the Limit

Comparison Test, let $a_k = \dfrac{1}{2k+1}$ and $b_k = \dfrac{1}{k}$.

$\displaystyle\lim_{k \to \infty} \dfrac{a_k}{b_k} = \lim_{k \to \infty} \left(\dfrac{1}{2k+1} \right)\left(\dfrac{k}{1} \right)$

Then $\quad = \displaystyle\lim_{k \to \infty} \dfrac{k}{2k+1}$

$= \dfrac{1}{2} > 0$

and since $\displaystyle\sum_{k=1}^{\infty} \dfrac{1}{k}$ is the divergent harmonic

series, then $\displaystyle\sum_{k=1}^{\infty} \dfrac{1}{2k+1}$ diverges.

63. If $x > 1$, $\zeta(x) = \sum\limits_{k=1}^{\infty} \dfrac{1}{k^x}$ is a convergent p-series.

If $x \leq 1$, $\zeta(x)$ is a divergent p-series.

65. $\sum\limits_{k=1}^{10} \dfrac{1}{k^4} = 1 + \dfrac{1}{2^4} + \dfrac{1}{3^4} + \dfrac{1}{4^4} + \dfrac{1}{5^4} + \dfrac{1}{6^4} + \dfrac{1}{7^4} + \dfrac{1}{8^4} + \dfrac{1}{9^4} + \dfrac{1}{10^4} \approx 1.0820$

$x = 4$

67. $\sum\limits_{k=1}^{7} \dfrac{1}{k^8} = 1 + \dfrac{1}{2^8} + \dfrac{1}{3^8} + \dfrac{1}{4^8} + \dfrac{1}{5^8} + \dfrac{1}{6^8} + \dfrac{1}{7^8} \approx 1.0041$

$x = 8$

Section 8.4

5. $\lim\limits_{k \to \infty} a_k = \lim\limits_{k \to \infty} \dfrac{3}{k} = 0$, and

$0 < a_{k+1} = \dfrac{3}{k+1} \leq \dfrac{3}{k} = a_k$

for all $k \geq 1$, so by the Alternating Series

Test, $\sum\limits_{k=1}^{\infty} (-1)^{k+1} \dfrac{3}{k}$ converges.

7. $\lim\limits_{k \to \infty} a_k = \lim\limits_{k \to \infty} \dfrac{2}{k^2} = 0$ and

$\dfrac{a_{k+1}}{a_k} = \dfrac{2}{(k+1)^2} \cdot \dfrac{k^2}{2} = \dfrac{2k^2}{2(k^2 + 2k + 1)} < 1$

so $a_{k+1} \leq a_k$ for all $k \geq 1$. Thus by the

Alternating Series Test, $\sum\limits_{k=1}^{\infty} (-1)^k \dfrac{2}{k^2}$ converges.

9. $\lim\limits_{k \to \infty} a_k = \lim\limits_{k \to \infty} \dfrac{k^2}{k+1} = \infty$ so by the kth-Term Test

for Divergence, $\sum\limits_{k=1}^{\infty} (-1)^{k+1} \dfrac{k^2}{k+1}$ diverges.

11. $\lim\limits_{k \to \infty} a_k = \lim\limits_{k \to \infty} \dfrac{k}{k^2 + 2} = 0$ and

$\dfrac{a_{k+1}}{a_k} = \dfrac{k+1}{(k+1)^2 + 2} \cdot \dfrac{k^2 + 2}{k}$

$= \dfrac{k^3 + k^2 + 2k + 2}{k^3 + 2k^2 + 3k} \leq 1$

for all $k \geq 1$ so $a_{k+1} < a_k$ for all $k \geq 1$. Thus by

the Alternating Series Test, $\sum\limits_{k=1}^{\infty} (-1)^k \dfrac{k}{k^2 + 2}$

converges.

13. $\lim\limits_{k \to \infty} a_k = \lim\limits_{k \to \infty} \dfrac{k}{2^k} = \lim\limits_{k \to \infty} \dfrac{1}{2^k \ln 2} = 0$

(by l'Hopital's Rule) and

$\dfrac{a_{k+1}}{a_k} = \dfrac{k+1}{2^{k+1}} \cdot \dfrac{2^k}{k} = \dfrac{k+1}{2k} \leq 1$ for all $k \geq 1$ so

$a_{k+1} \leq a_k$ for all $k \geq 1$. Thus by the Alternating

Series Test, $\sum\limits_{k=1}^{\infty} (-1)^{k+1} \dfrac{k}{2^k}$ converges.

15. $\lim\limits_{k \to \infty} a_k = \lim\limits_{k \to \infty} \dfrac{4^k}{k^2}$

$= \lim\limits_{k \to \infty} \dfrac{4^k \ln 4}{2k}$

$= \lim\limits_{k \to \infty} \dfrac{4^k (\ln 4)^2}{2}$

$= \infty$

(by l'Hopital's Rule) so by the kth-Term Test for

Divergence, $\sum\limits_{k=1}^{\infty} (-1)^k \dfrac{4^k}{k^2}$ diverges.

17. $\lim\limits_{k \to \infty} a_k = \lim\limits_{k \to \infty} \dfrac{2k}{k+1} = 2$, so by the kth-Term Test

for Divergence, $\sum\limits_{k=1}^{\infty} \dfrac{2k}{k+1}$ diverges.

19. $\lim\limits_{k \to \infty} a_k = \lim\limits_{k \to \infty} \dfrac{3}{\sqrt{k+1}} = 0$ and

$\dfrac{a_{k+1}}{a_k} = \dfrac{3}{\sqrt{k+2}} \cdot \dfrac{\sqrt{k+1}}{3} = \dfrac{\sqrt{k+1}}{\sqrt{k+2}} < 1$ for all

$k \geq 1$, so $a_{k+1} < a_k$ for all $k \geq 1$. Thus by the

Alternating Series Test, $\sum\limits_{k=1}^{\infty} (-1)^k \dfrac{3}{\sqrt{k+1}}$

converges.

21. $\lim\limits_{k\to\infty} a_k = \lim\limits_{k\to\infty}\dfrac{2}{k!}=0$ and

$\dfrac{a_{k+1}}{a_k}=\dfrac{2}{(k+1)!}\cdot\dfrac{k!}{2}=\dfrac{1}{k+1}<1$ for all $k\ge 1$ so

$a_{k+1}<a_k$ for all $k\ge 1$. Thus by the Alternating

Series Test, $\sum\limits_{k=1}^{\infty}(-1)^{k+1}\dfrac{2}{k!}$ converges.

23. $a_k=\dfrac{k!}{2^k}=\dfrac{1\cdot2\cdot3\cdots k}{2\cdot2\cdot2\cdots2}\ge\dfrac{1}{2}\cdot\dfrac{k}{2}=\dfrac{k}{4}$ and

$\lim\limits_{k\to\infty}=\dfrac{k}{4}=\infty$, so by the kth-Term Test for

Divergence $\sum\limits_{k=1}^{\infty}(-1)^k\dfrac{k!}{2^k}$ diverges.

25. $\lim\limits_{k\to\infty} a_k=\lim\limits_{k\to\infty}2e^{-k}=0$ and

$\dfrac{a_{k+1}}{a_k}=\dfrac{2e^{-(k+1)}}{2e^{-k}}=\dfrac{2e^{-1}}{2}=\dfrac{1}{e}<1$ for all $k\ge 0$, so

$a_{k+1}<a_k$ for all $k\ge 0$. Thus by the Alternating

Series Test, $\sum\limits_{k=0}^{\infty}(-1)^{k+1}2e^{-k}$ converges.

27. $\lim\limits_{k\to\infty} a_k=\lim\limits_{k\to\infty}\ln k=\infty$, so by the kth-Term Test

for Divergence, $\sum\limits_{k=2}^{\infty}(-1)^k\ln k$ diverges.

29. $\lim\limits_{k\to\infty} a_k=\lim\limits_{k\to\infty}\dfrac{1}{2^k}=0$ and

$\dfrac{a_{k+1}}{a_k}=\dfrac{1}{2^{k+1}}\cdot\dfrac{2^k}{1}=\dfrac{1}{2}<1$ for all $k\ge 0$, so

$a_{k+1}<a_k$ for all $k\ge 0$. Thus by the Alternating

Series Test, $\sum\limits_{k=0}^{\infty}(-1)^{k+1}\dfrac{1}{2^k}$ converges.

31. $|S-S_k|\le a_{k+1}=\dfrac{4}{(k+1)^3}\le.01$ so $400\le(k+1)^3$,

then $\sqrt[3]{400}\le k+1$, so $k\ge\sqrt[3]{400}-1\approx 6.37$. Thus
$k\ge 7$. So $S\approx S_7=3.61$.

33. $|S-S_k|\le a_{k+1}=\dfrac{k+1}{2^{k+1}}\le.01$

If $k=9$, $a_{10}=\dfrac{10}{2^{10}}\approx.00977<.01$

So $S\approx S_9=-.23$.

35. $|S-S_k|\le a_{k+1}=\dfrac{3}{(k+1)!}\le.01$

If $k=5$, $a_6=\dfrac{3}{6!}\approx.0042<.01$

So $S\approx S_5=1.10$.

37. $|S-S_k|\le a_{k+1}=\dfrac{4}{(k+1)^4}\le.01$, so

$400\le(k+1)^4$, then $\sqrt[4]{400}\le k+1$, so

$k\ge\sqrt[4]{400}-1\approx3.47$ so $k\ge 4$. Thus

$S\approx S_4=3.78$.

39. $|S-S_k|\le a_{k+1}=\dfrac{2}{k+1}<.0001$, so

$k+1\ge 20,000$ then $k\ge 19,999$. Thus $k=$
20,000.

41. $|S-S_k|\le a_{k+1}=\dfrac{2^{k+1}}{(k+1)!}\le.0001$

If $k=10$, $a_{11}=\dfrac{2^{11}}{11!}\approx.00005<.0001$.

So $k=10$, which is 11 terms (because k starts at
0).

43. $|S-S_k|\le a_{k+1}=\dfrac{(k+1)!}{(k+1)^{k+1}}\le.0001$

If $k=11$, $a_{12}=\dfrac{12!}{12^{12}}\approx.000054<.0001$

So $k=11$ which is 11 terms

45. If the derivative of a function $f(k)=a_k$ is
negative, it means that the function is decreasing,
i.e., each successive term is smaller than the one
before.

If $f(k)=a_k=\dfrac{k}{k^2+2}$, then

$f'(k)=\dfrac{-k^2+2}{(k^2+2)^2}<0$ for all $k\ge 2$ so a_k is

decreasing.

47. If k is odd, then a_k is positive and the series is the same as $\sum_{k=1}^{\infty} \frac{1}{2k-1}$, which is a divergent harmonic series (exercises 49 and 50, section 8.3). And if k is even, then a_k is negative and the series is the same as $-\sum_{k=1}^{\infty} \frac{1}{2k^2}$ which is a convergent p–series.

Thus the entire series diverges. Therefore positives diverge, negatives converge.

49. $\frac{3}{4} - \frac{3}{4}\left(\frac{3}{4}\right) + \frac{3}{4}\left(\frac{3}{4}\right)^2 - \cdots = \sum_{k=0}^{\infty} \left(\frac{3}{4}\right)\left(-\frac{3}{4}\right)^k$ which is a geometric series with $a = \frac{3}{4}$ and $|r| = \frac{3}{4} < 1$.

So $\sum_{k=0}^{\infty} \left(\frac{3}{4}\right)\left(-\frac{3}{4}\right)^k = \frac{\frac{3}{4}}{1+\frac{3}{4}} = \frac{3}{7}$. The person ends up $\frac{3}{7}$ of the distance from home.

51. The sum of the first million terms is 0.69314668; $\ln 2 \approx 0.69314718$.

Section 8.5

5. By the Ratio Test,

$$\lim_{k\to\infty}\left|\frac{a_{k+1}}{a_k}\right| = \lim_{k\to\infty} \frac{3}{(k+1)!} \cdot \frac{k!}{3}$$
$$= \lim_{k\to\infty}\frac{1}{k+1}$$
$$= 0 < 1$$

so $\sum_{k=0}^{\infty} (-1)^k \frac{3}{k!}$ converges absolutely.

7. By the Ratio Test,

$$\lim_{k\to\infty}\left|\frac{a_{k+1}}{a_k}\right| = \lim_{k\to\infty} \frac{2^{k+1}}{2^k}$$
$$= \lim_{k\to\infty} 2$$
$$= 2 > 1,$$

so $\sum_{k=0}^{\infty} (-1)^k 2^k$ diverges.

9. By the Alternating Series Test,

$$\lim_{k\to\infty} a_k = \lim_{k\to\infty}\frac{k}{k^2+1} = 0 \text{ and}$$

$$\frac{a_{k+1}}{a_k} = \frac{k+1}{(k+1)^2+1} \cdot \frac{k^2+1}{k}$$

$$= \frac{k^3+k^2+k+1}{k^3+2k^2+2k} < 1$$

for all $k \geq 1$, so $a_{k+1} < a_k$ for all $k \geq 1$, so the series converges.

But by the Limit Comparison Test, letting

$$a_k = \frac{k}{k^2+1} \text{ and } b_k = \frac{1}{k},$$

$$\lim_{k\to\infty}\frac{a_k}{b_k} = \lim_{k\to\infty}\frac{k^2}{k^2+1} = 1 > 0 \text{ and } \sum_{k=1}^{\infty}\frac{1}{k} \text{ is the}$$

divergent harmonic series. Therefore

$$\sum_{k=1}^{\infty}|a_k| = \sum_{k=1}^{\infty}\frac{k}{k^2+1} \text{ diverges. Thus}$$

$$\sum_{k=1}^{\infty}(-1)^{k+1}\frac{k}{k^2+1} \text{ converges conditionally.}$$

11. By the Ratio Test, $\lim_{k\to\infty}\left|\frac{a_{k+1}}{a_k}\right| = \lim_{k\to\infty}\frac{3^{k+1}}{(k+1)!}\cdot\frac{k!}{3^k}$

$$= \lim_{k\to\infty}\frac{3}{k+1}$$
$$= 0 < 1,$$

so $\sum_{k=0}^{\infty}(-1)^k\frac{3^k}{k!}$ converges absolutely.

13. $\lim_{k\to\infty} a_k = \lim_{k\to\infty}\frac{k}{2k+1} = \frac{1}{2}$, so by the kth-Term Test for Divergence, $\sum_{k=1}^{\infty}(-1)^{k+1}\frac{k}{2k+1}$ diverges.

15. Using the Ratio Test,

$$\lim_{k\to\infty}\left|\frac{a_{k+1}}{a_k}\right| = \lim_{k\to\infty}\frac{(k+1)2^{k+1}}{3^{k+1}}\cdot\frac{3^k}{k2^k}$$

$$= \lim_{k\to\infty}\frac{2(k+1)}{3k}$$

$$= \frac{2}{3} < 1,$$

so $\displaystyle\sum_{k=1}^{\infty}(-1)^k\frac{k2^k}{3^k}$ converges absolutely.

17. Using the Root Test,

$$\lim_{k\to\infty}\sqrt[k]{|a_k|} = \lim_{k\to\infty}\sqrt[k]{\left(\frac{4k}{5k+1}\right)^k}$$

$$= \lim_{k\to\infty}\frac{4k}{5k+1}$$

$$= \frac{4}{5} < 1$$

so $\displaystyle\sum_{k=1}^{\infty}\left(\frac{4k}{5k+1}\right)^k$ converges absolutely.

19. $\displaystyle\sum_{k=1}^{\infty}\frac{1}{k}$ is the divergent harmonic series; so $\displaystyle\sum_{k=1}^{\infty}\frac{-2}{k}$ diverges.

21. Using the Alternating Series Test,

$$\lim_{k\to\infty}a_k = \lim_{k\to\infty}\frac{\sqrt{k}}{k+1} = 0 \text{ and}$$

$$\frac{a_{k+1}}{a_k} = \frac{\sqrt{k+1}}{k+2}\cdot\frac{k+1}{\sqrt{k}} = \frac{(k+1)^{3/2}}{k^{3/2}+2k^{1/2}} < 1 \text{ for all}$$

$k \geq 1$, so $a_{k+1} < a_k$ for all $k \geq 1$ so the series converges.

But by the Limit Comparison Test, letting

$$a_k = \frac{\sqrt{k}}{k+1} \text{ and } b_k = \frac{1}{k^{1/2}},$$

$$\lim_{k\to\infty}\frac{a_k}{b_k} = \lim_{k\to\infty}\frac{\sqrt{k}}{k+1}\cdot\frac{k^{1/2}}{1} = \lim_{k\to\infty}\frac{k}{k+1} = 1 > 0,$$

and $\displaystyle\sum_{k=1}^{\infty}\frac{1}{k^{1/2}}$ is a divergent p-series $\left(p = \frac{1}{2} < 1\right)$.

Therefore, $\displaystyle\sum_{k=1}^{\infty}|a_k| = \sum_{k=1}^{\infty}\frac{\sqrt{k}}{k+1}$ diverges. So

$\displaystyle\sum_{k=1}^{\infty}(-1)^{k+1}\frac{\sqrt{k}}{k+1}$ converges conditionally.

23. Using the Ratio Test,

$$\lim_{k\to\infty}\left|\frac{a_{k+1}}{a_k}\right| = \lim_{k\to\infty}\frac{(k+1)^2}{e^{k+1}}\cdot\frac{e^k}{k^2}$$

$$= \lim_{k\to\infty}\frac{(k+1)^2}{ek^2}$$

$$= \frac{1}{e} < 1,$$

so $\displaystyle\sum_{k=1}^{\infty}\frac{k^2}{e^k}$ converges absolutely.

25. Using the Root Test,

$$\lim_{k\to\infty}\sqrt[k]{|a_k|} = \lim_{k\to\infty}\sqrt[k]{\left(\frac{e^3}{k^3}\right)^k} = \lim_{k\to\infty}\frac{e^3}{k^3} = 0 < 1, \text{ so}$$

$\displaystyle\sum_{k=2}^{\infty}\frac{e^{3k}}{k^{3k}}$ converges absolutely.

27. Since $|\sin k| \leq 1$ for all k, $\left|\frac{\sin k}{k^2}\right| = \frac{|\sin k|}{k^2} \leq \frac{1}{k^2}$,

and $\displaystyle\sum_{k=1}^{\infty}\frac{1}{k^2}$ is a convergent p-series $(p = 2 > 1)$

so by the Comparison Test, $\displaystyle\sum_{k=1}^{\infty}\left|\frac{\sin k}{k^2}\right|$ converges,

and by Theorem 5.1, $\displaystyle\sum_{k=1}^{\infty}\frac{\sin k}{k^2}$ converges absolutely.

29. Since $\cos k\pi = (-1)^k$ for all k,

$\left|\frac{\cos k\pi}{k}\right| = \left|\frac{(-1)^k}{k}\right| = \frac{1}{k}$ and $\displaystyle\sum_{k=1}^{\infty}\frac{1}{k}$ is the divergent

harmonic series, so $\displaystyle\sum_{k=1}^{\infty}\left|\frac{\cos k\pi}{k}\right|$ diverges by the

Comparison Test. But, using the Alternating

Series Test, $\displaystyle\lim_{k\to\infty}a_k = \lim_{k\to\infty}\frac{1}{k} = 0$ and

$$\frac{a_{k+1}}{a_k} = \frac{1}{k+1}\cdot\frac{k}{1} = \frac{k}{k+1} < 1 \text{ for all } k \geq 1, \text{ so}$$

$a_{k+1} < a_k$ for all $k \geq 1$, so $\displaystyle\sum_{k=1}^{\infty}\frac{(-1)^k}{k}$ converges.

Thus $\displaystyle\sum_{k=1}^{\infty}\frac{(-1)^k}{k} = \sum_{k=1}^{\infty}\frac{\cos k\pi}{k}$ converges

conditionally.

31. Using the Alternating Series Test,

$$\lim_{k \to \infty} a_k = \lim_{k \to \infty} \frac{1}{\ln k} = 0 \text{ and}$$

$$\frac{a_{k+1}}{a_k} = \frac{1}{\ln(k+1)} \cdot \frac{\ln k}{1} = \frac{\ln k}{\ln(k+1)} < 1 \text{ for all}$$

$k \ge 2$, so $a_{k+1} < a_k$ for all $k \ge 2$. So $\displaystyle\sum_{k=2}^{\infty} \frac{(-1)^k}{\ln k}$

converges. But by the Comparison Test, because

$\ln k < k$ for all $k \ge 2$, $\dfrac{1}{\ln k} > \dfrac{1}{k}$ for all $k \ge 2$

and $\displaystyle\sum_{k=2}^{\infty} \frac{1}{k}$ is the divergent harmonic series.

Therefore, $\displaystyle\sum_{k=2}^{\infty} |a_k| = \sum_{k=2}^{\infty} \frac{1}{\ln k}$ diverges. Thus

$\displaystyle\sum_{k=2}^{\infty} \frac{(-1)^k}{\ln k}$ converges conditionally.

33. $\displaystyle\sum_{k=1}^{\infty} |a_k| = \sum_{k=1}^{\infty} \frac{1}{k^{3/2}}$ which is a convergent

p-series $\left(p = \dfrac{3}{2} > 1 \right)$, so by Theorem 5.1,

$\displaystyle\sum_{k=1}^{\infty} \frac{(-1)^k}{k\sqrt{k}}$ converges absolutely.

35. Consider $\displaystyle\sum_{k=1}^{\infty} \frac{1}{k^k}$. Using the Root Test,

$$\lim_{k \to \infty} \sqrt[k]{|a_k|} = \lim_{k \to \infty} \sqrt[k]{\left(\frac{1}{k}\right)^k} = \lim_{k \to \infty} \frac{1}{k} = 0 < 1, \text{ so}$$

$\displaystyle\sum_{k=1}^{\infty} \frac{1}{k^k}$ converges absolutely, thus

$3\displaystyle\sum_{k=1}^{\infty} \frac{1}{k^k} = \sum_{k=1}^{\infty} \frac{3}{k^k}$ converges absolutely.

37. Using the Ratio Test,

$$\lim_{k \to \infty} \left| \frac{a_{k+1}}{a_k} \right| = \lim_{k \to \infty} \frac{(k+1)!}{4^{k+1}} \cdot \frac{4^k}{k!}$$

$$= \lim_{k \to \infty} \frac{k+1}{4}$$

$$= \infty > 1,$$

so $\displaystyle\sum_{k=1}^{\infty} (-1)^{k+1} \frac{k!}{4^k}$ diverges.

39. Using the Ratio Test,

$$\lim_{k \to \infty} \left| \frac{a_{k+1}}{a_k} \right| = \lim_{k \to \infty} \frac{(k+1)^{10}}{(2k+2)!} \cdot \frac{2k!}{k^{10}}$$

$$= \lim_{k \to \infty} \frac{(k+1)^{10}}{k^{10}(2k+1)(2k+2)}$$

$$= \lim_{k \to \infty} \frac{(k+1)^{10}}{4k^{12} + 6k^{11} + 2k^{10}}$$

$$= 0 < 1$$

so $\displaystyle\sum_{k=1}^{\infty} (-1)^{k+1} \frac{k^{10}}{(2k)!}$ converges absolutely.

41. Using the Ratio Test,

$$\lim_{k \to \infty} \left| \frac{a_{k+1}}{a_k} \right| = \lim_{k \to \infty} \left| \frac{(-2)^{k+1}(k+2)}{5^{k+1}} \cdot \frac{5^k}{(-2)^k(k+1)} \right|$$

$$= \lim_{k \to \infty} \frac{2(k+2)}{5(k+1)}$$

$$= \frac{2}{5} < 1$$

so $\displaystyle\sum_{k=0}^{\infty} \frac{(-2)^k(k+1)}{5^k}$ converges absolutely.

43. $\dfrac{\sqrt{8}}{9801} \displaystyle\sum_{k=0}^{0} \frac{(4k)!(1103 + 26390k)}{(k!)^4 396^{4k}} = \frac{\sqrt{8}}{9801}(1103) \approx .318309878 \approx \frac{1}{\pi}$

so $\pi \approx 3.14159273$

$\dfrac{\sqrt{8}}{9801} \displaystyle\sum_{k=0}^{1} \frac{(4k)!(1103 + 26390k)}{(k!)^4 396^{4k}} = \frac{\sqrt{8}}{9801}(1103) + \frac{\sqrt{8}}{9801} \cdot \frac{4!(27,493)}{396^4} \approx .318309886183791 \approx \frac{1}{\pi}$

For comparison, the value of $1/\pi$ to 15 places is 0.318309886183791, so two terms of the series give this value correct to 15 places.

45. Using the Ratio Test

$$\lim_{k \to \infty} \left| \frac{a_{k+1}}{a^k} \right| = \lim_{k \to \infty} \frac{(k+1)!}{(k+1)^{k+1}} \cdot \frac{k^k}{k!}$$

$$\lim_{k \to \infty} \frac{k^k}{(k+1)^k} = \lim_{k \to \infty} \left(\frac{k}{1+k} \right)^k = \frac{1}{e} < 1 \text{ so}$$

$$\sum_{k=1}^{\infty} \frac{k!}{k^k} \text{ converges absolutely.}$$

Section 8.6

5. $f(x) = \dfrac{2}{1-x} = 2\left(\dfrac{1}{1-x}\right) = 2\sum_{k=0}^{\infty} x^k = \sum_{k=0}^{\infty} 2x^k$ and

this is a geometric series that converges when $|x| < 1$, so $-1 < x < 1$, thus the interval of convergence is $(-1, 1)$ and $r = 1$.

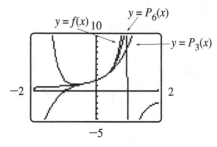

7. $f(x) = \dfrac{3}{1+x^2}$

$$= 3\left(\frac{1}{1-(-x^2)}\right)$$

$$= 3\sum_{k=0}^{\infty} (-x^2)^k$$

$$= \sum_{k=0}^{\infty} (-1)^k 3x^{2k}$$

and this is a geometric series that converges when $|-x^2| < 1$, so $x^2 < 1$, then $|x| < 1$ or $-1 < x < 1$, so the interval of convergence is $(-1, 1)$ and $r = 1$.

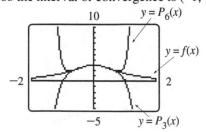

9. $f(x) = \dfrac{2x}{1-x^3}$

$$= 2x\left(\frac{1}{1-x^3}\right)$$

$$= 2x\sum_{k=0}^{\infty} (x^3)^k$$

$$= \sum_{k=0}^{\infty} 2x^{3k+1}$$

and this is geometric series that converges when $x^3 < 1$, so $|x| < 1$ or $-1 < x < 1$, so the interval of convergence is $(-1, 1)$ and r = 1.

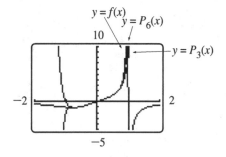

11. $f(x) = \dfrac{4}{1+4x}$

$$= 4\left(\frac{1}{1-(-4x)}\right)$$

$$= 4\sum_{k=0}^{\infty} (-4x)^k$$

$$= \sum_{k=0}^{\infty} (-1)^k 4^{k+1} x^k$$

which is a geometric series that converges when $|-4x| < 1$ so $-1 < -4x < 1$ or $1 > 4x > -1$, so the interval of convergence is $\left(-\dfrac{1}{4}, \dfrac{1}{4}\right)$ and $r = \dfrac{1}{4}$.

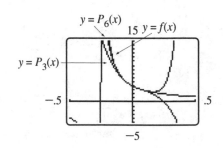

13.

$$f(x) = \frac{2}{4+x}$$

$$= \frac{\frac{1}{2}}{1+\frac{x}{4}}$$

$$= \frac{1}{2}\left(\frac{1}{1-\left(-\frac{x}{4}\right)}\right)$$

$$= \frac{1}{2}\sum_{k=0}^{\infty}\left(-\frac{x}{4}\right)^k$$

$$= \sum_{k=0}^{\infty}\frac{(-1)^k x^k}{2^{2k+1}}$$

which is a geometric series that converges when $\left|-\frac{x}{4}\right| < 1$ so $-1 < \frac{-x}{4} < 1$ or $1 > \frac{x}{4} > 1$, thus the interval of convergence is $(-4, 4)$ and $r = 4$.

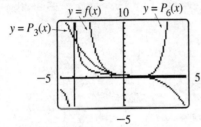

15. This is a geometric series

$$\sum_{k=0}^{\infty}(x+2)^k = \frac{1}{1-(x+2)} = \frac{-1}{1+x} = f(x)$$

which converges for $|x+2| < 1$ so $-1 < x+2 < 1$ or $-3 < x < -1$, thus the interval of convergence is $(-3, -1)$ and $r = 1$.

17. This is a geometric series

$$\sum_{k=0}^{\infty}(2x-1)^k = \frac{1}{1-(2x-1)} = \frac{1}{2-2x} = f(x)$$

which converges for $|2x-1| < 1$ so $-1 < 2x-1 < 1$ or $0 < x < 1$, thus the interval of convergence is $(0, 1)$ and $r = \frac{1}{2}$.

19. This is a geometric series

$$\sum_{k=0}^{\infty}(-1)^k\left(\frac{x}{2}\right)^k = \sum_{k=0}^{\infty}\left(\frac{-x}{2}\right)^k$$

$$= \frac{1}{1+\frac{x}{2}}$$

$$= \frac{2}{2+x}$$

$$= f(x)$$

which converges for $\left|\frac{-x}{2}\right| < 1$ so $-1 < \frac{-x}{2} < 1$ or $2 > x > -2$, thus the interval of convergence is $(-2, 2)$ and $r = 2$.

21. Using the Ratio Test,

$$\lim_{k\to\infty}\left|\frac{a_{k+1}}{a_k}\right| = \lim_{k\to\infty}\left|\frac{2^{k+1}k!(x-2)^{k+1}}{2^k(k+1)!(x-2)^k}\right|$$

$$= 2|x-2|\lim_{k\to\infty}\frac{1}{k+1}$$

$$= 0 \text{ and } 0 < 1 \text{ for all } x,$$

so the series converges absolutely for $x \in (-\infty, \infty)$. The interval of convergence is $(-\infty, \infty)$ and $r = \infty$.

23. Using the Ratio Test,

$$\lim_{k\to\infty}\left|\frac{a_{k+1}}{a_k}\right| = \lim_{k\to\infty}\left|\frac{(k+1)4^k x^{k+1}}{k4^{k+1}x^k}\right|$$

$$= \frac{|x|}{4}\lim_{k\to\infty}\frac{k+1}{k}$$

$$= \frac{|x|}{4} \text{ and } \frac{|x|}{4} < 1 \text{ when } |x| < 4$$

or $-4 < x < 4$. SO the series converges absolutely for $x \in (-4, 4)$. When $x = 4$, $\sum_{k=0}^{\infty}\frac{k}{4^k}4^k = \sum_{k=0}^{\infty}k$

and $\lim_{k\to\infty}k = \infty$, so the series diverges by the kth-Term Test for Divergence. When $x = -4$,

$$\sum_{k=0}^{\infty}\frac{k}{4^k}(-4)^k = \sum_{k=0}^{\infty}(-1)^k k, \text{ which diverges by the}$$

k^{th}-Term Test for Divergence. Thus the interval of convergence is $(-4, 4)$ and $r = 4$.

25. Using the Ratio Test,

$$\lim_{k\to\infty}\left|\frac{a_{k+1}}{a_k}\right| = \lim_{k\to\infty}\left|\frac{(-1)^{k+1}k3^k(x-1)^{k+1}}{(-1)^k(k+1)3^{k+1}(x-1)^k}\right|$$

$$= \frac{|x-1|}{3}\lim_{k\to\infty}\frac{k}{k+1}$$

$$= \frac{|x-1|}{3} \text{ and } \frac{|x-1|}{3} < 1$$

when $-3 < x-1 < 3$ or $-2 < x < 4$ so the series converges absolutely for $x \in (-2,4)$. When

$x = -2$, $\displaystyle\sum_{k=0}^{\infty}\frac{(-1)^k}{k3^k}(-3)^k = \sum_{k=0}^{\infty}\frac{1}{k}$ which is the

divergent harmonic series. When $x = 4$,

$$\sum_{k=0}^{\infty}\frac{(-1)^k}{k3^k}3^k = \sum_{k=0}^{\infty}\frac{(-1)^k}{k} \text{ which converges by the}$$

Alternating Series Test (example 4.2). So the interval of convergence is $(-2, 4]$ and $r = 3$.

27. Using the Ratio Test,

$$\lim_{k\to\infty}\left|\frac{a_{k+1}}{a_k}\right| = \lim_{k\to\infty}\left|\frac{(k+1)!(x+1)^{k+1}}{k!(x+1)^k}\right|$$

$$= \lim_{k\to\infty}\left[(k+1)|x+1|\right]$$

$$= \begin{cases} 0 \text{ if } x = -1 \\ \infty \text{ if } x \neq -1 \end{cases}$$

so this series converges absolutely for $x = -1$ and $r = 0$.

29. Using the Ratio Test,

$$\lim_{k\to\infty}\left|\frac{a_{k+1}}{a_k}\right| = \lim_{k\to\infty}\left|\frac{k(x-1)^{k+1}}{(k+1)(x-1)^k}\right|$$

$$= |x-1|\lim_{k\to\infty}\frac{k}{k+1}$$

$$= |x-1| \text{ and } |x-1| < 1 \text{ when}$$

$-1 < x-1 < 1$ or $0 < x < 2$, so the series converges absolutely for $x \in (0, 2)$.

When $x = 0$, $\displaystyle\sum_{k=0}^{\infty}\frac{(-1)^k}{k}$ which converges by the

Alternating Series Test (example 4.2). If $x = 2$,

$\displaystyle\sum_{k=1}^{\infty}\frac{1}{k}$ which is the divergent harmonic series. So

the interval of convergence is $[0, 2)$ and $r = 1$.

31. Using the Ratio Test,

$$\lim_{k\to\infty}\left|\frac{a_{k+1}}{a_k}\right| = \lim_{k\to\infty}\left|\frac{(k+1)^2(x-3)^{k+1}}{k^2(x-3)^k}\right|$$

$$= |x-3|\lim_{k\to\infty}\frac{(k+1)^2}{k^2}$$

$$= |x-3| < 1 \text{ when}$$

$-1 < x-3 < 1$ or $2 < x < 4$, so the series converges

absolutely for $x \in (2, 4)$. If $x = 2$, $\displaystyle\sum_{k=0}^{\infty}k^2(-1)^k$

and $\lim_{k\to\infty}k^2 = \infty$, so the series diverges by the

k^{th}-Term Test for Divergence. If $x = 4$, $\displaystyle\sum_{k=0}^{\infty}k^2$

and $\lim_{k\to\infty}k^2 = \infty$, so the series diverges by the

k^{th}-Term Test for Divergence. Thus, the interval of convergence is $(2, 4)$ and $r = 1$.

33. Using the Ratio Test,

$$\lim_{k\to\infty}\left|\frac{a_{k+1}}{a_k}\right| = \lim_{k\to\infty}\left|\frac{(k+1)!(2k)!x^{k+1}}{k!(2k+2)!x^k}\right|$$

$$= |x|\lim_{k\to\infty}\frac{k+1}{(2k+2)(2k+1)}$$

$$= 0 \text{ and } 0 < 1$$

for all x, so the series converges for all $x \in (-\infty, \infty)$. Thus the interval of convergence is $(-\infty, \infty)$ and $r = \infty$.

35. Using the Ratio Test,

$$\lim_{k\to\infty}\left|\frac{a_{k+1}}{a_k}\right| = \lim_{k\to\infty}\left|\frac{2^{k+1}k^2(x+2)^{k+1}}{2^k(k+1)^2(x+2)^k}\right|$$

$$= 2|x+2|\lim_{k\to\infty}\frac{k^2}{(k+1)^2}$$

$$= 2|x+2| \text{ and } 2|x+2| < 1$$

when $-\frac{1}{2} < x+2 < \frac{1}{2}$ or $-\frac{5}{2} < x < -\frac{3}{2}$, so the

series converges absolutely for $x \in \left(-\frac{5}{2}, -\frac{3}{2}\right)$. If

$x = -\frac{5}{2}$, then $\sum_{k=0}^{\infty}\frac{2^k}{k^2}\left(-\frac{1}{2}\right)^k = \sum_{k=0}^{\infty}\frac{(-1)^k}{k^2}$

and $\lim_{k\to\infty}\frac{1}{k^2} = 0$ and $\frac{a_{k+1}}{a_k} = \frac{k^2}{(k+1)^2} < 1$ for all

$k \geq 0$, so $a_{k+1} < a_k$ for all $k \geq 0$, so the series

converges by the Alternating Series Test. If

$x = -\frac{3}{2}$, then $\sum_{k=0}^{\infty}\frac{2^k}{k^2}\left(\frac{1}{2}\right)^k = \sum_{k=0}^{\infty}\frac{1}{k^2}$ which is a

convergent p-series $(p = 2 > 1)$. So the interval

of convergence is $\left[-\frac{5}{2}, -\frac{3}{2}\right]$ and $r = \frac{1}{2}$.

37. Using the Ratio Test,

$$\lim_{k\to\infty}\left|\frac{a_{k+1}}{a_k}\right| = \lim_{k\to\infty}\left|\frac{4^{k+1}k^{1/2}x^{k+1}}{4^k(k+1)^{1/2}x^k}\right|$$

$$= 4|x|\lim_{k\to\infty}\frac{k^{1/2}}{(k+1)^{1/2}}$$

$$= 4|x| \text{ and } 4|x| \leq 1 \text{ when}$$

$-\frac{1}{4} < x < \frac{1}{4}$, so the series converges absolutely

for

$x \in \left(-\frac{1}{4}, \frac{1}{4}\right)$. If $x = -\frac{1}{4}$, then

$\sum_{k=0}^{\infty}\frac{4^k}{\sqrt{k}}\left(-\frac{1}{4}\right)^k = \sum_{k=0}^{\infty}\frac{(-1)^k}{\sqrt{k}}$ and

$\lim_{k\to\infty}\frac{1}{\sqrt{k}} = 0$ and $\frac{a_{k+1}}{a_k} = \frac{\sqrt{k}}{\sqrt{k+1}} < 1$

for all $k \geq 0$, so $a_{k+1} < a_k$ for all $k \geq 0$, therefore

the series converges by the Alternating Series test.

If $x = \frac{1}{4}$, $\sum_{k=0}^{\infty}\frac{4^k}{\sqrt{k}}\left(\frac{1}{4}\right)^k \sum_{k=0}^{\infty}\frac{1}{\sqrt{k}}$ which is a

divergent p-series $\left(p = \frac{1}{2} < 1\right)$. Thus the interval

of convergence is $\left[-\frac{1}{4}, \frac{1}{4}\right)$ and $r = \frac{1}{4}$.

39. From #7, $\frac{3}{1+x^2} = \sum_{k=0}^{\infty}(-1)^k 3x^{2k}$ with $r = 1$

Integrating both sides gives

$$\int\frac{3}{1+x^2}dx = \sum_{k=0}^{\infty}3(-1)^k\int x^{2k}dx$$

$$3\tan^{-1}x = \sum_{k=0}^{\infty}\frac{3(-1)^k x^{2k+1}}{2k+1} + c$$

Taking $x = 0$,

$$3\tan^{-1}0 = \sum_{k=0}^{\infty}\frac{3(-1)^k(0)^{2k+1}}{2k+1} + c = c \text{ so that}$$

$c = 3\tan^{-1}(0) = 0$. Thus,

$$3\tan^{-1}x = \sum_{k=0}^{\infty}\frac{3(-1)^k x^{2k+1}}{2k+1} \text{ with } r = 1.$$

41. From #8, $\frac{2}{1-x^2} = \sum_{k=0}^{\infty}2x^{2k}$ with $r = 1$

Taking the derivative of both sides gives

$$\frac{d}{dx}\left[\frac{2}{1-x^2}\right] = \sum_{k=0}^{\infty}2\frac{d}{dx}\left[x^{2k}\right]$$

$$\frac{4x}{(1-x^2)^2} = \sum_{k=0}^{\infty}2\cdot 2kx^{2k-1}$$

$$\frac{2x}{(1-x^2)^2} = \sum_{k=0}^{\infty}2kx^{2k-1} \text{ with } r = 1.$$

43. From #10, $\dfrac{3x}{1+x^2} = \displaystyle\sum_{k=0}^{\infty} (-1)^k 3x^{2k+1}$ with $r = 1$

Integrating both sides gives

$$\int \dfrac{3x}{1+x^2}\,dx = \sum_{k=0}^{\infty} (-1)^k 3 \int x^{2k+1}\,dx$$

$$\dfrac{3}{2}\ln(1+x^2) = \sum_{k=0}^{\infty} \dfrac{(-1)^k 3x^{2k+2}}{2k+2} + c$$

$$\ln(1+x^2) = \sum_{k=0}^{\infty} \dfrac{(-1)^k x^{2k+2}}{k+1} + c.$$

Taking $x = 0$, $\ln(1) = \displaystyle\sum_{k=0}^{\infty} \dfrac{(-1)^k (0)^{2k+2}}{k+1} + c = c$ so

that $c = \ln(1) = 0$. Thus,

$$\ln(1+x^2) = \sum_{k=0}^{\infty} \dfrac{(-1)^k x^{2k+2}}{k+1} \text{ with } r = 1.$$

45. From #11, $\dfrac{4}{1+4x} = \displaystyle\sum_{k=0}^{\infty} (-1)^k 4^{k+1} x^k$ with $r = \dfrac{1}{4}$

Taking the derivative of both sides gives

$$\dfrac{d}{dx}\left[\dfrac{4}{1+4x}\right] = \sum_{k=0}^{\infty} (-1)^k 4^{k+1} \dfrac{d}{dx}\left[x^k\right]$$

$$\dfrac{-16}{(1+4x)^2} = \sum_{k=0}^{\infty} (-1)^k 4^{k+1} k x^{k-1}$$

$$\dfrac{1}{(1+4x)^2} = \sum_{k=0}^{\infty} (-1)^{k+1} k (4x)^{k-1} \text{ with } r = \dfrac{1}{4}.$$

47. Since $\left|\cos(k^3 x)\right| \le 1$ for all x, $\left|\dfrac{\cos(k^3 x)}{k^2}\right| \le \dfrac{1}{k^2}$ for

all x, and $\displaystyle\sum_{k=0}^{\infty} \dfrac{1}{k^2}$ is a convergent p–series, so by

the Comparison Test, $\displaystyle\sum_{k=0}^{\infty} \dfrac{\cos(k^3 x)}{k^2}$ converges

absolutely for all x. So the interval of convergence
is $(-\infty, \infty)$ and $r = \infty$. The series of derivatives is

$$\sum_{k=0}^{\infty} \dfrac{d}{dx}\left[\dfrac{\cos(k^3 x)}{k^2}\right] = \sum_{k=0}^{\infty} (-k)\sin(k^3 x) \text{ and}$$

$$\lim_{k\to\infty} (-k)\sin(k^3 x) = \begin{cases} 0 \text{ if } x = 0 \\ \infty \text{ if } x \ne 0 \end{cases} \text{ so}$$

$$\sum_{k=0}^{\infty} (-k)\sin(k^3(0)) = \sum_{k=0}^{\infty} 0 = 0, \text{ thus the series}$$

converges absolutely if $x = 0$.

49. Using the Ratio Test,

$$\lim_{k\to\infty}\left|\dfrac{a_{k+1}}{a_k}\right| = \lim_{k\to\infty}\left|\dfrac{e^{(k+1)x}}{e^{kx}}\right| = e^x \lim_{k\to\infty} 1 = e^x \text{ and}$$

$e^x < 1$ when $x < 0$, so the series converges
absolutely for all $x \in (-\infty, 0)$. When $x = 0$,

$$\sum_{k=0}^{\infty} e^0 = \sum_{k=0}^{\infty} 1 \text{ which diverges by the } k^{th}\text{-Term}$$

Test for Divergence because $\displaystyle\lim_{k\to\infty} 1 = 1$. So the

interval of convergence is $(-\infty, 0)$. The series of

derivatives is $\displaystyle\sum_{k=0}^{\infty} \dfrac{d}{dx}\left[e^{kx}\right] = \sum_{k=0}^{\infty} k e^{kx}$.

Using the Ratio Test,

$$\lim_{k\to\infty}\left|\dfrac{a_{k+1}}{a_k}\right| = \lim_{k\to\infty}\left|\dfrac{(k+1)e^{(k+1)x}}{k e^{kx}}\right|$$

$$= e^x \lim_{k\to\infty} \dfrac{k+1}{k}$$

$$= e^x \text{ and } e^x < 1 \text{ when } x < 0,$$

so the series converges absolutely for all

$x \in (-\infty, 0)$. When $x = 0$, $\displaystyle\sum_{k=0}^{\infty} k e^0 = \sum_{k=0}^{\infty} k$ which

diverges by the kth-Term Test for Divergence
because $\displaystyle\lim_{k\to\infty} k = \infty$. So the interval of

convergence is $(-\infty, 0)$.

51. Using the Ratio Test,

$$\lim_{k\to\infty}\left|\frac{a_{k+1}}{a_k}\right| = \lim_{k\to\infty}\left|\frac{(x-a)^{k+1}b^k}{b^{k+1}(x-a)^k}\right|$$

$$= \frac{|x-a|}{b}\lim_{k\to\infty}1$$

$$= \frac{|x-a|}{b} \text{ and } \frac{|x-a|}{b} < 1$$

when $-b < x - a < b$ or $a - b < x < a + b$. So the series converges absolutely for $x \in (a-b, a+b)$.

If $x = a - b$,

$$\sum_{k=0}^{\infty}\frac{(a-b-a)^k}{b^k} = \sum_{k=0}^{\infty}\frac{(-1)^k b^k}{b^k} = \sum_{k=0}^{\infty}(-1)^k \text{ which}$$

diverges by the kth-Term Test for Divergence because $\lim_{k\to\infty}1 = 1$. If $x = a + b$,

$$\sum_{k=0}^{\infty}\frac{(a+b-a)^k}{b^k} = \sum_{k=0}^{\infty}\frac{b^k}{b^k} = \sum_{k=0}^{\infty}1 \text{ which diverges}$$

by the kth-Term Test for Divergence because $\lim_{k\to\infty}1 = 1$. So the interval of convergence is $(a-b, a+b)$ and $r = b$.

53. If the radius of convergence of $\sum_{k=0}^{\infty}a_k x^k$ is r, then

$-r < x < r$. For any constant c,
$-r - c < x - c < r - c$. Thus, the radius of convergences of $\sum_{k=0}^{\infty}a_k(x-c)^k$ is

$$\frac{(r-c)-(-r-c)}{2} = r.$$

55. $f(x) = \dfrac{x+1}{(1-x)^2} = \dfrac{\frac{2x}{1-x}+1}{1-x} = \dfrac{2x}{(1-x)^2} + \dfrac{1}{1-x}$

$$\frac{1}{1-x} = \sum_{k=0}^{\infty}x^k$$

$$\frac{d}{dx}\left[\frac{1}{1-x}\right] = \sum_{k=0}^{\infty}\frac{d}{dx}[x^k]$$

$$\frac{1}{(1-x)^2} = \sum_{k=0}^{\infty}kx^{k-1}$$

$$\frac{2x}{(1-x)^2} = \sum_{k=0}^{\infty}2kx^k$$

So,

$$\frac{x+1}{(1-x)^2}$$

$$= \frac{2x}{(1-x)^2} + \frac{1}{1-x}$$

$$= \sum_{k=0}^{\infty}2kx^k + \sum_{k=0}^{\infty}x^k$$

$$= (2x + 4x^2 + 6x^3 + \ldots) + (1 + x + x^2 + x^3 + \ldots)$$

$$= 1 + 3x + 5x^2 + 7x^3 + \ldots$$

Since $\sum_{k=0}^{\infty}x^k$ converges for $|x| < 1$, $\dfrac{d}{dx}\left[\sum_{k=0}^{\infty}x^k\right]$

converges for $|x| < 1$, so $2x\dfrac{d}{dx}\left[\sum_{k=0}^{\infty}x^k\right]$

converges for $|x| < 1$. Hence

$$\frac{x+1}{(1-x)^2} = \sum_{k=0}^{\infty}x^k + \sum_{k=1}^{\infty}2kx^k \text{ converges for}$$

$|x| < 1$, so $r = 1$. If $x = \dfrac{1}{1000}$, then

$$\frac{1,001,000}{998,001} = 1.003005007$$

$$= 1 + \frac{3}{1000} + \frac{5}{(1000)^2} + \frac{7}{(1000)^3} + \ldots$$

57. If $x = 1, \ldots + 1 + 1 + 1 + 1 + 1 + \ldots \neq 0$.

$$\frac{1}{1-\frac{1}{x}} = \sum_{k=0}^{\infty}\left(\frac{1}{x}\right)^k \text{ is a geometric series which}$$

converges for $\dfrac{1}{|x|} < 1$ or $|x| > 1$.

$$\frac{x}{1-x} = \sum_{k=0}^{\infty}x^{k+1} \text{ is a geometric series which}$$

converges for $|x| < 1$. Euler's mistake was that there are no x's for which both series converge.

59. We would conjecture radii of convergence of 1/4, 4 and 2 for the series representing these three functions.

Section 8.7

7. $f(x) = \cos x, \quad f'(x) = -\sin x, \quad f''(x) = -\cos x, \quad f'''(x) = \sin x$
    $f(0) = 1, \qquad f'(0) = 0, \qquad f''(0) = -1, \qquad f'''(0) = 0$

$$\sum_{k=0}^{\infty} \frac{f^{(k)}(0)}{k!}(x-0)^k = 1 - \frac{1}{2}x^2 + \frac{1}{4!}x^4 - \frac{1}{6!}x^6 + \ldots = \sum_{k=0}^{\infty} \frac{(-1)^k}{(2k)!}x^{2k}$$

And, using the Ratio Test, $\lim_{k\to\infty} \left| \frac{a_{k+1}}{a_k} \right| = \lim_{k\to\infty} \left| \frac{(-1)^{k+1}(2k)!x^{2k+2}}{(-1)^k(2k+2)!x^{2k}} \right| = \left| x^2 \right| \lim_{k\to\infty} \frac{1}{(2k+1)(2k+2)} = 0$ and $0 < 1$ for

all x. So the interval of convergence is $(-\infty, \infty)$.

9. $f(x) = \sin x, \quad f'(x) = \cos x, \quad f''(x) = -\sin x, \quad f'''(x) = -\cos x$
    $f(0) = 0, \qquad f'(0) = 1, \qquad f''(0) = 0, \qquad f'''(0) = -1$

$$\sum_{k=0}^{\infty} \frac{f^{(k)}(0)}{k!}x^k = x - \frac{1}{3}x^3 + \frac{1}{5!}x^5 - \frac{1}{7!}x^7 \ldots = \sum_{k=0}^{\infty} \frac{(-1)^k x^{2k+1}}{(2k+1)!}$$

And, using the Ratio Test, $\lim_{k\to\infty} \left| \frac{a_{k+1}}{a_k} \right| = \lim_{k\to\infty} \left| \frac{(-1)^k(2k+1)!x^{2k+3}}{(-1)^k(2k+3)!x^{2k+1}} \right| = \left| x^2 \right| \lim_{k\to\infty} \frac{1}{(2k+2)(2k+3)} = 0$ and $0 < 1$ for

all x. So the interval of convergence is $(-\infty, \infty)$.

11. $f(x) = \ln(1+x), \quad f'(x) = \frac{1}{1+x}, \quad f''(x) = \frac{-1}{(1+x)^2} \quad f'''(x) = \frac{2}{(1+x)^3}$
    $f(0) = 0, \qquad f'(0) = 0, \qquad f''(0) = -1, \qquad f'''(0) = 2$

$$\sum_{k=0}^{\infty} \frac{f^{(k)}(0)}{k!}x^k = x - \frac{1}{2}x^2 + \frac{2!}{3!}x^3 - \frac{3!}{4!}x^4 + \ldots = \sum_{k=0}^{\infty} \frac{(-1)^k x^{k+1}}{k+1}$$

And, using the Ratio Test, $\lim_{k\to\infty} \left| \frac{a_{k+1}}{a_k} \right| = \lim_{k\to\infty} \left| \frac{(-1)^{k+1}(k+1)x^{k+2}}{(-1)^k(k+2)x^{k+1}} \right| = |x| \lim_{k\to\infty} \frac{k+1}{k+2} = |x|$ and $|x| < 1$ when

$-1 < x < 1$. So the series converges absolutely for all $x \in (-1, 1)$. When $x = 1$, $\sum_{k=0}^{\infty} \frac{(-1)^k(1)^{k+1}}{k+1} = \sum_{k=0}^{\infty} \frac{(-1)^k}{k+1}$ and

$\lim_{k\to\infty} \frac{1}{k+1} = 0$ and $\frac{a_{k+1}}{a_k} = \frac{k}{k+1} < 1$ for all $k \geq 0$ so $a_{k+1} > a_k$ for all $k \geq 0$, so the series converges by the

Alternating Series Test. When $x = -1$, $\sum_{k=0}^{\infty} \frac{(-1)^k(-1)^{k+1}}{k+1} = \sum_{k=0}^{\infty} \frac{-1}{k+1}$ which diverges by the Limit Comparison

Test, letting $a_k = \frac{1}{k+1}$ and $b_k = \frac{1}{k}$, then $\lim_{k\to\infty} \frac{a_k}{b_k} = \lim_{k\to\infty} \frac{k}{k+1} = 1 > 0$ and $\sum_{k=0}^{\infty} \frac{1}{k}$ is the divergent harmonic

series so $\sum_{k=1}^{\infty} \frac{1}{k+1}$ diverges, thus $-\sum_{k=1}^{\infty} \frac{1}{k+1}$ diverges. So the interval of convergence is $(-1, 1]$.

13. $f(x) = (1+x)^{-2}, \quad f'(x) = -2(1+x)^{-3}, \quad f''(x) = 6(1+x)^{-4}, \quad f'''(x) = -24(1+x)^{-5}$
$f(0) = 1, \qquad\qquad f'(0) = -2, \qquad\qquad f''(0) = 3!, \qquad\qquad f'''(0) = -4!$

$$\sum_{k=0}^{\infty} \frac{f^{(k)}(0)}{k!} x^k = 1 - 2x + \frac{3!}{2!}x^2 - \frac{4!}{3!}x^3 + \dots = \sum_{k=0}^{\infty} (-1)^k (k+1) x^k$$

And using the Ratio Test, $\lim\limits_{k \to \infty} \left| \dfrac{a_{k+1}}{a_k} \right| = \lim\limits_{k \to \infty} \left| \dfrac{(k+2)x^{k+1}}{(k+1)x^k} \right| = |x| \lim\limits_{k \to \infty} \dfrac{k+2}{k+1} = |x|$ and $|x| < 1$ when $-1 < x < 1$. So

the series converges absolutely for $x \in (-1, 1)$. When $x = -1$, $\sum\limits_{k=0}^{\infty} (k+1)$ and $\lim\limits_{k \to \infty} (k+1) = \infty$, so the series

diverges by the kth-Term Test for Divergence. When $x = 1$, $\sum\limits_{k=0}^{\infty} (-1)(k+1)^k$ and $\lim\limits_{k \to \infty} k+1 = \infty$, so the series

diverges by the kth-Term Test for Divergence. So the interval of convergence is $(-1, 1)$.

15. $f(x) = e^{x-1}, \quad f'(x) = e^{x-1}, \quad f''(x) = e^{x-1}$
$f(1) = 1, \qquad f'(1) = 1, \qquad f''(1) = 1$

$$\sum_{k=0}^{\infty} \frac{f^{(k)}(1)}{k!}(x-1)^k = 1 + (x-1) + \frac{1}{2!}(x-1)^2 + \dots = \sum_{k=0}^{\infty} \frac{(x-1)^k}{k!}. \text{ And, using the Ratio Test,}$$

$\lim\limits_{k \to \infty} \left| \dfrac{a_{k+1}}{a_k} \right| = \lim\limits_{k \to \infty} \left| \dfrac{k!(x-1)^{k+1}}{(k+1)!(x-1)^k} \right| = |x-1| \lim\limits_{k \to \infty} \dfrac{1}{k+1} = 0$, and $0 < 1$ for all x, so the series converges absolutely

for $x \in (-\infty, \infty)$. Thus the interval of convergence is $(-\infty, \infty)$.

17. $f(x) = \ln x, \quad f'(x) = \dfrac{1}{x}, \quad f''(x) = -\dfrac{1}{x^2}, \quad f'''(x) = \dfrac{2}{x^3}$
$f(e) = 1, \qquad f'(e) = \dfrac{1}{e}, \qquad f''(e) = -\dfrac{1}{e^2}, \qquad f'''(e) = \dfrac{2!}{e^3}$

$$\sum_{k=0}^{\infty} \frac{f^{(k)}(e)}{k!}(x-e)^k = 1 + \frac{1}{e}(x-e) - \frac{1}{e^2 2}(x-e)^2 + \frac{2!}{e^3 3!}(x-e)^3 + \dots = 1 + \sum_{k=1}^{\infty} (-1)^{k+1} \frac{e^{-k}}{k}(x-e)^k$$

Using the Ratio Test, $\lim\limits_{k \to \infty} \left| \dfrac{a_{k+1}}{a_k} \right| = \lim\limits_{k \to \infty} \left| \dfrac{e^{-k-1}k(x-e)^{k+1}}{e^{-k}(k+1)(x-e)^k} \right| = \dfrac{|x-e|}{e} \lim\limits_{k \to \infty} \dfrac{k}{k+1} = \dfrac{|x-e|}{e}$ and $\dfrac{|x-e|}{e} < 1$ when

$0 < x < 2e$. When $x = 0$, $\sum\limits_{k=1}^{\infty} \dfrac{(-1)^k e^{-k} (-e)^k}{k} = \sum\limits_{k=1}^{\infty} \left(-\dfrac{1}{k}\right)$ which is the divergent harmonic series. When $x = 2e$,

$\sum\limits_{k=1}^{\infty} \dfrac{(-1)^{k+1} e^{-k} e^k}{k} = \sum\limits_{k=1}^{\infty} \dfrac{(-1)^{k+1}}{k}$ converges (example 4.2). So the interval of convergence is $(0, 2e]$.

19. $f(x) = \dfrac{1}{x}, \quad f'(x) = \dfrac{-1}{x^2}, \quad f''(x) = \dfrac{2}{x^3}, \quad f'''(x) = \dfrac{-6}{x^4}$

$f(1) = 1, \qquad f'(1) = -1, \qquad f''(1) = 2!, \qquad f'''(1) = -3!$

$\displaystyle\sum_{k=0}^{\infty} \dfrac{f^{(k)}(1)}{k!}(x-1)^k = 1 - (x-1) + (x-1)^2 - (x-1)^3 + \ldots = \sum_{k=0}^{\infty} (-1)^k (x-1)^k$

Using the Ratio Test, $\displaystyle\lim_{k\to\infty}\left|\dfrac{a_{k+1}}{a_k}\right| = \lim_{k\to\infty}\left|\dfrac{(-1)^{k+1}(x-1)^{k+1}}{(-1)^k(x-1)^k}\right| = |x-1|\lim_{k\to\infty} 1 = |x-1|$ and $|x-1| < 1$ when $0 < x < 2$.

So the series converges absolutely for $x \in (0, 2)$. When $x = 0$, $\displaystyle\sum_{k=0}^{\infty}(-1)^k(-1)^k = \sum_{k=0}^{\infty} 1$ which diverges by the

k^{th}-Term Test for Divergence because $\displaystyle\lim_{k\to\infty} 1 = 1$. When $x = 2$, $\displaystyle\sum_{k=0}^{\infty}(-1)^k(1)^k = \sum_{k=0}^{\infty}(-1)^k$ which diverges by the

k^{th}-Term Test for Divergence. So the interval of convergence is $(0, 2)$.

21. $f(x) = \cos x$ about $c = 0$

$n = 5; \displaystyle\sum_{k=0}^{2}\dfrac{(-1)^k}{(2k)!}x^{2k} = 1 - \dfrac{x^2}{2!} + \dfrac{x^4}{4!}$ $\qquad n = 9; \displaystyle\sum_{k=0}^{4}\dfrac{(-1)^k}{(2k)!}x^{2k} = 1 - \dfrac{x^2}{2} + \dfrac{x^4}{4!} - \dfrac{x^6}{6!} + \dfrac{x^8}{8!}$

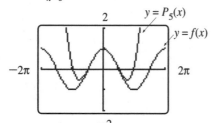
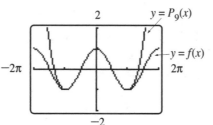

23. $f(x) = \sqrt{x}$ about $c = 1$

$f(x) = \sqrt{x}, \quad f'(x) = \dfrac{1}{2}x^{-1/2}, \quad f''(x) = \dfrac{-1}{4}x^{-3/2}, \quad f'''(x) = \dfrac{3}{8}x^{-5/2}$

$f(1) = 1, \qquad f'(1) = \dfrac{1}{2}, \qquad f''(1) = -\dfrac{1}{4}, \qquad f'''(1) = \dfrac{3}{8}$

$n = 3; \displaystyle\sum_{k=0}^{3}\dfrac{f^{(k)}(1)}{k!}(x-1)^k = 1 + \dfrac{1}{2}(x-1) - \dfrac{1}{8}(x-1)^2 + \dfrac{1}{16}(x-1)^3$

$n = 6; \displaystyle\sum_{k=0}^{6}\dfrac{f^{(k)}(1)}{k!}(x-1)^k = 1 + \dfrac{1}{2}(x-1) - \dfrac{1}{8}(x-1)^2 + \dfrac{1}{16}(x-1)^3 - \dfrac{5}{2^7}(x-1)^4 + \dfrac{7}{2^8}(x-1)^5 - \dfrac{21}{2^{10}}(x-1)^6$

$n = 3$ $\qquad\qquad\qquad\qquad\qquad\qquad\qquad n = 5$

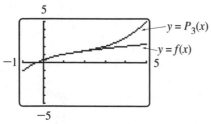
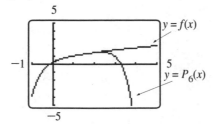

25. $f(x) = e^x$ about $c = 2$

$$f(x) = e^x, \quad f'(x) = e^x, \quad f''(x) = e^x$$
$$f(2) = e^2, \quad f'(2) = e^2, \quad f''(2) = e^2$$

$$n = 3; \quad \sum_{k=0}^{3} \frac{e^2}{k!}(x-2)^k = e^2\left[1 + (x-2) + \frac{(x-2)^2}{2} + \frac{(x-2)^3}{3!}\right]$$

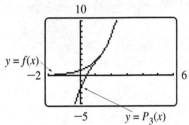

$$n = 6; \quad \sum_{k=0}^{6} \frac{e^2}{k!}(x-2)^k = e^2\left[1 + (x-2) + \frac{(x-2)^2}{2} + \frac{(x-2)^3}{3!} + \frac{(x-2)^4}{4!} + \frac{(x-2)^5}{5!} + \frac{(x-2)^6}{6!}\right]$$

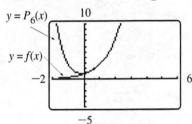

27. $f(x) = x^{-1/2}$ about $c = 4$

$$f(x) = x^{-1/2}, \quad f'(x) = -\frac{1}{2}x^{-3/2}, \quad f''(x) = \frac{3}{4}x^{-5/2}, \quad f'''(x) = -\frac{15}{8}x^{-7/8}$$
$$f(4) = \frac{1}{2}, \quad f'(4) = -\frac{1}{16}, \quad f''(4) = \frac{3}{2^7}, \quad f'''(x) = \frac{-15}{2^{10}}$$

$$n = 2; \quad \sum_{k=0}^{2} \frac{f^{(k)}(4)}{k!}(x-4)^k = \frac{1}{2} - \frac{1}{2^4}(x-4) + \frac{3}{2^8}(x-4)^2$$

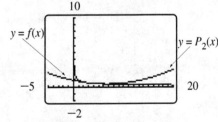

$$n = 4; \quad \sum_{k=0}^{4} \frac{f^{(k)}(4)}{k!}(x-4)^k = \frac{1}{2} - \frac{1}{2^4}(x-4) + \frac{3}{2^8}(x-4)^2 - \frac{5}{2^{11}}(x-4)^3 + \frac{35}{2^{16}}(x-4)^4$$

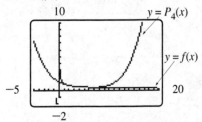

29. $R_n(x) = \dfrac{f^{(n+1)}(z)}{(n+1)!} x^{n+1}$. We have $f^{(n+1)}(z) = \begin{cases} \pm \cos z & \text{if } n \text{ is even} \\ \pm \sin z & \text{if } n \text{ is odd} \end{cases}$ so $\left| f^{(n+1)}(z) \right| \le 1$ for all n. So

$\left| R_n(x) \right| = \left| \dfrac{f^{(n+1)}(z)}{(n+1)!} x^{n+1} \right| \le \dfrac{|x|^{n+1}}{(n+1)!}$ so consider $\displaystyle\sum_{n=0}^{\infty} \dfrac{|x|^{n+1}}{(n+1)!}$. Using the Ratio Test, we have

$\displaystyle\lim_{k \to \infty} \left| \dfrac{a_{n+1}}{a_n} \right| = \lim_{n \to \infty} \dfrac{|x|^{n+2}(n+1)!}{|x|^{n+1}(n+2)!} = |x| \lim_{n \to \infty} \dfrac{1}{n+2} = 0$ and $0 < 1$ for all x. So $\displaystyle\sum_{n=0}^{\infty} \dfrac{|x|^{n+1}}{(n+1)!}$ converges absolutely

for all x. Thus, by the kth-Term Test for Divergence, $\displaystyle\lim_{n \to \infty} \dfrac{|x|^{n+1}}{(n+1)!} = 0$, and since

$\left| R_n(x) \right| \le \dfrac{|x|^{n+1}}{(n+1)!}$, $\displaystyle\lim_{n \to \infty} R_n(x) = 0$ for all x.

31. $R_n(x) = \dfrac{f^{(n+1)}(z)}{(n+1)!} x^{n+1}$. We have

$f^{(n+1)}(z) = \begin{cases} \dfrac{n!}{z^{n+1}} & \text{if } n \text{ is even} \\ \dfrac{n!}{z^{n+1}} & \text{if } n \text{ is odd} \end{cases}$ so

$\left| f^{(n+1)}(z) \right| = \dfrac{n!}{|z|^{n+1}}$ for all n. So

$\left| R_n(x) \right| = \dfrac{n!}{(n+1)!} \left| \dfrac{x-1}{z} \right|^{n+1} = \dfrac{1}{n+1} \left| \dfrac{x-1}{z} \right|^{n+1}$ and

$1 < z < 2$ so $\dfrac{1}{z} < \dfrac{1}{1} = 1$. So $\left| R_n(x) \right| < \dfrac{|x-1|^{n+1}}{n+1}$.

For $1 < x < 2$, the numerator $|x-1|^{n+1}$ is

bounded by 1 and the denominator tends to ∞

with n, so $\displaystyle\lim_{n \to \infty} R_n(x) = 0$.

33. a. Expand $f(x) = \ln x$ into a Taylor series about $c = 1$. Recall

$\ln(x) = \displaystyle\sum_{k=0}^{\infty} \dfrac{(-1)^{k+1}}{k}(x-1)^k$

(from example 7.6). So

$P_4(x)$

$= \displaystyle\sum_{k=0}^{4} \dfrac{f^{(k)}(1)}{k!}(x-1)^k$

$= (x-1) - \dfrac{1}{2}(x-1)^2 + \dfrac{1}{3}(x-1)^3 - \dfrac{1}{4}(x-1)^4$

letting $x = 1.05$ gives

$\ln(1.05) \approx P_4(1.05) = .04879$.

b. $|\text{Error}| = |\ln(1.05) - P_4(1.05)|$

$= |R_n(1.05)|$

$= \left| \dfrac{f^{(4+1)}(x)}{(4+1)!}(1.05-1)^{4+1} \right|$

$= \dfrac{4!|z|^{-5}(0.05)^5}{5!}$,

so because $1 < z < 1.05$, then $\dfrac{1}{z} < \dfrac{1}{1} = 1$.

Thus we have

$|\text{Error}| \dfrac{(.05)^5}{5z^5} < \dfrac{(.05)^5}{5(1)^5} = \dfrac{(.05)^5}{5}$.

c. From part (b) we have for $1 < z < 1.05$,

$|R_n(1.05)| = \left| \dfrac{f^{(n+1)}(z)}{(n+1)!}(1.05-1)^{n+1} \right|$

$= \dfrac{n!(.05)^{n+1}}{(n+1)!z^{n+1}}$

$= \dfrac{(.05)^{n+1}}{(n+1)z^{n+1}}$

$< \dfrac{(.05)^{n+1}}{n+1}$

since $1 < z < 1.05$ gives $\dfrac{1}{z} < \dfrac{1}{1} = 1$. So

$|R_n(1.05)| < \dfrac{(.05)^{n+1}}{n+1} < 10^{-10}$ if $n = 7$.

35. a. Expand $f(x) = \sqrt{x}$ into a Taylor series about $c = 1$. So, as shown in exercise 23,

$$\sqrt{x} = 1 + \sum_{k=1}^{\infty} \frac{(-1)^{k-1} 1 \cdot 3 \cdot 5 \cdots (2n-3)}{2^k k!}(x-1)^k$$

Then
$$P_4(x)$$
$$= \sum_{k=0}^{4} \frac{f^{(k)}(1)}{k!}(x-1)^k$$
$$= 1 + \frac{(x-1)}{2} - \frac{(x-1)}{8} + \frac{(x-1)^3}{16} - \frac{5(x-1)^4}{128}.$$

Letting $x = 1.1$ gives
$$\sqrt{1.1} \approx P_4(1.1) = 1.0488.$$

b. $|\text{Error}| = |\sqrt{1.1} - P_4(1.1)|$
$$= |R_4(1.1)|$$
$$= \left| \frac{f^{(4+1)}(z)}{(4+1)!}(1.1-1)^{4+1} \right|$$
$$= \frac{3 \cdot 5 \cdot 7(.1)^5}{2^5 \cdot 5! x^{9/2}}$$

Because $1 < z < 1.1$, $\frac{1}{z} < \frac{1}{1} = 1$, we have

$$|\text{Error}| = \frac{21(.1)^5}{2^5 4! z^{9/2}} < \frac{21(.1)^5}{2^5 4!} = \frac{7(.1)^5}{256}.$$

c. From part (b), we have for $1 < z < 1.1$,

$$|R_n(1.1)| = \left| \frac{f^{(n+1)}(z)}{(n+1)!}(1.1-1)^{n+1} \right|$$
$$= \frac{1 \cdot 3 \cdot 5 \cdots (2n-1)(.1)^{n+1}}{2^{n+1}(n+1)! z\left(\frac{2n+1}{2}\right)}$$
$$< \frac{1 \cdot 3 \cdot 5 \cdots (2n-1)(.1)^{n+1}}{2^{n+1}(n+1)!}$$

because $1 < z < 1.1$ gives $\frac{1}{z} < \frac{1}{1} = 1$. So

$$|R_n(1.1)| < \frac{1 \cdot 3 \cdot 5 \cdots (2n-1)(.1)^{n+1}}{2^{n+1}(n+1)!} < 10^{-10} \text{ if }$$

$n = 8$.

37. a. Expand $f(x) = e^x$ into Taylor series about $c = 0$. So $e^x = \sum_{k=0}^{\infty} \frac{x^k}{k!}$ (from Example 7.1).

Then

$$P_4(x) = \sum_{k=0}^{4} \frac{f^{(k)}(0)}{k!}x^k$$
$$= 1 + x + \frac{x^2}{2!} + \frac{x^3}{3!} + \frac{x^4}{4!}.$$

Letting $x = .1$ gives $e^{.1} \approx P_4(.1) = 1.10517$.

b. $|\text{Error}| = |e^{.1} - P_4(.1)|$
$$= |R_4(.1)|$$
$$= \left| \frac{f^{4+1)}(z)}{(4+1)!}(.1)^{4+1} \right|$$
$$= \frac{e^z(.1)^5}{5!}$$

Because $0 < z < .1$ gives $e^z < e^{.1} < 1.5$, we have $|\text{Error}| = \frac{e^z(.1)^5}{5!} < \frac{(1.5)(.1)^5}{5!}$.

c. From part (b), we have for $0 < z < .1$,

$$|R_n(.1)| = \left| \frac{f^{(n+1)}(z)}{(n+1)!}(.1)^{n+1} \right|$$
$$= \frac{e^z(.1)^{n+1}}{(n+1)!}$$
$$< \frac{(1.5)(.1)^{n+1}}{(n+1)!}$$

because $0 < z < .1$ gives $e^z < e^{.1} < 1.5$. So

$$|R_n(.1)| < \frac{1.5(.1)^{n+1}}{(n+1)!} < 10^{-10} \text{ if } n = 6.$$

39. Since $\sum_{k=0}^{\infty} \frac{1}{k!}x^k = e^x$ by replacing x with 2, we have $\sum_{k=0}^{\infty} \frac{(2)^k}{k!} = e^2$; e^x with $x = 2$.

41. Since $\sum_{k=0}^{\infty} \frac{(-1)^k x^{2k+1}}{(2k+1)!} = \sin x$, by replacing x with π, we have $\sum_{k=0}^{\infty} \frac{(-1)^k \pi^{2k+1}}{(2k+1)!} = \sin \pi = 0$;

$\sin x$ with $x = \pi$.

43. Since $\sum_{k=0}^{\infty}\frac{(-1)^k}{2k+1}x^{2k+1}=\tan^{-1}x$, by replacing x with 1, we have $\sum_{k=0}^{\infty}\frac{(-1)^k}{2k+1}=\tan^{-1}(1)=\frac{\pi}{4}$;

$\tan^{-1}(x)$ with $x=1$.

45. Since $e^x=\sum_{k=0}^{\infty}\frac{x^k}{k!}$ with $r=\infty$, by replacing x with $-2x$, we have $e^{-2x}=\sum_{k=0}^{\infty}\frac{(-1)^k(2x)^k}{k!}$ with

$r=\infty$.

47. Since $e^x=\sum_{k=0}^{\infty}\frac{x^k}{k!}$ with $r=\infty$, by replacing x with $-x^2$ and multiplying by x we have $xe^{-x^2}=\sum_{k=0}^{\infty}\frac{(-1)^k x^{2k+1}}{k!}$

with $r=\infty$.

49. Since $\sin x=\sum_{k=0}^{\infty}\frac{(-1)^k}{(2k+1)!}x^{2k+1}$ with $r=\infty$, by replacing x with x^2, we have

$\sin x^2=\sum_{k=0}^{\infty}\frac{(-1)^k x^{4k+2}}{(2k+1)!}$ with $r=\infty$.

51. Since $\cos x=\sum_{k=0}^{\infty}\frac{(-1)^k x^{2k}}{(2k)!}$ with $r=\infty$, by replacing x with $3x$, we have $\cos 3x=\sum_{k=0}^{\infty}\frac{(-1)^k 9^k x^{2k}}{(2k)!}$ with $r=\infty$.

53. If $f(x)=\begin{cases}e^{-1/x^2}&\text{if }x\neq 0\\0&\text{if }x=0\end{cases}$, then $f'(x)=\begin{cases}\frac{2}{x^3}e^{-1/x^2}&\text{if }x\neq 0\\0&\text{if }x=0\end{cases}$, so $f'(0)=0$, and

$f''(x)=\begin{cases}\frac{4}{x^6}e^{-1/x^2}-\frac{6}{x^4}e^{-1/x^2}&\text{if }x\neq 0\\0&\text{if }x=0\end{cases}$, so $f''(0)=0$.

55. $f(t)=\sum_{k=0}^{3}\frac{f^{(k)}(0)}{k!}t^k=10+10t+t^2-\frac{t^3}{3!}$

$f(2)=10+10(2)+(2)^2-\frac{(2)^3}{6}=\frac{98}{3}$ miles

57. $f(x)=e^x$ and $f'(x)=e^x$ so $f(c)=e^c=f'(c)$, so

$e^x=\sum_{k=0}^{\infty}\frac{f^{(k)}(c)}{k!}(x-c)^k=\sum_{k=0}^{\infty}\frac{e^c}{k!}(x-c)^k$

59. To generate the series, differentiate $(1+x)^r$ repeatedly. The series terminates if r is a positive integer, since all derivatives after the rth derivative will be 0. Otherwise the series is infinite.

61. $1+\sum_{k=1}^{\infty}\frac{\left(\frac{1}{2}\right)\left(-\frac{1}{2}\right)\cdots\left(\frac{1}{2}-k+1\right)}{k!}x^k$

63. $\cosh x=1+\frac{1}{2}x^2+\frac{1}{24}x^4+\frac{1}{720}x^6+\cdots$

$\sinh x=x+\frac{1}{6}x^3+\frac{1}{120}x^5+\frac{1}{5040}x^7+\cdots$

These match the cosine and sine series except that here all signs are positive.

Section 8.8

5. Using the first three terms of the sine series around the center $\pi/2$ we get $\sin 1.61 \approx 0.999231634426433$.

7. Using the first five terms of the cosine series around the center 0 we get $\cos 0.34 \approx 0.94275466553403$.

9. Using the first nine terms of the exponential series around the center 0 we get $e^{0.2} \approx 0.818730753079365$.

11. Since $\cos x^2 = \displaystyle\sum_{k=0}^{\infty} \frac{(-1)^k x^{4k}}{(2k)!}$

$$= 1 - \frac{x^4}{2!} + \frac{x^8}{4!} - \frac{x^{12}}{6!} + \frac{x^{16}}{8!}$$

$$= \ldots, \text{ then}$$

$$\lim_{x \to 0} \frac{\cos x^2 - 1}{x^4} = \lim_{x \to 0} \frac{\left(-\frac{x^4}{2!} + \frac{x^8}{4!} - \frac{x^{12}}{6!} \ldots\right) - 1}{x^4}$$

$$= \lim_{x \to 0}\left(-\frac{1}{2} + \frac{x^4}{4!} - \frac{x^8}{6!} + \ldots\right)$$

$$= -\frac{1}{2}.$$

13. Since $\ln x = \displaystyle\sum_{k=1}^{\infty} \frac{(-1)^k (x-1)^k}{k} = (x-1) - \frac{1}{2}(x-1)^2 + \frac{1}{3}(x-1)^3 - \ldots,$ then

$$\lim_{x \to 0} \frac{\ln x - (x-1)}{(x-1)^2} = \lim_{x \to 0} \frac{\left[(x-1) - \frac{1}{2}(x-1)^2 + \frac{1}{3}(x-1)^3 - \ldots\right] - (x-1)}{(x-1)^2}$$

$$= \lim_{x \to 0}\left(-\frac{1}{2} + \frac{1}{3}(x-1) - \frac{1}{4}(x-1)^2 + \ldots\right)$$

$$= -\frac{1}{2}.$$

15. Since $e^x = \displaystyle\sum_{k=0}^{\infty} \frac{x^k}{k!} = 1 + x + \frac{1}{2}x^2 + \frac{1}{3!}x^3 + \ldots,$ then

$$\lim_{x \to 0} \frac{e^x - 1}{x} = \lim_{x \to 0} \frac{\left(1 + x + \frac{x^2}{2} + \frac{x^3}{3!} + \ldots\right) - 1}{x} = \lim_{x \to 0}\left(1 + \frac{x}{2} + \frac{x^2}{3!} + \ldots\right) = 1.$$

17. Since $\sin x = \displaystyle\sum_{k=0}^{\infty} \frac{(-1)^k x^{2k+1}}{(2k+1)!}$, then when $n = 3$,

$$\sin x \approx x - \frac{x^3}{3!} + \frac{x^5}{5!} \text{ so}$$

$$\int_{-1}^{1} \frac{\sin x}{x} dx = \int_{-1}^{1}\left(1 - \frac{x^2}{6} + \frac{x^4}{120}\right)dx$$

$$= \left[x - \frac{x^3}{18} + \frac{x^5}{600}\right]_{-1}^{1}$$

$$= \frac{1703}{900} \approx 1.8922.$$

19. Since $e^{-x^2} = \displaystyle\sum_{k=0}^{\infty} \frac{(-1)^k x^{2k}}{k!}$, then when $n = 4$,

$$e^{-x^2} \approx 1 - x^2 + \frac{x^4}{2} - \frac{x^6}{6}, \text{ so}$$

$$\int_{-1}^{1} e^{-x^2} dx = \int_{-1}^{1}\left(1 - x^2 + \frac{x^4}{2} - \frac{x^6}{6}\right)dx$$

$$= \left[x - \frac{x^3}{3} + \frac{x^5}{10} - \frac{x^7}{42}\right]_{-1}^{1}$$

$$= \frac{52}{35} \approx 1.4857.$$

21. Since $\ln x = \sum_{k=1}^{\infty} \frac{(-1)^{k+1}}{k}(x-1)^k$, then when

$n = 4$,

$$\ln x \approx (x-1) - \frac{1}{2}(x-1)^2 + \frac{1}{3}(x-1)^3 - \frac{1}{4}(x-1)^4,$$

so

$$\int_1^2 \ln x\, dx$$

$$= \int_1^2 \left[(x-1) - \frac{1}{2}(x-1)^2 + \frac{1}{3}(x-1)^3 - \frac{1}{4}(x-1)^4 \right] dx$$

$$= \left[\frac{(x-1)^2}{2} - \frac{(x-1)^3}{6} + \frac{(x-1)^4}{12} - \frac{(x-1)^5}{20} \right]_1^2$$

$$= \frac{11}{30} \approx 0.3667.$$

23. $\left| \dfrac{a_{n+1}}{a_n} \right| \leq \dfrac{x^2}{2^2(n+1)(n+2)}$. Since this ratio tends

to 0, the radius of convergence for $J_1(x)$ is infinite.

25. We need the first neglected term to be less than 0.04. The kth term is bounded by

$$\frac{10^{2k+1}}{2^{2k+2}k!(k+2)!}, \text{ which is equal to } 0.0357 \text{ for}$$

$k = 12$. Thus we will need the terms up through $k = 11$, that is, the first 12 terms of the series.

27. $m(v) = \dfrac{m_o}{\sqrt{1 - \dfrac{v^2}{c^2}}}$ with $m(0) = m_o$, so

$$m'(v) = \frac{\dfrac{m_o}{c^2}v}{\left(1 - \dfrac{v^2}{c^2}\right)^{3/2}} \text{ with } m'(0) = 0, \text{ and}$$

$$m''(v) = \frac{\left(1 - \dfrac{v^2}{c^2}\right)\dfrac{m_o}{c^2} + \dfrac{3m_o}{c^4}\left(1 - \dfrac{v^2}{c^2}\right)^{1/2} v}{\left(1 - \dfrac{v^2}{c^2}\right)v^2}. \text{ To}$$

increase mass by 10%, we want $1 + \dfrac{1}{2c^2}v^2 - 1.1$,

then $m(v) = 1.1m_o$. So solving for v, we have

$$\frac{1}{2c^2}v^2 = .1 \text{ so } v^2 = .2c^2, \text{ thus}$$

$v = \sqrt{.2}c \approx 83,000$ miles per second .

29. $w(x) = \dfrac{mgR^2}{(R+x)^2}$ with $w(0) = mg$ and

$$w'(x) = \frac{-2mg}{(R+x)^3} \text{ with } w'(0) = \frac{-2mg}{R^3},$$

so $w(x) \approx \sum_{k=0}^{1} \dfrac{f^{(k)}(0)}{k!}x^k = mg\left(1 - \dfrac{2x}{R}\right).$

To reduce weight by 10%, we want $1 - \dfrac{2x}{R} = .9$,

then $w(x) = .9mg$. So solving for x, we have

$\dfrac{-2x}{R} = -.1$, so $2x = .1R$, thus $x = \dfrac{R}{20}$.

31. Since x is much smaller than R for high-altitude locations on Earth, we can neglect the last term in the numerator, which gives us the same estimate as in Exercise 29. You have to go out to an altitude of 200 miles before you weigh significantly less.

33. The first neglected term is negative, so this estimate is too large.

35. Use the first two terms of the series for tanh:

$\tanh x \approx x - \dfrac{1}{3}x^3$. Substitute $\sqrt{\dfrac{g}{40m}}\, t$ for x,

multiply by $\sqrt{40mg}$ and simplify. The result is

$$gt = \frac{g^2}{120m}t^3.$$

37. Using the series for $(1+x)^{3/2}$ around the center $x = 0$, we get

$S(d)$

$$\approx \frac{8\pi c^2}{3}\left(\frac{3}{32}\cdot\frac{d^2}{c^2} + \frac{3}{2048}\cdot\frac{d^4}{c^4} - \frac{1}{65536}\cdot\frac{d^6}{c^6} \right).$$

If we ignore the d^4 and d^6 terms, this simplifies

to $\dfrac{\pi d^2}{4}$.

Section 8.9

5. $f(x) = x$

$$a_0 = \frac{1}{\pi}\int_{-\pi}^{\pi} f(x)dx = \frac{1}{\pi}\int_{-\pi}^{\pi} x\,dx = \frac{1}{\pi}\left[\frac{x^2}{2}\right]_{-\pi}^{\pi} = 0$$

$$a_k = \frac{1}{\pi}\int_{-\pi}^{\pi} x\cos(kx)dx$$

$$= \frac{1}{\pi}\left[\frac{x}{k}\sin(kx) + \frac{1}{k^2}\cos(kx)\right]_{-\pi}^{\pi}$$

$$= \frac{1}{\pi}\left[\frac{\pi}{k}\sin(k\pi) + \frac{1}{k^2}\cos k\pi - \left(-\frac{\pi}{k}\sin(-k\pi)\right) + \frac{1}{k^2}\cos(-k\pi)\right]$$

$$= \frac{1}{\pi}\left(\frac{1}{k^2}\cos k\pi - \frac{1}{k^2}\cos(-k\pi)\right)$$

$$= 0$$

$$b_k = \frac{1}{\pi}\int_{-\pi}^{\pi} x\sin(kx)dx$$

$$= \frac{1}{\pi}\left[\frac{-x}{k}\cos kx + \frac{1}{k^2}\sin kx\right]_{-\pi}^{\pi}$$

$$= \frac{1}{\pi}\left[-\frac{\pi}{k}\cos(k\pi) + \frac{1}{k^2}\sin(k\pi) - \left(\frac{\pi}{2}\cos(-k\pi) + \frac{1}{k^2}\sin(-k\pi)\right)\right]$$

$$= \frac{1}{\pi}\left[\frac{-2\pi}{k}\cos(k\pi)\right]$$

$$= \frac{-2}{k}(-1)^k$$

so $f(x) = \sum_{k=1}^{\infty}(-1)^{k+1}\frac{2}{k}\sin(kx).$

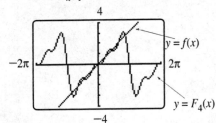

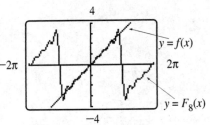

7. $f(x) = 2|x| = \begin{cases} -2x \text{ if } -\pi < x < 0 \\ 2x \text{ if } 0 < x < \pi \end{cases}$

$a_0 = \dfrac{1}{\pi} \displaystyle\int_{-\pi}^{\pi} f(x)dx$

$= \dfrac{1}{\pi} \displaystyle\int_{-\pi}^{0} (-2x)dx + \dfrac{1}{\pi} \displaystyle\int_{0}^{\pi} 2xdx$

$= \dfrac{1}{\pi}\left[-x^2\right]_{-\pi}^{0} + \dfrac{1}{\pi}\left[x^2\right]_{0}^{\pi}$

$= \dfrac{1}{\pi}\left[0 + \pi^2\right] + \left[\pi^2 - 0\right]$

$= 2\pi$

$a_k = \dfrac{1}{\pi} \displaystyle\int_{-\pi}^{\pi} f(x)\cos(kx)dx$

$= \dfrac{1}{\pi} \displaystyle\int_{-\pi}^{0} -2x\cos(kx)dx + \dfrac{1}{\pi} \displaystyle\int_{0}^{\pi} 2x\cos(kx)dx$

$= \dfrac{1}{\pi}\left[-\dfrac{2x}{k}\sin(kx) - \dfrac{2}{k^2}\cos(kx)\right]_{-\pi}^{0} + \dfrac{1}{\pi}\left[\dfrac{2x}{k}\sin(kx) + \dfrac{2}{k^2}\cos(kx)\right]_{0}^{\pi}$

$= \dfrac{1}{\pi}\left[0 - \dfrac{2}{k^2}\cos(0) - \left(\dfrac{2\pi}{k}\sin(-k\pi) - \dfrac{2}{k^2}\cos(-k\pi)\right)\right] + \dfrac{1}{\pi}\left[\dfrac{2\pi}{k}\sin(k\pi) + \dfrac{2}{k^2}\cos(k\pi) - \left(0 + \dfrac{2}{k^2}\cos(0)\right)\right]$

$= \dfrac{-2}{k^2\pi} + \dfrac{2}{k^2\pi}(-1)^k + \dfrac{2}{k^2\pi}(-1)^k - \dfrac{2}{k^2\pi}$

$= \dfrac{4}{k^2\pi}(-1)^k - \dfrac{4}{k^2\pi}$

$= \dfrac{4}{k^2\pi}[(-1)^k - 1]$

$= \begin{cases} 0 \text{ if } k \text{ is even} \\ \dfrac{-8}{k^2\pi} \text{ if } k \text{ is odd} \end{cases}$

So $a_{2k=0}$ and $a_{(2k-1)} = \dfrac{-8}{(2k-1)^2\pi}$.

$b_k = \dfrac{1}{\pi} \displaystyle\int_{-\pi}^{\pi} f(x)\sin(kx)dx$

$= \dfrac{1}{\pi} \displaystyle\int_{-\pi}^{0} -2x(kx)dx + \dfrac{1}{\pi} \displaystyle\int_{0}^{\pi} 2x\sin(kx)dx$

$= \dfrac{1}{\pi}\left[\dfrac{2x}{k}\cos(kx) - \dfrac{2}{k^2}\sin(kx)\right]_{-\pi}^{0} + \dfrac{1}{\pi}\left[\dfrac{-2x}{k}\cos(kx) + \dfrac{2}{k^2}\sin(kx)\right]_{0}^{\pi}$

$= \dfrac{1}{\pi}\left[0 - \dfrac{2}{k^2}\sin(0) - \left(\dfrac{-2\pi}{k}\cos(-k\pi) - \dfrac{2}{k^2}\sin(-k\pi)\right)\right] + \dfrac{1}{\pi}\left[\dfrac{-2\pi}{k}\cos(k\pi) + \dfrac{2}{k^2}\sin(k\pi) - \left(0 + \dfrac{2}{k^2}\sin(0)\right)\right]$

$= \dfrac{2}{k}\cos(k\pi) - \dfrac{2}{k}\cos k\pi$

$= 0$

So $f(x) = \dfrac{2\pi}{2} + \displaystyle\sum_{k=1}^{\infty} \dfrac{-8}{(2k-1)^2\pi}\cos(2k-1)x = \pi - \sum_{k=1}^{\infty} \dfrac{8}{(2k-1)^2\pi}\cos(2k-1)x$.

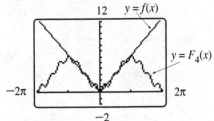

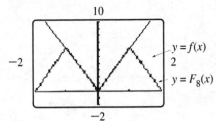

9. $f(x) = \begin{cases} 1 \text{ if } -\pi < x < 0 \\ -1 \text{ if } 0 < x < \pi \end{cases}$

$a_0 = \dfrac{1}{\pi}\displaystyle\int_{-\pi}^{0} 1\,dx + \dfrac{1}{\pi}\int_{0}^{\pi} (-1)\,dx = \dfrac{1}{\pi}[x]_{-\pi}^{0} + \dfrac{1}{\pi}[-x]_{0}^{\pi} = \dfrac{1}{\pi}(0-(-\pi)) + \dfrac{1}{\pi}(-\pi-0) = 0$

$a_k = \dfrac{1}{\pi}\displaystyle\int_{-\pi}^{0} \cos(kx)\,dx + \dfrac{1}{\pi}\int_{0}^{\pi}\big[-\cos(kx)\big]\,dx$

$= \dfrac{1}{\pi}\left[\dfrac{1}{k}\sin kx\right]_{-\pi}^{0} - \dfrac{1}{\pi}\left[\dfrac{1}{k}\sin kx\right]_{0}^{\pi}$

$= \dfrac{1}{\pi}\left[\dfrac{1}{k}\sin 0 - \dfrac{1}{k}\sin(-k\pi)\right] - \dfrac{1}{\pi}\left[\dfrac{1}{k}\sin k\pi - \dfrac{1}{k}\sin 0\right]$

$= 0$

$b_k = \dfrac{1}{\pi}\displaystyle\int_{\pi}^{0}\big[\sin(kx)\big]\,dx + \dfrac{1}{\pi}\int_{0}^{\pi} -\sin(kx)\,dx$

$= \dfrac{1}{\pi}\left[-\dfrac{1}{k}\cos kx\right]_{-\pi}^{0} - \dfrac{1}{\pi}\left[-\dfrac{1}{k}\cos kx\right]_{0}^{\pi}$

$= \dfrac{1}{\pi}\left[-\dfrac{1}{k}\cos 0 + \dfrac{1}{k}\cos(-k\pi)\right] - \dfrac{1}{\pi}\left[-\dfrac{1}{k}\cos k\pi + \dfrac{1}{k}\cos(0)\right]$

$= \dfrac{-2}{k\pi} + \dfrac{2}{k\pi}(-1)^k$

$= \dfrac{2}{k\pi}\left[(-1)^k - 1\right]$

$= \begin{cases} 0 \text{ if } k \text{ is even} \\ \dfrac{-4}{k\pi} \text{ if } k \text{ is odd} \end{cases}$

So $b_{2k} = 0$ and $b_{2k-1} = \dfrac{-4}{(2k-1)\pi}$. Thus, $f(x) = \displaystyle\sum_{k=1}^{\infty} \dfrac{-4}{(2k-1)\pi}\sin(2k-1)x$.

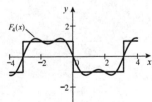

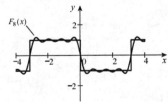

11. $f(x) = 3\sin 2x$ is already periodic on $[-2\pi, 2\pi]$

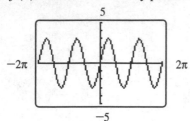

13. $f(x) = -x$ on $[-1, 1]$; so $l = 1$.

$$a_0 = \int_{-1}^{1}(-x)dx = \left[\frac{-x^2}{2}\right]_{-1}^{1} = \frac{-1}{2} - \frac{-1}{2} = 0$$

$$a_k = \int_{-1}^{1}(-x)\cos(k\pi x)dx$$

$$= \left[\frac{-x}{k\pi}\sin k\pi x - \frac{1}{(k\pi)^2}\cos k\pi x\right]_{-1}^{1}$$

$$= \frac{-1}{k\pi}\sin k\pi - \frac{1}{(k\pi)^2}\cos k\pi - \left[\frac{1}{k\pi}\sin(-k\pi) - \frac{1}{(k\pi)^2}\cos(-k\pi)\right]$$

$$= 0$$

$$b_k = \int_{-1}^{1}(-x)\sin(k\pi x)dx$$

$$= \left[\frac{x}{k\pi}\cos k\pi x - \frac{1}{(k\pi)^2}\sin k\pi x\right]_{-1}^{1}$$

$$= \frac{1}{k\pi}\cos k\pi - \frac{1}{(k\pi)^2}\sin k\pi - \left[\frac{-1}{k\pi}\cos(-k\pi) - \frac{1}{(k\pi)^2}\sin(-k\pi)\right]$$

$$= \frac{2}{k\pi}\cos k\pi$$

$$= \frac{2}{k\pi}(-1)^k$$

So $f(x) = \sum_{k=1}^{\infty}(-1)^k\frac{2}{k\pi}\sin k\pi x.$

15. $f(x) = x^2$ on $[-1, 1]$ so $l=1$

$$a_0 = \frac{1}{1}\int_{-1}^{1} x^2 dx = \frac{x^3}{3}\Bigg]_{-1}^{1} = \frac{1}{3} - \left(-\frac{1}{3}\right) = \frac{2}{3} \text{ and } \frac{a_0}{2} = \frac{1}{3}$$

$$a_k = \frac{1}{1}\int_{-1}^{1} x^2 \cos k\pi x \, dx$$

$$= \left[\frac{x^2}{k\pi}\sin k\pi x + \frac{2x}{(k\pi)^2}\cos k\pi x - \frac{2}{(k\pi)^3}\sin k\pi x\right]_{-1}^{1}$$

$$= \left[\frac{1}{k\pi}\sin k\pi + \frac{2}{(k\pi)^2}\cos k\pi - \frac{2}{(k\pi)^2}\sin k\pi - \left(\frac{1}{k\pi}\sin -k\pi - \frac{2}{(k\pi)^2}\cos -k\pi - \frac{2}{(k\pi)^3}\sin(-k\pi)\right)\right]$$

$$= \frac{4}{(k\pi)^2}\cos k\pi$$

$$= \frac{4}{(k\pi)^2}(-1)^k$$

$$b_k = \frac{1}{1}\int_{-1}^{1} x^2 \sin k\pi dx$$

$$= \left[\frac{x^2}{k\pi}\cos k\pi x + \frac{2x}{(k\pi)^2}\sin k\pi x + \frac{2}{(k\pi)^2}\cos k\pi x\right]_{-1}^{1}$$

$$= \left[\frac{-1}{k\pi}\cos k\pi + \frac{2}{(k\pi)^2}\sin k\pi + \frac{2}{(k\pi)^3}\cos k\pi - \left(-\frac{1}{k\pi}\cos(-k\pi) - \frac{2}{(k\pi)^2}\sin(-k\pi) + \frac{2}{(k\pi)^3}\cos(-k\pi)\right)\right]$$

$$= 0$$

So $f(x) = \frac{1}{3} + \sum_{k=1}^{\infty} \frac{(-1)^k 4}{(k\pi)^2}\cos k\pi x.$

17. $f(x) = \begin{cases} 0 \text{ if } -1 < x < 0 \\ x \text{ if } 0 < x < 1 \end{cases}$ so $l = 1$

$$a_0 = \int_{-1}^{1} f(x)\,dx = \int_{-1}^{0} 0\,dx + \int_{0}^{1} x\,dx = \left[\frac{x^2}{2}\right]_{0}^{1} = \frac{1}{2} \text{ and } \frac{a_0}{2} = \frac{1}{4}$$

$$a_k = \int_{-1}^{0} 0\,dx + \int_{0}^{1} x\cos(k\pi x)\,dx$$

$$= \left[\frac{x}{k\pi}\sin(k\pi x) + \frac{1}{(k\pi)^2}\cos(k\pi x)\right]_{0}^{1}$$

$$= \frac{1}{k\pi}\sin k\pi + \frac{1}{(k\pi)^2}\cos k\pi - 0 - \frac{1}{(k\pi)^2}\cos(0)$$

$$= \frac{1}{(k\pi)^2}[(-1)^k - 1]$$

$$= \begin{cases} 0 \text{ if } k \text{ is even} \\ \dfrac{-2}{(k\pi)^2} \text{ if } k \text{ is odd} \end{cases}$$

So $a_{2k} = 0$ and $a_{2k-1} = \dfrac{-2}{(2k-1)^2 \pi^2}$.

$$b_k = \int_{-1}^{0} 0 \, dx + \int_{0}^{1} x \sin(k\pi x) \, dx$$

$$= \left[\frac{-x}{k\pi} \cos(k\pi x) + \frac{1}{(k\pi)^2} \sin(k\pi x) \right]_0^1$$

$$= -\frac{1}{k\pi} \cos k\pi + \frac{1}{k\pi} \sin k\pi - 0 + \frac{1}{k\pi} \sin(0)$$

$$= \frac{-1}{k\pi} (-1)^k$$

So $f(x) = \dfrac{1}{4} + \displaystyle\sum_{k=1}^{\infty} \left[\dfrac{-2}{(2k-1)^2 \pi^2} \cos(2k-1)\pi x + \dfrac{(-1)^{k+1}}{k\pi} \sin k\pi x \right]$.

19. $f(x) = x$ on $[-2, 2]$

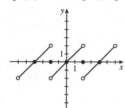

21. $f(x) = \begin{cases} -x \text{ if } -1 < x < 0 \\ 0 \text{ if } 0 < x < 1 \end{cases}$

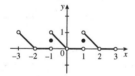

23. $f(x) = \begin{cases} -1 \text{ if } -2 < x < -1 \\ 0 \;\; \text{ if } -1 < x < 1 \\ 1 \;\; \text{ if } \;\;\; 1 < x < 2 \end{cases}$

25. $f(x) = x^2 = \dfrac{1}{3} + \displaystyle\sum_{k=1}^{\infty} (-1)^k \dfrac{4}{k^2 \pi^2} \cos k\pi x$

$$f(1) = 1 = \frac{1}{3} + \sum_{k=1}^{\infty} (-1)^k \frac{4}{k^2 \pi^2} (-1)^k$$

$$\frac{2}{3} = \sum_{k=1}^{\infty} \frac{4}{k^2 \pi^2} \text{ so } \frac{2}{3} = \frac{4}{\pi^2} \sum_{k=1}^{\infty} \frac{1}{k^2} \text{ so } \frac{\pi^2}{6} = \sum_{k=1}^{\infty} \frac{1}{k^2}$$

27. $f(x) = |x| = \begin{cases} -x \text{ if } -\pi \le x < 0 \\ x \;\; \text{ if } \;\; 0 \le x \le \pi \end{cases}$ gives us

$$f(x) = \frac{\pi}{2} - \sum_{k=1}^{\infty} \frac{4}{(2k-1)^2 \pi} \cos(2k-1)x$$

$$f(0) = 0 = \frac{\pi}{2} - \sum_{k=1}^{\infty} \frac{4}{(2k-1)^2 \pi} \cos 0$$

$$\frac{\pi}{2} = \sum_{k=1}^{\infty} \frac{4}{(2k-1)^2 \pi} = \frac{4}{\pi} \sum_{k=1}^{\infty} \frac{1}{(2k-1)^2}$$

$$\frac{\pi^2}{8} = \sum_{k=1}^{\infty} \frac{1}{(2k-1)^2}$$

29. If $f(x) = \cos x$, then
$f(-x) = \cos(-x) = \cos x = f(x)$ so $\cos x$ is even.
If $f(x) = \sin x$, then
$f(-x) = \sin(-x) = -\sin x = -f(x)$ so $\sin x$ is odd
If $f(x) = \cos x + \sin x$, then
$f(-x) = \cos(-x) + \sin(-x) = \cos x - \sin x \ne f(x)$
and $\cos x - \sin x \ne -f(x)$ so $\cos x + \sin x$ is neither
even nor odd.

31. If $f(x)$ is odd and $g(x) = f(x) \cos x$ then
$g(-x) = f(-x)\cos(-x) = -f(x)\cos x = -g(x)$ is odd.
Also if $h(x) = f(x)\sin x$ then
$$h(-x) = f(-x)\sin(-x)$$
$$= -f(x)(-\sin x)$$
$$= f(x)\sin x$$
$$= h(x)$$
so $h(x)$ is even.

33. If f is even and g is odd, then
$f(-x)g(-x) = f(x)[-g(x)] = -f(x)g(x)$, so $f \cdot g$ is odd.

35. We have the Fourier series expansion of f with period $2l$ is $f(x) = \dfrac{a_0}{2} + \sum_{k=1}^{\infty}\left[a_k \cos\left(\dfrac{k\pi x}{l}\right) + b_k \sin\left(\dfrac{k\pi x}{l}\right)\right]$.

Multiply both sides of this equation by $\cos\left(\dfrac{n\pi x}{l}\right)$ and integrate with respect to x on the interval $[-l,\, l]$ to get

$$\int_{-\ell}^{\ell} f(x)\cos\left(\frac{n\pi x}{\ell}\right)dx = \int_{-\ell}^{\ell}\frac{a_0}{2}\cos\left(\frac{n\pi x}{l}\right)dx + \sum_{k=1}^{\infty}\left[a_k \int_{-l}^{l}\cos\left(\frac{k\pi x}{l}\right)\cos\left(\frac{n\pi x}{l}\right)dx + b_k \int_{-l}^{l}\sin\left(\frac{k\pi x}{l}\right)\cos\left(\frac{n\pi x}{l}\right)dx\right]$$

Well $\int_{-l}^{l}\cos\left(\dfrac{n\pi x}{l}\right)dx = 0$ for all n. And, $\int_{-l}^{l}\sin\left(\dfrac{k\pi x}{l}\right)\cos\left(\dfrac{n\pi x}{l}\right)dx = \begin{cases} 0 & \text{if } n \neq k \\ l & \text{if } n = k \end{cases}$. So when $n = k$, we have

$\int_{-l}^{l} f(x)\cos\left(\dfrac{k\pi x}{l}\right)dx = a_k l$, so $a_k = \dfrac{1}{l}\int_{-l}^{l} f(x)\cos\left(\dfrac{k\pi x}{l}\right)dx$. Now multiply both sides of the original equation

by $\sin\left(\dfrac{n\pi x}{l}\right)$ and integrate on $[-l,\, l]$, we have

$$\int_{-l}^{l} f(x)\sin\left(\frac{n\pi x}{l}\right)dx = \int_{-l}^{l}\frac{a_0}{2}\sin\left(\frac{n\pi x}{l}\right)dx + \sum_{k=1}^{\infty}\left[a_k \int_{-l}^{l}\cos\left(\frac{k\pi x}{l}\right)\sin\left(\frac{n\pi x}{l}\right)dx + b_k \int_{-l}^{l}\sin\left(\frac{k\pi x}{l}\right)\sin\left(\frac{n\pi x}{l}\right)dx\right]$$

Well $\int_{-l}^{l}\sin\left(\dfrac{n\pi x}{l}\right)dx = 0$. And $\int_{-l}^{l}\cos\left(\dfrac{k\pi x}{l}\right)\sin\left(\dfrac{n\pi x}{l}\right)dx = 0$. Also $\int_{-l}^{l}\sin\left(\dfrac{k\pi x}{l}\right)\sin\left(\dfrac{n\pi x}{l}\right)dx = \begin{cases} 0 & \text{if } n \neq k \\ l & \text{if } n = k \end{cases}$.

So when $n = k$ we have $\int_{-l}^{l} f(x)\sin\left(\dfrac{k\pi x}{l}\right)dx = b_k l$. So $b_k = \dfrac{1}{l}\int_{-l}^{l} f(x)\sin\left(\dfrac{k\pi x}{l}\right)dx$.

37. $f(x) = x^3$

$f(-x) - (-x)^3 = -x^3$ so f is odd and the Fourier Series will contain only sine.

39. $f(x) = e^x$; $f(-x) = e^{-x} \neq f(x)$ and $e^{-x} \neq -f(x)$, so f is neither even nor odd, so the Fourier Series will contain both.

41. $f(x) = \begin{cases} 0 & \text{if } -1 < x < 0 \\ x & \text{if } \quad 0 < x < 1\end{cases}$

$f(-x) = -x = -f(x)$ if $x > 0$ or $x < 0$ and $f(-0) = 0 = f(x)$, so f is neither even nor odd so the Fourier Series will contain both.

43. Because $g(x)$ is an odd function, its series contains only sine terms, so $f(x)$ contains sine terms and the constant 1.

45. $f(x) = -$ on $[-4, 4]$

$f(x) = \begin{cases} -2 & \tfrac{1}{3} < |x| < 1 \\ 2 & |x| \le \tfrac{1}{3}\end{cases}$

$f(x) = \begin{cases} -2 & \tfrac{1}{4} < |x| < 1 \\ 2 & |x| \le \tfrac{1}{4}\end{cases}$

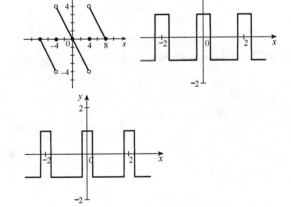

47. The cutoff frequency, n, corresponds to the partial sum $F_n(x)$ because in both, everything after the nth term is zero.

$f(x) = -x$ on $[-1, 1]$, so

$$f(x) = \sum_{k=1}^{\infty} \frac{(-1)^k 2}{k\pi} \sin k\pi x, \text{ thus}$$

$$F_2(x) = \frac{2}{\pi}[-\sin \pi x + \frac{1}{2}\sin 2\pi x]$$

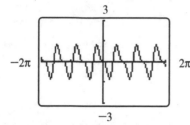

$$F_4(x)$$
$$= \frac{2}{\pi}\left[-\sin \pi x + \frac{1}{2}\sin 2\pi - \frac{1}{3}\sin 3\pi x + \frac{1}{4}\sin 4\pi x\right]$$

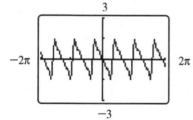

$$f(x) = \begin{cases} -2 & -1 < x \le 0 \\ 2 & 0 < x \le 1 \end{cases}, \text{ so}$$

$$f(x) = \sum_{k=1}^{\infty} \frac{8}{(2k-1)\pi} \sin[(2k-1)\pi x]$$

thus $F_2(x) = \frac{8}{\pi}[\sin \pi x + \frac{1}{3}\sin 3\pi x]$

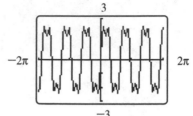

$$F_4(x)$$
$$= \frac{8}{\pi}\left[\sin \pi x + \frac{1}{3}\sin 3\pi x + \frac{1}{5}\sin 5\pi x + \frac{1}{7}\sin 7\pi x\right]$$

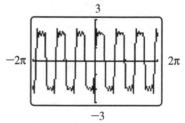

Because the wave is smoother.

49. The halo effect could be caused by an overshoot at the edges.

51. If $A = -1$ and $B = 4\pi$, then $-\sin 4\pi t = f(t)$ so $f(0) = 0$, $f\left(\frac{1}{3}\right) = \frac{\sqrt{3}}{2}$, $f\left(\frac{2}{3}\right) = \frac{\sqrt{3}}{2}$, $f(1) = 0$.

If $A = 1$ and $B = 2\pi$, then $\sin 2\pi t = f(t)$, so $f(0) = 0$, $f\left(\frac{1}{3}\right) = \frac{\sqrt{3}}{2}$, $f\left(\frac{2}{3}\right) = \frac{\sqrt{3}}{2}$, $f(1) = 0$.

Chapter 8 Review

1. See Example 1.3.

$$\lim_{n\to\infty}\frac{4}{3+n}=\lim_{n\to\infty}\frac{\frac{4}{n}}{\frac{3}{n}+1}=0 \text{ converges to } 0$$

3. See Example 1.8.

$$\lim_{n\to\infty}|a_n|=\lim_{n\to\infty}\frac{n}{n^2+4}=\lim_{n\to\infty}\frac{\frac{1}{n}}{1+\frac{4}{n^2}}=0$$

converges to 0

5. See Example 1.9.

$$0<\frac{4^n}{n!}=\frac{4\cdot4\cdot4\ldots4}{1\cdot2\cdot3\ldots n}<4\cdot\frac{4}{2}\cdot\frac{4}{3}\cdot\frac{4}{n}=\frac{128}{3n} \text{ and}$$

$$\lim_{n\to\infty}\frac{128}{3n}=0 \text{ , so by the Squeeze Theorem,}$$

$$\left\{\frac{4^n}{n!}\right\}_{n=1}^{\infty} \text{ converges to } 0.$$

7. See Remark 1.2.

$$\{\cos\pi n\}_{n=1}^{\infty}=\{-1,1,-1\ldots\} \text{ diverges}$$

For Exercises 9 through 19, see Sections 8.2 and 8.3.

9. diverges

11. can't tell

13. diverges

15. converges

17. converges

19. See Example 2.4.

$$\sum_{k=0}^{\infty}4\left(\frac{1}{2}\right)^k \text{ is a geometric series with } a=4 \text{ and}$$

$$|r|=\frac{1}{2}<1 \text{ so the series converges to}$$

$$\frac{a}{1-r}=\frac{4}{1-\frac{1}{2}}=8.$$

21. See Example 2.4.

$$\sum_{k=0}^{\infty}\left(\frac{1}{4}\right)^k \text{ is a geometric series with } a=1 \text{ and}$$

$$|r|=\frac{1}{4}<1 \text{ , so the series converges to}$$

$$\frac{a}{1-r}=\frac{1}{1-\frac{1}{4}}=\frac{4}{3}; \text{ converges to } \frac{4}{3}$$

23. See Example 4.6.

$$|S-S_k|\le a_{k+1}=\frac{k+1}{(k+1)^4+1}\le.01$$

when $k=4$, so $S\approx S_4\approx-0.40.$

25. See Example 2.6.

$$\lim_{k\to\infty}\frac{2k}{k+3}=\lim_{k\to\infty}\frac{2}{1+\frac{3}{k}}=2\ne0 \text{ so by the}$$

k^{th}-Term Test for Divergence, $\displaystyle\sum_{k=0}^{\infty}\frac{2k}{k+3}$

diverges.

27. See Example 4.3.

$$\lim_{k\to\infty}|a_k|=\lim_{k\to\infty}\frac{4}{(k+1)^{1/2}}=0 \text{ and}$$

$$\frac{a_{k+1}}{a_k}=\frac{4}{(k+2)^{1/2}}\cdot\frac{(k+1)^{1/2}}{4}=\frac{(k+1)^{1/2}}{(k+2)^{1/2}}<1$$

for all $k\ge0$, so $a_{k+1}<a_k$ for all $k\ge0$, so by

the Alternating Series Test, $\displaystyle\sum_{k=0}^{\infty}\frac{(-1)^x4}{\sqrt{k+1}}$

converges.

29. See Example 3.2.

$$\sum_{k=1}^{\infty}\frac{3}{k^{7/8}} \text{ is a divergent}$$

p-series $\left(p=\frac{7}{8}<1\right)$ diverges.

31. See Example 3.6.

Using the Limit Comparison Test, let $a_k = \dfrac{\sqrt{k}}{k^3 + 1}$

and $b_k = \dfrac{1}{k^{5/2}}$. Then

$$\lim_{k \to \infty} \frac{a_k}{b_k} = \lim_{k \to \infty} \frac{\sqrt{k}}{k^3 + 1} \cdot \frac{k^{5/2}}{1}$$

$$= \lim_{k \to \infty} \frac{k^3}{k^3 + 1} = 1 > 0$$

so because $\displaystyle\sum_{k=1}^{\infty} \frac{1}{k^{5/2}}$ is a convergent

p-series $\left(p = \dfrac{5}{2} > 1 \right)$, $\displaystyle\sum_{k=1}^{\infty} \frac{k}{\sqrt{k^3 + 1}}$ converges.

33. See Example 4.3.

Using the Alternating Series Test, $\displaystyle\lim_{k \to \infty} \frac{4^k}{k!} = 0$ (as

in Example 5) and $\dfrac{a_{k+1}}{a_k} = \dfrac{4^{k+1}}{(k+1)!} \cdot \dfrac{k!}{4} = \dfrac{4}{k+1} \le 1$

for $k \ge 3$ so $a_{k+1} \le a_k$ for $k \ge 3$, thus

$\displaystyle\sum_{k=1}^{\infty} (-1)^k \frac{4^k}{k!}$ converges.

35. See Example 4.2.

Using the Alternating Series Test,

$$\lim_{k \to \infty} |a_k| = \lim_{k \to \infty} \ln\left(1 + \frac{1}{k} \right) = \ln 1 = 0 \text{ and}$$

$$a_{k+1} = \ln\left(\frac{k+2}{k+1} \right) < \ln\left(\frac{k+1}{k} \right) = a_k \text{ for all } k \ge 1,$$

so $\displaystyle\sum_{k=1}^{\infty} (-1)^k \ln\left(1 + \frac{1}{k} \right)$ converges.

37. See Example 3.6.

Using the Limit Comparison Test, let

$a_k = \dfrac{2}{(k+3)^2}$ and $b_k = \dfrac{1}{k^2}$, so

$$\lim_{k \to \infty} \frac{a_k}{b_k} = \lim_{k \to \infty} \frac{2k^2}{(k+3)^2}$$

$$= \lim_{k \to \infty} \frac{2k^2}{k^2 + 6k + 9}$$

$$= 2 > 0$$

so because $\displaystyle\sum_{k=1}^{\infty} \frac{1}{k^2}$ is a convergent p-series

$p = 2 > 1$), $\displaystyle\sum_{k=1}^{\infty} \frac{2}{(k+3)^2}$ converges.

39. See Example 2.6.

$\dfrac{k!}{3^k} = \dfrac{1 \cdot 2 \cdot 3 \dots k}{3 \cdot 3 \cdot 3 \dots 3} > \dfrac{1}{3} \cdot \dfrac{2}{3} \cdot \dfrac{k}{3} = \dfrac{2k}{27}$, and

$\displaystyle\lim_{k \to \infty} \frac{2k}{27} = \infty$, so because $\dfrac{k!}{3^k} > \dfrac{2k}{27}$, $\displaystyle\lim_{k \to \infty} \frac{k!}{3^k} = \infty$

and by the kth-Term Test for Divergence, $\displaystyle\sum_{k=1}^{\infty} \frac{k!}{3^k}$

diverges.

41. See Example 3.3. Using the Comparison Test,

$e^{1/k} \le e$ for all $k \ge 1$, so $\dfrac{e^{1/k}}{k^2} \le \dfrac{e}{k^2}$. For all $k \ge$

1 and $\displaystyle\sum_{k=1}^{\infty} \frac{e}{k^2} = e \sum_{k=1}^{\infty} \frac{1}{k^2}$ which is a convergent p-

series ($p = 2 > 1$). Thus $\displaystyle\sum_{k=1}^{\infty} \frac{e^{1/k}}{k^2}$ converges.

43. See Example 5.4. Using the Ratio Test,

$$\lim_{k \to \infty} \left| \frac{a_{k+1}}{a_k} \right| = \lim_{k \to \infty} \left| \frac{4^{k+1} k!^2}{4^k (k+1)!^2} \right|$$

$$= \lim_{k \to \infty} \frac{4}{(k+1)^2} = 0 < 1$$

so the series converges absolutely. Thus $\displaystyle\sum_{k=1}^{\infty} \frac{4^k}{(k!)^2}$

converges.

45. See Example 5.2.

Using the Alternating Series Test,

$$\lim_{k \to \infty} |a_k| = \lim_{k \to \infty} \frac{k}{k^2 + 1} = 0 \text{ and}$$

$$\frac{a_{k+1}}{a_k} = \frac{(k+1)(k^2+1)}{k\left[(k+1)^2 + 1\right]} = \frac{k^3 + k^2 + k + 1}{k^3 + 2k^2 + 2k} < 1 \text{ for}$$

all $k \ge 1$ so $a_{k+1} < a_k$ for all $k \ge 1$, so the series

converges. But letting $a_k = \dfrac{k}{k^2 + 1}$ and $b_k = \dfrac{1}{k}$,

$\displaystyle\sum_{k=1}^{\infty} |a_k| = \sum_{k=1}^{\infty} \frac{k}{k^2 + 1}$ diverges using the Limit

Comparison Test, so

$\displaystyle\lim_{k \to \infty} \frac{a_k}{b_k} = \lim_{k \to \infty} \frac{k^2}{k^2 + 1} = 1 > 0$. Since $\displaystyle\sum_{k=1}^{\infty} \frac{1}{k}$ is the

divergent harmonic series, $\displaystyle\sum_{k=0}^{\infty} \frac{k}{k^2 + 1}$ diverges.

Thus $\displaystyle\sum_{k=1}^{\infty} \frac{(-1)^k k}{k^2 + 1}$ converges conditionally.

47. See Example 5.3.

$$\left|\frac{\sin k}{k^{3/2}}\right| = \frac{|\sin k|}{k^{3/2}} \le \frac{1}{k^{3/2}} \text{ because } |\sin k| \le 1 \text{ for all}$$

k. And $\displaystyle\sum_{k=1}^{\infty}\frac{1}{k^{3/2}}$ is a convergent p-series

$\left(p = \dfrac{3}{2} > 1\right)$. So by the Comparison Test,

$\displaystyle\sum_{k=1}^{\infty}\left|\frac{\sin k}{k^{3/2}}\right|$ converges, so $\displaystyle\sum_{k=1}^{\infty}\frac{\sin k}{k^{3/2}}$ converges

absolutely.

49. See Example 3.2.

If $p < 1$, then using the Limit Comparison Test we

have $a_k = \dfrac{2}{(3+k)^p}$ and $b_k = \dfrac{1}{k^p}$, so

$$\lim_{k\to\infty}\frac{a_k}{b_k} = \lim_{k\to\infty}\frac{2k^p}{(3+k)^p} = 2 > 0 \text{ but } \sum_{k=1}^{\infty}\frac{1}{k^p} \text{ is a}$$

divergent p-series ($p < 1$) so the series

$\displaystyle\sum_{k=1}^{\infty}\frac{2}{(3+k)^p}$ with $p < 1$ diverges. Likewise for

$p = 1$, by letting $b_k = \dfrac{1}{k}$ which is the divergent

harmonic series. If $p > 1$, however, letting

$b_k = \dfrac{1}{k^p}$, $\displaystyle\sum_{k=1}^{\infty}\frac{1}{k^p}$ is a convergent p-series

($p > 1$), so $\displaystyle\sum_{k=1}^{\infty}\frac{2}{(3+k)^p}$ converges when $p > 1$.

51. See Example 4.6.

$$|S - S_n| \le a_{n+1} = \frac{3}{(k+1)^2} \le 10^{-6}, \text{ so}$$

$3{,}000{,}000 \le (k+1)^2$, thus $1732.05 \le k+1$, so $1731.05 \le k$. Hence $k = 1732$ terms.

53. See Example 6.6.

$$f(x) = \frac{1}{4+x}$$

$$= \frac{1}{4}\left(\frac{1}{1-\left(-\frac{x}{4}\right)}\right)$$

$$= \frac{1}{4}\sum_{k=0}^{\infty}\left(-\frac{x}{4}\right)^k$$

$$= \sum_{k=0}^{\infty}\frac{(-1)^k x^k}{4^{k+1}}$$

which is a geometric series that converges when $\left|-\dfrac{x}{4}\right| < 1$, or $-4 < x < 4$. Thus $r = 4$.

55. See Example 6.6.

$$f(x) = \frac{3}{3+x^2} = \frac{1}{1-\left(-\frac{x^2}{3}\right)}$$

$$= \sum_{k=0}^{\infty}\left(-\frac{x^2}{3}\right)^k$$

$$= \sum_{k=0}^{\infty}\frac{(-1)^k x^{2k}}{3^k}$$

which is a geometric series that converges for

$\left|-\dfrac{x^2}{3}\right| < 1$ so $x^2 < 3$ or $|x| < \sqrt{3}$, so

$-\sqrt{3} < x < \sqrt{3}$. Thus, $r = \sqrt{3}$.

57. See pages Example 6.6.

From Exercise 53, $\dfrac{1}{4+x} = \displaystyle\sum_{k=0}^{\infty}\frac{(-1)^k x^k}{4^{k+1}}$ with

$r = 4$. By integrating both sides, we get

$$\int\frac{1}{4+x}\,dx = \sum_{k=0}^{\infty}\frac{(-1)^k}{4^{k+1}}\int x^k\,dx \text{ so}$$

$$\ln(4+x) = \sum_{k=0}^{\infty}\frac{(-1)^k x^{k+1}}{(k+1)4^{k+1}} + 4 \text{ with } r = 4.$$

59. See Example 6.2.

Using the Ratio Test, we have

$$\lim_{k\to\infty}\left|\frac{a_{k+1}}{a_k}\right| = \lim_{k\to\infty}\left|\frac{2x^{k+1}}{2x^k}\right| = |x|\lim_{k\to\infty}1 = |x| \text{ and}$$

$|x| < 1$ when $-1 < x < 1$. So the series converges absolutely for $x \in (-1, 1)$. When

$$x = -1, \sum_{k=0}^{\infty}(-1)^k 2(-1)^k = \sum_{k=1}^{\infty}2 \text{ and } \lim_{k\to\infty}2 = 2,$$

so the series diverges by the kth-Term Test for

Divergence. When $x = 1$, $\sum_{k=0}^{\infty}(-1)^k 2$ and

$\lim_{k\to\infty}2 = 2$, so the series diverges by the

k^{th}-Term Test for Divergence. So the interval of convergence is $(-1, 1)$.

61. See Example 6.3.

Using the Ratio Test, we have

$$\lim_{k\to\infty}\left|\frac{a_{k+1}}{a_k}\right| = \lim_{k\to\infty}\left|\frac{2kx^{k+1}}{2(k+1)x^k}\right| = |x|\lim_{k\to\infty}\frac{k}{k+1} = |x|$$

and $|x| < 1$ when $-1 < x < 1$. So the series converges absolutely for $x \in (-1, 1)$. When

$$x = -1, \sum_{k=0}^{\infty}(-1)^k \frac{2}{k}(-1)^k = \sum_{k=0}^{\infty}\frac{2}{k} \text{ which is the}$$

divergent harmonic series. When $x = 1$,

$$\sum_{k=0}^{\infty}(-1)^k \frac{2}{k} \text{ and } \lim_{k\to\infty}\frac{2}{k} = 0 \text{ and } \frac{a_{k+1}}{a_k} = \frac{k}{k+1} < 1$$

for all $k \geq 0$, so $a_{k+1} < a_k$ for all $k \geq 0$, so this series converges by the Alternating Series Test. Thus the interval of convergence is $(-1, 1]$.

63. See Example 6.2.

Using the Ratio Test, we have

$$\lim_{k\to\infty}\left|\frac{a_{k+1}}{a_k}\right| = \lim_{k\to\infty}\left|\frac{4k!(x-2)^{k+1}}{(k+1)!4(x-2)^k}\right|$$

$$= |x-2|\lim_{k\to\infty}\frac{1}{k+1}$$

$$= 0$$

and $0 < 1$ for all x, so the series converges absolutely for all $x \in (-\infty, \infty)$. Thus the interval of convergence is $(-\infty, \infty)$.

65. See Example 6.2.

Using the Ratio Test, we have

$$\lim_{k\to\infty}\left|\frac{a_{k+1}}{a_k}\right| = \lim_{k\to\infty}\left|\frac{3^{k+1}(x-2)^{k+1}}{3^k(x-2)^k}\right|$$

$$= 3|x-2|\lim_{k\to\infty}1$$

$$= 3|x-2|$$

and $3|x-2| < 1$ when $-\frac{1}{3} < x - 2 < \frac{1}{3}$ or

$\frac{5}{3} < x < \frac{7}{3}$, so the series converges for all

$x \in \left(\frac{5}{3}, \frac{7}{3}\right)$.

When $x = \frac{5}{3}$, $\sum_{k=0}^{\infty}3^k\left(-\frac{1}{3}\right)^k = \sum_{k=0}^{\infty}(-1)^k$ which

diverges by the kth-Term Test for Divergence

because $\lim_{k\to\infty}1 = 1$. When $x = \frac{7}{3}$,

$$\sum_{k=1}^{\infty}3^k\left(\frac{1}{3}\right)^k = \sum_{k=1}^{\infty}1 \text{ which diverges by the } k^{th}-$$

Term Test for Divergence because $\lim_{k\to\infty}1 = 1$. So

the interval of convergence is $\left(\frac{5}{3}, \frac{7}{3}\right)$.

67. See Example 7.1.

$f(x) = \sin x$ about $c = 0$.

$f(x) = \sin x, f'(x) = \cos x, f''(x) = -\sin x,$

$f'''(x) = -\cos x,$

$f(0) = 0, f'(0) = 1, f''(0) = 0, f'''(0) = -1$, so

$$\sum_{k=0}^{\infty}\frac{f^{(k)}(0)}{k!}x^k = x - \frac{1}{3!}x^3 + \frac{1}{5!}x^5 - \ldots$$

$$= \sum_{k=0}^{\infty}\frac{(-1)^k x^{2k+1}}{(2k+1)!}$$

69. See Example 7.2.

$f(x) = \ln x$ about $c = 1$.

$f(x) = \ln x, \; f'(x) = \dfrac{1}{x}, \; f''(x) = -\dfrac{1}{x^2}, \; f'''(x) = \dfrac{2}{x^3}, \; f^{(4)}(x) = -\dfrac{6}{x^4}$

$f(1) = 0, \; f'(1) = 1, \; f''(1) = -1, \; f'''(1) = 2!, \; f^{(4)}(1) = -3!$

$P_4(x) = \displaystyle\sum_{k=0}^{4} \dfrac{f^{(k)}(1)}{k!}(x-1)^k = (x-1) - \dfrac{1}{2}(x-1)^2 + \dfrac{1}{3}(x-1)^3 - \dfrac{1}{4}(x-1)^4$

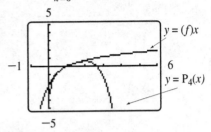

71. See Examples 7.6 and 7.7.

From Exercise 69,

$P_4(x) = (x-1) - \dfrac{1}{2}(x-1)^2 + \dfrac{1}{3}(x-1)^3 - \dfrac{1}{4}(x-1)^4$

Letting $x = 1.2$ gives $\ln(1.2) \approx P_4(1.2) = .1823$

Also, $\left| R_n(1.2) \right| = \left| \dfrac{f^{(n+1)}(z)}{(n+1)!}(1.2-1)^{n+1} \right| = \dfrac{n!(.2)^{n+1}}{(n+1)!\left|z\right|^{n+1}} = \dfrac{(.2)^{n+1}}{(n+1)z^{n+1}} < \dfrac{(.2)^{n+1}}{n+1}$ because $1 < z < 1.2$, so

$\dfrac{1}{z} < \dfrac{1}{1} = 1$. So $\left| R_n(1.2) \right| < \dfrac{(.2)^{n+1}}{n+1} < 10^{-8}$ if $n = 10$.

73. See Example 7.8.

Since $e^x = \displaystyle\sum_{k=0}^{\infty} \dfrac{x^k}{k!}$ with $r = \infty$, by replacing x with

$-3x^2$, we have

$e^{-3x^2} = \displaystyle\sum_{k=0}^{\infty} \dfrac{(-3x^2)^k}{k!} = \sum_{k=0}^{\infty} \dfrac{(-1)^k 3^k x^{2k}}{k!}$ with

75. See Example 7.8.

Since $\tan^{-1} x = \displaystyle\sum_{k=0}^{\infty} \dfrac{(-1)^k}{2k+1}x^{2k+1}$,

$\tan^{-1} x \approx x - \dfrac{1}{3}x^3 + \dfrac{1}{5}x^5 - \dfrac{1}{7}x^7 + \dfrac{1}{9}x^9$, so

$\displaystyle\int_0^1 \tan^{-1} x\,dx = \int_0^1 \left(x - \dfrac{1}{3}x^3 + \dfrac{1}{5}x^5 - \dfrac{1}{7}x^7 + \dfrac{1}{9}x^9 \right)dx$

$= \left[\dfrac{x^2}{2} - \dfrac{x^4}{12} + \dfrac{x^6}{30} - \dfrac{x^8}{58} + \dfrac{x^{10}}{90} \right]$

$= \dfrac{1117}{2520} \approx 0.4433.$

77. See Example 9.3.

$f(x) = x$, $-2 \le x \le 2$, so $l = 2$

$$a_0 = \frac{1}{2}\int_{-2}^{2} x\,dx = \frac{1}{2}\left[\frac{x^2}{2}\right]_{-2}^{2} = \frac{1}{2}\left[\frac{4}{2} - \frac{4}{2}\right] = 0$$

$$a_k = \frac{1}{2}\int_{-2}^{2} x\cos\left(\frac{k\pi x}{2}\right)dx$$

$$= \frac{1}{2}\left[\frac{2x}{k\pi}\sin\frac{k\pi x}{2} + \frac{4}{(k\pi)^2}\cos\frac{k\pi x}{2}\right]_{-2}^{2}$$

$$= \frac{1}{2}\left(\frac{4}{k\pi}\sin k\pi + \frac{4}{(k\pi)^2}\cos k\pi - \left[-\frac{4}{k\pi}\sin(-k\pi) + \frac{4}{(k\pi)^2}\cos(-k\pi)\right]\right)$$

$$= 0$$

$$b_k = \frac{1}{2}\int_{-2}^{2} x\sin\left(\frac{k\pi x}{2}\right)dx$$

$$= \frac{1}{2}\left[-\frac{2x}{k\pi}\cos\frac{k\pi x}{2} + \frac{4}{(k\pi)^2}\sin\frac{k\pi x}{2}\right]_{-2}^{2}$$

$$= \frac{1}{2}\left(-\frac{4}{k\pi}\cos k\pi + \frac{4}{(k\pi)^2}\sin k\pi - \left[\frac{4}{k\pi}\cos(-k\pi) + \frac{4}{(k\pi)^2}\sin(-k\pi)\right]\right)$$

$$= \frac{1}{2}\left(-\frac{8}{k\pi}\cos k\pi\right)$$

$$= -\frac{4}{k\pi}(-1)^k$$

So $f(x) = \displaystyle\sum_{k=1}^{\infty} \frac{(-1)^{k+1}4}{k\pi}\sin\left(\frac{k\pi x}{2}\right)$.

79. See Theorem 9.1.

$f(x) = x^2$, $-1 \le x \le 1$

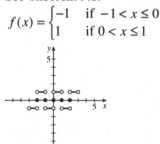

81. See Theorem 9.1.

$$f(x) = \begin{cases} -1 & \text{if } -1 < x \le 0 \\ 1 & \text{if } 0 < x \le 1 \end{cases}$$

83. See Example 2.4.

$\frac{1}{2} + \frac{1}{8} + \frac{1}{32} + \ldots = \displaystyle\sum_{k=0}^{\infty} \frac{1}{2}\left(\frac{1}{4}\right)^k$ which is a geometric series with $a = \frac{1}{2}$ and $|r| = \frac{1}{4} < 1$, so it converges to

$$\left(\frac{\frac{1}{2}}{1-\frac{1}{4}}\right) = \left(\frac{\frac{1}{2}}{\frac{3}{4}}\right) = \frac{2}{3}.$$

85. See Theorem 1.4.

$a_{n+1} = a_n + a_{n-1}$. Dividing by a_n gives $\dfrac{a_{n+1}}{a_n} = 1 + \dfrac{a_{n-1}}{a_n}$. $\dfrac{a_{n+1}}{a_n}$ is decreasing and bounded below so we know

that it has a limit r. Thus $\lim\limits_{n\to\infty} \dfrac{a_{n+1}}{a_n} = r$, and $\lim\limits_{n\to\infty} \dfrac{a_n}{a_n} = \lim\limits_{n\to\infty} 1 = 1 = \dfrac{r}{r} = \dfrac{\lim\limits_{n\to\infty}\frac{a_{n+1}}{a_n}}{r}$, so $\lim\limits_{n\to\infty} \dfrac{a_{n-1}}{a_n} = \dfrac{\lim\limits_{n\to\infty}\frac{a_n}{a_n}}{r} = \dfrac{1}{r}$.

So we have the equation $r = 1 + \dfrac{1}{r}$. Solving this for r gives us, $r^2 = r + 1$, so $r^2 - r - 1 = 0$, so

$r = \dfrac{1 \pm \sqrt{1 - 4(1)(-1)}}{2(1)} = \dfrac{1 \pm \sqrt{5}}{2}$ and since $r > 0$, $r = \dfrac{1 + \sqrt{5}}{2}$.

87. $a_1 = 1 + \dfrac{1}{1} = 2, a_2 = 1 + \dfrac{1}{2} = 1.5,$

$a_3 = 1 + \dfrac{1}{1.5} = 1.67$

$a_4 = 1 + \dfrac{1}{1.67} = 1.6, a_5 = 1 + \dfrac{1}{1.6} = 1.625,$

$a_6 = 1 + \dfrac{1}{1.625} = 1.615, \ldots$

So the sequence $a_n = 1 + \dfrac{1}{a_{n-1}}$ seems to be

converging to 1.618.

$a_1 = 1 + \dfrac{1^2}{2} = 1.5, a_2 = 1 + \dfrac{1^2}{2 + \frac{3^2}{2}} = 1.154,$

$a_3 = 1 + \dfrac{1^2}{2 + \dfrac{3^2}{2 + \frac{5^2}{2}}} = 1.382,$

$a_4 = 1 + \dfrac{1^2}{2 + \dfrac{3^2}{2 + \frac{5^2}{2 + \frac{7^2}{2}}}} = 1.198$

This sequence seems to be converging

to $\dfrac{4}{\pi} = 1.273$.

Chapter 9

Section 9.1

5.

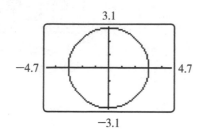

$$\begin{cases} x = 3\cos t \\ y = 3\sin t \end{cases}$$

$$\begin{cases} x^2 = 9\cos^2 t \\ y^2 = 9\sin^2 t \end{cases}$$

$$x^2 + y^2 = 9(\sin^2 t + \cos^2 t)$$

$$x^2 + y^2 = 9$$

7.

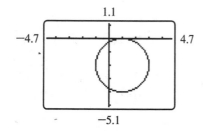

$$\begin{cases} x = 1 + 2\cos t \\ y = -2 + 2\sin t \end{cases}$$

$$\begin{cases} x - 1 = 2\cos t \\ y + 2 = 2\sin t \end{cases}$$

$$\begin{cases} (x-1)^2 = 4\cos^2 t \\ (y+2)^2 = 4\sin^2 t \end{cases}$$

$$(x-1)^2 + (y+2)^2 = 4(\sin^2 t + \cos^2 t)$$

$$(x-1)^2 + (y+2)^2 = 4$$

9.

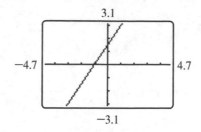

$$\begin{cases} x = -1 + 2t \\ y = 3t \end{cases}$$

$$t = \frac{y}{3}$$

$$x = -1 + 2\left(\frac{y}{3}\right)$$

$$\frac{2}{3}y = x + 1$$

$$y = \frac{3}{2}x + \frac{3}{2}$$

11.

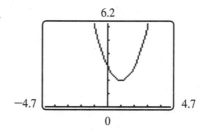

$$\begin{cases} x = 1 + t \\ y = t^2 + 2 \end{cases}$$

$$t = x - 1$$

$$y = (x-1)^2 + 2$$

$$y = x^2 - 2x + 3$$

13.

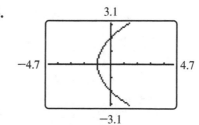

$$\begin{cases} x = t^2 - 1 \\ y = 2t \end{cases}$$

$$t = \frac{y}{2}$$

$$x = \left(\frac{y}{2}\right)^2 - 1 = \frac{1}{4}y^2 - 1$$

15.

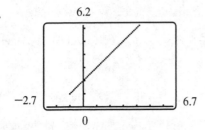

$$\begin{cases} x = t^2 - 1 \\ y = t^2 + 1 \end{cases}$$

$$y - x = t^2 + 1 - (t^2 - 1)$$

$$y = x + 2,\ x \ge -1$$

17.

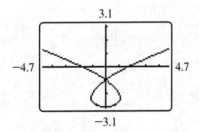

19.

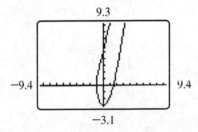

21.

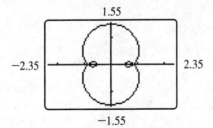

23.

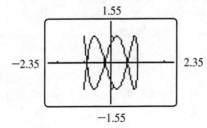

25.

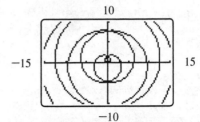

27.

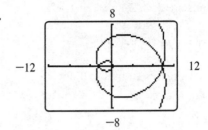

29.

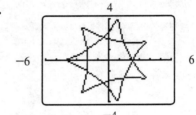

31.

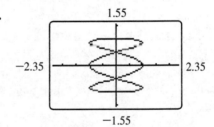

33.

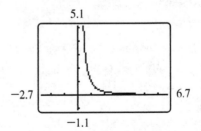

35. C. Since $(x+1)^2 = y$, with $x \ge -1$, this is part of an upward-opening parabola.

37. B. x is bounded below by -1. y is bounded below by -1 and above by 1. The graph will have x-intercepts at $t = n\pi$ or $x = (n\pi)^2 - 1$.

39. A. x and y both oscillate between -1 and 1, but with different periods.

41. Use $0 \le t \le 1$, with $t = 0$ corresponding to $(0, 1)$; then $x = 0 + bt$, $y = 1 + dt$. With $t = 1$, $x = 3 = 0 + b$, $y = 4 = 1 + d$. So $b = 3$, $d = 3$, the equations are $x = 3t$, $y = 1 + 3t$.

43. Use $0 \le t \le 1$, with $t = 0$ corresponding to $(-2, 4)$; then $x = -2 + bt$, $y = 4 + dt$. With $t = 1$, $x = 6 = -2 + b$, $y = 1 = 4 + d$. So $b = 8$, $d = -3$, the equations are $x = -2 + 8t$, $y = 4 - 3t$.

45. $x = t$, $y = t^2 + 1$ for $1 \le t \le 2$.

47. Since x decreases from 2 to 0, let
$x = 2 - t$, $y = 2 - (2 - t)^2 = -t^2 + 4t - 2$ for $0 \le t \le 2$.

49. Use $x = a + b\cos t$, $y = c + d\sin t$, with $0 \le t \le 2\pi$. The center is at $(2, 1)$, therefore $a = 2$ and $c = 1$. The radius is 3, therefore $b = d = 3$. So the equations are $x = 2 + 3\cos t$, $y = 1 + 3\sin t$.

51. Solve $\begin{cases} t = 1 + s \\ t^2 - 1 = 4 - s \end{cases}$

$s = t - 1$, so
$t^2 - 1 = 4 - (t - 1) \Rightarrow t^2 + t - 6 = 0 \Rightarrow t = -3$ or
$t = 2$. So either $t = -3$ and $s = -4$, in which case $x = -3$ and $y = 8$, or $t = 2$ and $s = 1$, in which case $x = 2$ and $y = 3$. So the intersection points are $(-3, 8)$ and $(2, 3)$.

53. Solve $\begin{cases} t + 3 = 1 + s \\ t^2 = 2 - s \end{cases}$

$s = t + 2$, so $t^2 = 2 - (t + 2) \Rightarrow t^2 + t = 0 \Rightarrow t = 0$ or $t = -1$. So either $t = 0$ and $s = 2$, in which case $x = 3$ and $y = 0$, or $t = -1$ and $s = 1$, in which case $x = 2$ and $y = 1$. So the intersection points are $(3, 0)$ and $(2, 1)$.

55. Solve $100t = 500 - 500(t - 2)$;
$100t = 1500 - 500t \Rightarrow t = \dfrac{15}{6}$. So at $t = \dfrac{5}{2}$, the missiles have the same x-coordinate.

For the target missile, the y-coordinate is
$80\left(\dfrac{15}{6}\right) - 16\left(\dfrac{15}{6}\right)^2 = 100$, and for the interceptor missile, the y-coordinate is
$208\left(\dfrac{15}{6} - 2\right) - 16\left(\dfrac{15}{6} - 2\right)^2 = 100$. Since the y-coordinates are equal when x-coordinates are equal, the interceptor missile hits its target.

57. For the interceptor missile, $y = 0$ at $t = $ time delay, and here $y = 0$ at $t = 2$. To find a time delay d that leads to missile collision, solve
$\begin{cases} 100t = 500 - 200(t - d) \\ 80t - 16t^2 = 80(t - d) - 16(t - d)^2. \end{cases}$

The first equation produces $d = \dfrac{3}{2}t - \dfrac{5}{2}$, and then from the second equation
$80t - 16t^2 = 80\left[t - \left(\dfrac{3}{2}t - \dfrac{5}{2}\right)\right] - 16\left[t - \left(\dfrac{3}{2}t - \dfrac{5}{2}\right)\right]^2$
$\Rightarrow 80t - 16t^2 = 80\left(-\dfrac{1}{2}t + \dfrac{5}{2}\right) - 16\left(-\dfrac{1}{2}t + \dfrac{5}{2}\right)^2$
$\Rightarrow 80t - 16t^2 = -40t + 200 - 4t^2 + 40t - 100$
$\Rightarrow 12t^2 - 80t + 100 = 0$
$\Rightarrow 3t^2 - 20t + 25 = 0$
$\Rightarrow (3t - 5)(t - 5) = 0$
$\Rightarrow t = 5$ or $t = \dfrac{5}{3}$

If $t = 5$, $d = \dfrac{3}{2}(5) - \dfrac{5}{2} = 5$ and if $t = \dfrac{5}{3}$,
$d = \dfrac{3}{2}\left(\dfrac{5}{3}\right) - \dfrac{5}{2} = 0$. So the delay time must be either 0 or 5. In the first case, the interceptor must be launched at the same instant as the enemy missile—which allows no time at all to spot the enemy launch and then respond—or the interceptor missile must be launched at the instant the enemy missile arrives at its target—which is too late to do any good. Better plot another interceptor trajectory.

59. The wave spreads in a circle, with radius t at time t.

61.

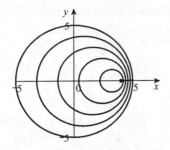

$x = \cos t$

$y = \sin 2t$

$\sin 2t = 2\cos t \sin t$

$\sin^2 2t = 4\cos^2 t \sin^2 t$

$\qquad = 4\cos^2 t(1 - \cos^2 t)$

Therefore $y^2 = 4x^2(1 - x^2)$

$\qquad\qquad$ or $y = \pm 2x\sqrt{1 - x^2}$

63.

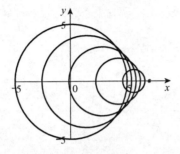

69.

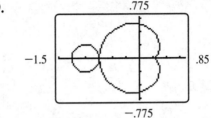

65. The shock wave will be a cone with its vertex at the location of the airplane.

67.

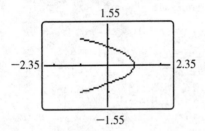

$x = \cos 2t$

$y = \sin t$

$\cos 2t = \cos^2 t - \sin^2 t$

$\qquad = 1 - \sin^2 t - \sin^2 t = 1 - 2\sin^2 t$

Therefore $x = 1 - 2y^2$.

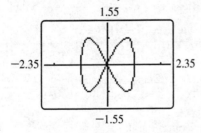

Section 9.2

5. $\dfrac{dy}{dx} = \dfrac{y'(t)}{x'(t)} = \dfrac{3t^2 - 1}{2t}$

$\quad$ **a.** $\dfrac{dy}{dx}(-1) = \dfrac{3(-1)^2 - 1}{2(-1)} = \dfrac{2}{-2} = -1$

$\quad$ **b.** $\dfrac{dy}{dx}(1) = \dfrac{3(1)^2 - 1}{2(1)} = \dfrac{2}{2} = 1$

$\quad$ **c.** $x = -2$ and $y = 0$ at $t = 0$, and

$\qquad \dfrac{dy}{dx}(0) = \dfrac{3(0)^2 - 1}{2(0)} = \dfrac{-1}{0}$, which is undefined.

7. $\dfrac{dy}{dx} = \dfrac{y'(t)}{x'(t)} = \dfrac{3\cos t}{-2\sin t} = -\dfrac{3}{2}\cot t$

$\quad$ **a.** $\dfrac{dy}{dx}\left(\dfrac{\pi}{4}\right) = -\dfrac{3}{2}\cot\dfrac{\pi}{4} = -\dfrac{3}{2}$

$\quad$ **b.** $\dfrac{dy}{dx}\left(\dfrac{\pi}{2}\right) = -\dfrac{3}{2}\cot\dfrac{\pi}{2} = 0$

$\quad$ **c.** $x = 0$ and $y = 3$ at $t = \dfrac{\pi}{2} + 2\pi n$, and

$\qquad \dfrac{dy}{dx}\left(\dfrac{\pi}{2} + 2\pi n\right) = -\dfrac{3}{2}\cot\left(\dfrac{\pi}{2} + 2\pi n\right) = 0$

9. $\dfrac{dy}{dx} = \dfrac{y'(t)}{x'(t)} = \dfrac{4\cos 4t}{-2\sin 2t} = -\dfrac{2\cos 4t}{\sin 2t}$

a. $\dfrac{dy}{dx}\left(\dfrac{\pi}{4}\right) = -\dfrac{2\cos \pi}{\sin \frac{\pi}{2}} = -\dfrac{-2}{1} = 2$

b. $\dfrac{dy}{dx}\left(\dfrac{\pi}{2}\right) = -\dfrac{2\cos 2\pi}{\sin \pi} = -\dfrac{2}{0}$, which is undefined.

c. $x = \dfrac{\sqrt{2}}{2}$ and $y = 1$ at $t = \dfrac{\pi}{8} + n\pi$ and

$\dfrac{dy}{dx}\left(\dfrac{\pi}{8} + n\pi\right) = -\dfrac{2\cos \frac{\pi}{2}}{\sin \frac{\pi}{4}} = -\dfrac{0}{\sqrt{2}/2} = 0$

11.

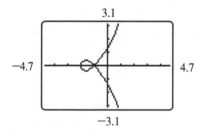

$\dfrac{dy}{dx} = \dfrac{y'(t)}{x'(t)} = \dfrac{3t^2 - 1}{2t}$

$x = -1$ and $y = 0$ at $t = \pm 1$.

$\dfrac{dy}{dx}(-1) = \dfrac{3(-1)^2 - 1}{2(-1)} = \dfrac{2}{-2} = -1$

$\dfrac{dy}{dx}(1) = \dfrac{3(1)^2 - 1}{2(1)} = \dfrac{2}{2} = 1$

13. $\dfrac{dy}{dx} = \dfrac{y'(t)}{x'(t)} = \dfrac{4\cos t}{-2\sin t} = -2\cot t$

a. Horizontal tangent: $\dfrac{dy}{dx} = -2\cot t = 0$

$\Rightarrow \cot t = 0 \Rightarrow t = \dfrac{\pi}{2} + n\pi$

$\Rightarrow x = 2\cos\left(\dfrac{\pi}{2} + n\pi\right)$,

$y = 4\sin\left(\dfrac{\pi}{2} + n\pi\right) \Rightarrow (0, -4), (0, 4)$

b. Vertical tangent: $\dfrac{dy}{dx}$ is undefined $\Rightarrow -2\cot t$ is undefined.
$\Rightarrow \cot t$ is undefined

$\Rightarrow t = n\pi \Rightarrow x = 2\cos n\pi,$
$y = 4\sin n\pi \Rightarrow (-2, 0), (2, 0)$

15. $\dfrac{dy}{dx} = \dfrac{y'(t)}{x'(t)} = \dfrac{4\cos 4t}{-2\sin 2t} = -\dfrac{2\cos 4t}{\sin 2t}$

a. Horizontal tangent:

$\dfrac{dy}{dx} = -\dfrac{2\cos 4t}{\sin 2t} = 0 \Rightarrow \cos 4t = 0 \Rightarrow$

$4t = \dfrac{\pi}{2} + n\pi \Rightarrow t = \dfrac{\pi}{8} + \dfrac{n\pi}{4}$

$\Rightarrow x = \cos\left(\dfrac{\pi}{4} + \dfrac{n\pi}{2}\right),$

$y = \sin\left(\dfrac{\pi}{2} + n\pi\right) \Rightarrow \left(\dfrac{\sqrt{2}}{2}, 1\right), \left(-\dfrac{\sqrt{2}}{2}, -1\right),$

$\left(-\dfrac{\sqrt{2}}{2}, 1\right), \left(\dfrac{\sqrt{2}}{2}, -1\right)$

b. Vertical tangent: $\dfrac{dy}{dx} = -\dfrac{2\cos 4t}{\sin 2t}$ is undefined

$\Rightarrow \sin 2t = 0 \Rightarrow 2t = n\pi \Rightarrow t = \dfrac{n\pi}{2}$

$\Rightarrow x = \cos n\pi,$
$y = \sin 2n\pi \Rightarrow (1, 0), (-1, 0)$

17. $\dfrac{dy}{dx} = \dfrac{y'(t)}{x'(t)} = \dfrac{4t^3 - 4}{2t} = \dfrac{2(t^3 - 1)}{t}$

a. Horizontal tangent:

$\dfrac{dy}{dx} = \dfrac{2(t^3 - 1)}{t} = 0 \Rightarrow t^3 - 1 = 0$

$\Rightarrow t = 1 \Rightarrow x = 0; \; y = -3 \Rightarrow (0, -3)$

b. Vertical tangent: $\dfrac{dy}{dx} = \dfrac{2(t^3 - 1)}{t}$ is undefined

$\Rightarrow t = 0 \Rightarrow x = -1, y = 0$
$\Rightarrow (-1, 0)$

19. $\dfrac{dy}{dx} = \dfrac{y'(t)}{x'(t)} = \dfrac{2\cos t - 2\sin 2t}{-2\sin t + 2\cos 2t} = \dfrac{\cos t - \sin 2t}{\cos 2t - \sin t}$

a. Horizontal tangent:

$\dfrac{dy}{dx} = \dfrac{\cos t - \sin 2t}{\cos 2t - \sin t} = 0 \Rightarrow \cos t - \sin 2t = 0$

$\Rightarrow \cos t = \sin 2t \Rightarrow \cos t = 2\sin t \cos t$

$\Rightarrow \cos t = 0$ or $\sin t = \dfrac{1}{2}$.

Since $\sin t = \dfrac{1}{2} \Rightarrow t = \dfrac{\pi}{2} \pm \dfrac{\pi}{3} + 2n\pi$, which causes both $y'(t)$ and $x'(t)$ to equal zero, we must consider the limit as $t \to \dfrac{\pi}{6}, t \to \dfrac{5\pi}{6}$

$$\lim_{t \to \frac{\pi}{6}} \left(\frac{\cos t - \sin 2t}{\cos 2t - \sin t} \right)$$

$$= \lim_{t \to \frac{\pi}{6}} \left(\frac{-\sin t - 2\cos 2t}{-2\sin 2t - \cos t} \right)$$

by L'Hôpital's Rule

$$= \frac{-\frac{1}{2} - 2\left(\frac{1}{2}\right)}{-2\left(\frac{\sqrt{3}}{2}\right) - \frac{1}{2}}$$

$$= -\frac{3}{2\sqrt{3} - 1}$$

$\neq 0$

$$\lim_{t \to \frac{5\pi}{6}} \left(\frac{\cos t - \sin 2t}{\cos 2t - \sin t} \right)$$

$$= \lim_{t \to \frac{5\pi}{6}} \left(\frac{-\sin t - 2\cos 2t}{-2\sin 2t - \cos t} \right)$$

$$= \frac{-\frac{1}{2} + 2\left(\frac{1}{2}\right)}{-2\left(\frac{\sqrt{3}}{2}\right) - \frac{1}{2}}$$

$$= \frac{1}{-2\sqrt{3} - 1}$$

$\neq 0$

Therefore, the only horizontal tangent line is

at $\cos t = 0 \Rightarrow t = \dfrac{\pi}{2} + n\pi \Rightarrow x = 0,$

$y = 1 \Rightarrow (0,1).$

b. Vertical tangent: $\dfrac{dy}{dx} = \dfrac{\cos t - \sin 2t}{\cos 2t - \sin t}$ is

undefined

$\Rightarrow \cos 2t - \sin t = 0 \Rightarrow \cos 2t = \sin t$

$\Rightarrow 1 - 2\sin^2 t = \sin t \Rightarrow 2\sin^2 t + \sin t - 1 = 0$

$\Rightarrow \sin t = \dfrac{-1 \pm \sqrt{1^2 - 4(2)(-1)}}{2(2)}$

$\Rightarrow \sin t = -1$ or $\sin t = \dfrac{1}{2} \Rightarrow t = \dfrac{3\pi}{2} + 2n\pi$

or $t = \dfrac{\pi}{2} \pm \dfrac{\pi}{3} + 2n\pi.$

The latter leads to $y'(t)$ and $x'(t)$ both equal to zero, and as we found in part a, the slope is defined and not zero.

So $t = \dfrac{3\pi}{2} + 2n\pi \Rightarrow$

$x = 2\cos\left(\dfrac{3\pi}{2} + 2n\pi\right) + \sin(3\pi + 4n\pi),$

$y = 2\sin\left(\dfrac{3\pi}{2} + 2n\pi\right) + \cos(3\pi + 4n\pi)$

$\Rightarrow (0, -3)$

The only vertical tangent is at $(0, -3).$

21. $x' = -2\sin t, \; y' = 2\cos t$

a. At $t = 0$, $x' = -2\sin 0 = 0$, $y' = 2\cos 0 = 2.$

Speed $= \sqrt{0^2 + 2^2} = 2$. Motion is up.

b. At $t = \dfrac{\pi}{2}$, $x' = -2\sin\dfrac{\pi}{2} = -2,$

$y' = 2\cos\dfrac{\pi}{2} = 0.$

Speed $= \sqrt{(-2)^2 + 0^2} = 2$. Motion is to the left.

23. $x' = 20, \; y' = -2 - 32t$

a. At $t = 0$, $x' = 20$, $y' = -2.$

Speed $= \sqrt{20^2 + (-2)^2} = 2\sqrt{101}.$

Motion is to the right and slightly down.

b. At $t = 2$, $x' = 20$, $y' = -2 - 64 = -66.$

Speed $= \sqrt{20^2 + (-66)^2} = 2\sqrt{1189}.$

Motion is to the right and down.

25. $x' = -4\sin 2t + 5\cos 5t, \; y' = 4\cos 2t - 5\sin 5t$

a. At $t = 0$, $x' = -4\sin 0 + 5\cos 0 = 5,$

$y' = 4\cos 0 - 5\sin 0 = 4.$

Speed $= \sqrt{5^2 + 4^2} = \sqrt{41}.$

Motion is to the right and up.

b. At $t = \dfrac{\pi}{2}$, $x' = -4\sin\pi + 5\cos\dfrac{5\pi}{2} = 0,$

$y' = 4\cos\pi - 5\sin\dfrac{5\pi}{2} = -9.$

Speed $= \sqrt{0^2 + (-9)^2} = 9$. Motion is down.

27. $x(t) = 3\cos t,\ y'(t) = 2\cos t$

$$A = \int_c^d x(t)\,y'(t)\,dt$$

$$= \int_0^{2\pi} (3\cos t)(2\cos t)\,dt$$

$$= \int_0^{2\pi} 6\cos^2 t\,dt$$

$$= \left[\frac{3}{2}(2t + \sin 2t)\right]_0^{2\pi}$$

$$= 6\pi$$

29. $x(t) = \dfrac{1}{2}\cos t - \dfrac{1}{4}\cos 2t,\ y'(t) = \dfrac{1}{2}\cos t - \dfrac{1}{2}\cos 2t$

$$A = \int_c^d x(t)\,y'(t)\,dt$$

$$= \int_0^{2\pi}\left(\frac{1}{2}\cos t - \frac{1}{4}\cos 2t\right)\left(\frac{1}{2}\cos t - \frac{1}{2}\cos 2t\right)dt$$

$$= \int_0^{2\pi}\frac{1}{4}\cos^2 t + \frac{1}{8}\cos^2 2t - \frac{3}{8}\cos t\cos 2t\,dt$$

$$= \left[\frac{1}{8}t + \frac{1}{16}\sin 2t + \frac{1}{16}t + \frac{1}{4}\sin 4t - \frac{3}{16}\sin t - \frac{1}{16}\sin 3t\right]_0^{2\pi}$$

$$= \frac{1}{64}\left[12t - 12\sin t + 4\sin 2t - 4\sin 3t + \sin 4t\right]_0^{2\pi}$$

$$= \frac{3\pi}{8}$$

31. $y(t) = \sin 2t,\ x'(t) = -\sin t$

$$A = \int_c^d y(t)\,x'(t)\,dt$$

$$= \int_{\frac{\pi}{2}}^{\frac{3\pi}{2}} (\sin 2t)(-\sin t)\,dt$$

$$= -\int_{\frac{\pi}{2}}^{\frac{3\pi}{2}} \sin t\sin 2t\,dt$$

$$= -\left[\frac{1}{2}\sin t - \frac{1}{6}\sin 3t\right]_{\frac{\pi}{2}}^{\frac{3\pi}{2}}$$

$$= \frac{4}{3}$$

33. $y(t) = t^2 - 3,\ x'(t) = 3t^2 - 4$

$$A = \int_c^d y(t)\,x'(t)\,dt = \int_{-2}^2 (t^2 - 3)(3t^2 - 4)\,dt$$

$$= \int_{-2}^2 (3t^4 - 13t^2 + 12)\,dt$$

$$= \left[\frac{3t^5}{5} - \frac{13t^3}{3} + 12t\right]_{-2}^2$$

$$= \frac{256}{15}$$

35. Crossing x-axis $\Rightarrow y = 0 \Rightarrow t = n\pi$

$x' = -4\cos t\sin t - 2\sin t,$

$y' = 2\sin^2 t + 2(1 - \cos t)\cos t$

At $t = n\pi$, n even, $(x, y) = (3, 0)$ and $x' = 0$, $y' = 0$, speed $= 0$.

At $t = n\pi$, n odd, $(x, y) = (-1, 0)$ and $x' = 0$, $y' = -4$, speed $= 4$.

37. The $3t$ and $5t$ indicate a ratio of 5–to–3.

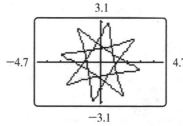

308

39. Use $x = 2\cos t + \sin 3t$, $y = 2\sin t + \cos 3t$.

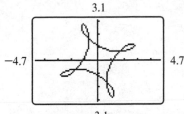

$x' = -2\sin t + 3\cos 3t$, $y' = 2\cos t - 3\sin 3t$

$s(t) = \sqrt{(-2\sin t + 3\cos 3t)^2 + (2\cos t - 3\sin 3t)^2}$

$\quad = \sqrt{4\sin^2 t - 12\sin t \cos 3t + 9\cos^2 3t + 4\cos^2 t - 12\cos t \sin 3t + 9\sin^2 3t}$

$\quad = \sqrt{4 + 9 - 12(\sin t \cos 3t + \cos t \sin 3t)}$

$\quad = \sqrt{13 - 12\sin 4t}$

Minimum speed $= \sqrt{13 - 12} = 1$; maximum speed $= \sqrt{13 + 12} = 5$.

41. $x' = 4\cos 4t$, $y' = 4\sin 4t$ and $s(t) = \sqrt{(4\cos 4t)^2 + (4\sin 4t)^2} = \sqrt{16(\sin^2 4t + \cos^2 4t)} = 4$

The slope of the tangent line is $\dfrac{y'}{x'} = \tan 4t$, and the slope of the origin-to-object line is $\dfrac{y}{x} = -\cot 4t$, and the product of the two slopes is -1.

43. $\dfrac{y(t)}{x(t)}$ is the slope of the origin-to-object line. If the slope is decreasing, the object is moving clockwise; and if the slope is increasing, the object is moving counterclockwise.

Section 9.3

5. $x'(t) = -2\sin t$; $y'(t) = 4\cos t$

t ranges from 0 to 2π.

$s = \displaystyle\int_0^{2\pi} \sqrt{(-2\sin t)^2 + (4\cos t)^2}\, dt$

$\quad = \displaystyle\int_0^{2\pi} \sqrt{4\sin^2 t + 16\cos^2 t}\, dt$

$\quad \approx 19.3769$

7. $x'(t) = 2\sin t$; $y'(t) = 3\cos t$

t ranges from 0 to 2π.

$s = \displaystyle\int_0^{2\pi} \sqrt{(2\sin t)^2 + (3\cos t)^2}\, dt$

$\quad = \displaystyle\int_0^{2\pi} \sqrt{4\sin^2 t + 9\cos^2 t}\, dt$

$\quad \approx 15.8654$

9. $x'(t) = 3t^2 - 4$, $y'(t) = 2t$

$s = \displaystyle\int_{-2}^{2} \sqrt{(3t^2 - 4)^2 + (2t)^2}\, dt$

$\quad \approx 15.6940$

11. $x'(t) = -2\sin 2t$; $y'(t) = 4\cos 4t$

t ranges from 0 to π.

$s = \displaystyle\int_0^{\pi} \sqrt{(-2\sin 2t)^2 + (4\cos 4t)^2}\, dt$

$\quad \approx \displaystyle\int_0^{\pi} \sqrt{4\sin^2 2t + 16\cos^2 4t}\, dt$

$\quad = 9.42943$

13. $x'(t) = \cos t - t\sin t$; $y'(t) = t\cos t + \sin t$

t ranges from -1 to 1.

$s = \displaystyle\int_{-1}^{1} \sqrt{(t\cos t + \sin t)^2 + (\cos t - t\sin t)^2}\, dt$

$\quad \approx 2.29559$

15. $x'(t) = 2\cos t \cos 2t - \sin t \sin 2t;\ y'(t) = 2\cos 2t \sin t + \cos t \sin 2t$

t ranges from 0 to $\dfrac{\pi}{2}$.

$s = \displaystyle\int_0^{\frac{\pi}{2}} \sqrt{(2\cos 2t \sin t + \cos t \sin 2t)^2 + (2\cos t \cos 2t - \sin t \sin 2t)^2}\ dt$

≈ 2.42211

17. $x'(t) = \cos t;\ y'(t) = \pi \cos \pi t$

t ranges from 0 to π.

$s = \displaystyle\int_0^{\pi} \sqrt{(\cos t)^2 + (\pi \cos \pi t)^2}\ dt$

$= \displaystyle\int_0^{\pi} \sqrt{\cos^2 t + \pi^2 \cos^2 \pi t}\ dt$

≈ 6.91395

19. At $t = 0$: $x = \pi(0) = 0,\ y = 2\sqrt{0} = 0$

At $t = 1$: $x = \pi(1) = \pi,\ y = 2\sqrt{1} = 2$

$x'(t) = \pi,\ y'(t) = \dfrac{1}{\sqrt{t}}$

$\text{Time} = \displaystyle\int_0^1 k \sqrt{\dfrac{\pi^2 + \left(\frac{1}{\sqrt{t}}\right)^2}{2\sqrt{t}}}\ dt$

$= k \displaystyle\int_0^1 \sqrt{\dfrac{\pi^2 t + 1}{2t^{3/2}}}\ dt$

$\approx 4.4864k$

21. At $t = 0$: $x = -\dfrac{1}{2}\pi(\cos 0 - 1) = 0,$

$y = 2(0) + \dfrac{7}{10}\sin 0 = 0$

At $t = 1$: $x = -\dfrac{1}{2}\pi(\cos \pi - 1) = \pi,$

$y = 2(1) + \dfrac{7}{10}\sin \pi = 2$

$x'(t) = \dfrac{1}{2}\pi^2 \sin \pi t,\ y'(t) = 2 + \dfrac{7\pi}{10}\cos \pi t$

Time

$= \displaystyle\int_0^1 k \sqrt{\dfrac{\left[\left(\frac{1}{2}\right)\pi^2 \sin \pi t\right]^2 + \left[2 + \left(\frac{7\pi}{10}\right)\cos \pi t\right]^2}{2t + \left(\frac{7}{10}\right)\sin \pi t}}\ dt$

$\approx 4.45691k$

23. For $x = \pi t,\ y = 2\sqrt{t},\ \dfrac{dy}{dx} = \dfrac{y'}{x'} = \dfrac{\frac{1}{\sqrt{t}}}{\pi}$ is undefined

at $t = 0$.

$s = \displaystyle\int_0^1 \sqrt{\left(\dfrac{1}{\sqrt{t}}\right)^2 + \pi^2}\ dt$

$= \displaystyle\int_0^1 \sqrt{\pi^2 + \dfrac{1}{t}}\ dt$

≈ 3.8897

25. For $x = -\dfrac{1}{2}\pi(\cos \pi t - 1),\ y = 2t + \dfrac{7}{10}\sin \pi t$,

$\dfrac{dy}{dx} = \dfrac{y'}{x'} = \dfrac{2 + \left(\frac{7\pi}{10}\right)\cos \pi t}{\left(\frac{1}{2}\right)\pi^2 \sin \pi t}$ is undefined at $t = 0$.

$s = \displaystyle\int_0^1 \sqrt{\left(\dfrac{1}{2}\pi^2 \sin \pi t\right)^2 + \left(2 + \dfrac{7}{10}\pi \cos \pi t\right)^2}\ dt$

≈ 4.06632

27. $x'(t) = 2t;\ y'(t) = 3t^2 - 4$

$A = 2\pi \displaystyle\int_{-2}^{0} \left| t^3 - 4t \right| \sqrt{(2t)^2 + (3t^2 - 4)^2}\ dt$

≈ 85.8228

29. $x'(t) = 2t;\ y'(t) = 3t^2 - 4$

$A = 2\pi \displaystyle\int_{-1}^{1} \left| t^2 - 1 \right| \sqrt{(2t)^2 + (3t^2 - 4)^2}\ dt$

≈ 29.6956

31. $x'(t) = 3t^2 - 4;\ y'(t) = 2t$

$A = 2\pi \displaystyle\int_{0}^{2} \left| t^3 - 4t \right| \sqrt{(3t^2 - 4)^2 + (2t)^2}\ dt$

≈ 85.8228

Section 9.4

5. $r = 2,\ \theta = 0$

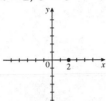

$x = 2 \cos 0 = 2$
$y = 2 \sin 0 = 0$
Rectangular representation: $(2, 0)$

7. $r = -2,\ \theta = \pi$

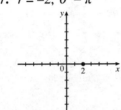

$x = -2 \cos \pi = 2$
$y = -2 \sin \pi = 0$
Rectangular representation: $(2, 0)$

9. $r = 3,\ \theta = -\pi$

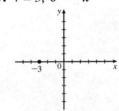

$x = 3 \cos(-\pi) = -3$
$y = 3 \sin(-\pi) = 0$
Rectangular representation: $(-3, 0)$

11. $r = \sqrt{2^2 + (-2)^2} = 2\sqrt{2}$

$\tan \theta = \dfrac{-2}{2} = -1$ in Quadrant IV,

so $\theta = -\dfrac{\pi}{4} + 2\pi n.$

Polar representation: $\left(2\sqrt{2}, -\dfrac{\pi}{4} + 2\pi n \right)$, or else

$\left(-2\sqrt{2}, \dfrac{3\pi}{4} + 2\pi n \right)$

13. $r = \sqrt{0^2 + 3^2} = 3$

θ is straight up, so $\theta = \dfrac{\pi}{2} + 2\pi n.$

Polar representation: $\left(3, \dfrac{\pi}{2} + 2\pi n \right)$, or else

$\left(-3, \dfrac{3\pi}{2} + 2\pi n \right)$

15. $r = \sqrt{2^2 + (-1)^2} = \sqrt{5}$

$\tan \theta = \dfrac{-1}{2} = -\dfrac{1}{2}$ in Quadrant IV, so

$\theta = -\tan^{-1}\left(\dfrac{1}{2} \right) + 2\pi n$

Polar representation: $\left(\sqrt{5}, -\tan^{-1}\left(\dfrac{1}{2} \right) + 2\pi n \right),$

or else $\left(-\sqrt{5}, -\tan^{-1}\left(\dfrac{1}{2} \right) + (2n+1)\pi \right)$

17. $r = \sqrt{(-\sqrt{3})^2 + 1^2} = 2$

$\tan \theta = \dfrac{1}{-\sqrt{3}} = -\dfrac{1}{\sqrt{3}}$ in Quadrant II, so

$\theta = \dfrac{5\pi}{6} + 2\pi n$

Polar representation: $\left(2, \dfrac{5\pi}{6} + 2\pi n \right)$, or else

$\left(-2, -\dfrac{\pi}{6} + 2\pi n \right)$

19. $r = 2;\ \theta = -\dfrac{\pi}{3}$

$x = 2\cos\left(-\dfrac{\pi}{3} \right) = 1$

$y = 2\sin\left(-\dfrac{\pi}{3} \right) = -\sqrt{3}$

Rectangular representation: $\left(1, -\sqrt{3} \right)$

21. $r = 0;\ \theta = 3$
$x = 0 \cos 3 = 0$
$y = 0 \sin 3 = 0$
Rectangular representation: $(0, 0)$

23. $r = 3; \theta = \dfrac{\pi}{8}$

$x = 3\cos\left(\dfrac{\pi}{8}\right)$

$y = 3\sin\left(\dfrac{\pi}{8}\right)$

Rectangular representation:

$\left(3\cos\left(\dfrac{\pi}{8}\right), 3\sin\left(\dfrac{\pi}{8}\right)\right) \approx (2.77, 1.15)$

25. $r = -3; \theta = 1$

$x = -3\cos 1$

$y = -3\sin 1$

Rectangular representation:

$(-3\cos 1, -3\sin 1) \approx (-1.62, -2.52)$

27.

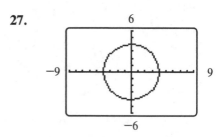

This is a circle centered at the origin with radius 4.

$x^2 + y^2 = 4^2$

$x^2 + y^2 = 16$

or $r = 4$

$r^2 = 16$

$x^2 + y^2 = 16$

29.

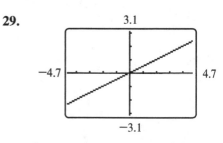

This is a line through the origin with slope

$\tan\dfrac{\pi}{6} = \dfrac{1}{\sqrt{3}} = \dfrac{y}{x}$

$y = \dfrac{1}{\sqrt{3}} x$

31.

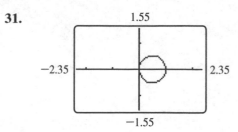

This is a circle centered at $\left(\dfrac{1}{2}, 0\right)$ with radius

$\dfrac{1}{2}$.

$\left(x - \dfrac{1}{2}\right)^2 + y^2 = \left(\dfrac{1}{2}\right)^2$

$\left(x - \dfrac{1}{2}\right)^2 + y^2 = \dfrac{1}{4}$

or

$r = \cos\theta$

$r^2 = r\cos\theta$

$x^2 + y^2 = x$

The two equations are equivalent.

33.

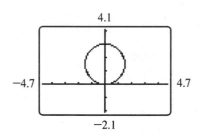

This is a circle centered at $\left(0, \dfrac{3}{2}\right)$ with radius $\dfrac{3}{2}$.

$x^2 + \left(y - \dfrac{3}{2}\right)^2 = \left(\dfrac{3}{2}\right)^2$

$x^2 + \left(y - \dfrac{3}{2}\right)^2 = \dfrac{9}{4}$

or

$r = 3\sin\theta$

$r^2 = 3r\sin\theta$

$x^2 + y^2 = 3y$

The two equations are equivalent.

35.

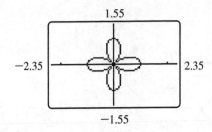

$r = 0$ when $2\theta = \dfrac{\pi}{2} + n\pi$ or $\theta = \dfrac{k\pi}{4}$ (k odd).

$0 \le \theta \le 2\pi$

37.

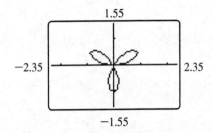

$r = 0$ when $3\theta = n\pi$ or $\theta = \dfrac{n\pi}{3}$.

$0 \le \theta \le \pi$

39.

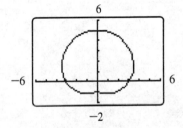

$r = 0$ when $3 + 2\sin\theta = 0 \Rightarrow \sin\theta = -\dfrac{3}{2} \Rightarrow$ never.

$0 \le \theta \le 2\pi$

41.

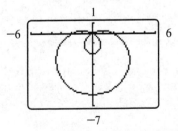

$r = 0$ when $\sin\theta = \dfrac{1}{2} \Rightarrow \theta = \dfrac{\pi}{6} + 2\pi n$ or

$\theta = \dfrac{5\pi}{6} + 2\pi n$.

$0 \le \theta \le 2\pi$

43.

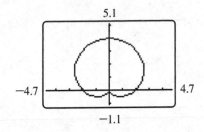

$r = 0$ when $\sin\theta = -1 \Rightarrow \theta = \dfrac{3\pi}{2} + 2\pi n$

$0 \le \theta \le 2\pi$

45.

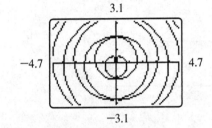

$r = 0$ when $\theta = 0$.

$-\infty < \theta < \infty$

47.

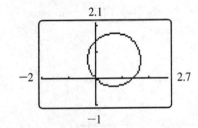

$r = 0$ when $\cos\left(\theta - \dfrac{\pi}{4}\right) = 0 \Rightarrow \theta - \dfrac{\pi}{4} = \dfrac{\pi}{2} + n\pi$.

$\Rightarrow \theta = \dfrac{3\pi}{4} + n\pi$

$0 \le \theta \le \pi$

49.

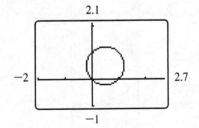

$r = 0$ when $\cos\theta = -\sin\theta$

$\Rightarrow \tan\theta = -1 \Rightarrow \theta = \dfrac{3\pi}{4} + n\pi$.

$0 \le \theta \le \pi$

51.

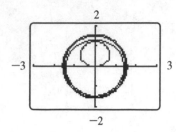

$r = 0$ when $2\theta = 0 \Rightarrow \theta = 0$.
$-\infty < \theta < \infty$

53.

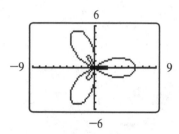

$r = 0$ when $\cos 3\theta = -\dfrac{1}{2}$

$\Rightarrow 3\theta = \dfrac{2\pi}{3} + 2\pi n$ or $3\theta = \dfrac{4\pi}{3} + 2\pi n$

$\Rightarrow \theta = \dfrac{2\pi}{9} + \dfrac{2\pi n}{3}$ or $\theta = \dfrac{4\pi}{9} + \dfrac{2\pi n}{3}$.
$0 \le \theta \le 2\pi$

55.

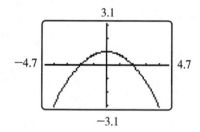

$r = 0$ when $\dfrac{2}{1 + \sin\theta} = 0 \Rightarrow$ never.

$-\dfrac{\pi}{2} < \theta < \dfrac{3\pi}{2}$

57.

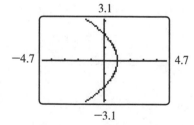

$r = 0$ when $\dfrac{2}{1 + \cos\theta} = 0 \Rightarrow$ never.

$-\pi < \theta < \pi$

59.

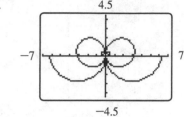

$r = 0$ when $\theta \cos\theta = 0$
$\Rightarrow \theta = 0$ or $\cos\theta = 0$
$\Rightarrow \theta = 0, \pm\dfrac{\pi}{2}, \pm\dfrac{3\pi}{2}$.

61. A circle with radius $\dfrac{a}{2}$ and center $\left(\dfrac{a}{2}, 0\right)$

63. A rose with n leaves (if n is odd) or $2n$ leaves (if n is even).

65. $y^2 - x^2 = 4$

$r^2 \sin^2\theta - r^2 \cos^2\theta = 4$

$r^2(-\cos 2\theta) = 4$

$r = \pm\sqrt{\dfrac{4}{-\cos 2\theta}}$

$= \pm 2\sqrt{-\sec 2\theta}, \dfrac{\pi}{4} < 0 < \dfrac{3\pi}{4}$

67. $x^2 + y^2 = 16$

$r^2 = 16$

$r = \pm 4$

$r = 4$ (since $r = -4$ duplicates)

69. $y = 3$

$r \sin\theta = 3$

$r = \dfrac{3}{\sin\theta}$

$r = 3 \csc\theta$

71. The graph consists of wide, overlapping "petals" on a flower. $0 \le \theta \le 24\pi$ produces a complete graph, with 24 petals. For larger domains the graph repeats.

73. A "line" that is tangent to the hole is the outside boundary for a successful putt. A tangent line to a circle must be perpendicular to a radius. Thus, a triangle formed by the points $(0, 0)$, $(d, 0)$ and the point of tangency has a right angle at the point of tangency, a hypotenuse of length d and the leg opposite the acute angle of length h. Thus any path with acute angle A smaller than the acute angle of this triangle must have $-\dfrac{h}{d} < \sin A < \dfrac{h}{d}$.

Thus $-\sin^{-1}\left(\dfrac{h}{d}\right) < A < \sin^{-1}\left(\dfrac{h}{d}\right)$.

75. $r_2(A)$ is given:

$$r_2(A) = d + b\left(1 - \left(\frac{A}{\sin^{-1}(h/d)}\right)^2\right).$$

$r_1(A)$ was found in exercise 74:

$$r_1(A) = d\cos A - \sqrt{d^2\cos^2 A - (d^2 - h^2)}$$

The constants A_1 and A_2 were found in Exercise 73:

$$A_1 = -\sin^{-1}\left(\frac{h}{d}\right) \text{ and } A_2 = \sin^{-1}\left(\frac{h}{d}\right).$$

Section 9.5

5. $x = r\cos\theta = \sin 3\theta \cos\theta$; $y = r\sin\theta = \sin 3\theta \sin\theta$

$$\frac{dx}{d\theta} = 3\cos\theta\cos 3\theta - \sin\theta\sin(3\theta); \quad \frac{dy}{d\theta} = 3\cos 3\theta\sin\theta + \cos\theta\sin 3\theta$$

At $\theta = \dfrac{\pi}{3}$, $\dfrac{dy}{dx} = \dfrac{\frac{dy}{d\theta}}{\frac{dx}{d\theta}} = \dfrac{3\cos\pi\sin\left(\frac{\pi}{3}\right) + \cos\left(\frac{\pi}{3}\right)\sin\pi}{3\cos\left(\frac{\pi}{3}\right)\cos\pi - \sin\left(\frac{\pi}{3}\right)\sin\pi} = \dfrac{-\frac{1}{2}\left(3\sqrt{3}\right)}{-\frac{3}{2}} = \sqrt{3}.$

7. $x = r\cos\theta = \cos 2\theta \cos\theta$; $y = r\sin\theta = \cos 2\theta \sin\theta$

$$\frac{dx}{d\theta} = -\cos 2\theta\sin\theta - 2\cos\theta\sin 2\theta; \quad \frac{dy}{d\theta} = \cos\theta\cos 2\theta - 2\sin\theta\sin 2\theta$$

At $\theta = 0$, $\dfrac{dy}{dx} = \dfrac{\frac{dy}{d\theta}}{\frac{dx}{d\theta}} = \dfrac{\cos 0\cos 0 - 2\sin 0\sin 0}{-\cos 0\sin 0 - 2\cos 0\sin 0} = \dfrac{1}{0}$, which is undefined.

9. $x = r\cos\theta = 3\sin\theta \cos\theta$; $y = r\sin\theta = 3\sin\theta \sin\theta$

$$\frac{dx}{d\theta} = 3\cos^2\theta - 3\sin^2\theta; \quad \frac{dy}{d\theta} = 6\cos\theta\sin\theta$$

At $\theta = 0$, $\dfrac{dy}{dx} = \dfrac{\frac{dy}{d\theta}}{\frac{dx}{d\theta}} = \dfrac{6\cos 0\sin 0}{3\cos^2 0 - 3\sin^2 0} = \dfrac{0}{3} = 0.$

11. $x = r\cos\theta = \sin 4\theta \cos\theta$; $y = r\sin\theta = \sin 4\theta \sin\theta$

$$\frac{dx}{d\theta} = 4\cos\theta\cos 4\theta - \sin\theta\sin 4\theta; \quad \frac{dy}{d\theta} = 4\cos 4\theta\sin\theta + \cos\theta\sin 4\theta$$

At $\theta = \dfrac{\pi}{4}$, $\dfrac{dy}{dx} = \dfrac{\frac{dy}{d\theta}}{\frac{dx}{d\theta}} = \dfrac{4\cos\pi\sin\left(\frac{\pi}{4}\right) + \cos\left(\frac{\pi}{4}\right)\sin\pi}{4\cos\left(\frac{\pi}{4}\right)\cos\pi - \sin\left(\frac{\pi}{4}\right)\sin\pi} = \dfrac{-2\sqrt{2}}{-2\sqrt{2}} = 1.$

13. $x = r\cos\theta = \cos 3\theta \cos\theta$; $y = r\sin\theta = \cos 3\theta \sin\theta$

$$\frac{dx}{d\theta} = -\cos 3\theta\sin\theta - 3\cos\theta\sin 3\theta; \quad \frac{dy}{d\theta} = \cos\theta\cos 3\theta - 3\sin\theta\sin 3\theta$$

At $\theta = \dfrac{\pi}{6}$, $\dfrac{dy}{dx} = \dfrac{\frac{dy}{d\theta}}{\frac{dx}{d\theta}} = \dfrac{\cos\left(\frac{\pi}{6}\right)\cos\left(\frac{\pi}{2}\right) - 3\sin\left(\frac{\pi}{6}\right)\sin\left(\frac{\pi}{2}\right)}{-\cos\left(\frac{\pi}{2}\right)\sin\left(\frac{\pi}{6}\right) - 3\cos\left(\frac{\pi}{6}\right)\sin\left(\frac{\pi}{2}\right)} = \dfrac{-\frac{3}{2}}{-\frac{1}{2}\left(3\sqrt{3}\right)} = \dfrac{1}{\sqrt{3}}.$

15. $|r|$ is at a maximum when $\sin 3\theta = \pm 1 \Rightarrow 3\theta = \pm\dfrac{\pi}{2} + 2\pi n \Rightarrow \theta = \pm\dfrac{\pi}{6} + \dfrac{2\pi n}{3}$.

$x = \sin 3\theta \cos\theta; \; y = \sin 3\theta \sin\theta$

$\dfrac{dx}{d\theta} = 3\cos\theta\cos 3\theta - \sin\theta\sin 3\theta; \; \dfrac{dy}{d\theta} = 3\cos 3\theta\sin\theta + \cos\theta\sin 3\theta$

$\dfrac{dy}{dx} = \dfrac{\frac{dy}{d\theta}}{\frac{dx}{d\theta}} = \dfrac{3\cos 3\theta\sin\theta + \cos\theta\sin 3\theta}{3\cos\theta\cos 3\theta - \sin\theta\sin 3\theta}$

For $\theta = \dfrac{\pi}{6}$: $(x,\, y) = \left(\sin\left(\dfrac{\pi}{2}\right)\cos\left(\dfrac{\pi}{6}\right),\; \sin\left(\dfrac{\pi}{2}\right)\sin\left(\dfrac{\pi}{6}\right) \right) = \left(\dfrac{\sqrt{3}}{2}, \dfrac{1}{2} \right)$.

The line from the origin has slope $\dfrac{\frac{1}{2}}{\frac{\sqrt{3}}{2}} = \dfrac{1}{\sqrt{3}}$, and the tangent line has slope

$\dfrac{3\cos\left(\frac{\pi}{2}\right)\sin\left(\frac{\pi}{6}\right) + \cos\left(\frac{\pi}{6}\right)\sin\left(\frac{\pi}{2}\right)}{3\cos\left(\frac{\pi}{6}\right)\cos\left(\frac{\pi}{2}\right) - \sin\left(\frac{\pi}{6}\right)\sin\left(\frac{\pi}{2}\right)} = \dfrac{\frac{\sqrt{3}}{2}}{-\frac{1}{2}} = -\sqrt{3}$

For $\theta = -\dfrac{\pi}{6}$: $(x,\, y) = \left(\sin\left(-\dfrac{\pi}{2}\right)\cos\left(-\dfrac{\pi}{6}\right),\; \sin\left(-\dfrac{\pi}{2}\right)\sin\left(-\dfrac{\pi}{6}\right) \right) = \left(-\dfrac{\sqrt{3}}{2}, \dfrac{1}{2} \right)$.

The line from the origin has slope $\dfrac{\frac{1}{2}}{-\frac{\sqrt{3}}{2}} = -\dfrac{1}{\sqrt{3}}$, and the tangent line has slope

$\dfrac{3\cos\left(-\frac{\pi}{2}\right)\sin\left(-\frac{\pi}{6}\right) + \cos\left(-\frac{\pi}{6}\right)\sin\left(-\frac{\pi}{2}\right)}{3\cos\left(-\frac{\pi}{6}\right)\cos\left(-\frac{\pi}{2}\right) - \sin\left(-\frac{\pi}{6}\right)\sin\left(-\frac{\pi}{2}\right)} = \dfrac{-\frac{\sqrt{3}}{2}}{-\frac{1}{2}} = \sqrt{3}$.

For $\theta = \dfrac{\pi}{2}$: $(x,\, y) = \left(\sin\left(\dfrac{3\pi}{2}\right)\cos\left(\dfrac{\pi}{2}\right),\; \sin\left(\dfrac{3\pi}{2}\right)\sin\left(\dfrac{\pi}{2}\right) \right) = (0,\, -1)$

The line from the origin is vertical, and the tangent line has slope

$\dfrac{3\cos\left(\frac{3\pi}{2}\right)\sin\left(\frac{\pi}{2}\right) + \cos\left(\frac{\pi}{2}\right)\sin\left(\frac{3\pi}{2}\right)}{3\cos\left(\frac{\pi}{2}\right)\cos\left(\frac{3\pi}{2}\right) - \sin\left(\frac{\pi}{2}\right)\sin\left(\frac{3\pi}{2}\right)} = \dfrac{0}{1} = 0$.

17. $|r|$ is at a maximum when $\sin 2\theta = -1 \Rightarrow 2\theta = -\dfrac{\pi}{2} + 2\pi n \Rightarrow \theta = -\dfrac{\pi}{4} + n\pi$.

$x = r\cos\theta = (2 - 4\sin 2\theta)\cos\theta; \; y = r\sin\theta = (2 - 4\sin 2\theta)\sin\theta$

$\dfrac{dx}{d\theta} = -8\cos\theta\cos 2\theta - \sin\theta(2 - 4\sin 2\theta); \; \dfrac{dy}{d\theta} = \cos\theta(2 - 4\sin 2\theta) - 8\cos 2\theta\sin\theta$

$\dfrac{dy}{dx} = \dfrac{\frac{dy}{d\theta}}{\frac{dx}{d\theta}} = \dfrac{\cos\theta(2 - 4\sin 2\theta) - 8\cos 2\theta\sin\theta}{-8\cos\theta\cos 2\theta - \sin\theta(2 - 4\sin 2\theta)}$

At $\theta = \dfrac{3\pi}{4}$: $(x,\, y) = \left(\left(2 - 4\sin\left(\dfrac{3\pi}{2}\right)\right)\cos\left(\dfrac{3\pi}{4}\right),\; \left(2 - 4\sin\left(\dfrac{3\pi}{2}\right)\right)\sin\left(\dfrac{3\pi}{4}\right) \right) = \left(-3\sqrt{2},\, 3\sqrt{2} \right)$.

The line from the origin has slope $\dfrac{3\sqrt{2}}{-3\sqrt{2}} = -1$, and the tangent line has slope

$$\frac{\cos\left(\frac{3\pi}{4}\right)\left(2-4\sin\left(\frac{3\pi}{2}\right)\right)-8\cos\left(\frac{3\pi}{2}\right)\sin\left(\frac{3\pi}{4}\right)}{-8\cos\left(\frac{3\pi}{4}\right)\cos\left(\frac{3\pi}{2}\right)-\sin\left(\frac{3\pi}{4}\right)\left(2-4\sin\frac{3\pi}{2}\right)} = \frac{\left(-\frac{1}{\sqrt{2}}\right)(6)-8(0)\left(\frac{1}{\sqrt{2}}\right)}{-8\left(-\frac{1}{\sqrt{2}}\right)(0)-\left(\frac{1}{\sqrt{2}}\right)(6)} = 1.$$

At $\theta = \dfrac{7\pi}{4}$: $(x,\ y) = \left(\left(2-4\sin\left(\dfrac{7\pi}{2}\right)\right)\cos\left(\dfrac{7\pi}{4}\right),\ \left(2-4\sin\left(\dfrac{7\pi}{2}\right)\right)\sin\left(\dfrac{7\pi}{4}\right)\right) = \left(3\sqrt{2},\ -3\sqrt{2}\right).$

The line from the origin has slope $\dfrac{-3\sqrt{2}}{3\sqrt{2}} = -1$, and the tangent line has slope

$$\frac{\cos\left(\frac{7\pi}{4}\right)\left(2-4\sin\left(\frac{7\pi}{2}\right)\right)-8\cos\left(\frac{7\pi}{2}\right)\sin\left(\frac{7\pi}{4}\right)}{-8\cos\left(\frac{7\pi}{4}\right)\cos\left(\frac{7\pi}{2}\right)-\sin\left(\frac{7\pi}{4}\right)\left(2-4\sin\left(\frac{7\pi}{2}\right)\right)} = \frac{\left(\frac{1}{\sqrt{2}}\right)(6)-8(0)\left(-\frac{1}{\sqrt{2}}\right)}{-8\left(\frac{1}{\sqrt{2}}\right)(0)-\left(-\frac{1}{\sqrt{2}}\right)(6)} = 1.$$

19.

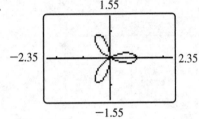

Use $\cos 3\theta = 0$ to find the appropriate range of θ: $-\dfrac{\pi}{6} \le \theta \le \dfrac{\pi}{6}$.

$$A = \int_{-\pi/6}^{\pi/6}\frac{1}{2}\cos^{2}3\theta\ d\theta = \frac{1}{4}\int_{-\pi/6}^{\pi/6}(1+\cos 6\theta)d\theta = \frac{1}{4}\left[\theta+\frac{1}{6}\sin 6\theta\right]_{-\pi/6}^{\pi/6} = \frac{\pi}{12}$$

21.

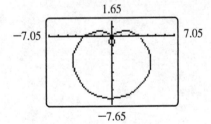

Use $3-4\sin\theta = 0$ to find the appropriate range of θ: $\sin^{-1}\left(\dfrac{3}{4}\right) = 0.8481 \le \theta \le \pi - \sin^{-1}\left(\dfrac{3}{4}\right) = 2.2935.$

$$A = \int_{0.8481}^{2.2935}\frac{1}{2}(3-4\sin\theta)^{2}\ d\theta = \int_{0.8481}^{2.2935}\frac{1}{2}(9-24\sin\theta+16\sin^{2}\theta)d\theta$$

$$= \int_{0.8481}^{2.2935}\frac{1}{2}\left[9-24\sin\theta+8(1-\cos 2\theta)\right]d\theta$$

$$= 0.3806$$

23.

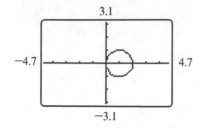

Use $2\cos\theta = 0$ to find the appropriate range of $\theta : 0 \le \theta \le \pi$.

$$A = \int_0^\pi 2\cos^2\theta \, d\theta$$

$$= \int_0^\pi (1 + \cos 2\theta) \, d\theta$$

$$= \left[\theta + \frac{1}{2}\sin 2\theta\right]_0^\pi$$

$$= \pi$$

25.

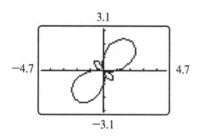

Use $1 + 2\sin 2\theta = 0$ to find the appropriate range of $\theta : \frac{7\pi}{12} \le \theta \le \frac{11\pi}{12}$.

$$A = \int_{7\pi/12}^{11\pi/12} \frac{1}{2}(1 + 2\sin 2\theta)^2 \, d\theta$$

$$= \int_{7\pi/12}^{11\pi/12} \frac{1}{2}(1 + 4\sin 2\theta + 4\sin^2 2\theta) \, d\theta$$

$$= \int_{7\pi/12}^{11\pi/12} \frac{1}{2}\left[1 + 4\sin 2\theta + 2(1 - \cos 4\theta)\right] d\theta$$

$$= \int_{7\pi/12}^{11\pi/12} \left[\frac{3}{2} + 2\sin 2\theta - \cos 4\theta\right] d\theta$$

$$= \left[\frac{3}{2}\theta - \cos 2\theta - \frac{1}{4}\sin 4\theta\right]_{7\pi/12}^{11\pi/12}$$

$$= \frac{\pi}{2} - \frac{3\sqrt{3}}{4}$$

$$\approx 0.2718$$

27.

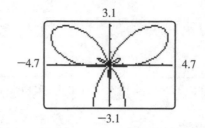

Use $2 + 3\sin 3\theta = 0$ to find the appropriate range of θ:

$$\frac{\pi}{3} + \frac{1}{3}\sin^{-1}\left(\frac{2}{3}\right) = 1.2904 \le \theta \le \frac{2\pi}{3} - \frac{1}{3}\sin^{-1}\left(\frac{2}{3}\right).$$

$$= 1.8512.$$

$$A = \int_{1.2904}^{1.8512} \frac{1}{2}(2 + 3\sin 3\theta)^2 \, d\theta = 0.1470$$

29.

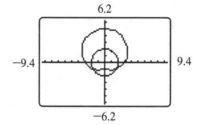

Use $3 + 2\sin\theta = 2$ to find the appropriate range of $\theta : -\frac{\pi}{6} \le \theta \le \frac{7\pi}{6}$.

$$A = \int_{-\pi/6}^{7\pi/6} \frac{1}{2}(3 + 2\sin\theta)^2 \, d\theta - \int_{-\pi/6}^{7\pi/6} \frac{1}{2}(2)^2 \, d\theta$$

$$= \int_{-\pi/6}^{7\pi/6} \frac{1}{2}(9 + 12\sin\theta + 4\sin^2\theta) \, d\theta - \int_{-\pi/6}^{7\pi/6} 2 \, d\theta$$

$$= \int_{-\pi/6}^{7\pi/6} \left(\frac{1}{2}\left[9 + 12\sin\theta + 2(1 - \cos 2\theta)\right] - 2\right) d\theta$$

$$= \int_{-\pi/6}^{7\pi/6} \left[\frac{7}{2} + 6\sin\theta - \cos 2\theta\right] d\theta$$

$$= \left[\frac{7}{2}\theta - 6\cos\theta - \frac{1}{2}\sin 2\theta\right]_{-\pi/6}^{7\pi/6}$$

$$= \frac{14\pi}{3} + \frac{11\sqrt{3}}{2}$$

$$\approx 24.1870$$

31.

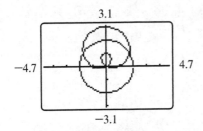

To exclude the loop, find half the area in question and double the result. For the limits along $r = 1 + 2\sin\theta$, use $1 + 2\sin\theta = 2$ and $1 + 2\sin\theta = 0$ to find $\dfrac{5\pi}{6} \le \theta \le \dfrac{7\pi}{6}$.

For the limits along $r = 2$, use $\dfrac{5\pi}{6} \le \theta \le \dfrac{3\pi}{2}$. So

$$A = 2\left\{ \int_{5\pi/6}^{3\pi/2} \frac{1}{2}(2)^2\, d\theta - \int_{5\pi/6}^{7\pi/6} \frac{1}{2}(1 + 2\sin\theta)^2\, d\theta \right\}$$

$$= \int_{5\pi/6}^{3\pi/2} 4\, d\theta - \int_{5\pi/6}^{7\pi/6} \left(1 + 4\sin\theta + 4\sin^2\theta\right) d\theta$$

$$= \frac{8\pi}{3} - \int_{5\pi/6}^{7\pi/6} \left[1 + 4\sin\theta + 2(1 - \cos 2\theta)\right] d\theta$$

$$= \frac{8\pi}{3} - \int_{5\pi/6}^{7\pi/6} \left[3 + 4\sin\theta - 2\cos 2\theta\right] d\theta$$

$$= \frac{8\pi}{3} - \left[3\theta - 4\cos\theta - \sin 2\theta\right]_{5\pi/6}^{7\pi/6}$$

$$= \frac{8\pi}{3} - \left(\pi - \sqrt{3}\right)$$

$$= \frac{5\pi}{3} + \sqrt{3}$$

$$\approx 6.9680$$

33.

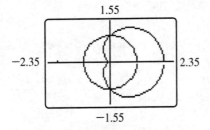

The area is a semicircle plus two "lobes:

$$A = \frac{12}{2}\pi(1)^2 + 2\int_{\pi/2}^{\pi} \frac{1}{2}(1 + \cos\theta)^2\, d\theta$$

$$= \frac{\pi}{2} + \int_{\pi/2}^{\pi} \left(1 + 2\cos\theta + \cos^2\theta\right) d\theta$$

$$= \frac{\pi}{2} + \int_{\pi/2}^{\pi} \left[1 + 2\cos\theta + \frac{1}{2}(1 + \cos 2\theta)\right] d\theta$$

$$= \frac{\pi}{2} + \int_{\pi/2}^{\pi} \left[\frac{3}{2} + 2\cos\theta + \frac{1}{2}\cos 2\theta\right] d\theta$$

$$A = \frac{\pi}{2} + \left[\frac{3}{2}\theta + 2\sin\theta + \frac{1}{4}\sin 2\theta\right]_{\pi/2}^{\pi}$$

$$= \frac{\pi}{2} + \left(\frac{3\pi}{4} - 2\right)$$

$$= \frac{5\pi}{4} - 2$$

$$\approx 1.9270$$

35.

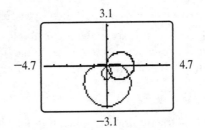

Let $f(\theta) = r = 1 - 2\sin\theta$ and $g(\theta) = r = 2\cos\theta$. Then intersections occur in each of the cases $f(\theta_1) = g(\theta_2) = 0$, $f(\theta) = g(\theta)$ and $f(\theta) = -g(\theta + \pi)$.

First, notice that $f(\theta) = 0$ when $\theta = \dfrac{\pi}{6}$ and $g(\theta) = 0$ when $\theta = \dfrac{\pi}{2}$. So the curves intersect at the origin.

Second, by drawing a *rectangular* plot of the two curves $r = f(\theta)$ and $r = g(\theta)$ in the interval $0 \le \theta \le 2\pi$, we find two intersection points at $\theta \approx 1.9948274$ and $\theta \approx 5.8591543$.

Third, since $-g(\theta + \pi) = -2\cos(\theta + \pi) = 2\cos\theta = g(\theta)$, no additional intersections occur in the case $f(\theta) = -g(\theta + \pi)$. The corresponding rectangular points on the *polar* plot are found using $(r\cos\theta,\ r\sin\theta)$ for each of these θ-values.

Hence, there are three intersection points: $(0, 0)$, $(0.3386, -0.75)$, $(1.6614, -0.75)$.

37.

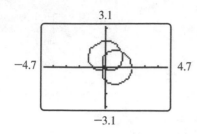

3.1

−4.7 ———————————— 4.7

−3.1

Let $f(\theta) = r = 1 + \sin\theta$ and $g(\theta) = 1 + \cos\theta$. The obvious symmetry in the polar plot implies that these curves intersect only along the line $y = x$. In particular, notice that $f(\theta) = 0$ when

$\theta = \dfrac{3\pi}{2}$ and $g(\theta) = 0$ when $\theta = \pi$. So the curves

intersect at the origin. The remaining intersection

points occur when $\theta = \dfrac{\pi}{4}$ and $\theta = \dfrac{5\pi}{4}$. The

corresponding rectangular points on *polar* plot are found using $(r\cos\theta,\ r\sin\theta)$ for each of these θ-values. Hence, there are three points of intersection:

39. $r = 2 - 2\sin\theta,\ r' = -2\cos\theta$

$$s = \int_0^{2\pi} \sqrt{(-2\cos\theta)^2 + (2 - 2\sin\theta)^2}\ d\theta$$

$$= \int_0^{2\pi} \sqrt{4\cos^2\theta + 4 - 8\sin\theta + 4\sin^2\theta}\ d\theta$$

$$= \int_0^{2\pi} \sqrt{8 - 8\sin\theta}\ d\theta$$

$$= \int_0^{2\pi} \sqrt{8}\sqrt{1 - \cos\left(\theta - \dfrac{\pi}{2}\right)}\ d\theta$$

$$= \int_0^{2\pi} \sqrt{8}\cdot\sqrt{2}\cdot\left|\sin\left(\dfrac{\theta}{2} - \dfrac{\pi}{4}\right)\right|\ d\theta$$

$$= 4\int_0^{\pi/2} -\sin\left(\dfrac{\theta}{2} - \dfrac{\pi}{4}\right)d\theta + 4\int_{\pi/2}^{2\pi}\sin\left(\dfrac{\theta}{2} - \dfrac{\pi}{4}\right)d\theta$$

$$= 8\left[\cos\left(\dfrac{\theta}{2} - \dfrac{\pi}{4}\right)\right]_0^{\pi/2} + (-8)\left[\cos\left(\dfrac{\theta}{2} - \dfrac{\pi}{4}\right)\right]_{\pi/2}^{2\pi}$$

$$= 16$$

41. $r = \sin 3\theta,\ r' = 3\cos 3\theta$

$$s = \int_0^\pi \sqrt{(3\cos 3\theta)^2 + (\sin 3\theta)^2}\ d\theta$$

$$= \int_0^\pi \sqrt{9\cos^2 3\theta + \sin^2 3\theta}\ d\theta$$

$$= \int_0^\pi \sqrt{1 + 8\cos^2 3\theta}\ d\theta$$

$$= 6.68245$$

43. $r = 1 + 2\sin 2\theta,\ r' = 4\cos 2\theta$

$$s = \int_0^{2\pi} \sqrt{(4\cos 2\theta)^2 + (1 + 2\sin 2\theta)^2}\ d\theta$$

$$= \int_0^{2\pi} \sqrt{16\cos^2 2\theta + 1 + 4\sin 2\theta + 4\sin^2 2\theta}\ d\theta$$

$$= \int_0^{2\pi} \sqrt{5 + 12\cos^2 2\theta + 4\sin 2\theta}\ d\theta$$

$$\approx 20.0158$$

45. For a depth of 1.4, solve
$2 = -0.6\csc\theta \Rightarrow \theta_1 = 3.4463,\ \theta_2 = 5.9785$.

$$\text{Area} = \int_{\theta_1}^{\theta_2}\dfrac{1}{2}[2]^2\ d\theta - \int_{\theta_1}^{\theta_2}\dfrac{1}{2}[-0.6\csc\theta]^2\ d\theta$$

$$= \left[2\theta + 0.18\cot\theta\right]_{3.4463}^{5.9785}$$

$$\approx 3.91969$$

So the percent remaining is about

$$\dfrac{3.91969}{4\pi}\times 100 \approx 31.2\%.$$

47. For a depth of 2.4, solve
$2 = 0.4\csc\theta \Rightarrow \theta_1 = 0.2014,\ \theta_2 = 2.9402$.

$$\text{Area} = \pi(2)^2 - \int_{\theta_1}^{\theta_2}\dfrac{1}{2}[2]^2\ d\theta + \int_{\theta_1}^{\theta_2}\dfrac{1}{2}[0.4\csc\theta]^2\ d\theta$$

$$= 4\pi + \left[-2\theta - 0.08\cot\theta\right]_{0.2014}^{2.9402}$$

$$\approx 7.87245$$

So the percent remaining is about

$$\dfrac{7.87245}{4\pi}\times 100 \approx 62.6\%.$$

49. $r = \sin 3\theta$ equals zero for

$\theta = 0,\ \theta = \dfrac{\pi}{3}$, and $\theta = \dfrac{2\pi}{3}$.

$x = \sin 3\theta \cos\theta,\ y = \sin 3\theta \sin\theta$

$\Rightarrow x' = 3\cos 3\theta \cos\theta - \sin 3\theta \sin\theta$,

$y' = 3\cos 3\theta \sin\theta + \sin 3\theta \cos\theta$

$\Rightarrow \dfrac{dy}{dx} = \dfrac{y'}{x'} = \dfrac{3\cos 3\theta \sin\theta + \sin 3\theta \cos\theta}{3\cos 3\theta \cos\theta - \sin 3\theta \sin\theta}$.

Then $\left.\dfrac{dy}{dx}\right|_{\theta=0} = 0$, $\left.\dfrac{dy}{dx}\right|_{\theta=\pi/3} = \dfrac{-3\frac{\sqrt{3}}{2}}{-\frac{3}{2}} = \sqrt{3}$,

and $\left.\dfrac{dy}{dx}\right|_{\theta=2\pi/3} = \dfrac{3\frac{\sqrt{3}}{2}}{-\frac{3}{2}} = -\sqrt{3}$.

Three different parts of the graph pass through the origin at three different angles.

51. If for $r = f(\theta),\ s = \int_a^b \sqrt{f'^2 + f^2}\ d\theta = L$,

then for $r = cf(\theta),\ s = \int_a^b \sqrt{(cf')^2 + (cf)^2}\ d\theta$

$$= |c|\int_a^b \sqrt{f'^2 + f^2}\ d\theta = |c|L.$$

Section 9.6

5. The vertex is halfway between the directrix and the focus, at $(0, 0)$, and the parabola opens down. From the vertex, the focus is shifted vertically by $\frac{1}{4a} = -1$ unit, so $a = -\frac{1}{4}$ and $y = -\frac{1}{4}x^2$.

7. The vertex is halfway between the directrix and the focus, at $(2, 0)$, and the parabola opens right. From the vertex, the focus is shifted horizontally by $\frac{1}{4a} = 1$ unit, so $a = \frac{1}{4}$ and $x = \frac{1}{4}y^2 + 2$.

9. The center is the midpoint of the foci, in this case $(0, 3)$. The foci are shifted $c = 2$ units from the center. The vertices are $a = 4$ units from the center. From $c^2 = a^2 - b^2$ we get $b^2 = 16 - 4 = 12$. The major axis is parallel to the y-axis, so a^2 is the divisor of the y-term. The equation is $\frac{x^2}{12} + \frac{(y-3)^2}{16} = 1$.

11. The center is the midpoint of the foci, in this case $(4, 1)$. The foci are shifted $c = 2$ units from the center. The vertices are shifted $a = 4$ units from the center. From $c^2 = a^2 - b^2$ we get $b^2 = 16 - 4 = 12$. The major axis is parallel to the x-axis, so a^2 is the divisor of the x-term. The equation is $\frac{(x-4)^2}{16} + \frac{(y-1)^2}{12} = 1$.

13. The center is the midpoint of the foci, in this case $(2, 0)$. The foci are shifted $c = 2$ units from the center. The vertices are shifted $a = 1$ unit from the center. From $b^2 = c^2 - a^2$ we get $b^2 = 4 - 1 = 3$. The major axis is parallel to the x-axis, so a^2 is the divisor of the x-term. The equation is $\frac{(x-2)^2}{1} - \frac{y^2}{3} = 1$.

15. The center is the midpoint of the foci, in this case $(2, 4)$. The foci are shifted $c = 2$ units from the center. The vertices are shifted $a = 1$ unit from the center. From $b^2 = c^2 - a^2$ we get $b^2 = 4 - 1 = 3$. The major axis is parallel to the y-axis, so a^2 is the divisor if the y-term. The equation is $\frac{(y-4)^2}{1} - \frac{(x-2)^2}{3} = 1$.

17. $y = 2(x+1)^2 - 1$
This is an upwards-opening parabola. The vertex is at $(h, k) = (-1, -1)$. The focus is $\frac{1}{4a} = \frac{1}{8}$ units from the vertex, at $\left(-1, -\frac{7}{8}\right)$. The directrix is also $\frac{1}{8}$ units from the vertex at $y = -\frac{9}{8}$.

19. $\frac{(x-1)^2}{4} + \frac{(y-2)^2}{9} = 1$.
This is an ellipse. The center is at $(h, k) = (1, 2)$. Here $a = 3$, so the vertices are 3 vertical units from the center $(1, -1)$ and $(1, 5)$, and the foci are $c = \sqrt{a^2 - b^2} = \sqrt{5}$ vertical units from the center at $(1, 2 - \sqrt{5})$ and $(1, 2 + \sqrt{5})$

21. $\frac{(x-1)^2}{9} - \frac{y^2}{4} = 1$.
This is a horizontally-opening hyperbola. The center is at $(h, k) = (1, 0)$. Here $a = 3$, so the vertices are 3 horizontal units from the center at $(-2, 0)$ and $(4, 0)$, and the foci are $c = \sqrt{a^2 + b^2} = \sqrt{13}$ horizontal units from the center at $(1 - \sqrt{13}, 0)$ and $(1 + \sqrt{13}, 0)$.

23. $\frac{(y+1)^2}{16} - \frac{(x+2)^2}{4} = 1$.
This is a vertically-opening hyperbola. The center is at $(h, k) = (-2, -1)$. Here $a = 4$, so the vertices are 4 vertical units from the center at $(-2, -5)$ and $(-2, 3)$, and the foci are $c = \sqrt{a^2 + b^2} = 2\sqrt{5}$ vertical units from the center at $(-2, -1 - 2\sqrt{5})$ and $(-2, -1 + 2\sqrt{5})$.

25. $(x-2)^2 + 9y^2 = 9$

$\dfrac{(x-2)^2}{9} + y^2 = 1.$

This is an ellipse. The center is at $(h, k) = (2, 0)$. Here $a = 3$, so the vertices are 3 horizontal units from the center at $(-1, 0)$ and $(5, 0)$, and the foci are $c = \sqrt{a^2 - b^2} = 2\sqrt{2}$ horizontal units from the center at $(2 - 2\sqrt{2}, 0)$ and $(2 + 2\sqrt{2}, 0)$.

27. $(x+1)^2 - 4(y-2) = 16$

$(x+1)^2 - 4y - 8 = 0$

$y = \dfrac{1}{4}(x+1)^2 - 2$

This is an upwards-opening parabola. The vertex is at $(h, k) = (-1, -2)$. The focus is $\dfrac{1}{4a} = 1$ unit from the vertex, at $(-1, -1)$. The directrix is also 1 unit from the focus, at $y = -3$.

29. This is a parabola with focus $(2, 1)$ and directrix $y = -3$. The vertex is halfway between the directrix and the focus, at $(2, -1)$, and the parabola opens up. From the vertex, the focus is shifted vertically by $\dfrac{1}{4a} = 2$ units, so $a = \dfrac{1}{8}$ and

$y = \dfrac{1}{8}(x-2)^2 - 1.$

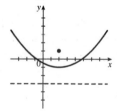

31. This is an ellipse with center at $(2, 2)$. Each focus is $c = 2$ units from the center, and the sum of distances is $2a = 8$, so $a = 4$. Then $b^2 = a^2 - c^2 = 16 - 4 = 12$, and the equation is

$\dfrac{(x-2)^2}{16} + \dfrac{(y-2)^2}{12} = 1.$

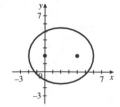

33. This is a vertically–opening hyperbola with center at $(0, 1)$. Each focus is $c = 3$ units from the center, and the difference of distances is $2a = 4$, so $a = 2$. Then $b^2 = c^2 - a^2 = 9 - 4 = 5$, and the equation is $\dfrac{(y-1)^2}{4} - \dfrac{x^2}{5} = 1.$

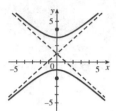

35. The vertex is at $(0, 0)$. The light bulb should be placed at the focus, $\dfrac{1}{4a} = \dfrac{1}{4(4)} = \dfrac{1}{16}$ units to the right of the vertex, at $\left(\dfrac{1}{16}, 0\right)$.

37. The vertex is at $(0, 0)$. The microphone should be placed at the focus, $\dfrac{1}{4a} = \dfrac{1}{4(2)} = \dfrac{1}{8}$ units up from the vertex at $\left(0, \dfrac{1}{8}\right)$.

39. The kidney stone should be at one focus and the transducer at the other, a distance of

$2c = 2\sqrt{a^2 - b^2} = 2\sqrt{124 - 24} = 20$ inches apart.

41. The center is at $(0, -4)$, and the foci are each $c = \sqrt{a^2 + b^2} = \sqrt{1 + 15} = 4$ units from the center, at $(0, 0)$ and $(0, -8)$. Light traveling along $y = cx$ toward $(0, 0)$ will be reflected toward $(0, -8)$.

43. The center is at $(0, 0)$, and the foci are each $c = \sqrt{a^2 + b^2} = \sqrt{3 + 1} = 2$ units from the center, at $(-2, 0)$ and $(2, 0)$. Light traveling along $y = c(x - 2)$ toward $(2, 0)$ will be reflected toward $(-2, 0)$.

45. The center is at $(0, 0)$, and the foci are each $c = \sqrt{a^2 - b^2} = \sqrt{400 - 100} = \sqrt{300}$ units from the center, at $(-\sqrt{300}, 0)$ and $(\sqrt{300}, 0)$. The desks should be at the foci.

47. Start with $d = at^2 + bt + c$. Since $d = 0$ when $t = 0$, it follows that $c = 0$. From the given information, $28 = a(2)^2 + b(2)$ and

$48 = a(4)^2 + b(4)$. Solving the system $4a + 2b = 28$, $16a + 4b = 48$ yields $a = -1$, $b = 16$. So $d = -t^2 + 16t$.
Since $d = t(16 - t)$, the object's distance from the spectator is 0 when $t = 16$ seconds. The object has come back towards the spectator, so it must be a boomerang.

Section 9.7

3. $d = 2, e = .6$
$$r = \frac{ed}{e\cos\theta + 1} = \frac{1.2}{.6\cos\theta + 1}$$
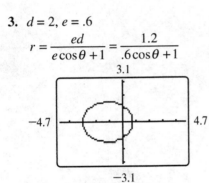

5. $d = 2, e = 1$
$$r = \frac{ed}{e\cos\theta + 1} = \frac{2}{\cos\theta + 1}$$
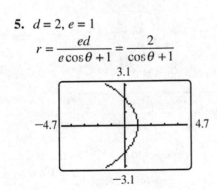

7. $d = 2, e = .6$
$$r = \frac{ed}{e\sin\theta + 1} = \frac{1.2}{.6\sin\theta + 1}$$
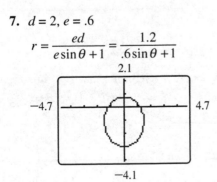

9. $d = 2, e = 1$
$$r = \frac{ed}{e\sin\theta + 1} = \frac{2}{\sin\theta + 1}$$
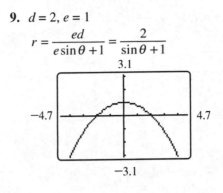

11. $d = -2, e = .4$
$$r = \frac{ed}{e\cos\theta - 1} = \frac{-.8}{.4\cos\theta - 1}$$
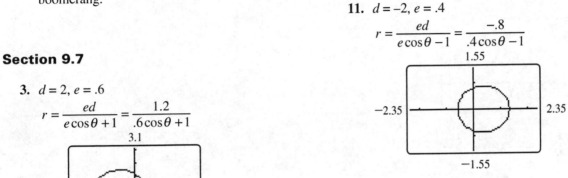

13. $d = -2, e = 2$
$$r = \frac{ed}{e\cos\theta - 1} = \frac{-4}{2\cos\theta - 1}$$
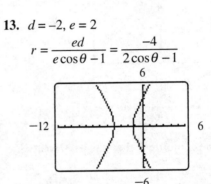

15. $d = -2, e = .4$
$$r = \frac{ed}{e\sin\theta - 1} = \frac{-.8}{.4\sin\theta - 1}$$
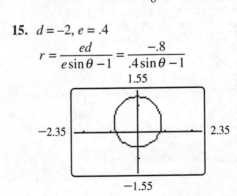

17. $d = -2,\ e = 1$

$$r = \frac{ed}{e\sin\theta - 1} = \frac{-2}{\sin\theta - 1}$$

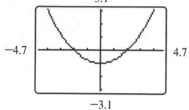

19.

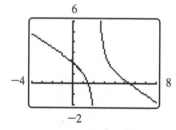

The graph is a hyperbola that has been rotated.

21.

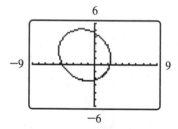

The graph is an ellipse that has been rotated.

23.

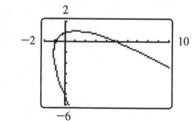

The graph is a parabola that has been rotated.

25. The graph is an ellipse with center at $(-1, 1)$, major axis parallel to the x-axis, $a = 3$, and $b = 2$.

$$\begin{cases} x = 3\cos t - 1 \\ y = 2\sin t + 1 \end{cases} \quad 0 \le t \le 2\pi$$

27. The graph is a hyperbola with center at $(-1, 0)$, vertices on a horizontal line, $a = 4$, and $b = 3$.

Left half: $\begin{cases} x = -4\cosh t - 1 \\ y = 3\sinh t \end{cases}$

Right half: $\begin{cases} x = 4\cosh t - 1 \\ y = 3\sinh t \end{cases}$

29. $\dfrac{x^2}{4} + y = 1$

$$y = -\frac{x^2}{4} + 1$$

$$\begin{cases} x = t \\ y = -\dfrac{1}{4}t^2 + 1 \end{cases}$$

31. Area for $0 \le \theta \le \dfrac{\pi}{2}$:

$$A = \frac{1}{2}\int_0^{\pi/2} \left(\frac{2}{\sin\theta + 2}\right)^2 d\theta \approx .4728$$

Area for $\dfrac{3\pi}{2} \le \theta \le 4.953$:

$$A = \frac{1}{2}\int_{3\pi/2}^{4.953} \left(\frac{2}{\sin\theta + 2}\right)^2 d\theta \approx .4722$$

Arc length for $0 \le \theta \le \dfrac{\pi}{2}$:

$$s = \int_0^{\pi/2} \sqrt{\frac{4\cos^2\theta}{(\sin\theta + 2)^4} + \frac{4}{(\sin\theta + 2)^2}}\,d\theta \approx 1.2661$$

Arc length for $\dfrac{3\pi}{2} \le \theta \le 4.953$:

$$s = \int_{3\pi/2}^{4.953} \sqrt{\frac{4\cos^2\theta}{(\sin\theta + 2)^4} + \frac{4}{(\sin\theta + 2)^2}}\,d\theta \approx .4810$$

The average speed for $0 \le \theta \le \dfrac{\pi}{2}$ is about 2.6 times as fast as the average speed for $\dfrac{3\pi}{2} \le \theta \le 4.953$.

33. With $d < 0$, Eq.(7.1) becomes

$$\sqrt{x^2 + y^2} = e(x - d)\ \text{ or, in polar coordinates,}$$

$r = e(r\cos\theta - d) \Rightarrow r(e\cos\theta - 1) = ed \Rightarrow$

$$r = \frac{ed}{e\cos\theta - 1}.$$

35. With directrix $y = d < 0$, Eq.(7.1) becomes

$$\sqrt{x^2 + y^2} = e(y - d) \text{ or, in polar coordinates,}$$

$$r = e(r\sin\theta - d) \Rightarrow r(e\sin\theta - 1) = ed \Rightarrow$$

$$r = \frac{ed}{e\sin\theta - 1}.$$

Chapter 9 Review

1. See Section 9.1.

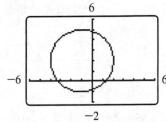

$$\begin{cases} x = -1 + 3\cos t \\ y = 2 + 3\sin t \end{cases} \quad \begin{cases} x + 1 = 3\cos t \\ y - 2 = 3\sin t \end{cases}$$

Since $(3\cos t)^2 + (3\sin t)^2 = 9$, a corresponding

x-y equation is $(x+1)^2 + (y-2)^2 = 9$.

3. See Section 9.1

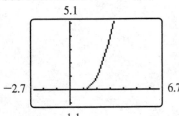

$$\begin{cases} x = t^2 + 1 \\ y = t^4 \end{cases}$$

Note that $t^2 = x - 1$; substitute this expression in the equation for y:

$$y = t^4 = (t^2)^2 = (x-1)^2 = x^2 - 2x + 1.$$

A corresponding x-y equation is $y = x^2 - 2x + 1$, for $x \geq 1$.

5. See Example 1.3.

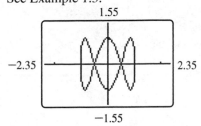

7. See Example 1.3.

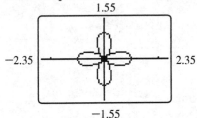

9. See Example 1.4.

C

11. See Example 1.4.

B

13. See Example 1.5.

$$\begin{cases} x = a + bt \\ y = c + dt \end{cases}$$

Let $(2, 1)$ correspond to $t = 0$, so $a = 2$ and $c = 1$. Let $(4, 7)$ correspond to $t = 1$, so $2 + b = 4$ and $1 + d = 7$, whence $b = 2$ and $d = 6$. Thus $x = 2 + 2t$, $y = 1 + 6t$, for $0 \leq t \leq 1$. (Other answers are possible.)

15. See Example 2.1.

Note that $\dfrac{dx}{dt} = 3t^2 - 3$ and $\dfrac{dy}{dt} = 2t - 1$.

a. $\left.\dfrac{dy}{dx}\right|_{t=0} = \dfrac{\left.\dfrac{dy}{dt}\right|_{t=0}}{\left.\dfrac{dx}{dt}\right|_{t=0}} = \dfrac{2(0) - 1}{3(0)^2 - 3} = \dfrac{1}{3}$

b. $\left.\dfrac{dy}{dx}\right|_{t=1} = \dfrac{\left.\dfrac{dy}{dt}\right|_{t=1}}{\left.\dfrac{dx}{dt}\right|_{t=1}} = \dfrac{2(1) - 1}{3(1)^2 - 3} = \dfrac{1}{0}$ (undefined)

The tangent line at this point is vertical.

c. The curve passes through $(2, 3)$ at $t = -1$ and at $t = 2$. At $t = -1$:

$$\left.\dfrac{dy}{dx}\right|_{t=-1} = \dfrac{\left.\dfrac{dy}{dt}\right|_{t=-1}}{\left.\dfrac{dx}{dt}\right|_{t=-1}} = \dfrac{2(-1) - 1}{3(-1)^2 - 3}$$

$$= \dfrac{-3}{0} \text{ (undefined)}$$

The tangent line at $t = -1$ is vertical.

At $t = 2$: $\left. \dfrac{dy}{dx} \right|_{t=2} = \dfrac{\left. \dfrac{dy}{dt} \right|_{t=2}}{\left. \dfrac{dx}{dt} \right|_{t=2}} = \dfrac{2(2)-1}{3(2)^2 - 3} = \dfrac{1}{3}$

17. See Example 2.3.

Note that $x'(t) = \dfrac{dx}{dt} = 3t^2 - 3$ and

$y'(t) = \dfrac{dy}{dt} = 2t + 2$. The velocity at

$t = 0$ is given by $x'(0) = -3$ and $y'(0) = 2$. The

speed at $t = 0$ is $\sqrt{(-3)^2 + 2^2} = \sqrt{13}$. The object is

moving toward the left and upward.

19. See Example 2.4.

For $0 \le t \le 2\pi$, the curve is traced clockwise.

$A = \displaystyle\int_0^{2\pi} y(t)x'(t)\,dt = \int_0^{2\pi} (2\cos t)(3\cos t)\,dt$

$= \displaystyle\int_0^{2\pi} 6\cos^2 t\,dt$

$= \displaystyle\int_0^{2\pi} 3(1 + \cos 2t)\,dt$

$= \left(3t + \dfrac{3}{2}\sin 2t \right) \Big|_0^{2\pi}$

$= 6\pi$

21. See Example 2.4.

For $-1 \le t \le 1$, the curve is traced counter clockwise. Note that the curve does not intersect itself and both endpoints are $(\cos 2, 0)$, so the requirements of Theorem 2.2 are satisfied.

$A = \displaystyle\int_{-1}^{1} x(t)\,y'(t)\,dt$

$= \displaystyle\int_{-1}^{1} (\cos 2t)(\pi \cos \pi t)\,dt$

$= \dfrac{4\pi \sin 2}{\pi^2 - 4}$

≈ 1.9467

We evaluate the integral using a CAS.

23. See Example 3.1.

$s = \displaystyle\int_{-1}^{1} \sqrt{\left[\dfrac{dx}{dt} \right]^2 + \left[\dfrac{dy}{dt} \right]^2}\,dt$

$= \displaystyle\int_{-1}^{1} \sqrt{[-2\sin 2t]^2 + [\pi \cos \pi t]^2}\,dt$

$= \displaystyle\int_{-1}^{1} \sqrt{4\sin^2 2t + \pi^2 \cos^2 \pi t}\,dt$

≈ 5.249

We approximated the integral numerically.

25. See Example 3.1.

The curve is traced once for $-\dfrac{\pi}{2} \le t \le \dfrac{\pi}{2}$.

$s = \displaystyle\int_{-\pi/2}^{\pi/2} \sqrt{\left[\dfrac{dx}{dt} \right]^2 + \left[\dfrac{dy}{dt} \right]^2}\,dt$

$= \displaystyle\int_{-\pi/2}^{\pi/2} \sqrt{[-4\sin 4t]^2 + [5\cos 5t]^2}\,dt$

$= \displaystyle\int_{-\pi/2}^{\pi/2} \sqrt{16\sin^2 4t + 25\cos^2 5t}\,dt$

≈ 13.593

We approximated the integral numerically.

27. See Example 3.4.

Surface Area $= \displaystyle\int_{-1}^{1} 2\pi(\text{radius})(\text{arc length})\,dt$

$= \displaystyle\int_{-1}^{1} 2\pi |y| \sqrt{[g'(t)]^2 + [h'(t)]^2}\,dt$

$= \displaystyle\int_{-1}^{1} 2\pi |t^4 - 4t| \sqrt{[3t^2 - 4]^2 + [4t^3 - 4]^2}\,dt$

≈ 128.075

We approximated the integral numerically. (Note that the absolute value is essential, since there are portions of the curve above and below the x-axis.)

29. See Example 4.7.

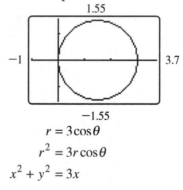

$r = 3\cos\theta$

$r^2 = 3r\cos\theta$

$x^2 + y^2 = 3x$

31. See Section 9.4.

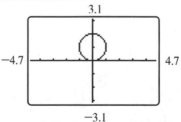

$r = 2\sin\theta$

$r = 0$ when $2\sin\theta = 0 : \theta = n\pi$

One copy of the graph is produced for $0 \le \theta \le \pi$.

33. See Section 9.4.

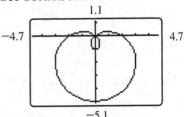

$r = 2 - 3\sin\theta$

$r = 0$ when $\sin\theta = \dfrac{2}{3}$:

$\theta = \sin^{-1}\dfrac{2}{3} + 2\pi n,\ \pi - \sin^{-1}\dfrac{2}{3} + 2\pi n$

One copy of the graph is produced for $0 \le \theta \le 2\pi$.

35. See Section 9.4.

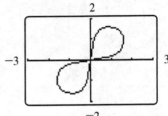

$r^2 = 4\sin 2\theta$

$r = 0$ when $\sin 2\theta = 0$, or $\theta = \dfrac{\pi}{2}n$.

Using $r = \pm 2\sqrt{\sin 2\theta}$, one copy of the graph is produced for $0 \le \theta \le \dfrac{\pi}{2}$.

37. See Section 9.4.

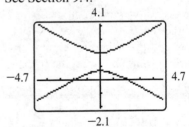

$r = \dfrac{2}{1 + 2\sin\theta}$

Note that $r \ne 0$; there are no values of θ for which $r = 0$. A complete copy of the graph is generated for $0 \le \theta \le 2\pi$ $\left(\theta \ne \dfrac{7\pi}{6}, \theta \ne \dfrac{11\pi}{6}\right)$.

39. See Example 4.6.

$$x^2 + y^2 = 9$$
$$(r\cos\theta)^2 + (r\sin\theta)^2 = 9$$
$$r^2 = 9$$
$$r = \pm 3$$

The answer may be given as $r = 3$, since this is sufficient to generate the entire graph.

41. See Example 5.1.

$r = \cos 3\theta$

$y = r\sin\theta = \cos 3\theta \sin\theta$

$x = r\cos\theta = \cos 3\theta \cos\theta$

$$\dfrac{dy}{dx}(\theta) = \dfrac{\dfrac{dy}{d\theta}}{\dfrac{dx}{d\theta}} = \dfrac{(-3\sin 3\theta)\sin\theta + \cos 3\theta(\cos\theta)}{(-3\sin 3\theta)\cos\theta - \cos 3\theta(\sin\theta)}$$

$$\dfrac{dy}{dx}\left(\dfrac{\pi}{6}\right) = \dfrac{-3\sin\left(\dfrac{\pi}{2}\right)\sin\left(\dfrac{\pi}{6}\right) + \cos\left(\dfrac{\pi}{2}\right)\cos\left(\dfrac{\pi}{6}\right)}{-3\sin\left(\dfrac{\pi}{2}\right)\cos\left(\dfrac{\pi}{6}\right) - \cos\left(\dfrac{\pi}{2}\right)\sin\left(\dfrac{\pi}{6}\right)}$$

$$= \dfrac{-3(1)\left(\dfrac{1}{2}\right) + (0)\left(\dfrac{\sqrt{3}}{2}\right)}{-3(1)\left(\dfrac{\sqrt{3}}{2}\right) - (0)\left(\dfrac{1}{2}\right)}$$

$$= \dfrac{1}{\sqrt{3}}$$

43. See Example 5.3.

$$A = \int_0^{\pi/5} \dfrac{1}{2} r^2\, d\theta$$

$$= \int_0^{\pi/5} \dfrac{1}{2}\sin^2 5\theta\, d\theta$$

$$= \int_0^{\pi/5} \dfrac{1}{4}(1 - \cos 10\theta)\, d\theta$$

$$= \dfrac{1}{4}\left(\theta - \dfrac{1}{10}\sin 10\theta\right)\Bigg|_0^{\pi/5}$$

$$= \dfrac{\pi}{20} \approx .157$$

45. See Example 5.4.

$$A = \int_{\pi/6}^{5\pi/6} \frac{1}{2} r^2 \, d\theta$$

$$= \int_{\pi/6}^{5\pi/6} \frac{1}{2}(1 - 2\sin\theta)^2 \, d\theta$$

$$= \int_{\pi/6}^{5\pi/6} \left(\frac{1}{2} - 2\sin\theta + 2\sin^2\theta \right) d\theta$$

$$= \int_{\pi/6}^{5\pi/6} \left[\frac{1}{2} - 2\sin\theta + (1 - \cos 2\theta) \right] d\theta$$

$$= \int_{\pi/6}^{5\pi/6} \left(\frac{3}{2} - 2\sin\theta - \cos 2\theta \right) d\theta$$

$$= \left(\frac{3}{2}\theta + 2\cos\theta - \frac{1}{2}\sin 2\theta \right) \Big|_{\pi/6}^{5\pi/6}$$

$$= \left(\frac{5\pi}{4} - \sqrt{3} + \frac{\sqrt{3}}{4} \right) - \left(\frac{\pi}{4} + \sqrt{3} - \frac{\sqrt{3}}{4} \right)$$

$$= \pi - \frac{3}{2}\sqrt{3}$$

$$\approx .543$$

47. See Example 5.5

Note that $r \geq 0$ for both curves and

$$1 + \cos\theta \leq 1 + \sin\theta \text{ for } \frac{\pi}{4} \leq r \leq \frac{5\pi}{4}.$$

Even though a graph of the polar functions looks somewhat complicated, the desired area can be found by integrating over this interval.

Area inside $r = 1 + \sin\theta$ on this interval:

$$A_1 = \int_{\pi/4}^{5\pi/4} \frac{1}{2}(1 + \sin\theta)^2 \, d\theta$$

$$= \int_{\pi/4}^{5\pi/4} \frac{1}{2}(1 + 2\sin\theta + \sin^2\theta) \, d\theta$$

$$= \int_{\pi/4}^{5\pi/4} \left[\frac{1}{2} + \sin\theta + \frac{1}{4}(1 - \cos 2\theta) \right] d\theta$$

$$= \int_{\pi/4}^{5\pi/4} \left(\frac{3}{4} + \sin\theta - \frac{1}{4}\cos 2\theta \right) d\theta$$

$$= \left(\frac{3}{4}\theta - \cos\theta - \frac{1}{8}\sin 2\theta \right) \Big|_{\pi/4}^{5\pi/4}$$

$$= \left(\frac{15}{16}\pi + \frac{\sqrt{2}}{2} - \frac{1}{8} \right) - \left(\frac{3}{16}\pi - \frac{\sqrt{2}}{2} - \frac{1}{8} \right)$$

$$= \frac{3}{4}\pi + \sqrt{2}$$

Area inside $r = 1 + \cos\theta$ on this interval:

$$A_2 = \int_{\pi/4}^{5\pi/4} \frac{1}{2}(1 + \cos\theta)^2 \, d\theta$$

$$= \int_{\pi/4}^{5\pi/4} \frac{1}{2}(1 + 2\cos\theta + \cos^2\theta) \, d\theta$$

$$= \int_{\pi/4}^{5\pi/4} \left[\frac{1}{2} + \cos\theta + \frac{1}{4}(1 + \cos 2\theta) \right] d\theta$$

$$= \int_{\pi/4}^{5\pi/4} \left(\frac{3}{4} + \cos\theta + \frac{1}{4}\cos 2\theta \right) d\theta$$

$$= \left(\frac{3}{4}\theta + \sin\theta + \frac{1}{8}\sin 2\theta \right) \Big|_{\pi/4}^{5\pi/4}$$

$$= \left(\frac{15}{16}\pi - \frac{\sqrt{2}}{2} + \frac{1}{8} \right) - \left(\frac{3}{16}\pi + \frac{\sqrt{2}}{2} + \frac{1}{8} \right)$$

$$= \frac{3}{4}\pi - \sqrt{2}$$

Desired area (difference):

$$A = \left(\frac{3}{4}\pi + \sqrt{2} \right) - \left(\frac{3}{4}\pi - \sqrt{2} \right) = 2\sqrt{2} \approx 2.828$$

49. See Example 5.8.

$$r = f(\theta) = 3 - 4\sin\theta$$

$$s = \int_0^{2\pi} \sqrt{[f'(\theta)]^2 + [f(\theta)]^2} \, d\theta$$

$$= \int_0^{2\pi} \sqrt{[-4\cos\theta]^2 + [3 - 4\sin\theta]^2} \, d\theta$$

$$= \int_0^{2\pi} \sqrt{16\cos^2\theta + 9 - 24\sin\theta + 16\sin^2\theta} \, d\theta$$

$$= \int_0^{2\pi} \sqrt{25 - 24\sin\theta} \, d\theta$$

$$\approx 28.814$$

We approximated the integral numerically.

51. See Example 6.1.

The vertex is midway between the focus $(1, 2)$ and the directrix $y = 0$, so the vertex is $(1, 1)$. The parabola opens upward because the focus is above the directrix. From the vertex, the focus is shifted upward by $\frac{1}{4a} = 1$ unit, so $a = \frac{1}{4}$. An equation is given by $y = a(x - b)^2 + c$ where $a = \frac{1}{4}$, $b = 1$, and $c = 1$:

$$y = \frac{1}{4}(x - 1)^2 + 1.$$

53. See Example 6.12.

The center is the midpoint of the foci, namely $(2, 2)$. The foci are $c = 2$ units from the center and the vertices are $a = 1$ unit from the center. Then $b^2 = c^2 - a^2 = 3$. The center, vertices, and foci lie on a vertical line, so the equation has the form

$$\frac{(y - y_0)^2}{a^2} - \frac{(x - x_0)^2}{b^2} = 1.$$

An equation of the hyperbola is

$$\frac{(y - 2)^2}{1} - \frac{(x - 2)^2}{3} = 1.$$

55. See Theorem 6.3.

$$\frac{(x + 1)^2}{9} + \frac{(y - 3)^2}{25} = 1$$

Since $25 > 9$, this is an ellipse in the form

$$\frac{(x - x_0)^2}{b^2} + \frac{(y - y_0)^2}{a^2} = 1 \text{ with } x_0 = -1, \ y_0 = 3,$$

$a = 5$, and $b = 3$. $c = \sqrt{a^2 - b^2} = 4$.

Vertices: $(x_0, \ y_0 \pm a) = (-1, -2)$ and $(-1, 8)$

Foci: $(x, \ y_0 \pm c) = (-1, -1)$ and $(-1, 7)$

57. See Theorem 6.1.

$$(x - 1)^2 + y = 4$$
$$y = -(x - 1)^2 + 4$$

This is a parabola in the form $y = a(x - b)^2 + c$ with $a = -1$, $b = 1$, and $c = 4$. The focus and directrix are shifted by $\dfrac{1}{4a} = -\dfrac{1}{4}$.

Vertex: $(b, c) = (1, 4)$

Focus: $\left(b, c + \dfrac{1}{4a}\right) = \left(1, \dfrac{15}{4}\right)$

Directrix: $y = c - \dfrac{1}{4a}$ or $y = \dfrac{17}{4}$

59. See Example 6.5.

$$y = \frac{1}{2}x^2$$

This is a parabola in the form $y = a(x - b)^2 + c$ with $a = \dfrac{1}{2}$, $b = 0$, and $c = 0$. The focus and directrix are shifted by $\dfrac{1}{4a} = \dfrac{1}{2}$. The microphone should be placed at the focus, $\left(0, \dfrac{1}{2}\right)$.

61. See Theorem 7.2.

$d = 3, e = .8$

$$r = \frac{ed}{e\cos\theta + 1} = \frac{2.4}{.8\cos\theta + 1}$$

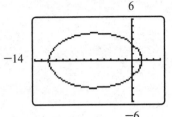

63. See Theorem 7.2.

$d = 2, e = 1.4$

$$r = \frac{ed}{e\sin\theta + 1} = \frac{2.8}{1.4\sin\theta + 1}$$

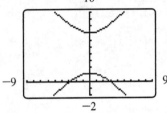

65. See Example 7.3.

$$\frac{(x + 1)^2}{9} + \frac{(y - 3)^2}{25} = 1$$

The graph is an ellipse with vertical major axis, $a = 5$, and $b = 3$.

$$\begin{cases} x = 3\cos t - 1 \\ y = 5\sin t + 3 \end{cases} \quad 0 \le t \le 2\pi$$

Chapter 10

Section 10.1

5.

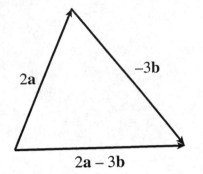

7.

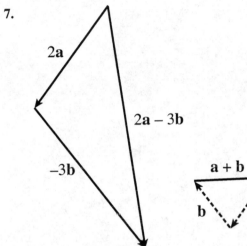

9. $a + b = \langle 2, 4 \rangle + \langle 3, -1 \rangle$
$= \langle 2 + 3, 4 - 1 \rangle$
$= \langle 5, 3 \rangle$

$a - 2b = \langle 2, 4 \rangle - 2\langle 3, -1 \rangle$
$= \langle 2, 4 \rangle - \langle 6, -2 \rangle$
$= \langle 2 - 6, 4 + 2 \rangle$
$= \langle -4, 6 \rangle$

$3a = 3\langle 2, 4 \rangle = \langle 3 \cdot 2, 3 \cdot 4 \rangle = \langle 6, 12 \rangle$

$5b - 2a = 5\langle 3, -1 \rangle - 2\langle 2, 4 \rangle$
$= \langle 15, -5 \rangle - \langle 4, 8 \rangle$
$= \langle 11, -13 \rangle$

$\|5b - 2a\| = \|\langle 11, -13 \rangle\| = \sqrt{11^2 + (-13)^2} = \sqrt{290}$

11. $a + b = (i + 2j) + (3i - j) = 4i + j$
$a - 2b = (i + 2j) - 2(3i - j)$
$= (i + 2j) - (6i - 2j)$
$= -5i + 4j$
$3a = 3(i + 2j) = 3i + 6j$
$5b - 2a = 5(3i - j) - 2(i + 2j)$
$= (15i - 5j) - (2i + 4j)$
$= 13i - 9j$
$\|5b - 2a\| = \|13i - 9j\|$
$= \sqrt{13^2 + (-9)^2}$
$= \sqrt{250}$
$= 5\sqrt{10}$

13. $a - b = \langle -2, 3 \rangle - \langle 1, 0 \rangle = \langle -2 - 1, 3 - 0 \rangle = \langle -3, 3 \rangle$
$-4b = -4\langle 1, 0 \rangle = \langle -4 \cdot 1, -4 \cdot 0 \rangle = \langle -4, 0 \rangle$
$3a + b = 3\langle -2, 3 \rangle + \langle 1, 0 \rangle$
$= \langle -6, 9 \rangle + \langle 1, 0 \rangle$
$= \langle -5, 9 \rangle$
$4a = 4\langle -2, 3 \rangle = \langle -8, 12 \rangle$
$\|4a\| = \|\langle -8, 12 \rangle\| = \sqrt{(-8)^2 + 12^2} = \sqrt{208} = 4\sqrt{13}$

15. $a - b = (i + 2j) - (3i - j) = -2i + 3j$
$-4b = -4(3i - j) = -12i + 4j$
$3a + b = 3(i + 2j) + (3i - j)$
$= (3i + 6j) + (3i - j)$
$= 6i + 5j$
$4a = 4(i + 2j) = 4i + 8j$
$\|4a\| = \|4i + 8j\| = \sqrt{4^2 + 8^2} = \sqrt{80} = 4\sqrt{5}$

17.

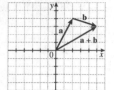 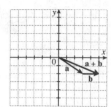

19. Notice that $-2\mathbf{a} = \mathbf{b}$. $\mathbf{a}$ and $\mathbf{b}$ are parallel.

21. If $\mathbf{a} = c\mathbf{b}$, then $\langle -2, 3 \rangle = c\langle 4, 6 \rangle$. That is,

$-2 = 4c\ \left(c = -\dfrac{1}{2} \right)$ and $3 = 6c\ \left(c = \dfrac{1}{2} \right)$. This is

a contradiction, so $\mathbf{a}$ and $\mathbf{b}$ are not parallel.

23. Notice that $3\mathbf{a} = \mathbf{b}$. $\mathbf{a}$ and $\mathbf{b}$ are parallel.

25. If $\mathbf{a} = c\mathbf{b}$, then $-5\mathbf{i} = c(10\mathbf{i} - 2\mathbf{j})$. That is,

$-5 = 10c\ \left(c = -\dfrac{1}{2} \right)$ and $0 = -2c\ (c = 0)$. This

is a contradiction, so $\mathbf{a}$ and $\mathbf{b}$ are not parallel.

27. $\overrightarrow{AB} = \langle 5 - 2,\ 4 - 3 \rangle = \langle 3, 1 \rangle$

29. $\overrightarrow{AB} = \langle 1 - 4,\ 0 - 3 \rangle = \langle -3, -3 \rangle$

31. $\overrightarrow{AB} = \langle 1 - (-1),\ -1 - 2 \rangle = \langle 2, -3 \rangle$

33. $\overrightarrow{AB} = \langle 0 - 2,\ -3 - 0 \rangle = \langle -2, -3 \rangle$

35. a. $\|\langle 4, -3 \rangle\| = \sqrt{4^2 + (-3)^2} = \sqrt{25} = 5$

$\mathbf{u} = \dfrac{1}{5}\langle 4, -3 \rangle = \left\langle \dfrac{4}{5}, -\dfrac{3}{5} \right\rangle$

b. $\langle 4, 1 \rangle = 5\left\langle \dfrac{4}{5}, -\dfrac{3}{5} \right\rangle$

37. a. $\|2\mathbf{i} - 4\mathbf{j}\| = \sqrt{2^2 + (-4)^2} = \sqrt{20} = 2\sqrt{5}$

$\mathbf{u} = \dfrac{1}{2\sqrt{5}}(2\mathbf{i} - 4\mathbf{j}) = \dfrac{1}{\sqrt{5}}\mathbf{i} - \dfrac{2}{\sqrt{5}}\mathbf{j}$

b. $2\mathbf{i} - 4\mathbf{j} = 2\sqrt{5}\left\langle \dfrac{1}{\sqrt{5}}, -\dfrac{2}{\sqrt{5}} \right\rangle$

39. a. $\|4\mathbf{i}\| = \sqrt{4^2} = 4$

$\mathbf{u} = \dfrac{1}{4}(4\mathbf{i}) = \mathbf{i}$

b. $4\mathbf{i} = 4\langle 1, 0 \rangle$

41. a. $\mathbf{v} = \langle 5 - 2,\ 2 - 1 \rangle = \langle 3, 1 \rangle$

$\|\mathbf{v}\| = \sqrt{3^2 + 1^2} = \sqrt{10}$

$\mathbf{u} = \dfrac{1}{\sqrt{10}}\langle 3, 1 \rangle = \left\langle \dfrac{3}{\sqrt{10}}, \dfrac{1}{\sqrt{10}} \right\rangle$

b. $3\mathbf{i} + \mathbf{j} = \sqrt{10}\left\langle \dfrac{3}{\sqrt{10}}, \dfrac{1}{\sqrt{10}} \right\rangle$

43. a. $\mathbf{v} = \langle 2 - 5,\ 3 - (-1) \rangle = \langle -3, 4 \rangle$

$\|\mathbf{v}\| = \sqrt{(-3)^2 + 4^2} = \sqrt{25} = 5$

$\mathbf{u} = \dfrac{1}{5}\langle -3, 4 \rangle = \left\langle -\dfrac{3}{5}, \dfrac{4}{5} \right\rangle$

b. $\langle -3, 4 \rangle = 5\left\langle -\dfrac{3}{5}, \dfrac{4}{5} \right\rangle$

45. $\|\mathbf{v}\| = \sqrt{3^2 + 4^2} = \sqrt{25} = 5$

$\mathbf{w} = \dfrac{3}{5}\mathbf{v} = \dfrac{9}{5}\mathbf{i} + \dfrac{12}{5}\mathbf{j}$

47. $\|\mathbf{v}\| = \sqrt{2^2 + 5^2} = \sqrt{29}$

$\mathbf{w} = \dfrac{29}{\sqrt{29}}\mathbf{v} = \sqrt{29}\mathbf{v} = \langle 2\sqrt{29}, 5\sqrt{29} \rangle$

49. $\|\mathbf{v}\| = \sqrt{3^2 + 0^2} = 3$

$\mathbf{w} = \dfrac{4}{3}\mathbf{v} = \langle 4, 0 \rangle$

51. $\mathbf{g} = \langle 0, -150 \rangle;\ \mathbf{r} = \langle 20, 140 \rangle$

$\mathbf{g} + \mathbf{r} = \langle 20, -10 \rangle$

Net force is 10 pounds down, 20 pounds right.

53. $\mathbf{g} = \langle 0, -200 \rangle;\ \mathbf{g} + \mathbf{r} = \langle 30, -10 \rangle$

$\mathbf{r} = \langle 30, -10 \rangle - \langle 0, -200 \rangle = \langle 30, 190 \rangle$

Air resistance is 190 pounds up, 30 pounds right.

55. Net force

$= \langle -164, 115 \rangle + \langle 177, 177 \rangle + \langle 0, -275 \rangle$

$= \langle 13, 17 \rangle$

Crate will move right and up.

57. $\mathbf{v} = \langle x, y \rangle;\ \mathbf{w} = \langle 30, -20 \rangle$

$\mathbf{v} + \mathbf{w} = \langle c, 0 \rangle$ where $c < 0$

$y + (-20) = 0$, so $y = 20$

$\|\mathbf{v}\| = 300$, so $\sqrt{x^2 + y^2} = 300$

$$x^2 + 400 = 90,000$$
$$x^2 = 89,600$$
$$x = -\sqrt{89,600} = -80\sqrt{14}$$

(we take the negative square root to have the plane moving westward)

$\mathbf{v} = \left\langle -80\sqrt{14}, \, 20 \right\rangle$ This points left and up, or

northwest, at an angle of $\tan^{-1} \dfrac{20}{80\sqrt{14}} \approx 3.8°$

north of west.

59. $\mathbf{v} = \left\langle x, \, y \right\rangle; \; \mathbf{w} = \left\langle -20, \, 30 \right\rangle$

$\mathbf{v} + \mathbf{w} = \left\langle 0, \, c \right\rangle$ where $c > 0$

$x + (-20) = 0$, so $x = 20$

$\|\mathbf{v}\| = 400$, so $\sqrt{x^2 + y^2} = 400$

$400 + y^2 = 160,000$
$$y^2 = 159,600$$
$$y = \sqrt{159,600} = 20\sqrt{399}$$

(we take the positive square root to have the plane moving northward)

$\mathbf{v} = \left\langle 20, \, 20\sqrt{399} \right\rangle$

This points right and up, or northeast, at an angle

of $\tan^{-1} \dfrac{20}{20\sqrt{399}} \approx 2.9°$ east of north.

61. The paper travels at velocity $\left\langle -50, 10 \right\rangle$. If the porch is at $(0, 0)$, then the paperboy throws from $(50, c)$. To travel 50 feet in the left direction, it takes 1 second. Thus in the street direction, the paper travels 10 feet. Hence the paperboy should be 10 ft up the road.

63. The force of the water is $\left\langle -200, 0 \right\rangle$ (magnitude of 200 lb in the direction $\left\langle -1, 0 \right\rangle$). The weight is $\left\langle 0, -20 \right\rangle$. Thus the force required to hold the hose horizontal is $\left\langle 200, 20 \right\rangle$, or

$\sqrt{200^2 + 20^2} = 20\sqrt{101}$ pounds at

$\tan^{-1} \dfrac{20}{200} \approx 5.7°$ to the horizontal.

65. The largest possible magnitude occurs when $\mathbf{a}$ and $\mathbf{b}$ are in the same direction. Thus the magnitude of $\mathbf{a} + \mathbf{b}$ is the sum of the magnitudes or 7.

The smallest possible magnitude occurs when $\mathbf{a}$ and $\mathbf{b}$ are in opposite directions. Thus the magnitude of $\mathbf{a} + \mathbf{b}$ is the absolute value of the difference or 1.

Suppose $\mathbf{a}$ and $\mathbf{b}$ are perpendicular. We can use a coordinate system. such that $\mathbf{a}$ lies in the direction $\left\langle 1, 0 \right\rangle$ and $\mathbf{b}$ lies in the direction $\left\langle 0, 1 \right\rangle$. Thus

$\mathbf{a} + \mathbf{b} = \left\langle 3, 4 \right\rangle$ and $\|\mathbf{a} + \mathbf{b}\| = \sqrt{3^2 + 4^2} = 5$.

67. $\mathbf{a} = \left\langle a_1, a_2 \right\rangle, \; \mathbf{b} = \left\langle b_1, b_2 \right\rangle, \; \mathbf{c} = \left\langle c_1, c_2 \right\rangle$

$$\begin{aligned} \mathbf{a} + (\mathbf{b} + \mathbf{c}) &= \left\langle a_1, a_2 \right\rangle + \left(\left\langle b_1, b_2 \right\rangle + \left\langle c_1, c_2 \right\rangle \right) \\ &= \left\langle a_1, a_2 \right\rangle + \left\langle b_1 + c_1, b_2 + c_2 \right\rangle \\ &= \left\langle a_1 + b_1 + c_1, a_2 + b_2 + c_2 \right\rangle \\ &= \left\langle a_1 + b_1, a_2 + b_2 \right\rangle + \left\langle c_1, c_2 \right\rangle \\ &= \left(\left\langle a_1, a_2 \right\rangle + \left\langle b_1, b_2 \right\rangle \right) + \left\langle c_1, c_2 \right\rangle \\ &= (\mathbf{a} + \mathbf{b}) + \mathbf{c} \end{aligned}$$

69. $\mathbf{a} + \mathbf{b} = \left\langle 2,3 \right\rangle + \left\langle 1,4 \right\rangle = \left\langle 3,7 \right\rangle$

$\|\mathbf{a} + \mathbf{b}\| = \sqrt{3^2 + 7^2} = \sqrt{58}$

$\|\mathbf{a}\| = \sqrt{2^2 + 3^2} = \sqrt{13}$

$\|\mathbf{b}\| = \sqrt{1^2 + 4^2} = \sqrt{17}$

$\|\mathbf{a} + \mathbf{b}\| < \|\mathbf{a}\| + \|\mathbf{b}\|$

For two other choices of a and b that are not in the same direction, $\|\mathbf{a} + \mathbf{b}\| < \|\mathbf{a}\| + \|\mathbf{b}\|$.

The sketch in Figure 10.6 illustrates an example of the triangle inequality, $\|\mathbf{a} + \mathbf{b}\| \leq \|\mathbf{a}\| + \|\mathbf{b}\|$.

71. $\|\mathbf{a} + \mathbf{b}\| = \|\mathbf{a}\| + \|\mathbf{b}\|$ when $\mathbf{a}$ and $\mathbf{b}$ have the same direction or $\mathbf{a} = c\mathbf{b}$ where $c > 0$.

$\|\mathbf{a} + \mathbf{b}\|^2 = \|\mathbf{a}\|^2 + \|\mathbf{b}\|^2$ when the triangle formed by $\mathbf{a}$, $\mathbf{b}$, and $\mathbf{a} + \mathbf{b}$ is a right angle. This is when $\mathbf{a} \perp \mathbf{b}$.

Using Figure 10.7 and the Law of Cosines, we have $\|\mathbf{a} + \mathbf{b}\|^2 = \|\mathbf{a}\|^2 + \|\mathbf{b}\|^2 - 2\|\mathbf{a}\|\|\mathbf{b}\|\cos\theta$ where θ is the angle between $\mathbf{a}$ and $\mathbf{b}$ in the triangle

formed by $\mathbf{a}$, $\mathbf{b}$ and $\mathbf{a} + \mathbf{b}$.

Thus $\|\mathbf{a} + \mathbf{b}\|^2 > \|\mathbf{a}\|^2 + \|\mathbf{b}\|^2$ when the triangle is obtuse $(\theta > 90°)$,

$\|\mathbf{a} + \mathbf{b}\|^2 < \|\mathbf{a}\|^2 + \|\mathbf{b}\|^2$ when the triangle is acute $(\theta < 90°)$,

$\|\mathbf{a} + \mathbf{b}\|^2 = \|\mathbf{a}\|^2 + \|\mathbf{b}\|^2$ when the triangle is right $(\theta = 90°)$.

Section 10.2

5. a.

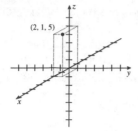

b.

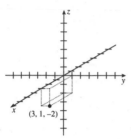

c.

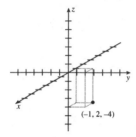

7.

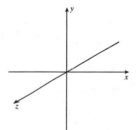

9. $d\{(2, 1, 2), (5, 5, 2)\}$

$= \sqrt{(5-2)^2 + (5-1)^2 + (2-2)^2}$

$= \sqrt{3^2 + 4^2 + 0^2}$

$= \sqrt{25}$

$= 5$

11. $d\{(-1, 0, 2), (1, 2, 3)\}$

$= \sqrt{[1-(-1)]^2 + (2-0)^2 + (3-2)^2}$

$= \sqrt{2^2 + 2^2 + 1^2}$

$= \sqrt{9}$

$= 3$

13. $d\{(0, 0, 0), (2, 3, 5)\}$

$= \sqrt{(2-0)^2 + (3-0)^2 + (5-0)^2}$

$= \sqrt{2^2 + 3^2 + 5^2}$

$= \sqrt{38}$

15. $\mathbf{a} + \mathbf{b} = \langle 2, 1, -2 \rangle + \langle 1, 3, 0 \rangle$

$= \langle 2+1, 1+3, -2+0 \rangle$

$= \langle 3, 4, -2 \rangle$

$\mathbf{a} - 3\mathbf{b} = \langle 2, 1, -2 \rangle - 3\langle 1, 3, 0 \rangle$

$= \langle 2, 1, -2 \rangle - \langle 3, 9, 0 \rangle$

$= \langle 2-3, 1-9, -2-0 \rangle$

$= \langle -1, -8, -2 \rangle$

$4\mathbf{a} + 2\mathbf{b} = 4\langle 2, 1, -2 \rangle + 2\langle 1, 3, 0 \rangle$

$= \langle 8, 4, -8 \rangle + \langle 2, 6, 0 \rangle$

$= \langle 8+2, 4+6, -8+0 \rangle$

$= \langle 10, 10, -8 \rangle$

$\|4\mathbf{a} + 2\mathbf{b}\| = \|\langle 10, 10, -8 \rangle\|$

$= \sqrt{10^2 + 10^2 + (-8)^2}$

$= \sqrt{264}$

$= 2\sqrt{66}$

17. $\mathbf{a} + \mathbf{b} = \langle -1, 0, 2 \rangle + \langle 4, 3, 2 \rangle$

$= \langle -1+4, 0+3, 2+2 \rangle$

$= \langle 3, 3, 4 \rangle$

$\mathbf{a} - 3\mathbf{b} = \langle -1, 0, 2 \rangle - 3\langle 4, 3, 2 \rangle$

$= \langle -1, 0, 2 \rangle - \langle 12, 9, 6 \rangle$

$= \langle -1-12, 0-9, 2-6 \rangle$

$= \langle -13, -9, -4 \rangle$

$4\mathbf{a} + 2\mathbf{b} = 4\langle -1, 0, 2 \rangle + 2\langle 4, 3, 2 \rangle$

$= \langle -4, 0, 8 \rangle + \langle 8, 6, 4 \rangle$

$= \langle -4+8, 0+6, 8+4 \rangle$

$= \langle 4, 6, 12 \rangle$

$\|4\mathbf{a} + 2\mathbf{b}\| = \|\langle 4, 6, 12 \rangle\|$

$= \sqrt{4^2 + 6^2 + 12^2}$

$= \sqrt{196}$

$= 14$

19. $\mathbf{a}+\mathbf{b} = (3\mathbf{i}-\mathbf{j}+4\mathbf{k})+(5\mathbf{i}+\mathbf{j}) = 8\mathbf{i}+4\mathbf{k}$

$\mathbf{a}-3\mathbf{b} = (3\mathbf{i}-\mathbf{j}+4\mathbf{k})-3(5\mathbf{i}+\mathbf{j})$

$\qquad = (3\mathbf{i}-\mathbf{j}+4\mathbf{k})-(15\mathbf{i}+3\mathbf{j})$

$\qquad = -12\mathbf{i}-4\mathbf{j}+4\mathbf{k}$

$4\mathbf{a}+2\mathbf{b} = 4(3\mathbf{i}-\mathbf{j}+4\mathbf{k})+2(5\mathbf{i}+\mathbf{j})$

$\qquad = (12\mathbf{i}-4\mathbf{j}+16\mathbf{k})+(10\mathbf{i}+2\mathbf{j})$

$\qquad = 22\mathbf{i}-2\mathbf{j}+16\mathbf{k}$

$\|4\mathbf{a}+2\mathbf{b}\| = \|22\mathbf{i}-2\mathbf{j}+16\mathbf{k}\|$

$\qquad = \sqrt{22^2+(-2)^2+16^2}$

$\qquad = \sqrt{744}$

$\qquad = 2\sqrt{186}$

21. a. $\|\langle 3,1,2\rangle\| = \sqrt{3^2+1^2+2^2} = \sqrt{14}$

$\pm\dfrac{1}{\sqrt{14}}\langle 3,1,2\rangle$

b. $\langle 3,1,3\rangle = \sqrt{14}\left\langle \dfrac{3}{\sqrt{14}}, \dfrac{1}{\sqrt{14}}, \dfrac{2}{\sqrt{14}}\right\rangle$

23. a. $\|\langle 2,-4,6\rangle\| = \sqrt{2^2+(-4)^2+6^2}$

$\qquad = \sqrt{56} = 2\sqrt{14}$

$\pm\dfrac{1}{2\sqrt{14}}\langle 2,-4,6\rangle = \pm\dfrac{1}{\sqrt{14}}\langle 1,-2,3\rangle$

b. $\langle 2,-4,6\rangle = 2\sqrt{14}\left\langle \dfrac{1}{\sqrt{14}}, \dfrac{-2}{\sqrt{14}}, \dfrac{3}{\sqrt{14}}\right\rangle$

25. a. $\|2\mathbf{i}-\mathbf{j}+2\mathbf{k}\| = \sqrt{2^2+(-1)^2+2^2} = \sqrt{9} = 3$

$\pm\dfrac{1}{3}(2\mathbf{i}-\mathbf{j}+2\mathbf{k}) = \pm\left(\dfrac{2}{3}\mathbf{i}-\dfrac{1}{3}\mathbf{j}+\dfrac{2}{3}\mathbf{k}\right)$

b. $2\mathbf{i}-\mathbf{j}+2\mathbf{k} = 3\left\langle \dfrac{2}{3}, -\dfrac{1}{3}, \dfrac{2}{3}\right\rangle$

27. a. $\mathbf{v} = \langle 3-1,2-2,1-3\rangle = \langle 2,0,-2\rangle$

$\|\mathbf{v}\| = \sqrt{2^2+0^2+(-2)^2} = \sqrt{8} = 2\sqrt{2}$

$\pm\dfrac{1}{\|\mathbf{v}\|}\mathbf{v} = \pm\dfrac{1}{2\sqrt{2}}\langle 2,0,-2\rangle$

$\qquad = \pm\dfrac{1}{\sqrt{2}}\langle 1,0,-1\rangle$

b. $\langle 2,0,-2\rangle = 2\sqrt{2}\left\langle \dfrac{1}{\sqrt{2}}, 0, -\dfrac{1}{\sqrt{2}}\right\rangle$

29. First find the unit vector in the same direction, and then scalar multiply by 6.

$\|\mathbf{v}\| = \sqrt{2^2+2^2+(-1)^2} = \sqrt{9} = 3$

$\mathbf{u} = \dfrac{1}{\|\mathbf{v}\|}\mathbf{v} = \dfrac{1}{3}\langle 2,2,-1\rangle = \left\langle \dfrac{2}{3}, \dfrac{2}{3}, -\dfrac{1}{3}\right\rangle$

$6\mathbf{u} = 6\left\langle \dfrac{2}{3}, \dfrac{2}{3}, -\dfrac{1}{3}\right\rangle = \langle 4,4,-2\rangle$

31. First find the unit vector in the same direction, and then scalar multiply by 2.

$\|\mathbf{v}\| = \sqrt{2^2+0^2+(-1)^2} = \sqrt{5}$

$\mathbf{u} = \dfrac{1}{\|\mathbf{v}\|}\mathbf{v} = \dfrac{1}{\sqrt{5}}\langle 2,0,-1\rangle$

$2\mathbf{u} = \dfrac{2}{\sqrt{5}}\langle 2,0,-1\rangle$

33. First find the unit vector in the same direction, and then scalar multiply by 4.

$\|\mathbf{v}\| = \sqrt{2^2+(-1)^2+3^2} = \sqrt{14}$

$\mathbf{u} = \dfrac{1}{\|\mathbf{v}\|}\mathbf{v} = \dfrac{1}{\sqrt{14}}(2\mathbf{i}-\mathbf{j}+3\mathbf{k})$

$4\mathbf{u} = \dfrac{4}{\sqrt{14}}(2\mathbf{i}-\mathbf{j}+3\mathbf{k})$

35. $(x-3)^2+(y-1)^2+(z-4)^2 = 2^2$

$(x-3)^2+(y-1)^2+(z-4)^2 = 4$

37. $(x-2)^2+(y-0)^2+[z-(-3)]^2 = 3^2$

$(x-2)^2+y^2+(z+3)^2 = 9$

39. $(x-\pi)^2+(y-1)^2+[z-(-3)]^2 = \left(\sqrt{5}\right)^2$

$(x-\pi)^2+(y-1)^2+(z+3)^2 = 5$

41. sphere, center $(1,0,-2)$, radius 2

43. $x^2+y^2-2y+z^2+4z = 4$

$x^2+(y^2-2y+1)+(z^2+4z+4) = 4+1+4$

$x^2+(y-1)^2+(z+2)^2 = 9$

sphere, center $(0,1,-2)$, radius 3

45. $x^2-2x+y^2+z^2-4z = 0$

$(x^2-2x+1)+y^2+(z^2-4z+4) = 1+4$

$(x-1)^2+y^2+(z-2)^2 = 5$

sphere, center $(1,0,2)$, radius $\sqrt{5}$

47. plane parallel to xz-plane

49. plane parallel to xy-plane

51. plane parallel to yz-plane

53. $y = 0$

55. $x = 0$

57. $\mathbf{a} = \langle a_1, a_2, a_3 \rangle, \mathbf{b} = \langle b_1, b_2, b_3 \rangle$

$$
\begin{aligned}
\mathbf{a} + \mathbf{b} &= \langle a_1, a_2, a_3 \rangle + \langle b_1, b_2, b_3 \rangle \\
&= \langle a_1 + b_1, a_2 + b_2, a_3 + b_3 \rangle \\
&= \langle b_1 + a_1, b_2 + a_2, b_3 + a_3 \rangle \\
&= \langle b_1, b_2, b_3 \rangle + \langle a_1, a_2, a_3 \rangle \\
&= \mathbf{b} + \mathbf{a}
\end{aligned}
$$

59. $\mathbf{a} = \langle a_1, a_2, a_3 \rangle, \mathbf{b} = \langle b_1, b_2, b_3 \rangle$

$$
\begin{aligned}
d(\mathbf{a} + \mathbf{b}) &= d\left(\langle a_1, a_2, a_3 \rangle + \langle b_1, b_2, b_3 \rangle\right) \\
&= d\langle a_1 + b_1, a_2 + b_2, a_3 + b_3 \rangle \\
&= \langle d(a_1 + b_1), d(a_2 + b_2), d(a_3 + b_3) \rangle \\
&= \langle da_1 + db_1, da_2 + db_2, da_3 + db_3 \rangle \\
&= \langle da_1, da_2, da_3 \rangle + \langle db_1, db_2, db_3 \rangle \\
&= d\langle a_1, a_2, a_3 \rangle + d\langle b_1, b_2, b_3 \rangle \\
&= d\mathbf{a} + d\mathbf{b}
\end{aligned}
$$

$$
\begin{aligned}
(d + e)\mathbf{a} &= (d + e)\langle a_1, a_2, a_3 \rangle \\
&= \langle (d + e)a_1, (d + e)a_2, (d + e)a_3 \rangle \\
&= \langle da_1 + ea_1, da_2 + ea_2, da_3 + ea_3 \rangle \\
&= \langle da_1, da_2, da_3 \rangle + \langle ea_1, ea_2, ea_3 \rangle \\
&= d\langle a_1, a_2, a_3 \rangle + e\langle a_1, a_2, a_3 \rangle \\
&= d\mathbf{a} + e\mathbf{a}
\end{aligned}
$$

61. $\overrightarrow{PQ} = \langle 4 - 2, 2 - 3, 2 - 1 \rangle = \langle 2, -1, 1 \rangle$

$\overrightarrow{QR} = \langle 8 - 4, 0 - 2, 4 - 2 \rangle = \langle 4, -2, 2 \rangle$

$2\overrightarrow{PQ} = \overrightarrow{QR}$ so P, Q and R are colinear.

63. The unit vector in the northeasterly direction is $\left\langle \dfrac{1}{\sqrt{2}}, \dfrac{1}{\sqrt{2}}, 0 \right\rangle$, so the vector for the wind force is $\left\langle \dfrac{150}{\sqrt{2}}, \dfrac{150}{\sqrt{2}}, 0 \right\rangle$. Combined with the weight of the helicopter, the total force acting on the helicopter is $\left\langle \dfrac{150}{\sqrt{2}}, \dfrac{150}{\sqrt{2}}, -1000 \right\rangle$. Thus to counter the force, the force needed is $\left\langle -\dfrac{150}{\sqrt{2}}, -\dfrac{150}{\sqrt{2}}, 1000 \right\rangle$ or 1000 pounds up, $\dfrac{150}{\sqrt{2}}$ pounds west, $\dfrac{150}{\sqrt{2}}$ pounds south.

Section 10.3

5. $\mathbf{a} \cdot \mathbf{b} = (3)(2) + (1)(4) = 10$

7. $\mathbf{a} \cdot \mathbf{b} = (0)(-2) + (-2)(4) = -8$

9. $\mathbf{a} \cdot \mathbf{b} = \langle 3, 1 \rangle \cdot \langle -2, 3 \rangle = (3)(-2) + (1)(3) = -3$

11. $\mathbf{a} \cdot \mathbf{b} = (2)(0) + (-1)(2) + (3)(4) = 10$

13. $\mathbf{a} \cdot \mathbf{b} = \langle 2, 0, -1 \rangle \cdot \langle 0, 4, -1 \rangle$
    $= (2)(0) + (0)(4) + (-1)(-1)$
    $= 1$

15. $\cos\theta = \dfrac{\mathbf{a} \cdot \mathbf{b}}{\|\mathbf{a}\|\|\mathbf{b}\|} = \dfrac{1}{\sqrt{13}\sqrt{2}} = \dfrac{1}{\sqrt{26}}$
    $\theta = \cos^{-1}\dfrac{1}{\sqrt{26}} \approx 1.37$

17. $\cos\theta = \dfrac{\mathbf{a} \cdot \mathbf{b}}{\|\mathbf{a}\|\|\mathbf{b}\|} = \dfrac{-8}{\sqrt{8}\sqrt{20}} = -\dfrac{2}{\sqrt{10}}$
    $\theta = \cos^{-1}\left(-\dfrac{2}{\sqrt{10}}\right) \approx 2.26$

19. $\cos\theta = \dfrac{\mathbf{a} \cdot \mathbf{b}}{\|\mathbf{a}\|\|\mathbf{b}\|} = \dfrac{-8}{\sqrt{26}\sqrt{9}} = -\dfrac{8}{\sqrt{234}}$
    $\theta = \cos^{-1}\left(-\dfrac{8}{\sqrt{234}}\right) \approx 2.12$

21. $\mathbf{a} \cdot \mathbf{b} = (2)(2) + (-1)(4) = 0$
    The vectors are orthogonal.

23. $\mathbf{a} \cdot \mathbf{b} = (4)(2) + (-1)(4) + (1)(4) = 8$
    The vectors are not orthogonal.

25. $\mathbf{a} \cdot \mathbf{b} = (6)(-1) + (2)(3) = 0$
    The vectors are orthogonal.

27. Answers may vary. Find a nonzero vector $\langle x, y \rangle$
    such that $\langle x, y \rangle \cdot \langle 2, -1 \rangle = 0$. For example,
    $\langle x, y \rangle = \langle 1, 2 \rangle$.

29. Answers may vary. Find a nonzero vector
    $\langle x, y, z \rangle$ such that $\langle x, y, z \rangle \cdot \langle 4, -1, 1 \rangle = 0$. For
    example, $\langle x, y, z \rangle = \langle 1, 4, 0 \rangle$.

31. Answers may vary. Find a nonzero vector
    $a\mathbf{i} + b\mathbf{j} + c\mathbf{k}$ such that
    $(a\mathbf{i} + b\mathbf{j} + c\mathbf{k}) \cdot (6\mathbf{i} + 2\mathbf{j} - \mathbf{k}) = 0$. For example,
    $a\mathbf{i} + b\mathbf{j} + c\mathbf{k} = \mathbf{j} + 2\mathbf{k}$.

33. $\text{comp}_b\,\mathbf{a} = \dfrac{\mathbf{a} \cdot \mathbf{b}}{\|\mathbf{b}\|} = \dfrac{10}{\sqrt{25}} = 2$
    $\text{proj}_b\,\mathbf{a} = \left(\dfrac{\mathbf{a} \cdot \mathbf{b}}{\|\mathbf{b}\|}\right)\dfrac{\mathbf{b}}{\|\mathbf{b}\|} = 2\dfrac{\langle 3, 4 \rangle}{5} = \left\langle \dfrac{6}{5}, \dfrac{8}{5} \right\rangle$

35. $\text{comp}_b\,\mathbf{a} = \dfrac{\mathbf{a} \cdot \mathbf{b}}{\|\mathbf{b}\|} = \dfrac{9}{\sqrt{25}} = \dfrac{9}{5}$
    $\text{proj}_b\,\mathbf{a} = \left(\dfrac{\mathbf{a} \cdot \mathbf{b}}{\|\mathbf{b}\|}\right)\dfrac{\mathbf{b}}{\|\mathbf{b}\|} = \dfrac{9}{5}\dfrac{(4\mathbf{i} - 3\mathbf{j})}{5} = \dfrac{9}{25}(4\mathbf{i} - 3\mathbf{j})$

37. $\text{comp}_b\,\mathbf{a} = \dfrac{\mathbf{a} \cdot \mathbf{b}}{\|\mathbf{b}\|} = \dfrac{6}{\sqrt{9}} = 2$
    $\text{proj}_b\,\mathbf{a} = \left(\dfrac{\mathbf{a} \cdot \mathbf{b}}{\|\mathbf{b}\|}\right)\dfrac{\mathbf{b}}{\|\mathbf{b}\|} = 2\dfrac{\langle 1, 2, 2 \rangle}{3} = \dfrac{2}{3}\langle 1, 2, 2 \rangle$

39. $\text{comp}_b\,\mathbf{a} = \dfrac{\mathbf{a} \cdot \mathbf{b}}{\|\mathbf{b}\|} = \dfrac{-8}{\sqrt{25}} = -\dfrac{8}{5}$
    $\text{proj}_b\,\mathbf{a} = \left(\dfrac{\mathbf{a} \cdot \mathbf{b}}{\|\mathbf{b}\|}\right)\dfrac{\mathbf{b}}{\|\mathbf{b}\|}$
    $= -\dfrac{8}{5}\dfrac{\langle 0, -3, 4 \rangle}{5}$
    $= -\dfrac{8}{25}\langle 0, -3, 4 \rangle$

41. $\mathbf{F} = 40\left\langle \cos\dfrac{\pi}{3}, \sin\dfrac{\pi}{3} \right\rangle$
    $= 40\left\langle \dfrac{1}{2}, \dfrac{\sqrt{3}}{2} \right\rangle$
    $= \left\langle 20, 20\sqrt{3} \right\rangle$
    $W = Fd = 20(5,280) = 105,600$ foot-pounds

43. In Exercise 42, more of the puller's force is in the
    direction of motion of the wagon, so more work is
    done. Since the force stays fixed at 40 lb in the
    direction of the handle, the lower handle angle
    represents more efficient use of the force.

45. $\mathbf{v} = \langle 24-0, 10-0 \rangle = \langle 24, 10 \rangle$

$F = \text{comp}_\mathbf{v}\langle 30, 20\rangle = \dfrac{\langle 30, 20\rangle \cdot \langle 24, 10\rangle}{\|\langle 24, 10\rangle\|}$

$d = \|\mathbf{v}\| = \|\langle 24, 10\rangle\|$

$W = Fd = \langle 30, 20\rangle \cdot \langle 24, 10\rangle = 920$ foot-pounds

47. a. false; for example, let $\mathbf{a} = \langle 1, 0\rangle$, $\mathbf{b} = \langle 0, 1\rangle$, and $\mathbf{c} = \langle 0, 2\rangle$. Then $\mathbf{a}\cdot\mathbf{b}=\mathbf{a}\cdot\mathbf{c}=0$, but $\mathbf{b} \ne \mathbf{c}$.

b. true; since $\mathbf{b} = \mathbf{c}$, $\mathbf{a}\cdot\mathbf{b} = \mathbf{a}\cdot\mathbf{c}$ by substitution.

c. true; $\mathbf{a}\cdot\mathbf{a} = \|\mathbf{a}\|\|\mathbf{a}\|\cos\theta$. Since $\theta = 0°$, $\cos\theta = 1$ so $\mathbf{a}\cdot\mathbf{a} = \|\mathbf{a}\|\|\mathbf{a}\|(1) = \|\mathbf{a}\|^2$.

d. false; for example, let $\mathbf{a} = \langle 3, 1\rangle$, $\mathbf{b} = \langle 2, 2\rangle$, and $\mathbf{c} = \langle 0, 1\rangle$. Then $\|\mathbf{a}\| = \sqrt{10} > \|\mathbf{b}\| = 2\sqrt{2}$, but $\mathbf{a}\cdot\mathbf{c} = 1$ and $\mathbf{b}\cdot\mathbf{c} = 2$ so $\mathbf{a}\cdot\mathbf{c}$ w $\mathbf{b}\cdot\mathbf{c}$.

e. false; consider $\mathbf{a} = \langle 0, 1\rangle$ and $\mathbf{b} = \langle 1, 0\rangle$.

49. $|\mathbf{a}\cdot\mathbf{b}| = \|\mathbf{a}\|\|\mathbf{b}\|\cos\theta = \|\mathbf{a}\|\|\mathbf{b}\||\cos\theta|$

If $|\mathbf{a}\cdot\mathbf{b}| = \|\mathbf{a}\|\|\mathbf{b}\|$ then $\cos\theta = \pm 1$ so the vectors are parallel (go in the same or opposite direction). $\mathbf{a} = c\mathbf{b}$.

51. Apply the triangle inequality to $\mathbf{a}-\mathbf{b}$ and $\mathbf{b}$.

$\|(\mathbf{a}-\mathbf{b})+\mathbf{b}\| \le \|\mathbf{a}-\mathbf{b}\| + \|\mathbf{b}\|$

$\|\mathbf{a}\| \le \|\mathbf{a}-\mathbf{b}\| + \|\mathbf{b}\|$

$\|\mathbf{a}\| - \|\mathbf{b}\| \le \|\mathbf{a}-\mathbf{b}\|$

53. $\mathbf{a} = \left\langle 0-\frac12, 0-\frac12, 0-\frac12\right\rangle = \left\langle -\frac12, -\frac12, -\frac12\right\rangle$

$\mathbf{b} = \left\langle 1-\frac12, 1-\frac12, 0-\frac12\right\rangle = \left\langle \frac12, \frac12, -\frac12\right\rangle$

$\mathbf{c} = \left\langle 1-\frac12, 0-\frac12, 1-\frac12\right\rangle = \left\langle \frac12, -\frac12, \frac12\right\rangle$

$\mathbf{d} = \left\langle 0-\frac12, 1-\frac12, 1-\frac12\right\rangle = \left\langle -\frac12, \frac12, \frac12\right\rangle$

$\|\mathbf{a}\| = \|\mathbf{b}\| = \|\mathbf{c}\| = \|\mathbf{d}\| = \dfrac{\sqrt{3}}{2}$

$\mathbf{a}\cdot\mathbf{b} = \mathbf{a}\cdot\mathbf{c} = \mathbf{a}\cdot\mathbf{d} = \mathbf{b}\cdot\mathbf{c} = \mathbf{b}\cdot\mathbf{d} = \mathbf{c}\cdot\mathbf{d} = -\frac14$

$\cos\theta = \dfrac{-\frac14}{\left(\frac{\sqrt3}{2}\right)\left(\frac{\sqrt3}{2}\right)} = -\frac13$

$\theta = \cos^{-1}\left(-\frac13\right) \approx 109.5°$ S

55. $\text{comp}_\mathbf{c}(\mathbf{a}+\mathbf{b}) = \dfrac{(\mathbf{a}+\mathbf{b})\cdot\mathbf{c}}{\|\mathbf{c}\|}$

$= \dfrac{\mathbf{a}\cdot\mathbf{b} + \mathbf{a}\cdot\mathbf{c}}{\|\mathbf{c}\|}$

$= \dfrac{\mathbf{a}\cdot\mathbf{b}}{\|\mathbf{c}\|} + \dfrac{\mathbf{a}\cdot\mathbf{c}}{\|\mathbf{c}\|}$

$= \text{comp}_\mathbf{c}\,\mathbf{a} + \text{comp}_\mathbf{c}\,\mathbf{b}$

57. Let $\mathbf{b}$ be the beam direction and $\mathbf{w}$ the wave direction. The component of the wave's force along the beam is

$\dfrac{\mathbf{b}\cdot\mathbf{w}}{|\mathbf{b}|} = \dfrac{\langle 10, 1, 5\rangle \cdot \langle 0, -200, 0\rangle\,\text{N}}{|\langle 10, 1, 5\rangle|} = -17.817\,\text{N}$

59. Let $\mathbf{b}$ be the vector representing the bank and $\mathbf{w}$ the vector representing the weight of the car (directed straight down). The component of the weight along the bank vector is

$\dfrac{\mathbf{b}\cdot\mathbf{w}}{|\mathbf{b}|} = \dfrac{\langle\cos 10°, \sin 10°\rangle \cdot \langle 0, -2000\rangle\,\text{lb}}{|\langle\cos 10°, \sin 10°\rangle|}$

$= \sin 10° \cdot (-2000\,\text{lb}) = -347.296\,\text{lb}$

The force is toward the inside of the curve.

61. $\mathbf{v}\cdot\mathbf{n} = \langle\cos\theta, \sin\theta\rangle \cdot \langle\sin\theta, -\cos\theta\rangle$

$= \cos\theta\sin\theta - \cos\theta\sin\theta$

$= 0$

The component of $\mathbf{w}$ along $\mathbf{v}$ is $-w\sin\theta$.
The component of $\mathbf{w}$ along $\mathbf{n}$ is $w\cos\theta$.

63. The angle is $\cos^{-1}\left(\dfrac{\mathbf{a}\cdot\mathbf{b}}{|\mathbf{a}|\cdot|\mathbf{b}|}\right)$.

$\cos^{-1}\left(\dfrac{50,000}{\sqrt{50,000}\cdot\sqrt{100,000}}\right) = \cos^{-1}\left(\dfrac{1}{\sqrt2}\right) = 45°$.

65. $\mathbf{s}\cdot\mathbf{p} = \langle 3000, 2000, 4000\rangle \cdot \langle 20, 15, 25\rangle$

$= (3000)(20) + (2000)(15) + (4000)(25)$

$= 190,000$

The monthly revenue is $190,000, which is given by $\mathbf{s}\cdot\mathbf{p}$.

Section 10.4

5. $\begin{vmatrix} 2 & 0 & -1 \\ 1 & 1 & 0 \\ -2 & -1 & 1 \end{vmatrix}$

$= (2)\begin{vmatrix} 1 & 0 \\ -1 & 1 \end{vmatrix} - (0)\begin{vmatrix} 1 & 0 \\ -2 & 1 \end{vmatrix} + (-1)\begin{vmatrix} 1 & 1 \\ -2 & -1 \end{vmatrix}$

$= (2)(1) - (0)(1) + (-1)(1)$

$= 1$

7. $\begin{vmatrix} 2 & 3 & -1 \\ 0 & 1 & 0 \\ -2 & -1 & 3 \end{vmatrix}$

$= (2)\begin{vmatrix} 1 & 0 \\ -1 & 3 \end{vmatrix} - (3)\begin{vmatrix} 0 & 0 \\ -2 & 3 \end{vmatrix} + (-1)\begin{vmatrix} 0 & 1 \\ -2 & -1 \end{vmatrix}$

$= (2)(3) - 3(0) + (-1)(2)$

$= 4$

9. $\langle 1, 2, -1 \rangle \times \langle 1, 0, 2 \rangle$

$= \begin{vmatrix} \mathbf{i} & \mathbf{j} & \mathbf{k} \\ 1 & 2 & -1 \\ 1 & 0 & 2 \end{vmatrix}$

$= \begin{vmatrix} 2 & -1 \\ 0 & 2 \end{vmatrix}\mathbf{i} - \begin{vmatrix} 1 & -1 \\ 1 & 2 \end{vmatrix}\mathbf{j} + \begin{vmatrix} 1 & 2 \\ 1 & 0 \end{vmatrix}\mathbf{k}$

$= 4\mathbf{i} - 3\mathbf{j} - 2\mathbf{k}$

$= \langle 4, -3, -2 \rangle$

11. $\langle 0, 1, 4 \rangle \times \langle -1, 2, -1 \rangle$

$= \begin{vmatrix} \mathbf{i} & \mathbf{j} & \mathbf{k} \\ 0 & 1 & 4 \\ -1 & 2 & -1 \end{vmatrix}$

$= \begin{vmatrix} 1 & 4 \\ 2 & -1 \end{vmatrix}\mathbf{i} - \begin{vmatrix} 0 & 4 \\ -1 & -1 \end{vmatrix}\mathbf{j} + \begin{vmatrix} 0 & 1 \\ -1 & 2 \end{vmatrix}\mathbf{k}$

$= -9\mathbf{i} - 4\mathbf{j} + \mathbf{k}$

$= \langle -9, -4, 1 \rangle$

13. $\langle -2, -1, 4 \rangle \times \langle 1, 0, 0 \rangle$

$= \begin{vmatrix} \mathbf{i} & \mathbf{j} & \mathbf{k} \\ -2 & -1 & 4 \\ 1 & 0 & 0 \end{vmatrix}$

$= \begin{vmatrix} -1 & 4 \\ 0 & 0 \end{vmatrix}\mathbf{i} - \begin{vmatrix} -2 & 4 \\ 1 & 0 \end{vmatrix}\mathbf{j} + \begin{vmatrix} -2 & -1 \\ 1 & 0 \end{vmatrix}\mathbf{k}$

$= 4\mathbf{j} + \mathbf{k}$

$= \langle 0, 4, 1 \rangle$

15. $(2\mathbf{i} - \mathbf{k}) \times (4\mathbf{j} + \mathbf{k})$

$= \langle 2, 0, -1 \rangle \times \langle 0, 4, 1 \rangle$

$= \begin{vmatrix} \mathbf{i} & \mathbf{j} & \mathbf{k} \\ 2 & 0 & -1 \\ 0 & 4 & 1 \end{vmatrix}$

$= \begin{vmatrix} 0 & -1 \\ 4 & 1 \end{vmatrix}\mathbf{i} - \begin{vmatrix} 2 & -1 \\ 0 & 1 \end{vmatrix}\mathbf{j} + \begin{vmatrix} 2 & 0 \\ 0 & 4 \end{vmatrix}\mathbf{k}$

$= 4\mathbf{i} - 2\mathbf{j} + 8\mathbf{k}$

$= \langle 4, -2, 8 \rangle$

17. $\langle 1, 0, 4 \rangle \times \langle 1, -4, 2 \rangle$

$= \begin{vmatrix} \mathbf{i} & \mathbf{j} & \mathbf{k} \\ 1 & 0 & 4 \\ 1 & -4 & 2 \end{vmatrix}$

$= \begin{vmatrix} 0 & 4 \\ -4 & 2 \end{vmatrix}\mathbf{i} - \begin{vmatrix} 1 & 4 \\ 1 & 2 \end{vmatrix}\mathbf{j} + \begin{vmatrix} 1 & 0 \\ 1 & -4 \end{vmatrix}\mathbf{k}$

$= 16\mathbf{i} + 2\mathbf{j} - 4\mathbf{k}$

$= \langle 16, 2, -4 \rangle$

$\|\langle 16, 2, -4 \rangle\| = \sqrt{16^2 + 2^2 + (-4)^2}$

$= \sqrt{276}$

$= 2\sqrt{69}$

$\pm \dfrac{1}{2\sqrt{69}}\langle 16, 2, -4 \rangle = \pm \dfrac{1}{\sqrt{69}}\langle 8, 1, -2 \rangle$

19. $\langle 2, -1, 0 \rangle \times \langle 1, 0, 3 \rangle$

$= \begin{vmatrix} \mathbf{i} & \mathbf{j} & \mathbf{k} \\ 2 & -1 & 0 \\ 1 & 0 & 3 \end{vmatrix}$

$= \begin{vmatrix} -1 & 0 \\ 0 & 3 \end{vmatrix}\mathbf{i} - \begin{vmatrix} 2 & 0 \\ 1 & 3 \end{vmatrix}\mathbf{j} + \begin{vmatrix} 2 & -1 \\ 1 & 0 \end{vmatrix}\mathbf{k}$

$= -3\mathbf{i} - 6\mathbf{j} + \mathbf{k}$

$= \langle -3, -6, 1 \rangle$

$\|\langle -3, -6, 1 \rangle\| = \sqrt{(-3)^2 + (-6)^2 + 1^2} = \sqrt{46}$

$\pm \dfrac{1}{\sqrt{46}}\langle -3, -6, 1 \rangle$

21. $(3\mathbf{i} - \mathbf{j}) \times (4\mathbf{j} + \mathbf{k})$

$= \langle 3, -1, 0 \rangle \times \langle 0, 4, 1 \rangle$

$= \begin{vmatrix} \mathbf{i} & \mathbf{j} & \mathbf{k} \\ 3 & -1 & 0 \\ 0 & 4 & 1 \end{vmatrix}$

$= \begin{vmatrix} -1 & 0 \\ 4 & 1 \end{vmatrix} \mathbf{i} - \begin{vmatrix} 3 & 0 \\ 0 & 1 \end{vmatrix} \mathbf{j} + \begin{vmatrix} 3 & -1 \\ 0 & 4 \end{vmatrix} \mathbf{k}$

$= -\mathbf{i} - 3\mathbf{j} + 12\mathbf{k} = \langle -1, -3, 12 \rangle$

$\| \langle -1, -3, 12 \rangle \| = \sqrt{(-1)^2 + (-3)^2 + 12^2} = \sqrt{154}$

$\pm \dfrac{1}{\sqrt{154}} \langle -1, -3, 12 \rangle$

23. $\mathbf{a} \times \mathbf{b} = \begin{vmatrix} \mathbf{i} & \mathbf{j} & \mathbf{k} \\ 1 & 0 & 4 \\ 2 & 0 & 1 \end{vmatrix}$

$= \begin{vmatrix} 0 & 4 \\ 0 & 1 \end{vmatrix} \mathbf{i} - \begin{vmatrix} 1 & 4 \\ 2 & 1 \end{vmatrix} \mathbf{j} + \begin{vmatrix} 1 & 0 \\ 2 & 0 \end{vmatrix} \mathbf{k}$

$= 7\mathbf{j} = \langle 0, 7, 0 \rangle$

$\| \mathbf{a} \times \mathbf{b} \| = 7, \; \| \mathbf{a} \| = \sqrt{17}, \; \| \mathbf{b} \| = \sqrt{5}$

$\sin \theta = \dfrac{\| \mathbf{a} \times \mathbf{b} \|}{\| \mathbf{a} \| \| \mathbf{b} \|} = \dfrac{7}{\sqrt{17}\sqrt{5}} = \dfrac{7}{\sqrt{85}}$

$\theta = \sin^{-1} \dfrac{7}{\sqrt{85}} \approx .86$

25. $\mathbf{a} \times \mathbf{b} = \begin{vmatrix} \mathbf{i} & \mathbf{j} & \mathbf{k} \\ 3 & 0 & 1 \\ 0 & 4 & 1 \end{vmatrix}$

$= \begin{vmatrix} 0 & 1 \\ 4 & 1 \end{vmatrix} \mathbf{i} - \begin{vmatrix} 3 & 1 \\ 0 & 1 \end{vmatrix} \mathbf{j} + \begin{vmatrix} 3 & 0 \\ 0 & 4 \end{vmatrix} \mathbf{k}$

$= -4\mathbf{i} - 3\mathbf{j} + 12\mathbf{k}$

$= \langle -4, -3, 12 \rangle$

$\| \mathbf{a} \times \mathbf{b} \| = 13, \; \| \mathbf{a} \| = \sqrt{10}, \; \| \mathbf{b} \| = \sqrt{17}$

$\sin \theta = \dfrac{\| \mathbf{a} \times \mathbf{b} \|}{\| \mathbf{a} \| \| \mathbf{b} \|} = \dfrac{13}{\sqrt{10}\sqrt{17}} = \dfrac{13}{\sqrt{170}}$

$\theta = \sin^{-1} \dfrac{13}{\sqrt{170}} \approx 1.49$

27. $P = (0, 1, 2), \; R = (3, 1, 1)$

$\overrightarrow{PQ} = \langle 1, 1, -2 \rangle, \; \overrightarrow{PR} = \langle 3, 0, -1 \rangle$

$\overrightarrow{PQ} \times \overrightarrow{PR} = \begin{vmatrix} \mathbf{i} & \mathbf{j} & \mathbf{k} \\ 1 & 1 & -2 \\ 3 & 0 & -1 \end{vmatrix}$

$= \begin{vmatrix} 1 & -2 \\ 0 & -1 \end{vmatrix} \mathbf{i} - \begin{vmatrix} 1 & -2 \\ 3 & -1 \end{vmatrix} \mathbf{j} + \begin{vmatrix} 1 & 1 \\ 3 & 0 \end{vmatrix} \mathbf{k}$

$= -\mathbf{i} - 5\mathbf{j} - 3\mathbf{k} = \langle -1, -5, -3 \rangle$

$d = \dfrac{\| \overrightarrow{PQ} \times \overrightarrow{PR} \|}{\| \overrightarrow{PR} \|}$

$= \dfrac{\| \langle -1, -5, -3 \rangle \|}{\| \langle 3, 0, -1 \rangle \|} = \dfrac{\sqrt{35}}{\sqrt{10}} = \sqrt{\dfrac{7}{2}} \approx 1.87$

29. $P = (2, 1, -1), \; R = (1, 1, 1)$

$\overrightarrow{PQ} = \langle 1, -3, 2 \rangle, \; \overrightarrow{PR} = \langle -1, 0, 2 \rangle$

$\overrightarrow{PQ} \times \overrightarrow{PR} = \begin{vmatrix} \mathbf{i} & \mathbf{j} & \mathbf{k} \\ 1 & -3 & 2 \\ -1 & 0 & 2 \end{vmatrix}$

$= \begin{vmatrix} -3 & 2 \\ 0 & 2 \end{vmatrix} \mathbf{i} - \begin{vmatrix} 1 & 2 \\ -1 & 2 \end{vmatrix} \mathbf{j} + \begin{vmatrix} 1 & -3 \\ -1 & 0 \end{vmatrix} \mathbf{k}$

$= -6\mathbf{i} - 4\mathbf{j} - 3\mathbf{k}$

$= \langle -6, -4, -3 \rangle$

$d = \dfrac{\| \overrightarrow{PQ} \times \overrightarrow{PR} \|}{\| \overrightarrow{PR} \|}$

$= \dfrac{\| \langle -6, -4, -3 \rangle \|}{\| \langle -1, 0, 2 \rangle \|}$

$= \dfrac{\sqrt{61}}{\sqrt{5}} = \sqrt{\dfrac{61}{5}} \approx 3.49$

31. $\| \tau \| = \| \mathbf{F} \| \| \mathbf{r} \| \sin \theta$

$= 20 \left(\dfrac{8}{12} \right) \sin \dfrac{\pi}{4}$

$= \dfrac{20\sqrt{2}}{3}$

≈ 9.4 foot-pounds

33. $\| \tau \| = \| \mathbf{F} \| \| \mathbf{r} \| \sin \theta$

$= 30 \left(\dfrac{8}{12} \right) \sin \dfrac{\pi}{6}$

$= 10$ foot-pounds

35. up

37. left

39. down, left

41. down

43. The ball rises.

45. The ball drops to the left.

47. The ball drops.

49. There is no effect.

51. The ball rises.

53. The ball rises (lands softly).

55. The ball rises to the right.

57. False. Let $\mathbf{a} = (1, 0, 0)$, $\mathbf{b} = (1, 0, 0)$, and
$\mathbf{c} = (2, 0, 0)$. Then $\mathbf{a} \times \mathbf{b} = \mathbf{a} \times \mathbf{c} = \mathbf{0}$ but $\mathbf{b} \neq \mathbf{c}$.

59. False. If $\mathbf{a} \neq \mathbf{0}$, $\|\mathbf{a}\|^2 \neq 0$ but $\mathbf{a} \times \mathbf{a} = \mathbf{0}$.

61. True. New torque $= \|2\mathbf{F}\|\|\mathbf{r}\|\sin\theta$
$$= 2\left(\|\mathbf{F}\|\|\mathbf{r}\|\sin\theta\right)$$
$$= 2 \times \text{original torque}$$

63. Consider the vectors to be in V_3.

$$\langle 2, 3, 0\rangle \times \langle 1, 4, 0\rangle = \begin{vmatrix} \mathbf{i} & \mathbf{j} & \mathbf{k} \\ 2 & 3 & 0 \\ 1 & 4 & 0 \end{vmatrix}$$

$$= \begin{vmatrix} 3 & 0 \\ 4 & 0 \end{vmatrix}\mathbf{i} - \begin{vmatrix} 2 & 0 \\ 1 & 0 \end{vmatrix}\mathbf{j} + \begin{vmatrix} 2 & 3 \\ 1 & 4 \end{vmatrix}\mathbf{k}$$

$$= 5\mathbf{k}$$
$$= \langle 0, 0, 5\rangle$$
$$\text{area} = \|\langle 0, 0, 5\rangle\| = 5$$

65. $\mathbf{a} = \langle 2, 3, -1\rangle$, $\mathbf{b} = \langle 3, -1, 4\rangle$

$$\mathbf{a} \times \mathbf{b} = \begin{vmatrix} \mathbf{i} & \mathbf{j} & \mathbf{k} \\ 2 & 3 & -1 \\ 3 & -1 & 4 \end{vmatrix}$$

$$= \begin{vmatrix} 3 & -1 \\ -1 & 4 \end{vmatrix}\mathbf{i} - \begin{vmatrix} 2 & -1 \\ 3 & 4 \end{vmatrix}\mathbf{j} + \begin{vmatrix} 2 & 3 \\ 3 & -1 \end{vmatrix}\mathbf{k}$$

$$= 11\mathbf{i} - 11\mathbf{j} - 11\mathbf{k}$$
$$= \langle 11, -11, -11\rangle$$

$$\text{area} = \frac{1}{2}\|\mathbf{a} \times \mathbf{b}\|$$

$$= \frac{1}{2}\sqrt{11^2 + (-11)^2 + (-11)^2}$$

$$= \frac{11\sqrt{3}}{2}$$

67. $\mathbf{a} = \langle 2, 1, 0\rangle$, $\mathbf{b} = \langle -1, 2, 0\rangle$, $\mathbf{c} = \langle 1, 1, 2\rangle$

$$\mathbf{c} \cdot (\mathbf{a} \times \mathbf{b}) = \begin{vmatrix} 1 & 1 & 2 \\ 2 & 1 & 0 \\ -1 & 2 & 0 \end{vmatrix}$$

$$= 1\begin{vmatrix} 1 & 0 \\ 2 & 0 \end{vmatrix} - 1\begin{vmatrix} 2 & 0 \\ -1 & 0 \end{vmatrix} + 2\begin{vmatrix} 2 & 1 \\ -1 & 2 \end{vmatrix}$$

$$= 1(0) - 1(0) + 2(5)$$
$$= 10$$

69. $\mathbf{i} \times (\mathbf{j} \times \mathbf{k}) = \mathbf{i} \times \mathbf{i} = \mathbf{0}$

71. $\mathbf{j} \times (\mathbf{j} \times \mathbf{i}) = \mathbf{j} \times (-\mathbf{k}) = -\mathbf{j} \times \mathbf{k} = -\mathbf{i}$

73. $\mathbf{i} \times (3\mathbf{k}) = 3(\mathbf{i} \times \mathbf{k}) = 3(-\mathbf{j}) = -3\mathbf{j}$

75. $\mathbf{a} = \langle 2, 3, 1\rangle$, $\mathbf{b} = \langle 1, 0, 2\rangle$, $\mathbf{c} = \langle 0, 3, -3\rangle$

$$\mathbf{c} \cdot (\mathbf{a} \times \mathbf{b}) = \begin{vmatrix} 0 & 3 & -3 \\ 2 & 3 & 1 \\ 1 & 0 & 2 \end{vmatrix}$$

$$= 0\begin{vmatrix} 3 & 1 \\ 0 & 2 \end{vmatrix} - 3\begin{vmatrix} 2 & 1 \\ 1 & 2 \end{vmatrix} + (-3)\begin{vmatrix} 2 & 3 \\ 1 & 0 \end{vmatrix}$$

$$= 0(6) - 3(3) + (-3)(-3)$$
$$= 0$$

Vectors are coplanar.

77. $\mathbf{a} = \langle 1, 0, -2\rangle$, $\mathbf{b} = \langle 3, 0, 1\rangle$, $\mathbf{c} = \langle 2, 1, 0\rangle$

$$\mathbf{c} \cdot (\mathbf{a} \times \mathbf{b}) = \begin{vmatrix} 2 & 1 & 0 \\ 1 & 0 & -2 \\ 3 & 0 & 1 \end{vmatrix}$$

$$= 2\begin{vmatrix} 0 & -2 \\ 0 & 1 \end{vmatrix} - 1\begin{vmatrix} 1 & -2 \\ 3 & 1 \end{vmatrix} + 0\begin{vmatrix} 1 & 0 \\ 3 & 0 \end{vmatrix}$$

$$= 2(0) - 1(7) + 0(0)$$
$$= -7$$

Vectors are not coplanar.

Section 10.5

5. a. $x - 1 = 2t,\ y - 2 = -t,\ z + 3 = 4t$
or $x = 1 + 2t,\ y = 2 - t,\ z = -3 + 4t$

b. $\dfrac{x-1}{2} = \dfrac{y-2}{-1} = \dfrac{z+3}{4}$

7. a. $\mathbf{a} = \langle 4 - 2,\ 0 - 1,\ 4 - 3 \rangle = \langle 2,\ -1,\ 1 \rangle$
$x - 2 = 2t,\ y - 1 = -t,\ z - 3 = t$
or $x = 2 + 2t,\ y = 1 - t,\ z - 3 = t$

b. $\dfrac{x-2}{2} = \dfrac{y-1}{-1} = \dfrac{z-3}{1}$

9. a. $\mathbf{a} = \langle -3,\ 0,\ 1 \rangle$
$x - 1 = -3t,\ y - 4 = 0,\ z - 1 = t$
or $x = 1 - 3t,\ y = 4,\ z = 1 + t$

b. $\dfrac{x-1}{-3} = \dfrac{z-1}{1},\ y = 4$

11. a. $\mathbf{a} = \langle 3,\ -4,\ 2 \rangle$
$x - 3 = 3t,\ y - 1 = -4t,\ z + 1 = 2t$
or $x = 3 + 3t,\ y = 1 - 4t,\ z = -1 + 2t$

b. $\dfrac{x-3}{3} = \dfrac{y-1}{-4} = \dfrac{z+1}{2}$

13. a. $\mathbf{a} = \langle 1,\ 0,\ 2 \rangle \times \langle 0,\ 2,\ 1 \rangle$

$= \begin{vmatrix} \mathbf{i} & \mathbf{j} & \mathbf{k} \\ 1 & 0 & 2 \\ 0 & 2 & 1 \end{vmatrix}$

$= \begin{vmatrix} 0 & 2 \\ 2 & 1 \end{vmatrix} \mathbf{i} - \begin{vmatrix} 1 & 2 \\ 0 & 1 \end{vmatrix} \mathbf{j} + \begin{vmatrix} 1 & 0 \\ 0 & 2 \end{vmatrix} \mathbf{k}$

$= -4\mathbf{i} - \mathbf{j} + 2\mathbf{k}$
$= \langle -4,\ -1,\ 2 \rangle$
$x - 2 = -4t,\ y = -t,\ z - 1 = 2t$
or $x = 2 - 4t,\ y = -t,\ z = 1 + 2t$

b. $\dfrac{x-2}{-4} = \dfrac{y}{-1} = \dfrac{z-1}{2}$

15. a. $\mathbf{a} = \langle 2,\ -1,\ 3 \rangle$
$x - 1 = 2t,\ y - 2 = -t,\ z + 1 = 3t$
or $x = 1 + 2t,\ y = 2 - t,\ z = -1 + 3t$

b. $\dfrac{x-1}{2} = \dfrac{y-2}{-1} = \dfrac{z+1}{3}$

17. $\mathbf{a}_1 = \langle -3,\ 4,\ 1 \rangle,\ \mathbf{a}_2 = \langle 2,\ -2,\ 1 \rangle$
$\mathbf{a}_1 \neq c\mathbf{a}_2$ so the lines are not parallel.
$\mathbf{a}_1 \cdot \mathbf{a}_2 = (-3)(2) + (4)(-2) + (1)(1) = -13$
$\|\mathbf{a}_1\| = \sqrt{26},\ \|\mathbf{a}_2\| = 3$

$\cos\theta = \dfrac{\mathbf{a}_1 \cdot \mathbf{a}_2}{\|\mathbf{a}_1\|\|\mathbf{a}_2\|} = \dfrac{-13}{3\sqrt{26}}$

$\theta = \cos^{-1}\dfrac{-13}{3\sqrt{26}} \approx 2.59$

19. $\mathbf{a}_1 = \langle 2,\ 0,\ 1 \rangle,\ \mathbf{a}_2 = \langle -1,\ 5,\ 2 \rangle$
$\mathbf{a}_1 \neq c\mathbf{a}_1$ so the lines are not parallel.
$\mathbf{a}_1 \cdot \mathbf{a}_2 = (2)(-1) + (0)(5) + (1)(2) = 0$
The lines are perpendicular.

21. $\mathbf{a}_1 = \langle 2,\ 4,\ -6 \rangle,\ \mathbf{a}_2 = \langle -1,\ -2,\ 3 \rangle$
$\mathbf{a}_1 = (-2)\mathbf{a}_2$ so the lines are parallel.

23. $\mathbf{a}_1 = \langle 1,\ 0,\ 2 \rangle,\ \mathbf{a}_2 = \langle 2,\ 2,\ 4 \rangle$
$\mathbf{a}_1 \neq c\mathbf{a}_2$ so the lines are not parallel.
Setting the y-coordinates equal, we get $2 = 2s$ or $s = 1$. Setting the x-coordinates equal and setting $s = 1$, we get $4 + t = 2 + 2$ or $2 + 2(1) = 4$. Solving for t yields $t = 0$. Setting the z-values equal, we get $3 + 2t = -1 + 4s$. This is satisfied when $t = 0$ and $s = 1$. Thus, the two lines intersect when $t = 0$ and $s = 1$.

25. $\mathbf{a}_1 = \langle 2,\ 0,\ -4 \rangle,\ \mathbf{a}_2 = \langle -1,\ 0,\ 2 \rangle$
$\mathbf{a}_1 = -2\mathbf{a}_2$ so the lines are parallel.

27. $2(x - 1) - (y - 3) + 5(z - 2) = 0$

29. $-3(x + 2) + 2z = 0$

31. $P = (2, 0, 3),\ Q = (1, 1, 0),\ R = (3, 2, -1)$
$\overrightarrow{PQ} = \langle -1,\ 1,\ -3 \rangle,\ \overrightarrow{PR} = \langle 1,\ 2,\ -4 \rangle$
$\mathbf{a} = \overrightarrow{PQ} \times \overrightarrow{PR}$

$= \begin{vmatrix} \mathbf{i} & \mathbf{j} & \mathbf{k} \\ -1 & 1 & -3 \\ 1 & 2 & -4 \end{vmatrix}$

$= \begin{vmatrix} 1 & -3 \\ 2 & -4 \end{vmatrix} \mathbf{i} - \begin{vmatrix} -1 & -3 \\ 1 & -4 \end{vmatrix} \mathbf{j} + \begin{vmatrix} -1 & 1 \\ 1 & 2 \end{vmatrix} \mathbf{k}$

$= 2\mathbf{i} - 7\mathbf{j} - 3\mathbf{k}$
$= \langle 2,\ -7,\ -3 \rangle$
$2(x - 2) - 7y - 3(z - 3) = 0$

33. $P = (-2, 2, 0)$, $Q = (-2, 3, 2)$, $R = (1, 2, 2)$

$\overrightarrow{PQ} = \langle 0, 1, 2 \rangle$, $\overrightarrow{PR} = \langle 3, 0, 2 \rangle$

$\mathbf{a} = \overrightarrow{PQ} \times \overrightarrow{PR}$

$= \begin{vmatrix} \mathbf{i} & \mathbf{j} & \mathbf{k} \\ 0 & 1 & 2 \\ 3 & 0 & 2 \end{vmatrix}$

$= \begin{vmatrix} 1 & 2 \\ 0 & 2 \end{vmatrix} \mathbf{i} - \begin{vmatrix} 0 & 2 \\ 3 & 2 \end{vmatrix} \mathbf{j} + \begin{vmatrix} 0 & 1 \\ 3 & 0 \end{vmatrix} \mathbf{k}$

$= 2\mathbf{i} + 6\mathbf{j} - 3\mathbf{k}$

$= \langle 2, 6, -3 \rangle$

$2(x+2) + 6(y-2) - 3z = 0$

35. $\mathbf{a} = \langle 3, -1, 2 \rangle$

$3(x-2) - (y-1) + 2(z+1) = 0$

37. $\mathbf{a} = \langle -2, 4, 0 \rangle$

$-2x + 4(y+2) = 0$

39. $\mathbf{a} = \langle 1, 1, 0 \rangle \times \langle 2, 1, -1 \rangle$

$= \begin{vmatrix} \mathbf{i} & \mathbf{j} & \mathbf{k} \\ 1 & 1 & 0 \\ 2 & 1 & -1 \end{vmatrix}$

$= \begin{vmatrix} 1 & 0 \\ 1 & -1 \end{vmatrix} \mathbf{i} - \begin{vmatrix} 1 & 0 \\ 2 & -1 \end{vmatrix} \mathbf{j} + \begin{vmatrix} 1 & 1 \\ 2 & 1 \end{vmatrix} \mathbf{k}$

$= -\mathbf{i} + \mathbf{j} - \mathbf{k}$

$= \langle -1, 1, -1 \rangle$

$-(x-1) + (y-2) - (z-1) = 0$

or $(x-1) - (y-2) + (z-1) = 0$

41.

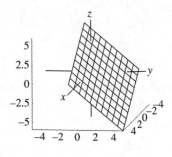

43.

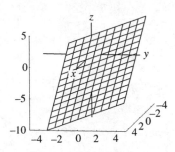

45.

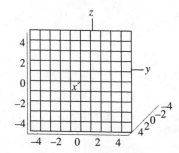

47.

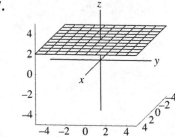

49.

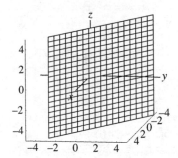

51.

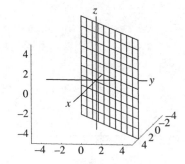

53. Solving both equations for z, we get
$z = -4 + 2x - y$ and $z = -3x + 2y$.

Setting these expressions for z equal, we get
$-4 + 2x - y = -3x + 2y$

Solving for y, we get $3y = 5x - 4$ or $y = \dfrac{5}{3}x - \dfrac{4}{3}$.

Solve for z in terms of x, we get
$$z = -3x + 2\left(\dfrac{5}{3}x - \dfrac{4}{3}\right) = \dfrac{1}{3}x - \dfrac{8}{3}.$$

Let $x = t$.

$x = t,\ y = \dfrac{5}{3}t - \dfrac{4}{3},\ z = \dfrac{1}{3}t - \dfrac{8}{3}$

55. Solving both equations for x, we get $x = \dfrac{1}{3} - \dfrac{4}{3}y$

and $x = 3 - y + z$.

Setting these expressions for x equal, we get
$\dfrac{1}{3} - \dfrac{4}{3}y = 3 - y + z$.

Solving for y, we get $\dfrac{1}{3}y = -z - \dfrac{8}{3}$ or

$y = -3z - 8$.

Solving for x in terms of z, we get
$x = 3 - (-3z - 8) + z = 4z + 11$.

Let $z = t$.

$x = 4t + 11,\ y = -3t - 8,\ z = t$

57. $\dfrac{\left|(2)(2) + (-1)(0) + (2)(1) + (-4)\right|}{\sqrt{2^2 + (-1)^2 + 2^2}} = \dfrac{2}{3}$

59. $\dfrac{\left|(1)(2) + (-1)(-1) + (1)(-1) + (-4)\right|}{\sqrt{1^2 + (-1)^2 + 1^2}} = \dfrac{2}{\sqrt{3}}$

61. $(2, 0, 0)$ is a point on the second plane.
$\dfrac{\left|(2)(2) + (-1)(0) + (-1)(0) + (-1)\right|}{\sqrt{2^2 + (-1)^2 + (-1)^2}} = \dfrac{3}{\sqrt{6}}$

63. Let (x_0, y_0, z_0) be a point on the second plane.
Thus $ax_0 + by_0 + cz_0 = d_2$. Using (5.8), we get
the distance between planes to be
$$\dfrac{\left|ax_0 + by_0 + cz_0 + (-d_1)\right|}{\sqrt{a^2 + b^2 + c^2}} = \dfrac{\left|d_2 - d_1\right|}{\sqrt{a^2 + b^2 + c^2}}.$$

65. Note that the lines intersect at $(4, 2, 3)$ when
$t = 0$ and $s = 1$.
$\mathbf{a} = \langle 1, 0, 2\rangle \times \langle 2, 2, 4\rangle$

$= \begin{vmatrix} \mathbf{i} & \mathbf{j} & \mathbf{k} \\ 1 & 0 & 2 \\ 2 & 2 & 4 \end{vmatrix}$

$= \begin{vmatrix} 0 & 2 \\ 2 & 4 \end{vmatrix}\mathbf{i} - \begin{vmatrix} 1 & 2 \\ 2 & 4 \end{vmatrix}\mathbf{j} + \begin{vmatrix} 1 & 0 \\ 2 & 2 \end{vmatrix}\mathbf{k}$

$= -4\mathbf{i} + 2\mathbf{k}$

$= \langle -4, 0, 2\rangle$

$-4(x - 4) + 2(z - 3) = 0$

67. The airplanes fly in straight lines.
Setting the x-coordinates equal, we get
$3 = 1 + 2s$. Solving for s, we get $s = 1$.
Setting the y-coordinates equal and setting
$s = 1$, we get $6 - 2t = 3 + 1$ or $t = 1$.
Setting the z-coordinates equal, we get
$3t + 1 = 2 + 2s$ which is satisfied when $t = 1$ and s
$= 1$.
Thus, the paths intersect. The planes collide
assuming $t = 1$ when $s = 1$.

Section 10.6

5.

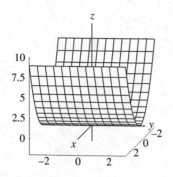

cylinder

7.

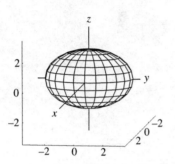

ellipsoid

9.

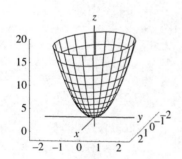

circular paraboloid

11.

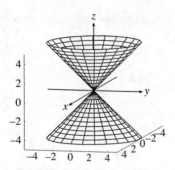

elliptic cone

13.

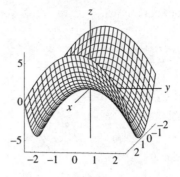

hyperbolic paraboloid

15.

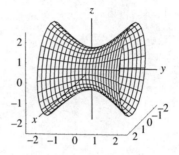

hyperboloid of one sheet

17.

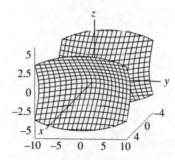

hyperboloid of two sheets

19.

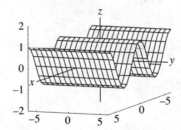

cylinder

21.

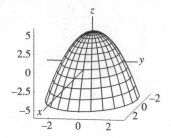

circular paraboloid

23.

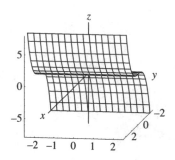

cylinder

25.

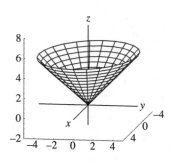

circular cone

27.

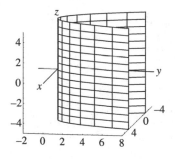

cylinder

29.

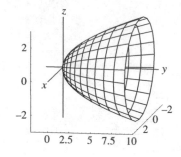

circular paraboloid

31.

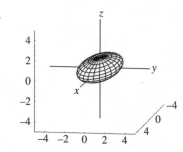

ellipsoid

33.

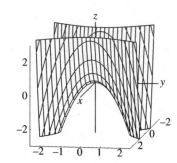

hyperbolic paraboloid

35.

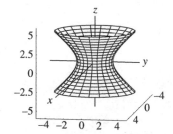

hyperboloid of one sheet

37.

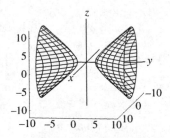

hyperboloid of two sheets

39.

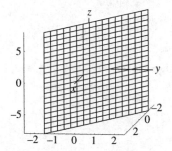

cylinder

41.

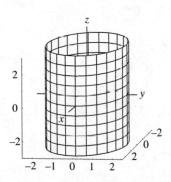

cylinder

43.

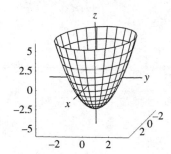

circular paraboloid

45.

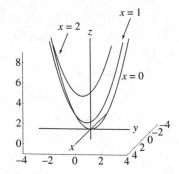

47.

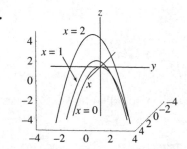

49. The traces in the planes $y = c$ are $z = x^2 - c^2$ which are parabolas that open up. In general, parabolas with the equation $z = x^2 - c^2$ have a minimum at $(0, -c^2)$ in the xz-plane. This can be shown by using calculus. The plane $x = 0$ in the picture would be halfway through the Saddle Dome, in the left-right direction going into the picture. When $x = 0$, $z = -y^2$ which has a maximum at $(0, 0)$. Thus the primary points of drainage are $(0, 1, -1)$ and $(0, -1, -1)$.

51. Substituting into the left side of the equation for the ellipsoid, we get

$$\frac{a^2 \sin^2 s \cos^2 t}{a^2} + \frac{b^2 \sin^2 s \sin^2 t}{b^2} + \frac{c^2 \cos^2 s}{c^2}$$

$$= \sin^2 s \left(\cos^2 t + \sin^2 t \right) + \cos^2 s$$

$$= \sin^2 s + \cos^2 s = 1$$

Since the coordinates satisfy the equation, the point lies on the elllipsoid.

53. Substituting into the equation for the cone, we get s^2 on the left side, and $s^2 \cos^2 t + s^2 \sin^2 t$ or s^2 on the right. Since the coordinates satisfy the equation, the point lies on the cone.

55. We use the identity $\cosh^2 x - \sinh^2 x = 1$.
Substituting into the left side of the equation for the hyperboloid we get
$$\cosh^2 s - \sinh^2 s \cos^2 t - \sinh^2 s \sin^2 t \text{ or}$$
$$\cosh^2 s - \sinh^2 s\left(\cos^2 t + \sin^2 t\right) \text{ which equals } 1.$$
Since the coordinates satisfy the equation, the point lies on the hyperboloid. Since $a > 0$, $x > 0$, so the point is on the right sheet.

57. **7)** $x = \sin s \cos t, \ y = 3\sin s \sin t, \ z = 2\cos s$

9) $x = \frac{1}{2}s\cos t, \ y = \frac{1}{2}s\sin t, \ z = s^2$

11) $x = \frac{1}{2}\sqrt{s^2}\cos t, \ y = \sqrt{s^2}\sin t, \ z = s$

For the graphs, see the answers for Exercises 7, 9, and 11.

59. One way of parametrizing the surface
$$z = 4 - x^2 - y^2 \text{ is}$$
$$x = s\cos t, \ y = s\sin t, \ z = 4 - s^2.$$

Chapter 10 Review

1. See Example 1.1.
$$\mathbf{a} + \mathbf{b} = \langle -2, 3 \rangle + \langle 1, 0 \rangle = \langle -2+1, 3+0 \rangle = \langle -1, 3 \rangle$$
$$4\mathbf{b} = 4\langle 1, 0 \rangle = \langle 4\cdot 1, 4\cdot 0 \rangle = \langle 4, 0 \rangle$$
$$2\mathbf{b} - \mathbf{a} = 2\langle 1, 0 \rangle - \langle -2, 3 \rangle$$
$$= \langle 2, 0 \rangle - \langle -2, 3 \rangle$$
$$= \langle 4, -3 \rangle$$
$$\|2\mathbf{b} - \mathbf{a}\| = \|\langle 4, -3 \rangle\| = \sqrt{4^2 + (-3)^2} = \sqrt{25} = 5$$

3. See Example 1.1 and page 793.
$$\mathbf{a} + \mathbf{b} = (10\mathbf{i} + 2\mathbf{j} - 2\mathbf{k}) + (-4\mathbf{i} + 3\mathbf{j} + 2\mathbf{k}) = 6\mathbf{i} + 5\mathbf{j}$$
$$4\mathbf{b} = 4(-4\mathbf{i} + 3\mathbf{j} + 2\mathbf{k}) = -16\mathbf{i} + 12\mathbf{j} + 8\mathbf{k}$$
$$2\mathbf{b} - \mathbf{a} = 2(-4\mathbf{i} + 3\mathbf{j} + 2\mathbf{k}) - (10\mathbf{i} + 2\mathbf{j} - 2\mathbf{k})$$
$$= (-8\mathbf{i} + 6\mathbf{j} + 4\mathbf{k}) - (10\mathbf{i} + 2\mathbf{j} - 2\mathbf{k})$$
$$= -18\mathbf{i} + 4\mathbf{j} + 6\mathbf{k}$$
$$\|2\mathbf{b} - \mathbf{a}\| = \|-18\mathbf{i} + 4\mathbf{j} + 6\mathbf{k}\|$$
$$= \sqrt{(-18)^2 + 4^2 + 6^2}$$
$$= \sqrt{376}$$
$$= 2\sqrt{94}$$

5. See Examples 1.2 and 3.4.
If $\mathbf{b} = c\mathbf{a}$, then $\langle 4, 5 \rangle = c\langle 2, 3 \rangle$. That is, $4 = 2c$ (so that $c = 2$) and $5 = 3c$ $\left(\text{so that } c = \frac{5}{3}\right)$. This is a contradiction, so $\mathbf{a}$ and $\mathbf{b}$ are not parallel.
$\mathbf{a} \cdot \mathbf{b} = \langle 2, 3 \rangle \cdot \langle 4, 5 \rangle = 8 + 15 = 23 \neq 0$ so $\mathbf{a}$ and $\mathbf{b}$ are not orthogonal. $\mathbf{a}$ and $\mathbf{b}$ are neither.

7. See Examples 1.2 and 3.4.
Observe that $\mathbf{b} = c\mathbf{a}$ when $c = -2$.
$\mathbf{a}$ and $\mathbf{b}$ are parallel.

9. See Example 1.3.
$$\overrightarrow{PQ} = \langle 2-3, \ -1-1, \ 1-(-2) \rangle = \langle -1, \ -2, \ 3 \rangle$$

11. See Example 1.4.
$$\|\langle 3, 6 \rangle\| = \sqrt{3^2 + 6^2} = \sqrt{45} = 3\sqrt{5}$$
$$\frac{1}{3\sqrt{5}}\langle 3, 6 \rangle = \frac{1}{\sqrt{5}}\langle 1, 2 \rangle$$

13. See Example 2.3.
$$\|10\mathbf{i} + 2\mathbf{j} - 2\mathbf{k}\| = \sqrt{10^2 + 2^2 + (-2)^2}$$
$$= \sqrt{108}$$
$$= 6\sqrt{3}$$
$$\frac{1}{6\sqrt{3}}(10\mathbf{i} + 2\mathbf{j} - 2\mathbf{k}) = \frac{1}{3\sqrt{3}}(5\mathbf{i} + \mathbf{j} - \mathbf{k})$$

15. See Example 1.4.
$$\langle 1-4, 1-1, 6-2 \rangle = \langle -3, 0, 4 \rangle$$
$$\|\langle -3, 0, 4 \rangle\| = \sqrt{(-3)^2 + 0^2 + 4^2} = \sqrt{25} = 5$$
$$\frac{1}{5}\langle -3, 0, 4 \rangle = \left\langle -\frac{3}{5}, 0, \frac{4}{5} \right\rangle$$

17. See Example 2.2.
$$d\{(0, -2, 2), (3, 4, 1)\}$$
$$= \sqrt{(3-0)^2 + [4-(-2)]^2 + (1-2)^2}$$
$$= \sqrt{3^2 + 6^2 + (-1)^2}$$
$$= \sqrt{46}$$

19. See the definition of magnitude and Example 2.3.

First find the unit vector in the direction of **v**.

$$\|\mathbf{v}\| = \sqrt{2^2 + (-2)^2 + (2)^2} = \sqrt{12} = 2\sqrt{3}$$

$$\frac{1}{\|\mathbf{v}\|}\mathbf{v} = \frac{1}{2\sqrt{3}}(2\mathbf{i} - 2\mathbf{j} + 2\mathbf{k}) = \frac{1}{\sqrt{3}}(\mathbf{i} - \mathbf{j} + \mathbf{k})$$

Multiply the unit vector by 2 to get a vector with magnitude 2.

$$\frac{2}{\sqrt{3}}(\mathbf{i} - \mathbf{j} + \mathbf{k}) = \frac{2}{\sqrt{3}}\mathbf{i} - \frac{2}{\sqrt{3}}\mathbf{j} + \frac{2}{\sqrt{3}}\mathbf{k}$$

21. See Example 1.6.

$$\mathbf{v} = \langle x, y \rangle$$

$$\mathbf{v} + \mathbf{w} = \langle x + 20, y - 80 \rangle = \langle c, 0 \rangle,$$

so $y - 80 = 0$ or $y = 80$

$$\|\mathbf{v}\| = \sqrt{x^2 + y^2} = \sqrt{x^2 + 6400} = 500$$

$$x^2 + 6400 = 250{,}000 \text{ or } x = \sqrt{243{,}600} = 20\sqrt{609}$$

(We take to positive square root to have the plane moving eastward.)

$$\mathbf{v} = \langle 20\sqrt{609}, 80 \rangle \text{ at an angle of}$$

$$\tan^{-1}\frac{80}{20\sqrt{609}} \approx 9.2° \text{ north of east}$$

23. See Example 2.4.

$$(x - 0)^2 + \left[y - (-2)\right]^2 + (z - 0)^2 = 6^2$$

$$x^2 + (y + 2)^2 + z^2 = 36$$

25. See Example 3.2.

$$\mathbf{a} \cdot \mathbf{b} = (2)(2) + (-1)(4) = 4 - 4 = 0$$

27. See Example 3.1.

$$\mathbf{a} \cdot \mathbf{b} = (3)(-2) + (1)(2) + (-4)(1) = -8$$

29. See Example 3.3.

$$\mathbf{a} \cdot \mathbf{b} = (3)(-1) + (2)(1) + (1)(2) = 1$$

$$\|\mathbf{a}\| = \sqrt{3^2 + 2^2 + 1^2} = \sqrt{14}$$

$$\|\mathbf{b}\| = \sqrt{(-1)^2 + 1^2 + 2^2} = \sqrt{6}$$

$$\cos\theta = \frac{\mathbf{a} \cdot \mathbf{b}}{\|\mathbf{a}\|\|\mathbf{b}\|} = \frac{1}{\sqrt{14}\sqrt{6}} = \frac{1}{2\sqrt{21}}$$

$$\theta = \cos^{-1}\frac{1}{2\sqrt{21}} \approx 1.46$$

31. See Example 3.5.

$$\mathbf{a} \cdot \mathbf{b} = (3)(1) + (1)(2) + (-4)(1) = 1$$

$$\|\mathbf{b}\| = \sqrt{1^2 + 2^2 + 1^2} = \sqrt{6}$$

$$\text{comp}_{\mathbf{b}}\,\mathbf{a} = \frac{\mathbf{a} \cdot \mathbf{b}}{\|\mathbf{b}\|} = \frac{1}{\sqrt{6}}$$

$$\text{proj}_{\mathbf{b}}\,\mathbf{a} = \left(\frac{\mathbf{a} \cdot \mathbf{b}}{\|\mathbf{b}\|}\right)\frac{\mathbf{b}}{\|\mathbf{b}\|}$$

$$= \left(\frac{1}{\sqrt{6}}\right)\frac{\mathbf{i} + 2\mathbf{j} + \mathbf{k}}{\sqrt{6}}$$

$$= \frac{1}{6}(\mathbf{i} + 2\mathbf{j} + \mathbf{k})$$

$$= \frac{1}{6}\mathbf{i} + \frac{1}{3}\mathbf{j} + \frac{1}{6}\mathbf{k}$$

33. See Example 4.3.

$$\mathbf{a} \times \mathbf{b} = \langle 1, -2, 1 \rangle \times \langle 2, 0, 1 \rangle$$

$$= \begin{vmatrix} \mathbf{i} & \mathbf{j} & \mathbf{k} \\ 1 & -2 & 1 \\ 2 & 0 & 1 \end{vmatrix}$$

$$= \begin{vmatrix} -2 & 1 \\ 0 & 1 \end{vmatrix}\mathbf{i} - \begin{vmatrix} 1 & 1 \\ 2 & 1 \end{vmatrix}\mathbf{j} + \begin{vmatrix} 1 & -2 \\ 2 & 0 \end{vmatrix}\mathbf{k}$$

$$= -2\mathbf{i} + \mathbf{j} + 4\mathbf{k}$$

$$= \langle -2, 1, 4 \rangle$$

35. See Example 4.3.

$$\mathbf{a} \times \mathbf{b} = (2\mathbf{j} + \mathbf{k}) \times (4\mathbf{i} + 2\mathbf{j} - \mathbf{k})$$

$$= \begin{vmatrix} \mathbf{i} & \mathbf{j} & \mathbf{k} \\ 0 & 2 & 1 \\ 4 & 2 & -1 \end{vmatrix}$$

$$= \begin{vmatrix} 2 & 1 \\ 2 & -1 \end{vmatrix}\mathbf{i} - \begin{vmatrix} 0 & 1 \\ 4 & -1 \end{vmatrix}\mathbf{j} + \begin{vmatrix} 0 & 2 \\ 4 & 2 \end{vmatrix}\mathbf{k}$$

$$= -4\mathbf{i} + 4\mathbf{j} - 8\mathbf{k}$$

37. See Theorem 4.2.

First find a vector orthogonal to **a** and **b**.

$$\mathbf{a} \times \mathbf{b} = (2\mathbf{i} + \mathbf{k}) \times (-\mathbf{i} + 2\mathbf{j} - \mathbf{k})$$

$$= \begin{vmatrix} \mathbf{i} & \mathbf{j} & \mathbf{k} \\ 2 & 0 & 1 \\ -1 & 2 & -1 \end{vmatrix}$$

$$= \begin{vmatrix} 0 & 1 \\ 2 & -1 \end{vmatrix}\mathbf{i} - \begin{vmatrix} 2 & 1 \\ -1 & -1 \end{vmatrix}\mathbf{j} + \begin{vmatrix} 2 & 0 \\ -1 & 2 \end{vmatrix}\mathbf{k}$$

$$= -2\mathbf{i} + \mathbf{j} + 4\mathbf{k}$$

Next find unit vectors parallel to $\mathbf{a} \times \mathbf{b}$.

$$\|\mathbf{a} \times \mathbf{b}\| = \sqrt{(-2)^2 + 1^2 + 4^2} = \sqrt{21}$$

$$\pm\frac{1}{\sqrt{21}}(-2\mathbf{i} + \mathbf{j} + 4\mathbf{k})$$

39. See Example 3.6.

To find the force along the direction given, compute the component of the force vector along $\langle 60-1,\, 22-0 \rangle = \langle 59,\, 22 \rangle$.

$$F = \frac{\langle 40,\, -30 \rangle \cdot \langle 59,\, 22 \rangle}{\|\langle 59,\, 22 \rangle\|}$$

$$d = \|\langle 59,\, 22 \rangle\|$$

$W = Fd$

$\quad = \langle 40,\, -30 \rangle \cdot \langle 59,\, 22 \rangle$

$\quad = 1700$ foot-pounds

41. See Example 4.5.

Let $P = (1, -1, 3)$ (when $t = 0$) and $R = (2, 1, 3)$ (when $t = 1$).

$\overrightarrow{PQ} = \langle 0,\, 0,\, -3 \rangle,\ \overrightarrow{PR} = \langle 1,\, 2,\, 0 \rangle$

$$\overrightarrow{PQ} \times \overrightarrow{PR} = \begin{vmatrix} \mathbf{i} & \mathbf{j} & \mathbf{k} \\ 0 & 0 & -3 \\ 1 & 2 & 0 \end{vmatrix}$$

$$= \begin{vmatrix} 0 & -3 \\ 2 & 0 \end{vmatrix}\mathbf{i} - \begin{vmatrix} 0 & -3 \\ 1 & 0 \end{vmatrix}\mathbf{j} + \begin{vmatrix} 0 & 0 \\ 1 & 2 \end{vmatrix}\mathbf{k}$$

$$= 6\mathbf{i} - 3\mathbf{j}$$

$$= \langle 6,\, -3,\, 0 \rangle$$

$$d = \frac{\|\overrightarrow{PQ} \times \overrightarrow{PR}\|}{\|\overrightarrow{PR}\|} = \frac{\sqrt{6^2 + (-3)^2 + 0^2}}{\sqrt{1^2 + 2^2 + 0^2}} = \frac{\sqrt{45}}{\sqrt{5}} = 3$$

43. See Example 4.4.

$$\langle 2,\, 0,\, 1 \rangle \times \langle 0,\, 1,\, -3 \rangle = \begin{vmatrix} \mathbf{i} & \mathbf{j} & \mathbf{k} \\ 2 & 0 & 1 \\ 0 & 1 & -3 \end{vmatrix}$$

$$= \begin{vmatrix} 0 & 1 \\ 1 & -3 \end{vmatrix}\mathbf{i} - \begin{vmatrix} 2 & 1 \\ 0 & -3 \end{vmatrix}\mathbf{j} + \begin{vmatrix} 2 & 0 \\ 0 & 1 \end{vmatrix}\mathbf{k}$$

$$= -\mathbf{i} + 6\mathbf{j} + 2\mathbf{k}$$

$$= \langle -1,\, 6,\, 2 \rangle$$

$$\|\langle 2,\, 0,\, 1 \rangle \times \langle 0,\, 1,\, -3 \rangle\| = \sqrt{(-1)^2 + 6^2 + 2^2}$$

$$= \sqrt{41}$$

45. See Example 4.7.

$$\|\boldsymbol{\tau}\| = \|\mathbf{F}\|\|\mathbf{r}\|\sin\theta = 50\left(\frac{6}{12}\right)\sin\frac{\pi}{6} = \frac{25}{2}$$

47. See Example 5.2.

a. $P = (2, -1, -3),\ Q = (0, 2, -3),$

$\overrightarrow{PQ} = \langle -2,\, 3,\, 0 \rangle$

$x - 2 = -2t,\ y + 1 = 3t,\ z + 3 = 0$

or $x = 2 - 2t,\ y = -1 + 3t,\ z = -3$

b. $\dfrac{x-2}{-2} = \dfrac{y+1}{3},\ z = -3$

49. See Example 5.1.

a. A vector parallel to the line is $\left\langle 2,\, \dfrac{1}{2},\, -3 \right\rangle$.

$x - 2 = 2t,\ y + 1 = \dfrac{1}{2}t,\ z - 1 = -3t$

or $x = 2 + 2t,\ y = -1 + \dfrac{1}{2}t,\ z = 1 - 3t$

b. $\dfrac{x-2}{2} = 2(y+1) = \dfrac{z-1}{-3}$

51. See Definition 5.1 (ii) (a).

$\mathbf{a} = \langle 1,\, 0,\, 2 \rangle,\ \mathbf{b} = \langle 2,\, 2,\, 4 \rangle$

$\mathbf{a} \cdot \mathbf{b} = (1)(2) + (0)(2) + (2)(4) = 10$

$\|\mathbf{a}\| = \sqrt{1^2 + 0^2 + 2^2} = \sqrt{5}$

$\|\mathbf{b}\| = \sqrt{2^2 + 2^2 + 4^2} = \sqrt{24}$

$\cos\theta = \dfrac{10}{\sqrt{5}\sqrt{24}} = \dfrac{10}{\sqrt{120}} = \dfrac{5}{\sqrt{30}}$

$\cos^{-1}\dfrac{5}{\sqrt{30}} \approx .42$

53. See Example 5.3.

$\mathbf{a} = \langle 2,\, 1,\, 4 \rangle,\ \mathbf{b} = \langle 0,\, 1,\, 1 \rangle$

The lines are not parallel since there is no c such that $\mathbf{a} = c\mathbf{b}$. Setting the x-components equal, we get $2t = 4$ so $t = 2$. Setting the y-components equal and setting $t = 2$, we get $3 + 2 = 4 + s$ so $s = 1$. Setting the z-components equal, we get $-1 + 4t = 3 + s$. This is not satisfied when $s = 1$ and $t = 2$. The lines are skew lines.

55. See Example 5.4.

$4(x + 5) + y - 2(z - 1) = 0$

57. See Example 5.5.

$P = (2, 1, 3)$, $Q = (2, -1, 2)$, $R = (3, 3, 2)$

$\overrightarrow{PQ} = \langle 0, -2, -1 \rangle$, $\overrightarrow{PR} = \langle 1, 2, -1 \rangle$

$$\overrightarrow{PQ} \times \overrightarrow{PR} = \begin{vmatrix} \mathbf{i} & \mathbf{j} & \mathbf{k} \\ 0 & -2 & -1 \\ 1 & 2 & -1 \end{vmatrix}$$

$$= \begin{vmatrix} -2 & -1 \\ 2 & -1 \end{vmatrix} \mathbf{i} - \begin{vmatrix} 0 & -1 \\ 1 & -1 \end{vmatrix} \mathbf{j} + \begin{vmatrix} 0 & -2 \\ 1 & 2 \end{vmatrix} \mathbf{k}$$

$$= 4\mathbf{i} - \mathbf{j} + 2\mathbf{k}$$

$$= \langle 4, -1, 2 \rangle$$

$$4(x - 2) - (y - 1) + 2(z - 3) = 0$$

59. See Example 6.4.

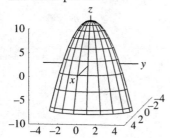

elliptic paraboloid

61. See page 836.

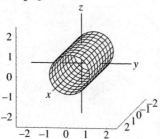

cylinder

63. See page 838.

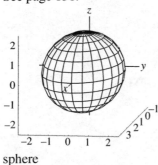

sphere

65. See page 836.

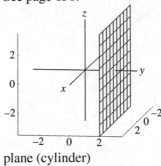

plane (cylinder)

67. See page 836.

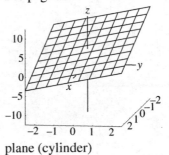

plane (cylinder)

69. See Example 6.6.

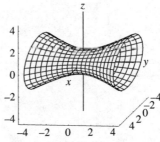

hyperboloid of one sheet

71. See Example 6.7.

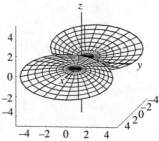

hyperboloid of two sheets.

351

Chapter 11

Section 11.1

5.

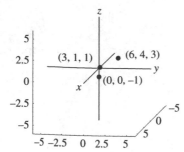

7.

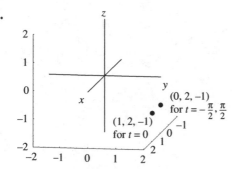

9.

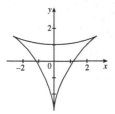

11.

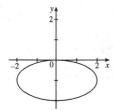

13.

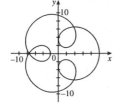

15.

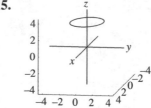

17.

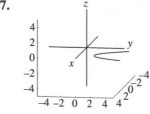

19.

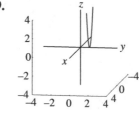

21.

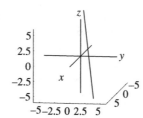

23.

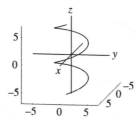

25.

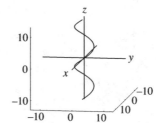

27.

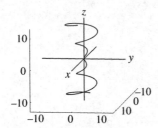

29.

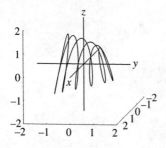

31.

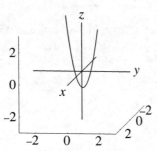

33.

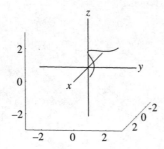

35. a. F

b. C

c. E

d. A

e. B

f. D

37.

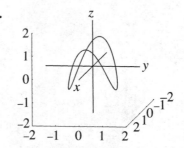

$$\mathbf{r}'(t) = \langle -\sin t, \cos t, -2\sin 2t \rangle$$
$$s = \int_0^{2\pi} \sqrt{(-\sin t)^2 + (\cos t)^2 + (-2\sin 2t)^2} \, dt$$
$$= \int_0^{2\pi} \sqrt{1 + 4\sin^2 2t} \, dt$$
$$\approx 10.54 \text{ by numerical integration}$$

39.

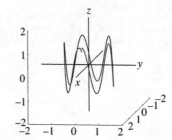

$$\mathbf{r}'(t) = \langle -\pi\sin \pi t, \, \pi\cos \pi t, \, -16\sin 16t \rangle$$
$$s = \int_0^2 \sqrt{[-\pi\sin \pi t]^2 + [\pi\cos \pi t]^2 + [-16\sin 16t]^2} \, dt$$
$$- \int_0^2 \sqrt{\pi^2 + 256\sin^2 16t} \, dt$$
$$\approx 21.56 \text{ by numerical integration}$$

41.

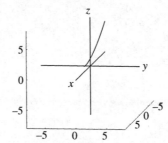

$$\mathbf{r}'(t) = \langle 1, \, 2t, \, 3t^2 \rangle$$
$$s = \int_0^2 \sqrt{1^2 + (2t)^2 + (3t^2)^2} \, dt$$
$$= \int_0^2 \sqrt{9t^4 + 4t^2 + 1} \, dt$$
$$\approx 9.57 \text{ by numerical integration}$$

43. $\cos 2t = \cos^2 t - \sin^2 t$

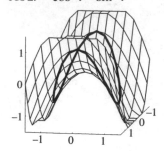

45. The two curves are identical, with the same endpoints. They are just parameterized using different t-values.

47. $g(t)$ and $h(t)$ are portions of
$$\mathbf{r}(t) = \left\langle t, t^2, t^2 \right\rangle, \quad -\infty < t < \infty.$$
$g(t) = \mathbf{r}(t)$ with $-1 \le t \le 1$, and
$h(t) = \mathbf{r}(t)$ with $0 \le t$.

49.

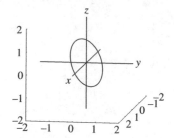

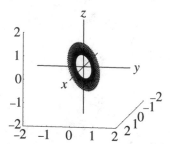

The graph is an ellipse in 3-space, and it is periodic with period 2π. However, for T much larger than 2π, the points plotted become too few and "jump" around the ellipse, causing the jagged lines.

Section 11.2

5. $\lim\limits_{t\to 0}\left\langle t^2-1,\ e^{2t},\ \sin t\right\rangle$

$=\left\langle \lim\limits_{t\to 0}(t^2-1),\ \lim\limits_{t\to 0}e^{2t},\ \lim\limits_{t\to 0}\sin t\right\rangle$

$=\left\langle -1,1,0\right\rangle$

7. $\lim\limits_{t\to 0}\left\langle \dfrac{\sin t}{t},\ \cos t,\ \dfrac{t+1}{t-1}\right\rangle$

$=\left\langle \lim\limits_{t\to 0}\dfrac{\sin t}{t},\ \lim\limits_{t\to 0}\cos t,\ \lim\limits_{t\to 0}\dfrac{t+1}{t-1}\right\rangle$

$=\left\langle 1,1,-1\right\rangle$

9. $\lim\limits_{t\to 0}\left\langle \ln t,\ \sqrt{t^2+1},\ t-3\right\rangle$

$=\left\langle \lim\limits_{t\to 0}\ln t,\ \lim\limits_{t\to 0}\sqrt{t^2+1},\ \lim\limits_{t\to 0}(t-3)\right\rangle$

which does not exist, because of the undefined
limit of the x-component.

11. $t\neq 1$, because $t=1$ is an excluded value for the x-component.

13. $t\neq \dfrac{n\pi}{2}$ (n odd), because the x-component is

undefined for $t=...,\ -\dfrac{3\pi}{2},\ -\dfrac{\pi}{2},\ \dfrac{\pi}{2},\ \dfrac{3\pi}{2},\ ...$

15. $t\geq 0$, because the y-component is undefined for $t<0$.

17. $\dfrac{d\mathbf{r}}{dt}=\left\langle \dfrac{d}{dt}(t^4),\ \dfrac{d}{dt}\left(\sqrt{t+1}\right),\ \dfrac{d}{dt}\left(\dfrac{3}{t^2}\right)\right\rangle$

$=\left\langle 4t^3,\ \dfrac{1}{2\sqrt{t+1}},\ -\dfrac{6}{t^3}\right\rangle$

19. $\dfrac{d\mathbf{r}}{dt}=\left\langle \dfrac{d}{dt}(\sin t),\ \dfrac{d}{dt}(\sin t^2),\ \dfrac{d}{dt}(\cos t)\right\rangle$

$=\left\langle \cos t,\ 2t\cos t^2,\ -\sin t\right\rangle$

21. $\dfrac{d\mathbf{r}}{dt}=\left\langle \dfrac{d}{dt}(e^{t^2}),\ \dfrac{d}{dt}(t^2),\ \dfrac{d}{dt}(\sec 2t)\right\rangle$

$=\left\langle 2te^{t^2},\ 2t,\ 2\sec 2t\tan 2t\right\rangle$

23. $\mathbf{r}'(t)=\left\langle -\sin t,\ \cos t\right\rangle$

$\mathbf{r}(0)=\left\langle 1,0\right\rangle;\ \mathbf{r}'(0)=\left\langle 0,1\right\rangle$

$\mathbf{r}\left(\dfrac{\pi}{2}\right)=\left\langle 0,1\right\rangle;\ \mathbf{r}'\left(\dfrac{\pi}{2}\right)=\left\langle -1,0\right\rangle$

$\mathbf{r}(\pi)=\left\langle -1,0\right\rangle;\ \mathbf{r}'(\pi)=\left\langle 0,-1\right\rangle$

25. $\mathbf{r}'(t)=\left\langle -\sin t,1,\cos t\right\rangle$

$\mathbf{r}(0)=\left\langle 1,0,0\right\rangle;\ \mathbf{r}'(0)=\left\langle 0,1,1\right\rangle$

$\mathbf{r}\left(\dfrac{\pi}{2}\right)=\left\langle 0,\dfrac{\pi}{2},1\right\rangle;\ \mathbf{r}'\left(\dfrac{\pi}{2}\right)=\left\langle -1,1,0\right\rangle$

$\mathbf{r}(\pi)=\left\langle -1,\pi,0\right\rangle;\ \mathbf{r}'(\pi)=\left\langle 0,1,-1\right\rangle$

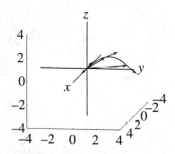

27. $\int\left\langle 3t-1,\sqrt{t}\right\rangle dt=\left\langle \int(3t-1)dt,\int\sqrt{t}\,dt\right\rangle$

$=\left\langle \dfrac{3}{2}t^2-t+c_1,\ \dfrac{2}{3}t^{3/2}+c_2\right\rangle$

$=\left\langle \dfrac{3}{2}t^2-t,\ \dfrac{2}{3}t^{3/2}\right\rangle+\mathbf{c}$

29. $\int\left\langle \cos 3t,\sin t,e^{4t}\right\rangle dt$

$=\left\langle \int\cos 3t\,dt,\int\sin t\,dt,\int e^{4t}\,dt\right\rangle$

$=\left\langle \dfrac{1}{3}\sin 3t+c_1,\ -\cos t+c_2,\ \dfrac{1}{4}e^{4t}+c_3\right\rangle$

$=\left\langle \dfrac{1}{3}\sin 3t,\ -\cos t,\ \dfrac{1}{4}e^{4t}\right\rangle+\mathbf{c}$

31. $\int \left\langle te^{t^2}, 3t\sin t, \dfrac{3t}{t^2+1}\right\rangle dt$

$= \left\langle \int te^{t^2}\,dt, \int 3t\sin t\,dt, \int \dfrac{3t}{t^2+1}\,dt\right\rangle$

$= \left\langle \dfrac{1}{2}e^{t^2}, 3\sin t - 3t\cos t, \dfrac{3}{2}\ln(t^2+1)\right\rangle + \mathbf{c}$

33. $\int_0^1 \left\langle t^2-1, 3t\right\rangle dt = \left\langle \int_0^1 (t^2-1)dt, \int_0^1 3t\,dt\right\rangle$

$= \left\langle \left[\dfrac{1}{3}t^3 - t\right]_0^1, \left[\dfrac{3t^2}{2}\right]_0^1\right\rangle$

$= \left\langle -\dfrac{2}{3}, \dfrac{3}{2}\right\rangle$

35. $\int_0^2 \left\langle \dfrac{4}{t+1}, e^{t-2}, te^t\right\rangle dt$

$= \left\langle \int_0^2 \dfrac{4}{t+1}\,dt, \int_0^2 e^{t-2}\,dt, \int_0^2 te^t\,dt\right\rangle$

$= \left\langle \left[4\ln|t+1|\right]_0^2, \left[e^{t-2}\right]_0^2, \left[e^t(t-1)\right]_0^2\right\rangle$

$= \left\langle 4\ln 3, 1 - \dfrac{1}{e^2}, e^2+1\right\rangle$

37. $\mathbf{r}'(t) = \left\langle -\sin t, \cos t\right\rangle$

$\mathbf{r}(t)\cdot\mathbf{r}'(t) = (\cos t)(-\sin t) + (\sin t)(\cos t) = 0$

$\mathbf{r}(t)$ and $\mathbf{r}'(t)$ are perpendicular for all t.

39. $\mathbf{r}'(t) = \left\langle 1, 1, 2t\right\rangle$

$\mathbf{r}(t)\cdot\mathbf{r}'(t) = (t)(1) + (t)(1) + (t^2-1)(2t) = 2t^3$

$\mathbf{r}(t)$ and $\mathbf{r}'(t)$ are perpendicular for $t = 0$.

49. $\dfrac{d}{dt}[f(t)\mathbf{r}(t)]$

$= \left\langle \dfrac{d}{dt}[f(t)x(t)], \dfrac{d}{dt}[f(t)y(t)], \dfrac{d}{dt}[f(t)z(t)]\right\rangle$

$= \left\langle f'(t)x(t) + f(t)x'(t), f'(t)y(t) + f(t)y'(t), f'(t)z(t) + f(t)z'(t)\right\rangle$

$= f'(t)\mathbf{r}(t) + f(t)\mathbf{r}'(t)$

41. In Exercises 37 and 38, for $\mathbf{r}(t)$ and $\mathbf{r}'(t)$ to be parallel there must be some number c such that $\mathbf{r}'(t) = c\mathbf{r}(t)$. With $\mathbf{r}(t) = \left\langle \cos t, \sin t\right\rangle$,

$\mathbf{r}'(t) = c\mathbf{r}(t)$ implies $-\sin t = c\cos t$ and

$\cos t = c\sin t$, so that $c = -\dfrac{\sin t}{\cos t} = \dfrac{\cos t}{\sin t}$. Then

$-\sin^2 t = \cos^2 t$ and $\sin^2 t + \cos^2 t = 0$, which is impossible.

With $\mathbf{r}(t) = \left\langle 2\cos t, \sin t\right\rangle$, $\mathbf{r}'(t) = c\mathbf{r}(t)$ implies

$-2\sin t = c(2\cos t)$ and $\cos t = c\sin t$ which in the same way leads to the impossible result that

$\sin^2 t + \cos^2 t = 0$.

43. $\mathbf{r}'(t)$ lies in the xy-plane when

$\dfrac{d}{dt}(t^3 - 3) = 3t^2 = 0 \Rightarrow t = 0$

45. $\mathbf{r}'(t)$ lies in the xy-plane when

$\dfrac{d}{dt}(\sin 2t) = 2\cos 2t = 0$, i.e. when

$2t = \dfrac{n\pi}{2}$ (n odd), and $t = \dfrac{n\pi}{4}$ (n odd)

47. $\dfrac{d}{dt}[c\mathbf{r}(t)] = \left\langle \dfrac{d}{dt}[c\cdot x(t)], \dfrac{d}{dt}[c\cdot y(t)], \dfrac{d}{dt}[c\cdot z(t)]\right\rangle$

$= \left\langle c\cdot x'(t), c\cdot y'(t), c\cdot z'(t)\right\rangle$

$= c\mathbf{r}'(t)$

51. False. If $\mathbf{r}(t)$ traces out, say, an ellipse, then $\mathbf{u}(t)$ traces out a circle. In such a case, $\mathbf{u}(t)$ and $\mathbf{u}'(t)$ are always perpendicular but $\mathbf{r}(t)$ and $\mathbf{r}'(t)$ are not always perpendicular.

53. If $\mathbf{r}(t)$ and $\mathbf{r}'(t)$ are orthogonal, then $\dfrac{d}{dt}\|\mathbf{r}(t)\|^2 = 2\mathbf{r}'(t)\cdot\mathbf{r}(t) = 0$, so $\|\mathbf{r}(t)\|^2$ is constant.

Since $\|\mathbf{r}(t)\|$ is nonnegative, this means $\|\mathbf{r}(t)\|$ is constant.

Section 11.3

5. $\mathbf{v}(t) = \mathbf{r}'(t)$

$$= \left\langle \frac{d}{dt}[5\cos(2t)], \frac{d}{dt}[5\sin(2t)] \right\rangle$$

$$= \langle -10\sin(2t), 10\cos(2t) \rangle$$

$\mathbf{a}(t) = \mathbf{v}'(t)$

$$= \left\langle \frac{d}{dt}[-10\sin(2t)], \frac{d}{dt}[10\cos(2t)] \right\rangle$$

$$= \langle -20\cos 2t, -20\sin 2t \rangle$$

7. $\mathbf{v}(t) = \mathbf{r}'(t)$

$$= \left\langle \frac{d}{dt}(25t), \frac{d}{dt}(-16t^2 + 15t + 5) \right\rangle$$

$$= \langle 25, -32t + 15 \rangle$$

$\mathbf{a}(t) = \mathbf{v}'(t)$

$$= \left\langle \frac{d}{dt}(25), \frac{d}{dt}(-32t + 15) \right\rangle$$

$$= \langle 0, -32 \rangle$$

9. $\mathbf{v}(t) = \mathbf{r}'(t)$

$$= \left\langle \frac{d}{dt}(4te^{-2t}), \frac{d}{dt}(2e^{-2t}), \frac{d}{dt}(-16t^2) \right\rangle$$

$$= \langle (4 - 8t)e^{-2t}, -4e^{-2t}, -32t \rangle$$

$\mathbf{a}(t) = \mathbf{v}'(t)$

$$= \left\langle \frac{d}{dt}[(4 - 8t)e^{-2t}], \frac{d}{dt}(-4e^{-2t}), \frac{d}{dt}(-32t) \right\rangle$$

$$= \left\langle (-16 + 16t)e^{-2t}, 8e^{-2t}, -32 \right\rangle$$

11. $\mathbf{r}(t) = \int \mathbf{v}(t)\, dt = \left\langle 10t, -16t^2 + 4t \right\rangle + \mathbf{c}$, and since
$\mathbf{r}(0) = \langle 3, 8 \rangle$, $\mathbf{c} = \langle 3, 8 \rangle$, so that
$\mathbf{r}(t) = \left\langle 10t + 3, -16t^2 + 4t + 8 \right\rangle$.

13. $\mathbf{v}(t) = \int \mathbf{a}(t)\, dt = \langle 0, -32t \rangle + \mathbf{c}_1$, and since
$\mathbf{v}(0) = \langle 5, 0 \rangle$, $\mathbf{c}_1 = \langle 5, 0 \rangle$, so that
$\mathbf{v}(t) = \langle 5, -32t \rangle$. Then
$\mathbf{r}(t) = \int \mathbf{v}(t)\, dt = \left\langle 5t, -16t^2 \right\rangle + \mathbf{c}_2$, and since
$\mathbf{r}(0) = \langle 0, 16 \rangle$, $\mathbf{c}_2 = \langle 0, 16 \rangle$, so that
$\mathbf{r}(t) = \left\langle 5t, -16t^2 + 16 \right\rangle$.

15. $\mathbf{r}(t) = \int \mathbf{v}(t)\, dt = \left\langle 10t, -3e^{-t}, -16t^2 + 4t \right\rangle + \mathbf{c}$, and
since $\mathbf{r}(0) = \langle 0, -6, 20 \rangle$, $\mathbf{c} = \langle 0, -3, 20 \rangle$, so that
$\mathbf{r}(t) = \left\langle 10t, -3e^{-t} - 3, -16t^2 + 4t + 20 \right\rangle$.

17. $\mathbf{v}(t) = \int \mathbf{a}(t)\, dt = \left\langle \frac{1}{2}t^2, 0, -16t \right\rangle + \mathbf{c}_1$, and since
$\mathbf{v}(0) = \langle 12, -4, 0 \rangle$, $\mathbf{c}_1 = \langle 12, -4, 0 \rangle$ and
$\mathbf{v}(t) = \left\langle \frac{1}{2}t^2 + 12, -4, -16t \right\rangle$. Then
$\mathbf{r}(t) = \int \mathbf{v}(t)\, dt = \left\langle \frac{1}{6}t^3 + 12t, -4t, -8t^2 \right\rangle + \mathbf{c}_2$, and
since $\mathbf{r}(0) = \langle 5, 0, 2 \rangle$, $\mathbf{c}_2 = \langle 5, 0, 2 \rangle$, so that
$\mathbf{r}(t) = \left\langle \frac{1}{6}t^3 + 12t + 5, -4t, -8t^2 + 2 \right\rangle$.

19. $\mathbf{v}(t) = \mathbf{r}'(t) = \langle -8\sin 2t, 8\cos 2t \rangle$
$\mathbf{a}(t) = \mathbf{v}'(t) = \langle -16\cos 2t, -16\sin 2t \rangle$
$\mathbf{F} = m\mathbf{a} = \langle -160\cos 2t, -160\sin 2t \rangle$ newtons

21. $\mathbf{v}(t) = \mathbf{r}'(t) = \langle -24\sin 4t, 24\cos 4t \rangle$
$\mathbf{a}(t) = \mathbf{v}'(t) = \langle -96\cos 4t, -96\sin 4t \rangle$
$\mathbf{F} = m\mathbf{a} = \langle -960\cos 4t, -960\sin 4t \rangle$ newtons

23. $\mathbf{v}(t) = \mathbf{r}'(t) = \langle -6\sin 2t, 10\cos 2t \rangle$
$\mathbf{a}(t) = \mathbf{v}'(t) = \langle -12\cos 2t, -20\sin 2t \rangle$
$\mathbf{F} = m\mathbf{a} = \langle -120\cos 2t, -200\sin 2t \rangle$

25. $\mathbf{v}(t) = \mathbf{r}'(t) = \langle 6t + 1, 3 \rangle$
$\mathbf{a}(t) = \mathbf{v}'(t) = \langle 6, 0 \rangle$
$\mathbf{F} = m\mathbf{a} = \langle 60, 0 \rangle$

27. Maximum altitude

$$= h + \frac{v_0^2 \sin^2 \theta}{2g}$$

$$= 0 + \frac{100^2 \sin^2\left(\frac{\pi}{3}\right)}{2(32)}$$

$$= \frac{1875}{16}$$

$$\approx 117 \text{ feet}$$

Time of impact comes from $0 = h + (v_0 \sin\theta)t - \frac{gt^2}{2} = 0 + 50\sqrt{3}\,t - 16t^2$

which leads to $t = \frac{25\sqrt{3}}{8}$

Then horizontal range

$$= (v_0 \cos\theta)t$$

$$= \left[100\cos\left(\frac{\pi}{3}\right)\right]\left(\frac{25\sqrt{3}}{8}\right)$$

$$= \frac{625\sqrt{3}}{4}$$

$$\approx 271 \text{ feet}$$

Because $h = 0$, impact speed = initial speed
$$= 100 \text{ ft/s}.$$

29. Maximum altitude $= h + \frac{v_0^2 \sin^2 \theta}{2g} = 10 + \frac{160^2 \sin^2\left(\frac{\pi}{4}\right)}{2(32)} = 210 \text{ feet}$

Time of impact comes from $0 = h + (v_0 \sin\theta)t - \frac{gt^2}{2} = 10 + 80\sqrt{2}\,t - 16t^2$

which leads to $t = \frac{10\sqrt{2} + \sqrt{210}}{4}$

Then horizontal range $= (v_0 \cos\theta)t = \left[160\cos\left(\frac{\pi}{4}\right)\right]\left(\frac{10\sqrt{2} + \sqrt{210}}{4}\right) = 400 + 40\sqrt{105} \approx 810 \text{ feet}$

Speed at impact $= \| \mathbf{v}(t) \| = \sqrt{(v_0 \cos\theta)^2 + (v_0 \sin\theta - gt)^2}$ with $t = \frac{10\sqrt{2} + \sqrt{210}}{4}$, which is

$$\sqrt{\left[160\cos\left(\frac{\pi}{4}\right)\right]^2 + \left[160\sin\left(\frac{\pi}{4}\right) - 32\left(\frac{10\sqrt{2} + \sqrt{210}}{4}\right)\right]^2} = 8\sqrt{410} \approx 162 \text{ ft/s}.$$

31. Maximum altitude $= h + \dfrac{v_0^2 \sin^2 \theta}{2g}$

$$= 10 + \dfrac{320^2 \sin^2\left(\frac{\pi}{4}\right)}{2(32)}$$

$$= 810 \text{ feet}$$

Time of impact comes from $0 = h + (v_0 \sin\theta)t - \dfrac{gt^2}{2} = 10 + 160\sqrt{2}\,t - 16t^2$

which leads to $t = \dfrac{20\sqrt{2} + 9\sqrt{10}}{4}$

Then horizontal range $= (v_0 \cos\theta)t$

$$= \left[320 \cos\left(\frac{\pi}{4}\right)\right]\left[\frac{20\sqrt{2} + 9\sqrt{10}}{4}\right]$$

$$= 1600 + 720\sqrt{5}$$

$$\approx 3210 \text{ feet}$$

Speed at impact $= \|\mathbf{v}(t)\|$

$$= \sqrt{(v_0 \cos\theta)^2 + (v_0 \sin\theta - gt)^2}$$

with $t = \dfrac{20\sqrt{2} + 9\sqrt{10}}{4}$,

which is $= \sqrt{\left[320\cos\left(\frac{\pi}{4}\right)\right]^2 + \left[320\sin\left(\frac{\pi}{4}\right) - 32\left(\frac{20\sqrt{2} + 9\sqrt{10}}{4}\right)\right]^2}$

$$= 8\sqrt{1610}$$

$$\approx 321 \text{ ft/s.}$$

33. Doubling the initial speed quadruples the horizontal distance the projectile travels before falling back to launch height.

35. Time of impact is found from

$0 = h + (v_0 \sin\theta)t - \dfrac{gt^2}{2}$, which $h = 0$ gives

impact at $t = \dfrac{2v_0 \sin\theta}{g}$, at which time the

horizontal distance traveled is

$(v_0 \cos\theta)t = \dfrac{v_0^2(2\sin\theta\cos\theta)}{g} = \dfrac{v_0^2 \sin 2\theta}{g}.$

37. Equation (3.8) becomes

$$\mathbf{r}(t) = \left\langle (120\cos 30°)t, 3 + (120\sin 30°)t - \frac{32t^2}{2}\right\rangle$$

$$= \left\langle 60\sqrt{3}\,t, 3 + 60t - 16t^2\right\rangle.$$

The x-component becomes 385 at

$t = \dfrac{385}{60\sqrt{3}} = \dfrac{77\sqrt{3}}{36} \approx 3.705$ sec, at which time the

y-component equals about

$3 + 60(3.705) - 16(3.705)^2 \approx 5.67$ feet. The ball does not clear the 6-foot wall and the hit is not a home run.

39. $\theta = 0$ and Equation (3.8) becomes

$$\mathbf{r}(t) = \left\langle (130\cos 0°)t, 6 + (130\sin 0°)t - \frac{32t^2}{2} \right\rangle$$

$$= \left\langle 130t, 6 - 16t^2 \right\rangle$$

The ball crosses home plate at $t = \frac{60}{130} = \frac{6}{13}$ sec, at which time the ball's height is

$$6 - 16\left(\frac{6}{13}\right)^2 \approx 2.59 \text{ feet.}$$

41. $\theta = 0$ and Equation (3.8) becomes

$$\mathbf{r}(t) = \left\langle (120\cos 0°)t, 8 + (120\sin 0°)t - \frac{32t^2}{2} \right\rangle$$

$$= \left\langle 120t, 8 - 16t^2 \right\rangle.$$

The ball is at the net when $t = \frac{39}{120} = 0.325$, and then its height is $8 - 16(0.325)^2 = 6.31$ feet, so it clears the 3-foot net. The ball is at the service line when $t = \frac{60}{120} = 0.5$, and then its height is $8 - 16(0.5)^2 = 4$ feet. Since the height is still positive at the service line, the ball flies over the line and the serve is "out."

43. Because the ball is launched from ground level, hang time = 2 × time to max. altitude.

55 mph $= \frac{242}{3}$ ft/s, so

$$\text{hang time} = 2\left(\frac{v_0\sin\theta}{g}\right)$$

$$= 2\left(\frac{\frac{242}{3}\sin 50°}{32}\right)$$

$$\approx 3.86 \text{ seconds.}$$

45. The northerly wind does not affect the eastward travel distance, which as calculated in exercise 27 is about 271 feet. Northward travel is found from

$$F = ma \Rightarrow a = \frac{F}{m} = \frac{8}{1} = 8, v = 8t, \text{ and } r = 4t^2.$$

Since t was found to be $\frac{25\sqrt{3}}{8}$, the northward travel distance is $4\left(\frac{25\sqrt{3}}{8}\right)^2 \approx 117$ feet. The landing point is $\langle 271, 117, 0 \rangle$.

47. Because $m = 1$, $\mathbf{a} = \mathbf{g} + \mathbf{w} + \mathbf{e}$. For $0 \le t \le 1$,

$\mathbf{a} = \langle 2t, 1, -8 \rangle$ and $\mathbf{v} = \int \mathbf{a}\, dt$

$= \langle t^2, t, -8t \rangle + \mathbf{c}_1$. Because $\mathbf{v}(0) = \langle 100, 0, 10 \rangle$,

for $0 \le t \le 1$, $\mathbf{v}(t) = \langle t^2 + 100, t, 10 - 8t \rangle$.

For $t > 1$, $\mathbf{a} = \langle 2t, 2, -8 \rangle$ and

$$\mathbf{v} = \int \mathbf{a}\, dt$$

$$= \langle t^2, 2t, -8t \rangle + \mathbf{c}_2$$

$$= \langle t^2 + a, 2t + b, -8t + c \rangle.$$

Since a physical object can't accelerate (or decelerate) instantaneously, at $t = 1$ it must be that $\langle t^2 + 100, t, 10 - 8t \rangle = \langle t^2 + a, 2t + b, -8t + c \rangle$.

This implies $a = 100$, $b = -1$, $c = 10$. However, $\mathbf{v}(t)$ is not differentiable because there is a stepwise jump (from 1 to 2) in the y-component of $\mathbf{v}'(t) = \mathbf{a}(t)$. This makes sense because there is a sudden change in applied force.

49. Centripetal acceleration is $\frac{v^2}{R}$, and weightless

feel implies $\frac{v^2}{R} = g$, so

$$v = \sqrt{gR} = \sqrt{3200} \approx 56.57 \text{ ft/s.}$$

51. $a = \frac{v^2}{R}$ and 900 km/hr = 250 m/s, so $5g = \frac{v^2}{R}$

and $R = \frac{v^2}{5g} = \frac{250^2}{5(9.8)} \approx 1275.5$ meters.

53. By Theorem 2.3,

$$\mathbf{L}'(t) = \frac{d}{dt}\left[m\mathbf{r}(t) \times \mathbf{v}(t) \right]$$

$$= m\left[\mathbf{r}'(t) \times \mathbf{v}(t) + \mathbf{r}(t) \times \mathbf{v}'(t) \right]$$

$$= m\left[\mathbf{v}(t) \times \mathbf{v}(t) + \mathbf{r}(t) \times \mathbf{a}(t) \right]$$

$$= m\left[\mathbf{0} + \mathbf{r}(t) \times \mathbf{a}(t) \right]$$

$$= m\mathbf{r}(t) \times \mathbf{a}(t)$$

Section 11.4

5. For radius $r = 2$, arc length $= 2\pi r = 4\pi$. So $0 \le s \le 4\pi$, and the parameterization is
$$x = 2\cos\left(\frac{s}{2}\right),\ y = 2\sin\left(\frac{s}{2}\right).$$

7. Arc length $= \sqrt{3^2 + 4^2} = 5$, so $0 \le s \le 5$, and the parameterization is $x = \frac{3}{5}s$, $y = \frac{4}{5}s$.

9. $\mathbf{r}'(t) = \langle 3, 2t \rangle$
$$T(t) = \frac{1}{\sqrt{9+4t^2}}\langle 3, 2t \rangle$$
$$T(0) = \frac{1}{\sqrt{9}}\langle 3, 0 \rangle = \langle 1, 0 \rangle$$
$$T(-1) = \frac{1}{\sqrt{13}}\langle 3, -2 \rangle$$
$$T(1) = \frac{1}{\sqrt{13}}\langle 3, 2 \rangle$$

11. $\mathbf{r}'(t) = (-3\sin t, 2\cos t)$
$$T(t) = \frac{1}{\sqrt{9\sin^2 t + 4\cos^2 t}}\langle -3\sin t, 2\cos t \rangle$$
$$T(0) = \frac{1}{\sqrt{4}}\langle 0, 2 \rangle - \langle 0, 1 \rangle$$
$$T\left(-\frac{\pi}{2}\right) = \frac{1}{\sqrt{9}}\langle 3, 0 \rangle = \langle 1, 0 \rangle$$
$$T\left(\frac{\pi}{2}\right) = \frac{1}{\sqrt{9}}\langle -3, 0 \rangle = \langle -1, 0 \rangle$$

13. $\mathbf{r}'(t) = \langle 3, -2\sin 2t, 2\cos 2t \rangle$
$$T(t) = \frac{\mathbf{r}'(t)}{\|\mathbf{r}'(t)\|}$$
$$= \frac{1}{\sqrt{13}}\langle 3, -2\sin 2t, 2\cos 2t \rangle$$
$$T(0) = \frac{1}{\sqrt{13}}\langle 3, 0, 2 \rangle$$
$$T(-\pi) = \frac{1}{\sqrt{13}}\langle 3, 0, 2 \rangle$$
$$T(\pi) = \frac{1}{\sqrt{13}}\langle 3, 0, 2 \rangle$$

15.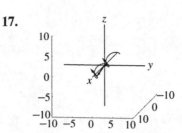

17.

19. $\mathbf{r}'(t) = \langle -2e^{-2t}, 2, 0 \rangle$ and $\mathbf{r}'(0) = \langle -2, 2, 0 \rangle$
$\mathbf{r}''(t) = \langle 4e^{-2t}, 0, 0 \rangle$ and $\mathbf{r}''(0) = \langle 4, 0, 0 \rangle$
$$\kappa = \frac{\|r'(0) \times r''(0)\|}{\|r'(0)\|^3}$$
$$= \frac{\|\langle 0, 0, -8 \rangle\|}{(2\sqrt{2})^3}$$
$$= \left(\frac{1}{2}\right)^{3/2}$$
$$\approx .3536$$

21. $\mathbf{r}'(t) = \langle 1, 2\cos(2t), 3 \rangle$ and $\mathbf{r}'(0) = \langle 1, 2, 3 \rangle$
$\mathbf{r}''(t) = \langle 0, -4\sin(2t), 0 \rangle$ and $\mathbf{r}''(0) = \langle 0, 0, 0 \rangle$
$$\kappa = \frac{\|r'(0) \times r''(0)\|}{\|r'(0)\|^3} = \frac{\|\langle 0, 0, 0 \rangle\|}{(\sqrt{14})^3} = 0$$

23. $f'(x) = 6x$
$f''(x) = 6$
$f'(1) = 6$
$f''(1) = 6$
$$\kappa = \frac{|f''(x)|}{\left\{1 + [f'(x)]^2\right\}^{3/2}}$$
$$= \frac{|6|}{(1+6^2)^{3/2}}$$
$$= \frac{6}{37\sqrt{37}}$$
$$\approx 0.0266$$

25. $f'(x) = \cos(x)$

$f''(x) = -\sin(x)$

$f'\left(\dfrac{\pi}{2}\right) = 0$

$f''\left(\dfrac{\pi}{2}\right) = -1$

$\kappa = \dfrac{|f''(x)|}{\left\{1 + [f'(x)]^2\right\}^{3/2}}$

$= \dfrac{|-1|}{(1 + 0^2)^{3/2}} = 1$

27. For $f(x) = \sin x$, $f'(x) = \cos x$ and

$f''(x) = -\sin x$. Then

$\left|f''\left(\dfrac{\pi}{2}\right)\right| = \left|f''\left(\dfrac{3\pi}{2}\right)\right| = 1$ and

$\left[f'\left(\dfrac{\pi}{2}\right)\right]^2 = \left[f'\left(\dfrac{3\pi}{2}\right)\right]^2 = 0$

So $\kappa = \dfrac{|f''(x)|}{\left\{1 + [f'(x)]^2\right\}^{3/2}}$ will be the same for

$x = \dfrac{\pi}{2}$ and $x = \dfrac{3\pi}{2}$. At $x = \pi$ the graph looks

(instantaneously) straight, so the curvature would

be smaller.

29.

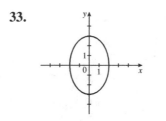

$\mathbf{r}'(t) = \left\langle -4\sin(2t),\ 4\cos(2t),\ 3 \right\rangle$ and

$\mathbf{r}''(t) = \left\langle -8\cos(2t),\ -8\sin(2t),\ 0 \right\rangle$

$\mathbf{r}'(0) = \left\langle 0, 4, 3 \right\rangle$ and $\mathbf{r}''(0) = \left\langle -8, 0, 0 \right\rangle$, so at $t = 0$

$\kappa = \dfrac{\|\mathbf{r}'(0) \times \mathbf{r}''(0)\|}{\|\mathbf{r}'0\|^3} = \dfrac{\|\langle 0, -24, 32 \rangle\|}{5^3} = \dfrac{8}{25}$

$\mathbf{r}'\left(\dfrac{\pi}{2}\right) = \left\langle 0, -4, 3 \right\rangle$ and

$\mathbf{r}''\left(\dfrac{\pi}{2}\right) = \left\langle 8, 0, 0 \right\rangle$, so at $t = \dfrac{\pi}{2}$

$\kappa = \dfrac{\left\|\mathbf{r}'\left(\frac{\pi}{2}\right) \times \mathbf{r}''\left(\frac{\pi}{2}\right)\right\|}{\left\|\mathbf{r}'\left(\frac{\pi}{2}\right)\right\|^3} = \dfrac{\|\langle 0, 24, 32 \rangle\|}{5^3} = \dfrac{8}{25}$

31.

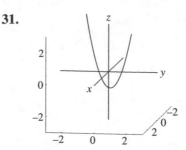

$\mathbf{r}'(t) = \left\langle 1, 1, 2t \right\rangle$ and $\mathbf{r}''(t) = \left\langle 0, 0, 2 \right\rangle$

$\mathbf{r}'(0) = \left\langle 1, 1, 0 \right\rangle$ and $\mathbf{r}''(0) = \left\langle 0, 0, 2 \right\rangle$, so at $t = 0$

$\kappa = \dfrac{\|\mathbf{r}'(0) \times \mathbf{r}''(0)\|}{\|\mathbf{r}'(0)\|^3} = \dfrac{\|\langle 2, -2, 0 \rangle\|}{\left(\sqrt{2}\right)^3} = 1$

$\mathbf{r}'(2) = \left\langle 1, 1, 4 \right\rangle$ and $\mathbf{r}''(2) = \left\langle 0, 0, 2 \right\rangle$, so at $t = 2$

$\kappa = \dfrac{\|\mathbf{r}'(0) \times \mathbf{r}''(0)\|}{\|\mathbf{r}'(0)\|^3} = \dfrac{\|\langle 2, -2, 0 \rangle\|}{\left(3\sqrt{2}\right)^3} = \dfrac{1}{27}$

33.

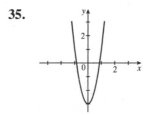

From inspection of the graph, the points of

maximum curvature are at $\left\langle 0, \pm 3 \right\rangle$ and the points

of minimum curvature are at $\left\langle \pm 2, 0 \right\rangle$.

35.

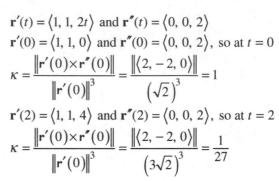

From inspection of the graph, the point of

maximum curvature is at $(0, -3)$ and there is no

point of minimum curvature.

37. $y' = 2e^{2x}$, $y'' = 4e^{2x}$

$$\lim_{x\to\infty} \kappa = \lim_{x\to\infty} \frac{\left|4e^{2x}\right|}{\left[1+(2e^{2x})^2\right]^{3/2}}$$

$$= \lim_{x\to\infty} \frac{(4^{2/3}e^{4x/3})^{3/2}}{(1+8e^{4x})^{3/2}}$$

$$= \lim_{x\to\infty} \left(\frac{4^{2/3}}{\frac{1}{e^{4x/3}}+8e^{8x/3}}\right)^{3/2} = 0$$

39. $y' = 3x^2$, $y'' = 6x$

$$\lim_{x\to\infty} \kappa = \lim_{x\to\infty} \frac{\left|6x\right|}{\left[1+(3x^2)^2\right]^{3/2}} = 0$$

41. As $x \to \infty$, the curve straightens out. So
$$\lim_{x\to\infty} \kappa = 0.$$

43. False; consider the spiral $\langle t\cos t,\ t\sin t\rangle$, $t \geq 0$:
the curvature steadily decreases with increasing t at every point, including local extrema.

45. True; the curves are the same, offset by c units.

47. With $\mathbf{s}(\theta) = \langle f(\theta)\cos\theta,\ f(\theta)\sin\theta,\ 0\rangle$ representing $r = f(\theta)$,

$\mathbf{s}'(\theta) = \langle f'(\theta)\cos\theta - f(\theta)\sin\theta,\ f'(\theta)\sin\theta + f(\theta)\cos\theta,\ 0\rangle$ and

$\mathbf{s}''(\theta) = \langle f''(\theta)\cos\theta - 2f'(\theta)\sin(\theta) - f(\theta)\cos\theta,\ f''(\theta)\sin\theta + 2f'(\theta)\cos\theta - f(\theta)\sin\theta,\ 0\rangle$.

$\mathbf{s}'(\theta) \times \mathbf{s}''(\theta)$

$= \left\langle 0, 0, 2[f'(\theta)\cos\theta]^2 - f(\theta)f''(\theta)\sin^2\theta + [f(\theta)\sin\theta]^2 + 2[f'(\theta)\sin\theta]^2 - f(\theta)f''(\theta)\cos^2\theta + [f(\theta)\cos\theta]^2 \right\rangle$

$= \left\langle 0, 0, 2[f'(\theta)]^2 - f(\theta)f''(\theta) + [f(\theta)]^2 \right\rangle$

Then $\kappa = \dfrac{\left\|\mathbf{s}'(t)\times\mathbf{s}''(t)\right\|}{\left\|\mathbf{s}'(t)\right\|^3} = \dfrac{\left|2[f'(\theta)]^2 - f(\theta)f''(\theta) + [f(\theta)]^2\right|}{\left\{[f'(\theta)]^2 + [f(\theta)]^2\right\}^{3/2}}$

49. $f(\theta) = \sin 3\theta$, $f'(\theta) = 3\cos 3\theta$,
$f''(\theta) = -9\sin 3\theta$ At $\theta = 0$, $f(0) = 0$,
$f'(0) = 3$, $f''(0) = 0$, and $\kappa = \dfrac{2}{|3|} = \dfrac{2}{3}$.

At $\theta = \dfrac{\pi}{6}$, $f\left(\dfrac{\pi}{6}\right) = 1$, $f'\left(\dfrac{\pi}{6}\right) = 0$,

$f''\left(\dfrac{\pi}{6}\right) = -9$, and

$\kappa = \dfrac{\left|2(0)^2 - (1)(-9) + (1)^2\right|}{\left[(0)^2 + (1)^2\right]^{3/2}} = \dfrac{10}{1} = 10.$

51. $f(\theta) = 3e^{2\theta}$, $f'(\theta) = 6e^{2\theta}$,
$f''(\theta) = 12e^{2\theta}$ At $\theta = 0$, $f(0) = 3$,
$f'(0) = 6$, $f''(0) = 12$, and
$\kappa = \dfrac{\left|2(6)^2 - (3)(12) + (3)^2\right|}{\left[(6)^2 + (3)^2\right]^{3/2}} = \dfrac{1}{\sqrt{45}}.$

At $\theta = 1$, $f(1) = 3e^2$, $f'(1) = 6e^2$,
$f''(1) = 12e^2$, and

$$\kappa = \frac{\left|2(6e^2)^2 - (3e^2)(12e^2) + (3e^2)^2\right|}{\left[(6e^2)^2 + (3e^2)^2\right]^{3/2}}$$

$$= \frac{45e^4}{(45e^4)^{3/2}} = \frac{1}{\sqrt{45}}e^{-2}.$$

53. Working through Example 4.5 for the general helix $\mathbf{r}(t) = \langle a\sin(t),\ a\cos(t),\ bt\rangle$

gives the curvature formula $\kappa = \dfrac{a}{a^2+b^2}$. For

$a = 2$ and $b = 0.4$, $\kappa = \dfrac{25}{52} \approx 0.481$.

The slower rate of climb of the helix produces a larger curvature than in Example 4.5

Section 11.5

5. $\mathbf{r}(t) = \left\langle t, t^2 \right\rangle$

$\mathbf{r}'(t) = \langle 1, 2t \rangle$ and $\langle \mathbf{r}'(t) \rangle = \sqrt{4t^2 + 1}$.

$\mathbf{T}(t) = -\dfrac{1}{\sqrt{4t^2 + 1}} \langle 1, 2t \rangle$

$\mathbf{T}'(t) = -\dfrac{4t}{(4t^2 + 1)^{3/2}} \langle 1, 2t \rangle + \dfrac{1}{\sqrt{4t^2 + 1}} \langle 0, 2 \rangle$

$\quad = \left\langle -\dfrac{4t}{(4t^2 + 1)^{3/2}}, \dfrac{2}{(4t^2 + 1)^{3/2}} \right\rangle$

and

$\| \mathbf{T}'(t) \| = \dfrac{2}{4t^2 + 1}$

$\mathbf{N}(t) = \dfrac{4t^2 + 1}{2} \left\langle -\dfrac{4t}{(4t^2 + 1)^{3/2}}, \dfrac{2}{(4t^2 + 1)^{3/2}} \right\rangle$

$\quad = \dfrac{1}{\sqrt{4t^2 + 1}} \langle -2t, 1 \rangle$

$\mathbf{T}(0) = \langle 1, 0 \rangle$ and $\mathbf{N}(0) = \langle 0, 1 \rangle$

$\mathbf{T}(1) = \dfrac{1}{\sqrt{5}} \langle 1, 2 \rangle$ and $\mathbf{N}(1) = \dfrac{1}{\sqrt{5}} \langle -2, 1 \rangle$

7. $\mathbf{r}(t) = \left\langle \cos 2t, \sin 2t \right\rangle$

$\mathbf{r}'(t) = \langle -2 \sin 2t, 2 \cos 2t \rangle$ and $\| \mathbf{r}'(t) \| = 2$.

$\mathbf{T}(t) = \dfrac{1}{2} \langle -2 \sin 2t, 2 \cos 2t \rangle$

$\quad = \langle -\sin 2t, \cos 2t \rangle$

$\mathbf{T}'(t) = \langle -2 \cos 2t, -2 \sin 2t \rangle$

and $\| \mathbf{T}'(t) \| = 2$

$\mathbf{N}(t) = \dfrac{1}{2} \langle -2 \cos 2t, -2 \sin 2t \rangle$

$\quad = \langle -\cos 2t, -\sin 2t \rangle$

$\mathbf{T}(0) = \langle 0, 1 \rangle$ and $\mathbf{N}(0) = \langle -1, 0 \rangle$

$\mathbf{T}\left(\dfrac{\pi}{4} \right) = \langle -1, 0 \rangle$ and $\mathbf{N}\left(\dfrac{\pi}{4} \right) = \langle 0, -1 \rangle$

9. $\mathbf{r}(t) = \left\langle \cos 2t, t, \sin 2t \right\rangle$

$\mathbf{r}'(t) = \langle -2 \sin 2t, 1, 2 \cos 2t \rangle$

and $\| \mathbf{r}'(t) \| = \sqrt{5}$.

$\mathbf{T}(t) = \dfrac{1}{\sqrt{5}} \langle -2 \sin 2t, 1, 2 \cos 2t \rangle$

$\mathbf{T}'(t) = \left\langle -\dfrac{4 \cos 2t}{\sqrt{5}}, 0, -\dfrac{4 \sin 2t}{\sqrt{5}} \right\rangle$

and $\| \mathbf{T}'(t) \| = \dfrac{4}{\sqrt{5}}$

$\mathbf{N}(t) = \dfrac{\sqrt{5}}{4} \left\langle -\dfrac{4 \cos 2t}{\sqrt{5}}, 0, -\dfrac{4 \sin 2t}{\sqrt{5}} \right\rangle$

$\quad = \langle -\cos 2t, 0, -\sin 2t \rangle$

$\mathbf{T}(0) = \dfrac{1}{\sqrt{5}} \langle 0, 1, 2 \rangle$ and $\mathbf{N}(0) = \langle -1, 0, 0 \rangle$

$\mathbf{T}\left(\dfrac{\pi}{2} \right) = \dfrac{1}{\sqrt{5}} \langle 0, 1, -2 \rangle$ and $\mathbf{N}\left(\dfrac{\pi}{2} \right) = \langle 1, 0, 0 \rangle$

11.　$\mathbf{r}(t) = \left\langle t, t^2 - 1, t \right\rangle$

$\mathbf{r}'(t) = \langle 1, 2t, 1 \rangle$ and $\| \mathbf{r}'(t) \| = \sqrt{4t^2 + 2}$.

$\mathbf{T}(t) = \dfrac{1}{\sqrt{4t^2 + 2}} \langle 1, 2t, 1 \rangle$

$\mathbf{T}'(t) = -\dfrac{4t}{(4t^2 + 2)^{3/2}} \langle 1, 2t, 1 \rangle + \dfrac{1}{\sqrt{4t^2 + 2}} \langle 0, 2, 0 \rangle = \left\langle -\dfrac{4t}{(4t^2 + 2)^{3/2}}, \dfrac{\sqrt{2}}{(2t^2 + 1)^{3/2}}, -\dfrac{4t}{(4t^2 + 2)^{3/2}} \right\rangle$

and $\| \mathbf{T}'(t) \| = \dfrac{\sqrt{2}}{2t^2 + 1}$

$\mathbf{N}(t) = \dfrac{2t^2 + 1}{\sqrt{2}} \left\langle -\dfrac{4t}{(4t^2 + 2)^{3/2}}, \dfrac{\sqrt{2}}{(2t^2 + 1)^{3/2}}, -\dfrac{4t}{(4t^2 + 2)^{3/2}} \right\rangle = \dfrac{1}{\sqrt{2t^2 + 1}} \langle -t, 1, -t \rangle$

$\mathbf{T}(0) = \dfrac{1}{\sqrt{2}} \langle 1, 0, 1 \rangle$ and $\mathbf{N}(0) = \langle 0, 1, 0 \rangle$

$\mathbf{T}(1) = \dfrac{1}{\sqrt{6}} \langle 1, 2, 1 \rangle$ and $\mathbf{N}(1) = \dfrac{1}{\sqrt{3}} \langle -1, 1, -1 \rangle$

13. The curve is the parabola $f(x) = x^2$. The principal unit normal vector at $\langle 0, 0 \rangle$ points upward along the y-axis: $\langle 0, 1 \rangle$. $f'(x) = 2x$ and $f''(x) = 2$, and so $f'(0) = 0$ and $f''(0) = 2$.

Then $\kappa = \dfrac{2}{\left[1 + 4(0)^2 \right]^{3/2}} = 2$

and $\rho = \dfrac{1}{2}$. The osculating circle is centered on the y-axis, $\dfrac{1}{2}$ unit up from the origin. It has

equation $x^2 + \left(y - \dfrac{1}{2} \right)^2 = \dfrac{1}{4}$.

15. The curve is a circle of radius 1, centered at the origin. Hence, at any point, the osculating circle is identical to the curve. It has equation $x^2 + y^2 = 1$.

17. $\mathbf{r}'(t) = \langle 8, 16 - 32t \rangle$

$\mathbf{a}(t) = \mathbf{r}''(t) = \langle 0, -32 \rangle$ and $\| \mathbf{a}(t) \| = 32$

$\dfrac{ds}{dt} = \| \mathbf{r}'(t) \|$

$= \sqrt{8^2 + (16 - 32t)^2}$

$= 8\sqrt{16t^2 - 16t + 5}$

$\dfrac{d^2 s}{dt^2} = \dfrac{d}{dt} \left(8\sqrt{16t^2 - 16t + 5} \right) = \dfrac{64(2t - 1)}{\sqrt{16t^2 - 16t + 5}}$

at $t = 0$:

$a_T = \dfrac{d^2 s}{dt^2} = \dfrac{64(-1)}{\sqrt{5}} = -\dfrac{64}{\sqrt{5}}$

$a_N = \sqrt{\| a \|^2 - a_T^2} = \sqrt{32^2 - \left(-\dfrac{64}{\sqrt{5}} \right)^2} = \dfrac{32}{\sqrt{5}}$

at $t = 1$:

$a_T = \dfrac{d^2 s}{dt^2} = \dfrac{64(1)}{\sqrt{16 - 16 + 5}} = \dfrac{64}{\sqrt{5}}$

$a_N = \sqrt{\| a \|^2 - a_T^2} = \sqrt{32^2 - \left(\dfrac{64}{\sqrt{5}} \right)^2} = \dfrac{32}{\sqrt{5}}$

19. $\mathbf{r}'(t) = \langle -2\sin 2t, \ 2t, \ 2\cos 2t \rangle$

$\mathbf{a}(t) = \mathbf{r}''(t)$

$= \langle -4\cos 2t, \ 2, \ -4\sin 2t \rangle$

and $\| \mathbf{a}(t) \| = 2\sqrt{5}$

$\dfrac{ds}{dt} = \| \mathbf{r}'(t) \|$

$= \sqrt{(-2\sin 2t)^2 + (2t)^2 + (2c\cos 2t)^2}$

$= 2\sqrt{t^2 + 1}$

$\dfrac{d^2 s}{dt^2} = \dfrac{d}{dt} \left(2\sqrt{t^2 + 1} \right) = \dfrac{2t}{\sqrt{t^2 + 1}}$

at $t = 0$:

$a_T = \dfrac{d^2 s}{dt^2} = \dfrac{2(0)}{\sqrt{0 + 1}} = 0$

$a_N = \sqrt{\| a \|^2 - a_T^2} = \sqrt{\left(2\sqrt{5} \right)^2 - 0^2} = 2\sqrt{5}$

at $t = \dfrac{\pi}{4}$:

$a_T = \dfrac{d^2 s}{dt^2} = \dfrac{2 \left(\dfrac{\pi}{4} \right)}{\sqrt{\left(\dfrac{\pi}{4} \right)^2 + 1}} = \dfrac{2\pi}{\sqrt{\pi^2 + 16}}$

$a_N = \sqrt{\| a \|^2 - a_T^2}$

$= \sqrt{\left(2\sqrt{5} \right)^2 - \left(\dfrac{2\pi}{\sqrt{\pi^2 + 16}} \right)^2}$

$= 4\sqrt{\dfrac{\pi^2 + 20}{\pi^2 + 16}}$

21. At $t = 0$, $a_T = 0$, so the object's speed is neither increasing nor decreasing. At $t = \dfrac{\pi}{4}$, $a_T = \dfrac{2\pi}{\sqrt{\pi^2 + 16}} \approx 1.2353 > 0$, so the object's speed is increasing.

23. $\mathbf{r}'(t) = \langle -a\sin t,\ a\cos t,\ b \rangle$

$\mathbf{a}(t) = \mathbf{r}''(t) = \langle -a\cos t,\ -a\sin t,\ 0 \rangle$

$\|\mathbf{a}(t)\| = \sqrt{(-a\cos t)^2 + (-a\sin t)^2 + 0^2} = a$

$\dfrac{ds}{dt} = \|\mathbf{r}'(t)\|$

$\qquad = \sqrt{(-a\sin t)^2 + (a\cos t)^2 + b^2}$

$\qquad = \sqrt{a^2 + b^2}$

$a_T = \dfrac{d^2 s}{dt^2} = \dfrac{d}{dt}\sqrt{a^2 + b^2} = 0$

$a_N = \sqrt{\|\mathbf{a}\|^2 - a_T^2} = \sqrt{a^2 - 0^2} = a$

25. $\mathbf{r}'(t) = \langle 1,\ 2,\ 2t \rangle$ and $\|\mathbf{r}'(t)\| = \sqrt{4t^2 + 5}$

$\mathbf{T}(t) = \dfrac{1}{\sqrt{4t^2 + 5}}\langle 1,\ 2,\ 2t \rangle$

$\mathbf{T}'(t)$

$= -\dfrac{4t}{\left(4t^2 + 5\right)^{3/2}}\langle 1,\ 2,\ 2t \rangle + \dfrac{1}{\sqrt{4t^2 + 5}}\langle 0,\ 0,\ 2 \rangle$

$= \dfrac{2}{\left(4t^2 + 5\right)^{3/2}}\langle -2t,\ -4t,\ 5 \rangle$

At $t = 0$,

$\mathbf{T}(0) = \dfrac{1}{\sqrt{0+5}}\langle 1,\ 2,\ 0 \rangle = \dfrac{1}{\sqrt{5}}\langle 1,\ 2,\ 0 \rangle,$

$\mathbf{T}'(0) = \dfrac{2}{5^{3/2}}\langle 0,\ 0,\ 5 \rangle = \dfrac{2}{\sqrt{5}}\langle 0,\ 0,\ 1 \rangle$

and $\|\mathbf{T}'(0)\| = \dfrac{2}{\sqrt{5}},$

$\mathbf{N}(0) = \dfrac{\mathbf{T}'(0)}{\|\mathbf{T}'(0)\|} = \langle 0,\ 0,\ 1 \rangle,$

$\mathbf{B}(0) = \mathbf{T}(0) \times \mathbf{N}(0)$

$\qquad = \dfrac{1}{\sqrt{5}}\big(\langle 1,\ 2,\ 0 \rangle \times \langle 0,\ 0,\ 1 \rangle\big)$

$\qquad = \dfrac{1}{\sqrt{5}}\langle 2,\ -1,\ 0 \rangle$

At $t = 1$,

$\mathbf{T}(1) = \dfrac{1}{\sqrt{4+5}}\langle 1,\ 2,\ 2 \rangle = \dfrac{1}{3}\langle 1,\ 2,\ 2 \rangle,$

$\mathbf{T}'(1) = \dfrac{2}{(4+5)^{3/2}}\langle -2,\ -4,\ 5 \rangle$

$\qquad = \dfrac{2}{27}\langle -2,\ -4,\ 5 \rangle$

and $\|\mathbf{T}'(1)\| = \dfrac{2\sqrt{5}}{9},$

$\mathbf{N}(1) = \dfrac{\mathbf{T}'(1)}{\|\mathbf{T}'(1)\|}$

$\qquad = \left(\dfrac{9}{2\sqrt{5}}\right)\left(\dfrac{2}{27}\right)\langle -2,\ -4,\ 5 \rangle$

$\qquad = \dfrac{1}{3\sqrt{5}}\langle -2,\ -4,\ 5 \rangle$

$\mathbf{B}(1) = \mathbf{T}(1) \times \mathbf{N}(1)$

$\qquad = \left(\dfrac{1}{3}\right)\left(\dfrac{1}{3\sqrt{5}}\right)\big(\langle 1,\ 2,\ 2 \rangle \times \langle -2,\ -4,\ 5 \rangle\big)$

$\qquad = \dfrac{1}{9\sqrt{5}}\langle 18,\ -9,\ 0 \rangle$

$\qquad = \dfrac{1}{\sqrt{5}}\langle 2,\ -1,\ 0 \rangle$

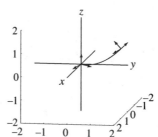

27. $\mathbf{r}'(t) = \langle -4\pi\sin\pi t,\ 4\pi\cos\pi t,\ 1\rangle$ and

$$\|\mathbf{r}'(t)\| = \sqrt{(-4\pi\sin\pi t)^2 + (4\pi\cos\pi t)^2 + 1^2}$$

$$= \sqrt{16\pi^2 + 1}$$

$$\mathbf{T}(t) = \frac{1}{\sqrt{16\pi^2+1}}\langle -4\pi\sin\pi t,\ 4\pi\cos\pi t,\ 1\rangle$$

$$\mathbf{T}'(t) = \frac{1}{\sqrt{16\pi^2+1}}\langle -4\pi^2\cos\pi t,\ -4\pi^2\sin\pi t,\ 0\rangle$$

At $t = 0$,

$$\mathbf{T}(0) = \frac{1}{\sqrt{16\pi^2+1}}\langle 0,\ 4\pi,\ 1\rangle$$

$$\mathbf{T}'(0) = \frac{1}{\sqrt{16\pi^2+1}}\langle -4\pi^2,\ 0,\ 0\rangle$$

$$\mathbf{N}(0) = \langle -1,\ 0,\ 0\rangle$$

$$\mathbf{B}(0) = \mathbf{T}(0)\times\mathbf{N}(0)$$

$$= \left(\frac{1}{\sqrt{16\pi^2+1}}\right)(\langle 0,\ 4\pi,\ 1\rangle\times\langle -1,\ 0,\ 0\rangle)$$

$$= \frac{1}{\sqrt{16\pi^2+1}}\langle 0,\ -1,\ 4\pi\rangle$$

At $t = 1$,

$$\mathbf{T}(1) = \frac{1}{\sqrt{16\pi^2+1}}\langle 0,\ -4\pi,\ 1\rangle$$

$$\mathbf{T}'(1) = \frac{1}{\sqrt{16\pi^2+1}}\langle 4\pi^2,\ 0,\ 0\rangle$$

$$\mathbf{N}(1) = \langle 1,\ 0,\ 0\rangle$$

$$\mathbf{B}(1) = \mathbf{T}(1)\times\mathbf{N}(1)$$

$$= \left(\frac{1}{\sqrt{16\pi^2+1}}\right)(\langle 0,\ -4\pi,\ 1\rangle\times\langle 1,\ 0,\ 0\rangle)$$

$$= \frac{1}{\sqrt{16\pi^2+1}}\langle 0,\ 1,\ 4\pi\rangle$$

29. True. $\dfrac{d\mathbf{T}}{ds} = \kappa\cdot\mathbf{N}$, and $\mathbf{N}$ and $\mathbf{T}$ are perpendicular.

31. True. $\mathbf{T}\cdot\mathbf{T} = \|\mathbf{T}\|^2 = 1^2 = 1$, which does not vary with S.

In exercises 33–35, $\mathbf{r}(t)$ moves in a circle at constant speed, and the tangential acceleration is zero. So

$$a_N = \sqrt{\|\mathbf{a}\|^2 - a_T^2} = \|\mathbf{a}\| = \|\mathbf{r}''(t)\|\ \text{and}$$

$$\mathbf{N}(t) = \frac{\mathbf{r}''(t)}{\|\mathbf{r}''(t)\|}.\ \text{Hence}\ \mathbf{F}_s(t) = ma_N\mathbf{N}(t) = 100\mathbf{r}''(t).$$

33. $\mathbf{F} = 100\dfrac{d^2}{dt^2}\langle 100\cos\pi t,\ 100\sin\pi t\rangle$

$$= 10{,}000\pi^2\langle -\cos\pi t,\ -\sin\pi t\rangle$$

35. $\mathbf{F} = 100\dfrac{d^2}{dt^2}\langle 100\cos 2\pi t,\ 100\sin 2\pi t\rangle$

$$= 40{,}000\pi^2\langle -\cos 2\pi t,\ -\sin 2\pi t\rangle$$

37. $\dfrac{20{,}000\pi^2}{10{,}000\pi^2} = 2$. The friction force doubles.

39. For $y = \cos x$, $\kappa = \dfrac{|-\cos x|}{\left[1 + (-\sin x)^2\right]^{3/2}}$ and

$$\rho = \frac{1}{\kappa} = \frac{\left[1 + (-\sin x)^2\right]^{3/2}}{|-\cos x|}.$$

At $x = 0$, $\rho = \dfrac{\left[1 + 0^2\right]^{3/2}}{|-1|} = 1$;

at $x = \dfrac{\pi}{4}$, $\rho = \dfrac{\left[1 + \left(-\frac{\sqrt{2}}{2}\right)^2\right]^{3/2}}{\left|-\frac{\sqrt{2}}{2}\right|} = \dfrac{\left(\frac{3}{2}\right)^{3/2}}{\left(\frac{1}{2}\right)^{1/2}} = \dfrac{3\sqrt{3}}{2}.$

Concavity: $y'' = -\cos x$.

$$y''(0) = -1\ \text{and}\ y''\left(\frac{\pi}{4}\right) = -\frac{\sqrt{2}}{2}$$

The absolute value of the concavity is greater at $x = 0$, where the curve bends more sharply and the approximating circle is smaller.

41. For the parametric curve $\mathbf{r}(t) = \langle t, t^2 \rangle$, the

curvature is $\dfrac{2}{\left(1+4t^2\right)^{3/2}}$, so the radius of the

osculating circle is $\dfrac{1}{2}\left(1+4t^2\right)^{3/2}$. At $\langle t, t^2 \rangle$ an

inward unit normal to the curve is

$\dfrac{1}{\sqrt{1+4t^2}}\langle -2t, 1 \rangle$, so the center of the osculating

circle is at

$\langle t, t^2 \rangle + \dfrac{1}{2}\left(1+4t^2\right)^{3/2} \cdot \dfrac{1}{\sqrt{1+4t^2}}\langle -2t, 1 \rangle$

or $\left\langle -4t^3, \dfrac{1}{2}+3t^2 \right\rangle$.

43. As outlined in the exercise.

45. $\dfrac{d\mathbf{B}}{dt} = \dfrac{d\mathbf{B}}{ds} \cdot \dfrac{ds}{dt}$ so $\dfrac{d\mathbf{B}}{ds}$ is a scalar multiple of

$\dfrac{d\mathbf{B}}{dt} = \mathbf{B}'$.

a. $\mathbf{B} \cdot \mathbf{B}' = 0$ because $\mathbf{B}$ is a unit vector.

b. $\mathbf{B}' = \dfrac{d}{dt}(\mathbf{T}\times\mathbf{N}) = \mathbf{T}'\times\mathbf{N} + \mathbf{T}\times\mathbf{N}'$

Since $\mathbf{N}$ is parallel to $\mathbf{T}'$, $\mathbf{T}'\times\mathbf{N} = \mathbf{0}$,

so $\mathbf{B}' = \mathbf{T}\times\mathbf{N}'$.

$\mathbf{T}\cdot\mathbf{B}' = \mathbf{T}\cdot(\mathbf{T}\times\mathbf{N}') = (\mathbf{T}\times\mathbf{T})\cdot\mathbf{N}' = \mathbf{0}\cdot\mathbf{N} = 0$.

Thus $\dfrac{d\mathbf{B}}{ds}$ is orthogonal to $\mathbf{T}$.

47. For a curve lying entirely in a plane parallel to the xy-plane, the binormal $\mathbf{B}$ will be parallel to the z-axis everywhere it is defined, so

$\dfrac{d\mathbf{B}}{ds} = \mathbf{0}$ and hence the scalar τ is 0.

49. a. $\mathbf{r}'(t) = \mathbf{T}(t)\cdot\|\mathbf{r}'(t)\| = \mathbf{T}(t)s'(t)$.

Differentiating again with respect to t,

$\mathbf{r}''(t) = s''(t)\mathbf{T}(t) + \mathbf{T}'(t)s'(t)$.

Since $\mathbf{T}' = \kappa s'(t)\mathbf{N}(t)$, this is equivalent to

$\mathbf{r}''(t) = s''(t)\mathbf{T}(t) + \kappa[s'(t)]^2\mathbf{N}(t)$.

b. Following the computation in (a), we have

$\mathbf{r}'(t) = \mathbf{T}(t)\cdot\|\mathbf{r}'(t)\| = \mathbf{T}(t)s'(t)$ and

$\mathbf{r}''(t) = s''(t)\mathbf{T}(t) + \kappa[s'(t)]^2\mathbf{N}(t)$.

$\mathbf{r}'(t)\times\mathbf{r}''(t)$

$= \mathbf{T}(t)s'(t)\times\left(s''(t)\mathbf{T}(t) + \kappa[s'(t)]^2\mathbf{N}(t)\right)$

$= s'(t)s''(t)\mathbf{T}(t)\times\mathbf{T}(t) + \kappa[s'(t)]^3\mathbf{T}(t)\times\mathbf{N}(t)$

$= \kappa[s'(t)]^3\mathbf{B}(t)$

c.

Differentiating $\mathbf{r}''(t) = s''(t)\mathbf{T}(t) + \kappa(t)[s'(t)]^2\mathbf{N}(t)$ with respect to t yields

$\mathbf{r}'''(t) = s'''(t)\mathbf{T}(t) + s''(t)\mathbf{T}'(t) + \kappa(t)2s'(t)s''(t)\mathbf{N}(t)$
$\qquad + \kappa(t)[s'(t)]^2\mathbf{N}'(t) + \kappa'(t)[s'(t)]^2\mathbf{N}(t)$

Using $\mathbf{T}' = \kappa(t)s'(t)\mathbf{N}(t)$ and, from Exercise 48(c), $\mathbf{N}'(t) = s'(t)\left(-\kappa(t)\mathbf{T}(t) + \tau(t)\mathbf{B}(t)\right)$, we have

$\mathbf{r}'''(t) = s'''(t)\mathbf{T}(t) + s''(t)\kappa(t)s'(t)\mathbf{N}(t)$
$\quad + \kappa(t)2s'(t)s''(t)\mathbf{N}(t)$
$\quad + \kappa(t)[s'(t)]^2\left(s'(t)(-\kappa(t)\mathbf{T}(t) + \tau(t)\mathbf{B}(t))\right)$
$\quad + \kappa'(t)[s'(t)]^2\mathbf{N}(t)$.

Collecting the coefficients of the vectors $\mathbf{T}$, $\mathbf{N}$, and $\mathbf{B}$, we get

$\mathbf{r}'''(t) = \left(s'''(t) - [\kappa(t)]^2[s'(t)]^3\right)\mathbf{T}(t)$
$\quad + \left(3\kappa(t)s'(t)s''(t) + \kappa'(t)[s'(t)]^2\right)\mathbf{N}(t)$
$\quad + \left(\kappa(t)\tau(t)[s'(t)]^3\right)\mathbf{B}(t)$.

d. Now when we take the dot product of our answer in (b) with the answer in (c), the only term that survives is the product of the coefficients of $\mathbf{B}$, namely

$\left(\kappa[s'(t)]^3\right)\left(\kappa\tau[s'(t)]^3\right)$,

and when we divide this by the squared magnitude of

$\mathbf{r}'(t)\times\mathbf{r}''(t)$, which is $\left(\kappa[s'(t)]^3\right)^2$,

we're left with the torsion, τ.

Chapter 11 Review

1. See Example 1.1.

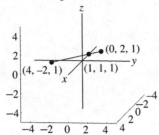

3. See Example 1.2.

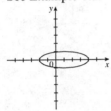

5. See Example 1.2.

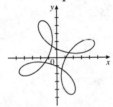

7. See Example 1.3.

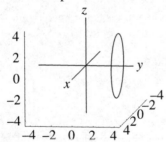

9. See Example 1.3.

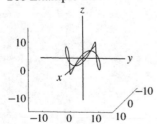

11. See Example 1.3.

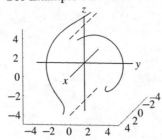

13. See Example 1.5.

   a. B

   b. C

   c. A

   d. F

   e. D

   f. E

15. See Example 1.6.

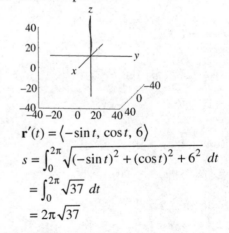

$$\mathbf{r}'(t) = \langle -\sin t,\ \cos t,\ 6 \rangle$$

$$s = \int_0^{2\pi} \sqrt{(-\sin t)^2 + (\cos t)^2 + 6^2}\ dt$$

$$= \int_0^{2\pi} \sqrt{37}\ dt$$

$$= 2\pi\sqrt{37}$$

17. See Example 2.1.

$$\lim_{t \to 1} \langle t^2 - 1,\ e^{2t},\ \cos \pi t \rangle$$

$$= \left\langle \lim_{t \to 1}(t^2 - 1),\ \lim_{t \to 1} e^{2t},\ \lim_{t \to 1} \cos \pi t \right\rangle$$

$$= \langle 0,\ e^2,\ -1 \rangle$$

19. See Example 2.3.

$\mathbf{r}(t) = \langle e^{4t},\ \ln t^2,\ 2t \rangle$ is continuous except where

$\ln t^2$, is undefined, namely at $t = 0$. So $\mathbf{r}(t)$ is

continuous for $t \neq 0$.

21. See Example 2.5.

$$\frac{d}{dt}\mathbf{r}(t) = \left\langle \frac{d}{dt}\sqrt{t^2+1},\ \frac{d}{dt}\sin 4t,\ \frac{d}{dt}\ln 4t \right\rangle$$

$$= \left\langle \frac{t}{\sqrt{t^2+1}},\ 4\cos 4t,\ \frac{1}{t} \right\rangle$$

23. See Example 2.7.

$$\int \left\langle e^{-4t},\ \frac{2}{t^3},\ 4t-1 \right\rangle dt$$

$$= \left\langle \int e^{-4t}\,dt,\ \int \frac{2}{t^3}\,dt,\ \int (4t-1)dt \right\rangle$$

$$= \left\langle -\frac{1}{4}e^{-4t},\ -t^{-2},\ 2t^2 - t \right\rangle + \mathbf{c}$$

25. See Example 2.8.

$$\int_0^1 \langle \cos \pi t,\ 4t,\ 2 \rangle\, dt$$

$$= \left\langle \int_0^1 \cos \pi t\ dt,\ \int_0^1 4t\ dt,\ \int_0^1 2\,dt \right\rangle$$

$$= \left\langle \left[\frac{1}{\pi}\sin \pi t\right]_0^1,\ \left[2t^2\right]_0^1,\ \left[2t\right]_0^1 \right\rangle$$

$$= \langle 0,\ 2,\ 2 \rangle$$

27. See Example 3.1.

$$\mathbf{v}(t) = \mathbf{r}'(t)$$

$$= \frac{d}{dt}\langle 4\cos 2t,\ 4\sin 2t,\ 4t \rangle$$

$$= \langle -8\sin 2t,\ 8\cos 2t,\ 4 \rangle$$

$$\mathbf{a}(t) = \mathbf{r}''(t)$$

$$= \frac{d}{dt}\langle -8\sin 2t,\ 8\cos 2t,\ 4 \rangle$$

$$= \langle -16\cos 2t,\ -16\sin 2t,\ 0 \rangle$$

29. See Example 3.2.

$$\mathbf{r}(t) = \int \mathbf{v}(t)\,dt = \left\langle t^2 + 4t,\ -16t^2 \right\rangle + \mathbf{c}$$

$$\mathbf{r}(0) = \langle 0,\ 0 \rangle + \mathbf{c} = \langle 2,\ 1 \rangle,\ \text{so}$$

$$\mathbf{r}(t) = \left\langle t^2 + 4t + 2,\ -16t^2 + 1 \right\rangle$$

31. See Example 3.2.

$$\mathbf{v}(t) = \int \mathbf{a}(t)\,dt = \langle 0,\ -32t \rangle + \mathbf{c}_1$$

$$\mathbf{v}(0) = \langle 0,\ 0 \rangle + \mathbf{c}_1 = \langle 4,\ 3 \rangle,\ \text{so}$$

$$\mathbf{v}(t) = \langle 0,\ -32t \rangle + \langle 4,\ 3 \rangle = \langle 4,\ -32t + 3 \rangle$$

$$\mathbf{r}(t) = \int \mathbf{v}(t)\,dt = \left\langle 4t,\ -16t^2 + 3t \right\rangle + \mathbf{c}_2$$

$$\mathbf{r}(0) = \langle 0,\ 0 \rangle + \mathbf{c}_2 = \langle 2,\ 6 \rangle,\ \text{so}$$

$$\mathbf{r}(t) = \left\langle 4t + 2,\ -16t^2 + 3t + 6 \right\rangle$$

33. See Example 3.2.and following remarks.

$$\mathbf{F} = m = 4\frac{d^2}{dt^2}\left\langle 12t,\ 12 - 16t^2 \right\rangle$$

$$= 4\langle 0,\ -32 \rangle$$

$$= \langle 0,\ -128 \rangle$$

35. See Example 3.4.

$$\text{Max. altitude} = h + \frac{v_0^2 \sin^2 \theta}{2g}$$

$$= 0 + \frac{80^2 \sin^2 \frac{\pi}{12}}{2(32)}$$

$$= \frac{80^2 \cdot \frac{1}{2}\left(1 - \cos\left(\frac{\pi}{6}\right)\right)}{2(32)}$$

$$= 25(2 - \sqrt{3})$$

$$\approx 6.70 \text{ feet}$$

Time of impact comes from solving

$$0 = h + (v_o \sin \theta)t - \frac{gt^2}{2}$$

$$0 = 0 + \left(80\sin\frac{\pi}{12}\right)t - \frac{32t^2}{2} \Rightarrow$$

$$t = \frac{80\sin\frac{\pi}{12}}{16} \approx 1.2941 \text{ seconds, at which time}$$

$$\text{horizontal range} = (v_o \cos \theta)t$$

$$\approx \left(80\cos\frac{\pi}{12}\right)(1.2941)$$

$$= 100 \text{ feet}$$

and speed at impact = launch speed
(because $h = 0$) which is 80 ft/s.

37. See Example 4.2.

$$\mathbf{r}'(t) = \left\langle -2e^{-2t},\ 2,\ 0 \right\rangle$$

At $t = 0$:

$$\mathbf{r}'(t) = \langle -2,\ 2,\ 0 \rangle \text{ and}$$

$$\mathbf{T}(0) = \frac{1}{2\sqrt{2}}\langle -2,\ 2,\ 0 \rangle = \frac{1}{\sqrt{2}}\langle -1,\ 1,\ 0 \rangle$$

At $t = 1$:

$$\mathbf{r}'(1) = \left\langle -2e^{-2},\ 2,\ 0 \right\rangle$$

$$\mathbf{T}(1) = \frac{1}{\sqrt{4e^{-4} + 4}}\left\langle -2e^{-2},\ 2,\ 0 \right\rangle$$

$$= \frac{1}{\sqrt{e^{-4} + 1}}\left\langle -e^{-2},\ 1,\ 0 \right\rangle$$

39. See Example 4.5.

$\mathbf{r}'(t) = \langle -\sin t, \cos t, \cos t \rangle$

$\mathbf{r}''(t) = \langle -\cos t, -\sin t, -\sin t \rangle$

At $t = 0$:

$\mathbf{r}'(0) = \langle 0, 1, 1 \rangle$ and $\mathbf{r}''(0) = \langle -1, 0, 0 \rangle$

$\mathbf{r}'(0) \times \mathbf{r}''(0) = \langle 0, -1, 1 \rangle$

and $\kappa = \dfrac{\| \mathbf{r}'(0) \times \mathbf{r}''(0) \|}{\| \mathbf{r}'(0) \|^3} = \dfrac{\sqrt{2}}{(\sqrt{2})^3} = \dfrac{1}{2}$

At $t = \dfrac{\pi}{4}$:

$\mathbf{r}'\left(\dfrac{\pi}{4}\right) = \left\langle -\dfrac{1}{\sqrt{2}}, \dfrac{1}{\sqrt{2}}, \dfrac{1}{\sqrt{2}} \right\rangle = \dfrac{1}{\sqrt{2}} \langle -1, 1, 1 \rangle$

$\mathbf{r}''\left(\dfrac{\pi}{4}\right) = \left\langle -\dfrac{1}{\sqrt{2}}, -\dfrac{1}{\sqrt{2}}, -\dfrac{1}{\sqrt{2}} \right\rangle$

$\qquad = \dfrac{1}{\sqrt{2}} \langle -1, -1, -1 \rangle$

$\mathbf{r}'\left(\dfrac{\pi}{4}\right) \times \mathbf{r}''\left(\dfrac{\pi}{4}\right)$

$= \left(\dfrac{1}{\sqrt{2}}\right)\left(\dfrac{1}{\sqrt{2}}\right)(\langle -1, 1, 1 \rangle \times \langle -1, -1, -1 \rangle)$

$= \dfrac{1}{2} \langle 0, -2, 2 \rangle$

$= \langle 0, -1, 1 \rangle$

and $\kappa = \dfrac{\left\| \mathbf{r}'\left(\frac{\pi}{4}\right) \times \mathbf{r}''\left(\frac{\pi}{4}\right) \right\|}{\left\| \mathbf{r}'\left(\frac{\pi}{4}\right) \right\|^3} = \dfrac{\sqrt{2}}{\left(\frac{\sqrt{3}}{\sqrt{2}}\right)^3} = \dfrac{4}{3\sqrt{3}}$

41. See Example 4.3.

Because the components of $\mathbf{r}(t)$ are linear functions of t, $\mathbf{r}(t)$ traces out a straight line. Its curvature everywhere is zero.

43. See Example 5.1.

$\mathbf{r}'(t) = \langle -\sin t, \cos t, \cos t \rangle$

$\| \mathbf{r}'(t) \| = \sqrt{(-\sin t)^2 + (\cos t)^2 + (\cos t)^2}$

$\qquad = \sqrt{1 + \cos^2 t}$

$\mathbf{T}(t) = \dfrac{1}{\sqrt{1 + \cos^2 t}} \langle -\sin t, \cos t, \cos t \rangle$

$\mathbf{T}'(t) = \dfrac{\sin t \cos t}{(1 + \cos^2 t)^{3/2}} \langle -\sin t, \cos t, \cos t \rangle$

$\qquad + \dfrac{1}{\sqrt{1 + \cos^2 t}} \langle -\cos t, -\sin t, -\sin t \rangle$

At $t = 0$:

$\mathbf{T}(0) = \dfrac{1}{\sqrt{1+1}} \langle 0, 1, 1 \rangle = \dfrac{1}{\sqrt{2}} \langle 0, 1, 1 \rangle$

$\mathbf{T}'(0) = \dfrac{0}{(1+1)^{3/2}} \langle 0, 1, 1 \rangle + \dfrac{1}{\sqrt{1+1}} \langle -1, 0, 0 \rangle$

$\qquad = \left\langle -\dfrac{1}{\sqrt{2}}, 0, 0 \right\rangle$

and $\mathbf{N}(0) = \langle -1, 0, 0 \rangle$

45. See Example 5.6.

$\mathbf{r}'(t) = \langle 2, 2t, 0 \rangle$ and $\| \mathbf{r}'(t) \| = 2\sqrt{t^2 + 1}$

$\mathbf{r}''(t) = \langle 0, 2, 0 \rangle$ and $\| \mathbf{r}''(t) \| = 2$

$\| \mathbf{a} \| = \| \mathbf{r}''(t) \| = 2$

$a_T = \dfrac{d^2 s}{dt^2} = \dfrac{d}{dt} \| \mathbf{r}'(t) \| = \dfrac{2t}{\sqrt{t^2 + 1}}$

At $t = 0$:

$a_T = \dfrac{2(0)}{\sqrt{0+1}} = 0$

$a_N = \sqrt{\| \mathbf{a} \|^2 - a_T^2} = \sqrt{2^2 - (0)^2} = 2$

At $t = 1$:

$a_T = \dfrac{2(1)}{\sqrt{1+1}} = \sqrt{2}$

$a_N = \sqrt{\| \mathbf{a} \|^2 - a_T^2} = \sqrt{2^2 - (\sqrt{2})^2} = \sqrt{4-2} = \sqrt{2}$

47. See Example 5.6 and Exercises 5.33–36.

$\mathbf{r}'(t) = \langle -480 \sin 6t, 480 \cos 6t \rangle$

and $\| \mathbf{r}'(t) \| = 480$

$\mathbf{r}''(t) = \langle -2880 \cos 6t, -2880 \sin 6t \rangle$

and $\| \mathbf{r}''(t) \| = 2880$

$a_T = 0$ because $\| \mathbf{r}'(t) \|$ is constant.

$a_N \mathbf{N}(t) = \mathbf{r}''(t)$ because $a_T = 0$.

$\mathbf{F}_s(t) = m a_N \mathbf{N}(t)$

$\qquad = 120 \langle -2880 \cos 6t, -2880 \sin 6t \rangle$

$\qquad = 345,600 \langle -\cos 6t, -\sin 6t \rangle$

Chapter 12

Section 12.1

5. The domain is the set of all values (x, y) except the line $y = -x$. $\{(x, y) | -x \neq y\}$

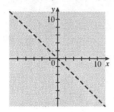

7. The domain is the half-plane given by $y > -x - 2$. $\{x, y | x + y + 2 > 0 \text{ or } y > -(x + 2)\}$

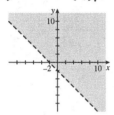

9. Range: $\{f(x, y) | f(x, y) \geq 0\}$

11. Range: $\{f(x, y) | -1 \leq f(x, y) \leq 1\}$

13. Range: $\{f(x, y) | f(x, y) \geq -1\}$

15. $f(1, 2) = 3, f(0, 3) = 3$

17. **a.** $R(150, 1000) = 312$
b. $R(150, 2000) = 331$
c. $R(150, 3000) = 350$
d. About 19 feet.

19. $f(x, y) = x^2 + y^2$

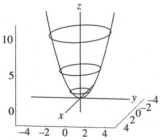

The traces give a graph that is representative of the function.

21. $f(x, y) = \sqrt{x^2 + y^2}$

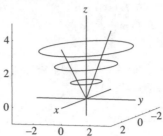

The traces give a graph that is representative of the function.

23.

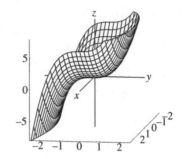

25.

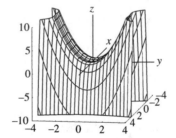

27.

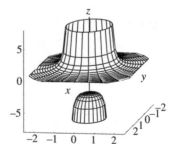

29.

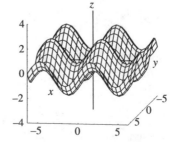

31.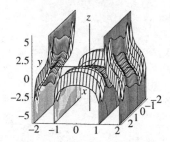

33.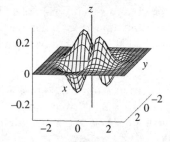

35.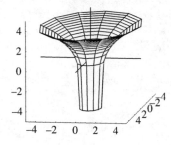

37.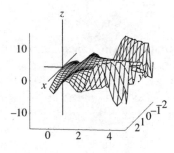

39.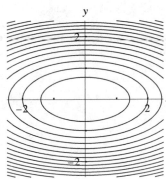

41.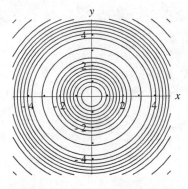

43.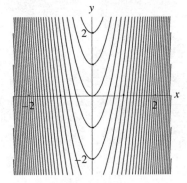

45.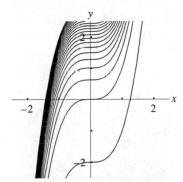

47.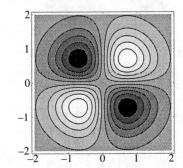

49.

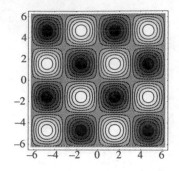

51. a. Surface B

   b. Surface D

   c. Surface A

   d. Surface F

   e. Surface C

   f. Surface E

53. a. Contour A

   b. Contour D

   c. Contour C

   d. Contour B

55.

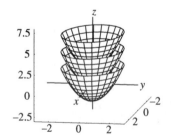

57. a. Looking from the positive x-axis you observe traces of $-y^2$ curves: Surface B

   b. Looking from the positive y-axis you observe traces of x^2 curves: Surface A

59. Since deformation of the square grid depends on changes in height, no deformation is observed from the z-axis.

61. The wave travels parallel to the line $y = x$. From (100, 100, 0) the waves fill in the vertical region between $z = -1$ and $z = 1$, so the top and bottom of the shape are straight lines. What the ends look like depends on how the plotting window is set.

63. For example: $(100, 100\sqrt{3}, 0)$

65. The stadium would be located in the upper left hand corner in the concentric ellipses. The line could be a highly used roadway and the other circular level curves could represent businesses along the roadway.

67. The point of maximum power would be at the smallest of the concentric ellipses. Power would increase away from the frame, and players would want to hit the ball at this spot to obtain maximum power.

69. Using a maximum of 4 for *HS,* 800 for *SATV* and *SATM,* we get *PGA* = 3.942 so that it is not possible to give a *PGA* of 4. Using a minimum of 0 for *HS*, 200 for *SATV* and *SATM*, we get *PGA* = −.57 so that it is possible to have a *PGA* that is negative. The *HS* variable is most important because of the weight carried by the term .708 *HS*. This term has a range from 0 to 2.832, a much larger range than the other terms.

71. Average speed is total distance over total time:
$$\frac{2d}{\dfrac{d}{x} + \dfrac{d}{y}} = \frac{2xy}{x + y}$$

Under the first scenario you would have to go 60 mph for the second 20 miles. Under the second scenario, you would have to go infinitely fast, because you have already used up all your time covering the first 20 miles.

73. The graphs have the same shape, but the images of the grid lines will be different.

75. The graphs have the same shape, but the images of the grid lines will be different.

Section 12.2

5. Since the limit of the denominator is not 0,
$$\lim_{(x,y)\to(1,3)}\left(\frac{x^2 y}{4x^2-y}\right)=\frac{3}{4-3}=3$$

7. Since the limit of the denominator is not 0,
$$\lim_{(x,y)\to(\pi,1)}\frac{\cos xy}{y^2+1}=\frac{-1}{1+1}=-\frac{1}{2}$$

9. Since the limit of the denominator is not 0,
$$\lim_{(x,y)\to(1,0,2)}\left(\frac{4xz}{y^2+z^2}\right)=\frac{8}{0+2^2}=2$$

11. Along $x=0$, $L_1=0$; along $y=0$, $L_2=3$, Therefore, L does not exist.

13. Along $x=0$, $L_1=0$; along $y=x$, $L_2=2$. Therefore, L does not exist.

15. Along $x=0$, $L_1=0$; along $x=y^2$, $L_2=1$. Therefore, L does not exist.

17. Along $x=0$, $L_1=0$; along $y^3=x$, $L_2=\frac{1}{2}$. Therefore, L does not exist.

19. Along $x=0$, $L_1=0$; along $y=x$, $L_2=\frac{1}{2}$. Therefore, L does not exist.

21. Along $x=1$, $L_1=0$; along $y=x+1$, $L_2=\frac{1}{2}$. Therefore, L does not exist.

23. Along $x=0$, $L_1=0$; along $x^2=y^2+z^2$, $L_2=\frac{3}{2}$. Therefore, L does not exist.

25. Along $y=-x$, L_1 does not exist. Therefore L does not exist.

27. If the limit exists, it must equal 0 since along $x=0$, $L_1=0$.
$$y^2\le x^2+y^2$$
$$|x|y^2\le|x|(x^2+y^2)$$
$$\left|\frac{xy^2}{x^2+y^2}-0\right|=\frac{|x|y^2}{x^2+y^2}\le|x|,\ (x,y)\ne(0,0)$$
Since $\lim\limits_{(x,y)\to(0,0)}|x|=0$ the limit must be 0 by theorem 2.1.

29. If the limit exists, it must equal 0 since along $x=0$, $L_1=0$.
$$2x^2\le 2x^2+y^2$$
$$|\sin y|2x^2\le(2x^2+y^2)|\sin y|$$
$$\left|\frac{2x^2\sin y}{2x^2+y^2}-0\right|=\frac{2x^2|\sin y|}{2x^2+y^2}\le|\sin y|,\ (x,y)\ne(0,0)$$
Since $\lim\limits_{(x,y)\to(0,0)}|\sin y|=0$, the limit must be 0 by theorem 2.1.

31. If the limit exists, it must equal 2 since along $x=0$, $L_1=2$.
$$\frac{x^3+4x^2+2y^2}{2x^2+y^2}=\frac{x^3}{2x^2+y^2}+2$$
$$x^2\le 2x^2+y^2$$
$$|x|x^2\le(2x^2+y^2)|x|$$
$$\left|\frac{x^3+4x^2+2y^2}{2x^2+y^2}-2\right|=\left|\frac{x^3}{2x^2+y^2}\right|=\frac{|x|x^2}{2x^2+y^2}\le|x|,$$
$(x,y)\ne(0,0)$
Since $\lim\limits_{(x,y)\to(0,0)}|x|=0$, the limit must be 2 by theorem 2.1.

33. If the limit exists, it must equal 0 since along $x=0$, $L_1=0$.
$$\frac{3x^3}{x^2+y^2+z^2}=3\left[\frac{x^3}{x^2+y^2+z^2}\right]$$
$$x^2\le x^2+y^2+z^2$$
$$|x|x^2\le(x^2+y^2+z^2)|x|$$
$$\left|\frac{x^3}{x^2+y^2+z^2}-0\right|=\frac{|x|x^2}{x^2+y^2+z^2}\le|x|,$$
$(x,y,z)\ne(0,0,0)$
Since $\lim\limits_{(x,y,z)\to(0,0,0)}|x|=0$, $\lim\limits_{(x,y,z)\to(0,0,0)}3|x|=0$, and the limit must equal 0 by theorem 2.1. Thus,
$$\lim_{(x,y,z)\to(0,0,0)}\frac{3x^3}{x^2+y^2+z^2}=0.$$

35. The density plot shows sharp color changes as we approach the origin.

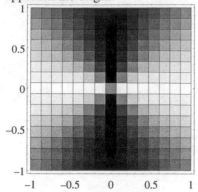

37. The density plot shows sharp color changes along the hyperbola $3y^2 - x^2 = c.$.

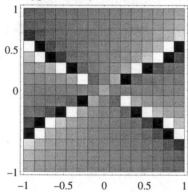

39. Since $4xy$ and $\sin 3x^2 y$ are continuous for all (x, y), the sum is continuous for all (x, y).

41. Since $9 - x^2 - y^2$ must be greater than or equal to zero to get real function values, $x^2 + y^2 \le 9$.
The function is continuous on
$\left\{ x \ y \middle| x^2 + y^2 \le 9 \right\}$.

43. Since the domain of the natural logarithm function is greater than zero, $x^2 - y < 3$.
The function is continuous on $\left\{ x, y \middle| x^2 - y < 3 \right\}$.

45. Since $\sin z$ is continuous for all x, y and $\dfrac{x^3}{y}$ is continuous except when $y = 0$, the sum is continuous on $\left\{ (x, y, z) \middle| y \ne 0 \right\}$.

47. Since $x^2 + y^2 + z^2 - 4$ must be greater than or equal to zero to get real function values, $x^2 + y^2 + z^2 \ge 4$. The function is continuous on $\left\{ x, y, z \middle| x^2 + y^2 + z^2 \ge 4 \right\}$.

49. $f(.1, \ .1) \approx .45 \qquad f(.01, \ .01) \approx .50$
$f(.001, \ .001) \approx .5 \qquad f(.1, \ -.1) \approx .56$
$f(.01, \ -.01) \approx .51 \qquad f(.001, \ -.001) \approx .5$
Along these paths, we would estimate the limit to be $\dfrac{1}{2}$.

51. True. If the limit exists and is equal to L then along any path, including $y = b$, the limit must be L.

53. False. The limit along particular paths being L does not imply that the limit along every path is L.

55. Along $x = 0$ the limit as $(x, y) \to (0, 0)$ is 0, but along $x = y^2$ the limit is $\dfrac{1}{2}$.

57. $\displaystyle\lim_{r \to 0} \left(\frac{r}{\sin r} \right) = 1$ by L'Hôpital's Rule.

59. $\displaystyle\lim_{r \to 0} \left(\frac{r^3 \cos \theta \sin^2 \theta}{r^2} \right) = \lim_{r \to 0} (r \cos \theta \sin^2 \theta) = 0$

Section 12.3

5. $f_x = 3x^2 - 4y^2, f_y = 4y^3 - 8xy$

7. $f_x = 2xe^y, f_y = x^2e^y - 4$

9. $f_x = x^2 y \cos xy + 2x \sin xy,$
    $f_y = x^3 \cos xy - 9y^2$

11. $f_x = 4 \cdot \dfrac{1}{y} e^{x/y} + \dfrac{y}{x^2} = \dfrac{4}{y} e^{x/y} + \dfrac{y}{x^2}$

    $f_y = 4\left(\dfrac{-x}{y^2}\right) e^{x/y} - \dfrac{1}{x} = -\dfrac{4x}{y^2} e^{x/y} - \dfrac{1}{x}$

13. $f_x = 12x^2 y^2 z + 3 \sin y,$
    $f_y = 8x^3 yz + 3x \cos y,$
    $f_z = 4x^3 y^2$

15. $f(x, y, z) = 2(x^2 + y^2 + z^2)^{-1/2}$
    $f_x = -(x^2 + y^2 + z^2)^{-3/2} \cdot 2x$
    $= -\dfrac{2x}{\sqrt{(x^2 + y^2 + z^2)^3}},$
    $f_y = -(x^2 + y^2 + z^2)^{-3/2} \cdot 2y$
    $= -\dfrac{2y}{\sqrt{(x^2 + y^2 + z^2)^3}},$
    $f_z = -(x^2 + y^2 + z^2)^{-3/2} \cdot 2z$
    $= -\dfrac{2z}{\sqrt{(x^2 + y^2 + z^2)^3}}$

17. $\dfrac{\partial f}{\partial x} = 3x^2 - 4y^2$

    $\dfrac{\partial f}{\partial y} = -8xy + 3$

    $\dfrac{\partial^2 f}{\partial x^2} = 6x$

    $\dfrac{\partial^2 f}{\partial y^2} = -8x$

    $\dfrac{\partial^2 f}{\partial y \partial x} = -8y$

19. $f_x = 4x^3 - 6xy^3$
    $f_{xx} = 12x^2 - 6y^3$
    $f_{xy} = -18xy^2$
    $f_{xyy} = -36xy$

21. $f_x = 3x^2 y^2$
    $f_{xx} = 6xy^2$
    $f_y = 2x^3 y - z \cos yz$
    $f_{yz} = yz \sin yz - \cos yz$
    $f_{xy} = 6x^2 y$
    $f_{xyz} = 0$

23. $f(x, y, z) = e^{2xy} - z^2 y^{-1} + xz \sin y$
    $f_x = 2ye^{2xy} + z \sin y$
    $f_{xx} = 4y^2 e^{2xy}$
    $f_y = 2xe^{2xy} + z^2 y^{-2} + xz \cos y$
    $f_{yy} = 4x^2 e^{2xy} - 2z^2 y^{-3} - xz \sin y$
    $\quad = 4x^2 e^{2xy} - \dfrac{2z^2}{y^3} - xz \sin y$
    $f_{yyz} = -\dfrac{4z}{y^3} - x \sin y$
    $f_{yyzz} = -\dfrac{4}{y^3}$

25. $f_w = 2wxy - ze^{wz}$
    $f_{ww} = 2xy - z^2 e^{wz}$
    $f_{wx} = 2wy$
    $f_{wxy} = 2w$
    $f_{wwx} = 2y$
    $f_{wwxy} = 2$
    $f_{wwxyz} = 0$

27. $f(x,1) = 3 - x^2$ is the curve that results when $f(x,$ $y)$ intersects the plane $y = 1$. $\frac{\partial f}{\partial x}(1,1)$ gives the slope of that curve that curve at $x = 1$, which is -2.

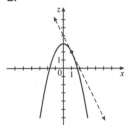

29. $f(1,y) = 3 - y^2$ is the curve that results when f intersect the plane $x = 1$. $\frac{\partial f}{\partial y}(1,1)$ gives the slope of that curve at $y = 1$, which is -2.

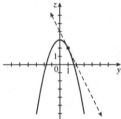

31. $f(2,y) = -y^2$ is the curve that results when f intersects the plane $x = 2$. $\frac{\partial f}{\partial y}(2,0)$ gives the slope of that curve at $y = 0$, which is 0.

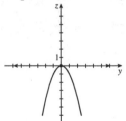

33. We have $\left(P + \dfrac{n^2 a}{V^2}\right)(V - nb) = nRT$

or $PV - Pnb + \dfrac{n^2 a}{V} - \dfrac{n^3 ab}{V^2} = nRT$

If we think of V as a function of T and treat R and P as constants, then differentiate with respect to T, we get

$$PV' - \frac{n^2 a}{V^2}V' + \frac{2V'n^3 ab}{V^3} = nR$$

$$PV^3 V' - Vn^2 aV' + 2V'n^3 ab = V^3 nR$$

$$V'(PV^3 - Vn^2 a + 2n^3 ab) = V^3 nR$$

$$\frac{\partial V}{\partial T} = \frac{V^3 nR}{PV^3 - Vn^2 a + 2n^3 ab}$$

35. $S = c\dfrac{L^4}{wh^3}$

$$\frac{\partial S}{\partial w} = -\frac{cL^4}{w^2 h^3} = -\frac{1}{w} \cdot \frac{cL^4}{wh^3} = -\frac{1}{w} \cdot S$$

37. $S = c\dfrac{L^3}{wh^4}$

$$\frac{\partial S}{\partial L} = \frac{3cL^2}{wh^4} = \frac{3S}{L}$$

$$\frac{\partial S}{\partial w} = -\frac{cL^3}{w^2 h^4} = -\frac{S}{w}$$

$$\frac{\partial S}{\partial h} = \frac{-4cL^3}{wh^5} = -\frac{4S}{h}$$

$$\frac{\Delta S}{S} \approx \frac{\frac{\partial S}{\partial L}\Delta L}{S} = \frac{3\Delta L}{L}$$

$$\frac{\Delta S}{S} \approx \frac{\frac{\partial S}{\partial w}\Delta w}{S} = -\frac{\Delta w}{w}$$

$$\frac{\Delta S}{S} \approx \frac{\frac{\partial S}{\partial h}\Delta h}{S} = -\frac{4\Delta h}{h}$$

The variable h has the greatest proportional effect.

39. $\dfrac{\partial f}{\partial x} = 2x, \dfrac{\partial f}{\partial y} = 2y$

$2x = 2y = 0$ at $(0, 0, 0)$

There are horizontal tangent lines to the curves in the $y = 0$ plane and the $x = 0$ plane at $(0, 0, 0)$.

41. $\dfrac{\partial f}{\partial x} = \cos x \sin y,\ \dfrac{\partial f}{\partial y} = \sin x \cos y$

$\cos x \sin y = 0$ at $y = n\pi,\ x = \dfrac{\pi}{2} + m\pi$

$\sin x \cos y = 0$ at $y = \dfrac{\pi}{2} + n\pi,\ x = m\pi$

At the points $\left(\dfrac{\pi}{2} + m\pi, \dfrac{\pi}{2} + n\pi \right)$ m, n integers,
there are horizontal tangent lines in the planes
$x = \dfrac{\pi}{2} + m\pi$ and $y = \dfrac{\pi}{2} + n\pi.$

43. $\dfrac{\partial f}{\partial x} \approx 4, \dfrac{\partial f}{\partial y} \approx 2$

45. $\dfrac{\partial f}{\partial x} \approx 1, \dfrac{\partial f}{\partial y} \approx -\dfrac{2}{3}$

47. $\dfrac{\partial c}{\partial t}(10,10) \approx 1.4, \dfrac{\partial c}{\partial s}(10,10) \approx -2.4$
You would expect that it would feel one degree
warmer when the temperature goes up one degree.

49. $\dfrac{\partial f}{\partial v}(170,3000) \approx \dfrac{44}{20} = 2.2$

$\dfrac{\partial f}{\partial \omega}(170,3000) \approx \dfrac{39}{2000} \approx 0.02$

51. $\dfrac{\partial f}{\partial x}(x,y,z) = \lim\limits_{h \to 0} \dfrac{f(x+h,y,z) - f(x,y,z)}{h}$

$\dfrac{\partial f}{\partial y}(x,y,z) = \lim\limits_{h \to 0} \dfrac{f(x,y+h,z) - f(x,y,z)}{h}$

$\dfrac{\partial f}{\partial z}(x,y,z) = \lim\limits_{h \to 0} \dfrac{f(x,y,z+h) - f(x,y,z)}{h}$

53. $f_n(x,t) = \sin n\pi x \cos n\pi ct$

$\dfrac{\partial f}{\partial x} = n\pi \cos n\pi x \cos n\pi ct$

$\dfrac{\partial^2 f}{\partial x^2} = -n^2\pi^2 \sin n\pi x \cos n\pi ct$

$\dfrac{\partial f}{\partial t} = \sin n\pi x (-n\pi c \sin n\pi ct)$

$\dfrac{\partial^2 f}{\partial t^2} = \sin n\pi x (-n^2\pi^2 c^2 \cos n\pi ct)$

$= c^2(-n^2\pi^2 \sin n\pi x \cos n\pi ct)$

$= c^2 \dfrac{\partial^2 f}{\partial x^2}$

55. $\dfrac{\partial V}{\partial I} = -\dfrac{5}{1+I}V$

$\dfrac{\partial V}{\partial T} = -\dfrac{.5}{1+.1(1-T)}V$

The inflation rate has a greater influence.

57. $\dfrac{\partial p}{\partial x} = \cos x \cos t$ gives the rate of change of
displacement with respect to distance along the
string at a particular time for $0 \le x \le L.$

$\dfrac{\partial p}{\partial t} = -\sin x \sin t$ gives the rate of change of
displacement with respect to time at a particular
distance along the string.

59.
$$G = H - TS$$

$$\frac{G}{T} = \frac{H}{T} - S, T \neq 0$$

$$\frac{\partial(G/T)}{\partial T} = -\frac{H}{T^2}$$

61. $\dfrac{\partial R}{\partial R_1}$

$$= \frac{R_2 R_3 (R_1 R_2 + R_1 R_3 + R_2 R_3) - R_1 R_2 R_3 (R_2 + R_3)}{(R_1 R_2 + R_1 R_3 + R_2 R_3)^2}$$

$$= \frac{R_2^2 R_3^2}{(R_1 R_2 + R_1 R_3 + R_2 R_3)^2}$$

$$= \left(\frac{R}{R_1}\right)^2$$

gives the rate of change of net resistance for fixed resistors R_2, R_3 and varying the resistance R_1.

$$\frac{\partial R}{\partial R_2} = \left(\frac{R}{R_2}\right)^2 \text{ and } \frac{\partial R}{\partial R_3} = \left(\frac{R}{R_3}\right)^2$$

since the function is symmetric with respect to $R_1, R_2,$ and R_3.

63. $P(100, 60, 15) = 400$

$$\frac{\partial P}{\partial t} = -T\frac{S}{t^2}, \text{ so } \frac{\partial P}{\partial t}(100, 60, 15) \approx -26.67.$$

So the population estimate would go *down* by about 27 animals.

65. $f_x(0, y) = \lim\limits_{h \to 0} \dfrac{f(0+h, y) - f(0, y)}{h}$

$$= \lim_{h \to 0} \frac{f(h, y) - f(0, y)}{h}$$

$$= \lim_{h \to 0} \frac{\frac{hy(h^2 - y^2)}{h^2 + y^2} - 0}{h}$$

$$= \lim_{h \to 0} \frac{y(h^2 - y^2)}{h + y^2} = -\frac{y^3}{y^2} = -y$$

$f_{xy}(0, 0) = \lim\limits_{k \to 0} \dfrac{f_x(0, 0+k) - f(0, 0)}{k}$

$$= \lim_{k \to 0} \frac{f_x(0, k) - f(0, 0)}{k}$$

$$= \lim_{k \to 0} \frac{-k - 0}{k}$$

$$= \lim_{k \to 0} -1$$

$$= -1$$

$f_y(x, 0) = \lim\limits_{k \to 0} \dfrac{f(x, 0+k) - f(x, 0)}{k}$

$$= \lim_{k \to 0} \frac{f(x, k) - f(x, 0)}{k}$$

$$= \lim_{k \to 0} \frac{\frac{xk(x^2 - k^2)}{x^2 + k^2} - 0}{k}$$

$$= \lim_{k \to 0} \frac{x(x^2 - k^2)}{x^2 + k^2}$$

$$= \frac{x^3}{x^2}$$

$$= x$$

$f_{yx}(0, 0) = \lim\limits_{h \to 0} \dfrac{f_y(0+h, 0) - f(0, 0)}{h}$

$$= \lim_{h \to 0} \frac{f_y(h, 0) - f(0, 0)}{h}$$

$$= \lim_{h \to 0} \frac{h}{h}$$

$$= \lim_{h \to 0} 1$$

$$= 1$$

f_{xy} and f_{yx} are not continuous at $(0, 0)$.

67. The slope of the curve at $x = x_0$ is $\dfrac{\partial f}{\partial x}(x_0, y_0)$ — the rate of change of $f(x, y)$ with respect to x when y is fixed at $y = y_0$, taken at the point $x = x_0$. The concavity of the curve at $x = x_0$ is

$$\frac{\partial^2 f}{\partial x^2}(x_0, y_0).$$

Section 12.4

5. $f_x = 2x, f_y = 2y$

$f_x(2,1) = 4, f_y(2,1) = 2$

at (2, 1, 4) the tangent plane is,
$4(x-2) + 2(y-1) + -(z-4) = 0$
$4x + 2y - z - 6 = 0$

7. $f_x = -2xe^{-x^2-y^2}, f_y = -2ye^{-x^2-y^2}$

$f_x(0,0) = 0, f_y(0,0) = 0$

at (0,0,1) the tangent plane is
$0(x-0) + 0(y-0) - (z-1) = 0$
$-z + 1 = 0$

9. $f_x = \cos x \cos y, f_y = -\sin x \sin y$

$f_x(0,\pi) = -1, f_y(0,\pi) = 0$

at (0, π, 0) the tangent plane is
$-1(x-0) + 0(y-\pi) - (z-0) = 0$
$-x - z = 0$

11. $f_x = 3x^2 - 2y, f_y = -2x$

$f_x(-2,3) = 6, f_y(-2,3) = 4$

at (−2, 3, 4) the tangent plane is
$6(x+2) + 4(y-3) - (z-4) = 0$
$6x + 4y - z + 4 = 0$

13. $f_x = \dfrac{x}{\sqrt{x^2+y^2}}, f_y = \dfrac{y}{\sqrt{x^2+y^2}}$

$f_x(-3,4) = -\dfrac{3}{5}, f_y(-3,4) = \dfrac{4}{5}$

at (−3, 4, 5) the tangent plane is

$-\dfrac{3}{5}(x+3) + \dfrac{4}{5}(y-4) - (z-5) = 0$

$-\dfrac{3}{5}x + \dfrac{4}{5}y - z = 0$

15. $f_x = \dfrac{4}{y}, f_y = -\dfrac{4x}{y^2}$

$f_x(1,2) = 2, f_y = (1,2) = -1$

at (1, 2, 2) the tangent plane is
$2(x-1) - 1(y-2) - (z-2) = 0$
$2x - y - z + 2 = 0$

17. $f_x = \dfrac{x}{\sqrt{x^2+y^2}}, f_y = \dfrac{y}{\sqrt{x^2+y^2}}$

$f_x(3,0) = 1, f_y(3,0) = 0$

$L(x,y) = 3 + 1(x-3) + 0(y-0) = x$

19. $f_x = \cos x \cos y, f_y = -\sin x \sin y$

$f_x(0,\pi) = -1, f_y(0,\pi) = 0$

$L(x,y) = 0 + -1(x-0) + 0(y-\pi) = -x$

21. $f_x = e^{y^2} - 4, f_y = 2xye^{y^2}$

$f_x(2,0) = -3, f_y(2,0) = 0$

$L(x,y) = -6 + -3(x-2) + 0(y-0) = -3x$

23. $f_x = 3x^2 z, f_y = z^2 \cos yz^2,$

$f_z = 2yz \cos yz^2 + x^3$

$f_x(-2,0,1) = 12, f_y(-2,0,1) = 1,$

$f_z = (-2,0,1) = -8$

$L(x,y,z) = -8 + 12(x+2) + 1(y-0) - 8(z-1)$

$= 12x + y - 8z + 24$

25. $f_w = 2wxy - yze^{wyz}, f_w(-2,3,1,0) = -12$

$f_x = w^2 y, f_x(-2,3,1,0) = 4$

$f_y = w^2 x - wze^{wyz}, f_y(-2,3,1,0) = 12$

$f_z = -wye^{wyz}, f_z(-2,3,1,0) = 2$

$L(w,x,y,z)$

$= 11 - 12(w+2) + 4(x-3) + 12(y-1)$

$+ 2(z-0) = -12w + 4x + 12y + 2z - 37$

27. From Exercise 17, $L(x,y) = x.$

$L(3,.1) = 3, f(3,.1) \approx 3.0017$

$L(3.1,0) = 3.1, f(3.1,0) = 3.1$

$L(3.1,-.1) = 3.1, f(3.1,-.1) \approx 3.1017$

29. From Exercise 19, $L(x,y) = -x.$

$L(0,3) = 0, f(0,3) = 0$

$L(.1,\pi) = -.1, f(.1,\pi) \approx -.0998$

$L(.1,3) = -.1, f(.1,3) \approx -.0988$

31. $S \approx 1.5552$

$\pm (.1728(.5) - .7776(-.2) - .7776(-.5))$

$= 1.5552 \pm .6307$

33. $g(9.9,930) = 4 + .3(-.1) - .004(30)$

$= 3.85$

35. $\dfrac{\partial g}{\partial s} \approx \dfrac{.03}{.3} = .1, \dfrac{\partial g}{\partial t} \approx \dfrac{-.02}{20} = -.001$

$g(10.2, 890) = 4 + .1(.2) - .001(-10)$
$\qquad = 4.03$

37. $f_x = 2y, f_y = 2x + 2y$

$\Delta z = 2(x + \Delta x)(y + \Delta y) + (y + \Delta y)^2 - 2xy - y^2$
$\quad = 2xy + 2x\Delta y + 2y\Delta x + 2\Delta x\Delta y + y^2 + 2y\Delta y + (\Delta y)^2 - 2xy - y^2$
$\quad = 2y\Delta x + (2x + 2y)\Delta y + 2\Delta y\Delta x + \Delta y\Delta y$

39. $f_x = 2x, f_y = 2y$

$\Delta z = (x + \Delta x)^2 + (y + \Delta y)^2 - x^2 - y^2$
$\quad = x^2 + 2x\Delta x + (\Delta x)^2 + y^2 + 2y\Delta y + (\Delta y)^2 - x^2 - y^2$
$\quad = 2x\Delta x + 2y\Delta y + \Delta x\Delta x + \Delta y\Delta y$

41. $f_x = 2x + 3y, f_y = 3x$

$\Delta z = (x + \Delta x)^2 + 3(x + \Delta x)(y + \Delta y) - x^2 - 3xy$
$\quad = x^2 + 2x\Delta x + (\Delta x)^2 + 3xy + 3x\Delta y + 3y\Delta x + \Delta x\Delta y - x^2 - 3xy$
$\quad = (2x + 3y)\Delta x + 3x\Delta y + \underbrace{(\Delta x)}_{\varepsilon_1}\Delta x + \underbrace{(\Delta x)}_{\varepsilon_2}\Delta y$

As $(\Delta x, \Delta y) \to (0, 0)$, $\varepsilon_1 = \Delta x$ and $\varepsilon_2 = \Delta x$ tend to 0.
Therefore $f(x, y)$ is differentiable.

43. $f_x = ye^x + \cos x, f_y = e^x$

$dz = (ye^x + \cos x)dx + e^x dy$

45. We will use Definition 4.1. Since the function value at the origin is 0,

$\Delta z = \dfrac{2(0 + \Delta x)(0 + \Delta y)}{(0 + \Delta x)^2 + (0 + \Delta y)^2} = \dfrac{2\Delta x\Delta y}{\Delta x^2 + \Delta y^2}$. Since

both partial derivatives are 0 at the origin, the function will be differentiable, according to Definition 4, if we can write

$\dfrac{2\Delta x\Delta y}{\Delta x^2 + \Delta y^2} = \varepsilon_1\Delta x + \varepsilon_2\Delta y$, where

$\varepsilon_1, \varepsilon_2 \to 0$ as $(\Delta x, \Delta y) \to (0, 0)$. But this is not possible since the expression on the left doesn't even have limit 0 as $(\Delta x, \Delta y) \to (0, 0)$.

47. To move from the $z = 1$ level curve to the $z = .95$ level curve you move .05 units up. Then $\dfrac{\partial f}{\partial y} = \dfrac{.05}{-.05} = -1$. Thus, $f(0, 0) = 1$,

$\dfrac{\partial f}{\partial y}(0,0) \approx -1$, and $\dfrac{\partial f}{\partial x}(0,0) \approx 2$,

$L(x, y) = 1 + 2(x - 0) - 1(y - 0) = 1 + 2x - y$

49. $f(0, 0) = 6$,

$\dfrac{\partial z}{\partial x}(0,0) \approx \dfrac{2}{.5} = 4, \dfrac{\partial z}{\partial y}(0,0) \approx \dfrac{2}{1} = 2$

$L(x, y) = 6 + 4x + 2y$

51. $f(0, 0) = 3$,

$\dfrac{\partial z}{\partial x}(0,0) \approx \dfrac{1}{1} = 1, \dfrac{\partial z}{\partial y} \approx \dfrac{-1}{1.5} = -\dfrac{2}{3}$

$L(x, y) = 3 + x - \dfrac{2}{3}y$

53. $\dfrac{\partial w}{\partial t} \approx 1.4, \dfrac{\partial w}{\partial s} \approx -2.4$

$L(t, s) = -9 + 1.4(t - 10) - 2.4(s - 10)$
$L(12, 13) = -13.4$

Section 12.5

5. $g'(t) = \dfrac{\partial f}{\partial x}\dfrac{dx}{dt} + \dfrac{\partial f}{\partial y}\dfrac{dy}{dt}$

   $\dfrac{\partial f}{\partial x} = 2xy, \dfrac{\partial f}{\partial y} = x^2 - \cos y$

   $\dfrac{dx}{dt} = \dfrac{t}{\sqrt{t^2+1}}, \dfrac{dy}{dt} = e^t$

   $g'(t) = 2e^t\sqrt{t^2+1}\,\dfrac{t}{\sqrt{t^2+1}} + \left[(t^2+1) - \cos e^t\right]e^t$

   $\qquad = (2t + t^2 + 1 - \cos e^t)e^t$

7. $\dfrac{\partial g}{\partial u} = \dfrac{\partial f}{\partial x}\dfrac{\partial x}{\partial u} + \dfrac{\partial f}{\partial y}\dfrac{\partial y}{\partial u}$

   $\dfrac{\partial f}{\partial x} = 8xy^3, \ \dfrac{\partial f}{\partial y} = 12x^2y^2$

   $\dfrac{\partial x}{\partial u} = 3u^2 - v\cos u, \ \dfrac{\partial y}{\partial u} = 8u$

   $\dfrac{\partial g}{\partial u} = 8(u^3 - v\sin u)(4u^2)^3(3u^2 - v\cos u) + 12(u^3 - v\sin u)^2(4u^2)^2 8u$

   $\qquad = 512u^6(u^3 - v\sin u)(3u^2 - v\cos u) + 1536u^5(u^3 - v\sin u)^2$

   $\dfrac{\partial g}{\partial v} = \dfrac{\partial f}{\partial x}\dfrac{\partial x}{\partial v} + \dfrac{\partial f}{\partial y}\dfrac{\partial y}{\partial v}$

   $\dfrac{\partial x}{\partial v} = -\sin u, \ \dfrac{\partial y}{\partial v} = 0$

   $\dfrac{\partial g}{\partial v} = 8(u^3 - v\sin u)(4u^2)^3(-\sin u) + 12(u^3 - v\sin u)^2(4u^2)^2 \cdot 0 = 512u^6\sin u(v\sin u - u^3)$

9. $g'(t) = \dfrac{\partial f}{\partial x}\dfrac{dx}{dt} + \dfrac{\partial f}{\partial y}\dfrac{dy}{dt} + \dfrac{\partial f}{\partial z}\dfrac{dz}{dt}$

11. $\dfrac{\partial g}{\partial u} = \dfrac{\partial f}{\partial x}\dfrac{\partial x}{\partial u} + \dfrac{\partial f}{\partial y}\dfrac{\partial y}{\partial u}$

   $\dfrac{\partial g}{\partial v} = \dfrac{\partial f}{\partial x}\dfrac{\partial x}{\partial v} + \dfrac{\partial f}{\partial y}\dfrac{\partial y}{\partial v}$

   $\dfrac{\partial g}{\partial w} = \dfrac{\partial f}{\partial x}\dfrac{\partial x}{\partial w} + \dfrac{\partial f}{\partial y}\dfrac{\partial y}{\partial w}$

13. $\dfrac{\partial P}{\partial k}(6,4) \approx 3.6889, \ \dfrac{\partial P}{\partial l}(6,4) \approx 16.6002$

   $k'(t) = .1, \ l'(t) = -.06$

   $g'(t) = 3.6889(.1) + 16.6002(-.06)$

   $\qquad \approx -.627$

15. $\dfrac{\partial P}{\partial k} = \dfrac{16}{3}k^{-2/3}l^{2/3}, \ \dfrac{\partial P}{\partial l} = \dfrac{32}{3}k^{1/3}l^{-1/3}$

   $\dfrac{\partial P}{\partial k}(4,3) \approx 4.4025, \ \dfrac{\partial P}{\partial l}(4,3) \approx 11.7401$

   $k'(t) = -.2, \qquad l'(t) = .08$

   $g'(t) = 4.4025(-.2) + 11.7401(.08)$

   $\qquad \approx .0587$

17. $I(t) = q(t)p(t)$

$$\frac{\partial q}{\partial t} = .05qt \text{ and } \frac{\partial p}{\partial t} = .03p(t)$$

$$\frac{dI}{dt} = \frac{\partial I}{\partial q}\frac{dq}{dt} + \frac{\partial I}{\partial p}\frac{dp}{dt}$$

$$= p(t)\frac{dq}{dt} + q(t)\frac{dp}{dt}$$

$$= p(t)[.05q(t)] + q(t)[.03p(t)]$$

$$= .08q(t)p(t)$$

$$= .08I(t)$$

Thus income increases at a rate of 8%.

19. $g'(t) = f_x x'(t) + f_y y'(t)$

$g''(t) = f_x x''(t) + x'(t)(f_{xx} x'(t) + f_{xy} y'(t)) + f_y y''(t) + y'(t)(f_{yx} x'(t) + f_{yy} y'(t))$

$\quad = x''(t)f_x + (y'(t))^2 f_{yy} + y''(t)f_y + 2f_{xy}x'(t)y'(t) + f_{xx}(x'(t))^2$

21. For convenience, we use subscripts and change the notation for the final answer.

$g_u = f_x x_u + f_y y_u$

$g_{uu} = f_x x_{uu} + x_u(f_{xx} x_u + f_{xy} y_u) + f_y y_{uu} + y_u(f_{yx} x_u + f_{yy} y_u)$

$\quad = f_x x_{uu} + f_y y_{uu} + f_{xx}(x_u)^2 + 2f_{xy}x_u y_u + f_{yy}(y_u)^2$

$\quad = \dfrac{\partial f}{\partial x}\dfrac{\partial^2 x}{\partial u^2} + \dfrac{\partial f}{\partial y}\dfrac{\partial^2 y}{\partial u^2} + \dfrac{\partial^2 f}{\partial x^2}\left(\dfrac{\partial x}{\partial u}\right)^2 + 2\dfrac{\partial^2 f}{\partial y \partial x}\dfrac{\partial x}{\partial u}\dfrac{\partial y}{\partial u} + \dfrac{\partial^2 f}{\partial y^2}\left(\dfrac{\partial y}{\partial u}\right)^2$

23. $F_x = 6xz$

$F_y = -3z$

$F_z = 3x^2 + 6z^2 - 3y$

$$\frac{\partial z}{\partial x} = -\frac{F_x}{F_z} = -\frac{2xz}{x^2 + 2z^2 - y}$$

$$\frac{\partial z}{\partial y} = -\frac{F_y}{F_z} = \frac{z}{x^2 + 2z^2 - y}$$

25. $F(x, y, z) = 3e^{xyz} - 4xz^2 + x\cos y - 2 = 0$

$F_x = 3yze^{xyz} - 4z^2 + \cos y$

$F_y = 3xze^{xyz} - x\sin y$

$F_z = 3xye^{xyz} - 8xz$

$$\frac{\partial z}{\partial x} = -\frac{F_x}{F_z} = \frac{4z^2 - 3yze^{xyz} - \cos y}{3xye^{xyz} - 8xz}$$

$$\frac{\partial z}{\partial y} = -\frac{F_y}{F_z} = \frac{x\sin y - 3xze^{xyz}}{3xye^{xyz} - 8xz}$$

27. $\dfrac{\partial f}{\partial \theta} = \dfrac{\partial f}{\partial x}\dfrac{\partial x}{\partial \theta} + \dfrac{\partial f}{\partial y}\dfrac{\partial y}{\partial \theta}$

$\quad = f_x(-r\sin\theta) + f_y(r\cos\theta)$

$\quad = -f_x r\sin\theta + f_y r\cos\theta$

29. $\dfrac{f_r}{r} = \dfrac{f_x \cos\theta}{r} + \dfrac{f_y \sin\theta}{r}$

$f_{rr} = f_{xx}\cos^2\theta + 2f_{xy}\cos\theta\sin\theta + f_{yy}\sin^2\theta$

$\dfrac{f_{\theta\theta}}{r^2} = f_{xx}\sin^2\theta - 2f_{xy}\cos\theta\sin\theta + f_{yy}\cos^2\theta$

$\qquad\qquad - \dfrac{f_x\cos(\theta)}{r} - \dfrac{f_y\cos(\theta)}{r}$

$f_{rr} + \dfrac{f_r}{r} + \dfrac{f_{\theta\theta}}{r^2} = f_{xx}(\cos^2\theta + \sin^2\theta)$

$\qquad\qquad\qquad + f_{yy}(\cos^2\theta + \sin^2\theta)$

$\qquad\qquad\quad = f_{xx} + f_{yy}$

31. $a = \dfrac{h(t)}{b(t)},$

$a' = \dfrac{b(t)h'(t) - h(t)\cdot b'(t)}{(b(t))^2} = \dfrac{b(t) - h(t)}{(b(t))^2}$

(since $h'(t) = b'(t) = 1$)

With 50 hits in 200 at bats,

$a = \dfrac{51}{201} \approx .254,$

$a = \dfrac{101}{401} \approx .252,$

$\dfrac{2(b(t) - h(t))}{4(b(t))^2} = \dfrac{1}{2} a'$ so a' is halved when b and h are both doubled.

Section 12.6

5. $f_x = 2x + 4y^2$, $f_y = 8xy - 5y^4$
$\nabla f = \left\langle 2x + 4y^2, \ 8xy - 5y^4 \right\rangle$

7. $f_x = xy^2 e^{xy^2} + e^{xy^2}$, $f_y = 2x^2 y e^{xy^2} - 2y\sin y^2$
$\nabla f = \left\langle e^{xy^2}(xy^2 + 1), \ 2y(x^2 e^{xy^2} - \sin y^2) \right\rangle$

9. $f_x = \dfrac{8}{y} e^{4x/y} - 2$, $f_y = -\dfrac{8x}{y^2} e^{4x/y}$
$f_x(2,-1) = -8e^{-8} - 2$, $f_y(2,-1) = -16e^{-8}$
$\nabla f(2,-1) = \left\langle -8e^{-8} - 2, \ -16e^{-8} \right\rangle$

11. $f_x = \dfrac{x}{\sqrt{x^2 + y^2}}$, $f_y = \dfrac{y}{\sqrt{x^2 + y^2}}$
$f_x(4,-3) = \dfrac{4}{5}$, $f_y(4,-3) = -\dfrac{3}{5}$
$\nabla f(4,-3) = \left\langle \dfrac{4}{5}, \ -\dfrac{3}{5} \right\rangle$

13. $f_x = 6xy + z\sin x$, $f_y = 3x^2$, $f_z = -\cos x$
$f_x(0,2,-1) = 0$, $f_y = (0,2,-1) = 0$,
$f_z = (0,2,-1) = -1$
$\nabla f(0,2,-1) = \left\langle 0, 0, -1 \right\rangle$

15. $\nabla f = \left\langle 2xy, \ x^2 + 8y \right\rangle$
$\nabla f(2,1) = \left\langle 4, 12 \right\rangle$
$\left\langle 4, 12 \right\rangle \cdot \left\langle \dfrac{1}{2}, \dfrac{\sqrt{3}}{2} \right\rangle = 2 + 6\sqrt{3}$

17. $\nabla f = \left\langle 2xy, \ x^2 + 8y \right\rangle$
$\nabla f(2,1) = \left\langle 4, 12 \right\rangle$
$\left\langle 4, 12 \right\rangle \cdot \left\langle \dfrac{1}{2}, \dfrac{-\sqrt{3}}{2} \right\rangle = 2 - 6\sqrt{3}$

19. $\nabla f = \left\langle \dfrac{x}{\sqrt{x^2 + y^2}}, \ \dfrac{y}{\sqrt{x^2 + y^2}} \right\rangle$
$\nabla f(3,-4) = \left\langle \dfrac{3}{5}, -\dfrac{4}{5} \right\rangle$
$\lVert 3, -2 \rVert = \sqrt{3^2 + (-2)^2} = \sqrt{13}$
$\left\langle \dfrac{3}{5}, -\dfrac{4}{5} \right\rangle \cdot \left\langle \dfrac{3}{\sqrt{13}}, -\dfrac{2}{\sqrt{13}} \right\rangle = \dfrac{17}{5\sqrt{13}}$

21. $\nabla f = \left\langle 8x e^{4x^2 - y}, \ -e^{4x^2 - y} \right\rangle$
$\nabla f(1,4) = \left\langle 8, -1 \right\rangle$
$\lVert -2, -1 \rVert = \sqrt{(-2)^2 + (-1)^2} = \sqrt{5}$
$\left\langle 8, -1 \right\rangle \cdot \left\langle -\dfrac{2}{\sqrt{5}}, -\dfrac{1}{\sqrt{5}} \right\rangle = -\dfrac{15}{\sqrt{5}} = -3\sqrt{5}$

23. $\nabla f = \left\langle -2\sin(2x - y), \sin(2x - y) \right\rangle$
$\nabla f(\pi, 0) = \left\langle 0, 0 \right\rangle$
$(\pi, 0)$ to $(2\pi, \pi)$ is $\left\langle \pi, \pi \right\rangle$
$\lVert \pi, \pi \rVert = \sqrt{\pi^2 + \pi^2} = \pi\sqrt{2}$
$\left\langle 0, 0 \right\rangle \cdot \left\langle \dfrac{1}{\sqrt{2}}, \dfrac{1}{\sqrt{2}} \right\rangle = 0$

25. $\nabla f = \left\langle 2x - 2y, -2x + 2y \right\rangle$
$\nabla f(-2,-1) = \left\langle -2, 2 \right\rangle$
$(-2,-1)$ to $(2,-3)$ is $\left\langle 4, -2 \right\rangle$
$\lVert 4, -2 \rVert = \sqrt{4^2 + (-2)^2} = 2\sqrt{5}$
$\left\langle -2, 2 \right\rangle \cdot \left\langle \dfrac{2}{\sqrt{5}}, -\dfrac{1}{\sqrt{5}} \right\rangle = -\dfrac{6}{\sqrt{5}}$

27. $\nabla f = \left\langle 3x^2 yz^2 - 4y, \ x^3 z^2 - 4x, \ 2x^3 yz \right\rangle$
$\nabla f(1,-1,2) = \left\langle -8, 0, -4 \right\rangle$
$\lVert 2, 0, -1 \rVert = \sqrt{2^2 + 0^2 + (-1)^2} = \sqrt{5}$
$\left\langle -8, 0, -4 \right\rangle \cdot \left\langle \dfrac{2}{\sqrt{5}}, 0, -\dfrac{1}{\sqrt{5}} \right\rangle = -\dfrac{12}{\sqrt{5}}$

29. $\nabla f = \left\langle y e^{xy+z}, x e^{xy+z}, e^{xy+z} \right\rangle$
$\nabla f(1,-1,1) = \left\langle -1, 1, 1 \right\rangle$
$\lVert 4, -2, 3 \rVert = \sqrt{4^2 + (-2)^2 + 3^2} = \sqrt{29}$
$\left\langle -1, 1, 1 \right\rangle \cdot \left\langle \dfrac{4}{\sqrt{29}}, -\dfrac{2}{\sqrt{29}}, \dfrac{3}{\sqrt{29}} \right\rangle = -\dfrac{3}{\sqrt{29}}$

31. $\nabla f = \left\langle 2x, -3y^2 \right\rangle$
$\nabla f(2,1) = \left\langle 4, -3 \right\rangle$
$\lVert \nabla f(2,1) \rVert = 5$
Max change is 5; in the direction $\mathbf{d} = \left\langle 4, -3 \right\rangle$.
Min change is -5; in the direction $\mathbf{d} = \left\langle -4, 3 \right\rangle$.

33. $\nabla f = \left\langle 4y^2 e^{4x}, 2y e^{4x} \right\rangle$
$\nabla f(0,-2) = \left\langle 16, -4 \right\rangle$
$\lVert \nabla f(0,-2) \rVert = \sqrt{272} = 4\sqrt{17}$
Max change is $4\sqrt{17}$; in direction $\mathbf{d} = \left\langle 16, -4 \right\rangle$.
Min change is $-4\sqrt{17}$; in direction $\mathbf{d} = \left\langle -16, 4 \right\rangle$.

35. $\nabla f = \left\langle \cos 3y, -3x\sin 3y \right\rangle$
$\nabla f(2,0) = \left\langle 1, 0 \right\rangle$
$\lVert \nabla f(2,0) \rVert = 1$
Max change is 1; in direction $\mathbf{d} = \left\langle 1, 0 \right\rangle$.
Min change is -1; in direction $\mathbf{d} = \left\langle -1, 0 \right\rangle$.

37. $\nabla f = \left\langle \dfrac{2x}{\sqrt{2x^2 - y}}, -\dfrac{1}{2\sqrt{2x^2 - y}} \right\rangle$

$\nabla f(3,2) = \left\langle \dfrac{3}{2}, -\dfrac{1}{8} \right\rangle \ \|\nabla f(3,2)\| = \sqrt{\dfrac{145}{64}} = \dfrac{\sqrt{145}}{8}$

Max change is $\dfrac{\sqrt{145}}{8}$; in direction $\mathbf{d} = \left\langle \dfrac{3}{2}, -\dfrac{1}{8} \right\rangle$.

Min change is $-\dfrac{\sqrt{145}}{8}$; in direction

$\mathbf{d} = \left\langle -\dfrac{3}{2}, \dfrac{1}{8} \right\rangle$.

39. $\nabla f = \left\langle 8xyz^3, 4x^2 z^3, 12x^2 yz^2 \right\rangle$

$\nabla f(1,2,1) = \langle 16, 4, 24 \rangle$

$|\nabla f(1,2,1)| = \sqrt{848} = 4\sqrt{53}$

Max change is $4\sqrt{53}$; in direction $\mathbf{d} = \langle 16, 4, 24 \rangle$.

Min change is $-4\sqrt{53}$; in direction

$\mathbf{d} = \langle -16, -4, -24 \rangle$.

41. For exercise 38 the gradient vectors are perpendicular to the level curves and the level curves here are circles. The vectors from the origin to any point will have the same direction as the gradient vector at that point. The same is true for exercise 40 except that the level surfaces are spheres.

43. For $f(x, y) = \sin(x + y)$,

$\nabla f = \langle \cos(x + y), \cos(x + y) \rangle$.

The sine wave travels in the direction of greatest increase and decrease along the surface, which is $\langle 1, 1 \rangle$ wherever the gradient is nonzero.

For $f(x, y) = \sin(2x - y)$,

$\nabla f = \langle 2\cos(2x - y), -\cos(2x - y) \rangle$.

Here the wave travels in the direction $\langle 2, -1 \rangle$.

45. $f(x, y, z) = x^2 + y^3 - z$

$\nabla f = \langle 2x, 3y^2, -1 \rangle$

$\nabla f(1, -1, 0) = \langle 2, 3, -1 \rangle$

$2(x - 1) + 3(y + 1) - z = 0$

$2x + 3y - z + 1 = 0$

47. $f(x, y, z) = x^2 + y^2 + z^2$

$\nabla f = \langle 2x, 2y, 2z \rangle$

$\nabla f(-1, 2, 1) = \langle -2, 4, 2 \rangle$

$-2(x + 1) + 4(y - 2) + 2(z - 1) = 0$

$-2x + 4y + 2z - 12 = 0$

49. $f_x = 4x - 4y \quad f_y = -4x + 4y^3$

$4x - 4y = 0$

$4x = 4y \qquad (2) \ -4x + 4y^3 = 0$

$(1) \ x = y$

Using (1) in (2) $\ -4x + 4x^3 = 0$

$-4x(1 - x^2) = 0$

$x = 0, \ x = \pm 1$

So $(0, 0, 0)$, $(1, 1, -1)$, and $(-1, -1, -1)$ are possible local extrema.

51. **53.**

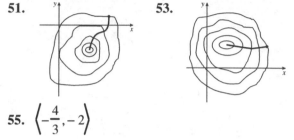

55. $\left\langle -\dfrac{4}{3}, -2 \right\rangle$

57. $f_x \approx \dfrac{2.2 - 1.8}{.1 - (-.1)} = 2, \ f_y \approx \dfrac{1.6 - 2.4}{.2 - (-.2)} \approx -2$

$\nabla f(0, 0) \approx \langle 2, -2 \rangle$

59. If the point on the mountain is (x_0, y_0, z_0) with positive x as east, positive y as north then the direction of steepest ascent will be

$\nabla f = \langle -\tan 10^\circ, \tan 6^\circ \rangle \approx \langle -.176, .105 \rangle$

The rise in that direction will be

$\tan^{-1}\left(|\nabla f|\right) \approx 11.60^\circ$.

61. $\nabla f = \langle -8x, -2y \rangle$

$-\nabla f(1, 2) = \langle 8, 4 \rangle$ is the direction of runoff.

63. $\nabla f = \left\langle -\dfrac{10e^{-z}}{x^3}, -\dfrac{5e^{-z}}{y^2}, -5e^{-z}(x^{-2} + y^{-1}) \right\rangle$

$-\nabla f(1, 4, 8) = \left\langle 10e^{-8}, \dfrac{5}{16}e^{-8}, \dfrac{25}{4}e^{-8} \right\rangle$

$\approx \langle .0033, \ .001, \ .0020 \rangle$

is the direction of most rapid temperature decrease.

65. $\dfrac{\partial g}{\partial G} \approx \dfrac{.04}{.05}, \ \dfrac{\partial g}{\partial s} \approx \dfrac{.06}{.2}, \ \dfrac{\partial g}{\partial t} \approx \dfrac{-.04}{10}$

direction of maximum increase is

$\nabla g = \langle .8, .3, -.004 \rangle$

67. $f_x = 3x^2 - 2y, \ f_y = -2x + 2y$

$f_{xx} = 6x \qquad f_{yy} = 2$

$\nabla^2 f(x, y) = 6x + 2$

Section 12.7

5. $f_x = 2xe^{-x^2}(y^2+1)$ $f_y = 2e^{-x^2}y$

$0 = 2xe^{-x^2}(y^2+1)$ $0 = 2e^{-x^2}y$

(1) $x = 0$ (2) $y = 0$

There is a critical point at $(0, 0)$.

$f_{xx} = 2e^{-x^2}(2x^2-1)(y^2+1)$, $f_{yy} = 2e^{-x^2}$,

$f_{xy} = -4xye^{-x^2}$

$D(0,0) = -2\cdot 2 - 0 = -4$ so the critical point is a saddle point.

7. $f_x = 3x^2 - 3y$ $f_y = -3x + 3y^2$

$0 = 3(x^2 - y)$ $0 = -3(x - y^2)$

(1) $x^2 = y$ (2) $y^2 = x$

Solving (1) and (2) simultaneously we have

$(x^2)^2 = x$

$x^4 - x = 0$

$x(x^3 - 1) = 0$

$x = 0, x = 1$

critical points at $(0, 0)$, $(1, 1)$.

$f_{xx} = 6x$, $f_{yy} = 6y$, $f_{xy} = -3$

$D(0,0) = 0\cdot 0 - (-3)^2 = -9 < 0$

saddle point at $(0, 0)$

$D(1,1) = 6\cdot 6 - (-3)^2 = 27 > 0$

$f_{xx}(1,1) = 6 > 0$

local minimum at $(1, 1)$

9. $f_x = 2xy + 2x$ $f_y = 2y + x^2 - 2$

$0 = 2xy + 2x$ (2) $0 = 2y + x^2 - 2$

(1) $0 = x(y + 1)$

Solving (1) and (2) simultaneously we have

$x = 0$ and $2y - 2 = 0 \Rightarrow y = 1$

$y = -1$ and $x^2 - 4 = 0 \Rightarrow x = \pm 2$

critical points at $(0, 1)$, $(2, -1)$, $(-2, -1)$

$f_{xx} = 2y + 2, f_{yy} = 2, f_{xy} = 2x$

$D(0,1) = (4)(2) - 0 = 8 > 0$

$f_{xx}(0,1) = 4 > 0$

local minimum at $(0, 1)$

$D(2,-1) = (0)(2) - (4)^2 = -16 < 0$

saddle point at $(2, -1)$

$D(-2,-1) = (0)(2) - (-4)^2 = -16 < 0$

saddle point at $(-2, -1)$

11. $f_x = -2xe^{-x^2-y^2}$ $f_y = -2ye^{-x^2-y^2}$

$0 = -2xe^{-x^2-y^2}$ $0 = -2ye^{-x^2-y^2}$

critical point at $(0, 0)$

$f_{xx} = 2(2x^2 - 1)e^{-x^2-y^2}$,

$f_{yy} = 2(2y^2 - 1)e^{-x^2-y^2}$,

$f_{xy} = 4xye^{-x^2-y^2}$

$D(0,0) = 2\cdot 2 - 4 = 4$, $f_{xx}(0,0) = -2$,

so at $(0, 0)$ there is a local maximum.

13. $f_x = 2x - \dfrac{4y}{y^2+1}$

$0 = \dfrac{2x(y^2+1) - 4y}{y^2+1}$

(1) $0 = \dfrac{2xy^2 + 2x - 4y}{y^2+1}$

$f_y = -\dfrac{(y^2+1)4x - 4xy(2y)}{(y^2+1)^2}$

$0 = \dfrac{4x(y^2-1)}{(1+y^2)^2}$

(2) $x = 0$, $y = \pm 1$

Solving (1) and (2) simultaneously we have

$x = 0 \Rightarrow y = 0$

$y = 1 \Rightarrow \dfrac{4x-4}{2} = 0 \Rightarrow x = 1$

$y = -1 \Rightarrow \dfrac{4x+4}{2} = 0 \Rightarrow x = -1$

critical points at $(0, 0)$, $(1, 1)$, $(-1, -1)$

Using a CAS:

$D(0,0) = -16 \Rightarrow$ saddle point at $(0,0)$

$D(1,1) = 4, f_{xx}(1,1) = 2 \Rightarrow$ local minimum at $(1,1)$

$D(-1,-1) = 4, f_x(-1,-1) = 2 \Rightarrow$

local minimum at $(-1, -1)$

15. $f_x = x(-2xe^{-x^2-y^2}) + e^{-x^2-y^2}$

$0 = e^{-x^2-y^2}(-2x^2+1)$

$2x^2 - 1 = 0$

$x = \pm\dfrac{1}{\sqrt{2}}$

$f_y = -2yxe^{-x^2-y^2}$

$0 = -2xye^{-x^2-y^2}$

critical points at $\left(\pm\dfrac{1}{\sqrt{2}}, 0\right)$

Using a CAS:

$D\left(\dfrac{1}{\sqrt{2}}, 0\right) = \dfrac{4}{e}, f_{xx}\left(\dfrac{1}{\sqrt{2}}, 0\right) = -2\sqrt{\dfrac{2}{e}}$

so there is a local maximum at $\left(\dfrac{1}{\sqrt{2}}, 0\right)$.

$D\left(-\dfrac{1}{\sqrt{2}}, 0\right) = \dfrac{4}{e}, f_{xx}\left(-\dfrac{1}{\sqrt{2}}, 0\right) = 2\sqrt{\dfrac{2}{e}}$

so there is a local minimum at $\left(-\dfrac{1}{\sqrt{2}}, 0\right)$.

17. $f_x = xy(-2xe^{-x^2-y^2}) + ye^{-x^2-y^2}$

$0 = ye^{-x^2-y^2}(-2x^2+1)$

$y(2x^2 - 1) = 0$

$x = \pm\dfrac{1}{\sqrt{2}}$ or $y = 0$

$f_y = xy(-2ye^{-x^2-y^2}) + xe^{-x^2-y^2}$

$xe^{-x^2-y^2}(-2y^2+1) = 0$

$x(-2y^2+1) = 0$

$y = \pm\dfrac{1}{\sqrt{2}}, x = 0$

critical points at $(0,0), \left(\pm\dfrac{1}{\sqrt{2}}, \pm\dfrac{1}{\sqrt{2}}\right)$

Using a CAS:

$D(0,0) = -1$, saddle point at $(0,0)$

$D\left(\dfrac{1}{\sqrt{2}}, \dfrac{1}{\sqrt{2}}\right) = \dfrac{4}{e^2}, f_{xx}\left(\dfrac{1}{\sqrt{2}}, \dfrac{1}{\sqrt{2}}\right) = -\dfrac{2}{e}$,

so there is a local maximum at $\left(\dfrac{1}{\sqrt{2}}, \dfrac{1}{\sqrt{2}}\right)$.

$D\left(-\dfrac{1}{\sqrt{2}}, -\dfrac{1}{\sqrt{2}}\right) = \dfrac{4}{e^2}, f_{xx}\left(\dfrac{1}{\sqrt{2}}, \dfrac{1}{\sqrt{2}}\right) = -\dfrac{2}{e}$,

so there is a local maximum at $\left(-\dfrac{1}{\sqrt{2}}, -\dfrac{1}{\sqrt{2}}\right)$.

$D\left(-\dfrac{1}{\sqrt{2}}, \dfrac{1}{\sqrt{2}}\right) = \dfrac{4}{e^2}, f_{xx}\left(-\dfrac{1}{\sqrt{2}}, \dfrac{1}{\sqrt{2}}\right) = \dfrac{2}{e}$,

so there is a local minimum at $\left(-\dfrac{1}{\sqrt{2}}, \dfrac{1}{\sqrt{2}}\right)$.

$D\left(\dfrac{1}{\sqrt{2}}, -\dfrac{1}{\sqrt{2}}\right) = \dfrac{4}{e^2}, f_{xx}\left(\dfrac{1}{\sqrt{2}}, -\dfrac{1}{\sqrt{2}}\right) 1 = \dfrac{2}{e}$,

so there is a local minimum at $\left(\dfrac{1}{\sqrt{2}}, -\dfrac{1}{\sqrt{2}}\right)$.

19. There are three critical points:
$(-2.836, -0.176), D = -22.994$, saddle
$(0.505, 0.989), D = -5.739$, saddle
$(2.821, 0.177), D = 22.265, f_{xx} = 3.969$,
     local minimum

21. There are seven critical points:
$(0, 0), D = 0$, not saddle, not local min or max
$(1, 0), D = 1.083, f_{xx} = -1.472$, local max
$(-1, 0), D = 1.083, f_{xx} = -1.472$, local max
$(0, 1.225), D = 2.076, f_{xx} = 1.266$, local min
$(0, -1.225), D = 0.613, f_{xx} = -0.374$, local max
$(0.839, -0.667), D = -0.567$, saddle
$(-0.839, -0.667), D = -0.567$, saddle

23. The residuals are:

$a + b - .6$

$4a + b - 2.0$

$9a + b - 6.4$

$16a + b - 21$

$f(a,b) = (a+b-.6)^2 + (4a+b-2.0)^2 + (9a+b-6.4)^2 + (16a+b-21)^2$

$\dfrac{\partial f}{\partial a} = 708a + 60b - 804.4$

$\dfrac{\partial f}{\partial b} = 60a + 8b - 60$

solving $\dfrac{\partial f}{\partial a} = \dfrac{\partial f}{\partial b} = 0$, we obtain

$a \approx 1.37, \ b \approx -2.80$

$y = 1.37x - 2.80$

25. The residuals are:

$b - 8910$

$2a + b - 8800$

$4a + b - 9040$

$6a + b - 9040$

$8a + b - 9050$

$f(a,b) = (b-8910)^2 + (2a+b-8800)^2 + (4a+b-9040)^2 + (6a+b-9040)^2 + (8a+b-9050)^2$

$\dfrac{\partial f}{\partial a} = 240a + 40b - 360,800$

$\dfrac{\partial f}{\partial b} = 40a + 10b - 89,680$

Solving $\dfrac{\partial f}{\partial a} = \dfrac{\partial f}{\partial b} = 0$, we obtain

$a = 26, \ b = 8864$

$y = 26x + 8864$

$y(12) = 9176$

27. The residuals are:

$68a + b - 160$

$70a + b - 172$

$70a + b - 184$

$71a + b - 180$

$f(a,b) = (68a+b-160)^2 + (70a+b-172)^2 + (70a+b-184)^2 + (71a+b-180)^2$

$\dfrac{\partial f}{\partial a} = 38,930a + 558b - 97,160$

$\dfrac{\partial f}{\partial b} = 558a + 8b - 1392$

Solving $\dfrac{\partial f}{\partial a} = \dfrac{\partial f}{\partial b} = 0$, we obtain

$a \approx 7.16,$

$b \approx -325.26$

$y = 7.16x - 325.26$

$y(80) = 248$ lbs

$y(60) = 104$ lbs

29. The residuals are:

$15a + b - 4.57$

$35a + b - 3.17$

$55a + b - 1.54$

$75a + b - .24$

$95a + b + 1.25$

$$f(a,b) = (15a + b - 4.57)^2 + (35a + b - 3.17)^2 + (55a + b - 1.54)^2 + (75a + b - .24)^2 + (95a + b + 1.25)^2$$

$$\frac{\partial f}{\partial a} = 38,250a + 550b - 326.9$$

$$\frac{\partial f}{\partial b} = 550a + 10b - 16.54$$

Solving $\dfrac{\partial f}{\partial b} = \dfrac{\partial f}{\partial a} = 0,$ we obtain

$a \approx -.07, b \approx 5.66$

$y = -.07x + 5.66$

a. $y(60) = 1.29$ points

b. $y(40) = 2.75$ points

31. $\nabla f = \left\langle 2y - 4x, 2x + 3y^2 \right\rangle$

$\nabla f(0,-1) = \langle -2, 3 \rangle$

$g(h) = f(0 - 2h, -1 + 3h)$

$g'(h) = -2(2(-1 + 3h) - 4(0 - 2h)) + 3(2(0 - 2h) + 3(-1 + 3h)^2)$

$g'(h) = 0$ when first

$h \approx .1604$ or $h = 1$

we arrive at $(-.3209, -.5185)$

$\nabla f(-.3209, -.5185) = \langle .2466, .1647 \rangle$

$g(h) = f(-.3209 + .2466h, -.5185 + .1647h)$

$g'(h) = .2466(2(-.5185 + .1647h) - 4(-.3209 + .2466h)) + .1647(2(-.3209 + .2466h) + 3(-.5185 + .1647h)^2)$

$g'(h) = 0$ first when $h = .5576$

We arrive at $(-.1833, -.4266)$.

33. $\nabla f = \left\langle 1 - 2xy^4, -4x^2y^3 + 2y \right\rangle$

$\nabla f(1,1) = \langle -1, -2 \rangle$

$g(h) = f(1 - h, 1 - 2h)$

$g'(h) = -(1 - 2(1-h)(1-2h)^4)$
$\quad\quad -2(-4(1-h)^2(1-2h)^3 + 2(1-2h))$

$g'(h) = 0$ first when $h = .0956$.

We arrive at $(.9044, .8088)$

$\nabla f(.9044, .8088) = \langle .2259, -.1134 \rangle$

$g(h) = f(.9044 + .2259h, .8088 - .1134h)$

$g'(h) =$

$\quad .2259(1 - 2(.9044 + .2259h)(.8088 - .1134h)^4)$
$\quad -.1134(-4(.9044 + .2259h)^2(.8088 - .1134h)^3$
$\quad +2(.8088 - .1134h))$

$g'(h) = 0$ first when $h \approx 10.5331$.

We arrive at $(3.2838, -.3856)$.

35. $\nabla f = \left\langle 2y - 4x, 2x + 3y^2 \right\rangle$

$\nabla f(0,0) = \langle 0,0 \rangle$

indicating that $(0, 0)$ is a critical point for f, so the gradient does not determine a direction for the first step.

37. In the interior:

$f_x = 2x - 3y \quad\quad f_y = 3 - 3x$

$(1)\, 0 = 2x - 3y \quad (2)\, 0 = 3(1 - x)$

Solving (1) and (2) simultaneously we obtain

$x = 1 \Rightarrow -3y + 2 = 0 \Rightarrow y = \frac{2}{3}$

$f\left(1, \frac{2}{3}\right) = 1$

Along $y = x$

$f(x,x) = g(x) = x^2 + 3x - 3x^2 = -2x^2 + 3x$

$g'(x) = -4x + 3$

$0 = -4x + 3$

$\frac{3}{4} = x = y$ which is included in the region.

$f\left(\frac{3}{4}, \frac{3}{4}\right) = \frac{9}{8}$

Along $y = 0$

$f(x, y) = h(x) = x^2$ with minimum at $(0,0)$

$f(0,0) = 0$

Along $x = 2$

$f(2, y) = h(y) = 4 + y^2 - 8y$

$h'(y) = 2y - 8$

$0 = 2y - 8$

$y = 4, x = 2$ which is not included in the region.
At the intersection points of the boundaries
$(0, 0), (2, 2), (2, 0)$, we have
$f(2, 2) = -2, f(2, 0) = 4, f(0, 0) = 0$
Finally, we compare the values of all the local extrema and boundary local extrema.
$f(2, 0) = 4$ is the absolute maximum and
$f(2, 2) = -2$ is the absolute minimum.

39. In the interior:

$f_x = 2x, f_y = 2y$

with critical point $(0, 0)$ at which $f(0, 0) = 0$
The function evaluated along the circle gives

$g(x) = x^2 + 4 - (x - 1)^2$

$g'(x) = 2x - 2(x - 1)$

$\quad\quad = 2$ Thus g is always increasing.

On the circle, the minimum value of x is $x = -1$
and the maximum value is $x = 3$.

$f(-1, 0) = 1$

$f(3, 0) = 9$

Comparing the values of the local extrema, we get:

$f(3, 0) = 9$ is the absolute maximum and
$f(0, 0) = 0$ is the absolute minimum.

41. Since $f_x = 2xy^2$ and $f_y = 2yx^2$, the critical

points would be where $x = 0$ and y is any real number or $y = 0$ and x is any real number, i.e., $(x, 0)$ or $(0, y)$.
Since $D(x, 0) = D(0, y) = 0$, no information is given about the types of extrema in the second derivative test. The function evaluated at points off the x- and y-axes will be greater than 0. Therefore, points along the x- and y-axes are all local minima.

43. $f(x, y) = x^2 + 2x + 1 + y^2 - 4y + 4 - 4$

$\quad\quad = (x + 1)^2 + (y - 2)^2 - 4$

Local minimum at $(-1, 2)$
$f(-1, 2) = -4$

45. $f(x) = (1 + k^3)x^3 - 3kx^2$

$\frac{d^2 f}{dx^2} = -6k$, which is positive for $k < 0$,

indicating a local minimum, and negative for $k > 0$, indicating a local maximum.

47. Substituting $y = kx$ we get

$f(x) = x^3(1+3x) - 2k^2x^2 - 2k^4x^4$

The second derivative of f with respect to

x evaluated at 0 is $-4k^2$. This doesnt

change sign, so all traces have a minimum

at 0, except possible for the trace

corresponding to $k = 0$.

In this case the trace reduces to

$z = x^3$ which has neither a max nor a min.

Thus there is no trace with a max at the origin

and the origin cannot be a saddle point.

49. $f(x,y,z) = xz - x + y^3 - 3y$

The partials with respect to x, y, and z are

$z - 1, 3(y^2 - 1)$, and x, so $(0,1,1)$ is a critical

point.

$f(\Delta x, 1 + \Delta y, 1 + \Delta z) - f(0,1,1)$

$= \Delta x \Delta z + 3\Delta y^2 + \Delta y^3$

So with $\Delta y = 0$, as we move away from the
critical point with $\Delta x \Delta z$ positive, f increases,
while with $\Delta x \Delta z$ negative, f decreses.
The critical point neither a local max nor a
local min.

51. False, the partials could be undefined at (a, b).

53. False, nothing is known about the function
between the two maxima.

55. Extrema at $\left(\pm\dfrac{\pi}{2}, \pm\dfrac{\pi}{2}\right)$, saddles at

$(\pm n\pi, \pm n\pi)$.

57. Extrema at $(1, 1)$ and $(-1, -1)$, saddle at $(0, 0)$.

59. Extrema at $(-.1, .1)$ and $(.1, .1)$, saddle at
$(0, 0)$.

61. $d(x,y) = \sqrt{(x-3)^2 + (y+2)^2 + (3-x^2-y^2)^2}$

minimize: $g(x,y) = (x-3)^2 + (y+2)^2 + (3-x^2-y^2)^2$

Here we will use polar coordinates.

$f(r,t) = (r\cos t - 3)^2 + (r\sin t + 2)^2 + (3-r^2)^2$

$= r^2\cos^2 t - 6r\cos t + 9 + r^2\sin^2 t + 4r\sin t + 4 + 9 - 6r^2 + r^4$

$= r^4 - 5r^2 - 6r\cos t + 4r\sin t + 22$

$f_r = 4r^3 - 10r - 6\cos t + 4\sin t$

$f_t = 6r\sin t + 4r\cos t$

$f_t = 0 \Rightarrow 6\sin t = -4\cos t$, since $r \neq 0$

$\tan t = -\dfrac{2}{3}$

or $t \approx -.5880$

substituting $t = -.5580$ in f_r we obtain,

$f_r = 4r^3 - 10r - 7.2111$

$0 = 4r^3 - 10r - 7.2111$

Solving numerically, we get $r \approx 1.8622$

$x = 1.8622\cos(-.5880) = 1.55$

$y = 1.8622\sin(-.5880) = -1.03$

$f_{rr} = 12r^2 - 10, f_{tt} = 6r\cos t - 4r\sin t$,

$f_{rt} = 6\sin t + 4\cos t$

$D(1.8622, -.5580) \approx 424.52 > 0$

$f_{rr}(1.8622, .5580) \approx 31.61 > 0$

Therefore, $(1.55, -1.03)$ is the point which minimizes the distance.

63. $d(x,y) = \sqrt{(x-2)^2 + (y-1)^2 + \left(-\sqrt{9-x^2-y^2}+3\right)^2}$

minimize: $g(x,y) = (x-2)^2 + (y-1)^2 + \left(-\sqrt{9-x^2-y^2}+3\right)^2$

using polar coordinates

$f(r,t) = (r\cos t - 2)^2 + (r\sin t - 1)^2 + \left(-\sqrt{9-r^2}+3\right)^2$

$\qquad = r^2\cos^2 t - 4r\cos t + 4 + r^2\sin^2 t - 2r\sin t + 1 + 9 - r^2 - 6\sqrt{9-r^2} + 9$

$\qquad = -4r\cos t - 2r\sin t + 23 - 6\sqrt{9-r^2}$

$\quad f_r = -4\cos t - 2\sin t + \dfrac{6r}{\sqrt{9-r^2}}$

$\quad f_t = 4r\sin t - 2r\cos t$

setting $f_t = 0$,

$4r\sin t = 2r\cos t$

$\tan t = \dfrac{1}{2}$ since $r \neq 0$.

$\quad t \approx .4636$

Substituting $t \approx .4636$ in f_r we obtain

$f_r = -4.4721 + \dfrac{6r}{\sqrt{9-r^2}}$

Setting $f_r = 0$,

$4.472 = \dfrac{6r}{\sqrt{9-r^2}}$

$20 = \dfrac{36r^2}{9-r^2}$

$180 = 56r^2$

$1.7928 \approx r$

$f_{rr} = \dfrac{\sqrt{9-r^2}\cdot 6 - 6r\dfrac{-r}{\sqrt{9-r^2}}}{9-r^2} = \dfrac{(9-r^2)6 + 6r^2}{\sqrt{(9-r^2)^3}} = \dfrac{54}{\sqrt{(9-r^2)^3}}$

$f_{tt} = 4r\cos t + 2r\sin t \qquad f_{rt} = 4\sin t - 2\cos t$

$D(1.7928, .4636) \approx 31.1091 > 0$

$f_{rr}(1.7928, .4636) \approx 3.8800 > 0$

Therefore, $x = 1.7928\cos .4636 \approx 1.60$

$\qquad\qquad y = 1.7928\sin .4636 \approx .8$

$\qquad\qquad z = -\sqrt{9 - 1.7928^2} \approx -2.4$

is the point that minimizes the distance.

(Use the negative value for z since the point is $(2, 1, -3)$.)

65. $\quad f_x = 5e^y - 5x^4 \qquad f_y = 5xe^y - 5e^{5y}$

$(1)\,0 = 5(e^y - x^4) \quad (2)\,0 = 5e^y(x - e^{4y})$

Solving (1) and (2) simultaneously, we obtain $x = 1, y = 0$.

$D(1, 0) = 375 \quad f_{xx} = (1,0) = -20$

Therefore $f(1, 0) = 3$ is a local maximum. Along the line $y = 0$ as $x \to -\infty$, $f(x,y) \to \infty$

Section 12.8

5. $f(x, y) = x^2 + y^2$
$g(x, y) = 3x - 4 - y$
$\nabla f = \langle 2x, 2y \rangle$
$\nabla g = \langle 3, -1 \rangle$
$\nabla f = \lambda \nabla g$
$2x = 3\lambda$
$2y = -\lambda$

eliminating λ we get $y = -\dfrac{x}{3}$.

$y = 3x - 4$

$-\dfrac{x}{3} = 3x - 4$

$x = \dfrac{6}{5}, \ y = -\dfrac{2}{5}$

7. $f(x, y) = (x-4)^2 + y^2$
$g(x, y) = 3 - 2x - y$
$\nabla f = \langle 2x - 8, 2y \rangle$
$\nabla g = \langle -2, -1 \rangle$
$\nabla f = \lambda \nabla g$
$2x - 8 = -2\lambda$
$2y = -\lambda$

eliminating λ we get $y = \dfrac{1}{2}x - 2$

$y = 3 - 2x$

$\dfrac{1}{2}x - 2 = 3 - 2x$

$x = 2, \ y = -1$

9. $f(x, y) = (x-3)^2 + y^2$
$g(x, y) = x^2 - y$
$\nabla f = \langle 2x - 6, \ 2y \rangle$
$\nabla g = \langle 2x, -1 \rangle$
$\nabla f = \lambda \nabla y$
$2x - 6 = 2\lambda x$
$2y = -\lambda$

eliminating λ we get $y = -\dfrac{x-3}{2x}$

$y = x^2$

$-\dfrac{x-3}{2x} = x^2$

$-x + 3 = 2x^3$

$2x^3 + x - 3 = 0$
$x = 1$ (by inspection), $y = 1$

11. $f(x, y) = (x-2)^2 + \left(y - \dfrac{1}{2} \right)^2$
$g(x, y) = x^2 - y$
$\nabla f = \langle 2x - 4, \ 2y - 1 \rangle$
$\nabla g = \langle 2x, -1 \rangle$
$\nabla f = \lambda \nabla g$
$2x - 4 = 2\lambda x$
$2y - 1 = -\lambda$

eliminating λ we get $y = \dfrac{1}{x}$

$y = x^2$

$\dfrac{1}{x} = x^2$

$1 = x^3$
$x = 1, y = 1$

13. $g(x, y) = x^2 + y^2 - 8 = 0$
$\nabla f = \langle 4y, 4x \rangle$
$\nabla g = \langle 2x, 2y \rangle$
$\nabla f = \lambda \nabla g$
$4y = 2\lambda x$
$4x = 2\lambda y$
eliminating λ we get $y = \pm x$

$g(x, x) = 2x^2 - 8, \ g(x, -x) = 2x^2 - 8$

$2x^2 = 8$

$x = \pm 2, \ y = \pm 2$

$f(2, 2) = f(-2, -2) = 16$ maxima
$f(-2, 2) = f(2, -2) = -16$ minima

15. $g(x, y) = x^2 + y^2 - 3 = 0$
$\nabla f = \langle 8xy, 4x^2 \rangle$
$\nabla g = \langle 2x, 2y \rangle$
$\nabla f = \lambda \nabla g$
$8xy = 2\lambda x$
$4x^2 = 2\lambda y$

eliminating λ we get $y = \pm \dfrac{1}{\sqrt{2}}x$

$g\left(x, \dfrac{1}{\sqrt{2}}x \right) = \dfrac{3x^2}{2} - 3, \ g\left(x, -\dfrac{1}{\sqrt{2}}x \right) = \dfrac{3x^2}{2} - 3$

$\dfrac{3x^2}{2} = 3$

$x = \pm\sqrt{2}, \ y = \pm 1$

$f(\sqrt{2}, 1) = f(-\sqrt{2}, 1) = 8$ maxima
$f(\sqrt{2}, -1) = f(-\sqrt{2}, -1) = -8$ minima

17. $g(x, y) = x^2 + y^2 - 2 = 0$

$\nabla f = \left\langle e^y, xe^y \right\rangle$

$\nabla g = \left\langle 2x, 2y \right\rangle$

$\nabla f = \lambda \nabla g$

$e^y = 2\lambda x$

$xe^y = 2\lambda y$

eliminating λ we get $y = x^2$

$\quad g(x, x^2) = x^4 + x^2 - 2$

$x^4 + x^2 - 2 = 0$

$\quad\quad x = \pm 1, \ y = 1$

$f(1, 1) = e$ maximum

$f(-1, 1) = -e$ minimum

19. $g(x, y) = x^2 + y^2 - 3 = 0$

$\nabla f = \left\langle 2xe^y, x^2 e^y \right\rangle$

$\nabla g = \left\langle 2x, 2y \right\rangle$

$\nabla f = \lambda \nabla g$

$2xe^y = 2\lambda x$

$x^2 e^y = 2\lambda y$

eliminating λ we get $y = \dfrac{x^2}{2}$ or $x = 0$ and $\lambda = 0$

if $x = 0, \ y = \pm\sqrt{3}$

$\quad g\left(x, \dfrac{x^2}{2}\right) = \dfrac{1}{4}x^4 + x^2 - 3$

$x^4 + 4x^2 - 12 = 0$

$\quad\quad x = \pm\sqrt{2}, \ y = \pm 1$

$f(\sqrt{2}, 1) = f(-\sqrt{2}, 1) = 2e$ maxima

$f(\sqrt{2}, 1) = f(-\sqrt{2}, -1) = 2e^{-1}$

$f\left(0, \pm\sqrt{3}\right) = 0$ minima

21. On the boundary $x^2 + y^2 = 8$:

$g(x, y) = x^2 + y^2 - 8$

$\nabla f = \left\langle 4y, 4x \right\rangle$

$\nabla g = \left\langle 2x, 2y \right\rangle$

$\nabla f = \lambda \nabla g$

$4y = 2\lambda x$

$4x = 2\lambda y$

eliminating λ we get $y = \pm x$

$g(x, \pm x) = 2x^2 - 8$

$\quad 2x^2 = 8$

$\quad\quad x = \pm 2, \ y = \pm 2$

$f(2, 2) = f(-2, -2) = 16$

$f(-2, 2) = f(2, -2) = -16$

In the interior $x^2 + y^2 < 8$:

$\quad f_x = 4y, \ f_y = 4x$

critical point is $(0, 0)$

$f(0, 0) = 0$

$f(2, 2) = f(-2, -2) = 16$ maxima

$f(-2, 2) = f(2, -2) = -16$ minima

23. On the boundary $x^2 + y^2 = 3$:

$g(x, y) = x^2 + y^2 - 3$

$\nabla f = \left\langle 8xy, 4x^2 \right\rangle$

$\nabla g = \left\langle 2x, 2y \right\rangle$

$\nabla f = \lambda \nabla g$

$8xy = 2\lambda x$

$4x^2 = 2\lambda y$

eliminating λ we get $y = \pm\dfrac{1}{\sqrt{2}}x$

$g\left(x, \pm\dfrac{1}{\sqrt{2}}x\right) = \dfrac{3x^2}{2} - 3$

$\quad\quad \dfrac{3x^2}{2} = 3$

$\quad\quad\quad x = \pm\sqrt{2}, \ y = \pm 1$

$f(\sqrt{2}, 1) = f(-\sqrt{2}, 1) = 8$

$f(\sqrt{2}, -1) = f(-\sqrt{2}, -1) = -8$

In the interior $x^2 + y^2 < 3$:

$\quad f_x = 8xy, \ f_y = 4x^2$

critical point at $(0, y)$

$f(0, y) = 0$

$f(\sqrt{2}, 1) = f(-\sqrt{2}, 1) = 8$ maxima

$f(\sqrt{2}, -1) = f(-\sqrt{2}, -1) = -8$ minima

25. $g(u,t) = u^2 t - 11,000$

$\nabla f = \left\langle (u-32)t, \frac{1}{2}t^2 \right\rangle \quad \nabla g = \left\langle u^2, 2ut \right\rangle$

$\nabla f = \lambda \nabla g$

$(u-32)t = \lambda u^2$

$\frac{1}{2}t^2 = 2\lambda ut$

eliminating λ we get $u = \frac{128}{3}$

$g\left(t, \frac{128}{3}\right) = 1820.44t - 11,000$

$1820.44t = 11,000$

$t \approx 6.04$

$z = \left(\frac{128}{3}, 6.04\right) \approx 195$

The maximum height is 195 feet.

27. $\nabla P = \langle 3, 6, 6 \rangle$ so no critical points in the interior

$g(x,y,z) = 2x^2 + y^2 + 4z^2 - 8800$

$\nabla g = \langle 4x, 2y, 8z \rangle$

$\nabla P = \lambda \nabla g$

$4\lambda x = 3 \quad 2\lambda y = 6 \quad 8\lambda z = 6$

$x = \frac{3}{4\lambda} \qquad y = \frac{3}{\lambda} \qquad z = \frac{3}{4\lambda}$

$g\left(\frac{3}{4\lambda}, \frac{3}{\lambda}, \frac{3}{4\lambda}\right) = \frac{9}{8\lambda^2} + \frac{9}{\lambda^2} + \frac{9}{4\lambda^2} - 8800$

$8800 = \frac{99}{8\lambda^2}$

$\lambda^2 = \frac{9}{6400}$

$\lambda = \frac{3}{80} = .0375$

$x = 20, y = 80, z = 20, P(20, 80, 20) = 660$

29. An increase of 1 unit in the constraint would give

us: $8801 = \frac{99}{8\lambda^2}$

$\lambda^2 = \frac{99}{8 \cdot (8801)} \approx .0014$

$\lambda \approx .037498$

$x = 20.0011, y = 81.0043, z = 20.0011$

$P(20.0011, 80.0043, 20.0011) = 660.0357$

The change in the profit is approximately λ.

31. $g(x,y) = xy - c$

$\nabla f = \langle 2, 2 \rangle$

$\nabla g = \langle y, x \rangle$

$\nabla f = \lambda \nabla g$

$2 = \lambda y, 2 = \lambda x$

$2 = \frac{2}{y}x$

$2y = 2x$

$x = y$

$f(x, y) = 2x + 2y$ is the perimeter of the rectangle with vertices (0, 0), (x, 0), (x, y), and (0, y).

33. $f(x,y) = y - x$

$g(x,y) = x^2 + y^2 - 1$

$\nabla f = \langle -1, 1 \rangle$

$\nabla g = \langle 2x, 2y \rangle$

$\nabla f = \lambda \nabla g$

$-1 = 2\lambda x \qquad 1 = 2\lambda y$

$-\frac{1}{2x} = \lambda \qquad 1 = \left(-\frac{1}{2x}\right)2y$

$1 = -\frac{y}{x}$

$x = -y$

$x^2 + y^2 = 1$

$2x^2 = 1$

$x = \pm\frac{1}{\sqrt{2}}$

$y = \pm\frac{1}{\sqrt{2}}$

$f\left(\frac{1}{\sqrt{2}}, \frac{1}{\sqrt{2}}\right) = f\left(-\frac{1}{\sqrt{2}}, -\frac{1}{\sqrt{2}}\right) = 0$

$f\left(\frac{1}{\sqrt{2}}, -\frac{1}{\sqrt{2}}\right) = -\sqrt{2}$

$f\left(-\frac{1}{\sqrt{2}}, \frac{1}{\sqrt{2}}\right) = \sqrt{2}$ is the maximum.

35. $\alpha + \beta$ represents the sum of the angles from due north to the hull of the sailboat, the remaining angle represents the angle from the hull to due east. Due north to due east is $\frac{\pi}{2}$.

$$f(\alpha,\beta,\theta) = \sin\alpha\sin\beta\sin\theta$$

$$g(\alpha,\beta,\theta) = \alpha + \beta + \theta - \frac{\pi}{2}$$

$$\nabla f = \langle \cos\alpha\sin\beta\sin\theta, \sin\alpha\cos\beta\sin\theta, \sin\alpha\sin\beta\cos\theta \rangle$$

$$\nabla g = \langle 1,1,1 \rangle$$

$$\nabla f = \lambda\nabla g$$

$$\cos\alpha\sin\beta\sin\theta = \lambda$$

$$\sin\alpha\cos\beta\sin\theta = \lambda$$

$$\sin\alpha\sin\beta\cos\theta = \lambda$$

$$\cos\alpha\sin\beta\sin\theta = \sin\alpha\cos\beta\sin\theta$$

$$\cos\alpha\sin\beta = \sin\alpha\cos\beta \quad \theta \neq 0$$

$$\tan\beta = \tan\alpha$$

$$\beta = \alpha \quad 0 < \alpha, \beta, \theta < \frac{\pi}{2}$$

$$\sin\alpha\cos\beta\sin\theta = \sin\alpha\sin\beta\cos\theta$$

$$\cos\beta\sin\theta = \sin\beta\cos\theta$$

$$\tan\theta = \tan\beta$$

$$\theta = \beta = \alpha$$

$$3\alpha = \frac{\pi}{2}$$

$$\alpha = \frac{\pi}{6} = \beta = \theta$$

$$f\left(\frac{\pi}{6},\frac{\pi}{6},\frac{\pi}{6}\right) = \frac{1}{8}$$

37. The constraint requires $y = -x$, so the function to be optimized is x^3, and this has neither a min nor a max at the origin.

39. $P(L, K) = 200L^{2/3}K^{1/3}$
$C(L, K) = 2L + 5K - 150$
Setting $\nabla P = \lambda\nabla C$ we get
$$\frac{400}{3}L^{-1/3}K^{1/3} = 2\lambda$$
$$\frac{200}{3}L^{2/3}K^{-2/3} = 5\lambda$$
Eliminating λ yields
$5K = L$, which together with
$2L + 5K = 150$ implies that
$L = 50$ and $K = 10$.

41. $C(L, K) = 25L + 100K$
$P(L, K) = 60L^{2/3}K^{1/3}$
Setting $\nabla C = \lambda\nabla P$ we get
$$25 = \lambda \cdot 40L^{-1/3}K^{1/3}$$
$$100 = \lambda \cdot 20L^{2/3}K^{-2/3}$$
Eliminating λ yields
$L = 8K$, which together with
$1920 = 60L^{2/3}K^{1/3}$ implies that
$1920 = 240K$, so $K = 8$ and $L = 64$.

43. $f(x,y,z) = x^2 + y^2 + z^2$
$g(x,y,z) = x + 2y + 3z - 6$
$h(x,y,z) = y + z$
Setting $\nabla f = \lambda \nabla g + \mu \nabla h$ yields the
five equations
$2x = \lambda$
$2y = 2\lambda + \mu$
$2z = 3\lambda + \mu$
$x + 2y + 3z - 6 = 0$
$y + z = 0$
Solving this system with a CAS yields
$\lambda = 8, \mu = -20, \langle x,y,z \rangle = \langle 4, -2, 2 \rangle$,
so $\langle 4, -2, 2 \rangle$ minimizes the distance.

45. $f(x,y,z) = xyz$
$g(x,y,z) = x + y + z - 4$
$h(x,y,z) = x + y - z$
Setting $\nabla f = \lambda \nabla g + \mu \nabla h$ yields the
five equations
$yz = \lambda + \mu$
$xz = \lambda + \mu$
$xy = \lambda - \mu$
$x + y + z - 4 = 0$
$x + y - z = 0$
Solving this system with a CAS yields
$\lambda = \frac{3}{2}, \mu = \frac{1}{2}, \langle x,y,z \rangle = \langle 1,1,2 \rangle$,
so $\langle 1,1,2 \rangle$ minimizes the function f.

47. These equations define intersecting cylinders with
the z-axis and the y-axis as axes. A sketch shows
that there are intersection points at $(\pm 1, 0, 0)$, and
since these are a distance 1 from the origin and all
points on the cylinders are at least 1 unit from the
origin, these must be the points that are closest to
the origin. Using the symmetry of the figure, it's
clear that the points of maximum distance from
the origin lie in the plane $x = 0$ at
$$\left(0, \pm \frac{1}{\sqrt{2}}, \pm \frac{1}{\sqrt{2}} \right).$$

Chapter 12 Review

1. See Examples 1.4 and 1.5.

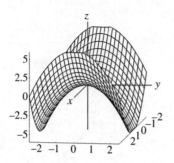

3. See Examples 1.4 and 1.5.

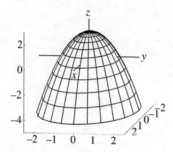

5. See Examples 1.4 and 1.5.

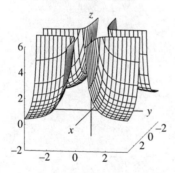

7. See Examples 1.4 and 1.5.

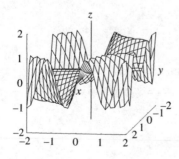

9. See Examples 1.4 and 1.5.

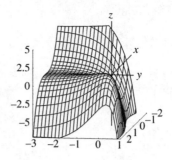

11. See Example 1.6.

a. For x = constant we get sine curves in the yz-plane with decreasing periods. The graph is bounded above by 1 and below by –1. Surface D.

b. Undefined at $y = 0$, this graph should oscillate near $(0, 0)$. Surface B.

c. At $x = 0$, there is a sine curve in the yz-plane. At $y = 0$, there is a sine curve in the xz-plane. Surface C.

d. As x gets large f will get large and for a fixed x we have a sine curve in the yz-plane. Surface A.

e. Undefined at $\left(\pm\dfrac{1}{\sqrt{2}}, 0 \right)$, this graph gets close the xy-plane away from the origin. Surface F.

f. Defined everywhere, this graph has a maximum at $(0, 0)$ of 4. Surface E.

13. See Example 1.9.

a. Contour C.

b. Contour A.

c. Contour D.

d. Contour B.

21. See Examples 2.8 and 2.9.

$3x^2 e^{4y}$ is continuous for all (x, y), $\dfrac{-3y}{x}$ is not continuous at $x = 0$ so the sum $3x^2 e^{4y} - \dfrac{3y}{x}$ is not continuous at $x = 0$.

23. See Examples 3.1 and 3.2.

$$f_x = \frac{4}{y} + xye^{xy} + e^{xy}$$

$$f_y = -\frac{4x}{y^2} + x^2 e^{xy}$$

25. See Examples 3.1 and 3.2.

$$f_x = 6xy\cos y - \frac{1}{2\sqrt{x}}$$

$$f_y = -3x^2 y\sin y + 3x^2 \cos y$$
$$= 3x^2 \cos y - 3x^2 y\sin y$$

27. See Examples 3.4 and 3.5.

$$\frac{\partial f}{\partial x} = e^x \sin y, \frac{\partial^2 f}{\partial x^2} = e^x \sin y$$

$$\frac{\partial f}{\partial y} = e^x \cos y, \frac{\partial^2 f}{\partial y^2} = -e^x \sin y$$

$$\frac{\partial^2 f}{\partial x^2} + \frac{\partial^2 f}{\partial y^2} = e^x \sin y - e^x \sin y = 0$$

29. See Section 12.3.

$$\frac{\partial f}{\partial x}(0,0) \approx \frac{1.6 - 2.4}{10 - (-10)} = -.04$$

$$\frac{\partial f}{\partial y}(0,0) \approx \frac{2.6 - 1.4}{10 - (-10)} = .06$$

31. See Example 4.3.

$$f_x = \frac{3xy}{\sqrt{x^2 + 5}}, f_y = 3\sqrt{x^2 + 5}$$

$$f_x(-2,5) = -10, f_y(-2,5) = 9, f(-2,5) = 45$$

$$L(x, y) = 45 - 10(x + 2) + 9(y - 5)$$
$$= -10x + 9y - 20$$

33. See Example 4.3.

$$f_x = \sec^2(x + 2y), f_y = 2\sec^2(x + 2y)$$

$$f_x = \left(\pi, \frac{\pi}{2}\right) = 1, f_y\left(\pi, \frac{\pi}{2}\right) = 2, f\left(\pi, \frac{\pi}{2}\right) = 0$$

$$L(x, y) = 0 + x - \pi + 2\left(y - \frac{\pi}{2}\right)$$

$$L(x, y) = x + 2y - 2\pi$$

35. See Example 3.5.

$$f_x = 8x^3 y + 6xy^2, f_{xx} = 24x^2 y + 6y^2$$
$$f_y = 2x^4 + 6x^2 y, f_{yy} = 6x^2$$
$$f_{xy} = 8x^3 + 12xy$$

37. See Examples 4.1 and 4.2.

$$f_x = 2xy + 2, f_y = x^2 - 2y$$

$$f_x(1,-1) = 0, f_y = (1,-1) = 3, f(1,-1) = 0$$

$$z = 3(y + 1)$$
$$z = 3y + 3$$

39. See Example 6.7.

$$\nabla f = \langle 2x + 2y, 2x + 2y, 2z \rangle$$

$$\nabla f(0,2,1) = \langle 4,4,2 \rangle$$

$$4(x - 0) + 4(y - 2) + 2(z - 1) = 0$$

$$4x + 4y + 2z - 10 = 0$$

41. See Example 5.1.

$$\frac{\partial f}{\partial x} = 2xy, \frac{\partial f}{\partial y} = x^2 + 2y$$

$$\frac{dx}{dt} = 4e^{4t}, \frac{dy}{dt} = \cos t$$

$$g'(t) = \frac{\partial f}{\partial x}\frac{dx}{dt} + \frac{\partial f}{\partial y}\frac{dy}{dt}$$

$$= 2e^{4t}\sin t(4e^{4t}) + ((e^{4t})^2 + 2\sin t)\cos t$$

$$= 8e^{8t}\sin t + (e^{8t} + 2\sin t)\cos t$$

43. See Theorem 5.1.

$$g'(t) = \frac{\partial f}{\partial x}\frac{dx}{dt} + \frac{\partial f}{\partial y}\frac{dy}{dt} + \frac{\partial f}{\partial z}\frac{dz}{dt} + \frac{\partial f}{\partial w}\frac{dw}{dt}$$

45. See Example 5.5.

$$f_x = 2x + 2y, f_z = 2z, f_y = 2x + 2y$$

$$\frac{\partial z}{\partial x} = -\frac{x + y}{z}, \frac{\partial z}{\partial y} = -\frac{x + y}{z}$$

47. See Definition 6.2.

$$f_x = 3\sin 4y - \frac{y}{2\sqrt{xy}}, f_y = 12x\cos 4y - \frac{x}{2\sqrt{xy}}$$

$$f_x(\pi, \pi) = 0 - \frac{1}{2}, f_y(\pi, \pi) = 12\pi - \frac{1}{2}$$

$$\nabla f(\pi, \pi) = \left\langle -\frac{1}{2}, 12\pi - \frac{1}{2}\right\rangle$$

49. See Examples 6.1 and 6.2.

$$\nabla f = \langle 3x^2 y, x^3 - 8y \rangle$$

$$\nabla f(-2,3) = \langle 36, -32 \rangle$$

$$\mathbf{u} = \left\langle \frac{3}{5}, \frac{4}{5}\right\rangle$$

$$\nabla f \cdot \mathbf{u} = \frac{108}{5} - \frac{128}{5} = -\frac{20}{5} = -4$$

51. See Examples 6.1 and 6.2.
$$\nabla f = \left\langle 3ye^{3xy}, 3xe^{3xy} - 2y \right\rangle$$
$$\nabla f(0,-1) = \langle -3, 2 \rangle$$
From (2, 3) to (3, 1) is $\langle 1, -2 \rangle$.
$$\mathbf{u} = \left\langle \frac{1}{\sqrt{5}}, -\frac{2}{\sqrt{5}} \right\rangle$$
$$\nabla f \cdot \mathbf{u} = -\frac{3}{\sqrt{5}} - \frac{4}{\sqrt{5}} = -\frac{7}{\sqrt{5}}$$

53. See Example 6.4.
$$\nabla f = \left\langle 3x^2 y, x^3 - 8y \right\rangle$$
$$\nabla f(-2,3) = \langle 36, -32 \rangle$$
maximum direction and rate:
$$\frac{\nabla f(-2,3)}{\|\nabla f(-2,3)\|} = \left\langle \frac{9}{\sqrt{145}}, -\frac{8}{\sqrt{145}} \right\rangle, \|\nabla f\| = 4\sqrt{145}$$
minimum direction and rate:
$$-\frac{\nabla f(-2,3)}{\|\nabla f(-2,3)\|} = \left\langle -\frac{9}{\sqrt{145}}, \frac{8}{\sqrt{145}} \right\rangle,$$
$$-\|\nabla f\| = -4\sqrt{145}$$

55. See Example 6.4.
$$\nabla f = \left\langle \frac{2x^3}{\sqrt{x^4 + y^4}}, \frac{2y^3}{\sqrt{x^4 + y^4}} \right\rangle$$
$$\nabla f(2,0) = \langle 4, 0 \rangle$$
maximum direction and rate:
$$\frac{\nabla f(2,0)}{\|\nabla f(2,0)\|} = \langle 1, 0 \rangle, \|\nabla f\| = 4$$
minimum direction and rate:
$$\frac{-\nabla f(2,0)}{\|\nabla f(2,0)\|} = \langle -1, 0 \rangle, -\|\nabla f\| = -4$$

57. See Example 6.6.
$$\nabla f = \langle -8x, -2 \rangle$$
$$\nabla f(2,1) = \langle -16, -2 \rangle$$
$$\|\nabla f(2,1)\| = 2\sqrt{65}$$
run off direction: $-\nabla f(2,1) = \langle 16, 2 \rangle$

59. See Examples 7.2 and 7.3.
$$f_x = 8x^3 - y^2 \quad f_y = -2xy + 4y$$
$$0 = 8x^3 - y^2 \quad 0 = -2y(x-2)$$
$$y = 0, x = 2$$
$$y = 0 \Rightarrow x = 0$$
$$x = 2 \Rightarrow y = \pm 8$$
Critical points at (0, 0), (2, 8), (2, –8)
$$f_{xx} = 24x^2, f_{yy} = -2x + 4, f_{xy} = -2y$$
$D(0, 0) = 0$ no information
$D(2,8) = -256 \Rightarrow$ saddle point at (2,8)
$D(2,-8) = -256 \Rightarrow$ saddle point at $(2, -8)$
Investigating the point (0, 0) further we rewrite
the function as $2x^4 + y^2(2 - x)$ and note that for
$x \le 2, 2 - x \ge 0$ so that the function
$$f(x, y) = 2x^4 - xy^2 + 2y^2 \ge 0 \text{ for } x \le 2.$$
Therefore, $f(0, 0) = 0$ is a local minimum.

61. See Examples 7.2 and 7.3.
$$f_x = 4y - 3x^2 \quad f_y = 4x - 4y$$
$$(1) \ 0 = 4y - 3x^2 \quad 0 = 4(x - y)$$
$$(2) \ x = y$$
Substituting (2) into (1) we obtain,
$$4x - 3x^2 = 0$$
$$x(4 - 3x) = 0$$
$$x = 0, x = \frac{4}{3}$$
Critical points at $(0,0), \left(\frac{4}{3}, \frac{4}{3} \right)$
$$f_{xx} = -6x, f_{yy} = -4, f_{xy} = 4$$
$D(0, 0) = -16$, saddle point at (0, 0)
$$D\left(\frac{4}{3}, \frac{4}{3} \right) = 16 > 0$$
$$f_{xx}\left(\frac{4}{3}, \frac{4}{3} \right) = -8 < 0,$$
$$f\left(\frac{4}{3}, \frac{4}{3} \right) = \frac{33}{27} \text{ is a local maximum.}$$

63. See Example 7.4.

The residuals are

$64a + b - 140$
$66a + b - 156$
$70a + b - 184$
$71a + b - 190$

$$g(a,b) = (64a + b - 140)^2 + (66a + b - 156)^2$$
$$+ (70a + b - 184)^2 + (71a + b - 190)^2$$

$$\frac{\partial g}{\partial a} = 36{,}786a + 542b - 91{,}252$$

$$\frac{\partial g}{\partial b} = 542a + 8b - 1340$$

Solving, $\dfrac{\partial g}{\partial a} = \dfrac{\partial g}{\partial b} = 0$, we obtain

$a \approx 7.129$, $b \approx -315.542$

$y = 7.129x - 315.542$

$y(74) \approx 212$ lbs

$y(60) \approx 112$ lbs

65. See Example 7.6.

$$f_x = 8x^3 - y^2 \quad f_y = -2xy + 4y$$

$$0 = 8x^3 - y^2 \quad 0 = -2y(x - 2)$$
$$y = 0, \, x = 2$$

$y = 0 \Rightarrow x = 0$

$x = 2 \Rightarrow y = \pm 8$

The only critical point in the region is $(0, 0)$

$f(0, 0) = 0$

Along $x = 4$ we have

$$f(4, y) = h(y) = 512 - 4y^2 + 2y^2$$
$$= 512 - 2y^2$$

$h'(y) = -4y$

$y = 0$ is the only critical value.

Along $x = 0$ we have

$$f(0, y) = l(y) = 2y^2$$

$l'(y) = 4y$

$y = 0$ is the only critical value.

Along $y = 0$ we have

$$f(x, 0) = k(x) = 2x^4$$

$k'(x) = 8x^3$

$x = 0$ is the only critical value.

Along $y = 2$ we have

$$f(x, 2) = g(x) = 2x^4 - 4x + 8$$

$$g'(x) = 8x^3 - 4$$

$$0 = 8x^3 - 4$$

$$\frac{1}{2} = x^3$$

$x = \dfrac{1}{\sqrt[3]{2}}$ is the only critical number.

$$f\left(\frac{1}{\sqrt[3]{2}}, 2\right) \approx 5.6189$$

At the intersection points of the boundaries we have

$f(4, 2) = 504$
$f(4, 0) = 512$
$f(0, 2) = 8$
$f(4, 0) = 512$ is the absolute maximum and
$f(0, 0) = 0$ is the absolute minimum.

67. See Example 8.2.

$\nabla f = \langle 1, 2 \rangle$

$\nabla g = \langle 2x, 2y \rangle$

$\nabla f = \lambda \nabla g$

$1 = 2\lambda x \quad 2 = 2\lambda y$

$\dfrac{1}{2x} = \lambda \quad 2 = 2\dfrac{1}{2x}y$

$\qquad\qquad\quad y = 2x$

$x^2 + y^2 = 5$

$x^2 + 4x^2 = 5$

$5x^2 = 5$

$x = \pm 1$

$x = 1, \, y = 2$

$x = -1, \, y = -2$

$f(1, 2) = 5$ maximum

$f(-1, -2) = -5$ minimum

69. See Example 8.2.

$\nabla f = \langle y, x \rangle$

$\nabla g = \langle 2x, 2y \rangle$

$\nabla f = \lambda \nabla g$

$y = 2\lambda x \qquad x = 2\lambda y$

$\dfrac{y}{2x} = \lambda \qquad x = 2\left(\dfrac{y}{2x}\right)y$

$y^2 = x^2$

$y = \pm x$

$x^2 + y^2 = 1$

$2x^2 = 1$

$x = \pm\dfrac{1}{\sqrt{2}}$

$f\left(-\dfrac{1}{\sqrt{2}}, -\dfrac{1}{\sqrt{2}}\right) = f\left(\dfrac{1}{\sqrt{2}}, \dfrac{1}{\sqrt{2}}\right) = \dfrac{1}{2}$ maxima

$f\left(-\dfrac{1}{\sqrt{2}}, \dfrac{1}{\sqrt{2}}\right) = f\left(\dfrac{1}{\sqrt{2}}, -\dfrac{1}{\sqrt{2}}\right) = -\dfrac{1}{2}$ minima

71. See Example 8.1.

$d(x, y) = \sqrt{(x-4)^2 + y^2}$

$f(x, y) = (x-4)^2 + y^2$

subject to $x^3 - y = 0$

$\nabla f = \langle 2x - 8, 2y \rangle$

$\nabla g = \langle 3x^2, -1 \rangle$

$\nabla f = \lambda \nabla g$

$2x - 8 = 3\lambda x^2, \quad 2y = -\lambda$

$2x - 8 = -6x^2 y$

$x - 4 = -3x^2 y$

$y = \dfrac{4 - x}{3x^2}$

$y = x^3$

$\dfrac{4 - x}{3x^2} = x^3$

$0 = 3x^5 + x - 4$

Using $x = 1$ (by inspection) we have $y = 1$.
(1, 1) is the closest point to (4, 0).

73. See Section 12.3.
For a given height, angular acceptance decreases as velocity increases due to the decreasing curvature of the path the ball follows. For a given velocity, the angular acceptance increases as height increases due to the increase in angle to the back of the service box.

Chapter 13

Section 13.1

5. $f(x, y) = x + 2y^2, 0 \le x \le 2, -1 \le y \le 1, n = 4$

The centers of the four squares are $\left(\frac{1}{2}, -\frac{1}{2}\right), \left(\frac{1}{2}, \frac{1}{2}\right), \left(\frac{3}{2}, -\frac{1}{2}\right)$ and $\left(\frac{3}{2}, \frac{1}{2}\right)$.

Since the four squares are the same size, $\Delta A_i = 1$, for each i.

$$V \approx \sum_{i=1}^{4} f(u_i, v_i) \Delta A_i$$

$$= f\left(\frac{1}{2}, -\frac{1}{2}\right)(1) + f\left(\frac{1}{2}, \frac{1}{2}\right)(1) + f\left(\frac{3}{2}, -\frac{1}{2}\right)(1)$$

$$+ f\left(\frac{3}{2}, \frac{1}{2}\right)(1)$$

$$= \left[\frac{1}{2} + 2\left(-\frac{1}{2}\right)^2\right] + \left[\frac{1}{2} + 2\left(\frac{1}{2}\right)^2\right] + \left[\frac{3}{2} + 2\left(-\frac{1}{2}\right)^2\right] + \left[\frac{3}{2} + 2\left(\frac{1}{2}\right)^2\right]$$

$$= 1 + 1 + 2 + 2$$

$$= 6$$

7. $f(x, y) = x + 2y^2, 0 \le x \le 2, -1 \le y \le 1, n = 16$

The centers of the sixteen squares are

$$\left(\frac{1}{4}, -\frac{3}{4}\right), \left(\frac{1}{4}, -\frac{1}{4}\right), \left(\frac{1}{4}, \frac{1}{4}\right), \left(\frac{1}{4}, \frac{3}{4}\right), \left(\frac{3}{4}, -\frac{3}{4}\right),$$

$$\left(\frac{3}{4}, -\frac{1}{4}\right), \left(\frac{3}{4}, \frac{1}{4}\right), \left(\frac{3}{4}, \frac{3}{4}\right), \left(\frac{5}{4}, -\frac{3}{4}\right), \left(\frac{5}{4}, -\frac{1}{4}\right),$$

$$\left(\frac{5}{4}, \frac{1}{4}\right), \left(\frac{5}{4}, \frac{3}{4}\right), \left(\frac{7}{4}, -\frac{3}{4}\right), \left(\frac{7}{4}, -\frac{1}{4}\right), \left(\frac{7}{4}, \frac{1}{4}\right)$$

and $\left(\frac{7}{4}, \frac{3}{4}\right)$.

Since the sixteen squares are the same,

$\Delta A_i = \frac{1}{4}$, for each i.

$$V \approx \sum_{i=1}^{16} f(u_i, v_i) \Delta A_i$$

$$= f\left(\frac{1}{4}, -\frac{3}{4}\right)\left(\frac{1}{4}\right) + f\left(\frac{1}{4}, -\frac{1}{4}\right)\left(\frac{1}{4}\right) + \ldots + f\left(\frac{7}{4}, \frac{3}{4}\right)\left(\frac{1}{4}\right)$$

$$= \frac{1}{4}\left(\frac{11}{8} + \frac{3}{8} + \frac{3}{8} + \frac{11}{8} + \frac{15}{8} + \frac{7}{8} + \frac{7}{8} + \frac{15}{8} + \frac{19}{8} + \frac{11}{8} + \frac{11}{8} + \frac{19}{8} + \frac{23}{8} + \frac{15}{8} + \frac{15}{8} + \frac{23}{8}\right)$$

$$= \frac{13}{2}$$

9. $f(x, y) = 3x - y, -1 \le x \le 1, 0 \le y \le 4, n = 4$

The upper right corners of the four rectangles are $(0, 2)$, $(0, 4)$, $(1, 2)$, and $(1, 4)$.

Since the rectangles are the same, $\Delta A_i = 2$, for each i.

$$V \approx \sum_{i=1}^{4} f(u_i, v_i) \Delta A_t$$
$$= f(0,2)(2) + f(0,4)(2) + f(1,2)(2) + f(1,4)(2)$$
$$= 2(-2 - 4 + 1 - 1)$$
$$= -12$$

11. $f(x, y) = 3x - y, 0 \le x \le 4, 0 \le y \le 2$

The centers of the areas are $\left(\frac{1}{2}, \frac{1}{2}\right), \left(\frac{1}{2}, \frac{3}{2}\right), \left(\frac{5}{2}, \frac{1}{2}\right)$ and $\left(\frac{5}{2}, \frac{3}{2}\right)$.

The areas are $A_1 = A_2 = 1, A_3 = A_4 = 3$.

$$V \approx \sum_{i=1}^{4} f(u_i, v_i) \Delta A_i$$
$$= f\left(\frac{1}{2}, \frac{1}{2}\right)(1) + f\left(\frac{1}{2}, \frac{3}{2}\right)(1) + f\left(\frac{5}{2}, \frac{1}{2}\right)(3) + f\left(\frac{5}{2}, \frac{3}{2}\right)(3)$$
$$= 1 \cdot 1 + 0 \cdot 1 + 7 \cdot 3 + 6 \cdot 3$$
$$= 40$$

13.
$$\iint_R (x^2 - 2y)\,dA = \int_0^2 \int_{-1}^1 (x^2 - 2y)\,dy\,dx$$
$$= \int_0^2 [x^2 y - y^2]_{y=-1}^{y=1}\,dx$$
$$= \int_0^2 2x^2\,dx = \left[\frac{2}{3}x^3\right]_0^2$$
$$= \frac{16}{3}$$

15.
$$\iint_R 4xe^{2y}\,dA = \int_2^4 \int_0^1 4xe^{2y}\,dy\,dx$$
$$= \int_2^4 [2xe^{2y}]_{y=0}^{y=1}\,dx$$
$$= \int_2^4 2x(e^2 - 1)\,dx = (e^2 - 1)[x^2]_2^4$$
$$= 12(e^2 - 1)$$

17.
$$\iint_R (1 - ye^{xy})\,dA = \int_0^3 \int_0^2 (1 - ye^{xy})\,dx\,dy$$
$$= \int_0^3 [x - e^{xy}]_{x=0}^{x=2}\,dy$$
$$= \int_0^3 (3 - e^{2y})\,dy$$
$$= \left[3y - \frac{1}{2}e^{2y}\right]_0^3$$
$$= \frac{19}{2} - \frac{1}{2}e^6$$

19.

21.

23.
$$\int_0^1 \int_0^{2x} (x + 2y)\,dy\,dx = \int_0^1 [xy + y^2]_{y=0}^{y=2x}\,dx$$
$$= \int_0^1 6x^2\,dx = 2[x^3]_0^1$$
$$= 2$$

25. $\int_0^2 \int_0^{4y} (x+2y)dxdy = \int_0^2 \left[\frac{1}{2}x^2 + 2xy\right]_{x=0}^{y=4y} dy$

$= \int_0^2 16y^2 dy$

$= \left[\frac{16}{3}y^3\right]_0^2 = \frac{128}{3}$

27. $\int_0^1 \int_0^{2y} (4x\sqrt{y} + y)dxdy = \int_0^1 [2x^2\sqrt{y} + xy]_{x=0}^{x=2y} dy$

$= \int_0^1 (8y^{5/2} + 2y^2)dy$

$= \left[\frac{16}{7}y^{7/2} + \frac{2}{3}y^3\right]_0^1$

$= \frac{62}{21}$

29. $\int_0^2 \int_0^{2y} e^{y^2} dxdy = \int_0^2 \left[xe^{y^2}\right]_{x=0}^{x=2y} dy$

$= \int_0^2 2ye^{y^2} dy$

$= \left[e^{y^2}\right]_0^2 = e^4 - 1$

31.

$\int_1^4 \int_0^{1/x} \cos xy\,dydx = \int_1^4 \left[\frac{1}{x}\sin xy\right]_{y=0}^{y=1/x} dx$

$= \int_1^4 \frac{1}{x}\sin 1\,dx$

$= \sin 1 \left[\ln|x|\right]_1^4$

$= \sin 1 (\ln 4)$

$= 2\ln 2 (\sin 1)$

33. $\int_0^1 \int_0^{2x} x^2 dydx = \int_0^1 \left[x^2 y\right]_{y=0}^{y=2x} dx$

$= \int_0^1 2x^3 dx = \left[\frac{1}{2}x^4\right]_0^1$

$= \frac{1}{2}$

$\int_0^1 \int_0^{2y} x^2 dxdy = \int_0^1 \left[\frac{1}{3}x^3\right]_{x=0}^{x=2y} dy$

$= \int_0^1 \frac{8}{3}y^3 dy$

$= \frac{2}{3}[y^4]_0^1 = \frac{2}{3}$

Therefore, $\int_0^1 \int_0^{2x} x^2 dydx \neq \int_0^1 \int_0^{2y} x^2 dxdy$.

35. $\int_0^3 \int_1^4 (x^2 + y^2)dydx = \int_0^3 \left[x^2 y + \frac{1}{3}y^3\right]_{y=1}^{y=4} dx$

$= \int_0^3 (3x^2 + 21)dx$

$= [x^3 + 21x]_0^3$

$= 90$

37. On the *xy*-plane, the region *R* lies between the parabola $y = x^2$ and the line $y = 1$. Thus,

$-1 \leq x \leq 1, x^2 \leq y \leq 1.$

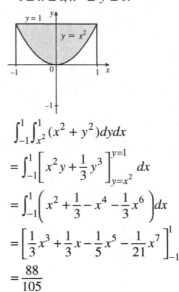

$\int_{-1}^1 \int_{x^2}^1 (x^2 + y^2)dydx$

$= \int_{-1}^1 \left[x^2 y + \frac{1}{3}y^3\right]_{y=x^2}^{y=1} dx$

$= \int_{-1}^1 \left(x^2 + \frac{1}{3} - x^4 - \frac{1}{3}x^6\right)dx$

$= \left[\frac{1}{3}x^3 + \frac{1}{3}x - \frac{1}{5}x^5 - \frac{1}{21}x^7\right]_{-1}^1$

$= \frac{88}{105}$

39. On the *xy*-plane, the region *R* lies between the parabola $x = 4 - y^2$ and the *y*-axis. Thus,

$0 \le x \le 4 - y^2, -2 \le y \le 2.$

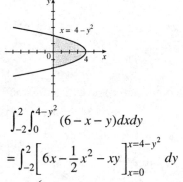

$$\int_{-2}^{2} \int_{0}^{4-y^2} (6 - x - y)\,dx\,dy$$

$$= \int_{-2}^{2} \left[6x - \frac{1}{2}x^2 - xy \right]_{x=0}^{x=4-y^2} dy$$

$$= \int_{-2}^{2} \left(-\frac{1}{2}y^4 + y^3 - 2y^2 - 4y + 16 \right) dy$$

$$= \left[-\frac{1}{10}x^5 + \frac{1}{4}y^4 - \frac{2}{3}y^3 - 2y^2 + 16y \right]_{-2}^{2}$$

$$= \frac{704}{15}$$

41. On the *xy*-plane, the region *R* lies between the *x*-axis and the line $y = x$. Thus, $0 \le x \le 2$, $0 \le y \le x$.

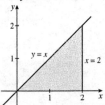

$$\int_{0}^{2} \int_{0}^{x} y^2\,dy\,dx = \int_{0}^{2} \left[\frac{1}{3}y^3 \right]_{y=0}^{y=2} dx = \int_{0}^{2} \frac{1}{3}x^3\,dx$$

$$= \left[\frac{1}{12}x^4 \right]_{0}^{2} = \frac{4}{3}$$

43.

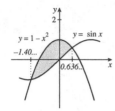

From the graph, y ranges from $\sin x$ to $1 - x^2$. For the outer limits of integration, we solve the equation $\sin x = 1 - x^2$ and obtain -1.40962 and 0.63673.

$$\iint\limits_R (2x - y)\,dA = \int_{-1.40962}^{0.63673} \int_{\sin x}^{1-x^2} (2x - y)\,dy\,dx$$

$$= \int_{-1.40962}^{0.63673} \left[2xy - \frac{1}{2}y^2 \right]_{y=\sin x}^{y=1-x^2} dx$$

$$= \int_{-1.40962}^{0.63673} \left(-\frac{1}{2}x^4 - 2x^3 + x^2 + 2x - \frac{1}{2} + \frac{1}{2}\sin^2 x - 2x\sin x \right) dx$$

$$= \left[-\frac{1}{10}x^5 - \frac{1}{2}x^4 + \frac{1}{3}x^3 + x^2 - \frac{1}{4}x + 2x\cos x - 2\sin x - \frac{1}{4}\sin x \cos x \right]_{-1.40962}^{0.63673}$$

$$\approx -1.5945$$

45.

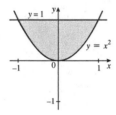

From the graph, y ranges from x^2 to 1 and $-1 \le x \le 1$. Using numerical procedures, we have

$$\iint\limits_R e^{x^2}\,dA = \int_{-1}^{1}\int_{x^2}^{1} e^{x^2}\,dy\,dx$$

$$= \int_{-1}^{1} \left[ye^{x^2} \right]_{y=x^2}^{y=1} dx$$

$$= \int_{-1}^{1} (1 - x^2)e^{x^2}\,dx$$

$$\approx 1.6697$$

47.

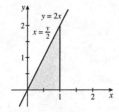

From the graph, $0 \le y \le 2x, 0 \le x \le 1$. Changing the order of the variables, we have

$0 \le y \le 2, \dfrac{y}{2} \le x \le 1.$

$\int_0^2 \int_{y/2}^1 f(x, y)dxdy$

49.

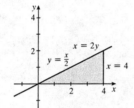

From the graph, $2y \le x \le 4, 0 \le y \le 2$. Changing the order of the variables, we have

$0 \le x \le 4,\ 0 \le y \le \dfrac{x}{2}$

$\int_0^4 \int_0^{x/2} f(x, y)dydx$

51.

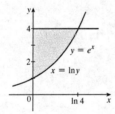

From the graph, $e^x \le y \le 4, 0 \le x \le \ln 4.$ Changing the order of the variables, we have
$0 \le x \le \ln y,\ 1 \le y \le 4.$

$\int_1^4 \int_0^{\ln y} f(x, y)dxdy$

53. $\displaystyle\int_0^2 \int_x^2 2e^{y^2} dydx = \int_0^2 \int_0^y 2e^{y^2} dxdy$

$$= \int_0^2 \left[2xe^{y^2} \right]_{x=0}^{x=y} dy$$

$$= \int_0^2 2ye^{y^2} dy$$

$$= \left[e^{y^2} \right]_0^2$$

$$= e^4 - 1$$

55. $\displaystyle\int_0^1\int_y^1 3xe^{x^3}\,dx\,dy = \int_0^1\int_0^x 3xe^{x^3}\,dy\,dx$

$$= \int_0^1\left[3xe^{x^3}y\right]_{y=0}^{y=x}dx$$

$$= \int_0^1 3x^2e^{x^3}\,dx$$

$$= \left[e^{x^3}\right]_0^1 = e-1$$

57. Answers will vary. The first integral cannot be evaluated.

59.

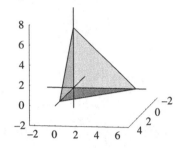

61.

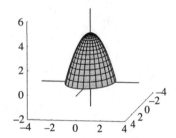

63. Answers will vary. In general, the integrals are integrals over different regions in the *xy*-plane.

Section 13.2

5. Limits of integration: $x^2 = 8 - x^2$

$$2x^2 = 8$$
$$x^2 = 4$$
$$x = \pm 2$$

$$A = \int_{-2}^{2} \int_{x^2}^{8-x^2} dy\,dx$$

$$= \int_{-2}^{2} [y]_{y=x^2}^{y=8-x^2} dx$$

$$= \int_{-2}^{2} (8 - 2x^2)\,dx$$

$$= \left[8x - \frac{2}{3}x^3 \right]_{-2}^{2} = \frac{64}{3}$$

7. Limits of integration: $\dfrac{y}{2} = 3 - y$

$$y = 2$$

$$A = \int_{0}^{2} \int_{y/2}^{3-y} dx\,dy = \int_{0}^{2} [x]_{x=y/2}^{x=3-y}$$

$$= \int_{0}^{2} \left(3 - \frac{3}{2}y \right) dy$$

$$= \left[3y - \frac{3}{4}y^2 \right]_{0}^{2} = 3$$

9. $A = \int_{0}^{1} \int_{x^2}^{\sqrt{x}} dy\,dx = \int_{0}^{1} [y]_{y=x^2}^{y=\sqrt{x}} dx$

$$= \int_{0}^{1} (\sqrt{x} - x^2)\,dx$$

$$= \left[\frac{2}{3}x^{3/2} - \frac{1}{3}x^3 \right]_{0}^{1} = \frac{1}{3}$$

11. Limits of integration will be the trace of

$$z = 6 - 2x - 3y, \text{ or } y = 2 - \frac{2}{3}x.$$

$$V = \int_{0}^{3} \int_{0}^{2-(2/3)x} (6 - 2x - 3y)\,dy\,dx$$

$$= \int_{0}^{3} \left[6y - 2xy - \frac{3}{2}y^2 \right]_{0}^{2-(2/3)x} dx$$

$$= \int_{0}^{3} \left(\frac{2}{3}x^2 - 4x + 6 \right) dx$$

$$= \left[\frac{2}{9}x^3 - 2x^2 + 6x \right]_{0}^{3} = 6$$

13. $V = \int_{-1}^{1} \int_{-1}^{1} (4 - x^2 - y^2)\,dx\,dy$

$$= \int_{-1}^{1} \left[4x - \frac{x^3}{3} - xy^2 \right]_{-1}^{1} dy$$

$$= \int_{-1}^{1} \left(\frac{22}{3} - 2y^2 \right) dy$$

$$= \left[\frac{22}{3}y - \frac{2}{3}y^3 \right]_{-1}^{1}$$

$$= \frac{40}{3}$$

15. $V = \int_{0}^{1} \int_{1}^{2} (1 - y)\,dx\,dy$

$$= \int_{0}^{1} [x - xy]_{1}^{2}\,dy$$

$$= \int_{0}^{1} (1 - y)\,dy$$

$$= \left[y - \frac{y^2}{2} \right]_{0}^{1}$$

$$= \frac{1}{2}$$

17. $V = \int_{0}^{1} \int_{0}^{1-x} (1 - y^2)\,dy\,dx$

$$= \int_{0}^{1} \left[y - \frac{1}{3}y^3 \right]_{y=0}^{y=1-x} dx$$

$$= \int_{0}^{1} \left(\frac{1}{3}x^3 - x^2 + \frac{2}{3} \right) dx$$

$$= \left[\frac{1}{12}x^4 - \frac{1}{3}x^3 + \frac{2}{3}x \right]_{0}^{1} = \frac{5}{12}$$

19. $V = \int_{-2}^{2} \int_{x^2}^{4} (x^2 + y^2)\,dy\,dx$

$$= \int_{-2}^{2} \left[x^2 y + \frac{1}{3}y^3 \right]_{y=x^2}^{y=4} dx$$

$$= \int_{-2}^{2} \left[-\frac{1}{3}x^6 - x^4 + 4x^2 + \frac{64}{3} \right] dx$$

$$= \left[-\frac{1}{21}x^7 - \frac{1}{5}x^5 + \frac{4}{3}x^3 + \frac{64}{3}x \right]_{-2}^{2}$$

$$= \frac{8576}{105}$$

21. Limits of integration:
$$y^2 - 2 = y$$
$$y^2 - y - 2 = 0$$
$$(y-2)(y+1) = 0$$
$$y = 2, -1$$

$$V = \int_{-1}^{2}\int_{y^2-2}^{y}(x+2)dxdy$$

$$= \int_{-1}^{2}\left[\frac{1}{2}x^2 + 2x\right]_{x=y^2-2}^{x=y} dy$$

$$= \int_{-1}^{2}\left(-\frac{1}{2}y^4 + \frac{1}{2}y^2 + 2y + 2\right)dy$$

$$= \left[-\frac{1}{10}y^5 + \frac{1}{6}y^3 + y^2 + 2y\right]_{-1}^{2}$$

$$= \frac{36}{5}$$

23. $V = \int_{0}^{2}\int_{0}^{4-x^2}\sqrt{x^2+y^2}\,dydx \approx 10.2753$

25. Limits of integration: $x + 2y = 4$
$$y = 2 - \frac{1}{2}x$$

$$V = \int_{0}^{4}\int_{0}^{2-x/2}e^{xy}dydx \approx 9.0032$$

27. $m = \int_{0}^{1}\int_{x^3}^{x^2}4dydx = \int_{0}^{1}[4y]_{y=x^3}^{y=x^2}dx$

$$= \int_{0}^{1}(4x^2 - 4x^3)dx$$

$$= \left[\frac{4}{3}x^3 - x^4\right]_{0}^{1} = \frac{1}{3}$$

$$M_y = \int_{0}^{1}\int_{x^3}^{x^2}4xdydx = \int_{0}^{1}[4xy]_{y=x^3}^{y=x^2}dx$$

$$= \int_{0}^{1}(4x^3 - 4x^4)dx$$

$$= \left[x^4 - \frac{4}{5}x^5\right]_{0}^{1} = \frac{1}{5}$$

$$\bar{x} = \frac{M_y}{m} = \frac{1/5}{1/3} = \frac{3}{5}$$

$$M_x = \int_{0}^{1}\int_{x^3}^{x^2}4ydydx = \int_{0}^{1}[2y^2]_{y=x^3}^{y=x^2}dx$$

$$= \int_{0}^{1}(2x^4 - 2x^6)dx$$

$$= \left[\frac{2}{5}x^5 - \frac{2}{7}x^7\right]_{0}^{1} = \frac{4}{35}$$

$$\bar{y} = \frac{M_x}{m} = \frac{4/35}{1/3} = \frac{12}{35}$$

29. $m = \int_{-1}^{1}\int_{y^2}^{1}(y^2 + x + 1)dxdy$

$$= \int_{-1}^{1}\left[xy^2 + \frac{1}{2}x^2 + x\right]_{x=y^2}^{x=1}dy$$

$$= \int_{-1}^{1}\left(-\frac{3}{2}y^4 + \frac{3}{2}\right)dy$$

$$= \left[-\frac{3}{10}y^5 + \frac{3}{2}y\right]_{-1}^{1} = \frac{12}{5}$$

$$M_y = \int_{-1}^{1}\int_{y^2}^{1}x(y^2 + x + 1)dxdy$$

$$= \int_{-1}^{1}\left[\frac{1}{2}x^2y^2 + \frac{1}{3}x^3 + \frac{1}{2}x^2\right]_{x=y^2}^{x=1}dy$$

$$= \int_{-1}^{1}\left(-\frac{5}{6}y^6 - \frac{1}{2}y^4 + \frac{1}{2}y^2 + \frac{5}{6}\right)dy$$

$$= \left[-\frac{5}{42}y^7 - \frac{1}{10}y^5 + \frac{1}{6}y^3 + \frac{5}{6}y\right]_{-1}^{1}$$

$$= \frac{164}{105}$$

$$\bar{x} = \frac{M_y}{m} = \frac{164/105}{12/5} = \frac{41}{63}$$

$$M_x = \int_{-1}^{1}\int_{y^2}^{1}y(y^2 + x + 1)dxdy$$

$$= \int_{-1}^{1}\left[xy^3 + \frac{1}{2}x^2y + xy\right]_{x=y^2}^{1}dy$$

$$= \int_{-1}^{1}\left(-\frac{3}{2}y^5 + \frac{3}{2}y\right)dy$$

$$= \left[-\frac{1}{4}y^6 + \frac{3}{4}y^2\right]_{-1}^{1} = 0$$

$$\bar{y} = \frac{M_x}{m} = 0$$

31. $m = \int_0^2 \int_{x^2}^4 x\,dy\,dx = \int_0^2 [xy]_{y=x^2}^{y=4}\,dx$

$\qquad = \int_0^2 (4x - x^3)\,dx$

$\qquad = \left[2x^2 - \frac{1}{4}x^4\right]_0^2 = 4$

$M_y = \int_0^2 \int_{x^2}^4 x \cdot x\,dy\,dx = \int_0^2 [x^2 y]_{y=x^2}^{y=4}\,dx$

$\qquad = \int_0^2 (4x^2 - x^4)\,dx$

$\qquad = \left[\frac{4}{3}x^3 - \frac{1}{5}x^5\right]_0^2 = \frac{64}{15}$

$\bar{x} = \frac{M_y}{m} = \frac{64/15}{4} = \frac{16}{15}$

$M_x = \int_0^2 \int_{x^2}^4 y \cdot x\,dy\,dx = \int_0^2 \left[\frac{1}{2}xy^2\right]_{y=x^2}^{y=4}\,dx$

$\qquad = \int_0^2 \left(8x - \frac{1}{2}x^5\right)dx$

$\qquad = \left[4x^2 - \frac{1}{12}x^6\right]_0^2 = \frac{32}{3}$

$\bar{y} = \frac{M_x}{m} = \frac{32/3}{4} = \frac{8}{3}$

33. In exercise 26, $\rho(x, y)$ is not symmetric about the x-axis.

35. If $\rho(-x, y) = \rho(x, y)$ then the center of mass is located on the y-axis.

37. $\int_1^2 \int_{x-1}^1 15000e^{-x^2-y^2}\,dy\,dx \approx 971.848$

39. $\int_0^{10} \int_0^4 20e^{-t/6}\,dt\,dx \approx 583.90$

41. mass $= \int_{-2}^2 \int_{x^2}^4 1\,dy\,dx \approx 10.667$

$I_y = \int_{-2}^2 \int_{x^2}^4 x^2\,dy\,dx \approx 8.533$

$I_x = \int_{-2}^2 \int_{x^2}^4 y^2\,dy\,dx \approx 73.143$

43. For the skater with extended arms, we add together the moments contributed by the central rectangle and the two rectangles representing the arms:

$Ie_y = \int_{-1}^1 \int_0^8 1 \cdot x^2\,dy\,dx + 2\int_1^3 \int_0^1 1 \cdot x^2\,dy\,dx$

$\qquad = \frac{68}{3}$

For the skater with arms raised we have a single rectangle with density 2:

$Ir_y = \int_{-1/2}^{1/2} \int_0^{10} 2 \cdot x^2\,dy\,dx = \frac{5}{3}$

Therefore the ratio of spin rates is

$\frac{68/3}{5/3} = 13.6$; the skater with raised arms spins 13.6 times faster.

45. Let s indicate the smaller ellipse and b the bigger one.

$Is_y = \int_{-4}^4 \int_{-(\sqrt{16-x^2})/2}^{(\sqrt{16-x^2})/2} 1 \cdot x^2\,dy\,dx \approx 100.53$

$Ib_y = \int_{-4}^4 \int_{-(\sqrt{36-x^2})/2}^{(\sqrt{36-x^2})/2} 1 \cdot x^2\,dy\,dx \approx 508.94$

47. $a = \int_{-2}^2 \int_{x^2}^4 dy\,dx = \int_{-2}^2 [y]_{y=x^2}^{y=4}\,dx$

$\qquad = \int_{-2}^2 (4 - x^2)\,dx$

$\qquad = \left[4x - \frac{1}{3}x^3\right]_{-2}^2 = \frac{32}{3}$

$\iint_R f(x, y)\,dA = \int_{-2}^2 \int_{x^2}^4 y\,dy\,dx$

$\qquad = \int_{-2}^2 \left[\frac{1}{2}y^2\right]_{y=x^2}^{y=4}\,dx$

$\qquad = \int_{-2}^2 \left(8 - \frac{1}{2}x^4\right)dx$

$\qquad = \left[8x - \frac{1}{10}x^5\right]_{-2}^2 = \frac{128}{5}$

average value $= \frac{1}{a}\iint_R f(x, y)\,dA = \frac{1}{32/3} \cdot \frac{128}{5}$

$\qquad = \frac{12}{5}$

49. They are the same.

51. Limits of integration: $x^2 - 4 = 3x$
$$x^2 - 3x - 4 = 0$$
$$(x-4)(x+1) = 0$$
$$x = 4, -1$$

$$a = \int_{-1}^{4}\int_{x^2-4}^{3x} dy\,dx = \int_{-1}^{4}[y]_{y=x^2-4}^{y=3x}\,dx$$
$$= \int_{-1}^{4}(3x - x^2 + 4)dx$$
$$= \left[-\frac{1}{3}x^3 + \frac{3}{2}x^2 + 4x\right]_{-1}^{4} = \frac{125}{6}$$

$$\iint_R f(x,y)dA = \int_{-1}^{4}\int_{x^2-4}^{3x}\sqrt{x^2+y^2}\,dy\,dx \approx 78.994$$

average value $= \dfrac{1}{a}\iint_R f(x,y)dA$

$$= \frac{1}{125/6}\cdot 78.994$$
$$\approx 3.792$$

53. Limits of integration: $x^2 = 8 - x^2$
$$2x^2 = 8$$
$$x = \pm 2$$

$$a = \int_{-2}^{2}\int_{x^2}^{8-x^2} dy\,dx = \int_{-2}^{2}[y]_{y=x^2}^{y=8-x^2}\,dx$$
$$= \int_{-2}^{2}(8 - 2x^2)dx$$
$$= \left[8x - \frac{2}{3}x^3\right]_{-2}^{2} = \frac{64}{3}$$

$$\iint_R f(x,y)dA = \int_{-2}^{2}\int_{x^2}^{8-x^2}[50+\cos(2x+y)]dy\,dx$$
$$\approx 1069.084$$

average temperature $= \dfrac{1}{a}\iint_R f(x,y)dA$

$$\approx \frac{1}{64/3}\cdot 1069.084$$
$$\approx 50.113$$

55. a. $\iint_R f(x,y)dA$ gives the total rainfall in the region R.

b. $\dfrac{\iint_R f(x,y)dA}{\iint_R 1\,dA}$ gives the average rainfall per unit area in the region R.

57.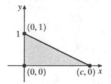

a. Without loss of generality, assume for simplicity that $c > 0$ as illustrated. The equation of the line containing $(0, 1)$ and $(c, 0)$ is $y = -\frac{1}{c}x + 1$.

$$m = \int_0^c\int_0^{-x/c+1} dy\,dx = \frac{1}{2}c$$
$$M_x = \int_0^c\int_0^{-x/c+1} y\,dy\,dx$$
$$= \int_0^c\left[\frac{1}{2}y^2\right]_{y=0}^{y=-x/c+1} dx$$
$$= \int_0^c\left(\frac{1}{2c^2}x^2 - \frac{1}{c}x + \frac{1}{2}\right)dx$$
$$= \left[\frac{1}{6c^2}x^3 - \frac{1}{2c}x^2 + \frac{1}{2}x\right]_0^c = \frac{1}{6}c$$

$$\overline{y} = \frac{M_x}{m} = \frac{c/6}{c/2} = \frac{1}{3}$$

Therefore $\overline{y}$ is independent of the constant c.

Section 13.3

5. $\int_0^{2\pi} \int_0^{3+2\sin\theta} r\,dr\,d\theta$

$= \int_0^{2\pi} \left[\frac{1}{2} r^2 \right]_{r=0}^{r=3+2\sin\theta} d\theta$

$= \frac{1}{2} \int_0^{2\pi} (9 + 12\sin\theta + 4\sin^2\theta)\,d\theta$

$= \frac{1}{2} \left[9\theta - 12\cos\theta + 4 \cdot \frac{1}{2}\left(\theta - \frac{1}{2}\sin 2\theta\right) \right]_0^{2\pi}$

$= 11\pi$

7. $\int_0^{\pi} \int_0^{2\sin\theta} r\,dr\,d\theta = \int_0^{\pi} \left[\frac{1}{2} r^2 \right]_{r=0}^{r=2\sin\theta} d\theta$

$= \frac{1}{2} \int_0^{\pi} 4\sin^2\theta\,d\theta$

$= \frac{1}{2} \cdot 4 \cdot \frac{1}{2}\left[\theta - \frac{1}{2}\sin 2\theta \right]_0^{\pi}$

$= \pi$

9. $\int_0^{\pi/3} \int_0^{\sin 3\theta} r\,dr\,d\theta = \int_0^{\pi/3} \left[\frac{1}{2} r^2 \right]_{r=0}^{r=\sin 3\theta} d\theta$

$= \frac{1}{2} \int_0^{\pi/3} \sin^2 3\theta\,d\theta$

$= \frac{1}{2} \cdot \frac{1}{3} \cdot \frac{1}{2} \left[3\theta - \frac{1}{2}\sin 6\theta \right]_0^{\pi/3}$

$= \frac{\pi}{12}$

11. Limits of integration: $2\sin 3\theta = 1$

$\sin 3\theta = \frac{1}{2}$

$3\theta = \frac{\pi}{6}, \frac{5\pi}{6}$

$\theta = \frac{\pi}{18}, \frac{5\pi}{18}$

$\int_{\pi/18}^{5\pi/18} \int_1^{2\sin 3\theta} r\,dr\,d\theta$

$= \int_{\pi/18}^{5\pi/18} \left[\frac{1}{2} r^2 \right]_{r=1}^{r=2\sin 3\theta} d\theta$

$= \frac{1}{2} \int_{\pi/18}^{5\pi/18} (4\sin^2 3\theta - 1)\,d\theta$

$= \frac{1}{2} \left[4 \cdot \frac{1}{3} \cdot \frac{1}{2}\left(3\theta - \frac{1}{2}\sin 6\theta\right) - \theta \right]_{\pi/18}^{5\pi/18}$

$= \frac{\sqrt{3}}{6} + \frac{\pi}{9}$

13. $\int_0^{2\pi} \int_0^3 r \cdot r\,dr\,d\theta = \int_0^{2\pi} \left[\frac{1}{3} r^3 \right]_{r=0}^{r=3} d\theta$

$= \int_0^{2\pi} 9\,d\theta$

$= [9\theta]_0^{2\pi} = 18\pi$

15. $\int_0^{2\pi} \int_0^2 e^{-r^2} \cdot r\,dr\,d\theta = \int_0^{2\pi} \left[-\frac{1}{2} e^{-r^2} \right]_{r=0}^{r=1} d\theta$

$= -\frac{1}{2} \int_0^{2\pi} (e^{-4} - 1)\,d\theta$

$= \pi - \pi e^{-4}$

17. $\int_0^{2\pi} \int_0^{2-\cos\theta} r\sin\theta \cdot r\,dr\,d\theta$

$= \int_0^{2\pi} \left[\frac{1}{3} r^3 \sin\theta \right]_{r=0}^{r=2-\cos\theta} d\theta$

$= \int_0^{2\pi} \left[\frac{1}{3}(2 - \cos\theta)^3 \sin\theta \right] d\theta$

$= \left[\frac{1}{3} \cdot \frac{1}{4}(2 - \cos\theta)^4 \right]_0^{2\pi}$

$= 0$

19. $\int_0^{2\pi} \int_0^3 r^2 \cdot r\,dr\,d\theta = \int_0^{2\pi} \left[\frac{1}{4} r^4 \right]_{r=0}^{r=3} d\theta$

$= \int_0^{2\pi} \frac{81}{4}\,d\theta$

$= \frac{81}{4} \cdot 2\pi = \frac{81\pi}{2}$

21. $\int_0^2 \int_0^x (x^2 + y^2)\,dy\,dx = \int_0^2 \left[x^2 y + \frac{1}{3} y^3 \right]_{y=0}^{y=x} dx$

$= \int_0^2 \frac{4}{3} x^3\,dx$

$= \left[\frac{1}{3} x^4 \right]_0^2 = \frac{16}{3}$

or

$\int_0^{\pi/4} \int_0^{2\sec\theta} r^2 \cdot r\,dr\,d\theta = \frac{16}{3}$

23. $V = \int_0^{2\pi} \int_0^3 r^2 \cdot r\,dr\,d\theta = \int_0^{2\pi} \left[\frac{1}{4} r^4 \right]_{r=0}^{r=3} d\theta$

$= \int_0^{2\pi} \frac{81}{4}\,d\theta$

$= \left[\frac{81}{4}\theta \right]_0^{2\pi} = \frac{81\pi}{2}$

415

25. $V = \int_0^{2\pi}\int_0^2 r \cdot r\,dr\,d\theta = \int_0^{2\pi}\left[\dfrac{1}{3}r^3\right]_{r=0}^{r=2}d\theta$

$\qquad = \int_0^{2\pi}\dfrac{8}{3}d\theta$

$\qquad = \left[\dfrac{8}{3}\theta\right]_0^{2\pi} = \dfrac{16\pi}{3}$

27. $V = \int_0^{2\pi}\int_0^{1/2}\sqrt{1-r^2}\cdot r\,dr\,d\theta$

$\qquad = \int_0^{2\pi}\left[-\dfrac{1}{3}(1-r^2)^{3/2}\right]_{r=0}^{r=1/2}d\theta$

$\qquad = \int_0^{2\pi}\left(-\dfrac{\sqrt{3}}{8}+\dfrac{1}{3}\right)d\theta$

$\qquad = \dfrac{8-3\sqrt{3}}{12}\pi$

29. $V = \int_0^6\int_0^{6-x}(6-x-y)\,dy\,dx$

$\qquad = \int_0^6\left[6y-xy-\dfrac{1}{2}y^2\right]_{y=0}^{y=6-x}dx$

$\qquad = \int_0^6\left(\dfrac{1}{2}x^2-6x+18\right)dx$

$\qquad = \left[\dfrac{1}{6}x^3-3x^2+18x\right]_0^6 = 36$

31. $V = \int_0^{\pi/4}\int_0^2(4-r^2)\cdot r\,dr\,d\theta$

$\qquad = \int_0^{\pi/4}\left[-\dfrac{1}{4}(4-r^2)^2\right]_{r=0}^{r=2}d\theta$

$\qquad = \int_0^{\pi/4}4\,d\theta$

$\qquad = [4\theta]_0^{\pi/4} = \pi$

33. $\int_0^{2\pi}\int_0^2 r \cdot r\,dr\,d\theta = \int_0^{2\pi}\left[\dfrac{1}{3}r^3\right]_{r=0}^{r=2}d\theta$

$\qquad = \int_0^{2\pi}\dfrac{8}{3}d\theta$

$\qquad = \left[\dfrac{8}{3}\theta\right]_0^{2\pi} = \dfrac{16\pi}{3}$

35. $\int_{-\pi/2}^{\pi/2}\int_0^2 e^{-r^2}\cdot r\,dr\,d\theta$

$\qquad = \int_{-\pi/2}^{\pi/2}\left[-\dfrac{1}{2}e^{-r^2}\right]_{r=0}^{r=2}d\theta$

$\qquad = \int_{-\pi/2}^{\pi/2}\left(\dfrac{1}{2}-\dfrac{1}{2}e^{-4}\right)d\theta$

$\qquad = \left(\dfrac{1}{2}-\dfrac{1}{2}e^{-4}\right)[\theta]_{-\pi/2}^{\pi/2}$

$\qquad = (1-e^{-4})\dfrac{\pi}{2}$

37. $\int_{\pi/4}^{\pi/2}\int_0^{\sqrt{8}} r^3 \cdot r\,dr\,d\theta = \int_{\pi/4}^{\pi/2}\left[\dfrac{1}{5}r^5\right]_{r=0}^{r=\sqrt{8}}d\theta$

$\qquad = \int_{\pi/4}^{\pi/2}\dfrac{64\sqrt{8}}{5}d\theta$

$\qquad = \dfrac{64\sqrt{8}}{5}[\theta]_{\pi/4}^{\pi/2} = \dfrac{16\pi\sqrt{8}}{5}$

39. $\int_0^{2\pi}\int_0^{1/4}\dfrac{1}{\pi}e^{-r^2}\cdot r\,dr\,d\theta$

$\qquad = \int_0^{2\pi}\left[-\dfrac{1}{2\pi}e^{-r^2}\right]_{r=0}^{r=1/4}d\theta$

$\qquad = \int_0^{2\pi}\dfrac{1}{2\pi}(1-e^{-1/16})d\theta$

$\qquad = \dfrac{1}{2\pi}(1-e^{-1/16})[\theta]_0^{2\pi}$

$\qquad = 1-e^{-1/16}\approx .06$

41. $\int_{9\pi/20}^{11\pi/20}\int_{15/4}^4\dfrac{1}{\pi}e^{-r^2}\cdot r\,dr\,d\theta$

$\qquad = \int_{9\pi/20}^{11\pi/20}\left[-\dfrac{1}{2\pi}e^{-r^2}\right]_{r=15/4}^{r=4}d\theta$

$\qquad = \int_{9\pi/20}^{11\pi/20}\dfrac{1}{2\pi}(e^{-225/16}-e^{-16})d\theta$

$\qquad = \dfrac{1}{2\pi}(e^{-225/16}-e^{-16})[\theta]_{9\pi/20}^{11\pi/20}$

$\qquad = \dfrac{1}{20}(e^{-225/16}-e^{-16})\approx 3.3\times10^{-8}$

43. $A = \int_{9\pi/20}^{11\pi/20}\int_{15/4}^4 r\,dr\,d\theta$

$\qquad = \int_{9\pi/20}^{11\pi/20}\left[\dfrac{1}{2}r^2\right]_{r=15/4}^{r=4}d\theta$

$\qquad = \int_{9\pi/20}^{11\pi/20}\dfrac{32}{31}d\theta$

$\qquad = \dfrac{31}{32}[\theta]_{9\pi/20}^{11\pi/20}$

$\qquad = \dfrac{31\pi}{320}$

45. $m = \int_0^\pi \int_0^{2\sin\theta} \frac{1}{r} \cdot r \, dr \, d\theta$

$\quad = \int_0^\pi [r]_{r=0}^{r=2\sin\theta} \, d\theta$

$\quad = \int_0^\pi 2\sin\theta \, d\theta$

$\quad = 2[-\cos\theta]_0^\pi = 4$

$M_y = \int_0^\pi \int_0^{2\sin\theta} \frac{1}{r} r\cos\theta \cdot r \, dr \, d\theta$

$\quad = \int_0^\pi \left[\frac{1}{2}r^2\cos\theta\right]_{r=0}^{r=2\sin\theta} d\theta$

$\quad = \int_0^\pi 2\sin^2\theta\cos\theta \, d\theta$

$\quad = \left[\frac{2}{3}\sin^3\theta\right]_0^\pi = 0$

$\bar{x} = \frac{M_y}{m} = 0$

$M_x = \int_0^\pi \int_0^{2\sin\theta} \frac{1}{r} r\sin\theta \cdot r \, dr \, d\theta$

$\quad = \int_0^\pi \left[\frac{1}{2}r^2\sin\theta\right]_{r=0}^{r=2\sin\theta} d\theta$

$\quad = \int_0^\pi 2\sin^3\theta \, d\theta$

$\quad = 2\left[\frac{1}{3}\cos^3\theta - \cos\theta\right]_0^\pi = \frac{8}{3}$

$\bar{y} = \frac{M_x}{m} = \frac{8/3}{4} = \frac{2}{3}$

47. $V = \int_0^{2\pi} \int_0^1 20,000 e^{-r^2} r \, dr \, d\theta$

$\quad = 20000\pi(1 - 1/e) \approx 39,717$

49. $I_y = \int_0^{2\pi} \int_0^R (r\cos\theta)^2 r \, dr \, d\theta = \frac{\pi R^4}{4}$

Since the moment of inertia is proportional to the fourth power of the radius, doubling the radius increases the moment by a factor of 16.

51. Using the area under $z(r,\theta) = \sqrt{R^2 - r^2}$ and above $z = 0$, and then doubling, we find the volume of a sphere of radius R as

$2\int_0^{2\pi} \int_0^R \sqrt{R^2 - r^2}\, r \, dr \, d\theta = \frac{4}{3}\pi R^3.$

Section 13.4

3.

$$S = \int_0^4 \int_0^x \sqrt{(2x)^2 + (2)^2 + 1}\, dydx$$

$$= \int_0^4 \int_0^x \sqrt{4x^2 + 5}\, dydx$$

$$= \int_0^4 \left[y\sqrt{4x^2 + 5} \right]_{y=0}^{y=x} dx$$

$$= \int_0^4 x\sqrt{4x^2 + 5}\, dx$$

$$= \frac{1}{12}\left[(4x^2 + 5)^{3/2} \right]_0^4$$

$$= \frac{1}{12}\left(69^{3/2} - 5^{3/2} \right) \approx 46.831$$

5.

$$S = \int_{-2}^2 \int_{-\sqrt{4-x^2}}^{\sqrt{4-x^2}} \sqrt{(-2x)^2 + (-2y)^2 + 1}\, dydx$$

$$= \int_0^{2\pi} \int_0^2 \sqrt{4r^2 + 1} \cdot r\, drd\theta$$

$$= \int_0^{2\pi} \frac{1}{12}\left[(4r^2 + 1)^{3/2} \right]_{r=0}^{r=2} d\theta$$

$$= \int_0^{2\pi} \frac{1}{12}(17^{3/2} - 1)d\theta$$

$$= \frac{1}{12}(17^{3/2} - 1)[\theta]_0^{2\pi}$$

$$= \frac{\pi}{6}(17^{3/2} - 1) \approx 36.177$$

7. $$S = \int_{-2}^2 \int_{-\sqrt{4-x^2}}^{\sqrt{4-x^2}} \sqrt{\left(\frac{x}{\sqrt{x^2 + y^2}} \right)^2 + \left(\frac{y}{\sqrt{x^2 + y^2}} \right)^2 + 1}\, dydx$$

$$= \int_{-2}^2 \int_{-\sqrt{4-x^2}}^{\sqrt{4-x^2}} \sqrt{2}\, dydx$$

$$= \int_0^{2\pi} \int_0^2 \sqrt{2} \cdot r\, drd\theta$$

$$= \int_0^{2\pi} \sqrt{2}\, \frac{1}{2}[r^2]_{r=0}^{r=2} d\theta$$

$$= \int_0^{2\pi} 2\sqrt{2}\, d\theta = 2\sqrt{2}[\theta]_0^{2\pi} = 4\pi\sqrt{2}$$

9. $$S = \int_{-2}^2 \int_{-\sqrt{4-x^2}}^{\sqrt{4-x^2}} \sqrt{(2xe^{x^2+y^2})^2 + (2ye^{x^2+y^2})^2 + 1}\, dydx$$

$$= \int_0^{2\pi} \int_0^2 \sqrt{4r^2 e^{2r^2} + 1} \cdot r\, drd\theta$$

Using numerical methods we obtain 583.769.

11. $$S = \int_0^6 \int_0^{2-x/3} \sqrt{(-1)^2 + (-3)^2 + 1}\, dydx = \int_0^6 \sqrt{11}[y]_{y=0}^{y=2-x/3} dx$$

$$= \int_0^6 \sqrt{11}\left(2 - \frac{1}{3}x \right) dx = \sqrt{11}\left[2x - \frac{1}{6}x^2 \right]_0^6 = 6\sqrt{11}$$

13. $$S = \int_0^4 \int_{x-4}^0 \sqrt{\left(\frac{1}{2} \right)^2 + \left(-\frac{1}{2} \right)^2 + 1}\, dydx = \int_0^4 \frac{\sqrt{6}}{2}[y]_{y=x-4}^0 dx$$

$$= \int_0^4 \frac{\sqrt{6}}{2}(4 - x)dx = \frac{\sqrt{6}}{2}\left[4x - \frac{1}{2}x^2 \right]_0^4 = 4\sqrt{6}$$

15. $$S = \int_0^1 \int_x^1 \sqrt{(2x)^2 + (2y)^2 + 1}\, dydx = \int_0^1 \int_x^1 \sqrt{4(x^2 + y^2) + 1}\, dydx$$
Using numerical methods we obtain 0.9308.

17. $$S = \int_{-2}^2 \int_{-2}^2 \sqrt{0^2 + (2y)^2 + 1}\, dydx = \int_{-2}^2 \int_{-2}^2 \sqrt{4y^2 + 1}\, dydx$$
Using numerical methods we obtain 37.174.

19.
$$S = \int_{-2}^{2}\int_{-\sqrt{4-x^2}}^{\sqrt{4-x^2}} \sqrt{\left(-\frac{x}{\sqrt{4-x^2-y^2}}\right)^2 + \left(-\frac{y}{\sqrt{4-x^2-y^2}}\right)^2 + 1}\ dy\,dx$$

$$= \int_{-2}^{2}\int_{-\sqrt{4-x^2}}^{\sqrt{4-x^2}} 2(4-x^2-y^2)^{-1/2}\,dy\,dx = \int_{0}^{2\pi}\int_{0}^{2} 2(4-r^2)^{-1/2}\cdot r\,dr\,d\theta$$

$$= \int_{0}^{2\pi}\left[-2\sqrt{4-r^2}\right]_{r=0}^{r=2} d\theta = \int_{0}^{2\pi} 4\,d\theta = 8\pi \approx 25.133$$

21. Exercise 7: The area A of the base R of the solid is
$A = \pi\cdot 2^2 = 4\pi$.
The surface area S is proportional to A:
$S = 4\pi\sqrt{2} = \sqrt{2}A$
Exercise 8: The area A of the base R of the solid is
$$A = \int_{-2}^{2}(4-x^2)\,dx = \left[4x-\frac{1}{3}x^3\right]_{-2}^{2} = \frac{32}{3}$$
The surface area S is proportional to A:
$S = \frac{32\sqrt{2}}{3} = \sqrt{2}A$

23. Exercise 13: $A = \frac{1}{2}(4)(4) = 8$, $S = 4\sqrt{6}$,
$$|\cos\theta| = \left|\frac{1}{\sqrt{6}}\langle 1,-1,-2\rangle\cdot\langle 0,0,1\rangle\right| = \sqrt{\frac{2}{3}};$$
$$S = \frac{A}{|\cos\theta|}$$

Exercise 14: $A = \frac{1}{2}(2)(4) = 4$, $S = \sqrt{21}$,
$$|\cos\theta| = \left|\frac{1}{\sqrt{21}}\langle 2,1,-4\rangle\cdot\langle 0,0,1\rangle\right| = \frac{4}{\sqrt{21}};$$
$$S = \frac{A}{|\cos\theta|}$$

25. Exercise 17:
$$L = \int_{-2}^{2}\sqrt{(2y)^2+1}\,dy = \int_{-2}^{2}\sqrt{4y^2+1}\,dy$$
and $A = \int_{-2}^{2}\int_{-2}^{2}\sqrt{0^2+(2y)^2+1}\,dy\,dx$
$$= 4\int_{-2}^{2}\sqrt{4y^2+1}\,dy$$
$$= 4L$$
Exercise 18:
$$L = \int_{-2}^{2}\sqrt{(-2x)^2+1}\,dx = \int_{-2}^{2}\sqrt{4x^2+1}\,dx$$
and $A = \int_{-2}^{2}\int_{-2}^{2}\sqrt{(-2x)^2+0^2+1}\,dy\,dx$
$$= 4\int_{-2}^{2}\sqrt{4x^2+1}\,dx$$
$$= 4L$$
So $A = 4L$.

Section 13.5

5. $\int_0^2\int_{-2}^2\int_0^2 (2x+y-z)\,dzdydx$

$= \int_0^2\int_{-2}^2\left[2xz+yz-\frac{1}{2}z^2\right]_{z=0}^{z=2}dydx$

$= \int_0^2\int_{-2}^2 (4x+2y-2)\,dydx$

$= \int_0^2 [4xy+y^2-2y]_{y=-2}^{y=2}\,dx$

$= \int_0^2 (16x-8)\,dx = [8x^2-8x]_0^2 = 16$

9. $\int_0^2\int_0^{1-x/2}\int_0^{2-x-2y} 4yz\,dzdydx = \int_0^2\int_0^{1-x/2}(2x^2y+8xy^2-8xy+8y^3-16y^2+8y)\,dydx$

$= \int_0^2\left(\frac{1}{24}x^4-\frac{1}{3}x^3+x^2-\frac{4}{3}x+\frac{2}{3}\right)dx = \frac{4}{15}$

11. $\int_0^2\int_0^{3-3x/2}\int_{3x+2y-6}^0 (3y^2-2z)\,dzdydx = \int_0^2\int_0^{3-3x/2}(9x^2-9xy^2+12xy-36x-6y^3+22y^2-24y+36)\,dydx$

$= \int_0^2\left(\frac{81}{32}x^4-\frac{99}{4}x^3+\frac{351}{4}x^2-135x+\frac{153}{2}\right)dx = \frac{171}{5}$

13. $\int_{-1}^1\int_{-\sqrt{1-x^2}}^{\sqrt{1-x^2}}\int_0^{1-x^2-y^2} 2xy\,dzdydx = \int_{-1}^1\int_{-\sqrt{1-x^2}}^{\sqrt{1-x^2}}(-2x^3y-2xy^3+2xy)\,dy = 0$

15. $\int_{-2}^2\int_{-2}^2\int_{-\sqrt{4-x^2}}^{\sqrt{4-x^2}}(x^2+y^2)y^2\,dzdydx = \int_{-2}^2\int_{-2}^2\left(\frac{4}{3}x^2y^2\sqrt{4-x^2}+\frac{8}{3}y^2\sqrt{4-x^2}\right)dydx$

$= \int_{-2}^2\left(\frac{64}{9}x^2\sqrt{4-x^2}+\frac{128}{9}\sqrt{4-x^2}\right)dx = \frac{128\pi}{3}$

17. $\int_{-2}^2\int_0^{4-y^2}\int_0^{10-2x-y} 15\,dzdxdy = \int_{-2}^2\int_0^{4-y^2}(-30x-15y+150)\,dxdy$

$= \int_{-2}^2 (-15y^4+15y^3-30y^2-60y+360)\,dy$

$= 1088$

7. $\int_2^3\int_0^1\int_{-1}^1 (\sqrt{y}-3z^2)\,dzdydx$

$= \int_2^3\int_0^1 2(\sqrt{y}-1)\,dydx$

$= \int_2^3\left(-\frac{2}{3}\right)dx = -\frac{2}{3}$

19.

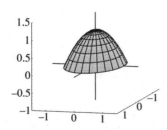

The integral equal 0 because of symmetry: the "positive" and "negative" parts of $f(x,y,z)=2xy$ cancel each other. If $f(x,y,z)=2x^2y$, the integral would still be 0, for the same reason.

21. $\int_0^2 \int_{-1}^1 \int_{x^2}^1 dz\,dx\,dy = \int_0^2 \int_{-1}^1 (1 - x^2)\,dx\,dy$

$$= \int_0^2 \frac{4}{3}\,dy = \frac{8}{3}$$

23. $\int_{-1}^1 \int_0^{1-y^2} \int_{2-z/2}^4 dx\,dz\,dy = \int_{-1}^1 \int_0^{1-y^2} \left(\frac{z}{2} + 2\right) dz\,dy$

$$= \int_{-1}^1 \left(\frac{1}{4}\,y^4 - \frac{5}{2}\,y^2 + \frac{9}{4}\right) dy$$

$$= \frac{44}{15}$$

25. $\int_{-2}^2 \int_{-\sqrt{4-y^2}}^{\sqrt{4-y^2}} \int_{y^2+z^2}^4 dx\,dz\,dy$

$$= \int_{-2}^2 \int_{-\sqrt{4-y^2}}^{\sqrt{4-y^2}} (4 - y^2 - z^2)\,dz\,dy$$

$$= \int_{-2}^2 \frac{4}{3} (4 - y^2)^{3/2}\,dy$$

$$= 8\pi$$

27. $\int_{-3}^3 \int_{-\sqrt{9-x^2}}^{\sqrt{9-x^2}} \int_{\sqrt{x^2+z^2}}^3 dy\,dz\,dx$

$$= \int_{-3}^3 \int_{-\sqrt{9-x^2}}^{\sqrt{9-x^2}} \left(3 - \sqrt{x^2 + z^2}\right) dz\,dx$$

$$= \int_{-3}^3 \left[\left(\frac{1}{2}\,x^2 \ln \frac{3 - \sqrt{9 - x^2}}{3 + \sqrt{9 - x^2}} + 3\sqrt{9 - x^2}\right)\right] dx$$

$$= 9\pi \approx 28.2743$$

29. $\int_{-2}^2 \int_0^{4-x^2} \int_0^{y+6} dz\,dy\,dx$

$$= \int_{-2}^2 \int_0^{4-x^2} (y + 6)\,dy\,dx$$

$$= \int_{-2}^2 \left(\frac{1}{2}\,x^4 - 10x^2 + 32\right) dx$$

$$= \frac{1216}{15}$$

31. $\int_{-1}^1 \int_{x^2}^1 \int_0^{3-x} dy\,dz\,dx = \int_{-1}^1 \int_{x^2}^1 (3 - x)\,dz\,dx$

$$= \int_{-1}^1 (x^3 - 3x^2 - x + 3)\,dx$$

$$= 4$$

33. $m = \int_{-2}^{2}\int_{-\sqrt{4-x^2}}^{\sqrt{4-x^2}}\int_{x^2+y^2}^{4} 4\,dz\,dy\,dx$

$\qquad = \int_{0}^{2\pi}\int_{0}^{2}\int_{r^2}^{4} 4\,dz \cdot r\,dr\,d\theta$

$\qquad = \int_{0}^{2\pi}\int_{0}^{2}(16r - 4r^3)\,dr\,d\theta$

$\qquad = \int_{0}^{2\pi}16\,d\theta = 32\pi$

$M_{yz} = \int_{-2}^{2}\int_{-\sqrt{4-x^2}}^{\sqrt{4-x^2}}\int_{x^2+y^2}^{4} 4x\,dz\,dy\,dx = 0$

$\bar{x} = \dfrac{M_{yz}}{m} = 0$

$M_{xz} = \int_{-2}^{2}\int_{-\sqrt{4-x^2}}^{\sqrt{4-x^2}}\int_{x^2+y^2}^{4} 4y\,dz\,dy\,dx = 0$

$\bar{y} = \dfrac{M_{xz}}{m} = 0$

$M_{xy} = \int_{-2}^{2}\int_{-\sqrt{4-x^2}}^{\sqrt{4-x^2}}\int_{x^2+y^2}^{4} 4z\,dz\,dy\,dx$

$\qquad = \int_{0}^{2\pi}\int_{0}^{2}\int_{r^2}^{4} 4z\,dz \cdot r\,dr\,d\theta$

$\qquad = \int_{0}^{2\pi}\int_{0}^{2}(32 - 2r^4)\,r\,dr\,d\theta$

$\qquad = \int_{0}^{2\pi}\dfrac{128}{3}\,d\theta = \dfrac{256\pi}{3}$

$\bar{z} = \dfrac{M_{xy}}{m} = \dfrac{256\pi/3}{32\pi} = \dfrac{8}{3}$

35. $\int_{0}^{6}\int_{0}^{2-x/3}\int_{0}^{6-x-3y}(10+x)\,dz\,dy\,dx$

$\qquad = \int_{0}^{6}\int_{0}^{2-x/3}(-x^2 - 3xy - 4x - 30y + 60)\,dy\,dx$

$\qquad = \int_{0}^{6}\left(\dfrac{1}{6}x^3 - \dfrac{1}{3}x^2 - 14x + 60\right)dx$

$\qquad = 138$

$M_{yz} = \int_{0}^{6}\int_{0}^{2-x/3}\int_{0}^{6-x-3y} x(10+x)\,dz\,dy\,dx = \dfrac{1116}{5}$

$\bar{x} = \dfrac{M_{yz}}{m} = \dfrac{1116/5}{138} = \dfrac{186}{115}$

$M_{xz} = \int_{0}^{6}\int_{0}^{2-x/3}\int_{0}^{6-x-3y} y(10+x)\,dz\,dy\,dx = \dfrac{336}{5}$

$\bar{y} = \dfrac{M_{xz}}{m} = \dfrac{336/5}{138} = \dfrac{56}{115}$

$M_{xy} = \int_{0}^{6}\int_{0}^{2-x/3}\int_{0}^{6-x-3y} z(10+x)\,dz\,dy\,dx = \dfrac{1008}{5}$

$\bar{z} = \dfrac{M_{xy}}{m} = \dfrac{1008/5}{138} = \dfrac{168}{115}$

37. The density function is symmetric about the
yz-plane in exercise 33 but not in exercise 34.

39. $\int_0^1 \int_0^{2-2y} \int_0^{2-x-2y} 4\,yzdzdxdy$

$= \int_0^2 \int_0^{2-x} \int_0^{1-x/2-z/2} 4\,yzdydzdx$

$= \int_0^1 \int_0^{2-2y} \int_0^{2-2y-z} 4\,yzdxdzdy = \dfrac{4}{15}$

41. $\int_0^2 \int_0^{4-2y} \int_0^{4-2y-z} dxdzdy$

$= \int_0^2 \int_0^{4-2y} \int_0^{4-2y-x} dzdxdy$

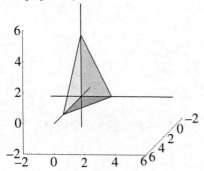

43. $\int_0^1 \int_0^{\sqrt{1-x^2}} \int_0^{\sqrt{1-x^2-y^2}} dzdydx$

$= \int_0^1 \int_0^{\sqrt{1-x^2}} \int_0^{\sqrt{1-x^2-z^2}} dydzdx$

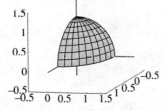

45. $\int_0^2 \int_0^{\sqrt{4-z^2}} \int_{x^2+z^2}^4 dydxdz = \int_0^2 \int_{x^2}^4 \int_0^{\sqrt{y-x^2}} dzdydx$

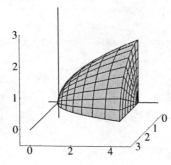

Section 13.6

5. $x^2 + y^2 = 16$
$$r^2 = 16$$
$$r = 4$$

7. $(x-2)^2 + y^2 = 4$
$$x^2 - 4x + 4 + y^2 = 4$$
$$x^2 + y^2 - 4x = 0$$
$$r^2 - 4r\cos\theta = 0$$
$$r(r - 4\cos\theta) = 0$$
$$r = 4\cos\theta$$

9. $z = x^2 + y^2$
$$z = r^2$$

11. $z = \cos(x^2 + y^2)$
$$z = \cos(r^2)$$

13.
$$y = x$$
$$r\sin\theta = r\cos\theta$$
$$\frac{\sin\theta}{\cos\theta} = 1$$
$$\tan\theta = 1$$
$$\theta = \frac{\pi}{4}$$

15. $\int_{-2}^{2}\int_{-\sqrt{4-x^2}}^{\sqrt{4-x^2}}\int_{\sqrt{x^2+y^2}}^{\sqrt{8-(x^2+y^2)}} f(x,y,z)\,dz\,dy\,dx$
$$= \int_{0}^{2\pi}\int_{0}^{2}\int_{r}^{\sqrt{8-r^2}} rf(r\cos\theta, r\sin\theta, z)\,dz\,dr\,d\theta$$

17. $\int_{-3}^{3}\int_{-\sqrt{9-x^2}}^{\sqrt{9-x^2}}\int_{0}^{9-(x^2+y^2)} f(x,y,z)\,dz\,dy\,dx$
$$= \int_{0}^{2\pi}\int_{0}^{3}\int_{0}^{9-r^2} rf(r\cos\theta, r\sin\theta, z)\,dz\,dr\,d\theta$$

19. $\int_{-2}^{2}\int_{-\sqrt{4-x^2}}^{\sqrt{4-x^2}}\int_{x^2+y^2}^{4} f(x,y,z)\,dz\,dy\,dx$
$$= \int_{0}^{2\pi}\int_{0}^{2}\int_{r^2}^{4} rf(r\cos\theta, r\sin\theta, z)\,dz\,dr\,d\theta$$

21. $\int_{-2}^{2}\int_{-\sqrt{4-x^2}}^{\sqrt{4-x^2}}\int_{0}^{4-(x^2+z^2)} f(x,y,z)\,dy\,dz\,dx$
$$= \int_{0}^{2\pi}\int_{0}^{2}\int_{0}^{4-r^2} rf(r\cos\theta, y, r\sin\theta)\,dy\,dr\,d\theta$$

23. $\int_{-1}^{1}\int_{-\sqrt{1-y^2}}^{\sqrt{1-y^2}}\int_{y^2+z^2}^{2-(y^2+z^2)} f(x,y,z)\,dx\,dz\,dy$
$$= \int_{0}^{2\pi}\int_{0}^{1}\int_{r^2}^{2-r^2} rf(x, r\cos\theta, r\sin\theta)\,dx\,dr\,d\theta$$

25. $\int_{0}^{2\pi}\int_{0}^{2}\int_{1}^{2} re^{r^2}\,dz\,dr\,d\theta = \int_{0}^{2\pi}\int_{0}^{2} re^{r^2}\,dr\,d\theta$
$$= \int_{0}^{2\pi}\frac{1}{2}(e^4 - 1)\,d\theta$$
$$= \pi(e^4 - 1)$$

27. $\int_{0}^{6}\int_{0}^{3-x/2}\int_{0}^{2-2y/3-x/3} (x+z)\,dz\,dy\,dx$
$$= \int_{0}^{6}\int_{0}^{3-x/2}\left(\begin{array}{c}-\dfrac{5}{18}x^2 - \dfrac{4}{9}xy + \dfrac{4}{3}x \\[4pt] + \dfrac{2}{9}y^2 - \dfrac{4}{3}y + 2\end{array}\right)dy\,dx$$
$$= \int_{0}^{6}\left(\frac{2}{27}x^3 - \frac{5}{6}x^2 + 2x + 2\right)dx = 12$$

29. $\int_{0}^{2\pi}\int_{0}^{\sqrt{2}}\int_{r}^{\sqrt{4-r^2}} rz\,dz\,dr\,d\theta = \int_{0}^{2\pi}\int_{0}^{\sqrt{2}} (2r - r^3)\,dr\,d\theta$
$$= \int_{0}^{2\pi} 1\,d\theta = 2\pi$$

31. $\int_{0}^{4}\int_{0}^{2-x/2}\int_{0}^{4-2y-x} (x+y)\,dz\,dy\,dx$
$$= \int_{0}^{4}\int_{0}^{2-x/2} (-x^2 - 3xy + 4x - 2y^2 + 4y)\,dy\,dx$$
$$= \int_{0}^{4}\left(\frac{5}{24}x^3 - \frac{3}{2}x^2 + 2x + \frac{8}{3}\right)dx = 8$$

33. $\int_{0}^{2\pi}\int_{0}^{3}\int_{0}^{r^2} re^z\,dz\,dr\,d\theta$
$$= \int_{0}^{2\pi}\int_{0}^{3} r(e^{r^2} - 1)\,dr\,d\theta$$
$$= \int_{0}^{2\pi}\frac{1}{2}(e^9 - 10)\,d\theta$$
$$= \pi(e^9 - 10)$$

35. The circle $x^2 + (y-1)^2 = 1$ is given in polar form
by $r = 2\sin\theta, 0 \le \theta \le \pi$; and the function $2x$
becomes $2r\cos\theta$.
$$\int_{0}^{\pi}\int_{0}^{2\sin\theta}\int_{0}^{r} 2r^2\cos\theta\,dz\,dr\,d\theta$$
$$= \int_{0}^{\pi}\int_{0}^{2\sin\theta} 2r^3\cos\theta\,dr\,d\theta$$
$$= \int_{0}^{\pi} 8\sin^4\theta\cos\theta\,dr\,d\theta = 0$$

37.

$$\int_{-1}^{1}\int_{-\sqrt{1-x^2}}^{\sqrt{1-x^2}}\int_{0}^{\sqrt{x^2+y^2}} 3z^2\,dz\,dy\,dx$$

$$=\int_{0}^{2\pi}\int_{0}^{1}\int_{0}^{r} 3rz^2\,dz\,dr\,d\theta$$

$$=\int_{0}^{2\pi}\int_{0}^{1} r^4\,dr\,d\theta$$

$$=\int_{0}^{2\pi}\frac{1}{5}\,d\theta=\frac{2\pi}{5}$$

39.

$$\int_{0}^{2}\int_{-\sqrt{4-y^2}}^{\sqrt{4-y^2}}\int_{\sqrt{x^2+y^2}}^{\sqrt{8-x^2-y^2}} 2\,dz\,dx\,dy$$

$$=\int_{0}^{\pi}\int_{0}^{2}\int_{r}^{\sqrt{8-r^2}} 2r\,dz\,dr\,d\theta$$

$$=\int_{0}^{\pi}\int_{0}^{2}(2r\sqrt{8-r^2}-2r^2)\,dr\,d\theta$$

$$=\int_{0}^{\pi}\frac{32}{3}(\sqrt{2}-1)\,d\theta=\frac{32\pi}{3}(\sqrt{2}-1)$$

41.

$$\int_{-3}^{3}\int_{-\sqrt{9-x^2}}^{\sqrt{9-x^2}}\int_{0}^{x^2+z^2}(x^2+z^2)\,dy\,dz\,dx$$

$$=\int_{0}^{2\pi}\int_{0}^{3}\int_{0}^{r^2} r^3\,dy\,dr\,d\theta$$

$$=\int_{0}^{2\pi}\int_{0}^{3} r^5\,dr\,d\theta=\int_{0}^{2\pi}\frac{243}{2}\,d\theta=243\pi$$

43.

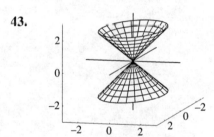

45.

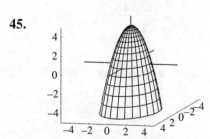

47.

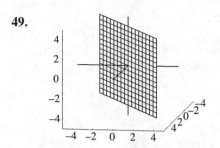

49.

51. $m=\int_{0}^{2\pi}\int_{0}^{4}\int_{r}^{4} r^2\,dz\,dr\,d\theta=\int_{0}^{2\pi}\int_{0}^{4}(4r^2-r^3)\,dr\,d\theta$

$$=\int_{0}^{2\pi}\frac{64}{3}\,d\theta=\frac{128\pi}{3}$$

$M_{yz}=\int_{0}^{2\pi}\int_{0}^{4}\int_{r}^{4} r\cos\theta\, r^2\,dz\,dr\,d\theta=0$

$\bar{x}=\dfrac{M_{yz}}{m}=0$

$M_{xz}=\int_{0}^{2\pi}\int_{0}^{4}\int_{r}^{4} r\sin\theta\, r^2\,dz\,dr\,d\theta=0$

$\bar{y}=\dfrac{M_{xz}}{m}=0$

$M_{xy}=\int_{0}^{2\pi}\int_{0}^{4}\int_{r}^{4} zr^2\,dz\,dr\,d\theta$

$$=\int_{0}^{2\pi}\int_{0}^{4}\frac{1}{2}(16r^2-r^4)\,dr\,d\theta$$

$$=\int_{0}^{2\pi}\frac{1024}{15}\,d\theta=\frac{2048\pi}{15}$$

$\bar{z}=\dfrac{M_{xy}}{m}=\dfrac{2048\pi/15}{128\pi/3}=\dfrac{16}{5}$

53. $m=\int_{0}^{\pi}\int_{0}^{2\sin\theta}\int_{r^2}^{4} 4r\,dz\,dr\,d\theta$

$$=\int_{0}^{\pi}\int_{0}^{2\sin\theta}(16r-4r^3)\,dr\,d\theta$$

$$=\int_{0}^{\pi}(16-16\cos^4\theta)\,d\theta=10\pi$$

$M_{yz}=\int_{0}^{\pi}\int_{0}^{2\sin\theta}\int_{r^2}^{4} 4r^2\cos\theta\,dz\,dr\,d\theta=0$

$\bar{x}=\dfrac{M_{yz}}{m}=0$

$M_{xz}=\int_{0}^{\pi}\int_{0}^{2\sin\theta}\int_{r^2}^{4} 4r^2\sin\theta\,dz\,dr\,d\theta=8\pi$

$\bar{y}=\dfrac{M_{xz}}{m}=\dfrac{8\pi}{10\pi}=\dfrac{4}{5}$

$M_{xy}=\int_{0}^{\pi}\int_{0}^{2\sin\theta}\int_{r^2}^{4} 4rz\,dz\,dr\,d\theta=\dfrac{76\pi}{3}$

$\bar{z}=\dfrac{M_{xy}}{m}=\dfrac{76\pi/3}{10\pi}=\dfrac{38}{15}$

Section 13.7

5. $x = \rho \sin\phi \cos\theta \quad y = \rho \sin\phi \sin\theta \quad z = \rho \cos\phi$
$\quad x = 4\sin 0 \cos\pi \quad y = 4\sin 0 \sin\pi \quad z = 4\cos 0$
$\quad x = 0 \qquad\qquad y = 0 \qquad\qquad z = 4$
$\quad (0, 0, 4)$

7. $x = 4\sin\dfrac{\pi}{2}\cos 0 \quad y = 4\sin\dfrac{\pi}{2}\sin 0 \quad z = 4\cos\dfrac{\pi}{2}$
$\quad x = 0 \qquad\qquad y = 0 \qquad\qquad z = 0$
$\quad (4, 0, 0)$

9. $x = 2\sin\dfrac{\pi}{4}\cos 0 \quad y = 2\sin\dfrac{\pi}{4}\sin 0 \quad z = 2\cos\dfrac{\pi}{4}$
$\quad x = \sqrt{2} \qquad\qquad y = 0 \qquad\qquad z = \sqrt{2}$
$\quad (\sqrt{2}, 0, \sqrt{2})$

11. $x = \sqrt{2}\sin\dfrac{\pi}{4}\cos\dfrac{\pi}{4} \quad y = \sqrt{2}\sin\dfrac{\pi}{4}\sin\dfrac{\pi}{4}$
$\quad x = \dfrac{\sqrt{2}}{2} \qquad\qquad y = \dfrac{\sqrt{2}}{2}$
$\quad z = \sqrt{2}\cos\dfrac{\pi}{4}$
$\quad z = 1$
$\quad \left(\dfrac{\sqrt{2}}{2}, \dfrac{\sqrt{2}}{2}, 1\right)$

13. $x^2 + y^2 + z^2 = 9$
$\qquad\qquad \rho^2 = 9$
$\qquad\qquad \rho = 3$

15. $\qquad y = x$
$\quad \rho\sin\theta = \rho\sin\phi\cos\theta$
$\qquad \sin\theta = \cos\theta$
$\qquad \tan\theta = 1$
$\qquad \theta = \dfrac{\pi}{4} \text{ or } \theta = \dfrac{5\pi}{4}$

17. $\qquad z = 2$
$\quad \rho\cos\phi = 2$

19. $\qquad z = \sqrt{3(x^2 + y^2)}$
$\qquad z^2 = 3(x^2 + y^2)$
$\quad (\rho\cos\phi)^2 = 3\left[\begin{array}{c}(\rho\sin\phi\cos\theta)^2 \\ + (\rho\sin\phi\sin\theta)^2\end{array}\right]$
$\quad \rho^2\cos^2\phi = 3\rho^2\sin^2\phi$
$\quad \left(\dfrac{\sin\phi}{\cos\phi}\right)^2 = \dfrac{1}{3}$
$\qquad \tan\phi = \dfrac{1}{\sqrt{3}}$
$\qquad \phi = \dfrac{\pi}{6}$

21.
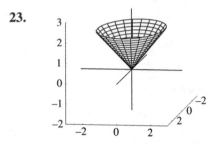

23.
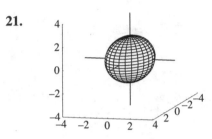

25.
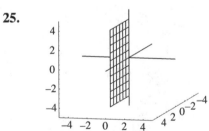

27.
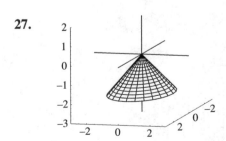

29.

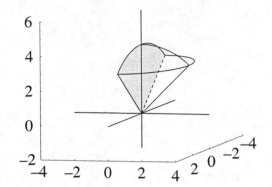

31.

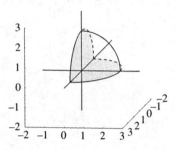

33.

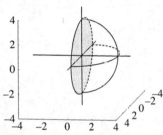

35.

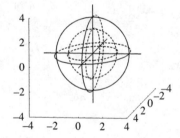

37. $\int_0^{2\pi} \int_0^{\pi/2} \int_0^2 e^{\rho^3} \rho^2 \sin\phi \, d\rho \, d\phi \, d\theta$

$= \int_0^{2\pi} \int_0^{\pi/2} \frac{1}{3}(e^8 - 1) \sin\phi \, d\phi \, d\theta$

$= \int_0^{2\pi} \frac{1}{3}(e^8 - 1) d\theta = \frac{2\pi}{3}(e^8 - 1)$

39. $\int_0^{2\pi} \int_0^{\pi/2} \int_0^{\sqrt{2}} \rho^7 \sin\phi \, d\rho \, d\phi \, d\theta$

$= \int_0^{2\pi} \int_0^{\pi/2} \left[\frac{1}{8}\rho^8 \sin\phi \right]_{\rho=0}^{\rho=\sqrt{2}} d\phi \, d\theta$

$= \int_0^{2\pi} \int_0^{\pi/2} 2\sin\phi \, d\phi \, d\theta$

$= \int_0^{2\pi} \left[-2\cos\phi \right]_0^{\pi/2} d\theta$

$= \int_0^{2\pi} 2 \, d\theta$

$= \left[2\theta \right]_0^{2\pi}$

$= 4\pi$

41. $\int_0^1 \int_1^2 \int_3^4 (x^2 + y^2 + z^2) dz \, dy \, dx = 15$

43. $\int_0^{2\pi} \int_0^2 \int_0^{4-r^2} r^3 dz \, dr \, d\theta = \frac{32\pi}{3}$

45. $\int_0^{2\pi} \int_0^{\pi/4} \int_0^{\sqrt{2}} \rho^3 \sin\phi \, d\rho \, d\phi \, d\theta = (2 - \sqrt{2})\pi$

47. $\int_0^{2\pi} \int_0^{\pi/4} \int_0^{4\cos\phi} \rho^2 \sin\phi \, d\rho \, d\phi \, d\theta = 8\pi$

49. $\int_0^{2\pi} \int_0^4 \int_r^4 r \, dz \, dr \, d\theta = \int_0^{2\pi} \int_0^4 (4r - r^2) dr \, d\theta$

$= \int_0^{2\pi} \frac{32}{3} d\theta$

$= \frac{64\pi}{3}$

51. $\int_{-1}^1 \int_{-1}^1 \int_0^{\sqrt{x^2+y^2}} dz \, dy \, dx \approx 3.061,$ using numeric methods.

53. $\int_0^{\pi/2} \int_0^{\pi/4} \int_0^2 \rho^2 \sin\phi \, d\rho \, d\phi \, d\theta$

$= \int_0^{\pi/2} \int_0^{\pi/4} \frac{8}{3} \sin\phi \, d\phi \, d\theta$

$= \int_0^{\pi/2} \frac{4}{3}(2 - \sqrt{2}) d\theta$

$= \frac{2\pi}{3}(2 - \sqrt{2})$

55. $\int_0^{2\pi} \int_0^2 \int_0^r r \, dz \, dr \, d\theta = \int_0^{2\pi} \int_0^2 r^2 dr \, d\theta$

$= \int_0^{2\pi} \frac{8}{3} d\theta$

$= \frac{16\pi}{3}$

57.

$$\int_{-\pi/2}^{\pi/2}\int_0^\pi\int_0^1 \rho^3 \sin\phi\, d\rho\, d\phi\, d\theta$$

$$=\int_{-\pi/2}^{\pi/2}\int_0^\pi \frac{1}{4}\sin\phi\, d\phi\, d\theta$$

$$=\int_{-\pi/2}^{\pi/2}\frac{1}{2}\, d\theta$$

$$=\frac{\pi}{2}$$

59.

$$\int_0^\pi\int_0^{\pi/4}\int_0^{\sqrt{8}} \rho^5 \sin\phi\, d\rho\, d\phi\, d\theta$$

$$=\int_0^\pi\int_0^{\pi/4}\frac{256}{3}\sin\phi\, d\phi\, d\theta$$

$$=\int_0^\pi\frac{128}{3}(2-\sqrt{2})\, d\theta$$

$$=\frac{128\pi}{3}(2-\sqrt{2})$$

$$M_{xz}=\int_0^{2\pi}\int_0^{\pi/4}\int_0^2 \rho^3 \sin^2\phi\sin\theta\, d\rho\, d\phi\, d\theta$$

$$=\int_0^{2\pi}\int_0^{\pi/4}4\sin^2\phi\sin\theta\, d\phi\, d\theta$$

$$=\int_0^{2\pi}\frac{\pi-2}{2}\sin\theta\, d\theta$$

$$=0$$

$$\overline{y}=\frac{M_{xz}}{m}=0$$

$$M_{xy}=\int_0^{2\pi}\int_0^{\pi/4}\int_0^2 \rho^3 \sin\phi\cos\phi\, d\rho\, d\phi\, d\theta$$

$$=\int_0^{2\pi}\int_0^{\pi/4}4\sin\phi\cos\phi\, d\phi\, d\theta$$

$$=\int_0^{2\pi}1\, d\theta$$

$$=2\pi$$

$$\overline{z}=\frac{M_{xy}}{m}=\frac{2\pi}{\frac{8\pi}{3}(2-\sqrt{2})}=\frac{3}{8}(2+\sqrt{2})$$

61.

$$m=\int_0^{2\pi}\int_0^{\pi/4}\int_0^2 \rho^2 \sin\phi\, d\rho\, d\phi\, d\theta$$

$$=\int_0^{2\pi}\int_0^{\pi/4}\frac{8}{3}\sin\phi\, d\phi\, d\theta$$

$$=\int_0^{2\pi}\frac{4}{3}(2-\sqrt{2})\, d\theta$$

$$=\frac{8\pi}{3}(2-\sqrt{2})$$

$$M_{yz}=\int_0^{2\pi}\int_0^{\pi/4}\int_0^2 \rho^3 \sin^2\phi\cos\theta\, d\rho\, d\phi\, d\theta$$

$$=\int_0^{2\pi}\int_0^{\pi/4}4\sin^2\phi\cos\theta\, d\phi\, d\theta$$

$$=\int_0^{2\pi}\frac{\pi-2}{2}\cos\theta\, d\theta$$

$$=0$$

$$\overline{x}=\frac{M_{yz}}{m}=0$$

Section 13.8

3. $\begin{aligned} y = 4x+2 &\rightarrow y-4x=2 \\ y = 4x+5 &\rightarrow y-4x=5 \end{aligned}\Bigg\}$ $\begin{aligned} u=y-4x \\ \\ v=y+2x \end{aligned}\Bigg\}$ $\rightarrow$ $\begin{cases} x=\dfrac{1}{6}(v-u) \\ \\ y=\dfrac{1}{3}(u+2v) \end{cases}$

$\begin{aligned} y = 3-2x &\rightarrow y+2x=3 \\ y = 1-2x &\rightarrow y+2x=1 \end{aligned}\Bigg\}$

$x=\dfrac{1}{6}(v-u),\ y=\dfrac{1}{3}(u+2v),\ 2\le u\le 5, 1\le v\le 3$

5. $\begin{aligned} y = 1-3x &\rightarrow y+3x=1 \\ y = 3-3x &\rightarrow y+3x=3 \end{aligned}\Bigg\}$ $\begin{aligned} u=y+3x \\ \\ v=y-x \end{aligned}\Bigg\}$ $\rightarrow$ $\begin{cases} x=\dfrac{1}{4}(u-v) \\ \\ y=\dfrac{1}{4}(u+3v) \end{cases}$

$\begin{aligned} y = x-1 &\rightarrow y-x=-1 \\ y = x-3 &\rightarrow y-x=-3 \end{aligned}\Bigg\}$

$x=\dfrac{1}{4}(u-v),\ y=\dfrac{1}{4}(u+3v),\ 1\le u\le 3, -3\le v\le -1$

7. $x=r\cos\theta,\ y=r\sin\theta, 1\le r\le 2, 0\le\theta\le\dfrac{\pi}{2}$

9. $x=r\cos\theta,\ y=r\sin\theta, 2\le r\le 3,\dfrac{\pi}{4}\le\theta\le\dfrac{3\pi}{4}$

11. $\begin{aligned} y = x^2 &\rightarrow y-x^2=0 \\ y = x^2+2 &\rightarrow y-x^2=2 \end{aligned}\Bigg\}$ $\begin{aligned} u=y-x^2 \\ \\ v=y+x^2 \end{aligned}\Bigg\}$ $\rightarrow$ $\begin{cases} x=\sqrt{\dfrac{1}{2}(v-u)} \\ \\ y=\dfrac{1}{2}(u+v) \end{cases}$

$\begin{aligned} y-3-x^2 &\rightarrow y+x^2=3 \\ y = 1-x^2 &\rightarrow y+x^2=1 \end{aligned}\Bigg\}$

$x=\sqrt{\dfrac{1}{2}(v-u)},\ y=\dfrac{1}{2}(u+v), 0\le u\le 2,\ 1\le v\le 3$

13. $\begin{aligned} y = e^x &\rightarrow y-e^x=0 \\ y = e^x+1 &\rightarrow y-e^x=1 \end{aligned}\Bigg\}$ $\begin{aligned} u=y-e^x \\ \\ v=y+e^x \end{aligned}\Bigg\}$ $\rightarrow$ $\begin{cases} x=\left(\ln\dfrac{1}{2}|v-u|\right) \\ \\ y=\dfrac{1}{2}(u+v) \end{cases}$

$\begin{aligned} y = 3-e^x &\rightarrow y+e^x=3 \\ y = 5-e^x &\rightarrow y+e^x=5 \end{aligned}\Bigg\}$

$x=\ln\left(\dfrac{1}{2}|v-u|\right),\ y=\dfrac{1}{2}(u+v), 0\le u\le 1, 3\le v\le 5$

15. $\dfrac{\partial(x,y)}{\partial(u,v)} = \begin{vmatrix} \frac{\partial x}{\partial u} & \frac{\partial x}{\partial v} \\ \frac{\partial y}{\partial u} & \frac{\partial y}{\partial v} \end{vmatrix} = \begin{vmatrix} -\frac{1}{6} & \frac{1}{6} \\ \frac{1}{3} & \frac{2}{3} \end{vmatrix} = -\dfrac{1}{6}$

$\displaystyle\iint\limits_{R}(y-4x)\,dA$

$\displaystyle= \int_1^3\int_2^5\left[\frac{1}{3}(u+2v)-4\left(\frac{1}{6}\right)(v-u)\right]\cdot\left|-\frac{1}{6}\right|du\,dv$

$\displaystyle= \frac{1}{6}\int_1^3\int_2^5 u\,du\,dv$

$\displaystyle= \frac{1}{12}\int_1^3 [u^2]_{u=2}^{u=5}\,dv$

$\displaystyle= \frac{7}{4}\int_1^3 dv$

$\displaystyle= \frac{7}{4}[v]_1^3$

$\displaystyle= \frac{7}{2}$

17. $\dfrac{\partial(x,y)}{\partial(u,v)} = \begin{vmatrix} \frac{1}{4} & -\frac{1}{4} \\ \frac{1}{4} & \frac{3}{4} \end{vmatrix} = \dfrac{1}{4}$

$\displaystyle\iint\limits_{R}(y+3x)^2\,dA$

$\displaystyle= \int_{-3}^{-1}\int_1^3\left[\frac{1}{4}(u+3v)+3\left(\frac{1}{4}\right)(u-v)\right]^2\left(\frac{1}{4}\right)du\,dv$

$\displaystyle= \frac{1}{4}\int_{-3}^{-1}\int_1^3 u^2\,du\,dv$

$\displaystyle= \frac{1}{12}\int_{-3}^{-1}[u^3]_1^3\,dv$

$\displaystyle= \frac{13}{6}\int_{-3}^{-1} dv$

$\displaystyle= \frac{13}{6}[v]_{-3}^{-1}$

$\displaystyle= \frac{13}{3}$

19. $\dfrac{\partial(x,y)}{\partial(r,\theta)} = \begin{vmatrix} \cos\theta & -r\sin\theta \\ \sin\theta & r\cos\theta \end{vmatrix} = r$

$\displaystyle\iint\limits_{R}x\,dA = \int_0^{\pi/2}\int_1^2 r\cos\theta(r)\,dr\,d\theta$

$\displaystyle= \frac{1}{3}\int_0^{\pi/2}[r^3\cos\theta]_{r=1}^{r=2}\,d\theta$

$\displaystyle= \frac{7}{3}\int_0^{\pi/2}\cos\theta\,d\theta$

$\displaystyle= \frac{7}{3}[\sin\theta]_0^{\pi/2}$

$\displaystyle= \frac{7}{3}$

21. $\displaystyle\iint\limits_{R}\frac{e^{y-4x}}{y+2x}\,dA$

$\displaystyle= \int_1^3\int_2^5\frac{e^{\frac{1}{3}(u+2v)-4\left(\frac{1}{6}\right)(v-u)}}{\frac{1}{3}(u+2v)+2\left(\frac{1}{6}\right)(v-u)}\left|-\frac{1}{6}\right|du\,dv$

$\displaystyle= \frac{1}{6}\int_1^3\int_2^5\frac{e^u}{v}\,du\,dv$

$\displaystyle= \frac{1}{6}\int_1^3\left[\frac{e^u}{v}\right]_{u=2}^{u=5}\,dv$

$\displaystyle= \frac{1}{6}\int_1^3\frac{e^5-e^2}{v}\,dv$

$\displaystyle= \frac{1}{6}(e^5-e^2)\left[\ln|v|\right]_1^3$

$\displaystyle= \frac{1}{6}(e^5-e^2)\ln 3$

23. $\displaystyle\iint\limits_{R}(x+y)\,dA$

$\displaystyle= \int_1^3\int_2^5\left[\frac{1}{3}(u+2v)+\frac{1}{6}(v-u)\right]\cdot\left|-\frac{1}{6}\right|du\,dv$

$\displaystyle= \frac{1}{6}\int_1^3\int_2^5\frac{1}{6}(u+5v)\,du\,dv$

$\displaystyle= \frac{1}{36}\int_1^3\left[\frac{1}{2}u^2+5uv\right]_2^5\,dv$

$\displaystyle= \frac{1}{36}\int_1^3\left(\frac{21}{2}+15v\right)dv$

$\displaystyle= \frac{1}{36}\left[\frac{21}{2}v+\frac{15}{2}v^2\right]_1^3$

$\displaystyle= \frac{9}{4}$

25. $x = ue^v,\ y = ue^{-v}$

$\dfrac{\partial(x,y)}{\partial(u,v)} = \begin{vmatrix} e^v & ue^v \\ e^{-v} & -ue^{-v} \end{vmatrix} = -2u$

27. $x = \dfrac{u}{v},\ y = v^2$

$\dfrac{\partial(x,y)}{\partial(u,v)} = \begin{vmatrix} \frac{1}{v} & -\frac{u}{v^2} \\ 0 & 2v \end{vmatrix} = 2$

29.
$$\left.\begin{array}{c}x+y+z=1\\x+y+z=2\end{array}\right\}\quad u=x+y+z$$
$$\left.\begin{array}{c}x+2y=0\\x+2y=1\end{array}\right\}\quad v=x+2y$$
$$\left.\begin{array}{c}y+z=2\\y+z=4\end{array}\right\}\quad w=y+z$$

$$\rightarrow\begin{cases}x=u-w\\[2mm]y=\dfrac{1}{2}(-u+v+w)\\[2mm]z=\dfrac{1}{2}(u-v+w)\end{cases}$$

$$x=u-w,\ y=\frac{1}{2}(-u+v+w),$$
$$z=\frac{1}{2}(u-v+w),$$
$$1\le u\le 2,\, 0\le v\le 1,\, 2\le w\le 4$$

31.
$$\frac{\partial(x,y,z)}{\partial(u,v,w)}\begin{vmatrix}1&0&-1\\-\frac{1}{2}&\frac{1}{2}&\frac{1}{2}\\\frac{1}{2}&-\frac{1}{2}&\frac{1}{2}\end{vmatrix}=\frac{1}{2}$$

$$\iiint\limits_{R}dA=\int_{2}^{4}\int_{0}^{1}\int_{1}^{2}\left|\frac{1}{2}\right|du\,dv\,dw$$
$$=\frac{1}{2}\int_{2}^{4}\int_{0}^{1}[u]_{1}^{2}\,dv\,dw$$
$$=\frac{1}{2}\cdot 1\int_{2}^{4}[v]_{0}^{1}\,dw$$
$$=\frac{1}{2}\cdot 1\cdot 1[w]_{2}^{4}$$
$$=\frac{1}{2}\cdot 1\cdot 1\cdot 2$$
$$=1$$

Chapter 13 Review

1. See Example 1.1.

$f(x, y) = 5x - 2y$, $1 \le x \le 3$, $0 \le y \le 1$, $n = 4$

The centers of the four rectangles are $\left(\dfrac{5}{4}, \dfrac{1}{2}\right), \left(\dfrac{7}{4}, \dfrac{1}{2}\right), \left(\dfrac{9}{4}, \dfrac{1}{2}\right)$, and $\left(\dfrac{11}{4}, \dfrac{1}{2}\right)$. Since the four rectangles are the

same size, $\Delta A_i = \dfrac{1}{2}$ for each i.

$$V \approx \sum_{i=1}^{4} f(u_i, v_i) \Delta A_i$$

$$= f\left(\frac{5}{4}, \frac{1}{2}\right)\left(\frac{1}{2}\right) + f\left(\frac{7}{4}, \frac{1}{2}\right)\left(\frac{1}{2}\right) + f\left(\frac{9}{4}, \frac{1}{2}\right)\left(\frac{1}{2}\right) + f\left(\frac{11}{4}, \frac{1}{2}\right)\left(\frac{1}{2}\right)$$

$$= \frac{1}{2}\left(\frac{21}{4} + \frac{31}{4} + \frac{41}{4} + \frac{51}{4}\right)$$

$$= 18$$

3. See Example 1.2.

$$\iint_R (4x + 9x^2 y^2)\, dA = \int_0^3 \int_1^2 (4x + 9x^2 y^2)\, dy\, dx$$

$$= \int_0^3 [4xy + 3x^2 y^3]_{y=1}^{y=2}\, dx$$

$$= \int_0^3 (4x + 21x^2)\, dx$$

$$= [2x^2 + 7x^3]_0^3$$

$$= 207$$

5. See Example 3.4.

$$\iint_R e^{-x^2 - y^2}\, dA = \int_0^{2\pi} \int_1^2 r e^{-r^2}\, dr\, d\theta$$

$$= -\frac{1}{2}\int_0^{2\pi} \left[e^{-r^2} \right]_1^2\, d\theta$$

$$= -\frac{1}{2}\int_0^{2\pi} (e^{-4} - e^{-1})\, d\theta$$

$$= -\frac{1}{2}(e^{-4} - e^{-1})[\theta]_0^{2\pi}$$

$$= (e^{-1} - e^{-4})\pi$$

7. See Example 1.3.

$$\int_{-1}^1 \int_{x^2}^{2x} (2xy - 1)\, dy\, dx$$

$$= \int_{-1}^1 [xy^2 - y]_{y=x^2}^{y=2x}\, dx$$

$$= \int_{-1}^1 (-x^5 + 4x^3 + x^2 - 2x)\, dx$$

$$= \left[-\frac{1}{6}x^6 + x^4 + \frac{1}{3}x^3 - x^2 \right]_{-1}^1$$

$$= \frac{2}{3}$$

9. See Example 3.4.

$$\iint_R xy\, dA = \int_0^\pi \int_0^{2\cos\theta} r\cos\theta \cdot r\sin\theta \cdot r\, dr\, d\theta$$

$$= \int_0^\pi \left[\frac{1}{4}r^4 \sin\theta \cos\theta \right]_{r=0}^{r=2\cos\theta}\, d\theta$$

$$= \int_0^\pi 4\sin\theta \cos^5\theta\, d\theta$$

$$= -\frac{2}{3}[\cos^6\theta]_0^\pi$$

$$= 0$$

11. See Example 1.4.

To find the limits of integration, we solve

$x^2 - 4 = \ln x$ using numerical methods and obtain

$x \approx 0.018, 2.187$.

$$\int_R \int 4xy\, dA = \int_{0.018}^{2.187} \int_{x^2-4}^{\ln x} 4xy\, dy\, dx \approx -19.917$$

13. See Example 5.5.

$$\int_0^1 \int_{-1}^1 \int_0^{1-x^2} dz\, dx\, dy = \int_0^1 \int_{-1}^1 (1 - x^2)\, dx\, dy$$

$$= \int_0^1 \frac{4}{3}\, dy$$

$$= \frac{4}{3}$$

15. See Examples 5.4, 6.1 and 6.5.

$$\int_0^{2\pi} \int_0^2 \int_{r^2}^{8-r^2} r\, dz\, dr\, d\theta = \int_0^{2\pi} \int_0^2 (8r - 2r^3)\, dr\, d\theta$$

$$= \int_0^{2\pi} 8\, d\theta$$

$$= 16\pi$$

17. See Example 5.4.

$$\int_0^8 \int_0^{4-x/2} \int_0^{8-x-2y} dz\, dy\, dx$$

$$= \int_0^8 \int_0^{4-x/2} (8-x-2y)dy\, dx$$

$$= \int_0^8 \left(\frac{1}{4}x^2 - 4x + 16\right)dx$$

$$= \frac{128}{3}$$

19. See Examples 5.4, 6.1 and 6.5.

$$\int_0^{2\pi}\int_0^4\int_r^4 r\, dz\, dr\, d\theta = \int_0^{2\pi}\int_0^4 (4r - r^2)dr\, d\theta$$

$$= \int_0^{2\pi}\frac{32}{3}d\theta$$

$$= \frac{64\pi}{3}$$

21. See Example 7.4.

$$\int_0^{2\pi}\int_0^{\pi/4}\int_0^2 \rho^2\sin\phi\, d\rho\, d\phi\, d\theta$$

$$= \int_0^{2\pi}\int_0^{\pi/4}\frac{8}{3}\sin\phi\, d\phi\, d\theta$$

$$= \int_0^{2\pi}\frac{4}{3}(2-\sqrt{2})d\theta$$

$$= \frac{8\pi}{3}(2-\sqrt{2})$$

23. See Examples 5.4, 6.1 and 6.5.

$$\int_0^{2\pi}\int_0^1\int_0^{6-r^2} r\, dz\, dr\, d\theta = \int_0^{2\pi}\int_0^1 (6r - r^3)dr\, d\theta$$

$$= \int_0^{2\pi}\frac{11}{4}d\theta = \frac{11\pi}{2}$$

25. See Example 1.7.

$$\int_0^4 \int_{\sqrt{y}}^2 f(x,y)dx\, dy$$

27. See Example 3.4.

$$\int_{-\pi/2}^{\pi/2}\int_0^2 2r\cos\theta \cdot r\, dr\, d\theta = \int_{-\pi/2}^{\pi/2}\frac{16}{3}\cos\theta\, d\theta = \frac{32}{3}$$

29. See Example 2.6.

$$m = \int_0^2\int_x^{2x} 2x\, dy\, dx = \int_0^2 2x^2 dx = \frac{16}{3}$$

$$M_y = \int_0^2\int_x^{2x} x\cdot 2x\, dy\, dx = \int_0^2 2x^3 dx = 8$$

$$\bar{x} = \frac{M_y}{m} = \frac{8}{\frac{16}{3}} = \frac{3}{2}$$

$$M_x = \int_0^2\int_x^{2x} y\cdot 2x\, dy\, dx = \int_0^2 3x^3 dx = 12$$

$$\bar{y} = \frac{M_x}{m} = \frac{12}{\frac{16}{3}} = \frac{9}{4}$$

31. See Example 5.6.

$$m = \int_{-1}^1\int_0^{1-x^2}\int_0^{2-z} 2\, dy\, dz\, dx$$

$$= \int_{-1}^1\int_0^{1-x^2}(4-2z)dz\, dx$$

$$= \int_{-1}^1 (-x^4 - 2x^2 + 3)dx$$

$$= \frac{64}{15}$$

$$M_{yz} = \int_{-1}^1\int_0^{1-x^2}\int_0^{2-z} x\cdot 2\, dy\, dz\, dx$$

$$= \int_{-1}^1\int_0^{1-x^2}(4x - 2xz)dz\, dx$$

$$= \int_{-1}^1 (-x^5 - 2x^3 + 3x)dx$$

$$= 0$$

$$\bar{x} = 0$$

$$M_{xz} = \int_{-1}^1\int_0^{1-x^2}\int_0^{2-z} y\cdot 2\, dy\, dz\, dx$$

$$= \int_{-1}^1\int_0^{1-x^2}(z-2)^2 dz\, dx$$

$$= \int_{-1}^1\left(-\frac{1}{3}x^6 - x^4 - x^2 + \frac{7}{3}\right)dx$$

$$= \frac{368}{105}$$

$$\bar{y} = \frac{M_{xz}}{m} = \frac{\frac{368}{105}}{\frac{64}{15}} = \frac{23}{28}$$

$$M_{xy} = \int_{-1}^1\int_0^{1-x^2}\int_0^{2-z} z\cdot 2\, dy\, dz\, dx$$

$$= \int_{-1}^1\int_0^{1-x^2}(4z - 2z^2)dz\, dx$$

$$= \int_{-1}^1\left(\frac{2}{3}x^6 - 2x^2 + \frac{4}{3}\right)dx$$

$$= \frac{32}{21}$$

$$\bar{z} = \frac{M_{xy}}{m} = \frac{\frac{32}{21}}{\frac{64}{15}} = \frac{5}{14}$$

33. See Example 2.1.

$$\int_0^1 \int_{\sqrt{y}}^{2-y} dx\, dy = \int_0^1 \left(2 - y - \sqrt{y}\right) dy = \frac{5}{6}$$

35. See Exercises 2.47–2.54.

$$a = \int_0^1 \int_x^{2x} dy\, dx = \int_0^1 x\, dx = \frac{1}{2}$$

$$\iint_R f(x, y)\, dA = \int_0^1 \int_x^{2x} x^2\, dy\, dx = \int_0^1 x^3\, dx = \frac{1}{4}$$

$$\text{Average value} = \frac{1}{a} \iint_R f(x, y)\, dA = \frac{1}{\frac{1}{2}} \cdot \frac{1}{4} = \frac{1}{2}$$

37. See Example 4.1.

$$\int_0^2 \int_x^2 \sqrt{2^2 + 4^2 + 1}\, dy\, dx = \int_0^2 \int_x^2 \sqrt{21}\, dy\, dx$$
$$= \int_0^2 \sqrt{21}(2 - x)\, dx$$
$$= 2\sqrt{21}$$

39. See Example 4.2.

$$\int_0^{2\sqrt{2}} \int_0^{\sqrt{8-x^2}} \sqrt{y^2 + x^2 + 1}\, dy\, dx$$
$$= \int_0^{\pi/2} \int_0^{\sqrt{8}} \sqrt{r^2 + 1} \cdot r\, dr\, d\theta$$
$$= \int_0^{\pi/2} \frac{26}{3}\, d\theta$$
$$= \frac{13\pi}{3}$$

41. See Example 4.2.

$$S = \iint_R \sqrt{[f_x(x, y)]^2 + [f_y(x, y)]^2 + 1}\, dA$$

$$= \iint_R \sqrt{\left(\frac{x}{\sqrt{x^2 + y^2}}\right)^2 + \left(\frac{y}{\sqrt{x^2 + y^2}}\right)^2 + 1}\, dA$$
$$= \int_0^{2\pi} \int_0^4 \sqrt{2}\, r\, dr\, d\theta = \int_0^{2\pi} \sqrt{2} \cdot \frac{1}{2}[r^2]_{r=0}^{r=4}\, d\theta$$
$$= \int_0^{2\pi} \frac{\sqrt{2}}{2} 16\, d\theta = 8\sqrt{2}[\theta]_0^{2\pi}$$
$$= 16\pi\sqrt{2}$$

43. See Example 5.1.

$$\int_0^2 \int_{-1}^1 \int_{-1}^1 z(x + y)\, dz\, dy\, dx = \int_0^2 \int_{-1}^1 0\, dy\, dx = 0$$

45. See Example 7.3.

$$\int_0^{2\pi} \int_0^{\pi/4} \int_0^2 \rho \cdot \rho^2 \sin\phi\, d\rho\, d\phi\, d\theta$$
$$= \int_0^{2\pi} \int_0^{\pi/4} 4\sin\phi\, d\phi\, d\theta$$
$$= \int_0^{2\pi} (4 - 2\sqrt{2})\, d\theta$$
$$= 4\pi(2 - \sqrt{2})$$

47. See Example 5.1.

$$\int_0^2 \int_x^2 \int_0^{6-x-y} f(x, y, z)\, dz\, dy\, dx$$

49. See Examples 6.4 and 7.3.

$$\int_0^{2\pi} \int_0^2 \int_0^{\sqrt{4-r^2}} f(x,\ y,\ z) \cdot r\, dz\, dr\, d\theta = \int_0^{2\pi} \int_0^{\pi/2} \int_0^2 f(\rho\sin\phi\cos\theta, \rho\sin\phi\sin\theta, \rho\cos\phi) \cdot \rho^2 \sin\phi\, d\rho\, d\phi\, d\theta$$

51. See Examples 6.5 and 6.6.

$$\int_{\pi/4}^{\pi/2} \int_0^{\sqrt{2}} \int_0^r e^z \cdot r\, dz\, dr\, d\theta = \int_{\pi/4}^{\pi/2} \int_0^{\sqrt{2}} r(e^r - 1)\, dr\, d\theta$$
$$= \int_{\pi/4}^{\pi/2} e^{\sqrt{2}}(\sqrt{2} - 1)\, d\theta$$
$$= \frac{\pi}{4} e^{\sqrt{2}}(\sqrt{2} - 1)$$

53. See Example 7.5.

$$\int_0^\pi \int_0^{\pi/4} \int_0^{\sqrt{2}} \rho \cdot \rho^2 \sin\phi \, d\rho \, d\phi \, d\theta$$

$$= \int_0^\pi \int_0^{\pi/4} \sin\phi \, d\phi \, d\theta$$

$$= \int_0^\pi \frac{1}{2}(2 - \sqrt{2}) d\theta$$

$$= \frac{\pi}{2}(2 - \sqrt{2})$$

55. See Sections 13.6 and 13.7

   a. $r\sin\theta = 3$

   b. $\rho \sin\phi \sin\theta = 3$

57. See Sections 13.6 and 13.7.

   a. $(r\cos\theta)^2 + (r\sin\theta)^2 + z^2 = 4$
$$r^2 + z^2 = 4$$

   b. $\rho^2 = 4$
$$\rho = 2$$

59. See Sections 13.6 and 13.7.

   a. $z = \sqrt{(r\cos\theta)^2 + (r\sin\theta)^2}$
$$z = \sqrt{r^2}$$
$$z = r$$

   b.

$$\rho\cos\phi = \sqrt{(\rho\sin\phi\cos\theta)^2 + (\rho\sin\phi\sin\theta)^2}$$
$$\rho\cos\phi = \rho\sin\phi$$
$$\frac{\sin\phi}{\cos\phi} = 1$$
$$\tan\phi = 1$$
$$\phi = \frac{\pi}{4}$$

61. See Section 13.6.

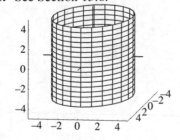

63. See Section 13.6.

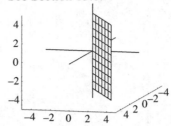

65. See Section 13.6.

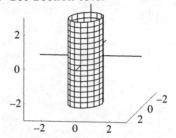

67. See Example 8.1.

$$\left.\begin{array}{r} y - 2x = -1 \\ y - 2x = 1 \end{array}\right\} \; u = y - 2x \\ \left.\begin{array}{r} y + 2x = 2 \\ y + 2x = 4 \end{array}\right\} \; v = y + 2x \quad \rightarrow \quad \left\{\begin{array}{l} x = \dfrac{1}{4}(v - u) \\ y = \dfrac{1}{2}(u + v) \end{array}\right.$$

$$x = \frac{1}{4}(v - u), \; y = \frac{1}{2}(u + v), \; -1 \le u \le 1, \; 2 \le v \le 4$$

69. See Example 8.5.

$$\frac{\partial(x, y)}{\partial(u, v)} = \begin{vmatrix} -\frac{1}{4} & \frac{1}{4} \\ \frac{1}{2} & \frac{1}{2} \end{vmatrix} = -\frac{1}{4}$$

$$\iint_R e^{y-2x} dA = \int_2^4 \int_{-1}^1 e^{\frac{1}{2}(u+v) - 2\left(\frac{1}{4}\right)(v-u)} \left| -\frac{1}{4} \right| du \, dv$$

$$= \frac{1}{4} \int_2^4 \int_{-1}^1 e^u \, du \, dv$$

$$= \frac{1}{4} \int_2^4 (e - e^{-1}) dv$$

$$= \frac{1}{2}(e - e^{-1})$$

71. See Definition 8.1. and Examples 8.4 and 8.5.

$$x = u^2 v, \; y = 4u + v^2$$

$$\frac{\partial(x, y)}{\partial(u, v)} = \begin{vmatrix} 2uv & u^2 \\ 4 & 2v \end{vmatrix} = 4uv^2 - 4u^2$$

Chapter 14

Section 14.1

5.

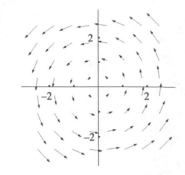

7.

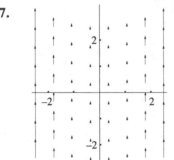

9.

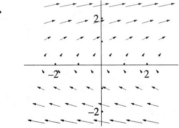

11.

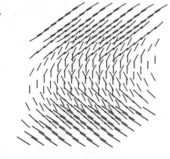

13.

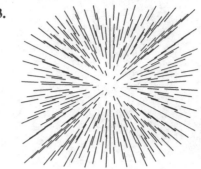

15. $F_1(x, y) = \dfrac{\langle x, y \rangle}{\sqrt{x^2 + y^2}}$: Graph D, because all

vectors point away from the origin and have the same length.

$F_2(x, y) = \langle x, y \rangle$: Graph B, because all vectors point away from the origin and the lengths are proportional to the distances from the origin.

$F_3(x, y) = \langle e^y, x \rangle$: Graph A, because the vectors point upward when $x > 0$ and downward when $x < 0$.

$F_4(x, y) = \langle e^y, y \rangle$: Graph C, because the vectors depend only on y.

17. $f(x, y) = x^2 + y^2$

$\dfrac{\partial f}{\partial x} = 2x$

$\dfrac{\partial f}{\partial y} = 2y$

$\nabla f(x, y) = \langle 2x, 2y \rangle$

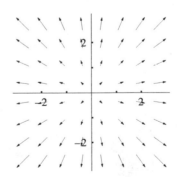

19. $f(x, y) = \sqrt{x^2 + y^2} = (x^2 + y^2)^{1/2}$

$\dfrac{\partial f}{\partial x} = \dfrac{1}{2}(x^2 + y^2)^{-1/2}(2x) = \dfrac{x}{\sqrt{x^2 + y^2}}$

$\dfrac{\partial f}{\partial x} = \dfrac{1}{2}(x^2 + y^2)^{-1/2}(2y) = \dfrac{x}{\sqrt{x^2 + y^2}}$

$\nabla f(x, y) = \dfrac{\langle x, y \rangle}{\sqrt{x^2 + y^2}}$

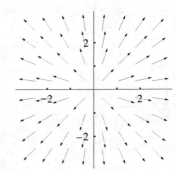

21. $f(x, y) = xe^{-y}$

$\dfrac{\partial f}{\partial x} = e^{-y}$

$\dfrac{\partial f}{\partial y} = -xe^{-y}$

$\nabla f(x, y) = \left\langle e^{-y}, -xe^{-y} \right\rangle$

23. $f(x, y, z) = \sqrt{x^2 + y^2 + z^2}$

$\qquad = (x^2 + y^2 + z^2)^{1/2}$

$\dfrac{\partial f}{\partial x} = \dfrac{1}{2}(x^2 + y^2 + z^2)^{-1/2}(2x)$

$\qquad = \dfrac{x}{\sqrt{x^2 + y^2 + z^2}}$

$\dfrac{\partial f}{\partial y} = \dfrac{1}{2}(x^2 + y^2 + z^2)^{-1/2}(2y)$

$\qquad = \dfrac{y}{\sqrt{x^2 + y^2 + z^2}}$

$\dfrac{\partial f}{\partial z} = \dfrac{1}{2}(x^2 + y^2 + z^2)^{-1/2}(2z)$

$\qquad = \dfrac{z}{\sqrt{x^2 + y^2 + z^2}}$

$\nabla f(x, y, z) = \dfrac{\langle x, y, z \rangle}{\sqrt{x^2 + y^2 + z^2}}$

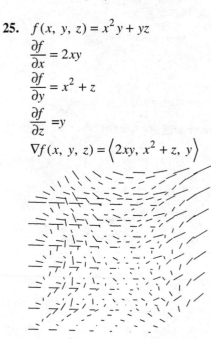

25. $f(x, y, z) = x^2 y + yz$

$\dfrac{\partial f}{\partial x} = 2xy$

$\dfrac{\partial f}{\partial y} = x^2 + z$

$\dfrac{\partial f}{\partial z} = y$

$\nabla f(x, y, z) = \left\langle 2xy, x^2 + z, y \right\rangle$

27. If $\nabla f(x, y) = \langle y, x \rangle$, then $\dfrac{\partial f}{\partial x} = y$ and $\dfrac{\partial f}{\partial y} = x$.

$f(x, y) = \displaystyle\int y \, dx = xy + g(y)$

$\dfrac{\partial f}{\partial y} = x + g'(y) = x$

Therefore, $g'(y) = 0$ and $g(y) = c$.

The vector field is conservative: $f(x, y) = xy + c$.

29. If $\nabla f(x, y) = \langle y, -x \rangle$, then $\dfrac{\partial f}{\partial x} = y$

and $\dfrac{\partial f}{\partial y} = -x$.

$f(x, y) = \displaystyle\int y \, dx = xy + g(y)$

$\dfrac{\partial f}{\partial y} = x + g'(y) = -x$, so $g'(y) = -2x$.

Since this is not possible, the vector field is not conservative.

31. If $\nabla f(x, y) = \left\langle x - 2xy, y^2 - x^2 \right\rangle$, then

$\dfrac{\partial f}{\partial x} = x - 2xy$ and $\dfrac{\partial f}{\partial y} = y^2 - x^2$.

$f(x, y) = \displaystyle\int (x - 2xy) dx = \dfrac{1}{2}x^2 - x^2 y + g'(y)$

$\dfrac{\partial f}{\partial y} = -x^2 + g'(y)$

$\qquad = y^2 - x^2$, so $g'(y)$

$\qquad = y^2$

$g(y) = \displaystyle\int y^2 dy = \dfrac{1}{3}y^3 + c$

The vector field is conservative;

$f(x, y) = \dfrac{1}{2}x^2 - x^2 y + \dfrac{1}{3}y^3 + c$.

33. If $\nabla f(x, y) = \langle y \sin xy, x \sin xy \rangle$, then

$\dfrac{\partial f}{\partial x} = y \sin xy$ and $\dfrac{\partial f}{\partial y} = x \sin xy$.

$f(x, y) = \displaystyle\int y \sin xy \, dx = -\cos xy + g(y)$

$\dfrac{\partial f}{\partial y} = x \sin xy + g'(y) = x \sin xy$

Therefore, $g'(y) = 0$ and $g(y) = c$.

The vector field is conservative;

$f(x, y) = -\cos xy + c$.

35. If $\nabla f(x, y, z) = \langle 4x - z, 3y + z, y - x \rangle$, then

$\dfrac{\partial f}{\partial x} = 4x - z, \dfrac{\partial f}{\partial y} = 3y + z,$

and $\dfrac{\partial f}{\partial z} = y - x$.

$f(x, y, z) = \displaystyle\int (4x - z) dx = 2x^2 - xz + g(y, z)$

$\dfrac{\partial f}{\partial y} = 0 - 0 + \dfrac{\partial g}{\partial y} = 3y + z$

$g(y, z) = \displaystyle\int (3y + z) dy = \dfrac{3}{2}y^2 + yz + h(z)$

$f(x, y, z) = 2x^2 - xz + \dfrac{3}{2}y^2 + yz + h(z)$

$\dfrac{\partial f}{\partial z} = -x + y + h'(z) = y - x,$

so $h'(z) = 0$ and $h(z) = c$.

The vector field is conservative:

$f(x, y, z) = 2x^2 - xz + \dfrac{3}{2}y^2 + yz + c$.

37. If $\nabla f(x, y, z) = \left\langle y^2 z^2 - 1, 2xyz^2, 4z^3 \right\rangle$ then

$\dfrac{\partial f}{\partial x} = y^2 z^2 - 1, \dfrac{\partial f}{\partial y} = 2xyz^2,$ and $\dfrac{\partial f}{\partial z} = 4z^3$.

$f(x, y, z) = \displaystyle\int (y^2 z^2 - 1) dx = xy^2 z^2 - x + g(y, z)$

$\dfrac{\partial f}{\partial y} = 2xyz^2 + \dfrac{\partial g}{\partial y} = 2xyz^2,$

so $\dfrac{\partial g}{\partial y} = 0$ and $g(y, z) = h(z)$.

$f(x, y, z) = xy^2 z^2 - x + h(z)$

$\dfrac{\partial f}{\partial z} = 2xy^2 z + h'(z) = 4z^3,$

so $h'(z) = 4z^3 - 2xy^2 z$.

Since this is not possible, the vector field is not conservative.

39. $\dfrac{dy}{dx} = \dfrac{\cos x}{2}$

$\displaystyle\int dy = \int \dfrac{1}{2}\cos x \, dx$

$y = \dfrac{1}{2}\sin x + c$

41. $\dfrac{dy}{dx} = \dfrac{3x^2}{2y}$

$2y\dfrac{dy}{dx} = 3x^2$

$\displaystyle\int 2y \, dy = \int 3x^2 dx$

$y^2 = x^3 + c$

43. $\quad \dfrac{dy}{dx} = \dfrac{xe^y}{y}$

$$ye^{-y}\frac{dy}{dx} = x$$

$$\int ye^{-y}dy = \int x\,dx$$

Integrate by parts: $\quad u = y \qquad dv = e^{-y}dy$

$$\qquad\qquad\qquad\qquad du = dy \qquad v = -e^{-y}$$

$$\int ye^{-y}dy = -ye^{-y} - \int(-e^{-y})dy$$

$$= -ye^{-y} - e^{-y} + c$$

$$-ye^{-y} - e^{-y} + c = \frac{1}{2}x^2 + c$$

$$(y+1)e^{-y} = -\frac{1}{2}x^2 + c$$

45. $\quad \dfrac{dy}{dx} = \dfrac{y^2+1}{y}$

$$\frac{y}{y^2+1}\frac{dy}{dx} = 1$$

$$\int \frac{y}{y^2+1}dy = \int dx$$

$$\frac{1}{2}\ln(y^2+1) = x + c$$

$$\ln(y^2+1) = 2x + 2c$$

$$y^2+1 = e^{2x+2c}$$

Since e^{2c} is an arbitrary positive constant, we may write $y^2+1 = ce^{2x}\ (c>0)$.

47. $\quad f(x,y,z) = \displaystyle\int_0^x f(u)du + \int_0^y g(u)du$

$$\qquad\qquad\qquad + \int_0^z h(u)du + c$$

49. Using the result of exercise 19,

$$\nabla r = \frac{\langle x,y\rangle}{\sqrt{x^2+y^2}} = \frac{\mathbf{r}}{r}.$$

51. $\quad f(x,y) = r^3 = (x^2+y^2)^{3/2}$

$$\frac{\partial f}{\partial x} = \frac{3}{2}(x^2+y^2)^{1/2}(2x) = 3x\sqrt{x^2+y^2}$$

$$\frac{\partial f}{\partial y} = \frac{3}{2}(x^2+y^2)^{1/2}(2y) = 3y\sqrt{x^2+y^2}$$

$$\nabla f(r^3) = \left\langle 3x\sqrt{x^2+y^2}, 3y\sqrt{x^2+y^2}\right\rangle$$

$$= 3\langle x,y\rangle\sqrt{x^2+y^2}$$

$$= 3r\mathbf{r}$$

53. If $\nabla f(x,y) = \dfrac{\langle 1,1\rangle}{r} = \dfrac{\langle 1,1\rangle}{\sqrt{x^2+y^2}}$, then

$$\frac{\partial f}{\partial x} = \frac{1}{\sqrt{x^2+y^2}} \text{ and } \frac{\partial f}{\partial y} = \frac{1}{\sqrt{x^2+y^2}}.$$

$$f(x,y) = \int \frac{1}{\sqrt{x^2+y^2}}dx = \ln\left|x+\sqrt{x^2+y^2}\right| + g(y)$$

$$\frac{\partial f}{\partial y} = \left(\frac{1}{x+\sqrt{x^2+y^2}}\right)\left(\frac{y}{\sqrt{x^2+y^2}}\right) + g'(y)$$

$$= \frac{1}{\sqrt{x^2+y^2}}$$

$$g'(y) = \frac{1}{\sqrt{x^2+y^2}} - \left(\frac{1}{x+\sqrt{x^2+y^2}}\right)\left(\frac{y}{\sqrt{x^2+y^2}}\right)$$

$$= \frac{x+\sqrt{x^2+y^2}-y}{\left(\sqrt{x^2+y^2}\right)\left(x+\sqrt{x^2+y^2}\right)}$$

This is not possible since $g'(y)$ cannot depend on x. Therefore, $\dfrac{\langle 1,1\rangle}{r}$ is not conservative.

55. The magnetic field is shown. (The wire is perpendicular to this page.)

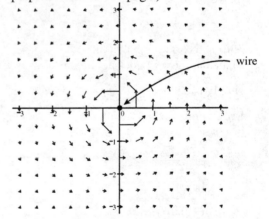

Section 14.2

5. $x = 1 + 2t, \ y = 2 + 3t; \ 0 \le t \le 1$

$$ds = \sqrt{[x'(t)]^2 + [y'(t)]^2} \, dt$$
$$= \sqrt{2^2 + 3^2} \, dt$$
$$= \sqrt{13} \, dt$$

$$\int_C 2x \, ds = \int_0^1 2(1 + 2t)\sqrt{13} \, dt$$
$$= 2\sqrt{13}(t + t^2)\Big|_0^1$$
$$= 4\sqrt{13}$$

7. $x = 5 - 4t, \ y = 2 - t; \ 0 \le t \le 1$

$$ds = \sqrt{[x'(t)]^2 + [y'(t)]^2} \, dt$$
$$= \sqrt{(-4)^2 + (-1)^2} \, dt$$
$$= \sqrt{17} \, dt$$

$$\int_C (3x + y) ds = \int_0^1 [3(5 - 4t) + (2 - t)]\sqrt{17} \, dt$$
$$= \sqrt{17} \int_0^1 (17 - 13t) dt$$
$$= \sqrt{17}\left(17t - \frac{13}{2}t^2\right)\Big|_0^1$$
$$= \frac{21}{2}\sqrt{17}$$

9. $x = 2t, \ y = 2 + 4t; \ 0 \le t \le 1$
$$dx = 2 \, dt$$

$$\int_C 2x \, dx = \int_C 2(2t)2 \, dt = 4t^2\Big|_0^1 = 4$$

Alternative solution method:
Since the integral does not depend on y, we may simply calculate $\int_0^2 2x \, dx = x^2\Big|_0^2 = 4$.
See Exercise 72.

11. $x = 2\cos t, \ y = 2\sin t; \ 0 \le t \le \dfrac{\pi}{2}$

$$ds = \sqrt{[x'(t)]^2 + [y'(t)]^2} \, dt$$
$$= \sqrt{(-2\sin t)^2 + (2\cos t)^2} \, dt$$
$$= 2 \, dt$$

$$\int_C 3x \, ds = \int_0^{\pi/2} 3(2\cos t)2 \, dt$$
$$= (12\sin t)\Big|_0^{\pi/2}$$
$$= 12$$

13. $x = 2\cos t, \ y = 2\sin t; \ 0 \le t \le \dfrac{\pi}{2}$

$$dx = -2\sin t \, dt$$

$$\int_C 2x \, dx = \int_0^{\pi/2} 2(2\cos t)(-2\sin t) dt$$
$$= 4\cos^2 t\Big|_0^{\pi/2}$$
$$= -4$$

Alternative solution method:
Since the integral does not depend on y, we may simply calculate $\int_2^0 2x \, dx = x^2\Big|_2^0 = -4$.
See Exercise 72.

15. $x = 2\sin t, \ y = \cos t; \ 0 \le t \le \pi$

$$dx = 2\cos t \, dt$$

$$\int_C 3y \, dx = \int_0^\pi (3\cos t)(2\cos t) dt$$
$$= \int_0^\pi 6\cos^2 t \, dt$$
$$= \int_0^\pi 3(1 + \cos 2t) dt$$
$$= \left(3t + \frac{3}{2}\sin 2t\right)\Big|_0^\pi$$
$$= 3\pi$$

17. $x = t, \ y = t^2; \ 0 \le t \le 2$

$$ds = \sqrt{[x'(t)]^2 + [y'(t)]^2} \, dt$$
$$= \sqrt{1^2 + (2t)^2} \, dt$$
$$= \sqrt{4t^2 + 1} \, dt$$

$$\int_C 3y \, ds = \int_0^2 3t^2 \sqrt{4t^2 + 1} \, dt$$
$$\approx 25.41$$

We approximated the definite integral numerically.

19. $x = -t, \ y = t^2, \ -2 \le t \le 0$

$$dx = -dt$$

$$\int_C 2x \, dx = \int_{-2}^0 2(-t)(-dt)$$
$$= t^2\Big|_{-2}^0$$
$$= -4$$

Alternative solution method:
Since the integral does not depend on y, we may simply calculate $\int_2^0 2x \, dx = x^2\Big|_2^0 = -4$.
See Exercise 72.

21. $x = t^2, y = t; 1 \le t \le 2$

$dx = 2t\ dt$

$\int_C 3y\ dx = \int_1^2 3t(2t)dt$

$\qquad = 2t^3 \Big|_1^2$

$\qquad = 14$

23. $C_1 : x = t, y = 0; 0 \le t \le 1$

$ds = \sqrt{[x'(t)]^2 + [y'(t)]^2}\ dt$

$\quad = \sqrt{1^2 + 0^2}\ dt$

$\quad = dt$

$\int_{C_1} 3x\ ds = \int_0^1 3t\ dt = \frac{3}{2}t^2 \Big|_0^1 = \frac{3}{2}$

$C_2 : x = \cos t, y = \sin t; 0 \le t \le \frac{\pi}{2}$

$ds = \sqrt{[x'(t)]^2 + [y'(t)]^2}\ dt$

$\quad = \sqrt{(-\sin t)^2 + (\cos t)^2}\ dt$

$\quad = dt$

$\int_{C_2} 3x\ ds = \int_0^{\pi/2} 3\cos t\ dt$

$\qquad = 3\sin t \Big|_0^{\pi/2}$

$\qquad = 3$

Total: $\int_C 3x\ ds = \frac{3}{2} + 3 = \frac{9}{2}$

25. $x = 1 + t, y = -2t, z = 1 + t; 0 \le t \le 1$

$ds = \sqrt{[x'(t)]^2 + [y'(t)]^2 + [z'(t)]^2}\ dt$

$\quad = \sqrt{1^2 + 2^2 + 1^2}\ dt$

$\quad = \sqrt{6}dt$

$\int_C 4z\ ds = \int_0^1 4(1+t)\sqrt{6}dt$

$\qquad = \sqrt{6}(4t + 2t^2)\Big|_0^1$

$\qquad = 6\sqrt{6}$

27. $x = t, y = t^2, z = 2; 1 \le t \le 2$

$\int_C 4(x-z)z\,dx = \int_1^2 4(t-2)(2)dt$

$\qquad = \int_1^2 (8t - 16)dt$

$\qquad = (4t^2 - 16t)\Big|_1^2$

$\qquad = -4$

29. $x = 3 + 2t, y = 1 + 3t; 0 \le t \le 1$

$\int_C \mathbf{F} \cdot d\mathbf{r} = \int_C 2x\ dx + 2y\ dy$

$\qquad = \int_0^1 [2(3 + 2t)(2) + 2(1 + 3t)(3)]dt$

$\qquad = \int_0^1 (26t + 18)dt$

$\qquad = (13t^2 + 18t)\Big|_0^1$

$\qquad = 31$

31. $x = 4\cos t, y = 4\sin t; 0 \le t \le \frac{\pi}{2}$

$\int_C \mathbf{F} \cdot d\mathbf{r}$

$= \int_C 2x\ dx + 2y\ dy$

$= \int_0^{\pi/2} [2(4\cos t)(-4\sin t) + 2(4\sin t)(4\cos t)]dt$

$= \int_0^{\pi/2} 0\ dt$

$= 0$

33. $x = t, y = t^2; 0 \le t \le 1$

$\int_C \mathbf{F} \cdot d\mathbf{r} = \int_C 2\ dx + x\ dy$

$\qquad = \int_0^1 [2(1) + t(2t)]dt$

$\qquad = \int_0^1 (2t^2 + 2)dt$

$\qquad = \left(\frac{2}{3}t^3 + 2t\right)\Big|_0^1$

$\qquad = \frac{8}{3}$

35. $C_1 : x - 0, y = t; 0 \le t \le 1$

$\int_{C_1} \mathbf{F} \cdot d\mathbf{r} = \int_{C_1} 3x\ dx + 2\ dy$

$\qquad = \int_0^1 (0 + 2)dt$

$\qquad = 2$

$C_2 : x = 4t, y = 1; 0 \le t \le 1$

$\int_{C_2} \mathbf{F} \cdot d\mathbf{r} = \int_{C_1} 3x\ dx + 2\ dy$

$\qquad = \int_0^1 [3(4t)(4) + 2(0)]dt$

$\qquad = 24t^2 \Big|_0^1$

$\qquad = 24$

Total: $\int_C \mathbf{F} \cdot d\mathbf{r} = 2 + 24 = 26$

441

37. $\int_C \mathbf{F} \cdot d\mathbf{r} = \int_C y\, dx + 0\, dy + z\, dz$

Note that $\int_C z\, dz = 0$, because the ending point is the same as the starting point, and obviously $\int_C 0\, dy = 0$. So, we need only find $\int_C y\, dx$. This term does not depend on z, so the portion from (0, 0, 0) to (2, 1, 2) will cancel with the portion from (2, 1, 0) to (0, 0, 0). Furthermore, is 0 or the portion from (2, 1, 2) to (2, 1, 0) because $dx = 0$. So, the total work done is 0.

39. $x = \cos t, \ y = \sin t, \ z = 2t; \ 0 \le t \le \dfrac{\pi}{2}$

$\int_C \mathbf{F} \cdot d\mathbf{r} = \int_C xy\, dx + 3z\, dy + 1\, dz$

$= \int_0^{\pi/2} \left[(\cos t)(\sin t)(-\sin t) + 3(2t)(\cos t) + 2 \right] dt$

$= \int_0^{\pi/2} \left(-\sin^2 t \cos t + 6t \cos t + 2 \right) dt$

$= \left(-\dfrac{1}{3}\sin^3 t - 6\sin t + 6t\sin t + 2t \right) \Big|_0^{\pi/2}$

$= \left(-\dfrac{1}{3} - 6 + 3\pi + \pi \right) - (0)$

$= 4\pi - \dfrac{19}{3}$

Note: We used integration by parts to find $\int 6t \cos t\, dt$.

41. The motion is in the same direction as the force field, so the work done by the force field is positive.

43. The motion is perpendicular to the direction of the force field, so the work done by the force field is zero.

45. Most of the motion is against the force field, so the work done by the force field is negative.

47. $x = t, \ y = t^2; \ 0 \le t \le 3$

$ds = \sqrt{[x'(t)]^2 + [y'(t)]^2}\, dt$

$= \sqrt{1^2 + (2t)^2}\, dt$

$= \sqrt{4t^2 + 1}\, dt$

$m = \int_C \rho\, ds = \int_C x\, ds$

$= \int_0^3 t\sqrt{4t^2 + 1}\, dt$

$= \dfrac{1}{12}(4t^2 + 1)^{3/2} \Big|_0^3$

$= \dfrac{1}{12}(37)^{3/2} - \dfrac{1}{12}$

≈ 18.67

49. $\bar{x} = \dfrac{1}{m}\int_C x\rho\, ds$

$= \dfrac{1}{m}\int_0^3 (t)(t)\sqrt{4t^2 + 1}\, dt$

$\approx \dfrac{1}{18.67}(41.59)$

≈ 2.227

$\bar{y} = \dfrac{1}{m}\int_C y\rho\, ds$

$= \dfrac{1}{m}\int_0^3 (t^2)(t)\sqrt{4t^2 + 1}\, dt$

$\approx \dfrac{1}{18.67}(99.41)$

≈ 5.324

51. $I = \int_C w^2 \rho\, ds$

$= \int_C x^2 x\, ds$

$= \int_0^3 t^3 \sqrt{4t^2 + 1}\, dt$

≈ 99.41

53. $I = \int_C w^2 \rho\, ds$

$= \int_C (9 - y)^2 x\, ds$

$= \int_0^3 (9 - t^2)^2 t\sqrt{4t^2 + 1}\, dt$

≈ 359.9

55. $x = \cos 2t, \ y = \sin 2t, \ z = t; \ 0 \le t \le \pi$

$ds = \sqrt{[x'(t)]^2 + [y'(t)]^2 + [z'(t)]^2}\, dt$

$= \sqrt{(-2\sin 2t)^2 + (2\cos 2t)^2 + (1)^2}\, dt$

$= \sqrt{5}\, dt$

$m = \int_C \rho\, ds$

$= \int_0^\pi t^2 \sqrt{5}\, dt$

$= \sqrt{5}\, \dfrac{t^3}{3} \Big|_0^\pi = \dfrac{\pi^3}{3}\sqrt{5}$

57. For Exercise 49:
The center of mass is (2.227, 5.324). Since $5.324 \neq 2.227^2$, this point is not on the rod.
For Exercise 50:
The center of mass is (.9478, 2.855).
Since $2.855 \neq 4 - .9478^2$, this point is not on the rod. You cannot balance an object on a point which is not part of the object.

59. $x = 2\cos t,\ y = 2\sin t;\ 0 \leq t \leq \dfrac{\pi}{2}$

$$ds = \sqrt{[x'(t)]^2 + [y'(t)]^2}\,dt$$
$$= \sqrt{(-2\sin t)^2 + (2\cos t)^2}\,dt$$
$$= 2\,dt$$

Surface area $= \displaystyle\int_C (\text{height})ds$
$$= \int_C (x^2 + y^2)\,ds$$
$$= \int_0^{\pi/2} [(2\cos t)^2 + (2\sin t)^2](2)\,dt$$
$$= \int_0^{\pi/2} 8\,dt = 4\pi$$

Alternative solution:
On $x^2 + y^2 = 4$, the height is $z = x^2 + y^2 = 4$.
Surface area $=$ (length of quarter circel)(height)
$$= \frac{1}{4}(\pi \cdot 2^2)(4) = 4\pi$$

61. $x = 2 - 4t,\ y = 0;\ 0 \leq t \leq 1$

$$ds = \sqrt{[x'(t)]^2 + [y'(t)]^2}\,dt$$
$$= \sqrt{(-4)^2 + 0^2}\,dt$$
$$= 4\,dt$$

Surface area $= \displaystyle\int_C (\text{height})ds$
$$= \int_C (4 - x^2 - y^2)\,ds$$
$$= \int_0^1 [4 - (2 - 4t)^2 - 0]4\,dt$$
$$= \int_0^1 (16t - 16t^2)4\,dt$$
$$= \left(32t^2 - \frac{64}{3}t^3 \right)\Big|_0^1$$
$$= \frac{32}{3}$$

63. Here is a simple geometric solution.
Height at (0, 0): $4 - 0 - 0 = 4$
Height at (0, 1): $4 - 0 - 1 = 3$
Height at (1, 1): $4 - 1 - 1 = 2$
Height at (1, 0): $4 - 1 - 0 = 3$
Clearly the average height is 3.
Surface area
$=$ (average height)(perimieter of square)
$= 3 \cdot 4 = 12$

65. The line integrals can be estimated using a modified trapezoid rule. Note that $\Delta s = \sqrt{(\Delta x)^2 + (\Delta y)^2}$, and $f\!*$ is the average of the f-values at the endpoints. This method is likely to underestimate the true value slightly, since we have assumed that the curve is piecewise linear and the actual curve is probably somewhat longer.

	dx	dy	ds	f*	f*dx	f*dy	f*ds
(0, 0) to (1, 0)	1	0	1	2.5	2.5	0	2.5
(1, 0) to (1, 1)	0	1	1	3.3	0	3.3	3.3
(1, 1) to (1.5, 1.5)	.5	.5	$\sqrt{.5}$	4	2	2	2.83
(1.5, 1.5) to (2, 2)	.5	.5	$\sqrt{.5}$	4.7	2.35	2.35	3.32
(2, 2) to (3, 2)	1	0	1	4.5	4.5	0	4.5
(3, 2) to (4, 1)	1	−1	$\sqrt{2}$	4	4	−4	5.66
				Totals →	15.35	3.65	22.11

a. $\displaystyle\int_C f\,ds \approx 22.11$ **b.** $\displaystyle\int_C f\,dx \approx 15.35$ **c.** $\displaystyle\int_C f\,dy \approx 3.65$

Section 14.3

5. $M_y = \dfrac{\partial}{\partial y}(2xy - 1) = 2x$

$N_x = \dfrac{\partial}{\partial x} x^2 = 2x$

Since $M_y = N_x$, the vector field is

conservative. If $\nabla f(x, y) = \mathbf{F}$, then

$\dfrac{\partial f}{\partial x} = 2xy - 1$ and $\dfrac{\partial f}{\partial y} = x^2.$

$f(x, y) = \int (2xy - 1)dx = x^2 y - x + g(y)$

$\dfrac{\partial f}{\partial y} = x^2 + g'(y) = x^2$

$g'(y) = 0$

$g(y) = c$

potential function: $f(x, y) = x^2 y - x + c$

7. $M_y = \dfrac{\partial}{\partial y}\left(\dfrac{1}{y} - 2x\right) = -\dfrac{1}{y^2}$

$N_x = \dfrac{\partial}{\partial x}\left(y - \dfrac{x}{y^2}\right) = -\dfrac{1}{y^2}$

Since $M_y = N_x$, the vector field is

conservative. If $\nabla f(x, y) = \mathbf{F}$, then

$\dfrac{\partial f}{\partial x} = \dfrac{1}{y} - 2x$ and $\dfrac{\partial f}{\partial y} = y - \dfrac{x}{y^2}.$

$f(x, y) = \int\left(\dfrac{1}{y} - 2x\right)dx = \dfrac{x}{y} - x^2 + g(y)$

$\dfrac{\partial f}{\partial y} = -\dfrac{x}{y^2} + g'(y) = y - \dfrac{x}{y^2}$

$g'(y) = y$

$g(y) = \int y\,dy = \dfrac{1}{2}y^2 + c$

potential function: $f(x, y) = \dfrac{x}{y} - x^2 + \dfrac{1}{2}y^2 + c$

9. $M_y = \dfrac{\partial}{\partial y}(e^{xy} - 1) = xe^{xy}$

$N_x = \dfrac{\partial}{\partial x}(xe^{xy}) = xye^{xy} - e^{xy}$

Since $M_y \neq N_x$, the vector field is not

conservative.

11. $M_y = \dfrac{\partial}{\partial y}(ye^{xy}) = xye^{xy} + e^{xy}$

$N_x = \dfrac{\partial}{\partial x}(xe^{xy} + \cos y) = xye^{xy} + e^{xy} + 0$

Since $M_y = N_x$, the vector field is

conservative. If $\nabla f(x, y) = \mathbf{F}$, then

$\dfrac{\partial f}{\partial x} = ye^{xy}$ and $\dfrac{\partial f}{\partial y} = xe^{xy} + \cos y.$

$f(x, y) = \int ye^{xy}dx = e^{xy} + g(y)$

$\dfrac{\partial f}{\partial y} = xe^{xy} + g'(y) = xe^{xy} + \cos y$

$g'(y) = \cos y$

$g(y) = \int \cos y\,dy = \sin y + c$

potential function: $f(x, y) = e^{xy} + \sin y + c$

13. If $\nabla f(x, y, z) = \mathbf{F}$, then $\dfrac{\partial f}{\partial x} = z^2 + 2xy,$

$\dfrac{\partial f}{\partial y} = x^2 + 1,$ and $\dfrac{\partial f}{\partial z} = 2xz - 3.$

$f(x, y, z) = \int (z^2 + 2xy)dx = xz^2 + x^2 y + g(y, z)$

$\dfrac{\partial f}{\partial y} = x^2 + \dfrac{\partial}{\partial y}g(y, z) = x^2 + 1$

$\dfrac{\partial}{\partial y}g(y, z) = 1$

$g(y, z) = \int 1\,dy = y + h(z)$

$f(x, y, z) = xz^2 + x^2 y + y + h(z)$

$\dfrac{\partial f}{\partial z} = 2xz + h'(z) = 2xz - 3$

$h'(z) = -3$

$h(z) = \int (-3)dz = -3z + c$

The vector field is conservative.
potential function:

$f(x, y, z) = xz^2 + x^2 y + y - 3z + c$

15. If $\nabla f(x, y, z) = \mathbf{F}$, then $\dfrac{\partial f}{\partial x} = y^2 z^2 + x$,

$\dfrac{\partial f}{\partial y} = y + 2xyz^2$, and $\dfrac{\partial f}{\partial z} = 2xy^2 z$.

$f(x, y, z) = \displaystyle\int (yz^2 + x)dx = xy^2 z^2 + \dfrac{1}{2}x^2 + g(y, z)$

$\dfrac{\partial f}{\partial y} = 2xyz^2 + \dfrac{\partial}{\partial y} g(y, z) = y + 2xyz^2$

$\dfrac{\partial}{\partial y} g(y, z) = y$

$g(y, z) = \displaystyle\int y\, dy = \dfrac{1}{2}y^2 + h(z)$

$f(x, y, z) = xy^2 z^2 + \dfrac{1}{2}x^2 + \dfrac{1}{2}y^2 + h(z)$

$\dfrac{\partial f}{\partial z} = 2xy^2 z + h'(z) = 2xy^2 z$

$h'(z) = 0$

$h(z) = c$

The vector field is conservative.
potential function:

$f(x, y, z) = xy^2 z^2 + \dfrac{1}{2}x^2 + \dfrac{1}{2}y^2 + c$

17. The integral is $\displaystyle\int_C \mathbf{F} \cdot d\mathbf{r}$ where

$\mathbf{F}(x, y) = \left\langle 2xy, x^2 - 1 \right\rangle$.

A potential function is $f(x, y) = x^2 y - y$, so
the line integral is independent of path.

$\displaystyle\int_C \mathbf{F} \cdot d\mathbf{r} = f(x, y)\Big|_{(1,0)}^{(3,1)}$

$= [(3)^2(1) - 1] - [(1)^2(0) - 0]$

$= 8$

19. The integral is $\displaystyle\int_C \mathbf{F} \cdot d\mathbf{r}$ where

$\mathbf{F}(x, y) = \left\langle ye^{xy}, xe^{xy} - 2y \right\rangle$.

A potential function is $f(x, y) = e^{xy} - y^2$, so
the line integral is independent of path.

$\displaystyle\int \mathbf{F} \cdot d\mathbf{r} = f(x, y)\Big|_{(1,0)}^{(0,4)}$

$= (e^{0 \cdot 4} - 4^2) - (e^{1 \cdot 0} - 0^2)$

$= -16$

21. The integral is $\displaystyle\int_C \mathbf{F} \cdot d\mathbf{r}$ where

$\mathbf{F}(x, y, z) = \left\langle z^2 + 2xy, x^2, 2xz \right\rangle$.

A potential function is $f(x, y, z) = xz^2 + x^2 y$,
so the line integral is independent of path.

$\displaystyle\int_C \mathbf{F} \cdot d\mathbf{r}$

$= f(x, y, z)\Big|_{(2,1,3)}^{(4,-1,0)}$

$= [(4)(0)^2 + (4)^2(-1)] - [(2)(3)^2 + (2)^2(1)]$

$= -38$

23. A potential function is

$f(x, y) = \dfrac{1}{3}x^3 + x + \dfrac{1}{4}y^4 - \dfrac{3}{2}y^2 + 2y$, so the

line integral is independent of path.

$\displaystyle\int_C \mathbf{F} \cdot d\mathbf{r} = f(x, y)\Big|_{(-4,0)}^{(4,0)}$

$= \left[\dfrac{1}{3}(4)^3 + 4 + 0\right] - \left[\dfrac{1}{3}(-4)^3 - 4 + 0\right]$

$= \dfrac{152}{3}$

25. A potential function is

$f(x, y) = \dfrac{1}{3}(x^3 + y^3 + z^3)$, so the line integral

is independent of path.

$\displaystyle\int_C \mathbf{F} \cdot d\mathbf{r} = f(x, y, z)\Big|_{(1,4,-3)}^{(1,4,3)}$

$= \dfrac{1}{3}(1^3 + 4^3 + 3^3) - \dfrac{1}{3}[1^3 + 4^3 + (-3)^3]$

$= 18$

27. A potential function is

$f(x, y, z) = \sqrt{x^2 + y^2 + z^2}$, so the line integral

is independent of path.

$\displaystyle\int_C \mathbf{F} \cdot d\mathbf{r} = f(x, y, z)\Big|_{(1,3,2)}^{(2,1,5)}$

$= \sqrt{2^2 + 1^2 + 5^2} - \sqrt{1^2 + 3^2 + 2^2}$

$= \sqrt{30} - \sqrt{14}$

29. $M_y(x, y) = \dfrac{\partial}{\partial y}(3x^2 y + 1) = 3x^2$

$N_x(x, y) = \dfrac{\partial}{\partial x}(3xy^2) = 3y^2$

Since $M_y \neq N_x$, the vector field is not conservative and the line integral is not independent of path.

$x = \cos t, y = -\sin t; 0 \le t \le \pi$

$\displaystyle\int_C \mathbf{F} \cdot d\mathbf{r} = \int_C (3x^2 y + 1)dx + 3xy^2 dy$

$\displaystyle = \int_0^\pi [(-3\cos^2 t \sin t + 1)(-\sin t) \\ \qquad + (3\cos t \sin^2 t)(-\cos t)]dt$

$\displaystyle = \int_0^\pi (-\sin t)dt$

$= \cos t \big|_0^\pi$

$= -2$

31. A potential function is $f(x, y) = e^{xy^2} - xy - y$, so the line integral is independent of path.

$\displaystyle\int_C \mathbf{F} \cdot d\mathbf{r} = f(x, y)\big|_{(2,3)}^{(3,0)}$

$\displaystyle = [e^{3 \cdot 0^2} - 0] - [e^{2 \cdot 3^2} - (2)(3) - 3]$

$= 10 - e^{18}$

33. A potential function is

$f(x, y) = \dfrac{x}{y} - \dfrac{1}{2}e^{2x} + y^2$, so the line integral is independent of path. Since C is a closed curve, $\displaystyle\int_C \mathbf{F} \cdot d\mathbf{r} = 0$.

35. Since $\mathbf{F}$ is constant, $M_y = N_x = 0$.

The vector field is conservative.

37. Consider $\displaystyle\int_C \mathbf{F} \cdot d\mathbf{r}$ where C is the unit circle, $x^2 + y^2 = 1$. This is a closed curve, but the line integral is clearly non zero. The vector field is not conservative.

39. The vector field appears to be $\mathbf{F}(x, y) = \langle 1, x \rangle$, so $M_y = 0$ and $N_x = 1$. Since $M_y \neq N_x$, the vector field is not conservative.

41. Possible answer:

$C_1 : x = -2 + t, y = 0; 0 \le t \le 4$

$\displaystyle\int_{C_1} y\,dx - x\,dy = \int_0^4 [0(1) - (-2 + t)(0)]dt = 0$

$C_2 : x = -\cos t, y = \sin t; 0 \le t \le \pi$

$\displaystyle\int_{C_2} y\,dx - x\,dy$

$\displaystyle = \int_0^\pi [(\sin t)(\sin t) - (-\cos t)(\cos t)]dt$

$\displaystyle = \int_0^\pi 1\,dt$

$= \pi$

The two paths from $(-2, 0)$ to $(2, 0)$ give different values, so the line integral is not independent of path.

43. Possible answer:

$C_1 : x = -2 + t, y = 2 - t; 0 \le t \le 2$

$\displaystyle\int_{C_1} y\,dx - 3\,dy = \int_0^2 [(2 - t)(1) - 3(-1)]dt$

$\displaystyle = \int_0^2 (5 - t)dt$

$\displaystyle = \left(5t - \frac{1}{2}t^2\right)\Big|_0^2$

$= 8$

$C_2 : x = -2 + t, y = 2 - \dfrac{1}{2}t^2; 0 \le t \le 2$

$\displaystyle\int_{C_2} y\,dx - 3\,dy = \int_0^2 \left[\left(2 - \frac{1}{2}t^2\right)(1) - 3(-t)\right]dt$

$\displaystyle = \int_0^2 \left(2 + 3t - \frac{1}{2}t^2\right)dt$

$\displaystyle = \left(2t + \frac{3}{2}t^2 - \frac{1}{6}t^3\right)\Big|_0^2$

$\displaystyle = -\frac{26}{3}$

The two paths from $(-2, 2)$ to $(0, 0)$ give different values, so the line integral is not independent of path.

45. False; this is true only if C is a closed curve.

47. True; the potential function has the same value at both endpoints (since the endpoints are the same), so the line integral is 0.

49. If $\nabla f(x, y) = \mathbf{F}(x, y),$ then

$$\frac{\partial f}{\partial x} = \frac{-y}{x^2 + y^2} \text{ and } \frac{\partial f}{\partial y} = \frac{x}{x^2 + y^2}.$$

$$f(x, y) = \int \frac{x}{x^2 + y^2} \, dy = \tan^{-1} \frac{y}{x} + g(x)$$

$$\frac{\partial f}{\partial y} = \frac{x}{x^2 + y^2} + g'(x) = \frac{-y}{x^2 + y^2},$$

so $g(x) = c$ and $f(x, y) = \tan^{-1} \frac{y}{x} + c.$

If $c = 0, f(x, y)$ is the polar angle θ, where

$-\frac{\pi}{2} < \theta < \frac{\pi}{2}.$ When defined as

$f(x, y) = \tan^{-1} \frac{y}{x},$ the domain of f is $x \neq 0,$ but

when defined continuously in terms of angles, f can be defined on any simply connected region which excludes the origin!

To calculate $\int_C \mathbf{F} \cdot d\mathbf{r}$ over the circle

$(x - 2)^2 + (y - 3)^2 = 1,$ note that the circle lies in a simply connected region (say, the first quadrant) which excludes the origin. Since a potential function exists on this region and the curve is closed, the integral is 0.

51. a. A disk; simply connected.

b. A ring; not simply connected.

Section 14.4

5. **a.** $x = \cos t, \, y = \sin t, \, 0 \le t \le 2\pi$

$$\oint_C (x^2 - y)dx - y^2 dy = \int_0^{2\pi} [(\cos^2 t - \sin t)(-\sin t) - (\sin^2 t)(\cos t)]dt$$

$$= \int_0^{2\pi} (\sin^2 t - \cos^2 t \sin t - \sin^2 t \cos t)dt$$

$$= \int_0^{2\pi} \left[\frac{1}{2}(1 - \cos 2t) - \cos^2 t \sin t - \sin^2 t \cos t \right]dt$$

$$= \left(\frac{1}{2}t - \frac{1}{4}\sin 2t + \frac{1}{3}\cos^3 t - \frac{1}{3}\sin^3 t \right)\Bigg|_0^{2\pi}$$

$$= \pi$$

b.   Let $M(x, y) = x^2 - y$ and $N(x, y) = y^2$.

$$\oint_C M dx + N dy = \iint_R \left(\frac{\partial N}{\partial x} - \frac{\partial M}{\partial y} \right)dA$$

$$= \iint_R [0 - (-1)]dA$$

$$= \text{Area of circle}$$

$$= \pi$$

7. **a.**   Left edge: $(0, 0)$ to $(0, 2)$, $x = 0, \, dx = 0$

$$\int_0^2 (-x^3)dy = \int_0^2 0 \, dy = 0$$

Top edge: $(0, 2)$ to $(2, 2)$, $y = 2, \, dy = 0$

$$\int_0^2 x^2 dx = \frac{1}{3}x^3 \Bigg|_0^2 = \frac{8}{3}$$

Right edge: $(2, 2)$ to $(2, 0)$, $x = 2, \, dx = 0$

$$\int_2^0 (-x^3)dy = \int_2^0 (-8)dy = 16$$

Bottom edge: $(2, 0)$ to $(0, 0)$, $y = 0, \, dy = 0$

$$\int_2^0 x^2 dx = \frac{1}{3}x^3 \Bigg|_2^0 = -\frac{8}{3}$$

Summing the above, the line integral is 16.

b.   Let $M(x, y) = x^2$ and $N(x, y) = -x^3$.

Note that the curve has negative orientation.

$$\oint_C M \, dx + N \, dy = -\iint_R \left(\frac{\partial N}{\partial x} - \frac{\partial M}{\partial y} \right)dA$$

$$= -\iint_R [-3x^2 + 0]dA$$

$$= \int_0^2 \int_0^2 3x^2 dx \, dy$$

$$= \int_0^2 [x^3]_{x=0}^{x=2} dy$$

$$= \int_0^2 8 \, dy$$

$$= 16$$

9. Let $M(x, y) = xe^{2x}$ and $N(x, y) = -3x^2 y$.

Note that the curve has positive orientation.

$$\oint_C M \, dx + N \, dy = \iint_R \left(\frac{\partial N}{\partial x} - \frac{\partial M}{\partial y} \right)dA$$

$$= \int_0^2 \int_0^3 (-6xy - 0)dx \, dy$$

$$= \int_0^2 [-3x^2 y]_{x=0}^{x=3} dy$$

$$= -\frac{27}{2}\Bigg|_0^2$$

$$= -54$$

11. Let $M(x, y) = \dfrac{x}{x^2 + 1} - y$ and

$N(x, y) = 3x - 4\tan y$. Note that the curve has positive orientation.

$$\oint_C M \, dx + N \, dy = \iint_R \left(\frac{\partial N}{\partial x} - \frac{\partial M}{\partial y} \right)dA$$

$$= \int_{-1}^1 \int_{x^2}^{2-x^2} [3 - (-1)]dy \, dx$$

$$= 4\int_{-1}^1 \int_{x^2}^{2-x^2} dy \, dx$$

$$= 4\int_{-1}^1 (2 - 2x^2)dx$$

$$= 4\left(2x - \frac{2}{3}x^3 \right)\Bigg|_{x=-1}^{x=1}$$

$$= \frac{32}{3}$$

13. Let $M(x, y) = \tan x - y^3$ and

$N(x, y) = x^3 - \sin y$. Note that the curve has positive orientation.

$$\int_C M\,dx + N\,dy = \iint_R \left(\frac{\partial N}{\partial x} - \frac{\partial M}{\partial y} \right) dA$$

$$= \iint_R (3x^2 + 3y^2)\,dA$$

$$= \int_0^{2\pi} \int_0^{\sqrt{2}} (3r^2)r\,dr\,d\theta$$

$$= \int_0^{2\pi} \left[\frac{3}{4} r^4 \right]_{r=0}^{r=\sqrt{2}} d\theta$$

$$= \int_0^{2\pi} 3\,d\theta$$

$$= 6\pi$$

15. Let $M(x, y) = x^3 - y$ and $N(x, y) = x + y^3$.
Note that the curve has positive orientation.
Since $y = x^2$ and $y = x$ intersect at $(0, 0)$ and
$(1, 1)$, the region extends from $x = 0$ to $x = 1$.

$$\oint_C \mathbf{F} \cdot d\mathbf{r} = \oint_C M\,dx + N\,dy$$

$$= \iint_R \left(\frac{\partial N}{\partial x} - \frac{\partial M}{\partial y} \right) dA$$

$$\int_0^1 \int_{x^2}^x [1 - (-1)]\,dy\,dx$$

$$= 2\int (x - x^2)\,dx$$

$$= 2\left(\frac{1}{2}x^2 - \frac{1}{3}x^3 \right)\Big|_{x=0}^{x=1}$$

$$= \frac{1}{3}$$

17. Let $M(x, y) = e^{x^2} - y$ and $N(x, y) = e^{2x} + y$.
Note that the curve has positive orientation.
Since $y = 1 - x^2$ and $y = 0$ intersect at $(\pm 1, 0)$,
the region extends from $x = -1$, to $x = 1$.

$$\oint_C \mathbf{F} \cdot d\mathbf{r}$$

$$= \oint_C M\,dx + N\,dy$$

$$= \iint_R \left(\frac{\partial N}{\partial x} - \frac{\partial M}{\partial y} \right) dA$$

$$= \int_{-1}^1 \int_0^{1-x^2} (2e^{2x} + 1)\,dy\,dx$$

$$= \int_{-1}^1 [(2e^{2x} + 1)y]_{y=0}^{y=1-x^2}\,dx$$

$$= \int_{-1}^1 (2e^{2x} + 1)(1 - x^2)\,dx$$

$$= \int_{-1}^1 (2e^{2x} + 1 - 2x^2e^{2x} - x^2)\,dx$$

$$= \left[e^{2x} + x - e^{2x}\left(x^2 - x + \frac{1}{2} \right) - \frac{1}{3}x^3 \right]_{-1}^1$$

$$= \left[e^2 + 1 - e^2\left(\frac{1}{2} \right) - \frac{1}{3} \right] - \left[e^{-2} - 1 - e^{-2}\left(\frac{5}{2} \right) + \frac{1}{3} \right]$$

$$= \frac{4}{3} + \frac{1}{2}e^2 + \frac{3}{2}e^{-2}$$

Note: To integrate $2x^2e^{2x}$, integrate by parts twice.

19. Let $M(x, y) = y^3 - \ln x$ and

$N(x, y) = \sqrt{y^2 + 1} + 3x$. Note that the curve has positive orientation. Since $x = y^2$ and $x = 4$ intersect at $(4, \pm 2)$, the region extends from $y = -2$ to $y = 2$.

$$\oint_C M\,dx + N\,dy$$

$$= \iint_R \left(\frac{\partial N}{\partial x} - \frac{\partial M}{\partial y} \right) dA$$

$$= \int_{-2}^2 \int_{y^2}^4 (3 - 3y^2)\,dx\,dy$$

$$= \int_{-2}^2 [3x - 3xy^2]_{x=y^2}^{x=4}\,dy$$

$$= \int_{-2}^2 [(12 - 12y^2) - (3y^2 - 3y^4)]\,dy$$

$$= \int_{-2}^2 (3y^4 - 15y^2 + 12)\,dy$$

$$= \left(\frac{3}{5}y^5 - 5y^3 + 12y \right)\Big|_{-2}^2$$

$$= \frac{32}{5}$$

21. The triangle lies in the plane $z = 2$, so $dz = 0$. Therefore, we can reduce the problem to two dimensions (in terms of x and y):

Evaluate $\oint_C x^2 dx + 2x\, dy$ where C is the triangle from (0, 0) to (2, 0) to (2, 2) to (0, 0).

Let $M(x, y) = x^2$ and $N(x, y) = 2x$. Note that the curve has positive orientation.

$$\oint_C M dx + N dy = \iint_R \left(\frac{\partial N}{\partial x} - \frac{\partial M}{\partial y} \right) dA$$

$$= \iint_R (2 - 0) dA$$

$$= 2(\text{Area of triangle})$$

$$= 2\left(\frac{1}{2} \cdot 2 \cdot 2 \right) = 4$$

23. Since $x^2 + z^2 = 1$ and $y = 0$ along the curve, we may use $\mathbf{F} = \left\langle x^3, e, x^2 \right\rangle$. Then

$\oint_C \mathbf{F} \cdot d\mathbf{r} = \oint_C x^3 dx + e\, dy + x^2 dz$. But $dy = 0$,

so we can reduce the problem to two dimensions (in terms of x and z):

Evaluate $\oint_C x^3 dx + x^2 dz$ where C is the circle

$x^2 + z^2 = 1$ oriented positively.

Let $M(x, z) = x^3$ and $N(x, z) = x^2$.

$$\oint_C M\, dx + N\, dz = \iint_R \left(\frac{\partial N}{\partial x} - \frac{\partial M}{\partial z} \right) dA$$

$$= \iint_R 2x\, dA$$

$$= \int_{-1}^{1} \int_{-\sqrt{1-z^2}}^{\sqrt{1-z^2}} 2x\, dx\, dz$$

$$= \int_{-1}^{1} [x^2]_{x=-\sqrt{1-z^2}}^{x=\sqrt{1-z^2}} dz$$

$$= \int_{-1}^{1} 0\, dz$$

$$= 0$$

25. $C = \{(x, y) \mid x = 2\cos t, \ y = 4\sin t, \ 0 \le t \le 2\pi\}$

$$A = \frac{1}{2} \oint_C x\, dy - y\, dx$$

$$= \frac{1}{2} \int_0^{2\pi} [(2\cos t)(4\cos t) - (4\sin t)(-2\sin t)] dt$$

$$= \int_0^{2\pi} 4(\cos^2 t + \sin^2 t) dt$$

$$= 8\pi$$

27. $C = \{(x, y) \mid x = \cos^3 t, \ y = \sin^3 t, \ 0 \le t \le 2\pi\}$

$$A = \frac{1}{2} \oint_C x\, dy - y\, dx$$

$$= \frac{1}{2} \int_0^{2\pi} \left[\begin{array}{l} (\cos^3 t)(3\sin^2 t \cos t) \\ \quad - (\sin^3 t)(-3\cos^2 t \sin t) \end{array} \right] dt$$

$$= \frac{3}{2} \int_0^{2\pi} (\cos^2 t \sin^2 t)(\cos^2 t + \sin^2 t) dt$$

$$= \frac{3}{2} \int_0^{2\pi} \cos^2 t \sin^2 t\, dt$$

$$= \frac{3}{8} \int_0^{2\pi} (2\cos t \sin t)^2 dt$$

$$= \frac{3}{8} \int_0^{2\pi} \sin^2 2t\, dt$$

$$= \frac{3}{16} \int_0^{2\pi} (1 - \cos 4t) dt$$

$$= \left(\frac{3}{16} t - \frac{3}{64} \sin 4t \right)\Big|_{t=0}^{t=2\pi}$$

$$= \frac{3\pi}{8}$$

29. $C_1 = \{(x, y) \mid x = t, \ y = t^2, \ -2 \le t \le 2\}$

$C_2 = \{(x, y) \mid x = 2 - t, \ y = 4, \ 0 \le t \le 4\}$

$$A = \frac{1}{2} \left(\int_{C_1} x\, dy - y\, dx \right) + \frac{1}{2} \left(\int_{C_2} x\, dy - y\, dx \right)$$

$$= \frac{1}{2} \int_{-2}^{2} [t(2t) - t^2] dt + \frac{1}{2} \int_0^{4} (0 - 4)(-dt)$$

$$= \frac{1}{2} \int_{-2}^{2} t^2 dt + \frac{1}{2} \int_0^{4} 4\, dt$$

$$= \frac{1}{2} \left[\frac{1}{3} t^3 \right]_{t=-2}^{t=2} + \frac{1}{2}(16)$$

$$= \frac{32}{3}$$

31. For $\bar{x}$, let $M(x,y) = 0$ and $N(x,y) = x^2$.

$$\frac{1}{2A}\oint_C x^2 dy = \frac{1}{2A}\oint_C M\,dx + N\,dy$$
$$= \frac{1}{2A}\iint_R \left(\frac{\partial N}{\partial x} - \frac{\partial M}{\partial y}\right)dA$$
$$= \frac{1}{2A}\iint_R 2x\,dA$$
$$= \frac{1}{A}\iint_R x\,dA$$
$$= \bar{x}$$

For $\bar{y}$, let $M(x,y) = y^2$ and $N(x,y) = 0$.

$$-\frac{1}{2A}\oint_C y^2 dx = -\frac{1}{2A}\oint_C M\,dx + N\,dy$$
$$= -\frac{1}{2A}\iint_R \left(\frac{\partial N}{\partial x} - \frac{\partial M}{\partial y}\right)dA$$
$$= -\frac{1}{2A}\iint_R (-2y)dA$$
$$= \frac{1}{A}\iint_R y\,dA$$
$$= \bar{y}$$

33. $x = t^3 - t,\; y = 1 - t^2,\; -1 \le t \le 1$

Note that the curve has positive orientation.

$$A = \frac{1}{2}\oint_C x\,dy - y\,dx$$
$$= \frac{1}{2}\int_{-1}^1 [(t^3 - t)(-2t) - (1 - t^2)(3t^2 - 1)]dt$$
$$= \frac{1}{2}\int_{-1}^1 (t^4 - 2t^2 + 1)dt$$
$$= \frac{1}{2}\left(\frac{1}{5}t^5 - \frac{2}{3}t^3 + t\right)\Big|_{-1}^1$$
$$= \frac{8}{15}$$

$$\bar{x} = \frac{1}{2A}\oint_C x^2 dy$$
$$= \frac{15}{16}\int_{-1}^1 (t^3 - t)^2(-2t)dt$$
$$= \frac{15}{16}\int_{-1}^1 (-2t^7 + 4t^5 - 2t^3)dt$$
$$= 0 \text{ (by symmetry)}$$

$$\bar{y} = -\frac{1}{2A}\oint_C y^2 dx$$
$$= -\frac{15}{16}\int_{-1}^1 (1 - t^2)^2(3t^2 - 1)dt$$
$$= \frac{15}{16}\int_{-1}^1 (-3t^6 + 7t^4 - 5t^2 + 1)dt$$
$$= \frac{15}{16}\left(-\frac{3}{7}t^7 + \frac{7}{5}t^5 - \frac{5}{3}t^3 + t\right)\Big|_{-1}^1$$
$$= \frac{4}{7}$$

35. Apply Green's Theorem twice: first, in terms of x and y and then, in terms of u and v. The absolute value is necessary because, even though C is chosen to be positively oriented in the xy-plane, it may turn out to be negatively oriented in the uv-plane.

$$\iint_R dA = \frac{1}{2}\oint_C x\,dy - y\,dx$$
$$= \frac{1}{2}\oint_C \left(x\frac{\partial y}{\partial u} - y\frac{\partial x}{\partial u}\right)du + \left(x\frac{\partial y}{\partial v} - y\frac{\partial x}{\partial v}\right)dv$$
$$= \frac{1}{2}\iint_S \left|\frac{\partial}{\partial v}\left(x\frac{\partial y}{\partial u} - y\frac{\partial x}{\partial u}\right) - \frac{\partial}{\partial u}\left(x\frac{\partial y}{\partial v} - y\frac{\partial x}{\partial v}\right)\right|du\,dv$$
$$= \frac{1}{2}\iint_S \left|\left(\frac{\partial x}{\partial v}\frac{\partial y}{\partial u} + x\frac{\partial^2 y}{\partial u\partial v} - \frac{\partial y}{\partial v}\frac{\partial x}{\partial u} - y\frac{\partial^2 x}{\partial u\partial v}\right) - \left(\frac{\partial x}{\partial u}\frac{\partial y}{\partial v} + x\frac{\partial^2 y}{\partial v\partial u}\right) - \frac{\partial y}{\partial u}\frac{\partial x}{\partial v} - y\frac{\partial^2 x}{\partial v\partial u}\right|du\,dv$$
$$= \frac{1}{2}\iint_S \left|2\frac{\partial x}{\partial v}\frac{\partial y}{\partial u} - 2\frac{\partial y}{\partial v}\frac{\partial x}{\partial u}\right|du\,dv$$
$$= \iint_S \left|\begin{vmatrix}\frac{\partial x}{\partial u} & \frac{\partial x}{\partial v} \\ \frac{\partial y}{\partial u} & \frac{\partial y}{\partial v}\end{vmatrix}\right|du\,dv$$
$$= \iint_S \left|\frac{\partial(x,y)}{\partial(u,v)}\right|du\,dv$$

37. Let $M(x, y) = \dfrac{x}{x^2 + y^2}$ and $N(x, y) = \dfrac{y}{x^2 + y^2}$.

Define C, C_1, and R as in example 4.5.

$$\int_C \mathbf{F} \cdot d\mathbf{r} - \int_{C_1} \mathbf{F} \cdot d\mathbf{r} = \iint_R \left(\frac{\partial N}{\partial x} - \frac{\partial M}{\partial y} \right) dA$$

$$= \iint_R \left[\frac{(0)(x^2 + y^2) - (y)(2x)}{(x^2 + y^2)^2} - \frac{(0)(x^2 + y^2) - (x)(2y)}{(x^2 + y^2)^2} \right] dA$$

$$= \iint_R 0 \, dA = 0$$

Therefore, the line integral is the same for only positive oriented simple closed curve containing the origin. So we may assume $x = \cos t$, $y = \sin t$, $0 \le t \le 2\pi$.

$$\int_C \mathbf{F} \cdot d\mathbf{r} = \int_C \frac{x}{x^2 + y^2} dx + \frac{y}{x^2 + y^2} dy$$

$$= \int_0^{2\pi} [(\cos t)(\cos t) + (\sin t)(-\sin t)] dt$$

$$= \int_0^{2\pi} \left(\cos^2 t - \sin^2 t \right) t \, dt$$

$$= \int_0^{2\pi} \cos 2t \, dt$$

$$= \frac{1}{2} \sin 2t \Big|_0^{2\pi} = 0$$

39. Let $M(x, y) = \dfrac{x^3}{x^4 + y^4}$ and $N(x, y) = \dfrac{y^3}{x^4 + y^4}$.

Define C, C_1, and R as in example 4.5.

$$\int_C \mathbf{F} \cdot d\mathbf{r} - \int_{C_1} \mathbf{F} \cdot d\mathbf{r} = \iint_R \left(\frac{\partial N}{\partial x} - \frac{\partial M}{\partial y} \right) dA$$

$$= \iint_R \left[\frac{(0)(x^4 + y^4) - (y^3)(4x^3)}{(x^4 + y^4)^2} - \frac{(0)(x^4 + y^4) - (x^3)(4y^3)}{(x^4 + y^4)^2} \right] dA$$

$$= \iint_R 0 \, dA = 0$$

Therefore, the line integral is the same for any positively oriented simply closed curve containing the origin. So we may assume $x = \cos t$, $y = \sin t$, $0 \le t \le 2\pi$.

$$\int_C \mathbf{F} \cdot d\mathbf{r} = \int_C \frac{x^3}{x^4 + y^4} dx + \frac{y^3}{x^4 + y^4} dy$$

$$= \int_0^{2\pi} \frac{(\cos^3 t)(-\sin t) + (\sin^3 t)(\cos t)}{\cos^4 t + \sin^4 t} dt$$

$$= \int_0^{2\pi} \frac{(\sin^2 t - \cos^2 t)(\cos t \sin t)}{\cos^4 t + \sin^4 t} dt$$

$$= \int_0^{2\pi} \frac{-\cos 2t \sin 2t}{2(\cos^4 t + \sin^4 t)} dt = 0$$

Note that the integral can be evaluated by symmetry or by numerical approximation.

41. Green 's Theorem assumes that the partial derivatives are defined in some open region which contains the region enclosed by C. Since for this field $\mathbf{F}$ the partials M_y and N_x aren't defined at the origin (indeed, $\mathbf{F}$ isn't defined at the origin), Green's Theorem says nothing about the integral of $\mathbf{F} \cdot d\mathbf{r}$ around a curve enclosing the origin.

Section 14.5

5. $\mathbf{F}(x,y,z) = \left\langle x^2, -3xy, 0 \right\rangle$

$$\text{curl } \mathbf{F} = \nabla \times \mathbf{F} = \begin{vmatrix} \mathbf{i} & \mathbf{j} & \mathbf{k} \\ \frac{\partial}{\partial x} & \frac{\partial}{\partial y} & \frac{\partial}{\partial z} \\ x^2 & -3xy & 0 \end{vmatrix}$$

$$= \left(\frac{\partial(0)}{\partial y} - \frac{\partial(-3xy)}{\partial z} \right)\mathbf{i} - \left(\frac{\partial(0)}{\partial x} - \frac{\partial(x^2)}{\partial z} \right)\mathbf{j} + \left(\frac{\partial(-3xy)}{\partial x} - \frac{\partial(x^2)}{\partial y} \right)\mathbf{k}$$

$$= (0-0)\mathbf{i} - (0-0)\mathbf{j} + (-3y-0)\mathbf{k}$$

$$= \left\langle 0, 0, -3y \right\rangle$$

$$\text{div } \mathbf{F} = \nabla \cdot \mathbf{F}$$

$$= \frac{\partial(x^2)}{\partial x} + \frac{\partial(-3xy)}{\partial y} + \frac{\partial(0)}{\partial z}$$

$$= 2x - 3x + 0$$

$$= -x$$

7. $\mathbf{F}(x,y,z) = \left\langle 2xz, 0, -3y \right\rangle$

$$\text{curl } \mathbf{F} = \nabla \times \mathbf{F}$$

$$= \begin{vmatrix} \mathbf{i} & \mathbf{j} & \mathbf{k} \\ \frac{\partial}{\partial x} & \frac{\partial}{\partial y} & \frac{\partial}{\partial z} \\ 2xz & 0 & -3y \end{vmatrix}$$

$$= \left(\frac{\partial(-3y)}{\partial y} - \frac{\partial(0)}{\partial z} \right)\mathbf{i} - \left(\frac{\partial(-3y)}{\partial x} - \frac{\partial(2xz)}{\partial z} \right)\mathbf{j} + \left(\frac{\partial(0)}{\partial x} - \frac{\partial(2xz)}{\partial y} \right)\mathbf{k}$$

$$= (-3-0)\mathbf{i} - (0-2x)\mathbf{j} + (0-0)\mathbf{k}$$

$$= \left\langle -3, 2x, 0 \right\rangle$$

$$\text{div } \mathbf{F} = \nabla \cdot \mathbf{F}$$

$$= \frac{\partial(2xz)}{\partial x} + \frac{\partial(0)}{\partial y} + \frac{\partial(-3y)}{\partial z}$$

$$= 2z + 0 + 0 = 2z$$

9. $\mathbf{F}(x,y,z) = \left\langle xy, yz, x^2 \right\rangle$

$$\text{curl } \mathbf{F} = \nabla \times \mathbf{F}$$

$$= \begin{vmatrix} \mathbf{i} & \mathbf{j} & \mathbf{k} \\ \frac{\partial}{\partial x} & \frac{\partial}{\partial y} & \frac{\partial}{\partial z} \\ xy & yz & x^2 \end{vmatrix}$$

$$= \left(\frac{\partial(x^2)}{\partial y} - \frac{\partial(yz)}{\partial z} \right)\mathbf{i} - \left(\frac{\partial(x^2)}{\partial x} - \frac{\partial(xy)}{\partial z} \right)\mathbf{j} + \left(\frac{\partial(yz)}{\partial x} - \frac{\partial(xy)}{\partial y} \right)\mathbf{k}$$

$$= (0-y)\mathbf{i} - (2x-0)\mathbf{j} + (0-x)\mathbf{k}$$

$$= \left\langle -y, -2x, -x \right\rangle$$

$$\text{div } \mathbf{F} = \nabla \cdot \mathbf{F}$$

$$= \frac{\partial(xy)}{\partial x} + \frac{\partial(yz)}{\partial y} + \frac{\partial(x^2)}{\partial z} = y + z + 0 = y + z$$

11. $\mathbf{F}(x, y, z) = \left\langle x^2, y - z, xe^y \right\rangle$

curl $\mathbf{F} = \nabla \times \mathbf{F}$

$$= \begin{vmatrix} \mathbf{i} & \mathbf{j} & \mathbf{k} \\ \dfrac{\partial}{\partial x} & \dfrac{\partial}{\partial y} & \dfrac{\partial}{\partial z} \\ x^2 & y - z & xe^y \end{vmatrix}$$

$$= \left(\frac{\partial(xe^y)}{\partial y} - \frac{\partial(y-z)}{\partial z} \right)\mathbf{i} - \left(\frac{\partial(xe^y)}{\partial x} - \frac{\partial(x^2)}{\partial z} \right)\mathbf{j} + \left(\frac{\partial(y-z)}{\partial x} - \frac{\partial(x^2)}{\partial y} \right)\mathbf{k}$$

$$= (xe^y + 1)\mathbf{i} - (e^y - 0)\mathbf{j} + (0 - 0)\mathbf{k}$$

$$= \left\langle xe^y + 1, -e^y, 0 \right\rangle$$

div $\mathbf{F} = \nabla \cdot \mathbf{F}$

$$= \frac{\partial(x^2)}{\partial x} + \frac{\partial(y-z)}{\partial y} + \frac{\partial(xe^y)}{\partial z}$$

$$= 2x + 1 + 0 = 2x + 1$$

13. $\mathbf{F}(x, y, z) = \left\langle 3yz, x^2, x\cos y \right\rangle$

curl $\mathbf{F} = \nabla \times \mathbf{F}$

$$= \begin{vmatrix} \mathbf{i} & \mathbf{j} & \mathbf{k} \\ \dfrac{\partial}{\partial x} & \dfrac{\partial}{\partial y} & \dfrac{\partial}{\partial z} \\ 3yz & x^2 & x\cos y \end{vmatrix}$$

$$= \left(\frac{\partial(x\cos y)}{\partial y} - \frac{\partial(x^2)}{\partial z} \right)\mathbf{i} - \left(\frac{\partial(x\cos y)}{\partial x} - \frac{\partial(3yz)}{\partial z} \right)\mathbf{j} + \left(\frac{\partial(x^2)}{\partial x} - \frac{\partial(3yz)}{\partial y} \right)\mathbf{k}$$

$$= (-x\sin y - 0)\mathbf{i} - (\cos y - 3y)\mathbf{j} + (2x - 3z)\mathbf{k}$$

$$= \left\langle -x\sin y, 3y - \cos y, 2x - 3z \right\rangle$$

div $\mathbf{F} = \nabla \cdot \mathbf{F}$

$$= \frac{\partial(3yz)}{\partial x} + \frac{\partial(x^2)}{\partial y} + \frac{\partial(x\cos y)}{\partial z}$$

$$= 0 + 0 + 0 = 0$$

15. $\mathbf{F}(x, y, z) = \left\langle 2xz,\ y + z^2,\ y^2 z \right\rangle$

curl $\mathbf{F} = \nabla \times \mathbf{F}$

$$= \begin{vmatrix} \mathbf{i} & \mathbf{j} & \mathbf{k} \\ \dfrac{\partial}{\partial x} & \dfrac{\partial}{\partial y} & \dfrac{\partial}{\partial z} \\ 2xz & y + z^2 & y^2 z \end{vmatrix}$$

$$= \left(\frac{\partial (y^2 z)}{\partial y} - \frac{\partial (y + z^2)}{\partial z} \right) \mathbf{i} - \left(\frac{\partial (y^2 z)}{\partial x} - \frac{\partial (2xz)}{\partial z} \right) \mathbf{j} + \left(\frac{\partial (y + z^2)}{\partial x} - \frac{\partial (2xz)}{\partial y} \right) \mathbf{k}$$

$$= (2yz - 2z)\mathbf{i} - (0 - 2x)\mathbf{j} + (0 - 0)\mathbf{k}$$

$$= \left\langle 2yz - 2z,\ 2x,\ 0 \right\rangle$$

div $\mathbf{F} = \nabla \cdot \mathbf{F}$

$$= \frac{\partial (2xz)}{\partial x} + \frac{\partial (y + z^2)}{\partial y} + \frac{\partial (y^2 z)}{\partial z}$$

$$= 2z + 1 + y^2$$

17. $\mathbf{F}(x, y, z) = \left\langle 2x,\ 2yz^2,\ 2y^2 z \right\rangle$

curl $\mathbf{F} = \nabla \times \mathbf{F}$

$$= \begin{vmatrix} \mathbf{i} & \mathbf{j} & \mathbf{k} \\ \dfrac{\partial}{\partial x} & \dfrac{\partial}{\partial y} & \dfrac{\partial}{\partial z} \\ 2x & 2yz^2 & 2y^2 z \end{vmatrix}$$

$$= \left(\frac{\partial (2y^2 z)}{\partial y} - \frac{\partial (2yz^2)}{\partial z} \right) \mathbf{i} - \left(\frac{\partial (2y^2 z)}{\partial x} - \frac{\partial (2x)}{\partial z} \right) \mathbf{j} + \left(\frac{\partial (2yz^2)}{\partial x} - \frac{\partial (2x)}{\partial y} \right) \mathbf{k}$$

$$= (4yz - 4yz)\mathbf{i} - (0 - 0)\mathbf{j} + (0 - 0)\mathbf{k}$$

$$= \left\langle 0, 0, 0 \right\rangle$$

Since the components of $\mathbf{F}$ have continuous partial derivatives throughout $\mathbf{R}^3$ and curl $\mathbf{F} = 0$, $\mathbf{F}$ is conservative.

div $\mathbf{F} = \nabla \cdot \mathbf{F}$

$$= \frac{\partial (2x)}{\partial x} + \frac{\partial (2yz^2)}{\partial y} + \frac{\partial (2y^2 z)}{\partial z}$$

$$= 2 + 2z^2 + 2y^2$$

Since div $\mathbf{F} \neq 0$, $\mathbf{F}$ is not incompressible.

19. $\mathbf{F}(x, y, z) = \left\langle 3yz, x^2, x\cos y \right\rangle$

curl $\mathbf{F} = \nabla \times \mathbf{F}$

$$= \begin{vmatrix} \mathbf{i} & \mathbf{j} & \mathbf{k} \\ \dfrac{\partial}{\partial x} & \dfrac{\partial}{\partial y} & \dfrac{\partial}{\partial z} \\ 3yz & x^2 & x\cos y \end{vmatrix}$$

$$= \left(\frac{\partial(x\cos y)}{\partial y} - \frac{\partial(x^2)}{\partial z} \right)\mathbf{i} - \left(\frac{\partial(x\cos y)}{\partial x} - \frac{\partial(3yz)}{\partial z} \right)\mathbf{j} + \left(\frac{\partial(x^2)}{\partial x} - \frac{\partial(3yz)}{\partial y} \right)\mathbf{k}$$

$$= (-x\sin y - 0)\mathbf{i} - (\cos y - 3y)\mathbf{j} + (2x - 3z)\mathbf{k}$$

$$= \left\langle -x\sin y, 3y - \cos y, 2x - 3z \right\rangle$$

Since curl $\mathbf{F} \neq \mathbf{0}$, $\mathbf{F}$ is not conservative.

div $\mathbf{F} = \nabla \cdot \mathbf{F}$

$$= \frac{\partial(3yz)}{\partial x} + \frac{\partial(x^2)}{\partial y} + \frac{\partial(x\cos y)}{\partial z}$$

$$= 0 + 0 + 0$$

$$= 0$$

Since div $\mathbf{F} = 0$, $\mathbf{F}$ is incompressible.

21. $\mathbf{F}(x, y, z) = \left\langle \sin z, e^{yz^2}z^2, 2e^{yz^2}yz + x\cos z \right\rangle$

curl $\mathbf{F} = \nabla \times \mathbf{F}$

$$= \begin{vmatrix} \mathbf{i} & \mathbf{j} & \mathbf{k} \\ \dfrac{\partial}{\partial x} & \dfrac{\partial}{\partial y} & \dfrac{\partial}{\partial z} \\ \sin z & z^2 e^{yz^2} & 2yze^{yx^2} + x\cos z \end{vmatrix}$$

$$= \left(\frac{\partial(2yze^{yz^2} + x\cos z)}{\partial y} - \frac{\partial(z^2 e^{yz^2})}{\partial z} \right)\mathbf{i} - \left(\frac{\partial(2yze^{yz^2} + x\cos z)}{\partial x} - \frac{\partial(\sin z)}{\partial z} \right)\mathbf{j} + \left(\frac{\partial(z^2 e^{yz^2})}{\partial x} - \frac{\partial(\sin z)}{\partial y} \right)\mathbf{k}$$

$$= [(2ze^{yz^2} + 2yze^{yz^2} \cdot z^2) - (2ze^{yz^2} + z^2 e^{yz^2} \cdot 2yz)]\mathbf{i} - (\cos z - \cos z)\mathbf{j} + (0 - 0)\mathbf{k}$$

$$= \left\langle 0, 0, 0 \right\rangle$$

Since the components of $\mathbf{F}$ have continuous partial derivatives throughout R^3 and curl $\mathbf{F} = \mathbf{0}$, $\mathbf{F}$ is conservative.

div $\mathbf{F} = \nabla \cdot \mathbf{F}$

$$= \frac{\partial(\sin z)}{\partial x} + \frac{\partial(z^2 e^{yz^2})}{\partial y} + \frac{\partial(2yze^{yz^2} + x\cos z)}{\partial z}$$

$$= 0 + z^4 e^{yz^2} + (2ye^{yz^2} + 2yze^{yz^2} \cdot 2yz - x\sin z)$$

$$= 2e^{yz^2}y + 4e^{yz^2}y^2 z^2 + e^{yz^2}z^4 - x\sin z$$

Since div $\mathbf{F} \neq 0$, $\mathbf{F}$ is not incompressible.

23. $\mathbf{F}(x, y, z) = \left\langle z^2 - 3e^{3x}y, z^2 - e^{3x}, 2z\sqrt{xy} \right\rangle$

curl $\mathbf{F} = \nabla \times \mathbf{F}$

$$= \begin{vmatrix} \mathbf{i} & \mathbf{j} & \mathbf{k} \\ \frac{\partial}{\partial x} & \frac{\partial}{\partial y} & \frac{\partial}{\partial z} \\ z^2 - 3e^{3x}y & z^2 - e^{3x} & 2z\sqrt{xy}z \end{vmatrix}$$

$$= \left(\frac{\partial(2z\sqrt{xy})}{\partial y} - \frac{\partial(z^2 - e^{3x})}{\partial z} \right)\mathbf{i} - \left(\frac{\partial(2z\sqrt{xy})}{\partial x} - \frac{\partial(z^2 - 3e^{3x}y)}{\partial z} \right)\mathbf{j} + \left(\frac{\partial(z^2 - e^{3x})}{\partial x} - \frac{\partial(z^2 - 3e^{3x}y)}{\partial y} \right)\mathbf{k}$$

$$= \left(\frac{xz}{\sqrt{xy}} - 2z \right)\mathbf{i} - \left(\frac{yz}{\sqrt{xy}} - 2z \right)\mathbf{j} + (-3e^{3x} + 3e^{3x})\mathbf{k}$$

$$= \left\langle -2z + \frac{xz}{\sqrt{xy}}, 2z - \frac{yz}{\sqrt{xy}}, 0 \right\rangle$$

Since curl $\mathbf{F} \neq \mathbf{0}$, $\mathbf{F}$ is not conservative.

div $\mathsf{F} = \nabla \cdot \mathbf{F}$

$$= \frac{\partial(z^2 - 3e^{3x}y)}{\partial x} + \frac{\partial(z^2 - e^{3x})}{\partial y} + \frac{\partial(2z\sqrt{xy})}{\partial z}$$

$$= -9ye^{3x} + 0 + 2\sqrt{xy}$$

$$= 2\sqrt{xy} - 9ye^{3x}$$

Sinve div $\mathbf{F} \neq 0$, $\mathbf{F}$ is not incompressible.

25. $\mathbf{F}(x, y, z) = \left\langle xy^2, 3xz, 4 - y^2z \right\rangle$

curl $\mathbf{F} = \nabla \times \mathbf{F}$

$$= \begin{vmatrix} \mathbf{i} & \mathbf{j} & \mathbf{k} \\ \frac{\partial}{\partial x} & \frac{\partial}{\partial y} & \frac{\partial}{\partial z} \\ xy^2 & 3xz & 4 - y^2z \end{vmatrix}$$

$$= \left(\frac{\partial(4 - y^2z)}{\partial y} - \frac{\partial(3xz)}{\partial z} \right)\mathbf{i} - \left(\frac{\partial(4 - y^2z)}{\partial x} - \frac{\partial(xy^2)}{\partial z} \right)\mathbf{j} + \left(\frac{\partial(3xz)}{\partial x} - \frac{\partial(xy^2)}{\partial y} \right)\mathbf{k}$$

$$= (-2yz - 3x)\mathbf{i} - (0 - 0)\mathbf{j} + (3z - 2xy)\mathbf{k}$$

$$= \left\langle -3x - 2yz, 0, -2xy + 3z \right\rangle$$

Since curl $\mathbf{F} \neq \mathbf{0}$, $\mathbf{F}$ is not conservative.

div $\mathbf{F} = \nabla \cdot \mathbf{F}$

$$= \frac{\partial(xy^2)}{\partial x} + \frac{\partial(3xz)}{\partial y} + \frac{\partial(4 - y^2z)}{\partial z}$$

$$= y^2 + 0 - y^2$$

$$= 0$$

Since div $\mathbf{F} = 0$, $\mathbf{F}$ is incompressible.

27. $\mathbf{F}(x, y, z) = \left\langle 4x, 3y^3, e^z \right\rangle$

    $\text{curl } \mathbf{F} = \nabla \times \mathbf{F}$

$$= \begin{vmatrix} \mathbf{i} & \mathbf{j} & \mathbf{k} \\ \dfrac{\partial}{\partial x} & \dfrac{\partial}{\partial y} & \dfrac{\partial}{\partial z} \\ 4x & 3y^3 & e^z \end{vmatrix}$$

$$= \left(\frac{\partial(e^z)}{\partial y} - \frac{\partial(3y^3)}{\partial z} \right)\mathbf{i} - \left(\frac{\partial(e^z)}{\partial x} - \frac{\partial(4x)}{\partial z} \right)\mathbf{j} + \left(\frac{\partial(3y^3)}{\partial x} - \frac{\partial(4x)}{\partial y} \right)\mathbf{k}$$

$$= (0 - 0)\mathbf{i} - (0 - 0)\mathbf{j} + (0 - 0)\mathbf{k}$$

$$= \left\langle 0, 0, 0 \right\rangle$$

Since the components of $\mathbf{F}$ have continuous partial derivatives throughout R^3 and curl $\mathbf{F} = \mathbf{0}$, $\mathbf{F}$ is conservative.

div $\mathbf{F} = \nabla \cdot \mathbf{F}$

$$= \frac{\partial(4x)}{\partial x} + \frac{\partial(3y^3)}{\partial y} + \frac{\partial(e^z)}{\partial z}$$

$$= 4 + 9y^2 + e^z$$

Since div $\mathbf{F} \neq 0$, $\mathbf{F}$ is not incompressible.

29. $\mathbf{F}(x, y, z) = \left\langle -2xy, -x^2 + z^2 \cos yz^2, 2yz \cos yz^2 \right\rangle$

    $\text{curl } \mathbf{F} = \nabla \times \mathbf{F}$

$$= \begin{vmatrix} \mathbf{i} & \mathbf{j} & \mathbf{k} \\ \dfrac{\partial}{\partial x} & \dfrac{\partial}{\partial y} & \dfrac{\partial}{\partial z} \\ -2xy & -x^2 + z^2 \cos yz^2 & 2yz \cos yz^2 \end{vmatrix}$$

$$= \left(\frac{\partial(2yz(\cos yz^2))}{\partial y} - \frac{\partial(-x^2 + z^2 \cos yz^2)}{\partial z} \right)\mathbf{i} - \left(\frac{\partial(2yz \cos yz^2)}{\partial x} - \frac{\partial(-2xy)}{\partial z} \right)\mathbf{j}$$

$$+ \left(\frac{\partial(-x^2 + z^2 \cos yz^2)}{\partial x} - \frac{\partial(-2xy)}{\partial y} \right)\mathbf{k}$$

$$= [(2z \cos yz^2 - 2yz^3 \sin yz^2) - (2z \cos yz^2 - 2yz^3 \sin yz^2)]\mathbf{i} - (0 - 0)\mathbf{j} + (-2x + 2x)\mathbf{k}$$

$$= \left\langle 0, 0, 0 \right\rangle$$

Since the components of $\mathbf{F}$ have continuous partial derivatives throughout R^3 and curl $\mathbf{F} = \mathbf{0}$, $\mathbf{F}$ is conservative.

div $\mathbf{F} = \nabla \cdot \mathbf{F}$

$$= \frac{\partial(-2xy)}{\partial x} + \frac{\partial(-x^2 + z^2 \cos yz^2)}{\partial y} + \frac{\partial(2yz \cos yz^2)}{\partial z}$$

$$= -2y - z^4 \sin yz^2 + 2y \cos yz^2 - 4y^2 z^2 \sin yz^2$$

Since div $\mathbf{F} \neq 0$, $\mathbf{F}$ is not incompressible.

31. a. $\nabla \cdot (\nabla f) = \nabla \cdot (\nabla \text{ scalar}) = \nabla \cdot \text{vector} = \text{scalar}$

    b. $\nabla \times (\nabla \cdot \mathbf{F}) = \nabla \times (\nabla \cdot \text{vector}) = \nabla \times \text{scalar} = \text{undefined}$

    c. $\nabla (\nabla \times \mathbf{F}) = \nabla (\nabla \times \text{vector}) = \nabla \text{vector} = \text{undefined}$

    d. $\nabla (\nabla \cdot \mathbf{F}) = \nabla (\nabla \cdot \text{vector}) = \nabla \text{scalar} = \text{vector}$

    e. $\nabla \times (\nabla f) = \nabla x (\nabla \text{scalar}) = \nabla \times \text{vector} = \text{vector}$

33. $\mathbf{r}(x, y, z) = \langle x, y, z \rangle$

$$\nabla \times \mathbf{r} = \begin{vmatrix} \mathbf{i} & \mathbf{j} & \mathbf{k} \\ \frac{\partial}{\partial x} & \frac{\partial}{\partial y} & \frac{\partial}{\partial z} \\ x & y & z \end{vmatrix}$$

$$= \left(\frac{\partial(z)}{\partial y} - \frac{\partial(y)}{\partial z} \right)\mathbf{i} - \left(\frac{\partial(z)}{\partial x} - \frac{\partial(x)}{\partial z} \right)\mathbf{j} + \left(\frac{\partial(y)}{\partial x} - \frac{\partial(x)}{\partial y} \right)\mathbf{k}$$

$$= (0 - 0)\mathbf{i} - (0 - 0)\mathbf{j} + (0 - 0)\mathbf{k}$$

$$= \langle 0, 0, 0 \rangle$$

$$= 0$$

$$\text{div } \mathbf{r} = \nabla \cdot \mathbf{r}$$

$$= \frac{\partial(x)}{\partial x} + \frac{\partial(y)}{\partial y} + \frac{\partial(z)}{\partial z}$$

$$= 1 + 1 + 1$$

$$= 3$$

35. The divergence is positive, because if we draw a box around P, the outflow (top and right sides) is greater than the inflow (bottom and left sides).

37. The divergence appears to be zero, because if we draw a box around P, the outflow (top and right sides) appears to be the same as the inflow (bottom and left sides). Note that the graph could plausibly represent $\mathbf{F}(x, y, z) = \langle -x, y \rangle$, (in the second quadrant, anyway) and $\nabla \cdot \mathbf{F} = 0$.

39. The divergence is negative, because if we draw a box around P, the outflow (mostly on the right side) is less than the inflow (mostly on the left side).

51. For $f(x, y, z) = \sqrt{x^2 + y^2 + z^2}$,

$$\nabla f = \left\langle \frac{x}{\sqrt{x^2 + y^2 + z^2}}, \frac{y}{\sqrt{x^2 + y^2 + z^2}}, \frac{z}{\sqrt{x^2 + y^2 + z^2}} \right\rangle \text{ and}$$

$$\nabla \cdot (\nabla f) = \frac{2}{\sqrt{x^2 + y^2 + z^2}}.$$

53. For $\mathbf{F}(x, y) = \langle x^2, y^2 - 4x \rangle$, div $\mathbf{F} = 2x + 2y$.

    a. At $(0, 0)$, div $\mathbf{F} = 0$ so the flow in and out is equal.

    b. At $(1, 0)$, div $\mathbf{F} = 2$, so the flow out is greater than the flow in.

55. $\mathbf{F}(x, y, z) = \langle x, y^3, z \rangle$

Section 14.6

5. Possible answer:
$x = x,\ y = y,\ z = 3x + 4y,$
$-\infty < x < \infty, -\infty < y < \infty$

7. Possible answer:
$x = \cos u \cosh v,\ y = \sin u \cosh v,$
$z = \sinh v,\ 0 \le u \le 2\pi, -\infty < v < \infty$

9. Possible answer:
$x = 2\cos\theta,\ y = 2\sin\theta,\ z = z,$
$0 \le \theta \le 2\pi, 0 \le z \le 2$

11. Possible answer:
$x = r\cos\theta,\ y = r\sin\theta,$
$z = 4 - r^2, 0 \le r \le 2, 0 \le \theta \le 2\pi$

13.

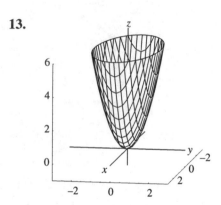

15.

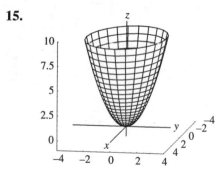

17.

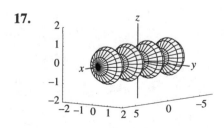

19.

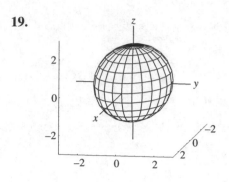

21. a. A

　 b. C

　 c. B

23. $z = f(x, y) = \sqrt{x^2 + y^2}$

$$\mathbf{n} = \langle f_x, f_y, -1 \rangle = \left\langle \frac{x}{\sqrt{x^2 + y^2}}, \frac{y}{\sqrt{x^2 + y^2}}, -1 \right\rangle$$

$$\|\mathbf{n}\| = \sqrt{\frac{x^2}{x^2 + y^2} + \frac{y^2}{x^2 + y^2} + 1} = \sqrt{2}$$

$$\iint_S dS = \iint_R \|\mathbf{n}\|\, dA$$
$$= \iint_R \sqrt{2}\, dA$$
$$= \left(\sqrt{2}\right)(\text{area of circle of radius } 2)$$
$$= 16\pi\sqrt{2}$$

25.　$z = f(x, y) = 6 - 3x - 2y$
$\mathbf{n} = \langle f_x, f_y, -1 \rangle = \langle -3, -2, -1 \rangle$
$\|\mathbf{n}\| = \sqrt{(-3)^2 + (-2)^2 + (-1)^2} = \sqrt{14}$
$\iint_S dS = \iint_R \|n\|\, dA$
$\qquad = \iint_R \sqrt{14}\, dA$
$\qquad = \left(\sqrt{14}\right)(\text{area of a circle of radius } 2)$
$\qquad = 4\pi\sqrt{14}$

27. $z = f(x, y) = \sqrt{x^2 + y^2}$

$$\mathbf{n} = \left\langle f_x, f_y, -1 \right\rangle = \left\langle \frac{x}{\sqrt{x^2 + y^2}}, \frac{y}{\sqrt{x^2 + y^2}}, -1 \right\rangle$$

$$\|\mathbf{n}\| = \sqrt{\frac{x^2}{x^2 + y^2} + \frac{y^2}{x^2 + y^2} + 1} = \sqrt{2}$$

$$\iint_S dS = \iint_R \|\mathbf{n}\| dA$$

$$= \iint_R \sqrt{2} dA$$

$$= \sqrt{2} \int_0^1 \int_x^1 dy \, dx$$

$$= \sqrt{2} \int_0^1 (1 - x) \, dx$$

$$= \sqrt{2} \left(x - \frac{1}{2}x^2 \right) \bigg|_0^1$$

$$= \frac{1}{2}\sqrt{2}$$

29. $z = f(x, y) = \sqrt{4 - x^2 - y^2}$

$$\mathbf{n} = \left\langle f_x, f_y, -1 \right\rangle$$

$$= \left\langle \frac{-x}{\sqrt{4 - x^2 - y^2}}, \frac{-y}{\sqrt{4 - x^2 - y^2}}, -1 \right\rangle$$

$$\|\mathbf{n}\| = \sqrt{\frac{x^2}{4 - x^2 - y^2} + \frac{y^2}{4 - x^2 - y^2} + 1}$$

$$= \sqrt{\frac{x^2 + y^2 + \left(4 - x^2 - y^2\right)}{4 - x^2 - y^2}}$$

$$= \frac{2}{\sqrt{4 - x^2 - y^2}}$$

$$\iint_S dS = \iint_R \|\mathbf{n}\| dA$$

$$= \iint_R \frac{2}{\sqrt{4 - x^2 - y^2}} dA$$

$$= \int_0^{2\pi} \int_0^2 \frac{2}{\sqrt{4 - r^2}} r \, dr \, d\theta$$

$$= \int_0^{2\pi} \left[-2\sqrt{4 - r^2} \right]_{r=0}^{r=\sqrt{3}}$$

$$= \int_0^{2\pi} 2 \, d\theta = 4\pi$$

Alternative method: Use spherical coordinates.

$(x, y, z) = (2\sin\phi\cos\theta, 2\sin\phi\sin\theta, 2\cos\theta)$

$dS = 4\sin\phi \, d\phi \, d\theta$

$$\iint_S dS = \int_0^{2\pi} \int_0^{\frac{\pi}{3}} 4\sin\phi \, d\phi \, d\theta$$

$$= \int_0^{2\pi} \left[-4\cos\phi \right]_{\phi=0}^{\phi=\pi/3} d\theta$$

$$= \int_0^{2\pi} (2) \, d\theta$$

$$= 4\pi$$

31. $\mathbf{n} = \left\langle 2, 3, -1 \right\rangle$

$$\|\mathbf{n}\| = \sqrt{2^2 + 3^2 + (-1)^2} = \sqrt{14}$$

$$\iint_S x \, dS = \iint_R x \|\mathbf{n}\| dA$$

$$= \int_1^3 \int_1^2 x\sqrt{14} \, dx \, dy$$

$$= \sqrt{14} \int_1^3 \left[\frac{1}{2}x^2 \right]_{x=1}^{x=2} dy$$

$$= \sqrt{14} \int_1^3 \frac{3}{2} dy$$

$$= \sqrt{14} \cdot 2 \cdot \frac{3}{2}$$

$$= 3\sqrt{14}$$

33. $\mathbf{n} = \left\langle 2x, 2y, -1 \right\rangle$

$$\|\mathbf{n}\| = \sqrt{4x^2 + 4y^2 + 1}$$

$$\iint_S y \, dS = \iint_R y \|\mathbf{n}\| dA$$

$$= \iint_R y\sqrt{4x^2 + 4y^2 + 1} \, dA$$

$$= \int_0^{2\pi} \int_0^2 (r\sin\theta)\sqrt{4r^2 + 1} \, r \, dr \, d\theta$$

$$= \int_0^{2\pi} \int_0^2 r^2 \sin\theta \sqrt{4r^2 + 1} \, dr \, d\theta$$

$$= \int_0^2 \left(r^2\sqrt{4r^2 + 1} \right) \left(\int_0^{2\pi} \sin\theta \, d\theta \right) dr$$

$$= \int_0^2 \left(r^2\sqrt{4r^2 + 1} \right) (0) \, dr$$

$$= 0$$

35. $\mathbf{n} = \langle -2x, -2y, -1 \rangle$

$\|\mathbf{n}\| = \sqrt{4x^2 + 4y^2 + 1}$

$\displaystyle\iint_S \left(x^2 + y^2\right) dS$

$\displaystyle = \iint_S \left(x^2 + y^2\right)\|\mathbf{n}\| \, dA$

$\displaystyle = \iint_S \left(x^2 + y^2\right)\sqrt{4x^2 + 4y^2 + 1} \, dA$

$\displaystyle = \int_0^{2\pi} \int_0^2 r^2 \sqrt{4r^2 + 1} \; r \, dr \, d\theta$

$\displaystyle = \int_0^{2\pi} \int_0^2 r^3 \sqrt{4r^2 + 1} \; dr \, d\theta$

$\displaystyle = 2\pi \int_0^2 r^3 \sqrt{4r^2 + 1} \; dr$

Integrate by

parts: $u = r^2 \qquad dv = r\sqrt{4r^2 = 1}\, dr$

$\qquad du = 2r\,dr \qquad v = \dfrac{1}{12}\left(4r^2 + 1\right)^{3/2}$

$\displaystyle \int r^3 \sqrt{4r^2 + 1}$

$\displaystyle = \frac{r^2}{12}\left(4r^2 + 1\right)^{3/2} - \int \frac{1}{6} r \left(4r^2 + 1\right)^{3/2} dr$

$\displaystyle = \frac{r^2}{12}\left(4r^2 + 1\right)^{3/2} - \frac{1}{120}\left(4r^2 + 1\right)^{5/2} + c$

$\displaystyle = \left[\frac{r^2}{12}\left(4r^2 + 1\right) - \frac{1}{120}\left(4r^2 + 1\right)^2\right]\sqrt{4r^2 + 1} + c$

$\displaystyle = \frac{1}{120}\left(24r^4 + 2r^2 - 1\right)\sqrt{4r^2 + 1} + c$

$\displaystyle \iint_S \left(x^2 + y^2\right) dS$

$\displaystyle = 2\pi \int_0^2 r^3 \sqrt{4r^2 + 1}\, dr$

$\displaystyle = \frac{\pi}{60}\left(24r^4 + 2r^2 - 1\right)\sqrt{4r^2 + 1}\,\Big|_{r=0}^{r=2}$

$\displaystyle = \frac{\pi}{60}\left(1 + 391\sqrt{17}\right)$

37. $\mathbf{n} = \left\langle \dfrac{x}{\sqrt{x^2 + y^2}}, \dfrac{y}{\sqrt{x^2 + y^2}}, -1 \right\rangle$

$\|\mathbf{n}\| = \sqrt{\dfrac{x^2}{x^2 + y^2} + \dfrac{y^2}{x^2 + y^2} + 1}$

$\qquad = \sqrt{2}$

$\displaystyle \iint_S z \, dS = \iint_R \sqrt{x^2 + y^2}\,\sqrt{2}\, dA$

$\displaystyle = \int_0^{2\pi} \int_0^4 r\sqrt{2}\, r \, dr \, d\theta$

$\displaystyle = \sqrt{2} \int_0^{2\pi} \int_0^4 r^2 \, dr \, d\theta$

$\displaystyle = \sqrt{2} \int_0^{2\pi} \left[\frac{1}{3} r^3\right]_0^4 d\theta$

$\displaystyle = \sqrt{2} \int_0^{2\pi} \frac{64}{3}\, d\theta$

$\displaystyle = \frac{128}{3} \pi \sqrt{2}$

39.   Use spherical coordinates.

$(x, y, z) = (2\sin\phi\cos\theta, 2\sin\phi\sin\theta, 2\cos\phi)$

$dS = 4\sin\phi\, d\phi\, d\theta$

Note that $x^2 + y^2 + z^2 = 4$.

$\displaystyle \iint_S \left(x^2 + y^2 + z^2\right) dS = \int_0^{2\pi} \int_0^{\frac{\pi}{2}} (4)(4\sin\phi)\, d\phi\, d\theta$

$\displaystyle = \int_0^{2\pi} \left[-16\cos\phi\right]_{\phi=0}^{\phi=\frac{\pi}{2}} d\theta$

$\displaystyle = \int_0^{2\pi} 16\, d\theta$

$\displaystyle = 32\pi$

41. $\mathbf{n} = \dfrac{\langle 2x, 2y, 1 \rangle}{\sqrt{4x^2 + 4y^2 + 1}}$

$dS = \sqrt{4x^2 + 4y^2 + 1}\, dA$

$\displaystyle \iint_S \mathbf{F} \cdot \mathbf{n}\, dS = \iint_R \langle x, y, z \rangle \cdot \dfrac{\langle 2x, 2y, 1 \rangle}{\sqrt{4x^2 + 4y^2 + 1}} \sqrt{4x^2 + 4y^2 + 1}\, dA$

$\displaystyle \qquad = \iint_R \left(2x^2 + 2y^2 + z \right) dA$

$\displaystyle \qquad = \int_0^{2\pi} \int_0^2 \left[2r^2 + \left(4 - r^2 \right) \right] r\, dr\, r\theta$

$\displaystyle \qquad = \int_0^{2\pi} \int_0^2 (r^3 + 4r)\, dr\, d\theta$

$\displaystyle \qquad = \int_0^{2\pi} \left[\frac{1}{4} r^4 + 2r^2 \right]_{r=0}^{r=2} d\theta$

$\displaystyle \qquad = \int_0^{2\pi} 12\, d\theta$

$\qquad = 24\pi$

43. $\mathbf{n} = \left\langle \dfrac{x}{\sqrt{2\left(x^2 + y^2\right)}}, \dfrac{y}{\sqrt{2\left(x^2 + y^2\right)}}, -\dfrac{1}{\sqrt{2}} \right\rangle$

$dS = \sqrt{2}\, dA$

$\displaystyle \iint_S \mathbf{F} \cdot \mathbf{n}\, dA = \iint_R \langle y, -x, z \rangle \cdot \left\langle \dfrac{x}{\sqrt{2\left(x^2 + y^2\right)}}, \dfrac{y}{\sqrt{2\left(x^2 + y^2\right)}}, -\dfrac{1}{\sqrt{2}} \right\rangle \sqrt{2}\, dA$

$\displaystyle \qquad = \iint_R \left(\dfrac{xy}{\sqrt{2\left(x^2 + y^2\right)}} - \dfrac{xy}{\sqrt{2\left(x^2 + y^2\right)}} - \dfrac{z}{\sqrt{2}} \right) \sqrt{2}\, dA$

$\displaystyle \qquad = \iint_R (-z)\, dA$

$\displaystyle \qquad = -\int_0^{2\pi} \int_0^3 (r)\, r\, dr\, d\theta$

$\displaystyle \qquad = -\int_0^{2\pi} \left[\frac{1}{3} r^3 \right]_0^3 d\theta$

$\displaystyle \qquad = -\int_0^{2\pi} 9\, d\theta$

$\qquad = -18\pi$

45. Back surface ($x = 0$):

$\mathbf{F}\cdot\mathbf{n} = \left\langle xy, y^2, z\right\rangle\cdot\left\langle -1, 0, 0\right\rangle = -xy = 0$

Left surface ($y = 0$):

$\mathbf{F}\cdot\mathbf{n} = \left\langle xy, y^2, z\right\rangle\cdot\left\langle 0, -1, 0\right\rangle = -y^2 = 0$

Bottom surface ($z = 0$):

$\mathbf{F}\cdot\mathbf{n} = \left\langle xy, y^2, z\right\rangle\cdot\left\langle 0, 0, -1\right\rangle = -z = 0$

Front surface ($x = 1$):

$\mathbf{F}\cdot\mathbf{n} = \left\langle xy, y^2, z\right\rangle\cdot\left\langle 1, 0, 0\right\rangle = xy = y$

$\int_0^1\int_0^1 y\,dy\,dz = \int_0^1 \frac{1}{2}dz = \frac{1}{2}$

Right surface ($y = 1$):

$\mathbf{F}\cdot\mathbf{n} = \left\langle xy, y^2, z\right\rangle\cdot\left\langle 0, 1, 0\right\rangle = y^2 = 1$

$\int_0^1\int_0^1 1\,dx\,dz = 1$

Top surface ($z = 1$):

Therefore, $\iint_S \mathbf{F}\cdot\mathbf{n}\,dS = \frac{1}{2}+1+1 = \frac{5}{2}$.

47. Note that on the base ($z = 0$),

$\mathbf{F}\cdot\mathbf{n} = \left\langle 1, 0, z\right\rangle\cdot\left\langle 0, 0, -1\right\rangle = -z = 0.$

Therefore, we need only consider the portion of $z = 1-x^2-y^2$ above the xy-plane.

$\mathbf{n} = \dfrac{\left\langle 2x, 2y, 1\right\rangle}{\sqrt{4x^2, 4y^2+1}}$

$dS = \sqrt{4x^2+4y^2+1}\,dA$

$\iint_S \mathbf{F}\cdot\mathbf{n}\,dS$

$= \iint_R \left\langle 1, 0, z\right\rangle\cdot\dfrac{\left\langle 2x, 2y, 1\right\rangle}{\sqrt{4x^2+4y^2+1}}\sqrt{4x^2+4y^2+1}\,dA$

$= \iint_R (2x+z)\,dA$

$= \int_0^{2\pi}\int_0^1 \left(2r\sin\theta+1-r^2\right)r\,dr\,d\theta$

$= \int_0^{2\pi}\int_0^1 \left(2r^2\sin\theta+r-r^3\right)dr\,d\theta$

$= \int_0^{2\pi}\left[\frac{2}{3}r^3\sin\theta+\frac{1}{2}r^2-\frac{1}{4}r^3\right]_{r=0}^{r=1}d\theta$

$= \int_0^{2\pi}\left(\frac{2}{3}\sin\theta+\frac{1}{4}\right)d\theta$

$= \left(-\frac{2}{3}\cos\theta+\frac{1}{4}\right)\Big|_{\theta=0}^{\theta=2\pi}$

$= \frac{\pi}{2}$

49. $\mathbf{n} = \dfrac{\left\langle 1,1,1\right\rangle}{\sqrt{3}}$

$dS = \sqrt{3}\,dA = \sqrt{3}\,dx\,dy$

$\iint_S \mathbf{F}\cdot\mathbf{n}\,dS = \int_0^1\int_0^1\left\langle yx, 1, x\right\rangle\cdot\dfrac{\left\langle 1,1,1\right\rangle}{\sqrt{3}}\sqrt{3}\,dx\,dy$

$= \int_0^1\int_0^1\left\langle yx+1+x\right\rangle dx\,dy$

$= \int_0^1\left[\frac{1}{2}x^2y+x+\frac{1}{2}x^2\right]_{x=0}^{x=1}dy$

$= \int_0^1\left(\frac{1}{2}y+\frac{3}{2}\right)dy$

$= \left(\frac{1}{4}y^2+\frac{3}{2}y\right)\Big|_{y=0}^{y=1}$

$= \frac{7}{4}$

51. Note that the upper and lower surfaces intersect where $\sqrt{8 - x^2 - y^2} = \sqrt{x^2 + y^2}$,

which gives $x^2 + y^2 = 4$, or $r = 2$. For the upper surface S_1:

$$n = \frac{\langle x, y, z \rangle}{\sqrt{x^2 + y^2 + z^2}} = \frac{\langle x, y, z \rangle}{\sqrt{8}}$$

$$dS = \sqrt{(f_x)^2 + (f_y)^2 + 1}\, dA$$

$$= \sqrt{\left(\frac{x}{\sqrt{8 - x^2 - y^2}} \right)^2 + \left(\frac{y}{\sqrt{8 - x^2 - y^2}} \right)^2 + 1}\, dA$$

$$= \sqrt{\frac{8}{8 - x^2 - y^2}}\, dA$$

$$\iint_S \mathbf{F} \cdot \mathbf{n}\, dS = \iint_R \langle y, 0, 2 \rangle \cdot \frac{\langle x, y, z \rangle}{\sqrt{8}} \sqrt{\frac{8}{8 - x^2 - y^2}}\, dA$$

$$= \iint_R \frac{xy}{\sqrt{8 - x^2 - y^2}}\, dA + \iint \frac{2\sqrt{8 - x^2 - y^2}}{\sqrt{8 - x^2 - y^2}}\, dA$$

$$= 0 + \iint_R 2\, dA$$

$$= 2\,(\text{area of circle of radius 2})$$

$$= 8\pi$$

Note that the first integral is 0 because $\dfrac{xy}{\sqrt{8 - x^2 - y^2}}$ is an odd function of x (or of y).

For the lower surface of S_2:

$$n = \left\langle \frac{x}{\sqrt{2(x^2 + y^2)}}, \frac{y}{\sqrt{2(x^2 + y^2)}}, -\frac{1}{\sqrt{2}} \right\rangle$$

$$dS = \sqrt{2}\, dA$$

$$\iint_{S_2} \mathbf{F} \cdot \mathbf{n}\, dA = \iint_R \langle y, 0, 2 \rangle \cdot \left\langle \frac{x}{\sqrt{2(x^2 + y^2)}}, \frac{y}{\sqrt{2(x^2 + y^2)}}, -\frac{1}{\sqrt{2}} \right\rangle \sqrt{2}\, dA$$

$$= \iint_R \left(\frac{xy}{\sqrt{x^2 + y^2}} - 2 \right) dA$$

$$= \iint \frac{xy}{\sqrt{x^2 + y^2}}\, dA - \iint_R 2\, dA$$

$$= 0 - \iint_R 2\, dA$$

$$= -2\,(\text{area of circle of radius 2})$$

$$= -8\pi$$

Again, the first integral is 0 because $\dfrac{xy}{\sqrt{x^2 + y^2}}$ is an odd function of x (or of y).

Combine the results above: $\iint_S \mathbf{F} \cdot \mathbf{n}\, dS = 8\pi - 8\pi = 0$

53. $z = 6 - 3x - 2y$

$\mathbf{n} = \langle -3, -2, -1 \rangle$

$dS = \sqrt{(-3)^2 + (-2)^2 + (-1)^2} \; dA = \sqrt{14} \; dA$

$m = \iint_S \rho \; dS$

$\quad = \iint_R (x^2 + 1)\sqrt{14} \; dA$

$\quad = \int_0^{2\pi} \int_0^2 (r^2 \cos^2 \theta + 1)\sqrt{14} \; r \; dr \; d\theta$

$\quad = \sqrt{14} \int_0^{2\pi} \int_0^2 (r^3 \cos^2 \theta + r) \; dr \; d\theta$

$\quad = \sqrt{14} \int_0^{2\pi} \left[\frac{1}{4} r^4 \cos^2 \theta + \frac{1}{2} r^2 \right]_{r=0}^{r=2} d\theta$

$\quad = \sqrt{14} \int_0^{2\pi} (4\cos^2 \theta + 2) d\theta$

$\quad = \sqrt{14} \int_0^{2\pi} (4 + 2\cos 2\theta) d\theta$

$\quad = \sqrt{14}(4\theta + \sin 2\theta)\Big|_0^{2\pi}$

$\quad = 8\pi\sqrt{14}$

$\bar{x} = \frac{1}{m} \iint_S x\rho \; dS$

$\quad = \frac{1}{m} \iint_R (x)(x^2 + 1)\left(\sqrt{14}\right) dA$

$\quad = 0 \text{ (by symmetry)}$

$\bar{y} = \frac{1}{m} \iint_S y\rho \; dS$

$\quad = \frac{1}{m} \iint_R (y)(x^2 + 1)\left(\sqrt{14}\right) dA$

$\quad = 0 \text{ (by symmetry)}$

$\bar{z} = \frac{1}{m} \iint_S z\rho \; dS$

$\quad = \frac{1}{m} \iint_R (6 - 3x - 2y)(x^2 + 1)\sqrt{14} dS$

$\quad = \frac{6}{m} \iint_R (x^2 + 1)\sqrt{14} dS + \frac{1}{m} \iint (-3x - 2y)(x^2 + 1)\sqrt{14} dS$

$\quad = \frac{6}{m}(m) + 0$

$\quad = 6$

Note: the first integral is m because it is the same as the integral we evaluated to find m; the second integral is 0 by symmetry.

55. Use spherical coordinates. Note that ρ means density in this problem; the radius is always 1 here.

$(x, y, z) = (\sin\phi\cos\theta, \sin\phi\sin\theta, \cos\phi)$

$\rho = 1 + x = 1 + \sin\phi\ \cos\theta$

$dS = \sin\phi\ d\phi\ d\theta$

$m = \iint_S \rho\ dS$

$\quad = \int_0^{2\pi}\int_0^{\pi/2}(1+\sin\phi\ \cos\theta)\sin\phi\ d\phi\ d\theta$

$\quad = \int_0^{2\pi}\int_0^{\pi/2}(\sin\phi+\sin^2\phi\ \cos\theta)d\phi\ d\theta$

$\quad = \int_0^{2\pi}\int_0^{\pi/2}\left(\sin\phi+\frac{1}{2}\cos\theta-\frac{1}{2}\cos 2\phi\cos\theta\right)d\phi d\theta$

$\quad = \int_0^{2\pi}\left[-\cos\phi+\frac{1}{2}\phi\cos\theta-\frac{1}{4}\sin 2\phi\cos\theta\right]_{\phi=0}^{\phi=\pi/2}d\theta$

$\quad = \int_0^{2\pi}\left(\frac{\pi}{4}\cos\theta+1\right)d\theta$

$\quad = \left(\frac{\pi}{4}\sin\theta+\theta\right)\Big|_{\theta=0}^{\theta=2\pi}$

$\quad = 2\pi$

$\bar{x} = \frac{1}{m}\iint_S x\rho\ dS$

$\quad = \frac{1}{2\pi}\int_0^{2\pi}\int_0^{\pi/2}(\sin\phi\ \cos\theta)(1+\sin\phi\ \cos\theta)\sin\phi\ d\phi\ d\theta$

$\quad = \frac{1}{2\pi}\int_0^{2\pi}\int_0^{\pi/2}(\sin^2\phi\ \cos\theta+\sin^3\phi\cos^2\theta)d\phi\ d\theta$

$\quad = \frac{1}{2\pi}\int_0^{\pi/2}\int_0^{2\pi}\left(\sin^2\phi\ \cos\theta+\frac{1}{2}\sin^3\phi+\frac{1}{2}\sin^3\phi\ \cos 2\theta\right)d\theta\ d\phi$

$\quad = \frac{1}{2\pi}\int_0^{\pi/2}\left[\sin^2\phi\ \sin\theta+\frac{1}{2}\theta\sin^3\phi+\frac{1}{4}\sin^3\phi\sin 2\theta\right]_{\theta=0}^{\theta=2\pi}d\phi$

$\quad = \frac{1}{2\pi}\int_0^{\pi/2}\pi\sin^3\phi\ d\phi$

$\quad = \frac{1}{2}\int_0^{\pi/2}(\sin\phi-\cos^2\phi\sin\phi)d\phi$

$\quad = \frac{1}{2}\left(-\cos\phi+\frac{1}{3}\cos^3\phi\right)\Big|_0^{\pi/2}$

$\quad = \frac{1}{3}$

$\bar{y} = \frac{1}{m}\iint_S y\rho\ dS$

$\quad = \frac{1}{m}\iint_S (y)(1+x)dS$

$\quad = 0$ (by symmetry)

$$\overline{z} = \frac{1}{m}\iint_S z\rho \; dS$$

$$= \frac{1}{2\pi}\int_0^{2\pi}\int_0^{\pi/2}(\cos\phi)(1+\sin\phi\cos\theta)\sin\phi \; d\phi \; d\theta$$

$$= \frac{1}{2\pi}\int_0^{2\pi}\int_0^{\pi/2}(\cos\phi\sin\phi+\sin^2\phi\cos\phi\cos\theta)d\phi \; d\theta$$

$$= \frac{1}{2\pi}\int_0^{2\pi}\left[\frac{1}{2}\sin^2\phi+\frac{1}{3}\sin^3\phi\cos\theta\right]_{\phi=0}^{\phi=\pi/2} d\theta$$

$$= \frac{1}{2\pi}\int_0^{2\pi}\left(\frac{1}{2}+\frac{1}{3}\cos\theta\right)d\theta$$

$$= \frac{1}{2\pi}\left(\frac{1}{2}\theta+\frac{1}{3}\sin\theta\right)\Big|_{\theta=0}^{\theta=2\pi}$$

$$= \frac{1}{2\pi}\left(\frac{1}{2}\cdot 2\pi\right)$$

$$= \frac{1}{2}$$

57. $\iint_S g(x,\; y,\; z)dS = \iint_R g(f(y,\; z),\; y,z)\sqrt{(f_y)^2+(f_z)^2+1}\; dA$ where S is given by $f(y,z)$ for (y,z) in region R.

59. $x = f(y,\; z) = \sqrt{1-y^2}$

$$\|\mathbf{n}\| = \sqrt{(f_y)^2+(f_z)^2+1}$$

$$= \sqrt{\left(\frac{-y}{\sqrt{1-y}}\right)^2+(0)^2+1}$$

$$= \sqrt{\frac{y^2}{1-y^2}+1}$$

$$= \frac{1}{\sqrt{1-y^2}}$$

$$\iint_S z\; dS = \iint_R (z)\left(\frac{1}{\sqrt{1-y^2}}\right)dA$$

$$= \int_{-1}^1\int_1^2 \frac{z}{\sqrt{1-y^2}}dz\; dy$$

$$= \int_{-1}^1\left[\frac{1}{\sqrt{1-y^2}}\left(\frac{1}{2}z^2\right)\right]_{z=1}^{z=2}dy$$

$$= \frac{3}{2}\int_{-1}^1\frac{1}{\sqrt{1-y^2}}dy$$

$$= \frac{3}{2}(\sin^{-1}y)\Big|_{y=-1}^{y=1}$$

$$= \frac{3}{2}\pi$$

61. $x = f(y, z) = 9 - y^2 - z^2$

$$\|\mathbf{n}\| = \sqrt{(f_y)^2 + (f_y)^2 + 1}$$
$$= \sqrt{(-2y)^2 + (-2z)^2 + 1}$$
$$= \sqrt{4y^2 + 4z^2 + 1}$$

$$\iint (y^2 + z^2)\, dS$$
$$= \iint_R (y^2 + z^2)\sqrt{4y^2 + 4x^2 + 1}\ dA$$
$$= \int_0^{2\pi}\int_0^3 r^2\sqrt{4r^2 + 1}\ r\ dr\ d\theta$$
$$= 2\pi\int_0^3 r^3\sqrt{4r^2 + 1}\,dr$$
$$= 2\pi \cdot \frac{1}{120}(24r^4 + 2r^2 - 1)\sqrt{4r^2 + 1}\ \Big|_{r=0}^{r=3}$$
$$= \frac{\pi}{60}(1 + 1961\sqrt{37})$$
$$\approx 198.8\pi$$

Note: $x = f(y, z) = \sqrt{4 - y^2 - z^2}$

63. $y = f(x, z) = x^2 + z^2$

$$\|\mathbf{n}\| = \sqrt{(f_x)^2 + (f_z)^2 + 1}$$
$$= \sqrt{(2x)^2 + (2x)^2 + 1}$$
$$= \sqrt{4x^2 + 4x^2 + 1}$$

$$\iint x^2\, dS$$
$$= \iint_R x^2\sqrt{4x^2 + 4z^2 + 1}\ dA$$
$$= \int_0^{2\pi}\int_0^1 r^2\cos^2\theta\sqrt{4r^2 + 1}\,r\ dr\,d\theta$$
$$= \int_0^1 r^3\sqrt{4r^2 + 1}\left(\int_0^{2\pi}\cos^2\theta\ d\theta\right)dr$$
$$= \int_0^1 r^3\sqrt{4r^2 + 1}\left[\int_0^{2\pi}\left(\frac{1}{2} + \frac{1}{2}\cos 2\theta\right)d\theta\right]dr$$
$$= \int_0^1 r^3\sqrt{4r^2 + 1}\left[\frac{1}{2}\theta + \frac{1}{4}\sin 2\theta\right]_{\theta=0}^{\theta=2\pi}dr$$
$$= \pi\int_0^1 r^3\sqrt{4r^2 + 1}\ dr$$
$$= \pi \cdot \frac{1}{120}(24r^4 + 2r^2 - 1)\sqrt{4r^2 + 1}\ \Big|_{r=0}^{r=1}$$
$$= \frac{\pi}{120}(25\sqrt{5} + 1)$$
$$\approx .474\pi$$

Note: $\int r^3\sqrt{4r^2 + 1}$ can be evaluated using integration by parts, as shown in the solution to exercise 35.

65. $y = f(x, z) = 1 - x^2$

$$\|\mathbf{n}\| = \sqrt{(f_x)^2 + (f_z)^2 + 1}$$
$$= \sqrt{(-2x)^2 + (0)^2 + 1}$$
$$= \sqrt{4x^2 + 1}$$

$$\iint_S 4\,dS = \iint_R 4\sqrt{4x^2 + 1}\,dA$$
$$= \int_{=1}^{1}\int_0^2 4\sqrt{4x^2 + 1}\,dz\,dx$$
$$= \int_{-1}^{1} 8\sqrt{4x^2 + 1}\,dx$$
$$\approx 23.66$$

Note: We approximated the integral numerically.

67. The flux integral over any portion of the cone is zero because **F** is always perpendicular to the normal vector **m** (and hence to the normal unit vector **n**).

$\mathbf{F} \cdot \mathbf{n}$

$$= \langle x, y, z\rangle \cdot \left\langle \frac{x}{\sqrt{x^2 + y^2}}, \frac{y}{\sqrt{x^2 + y^2}}, -1 \right\rangle$$

$$= \left\langle x, y, \sqrt{x^2 + y^2}\right\rangle \cdot \left\langle \frac{x}{\sqrt{x^2 + y^2}}, \frac{y}{\sqrt{x^2 + y^2}}, -1 \right\rangle$$

$$= \frac{x^2}{\sqrt{x^2 + y^2}} + \frac{y^2}{\sqrt{x^2 + y^2}} - \sqrt{x^2 + y^2}$$

$$= \frac{x^2 + y^2}{\sqrt{x^2 + y^2}} - \sqrt{x^2 + y^2}$$

$$= 0$$

Section 14.7

5. Compute $\iint_{\partial Q} \mathbf{F} \cdot \mathbf{n}\, dS$:

Back surface $(x - 0)$:
$$\mathbf{F} \cdot \mathbf{n} = \left\langle 2xz,\ y^2,\ -xz \right\rangle \cdot \langle -1,\ 0,\ 0 \rangle$$
$$= 2xz$$
$$= 0$$

Left surface $(y = 0)$:
$$\mathbf{F} \cdot \mathbf{n} = \left\langle 2xz,\ y^2,\ -xz \right\rangle \cdot \langle 0,\ -1,\ 0 \rangle$$
$$= -y^2$$
$$= 0$$

Bottom surface $(z = 0)$:
$$\mathbf{F} \cdot \mathbf{n} = \left\langle 2xz,\ y^2,\ -xz \right\rangle \cdot \langle 0,\ 0,\ -1 \rangle$$
$$= xz$$
$$= 0$$

Front surface $(x = 1)$:
$$\mathbf{F} \cdot \mathbf{n} = \left\langle 2xz,\ y^2,\ -xz \right\rangle \cdot \langle 1,\ 0,\ 0 \rangle$$
$$= 2xz$$
$$= 2z$$
$$\int_0^1 \int_0^1 1\, dx\, dz = \int_0^1 1\, dz = 1$$

Top surface $(z = 1)$:
$$\mathbf{F} \cdot \mathbf{n} = \left\langle 2xz,\ y^2,\ -xz \right\rangle \cdot \langle 0,\ 0,\ 1 \rangle$$
$$= -xz$$
$$= -x$$
$$\int_0^1 \int_0^1 -x\, dx\, dy = \int_0^1 -\frac{1}{2}\, dy = -\frac{1}{2}$$

Summing the above results,
$$\iint_{\partial Q} \mathbf{F} \cdot \mathbf{n}\, dS = 1 + 1 - \frac{1}{2} = \frac{3}{2}$$

Compute $\iiint_Q \nabla \cdot \mathbf{F}(x,\ y,\ z)\, dV$:

$$\nabla \cdot \mathbf{F} = \nabla \left\langle 2xz,\ y^2,\ -xz \right\rangle$$
$$= 2z + 2y - z$$

$$\iiint_Q \nabla \cdot \mathbf{F}(x,\ y,\ z)\, dV$$
$$= \int_0^1 \int_0^1 \int_0^1 (2z + 2y - x)\, dx\, dy\, dz$$
$$= \int_0^1 \int_0^1 \left[2xz + 2xy - \frac{1}{2}x^2 \right]_{x=0}^{x=1} dy\, dz$$
$$= \int_0^1 \int_0^1 \left(2z + 2y - \frac{1}{2} \right) dy\, dz$$
$$= \int_0^1 \left[2yz + y^2 - \frac{1}{2}y \right]_{y=0}^{y=1} dz$$
$$= \int_0^1 \left(2z + \frac{1}{2}z \right) dz$$
$$= \left(z^2 + \frac{1}{2}z \right)\Big|_{z=0}^{z=1}$$
$$= \frac{3}{2}$$

7. Compute $\iint_{\partial Q} \mathbf{F} \cdot \mathbf{n}\, dS$:

On the bottom surface $(z = 0)$, $\mathbf{F} \cdot \mathbf{n} = \left\langle xz,\ xy,\ 2x^2 \right\rangle \cdot \left\langle 0,\ 0,\ -1 \right\rangle = -2z^2 = 0$, so we need only consider the top surface.

$z = f(x,\ y) = 1 - x^2 - y^2$

$\mathbf{n} = \dfrac{\langle 2x, 2y, 1 \rangle}{\sqrt{4x^2 + 4y^2 + 1}}$

$dS = \sqrt{4x^2 + 4y^2 + 1}\ dA$

$\iint_{\partial Q} \mathbf{F} \cdot \mathbf{n}\ dS$

$= \iint_R \left\langle xz,\ zy,\ 2z^2 \right\rangle \cdot \dfrac{\langle 2x,\ 2y,\ 1 \rangle}{\sqrt{4x^2 + 4y^2 + 1}} \sqrt{4x^2 + 4y^2 + 1}\ dA$

$= \iint_R (2x^2 z + 2y^2 z + 2z^2)\ dA$

$= \iint_R 2z(x^2 + y^2 + x)\, dA$

$= \iint_R 2(1 - x^2 - y^2)[x^2 + y^2 + (1 - x^2 - y^2)]\, dA$

$= \iint_R 2(1 - x^2 - y^2)\, dA$

$= \int_0^{2\pi} \int_0^1 (2 - 2r^2)r\, dr\, d\theta$

$= \int_0^{2\pi} \left[r^2 - \dfrac{1}{2} r^4 \right]_{r=0}^{r=1} d\theta$

$= \int_0^{2\pi} \dfrac{1}{2} d\theta$

$= \pi$

$\iiint_Q \nabla \cdot \mathbf{F}(x,\ y,\ z)\ dS$:

$\nabla \cdot \mathbf{F}(x,\ y,\ z) = \nabla \cdot \mathbf{F}\left\langle xz,\ zy,\ 2z^2 \right\rangle$

$\qquad\qquad\quad = z + z + 4z$

$\qquad\qquad\quad = 6z$

$\iiint_Q \nabla \cdot \mathbf{F}\ dS = \int_0^{2\pi} \int_0^1 \int_0^{1-r^2} (6z)r\ dz\ dr\ d\theta$

$\qquad\qquad\quad = \int_0^{2\pi} \int_0^1 \left[3rz^2 \right]_{z=0}^{z=1-r^2} d\, rd\theta$

$\qquad\qquad\quad = \int_0^{2\pi} \int_0^1 3r(1 - r^2)^2\ dr\ d\theta$

$\qquad\qquad\quad = \int_0^{2\pi} \int_0^1 (3r^5 - 6r^3 + 3r)dr d\theta$

$\qquad\qquad\quad = \int_0^{2\pi} \left[\dfrac{1}{2} r^6 - \dfrac{3}{2} r^4 + \dfrac{3}{2} r^2 \right]_{r=0}^{r=1} d\theta$

$\qquad\qquad\quad = \int_0^{2\pi} \dfrac{1}{2} d\theta$

$\qquad\qquad\quad = \pi$

9. $\nabla \cdot \mathbf{F}(x, y, z) = \nabla \cdot \left\langle 2x - y^2, 4xz - 2y, xy^3 \right\rangle$

$= 2 - 2 + 0$

$= 0$

$\iint_{\partial Q} \mathbf{F} \cdot \mathbf{n} \, dS = \iiint_Q \nabla \cdot \mathbf{F}(x, y, z) \, dV$

$= \iiint_Q 0 \, dV$

$= 0$

11. $\nabla \cdot \mathbf{F}(x, y, z) = \nabla \cdot \left\langle 4y^2, 3z - \cos x, z^3 - x \right\rangle$

$= 0 + 0 + 3z^2$

$= 3z^2$

$\iint_{\partial Q} \mathbf{F} \cdot \mathbf{n} \, dS = \iiint_Q \nabla \cdot \mathbf{F}(x, y, z) dV$

$= \int_{-1}^1 \int_{-1}^1 \int_{-1}^1 3z^2 dx \, dy \, dz$

$= \int_{-1}^1 \int_{-1}^1 6z^2 dy \, dz$

$= \int_{-1}^1 12z^2 \, dz$

$= 4z^3 \Big|_{z=-1}^{z=1}$

$= 8$

13. $\nabla \cdot \mathbf{F}(x, y, z) = \nabla \cdot \left\langle x^3, y^3 - z, xy^2 \right\rangle$

$= 3x^2 + 3y^2 + 0$

$= 3x^2 + 3y^2$

$\iint_{\partial Q} \mathbf{F} \cdot \mathbf{n} \, dS = \iiint_Q \nabla \cdot \mathbf{F}(x, y, z) \, dV$

$= \iiint_{Qd} 3(x^2 + y^2) dV$

$= \int_0^{2\pi} \int_0^2 \int_{r^2}^4 (3r^2) r \, dz \, dr \, d\theta$

$= \int_0^{2\pi} \int_0^2 [3r^3 z]_{z=r^2}^{z=4} dr d\theta$

$= \int_0^{2\pi} \int_0^2 (12r^3 - 3r^5) dr d\theta$

$= \int_0^{2\pi} \left[3r^4 - \frac{1}{2} r^6 \right]_{r=0}^{r=2} d\theta$

$= \int_0^{2\pi} 16 d\theta$

$= 32\pi$

15. $\nabla \cdot \mathbf{F}(x, y, z) = \nabla \cdot \left\langle y^3, x + z^2, z + y^2 \right\rangle$

$= 0 + 0 + 1$

$= 1$

$\iint_{\partial Q} \mathbf{F} \cdot \mathbf{n} \, dS = \iiint_Q \nabla \cdot \mathbf{F}(x, y, z) dV$

$= \iiint_Q 1 \, dV$

$=$ (volume of cone with radius 4

and height 4)

$= \frac{1}{3} \cdot (\pi \cdot 4^2)(4)$

$= \frac{64}{3} \pi$

17. $\nabla \cdot \mathbf{F}(x, y, z) = \nabla \cdot \left\langle x - y^3, x^2 \sin z, 3z \right\rangle$

$= 1 + 0 + 3$

$= 4$

$\iint_{\partial Q} \mathbf{F} \cdot \mathbf{n} \, dS = \iiint_Q \nabla \cdot \mathbf{F}(x, y, z) \, dV$

$= \iiint_Q 4 \, dV$

$= 4$ (volume of cylinder of radius 1

and height 1)

$= 4(\pi \cdot 1^2)(1)$

$= 4\pi$

19. $\nabla \cdot \mathbf{F}(x, y, z) = \nabla \cdot \mathbf{F} \left\langle x^3, y^3, z^3 \right\rangle$

$= 3x^2 + 3y^2 + 3z^2$

$\iint_{\partial Q} \mathbf{F} \cdot \mathbf{n} \, dS$

$= \iiint_Q \nabla \cdot \mathbf{F}(x, y, z) dV$

$= \iiint_Q 3(x^2 + y^2 + z^2) \, dV$

$= \int_0^{2\pi} \int_0^{\pi/2} \int_0^1 (3\rho^2) \rho^2 \sin\phi \, d\rho \, d\phi \, d\theta$

$= \int_0^{2\pi} \int_0^{\pi/2} \left[\frac{3}{5} \rho^5 \sin\phi \right]_{\rho=0}^{\rho=1} d\phi \, d\theta$

$= \int_0^{2\pi} \int_0^{\pi/2} \frac{3}{5} \sin\phi \, d\phi \, d\theta$

$= \int_0^{2\pi} \left[-\frac{3}{5} \cos\phi \right]_{\phi=0}^{\phi=\pi/2} d\theta$

$= \int_0^{2\pi} \frac{3}{5} d\theta$

$= \frac{6}{5} \pi$ or 1.2π

21. $\nabla \cdot \mathbf{F}(x, y, z) = \nabla \cdot \left\langle x^2, z^2 - x, y^3 \right\rangle$

$$= 2x + 0 + 0$$
$$= 2x$$
$$\iint_{\partial Q} \mathbf{F} \cdot \mathbf{n}\, dS = \iiint_Q \nabla \cdot \mathbf{F}(x, y, z)dV$$
$$= \iiint_Q 2x\, dV$$
$$= 0 \text{ (by symmetry)}$$

23. $\nabla \cdot \mathbf{F}(x, y, z) = \nabla \cdot \mathbf{F}\left\langle y^2, x^2 z, z^2 \right\rangle$

$$= 0 + 0 + 2z$$
$$\iint_Q \mathbf{F} \cdot \mathbf{n}\, dS = \iiint_Q \nabla \cdot \mathbf{F}(x, y, . z)dV$$
$$= \int_0^{2\pi} \int_0^1 \int_0^r (2z)r\, dz\, dr\, d\theta$$
$$= \int_0^{2\pi} \int_0^1 [z^2 r]_{z=0}^{z=r}\, dr\, d\theta$$
$$= \int_0^{2\pi} \int_0^1 r^3\, dr\, d\theta$$
$$= \int_0^{2\pi} \left[\frac{1}{4}r^4 \right]_{r=0}^{r=1}\, d\theta$$
$$= \int_0^{2\pi} \frac{1}{4}\, d\theta$$
$$= \frac{1}{2}\pi$$

25. $\nabla \cdot \mathbf{F}(x, y, z) = \nabla \cdot \left\langle z - y^3, 2y - \sin z, x^2 - z \right\rangle$

$$= 0 + 2 - 1$$
$$= 1$$
$$\iint_{\partial Q} \mathbf{F} \cdot \mathbf{n}\, dS = \iiint_Q \nabla \cdot \mathbf{F}(x, y, z)dV$$
$$= \iiint_Q 1\, dV$$
$$= \text{(volume of cylinder of radius 1}$$
$$\text{and height 1)}$$
$$= \pi(1)^2(1)$$
$$= \pi$$

27. $\nabla \cdot \mathbf{F}(x, y, z) = \nabla \cdot \left\langle x^3, y^3 - z, z^3 - y^2 \right\rangle$

$$= 3x^2 + 3y^2 + 3z^2$$

Use modified cylindrical coordinates: $x = x$, $y = r\cos\theta$, $z = r\sin\theta$, $dV = r\, dx\, dr\, d\theta$.

$$\iint_{\partial Q} \mathbf{F} \cdot \mathbf{n}\, dS$$
$$= \iiint_Q \nabla \cdot \mathbf{F}(x, y, z)dV$$
$$= \iiint_Q (3x^2 + 3y^2 + 3x^2)dV$$
$$= \int_0^{2\pi} \int_0^2 \int_{r^2}^4 (3x^2 + 3r^2)r\, dx\, dr\, d\theta$$
$$= \int_0^{2\pi} \int_0^2 \left[x^3 r + 3r^3 x \right]_{x=r^2}^{x=4}\, dr\, d\theta$$
$$= \int_0^{2\pi} \int_0^2 (64r + 12r^3 - r^7 - 3r^5)\, dr\, d\theta$$
$$= \int_0^{2\pi} \left[32r^2 + 3r^4 - \frac{1}{8}r^8 - \frac{1}{2}r^6 \right]_{r=0}^{r=2}\, d\theta$$
$$= \int_0^{2\pi} 112\, d\theta$$
$$= 224\pi$$

29. $\nabla \cdot \mathbf{F}(x,\ y,\ z) = \nabla \cdot \left\langle y^2 x,\ 4x^2 \sin z,\ 3 \right\rangle$

$\qquad\qquad = y^2 + 0 + 0$

$\qquad\qquad = y^2$

$\iint_{\partial Q} \mathbf{F} \cdot \mathbf{n} = \iiint_Q \nabla \cdot \mathbf{F}(x,\ y,\ z)\,dV$

$\qquad = \int_0^3 \int_0^{2-2/3y} \int_0^{6-3x-2y} y^2 dz\ dx\ dy$

$\qquad = \int_0^3 \int_0^{2-2/3y} \left[y^2 z \right]_{z=0}^{z=6-3x-2y} dx\ dy$

$\qquad = \int_0^3 \int_0^{2-2/3y} (6y^2 - 3xy^2 - 2y^3)dx\ dy$

$\qquad = \int_0^3 \left[6xy^2 - \frac{3}{2}x^2 y^2 - 2xy^3 \right]_{x=0}^{x=2-2/3y} dy$

$\qquad = \int_0^3 \left[6\left(2-\frac{2}{3}y\right)y^2 - \frac{3}{2}\left(2-\frac{2}{3}y^2\right)y^2 - 2\left(2-\frac{2}{3}y\right)y^3 \right] dy$

$\qquad = \int_0^3 \left(\frac{2}{3}y^4 - 4y^3 + 6y^2 \right) dy$

$\qquad = \left(\frac{2}{15}y^5 - y^4 + 2y^3 \right)\Big|_{y=0}^{y=3}$

$\qquad = \frac{27}{5}$

31. $\nabla \cdot \mathbf{F}(x,\ y,z) = \nabla \cdot \left\langle x^2,\ y^3,\ x^3 y^2 \right\rangle$

$\qquad\qquad = 2x + 3y^2 + 0$

$\qquad\qquad = 2x + 3y^2$

$\iint_{\partial Q} \mathbf{F} \cdot \mathbf{n}\ dS = \iiint_Q \nabla \cdot \mathbf{F}(x,\ y,\ z)dV$

$\qquad = \int_{-2}^2 \int_{-2}^2 \int_{-3}^{1-x^2} (2x + 3y^2)dz\ dy\ dx$

$\qquad = \int_{-2}^2 \int_{-2}^2 \left[(2x+3y^2)z \right]_{z=-3}^{z=1-x^2} dy\ dx$

$\qquad = \int_{-2}^2 \int_{-2}^2 (2x+3y^2)(4-x^2)dy\ dx$

$\qquad = \int_{-2}^2 \left[2xy + y^3)(4-x^2) \right]_{y=-2}^{y=2} dx$

$\qquad = \int_{-2}^2 (8x+16)(4-x^2)dx$

$\qquad = \int_{-2}^2 (-8x^3 - 16x^2 + 32x + 64)dx$

$\qquad = \left(-2x^4 - \frac{16}{3}x^3 16x^2 + 64x \right)\Big|_{x=-2}^{x=2}$

$\qquad = \frac{512}{3}$

Section 14.8

5. Compute $\int_{\partial S} \mathbf{F}(x, y, z) \cdot d\mathbf{r}$:

$x = \cos t,\ y = 2\sin t,\ z = 0,\ 0 \le t \le 2\pi$

$$\int_{\partial S} \mathbf{F}(x, y, z) \cdot dr = \int_0^{2\pi} zx\ dx + 2y\ dy + z^3\ dz$$

$$= \int_0^{2\pi} [(0)(\cos t)(-2\sin t) + 2(2\sin t)(2\cos t) + (0^3)(0)]dt$$

$$= \int_0^{2\pi} 8\sin t \cos t\ dt$$

$$= 4\sin^2 t \Big|_0^{2\pi}$$

$$= 0$$

Compute $\iint_S (\nabla \times \mathbf{F}) \cdot \mathbf{n}\ dS$:

$$\nabla \times \mathbf{F} = \begin{vmatrix} \mathbf{i} & \mathbf{j} & \mathbf{k} \\ \dfrac{\partial}{\partial x} & \dfrac{\partial}{\partial y} & \dfrac{\partial}{\partial z} \\ xz & 2y & z^3 \end{vmatrix}$$

$$= \left(\frac{\partial}{\partial y}(z^3) - \frac{\partial}{\partial z}(2y) \right)\mathbf{i} - \left(\frac{\partial}{\partial x}(z^3) - \frac{\partial}{\partial z}(xz) \right)\mathbf{j} + \left(\frac{\partial}{\partial x}(2y) - \frac{\partial}{\partial y}(xz) \right)\mathbf{k}$$

$$= (0 - 0)\mathbf{i} - (0 - x)\mathbf{j} + (0 - 0)\mathbf{k}$$

$$= \langle 0, x, 0 \rangle$$

$$\mathbf{n} = \frac{\langle 2x,\ 2y,\ 1 \rangle}{\sqrt{4x^2 + 4y^2 + 1}}$$

$$dS = \sqrt{4x^2 + 4y^2 + 1}\ dA$$

$$\iint_S (\nabla \times \mathbf{F}) \cdot \mathbf{n}\ dS = \iint_S \langle 0,\ x,\ 0 \rangle \frac{\langle 2x,\ 2y,\ 1 \rangle}{\sqrt{4x^2 + 4y^2 + 1}} \sqrt{4x^2 + 4y^2 + 1}\ dA$$

$$= \iint_S 2xy\ dA$$

$$= 0 \text{ (by symmetry)}$$

7. Compute $\int_{\partial S} \mathbf{F}(x,\ y,\ z) \cdot d\mathbf{r}$:

$x = 2\cos t,\ y = 2\sin t,\ z = 0$

$$\int_{\partial S} \mathbf{F}(x,\ y,\ z)d\mathbf{r} = \int_{\partial S}(2x - y)dx + (yz^2)dy + (y^2 z)dz$$

$$= \int_0^{2\pi}[(2\cos t - 2\sin t)(-2\sin t) + (0)(2\cos t) + (0)(0)]dt$$

$$= \int_0^{2\pi}[-4\cos t \sin t + 4\sin^2 t]dt$$

$$= \int_0^{2\pi}[-4\cos t \sin t + 2 - 2\cos 2t]dt$$

$$= (2\cos^2 t + 2t - \sin 2t)\Big|_0^{2\pi}$$

$$= 4\pi$$

Compute $\iint_S (\nabla \times \mathbf{F}) \cdot \mathbf{n}\ dS$:

$$\nabla \times \mathbf{F} = \begin{vmatrix} \mathbf{i} & \mathbf{j} & \mathbf{k} \\ \dfrac{\partial}{\partial x} & \dfrac{\partial}{\partial y} & \dfrac{\partial}{\partial z} \\ 2x - y & yz^2 & y^2 z \end{vmatrix}$$

$$= \left(\frac{\partial}{\partial y}(y^2 z) - \frac{\partial}{\partial z}(yz^2)\right)\mathbf{i} - \left(\frac{\partial}{\partial x}(y^2 z) - \frac{\partial}{\partial z}(2x - y)\right)\mathbf{j} + \left(\frac{\partial}{\partial x}(yz^2) - \frac{\partial}{\partial y}(2x - y)\right)\mathbf{k}$$

$$= (2yz - 2yz)\mathbf{i} - (0 - 0)\mathbf{j} + (0 + 1)\mathbf{k}$$

$$= \langle 0,\ 0,\ 1 \rangle$$

$$\mathbf{n} = \frac{\langle x,\ y,\ z \rangle}{2}$$

$$(\nabla \times \mathbf{F}) \cdot \mathbf{n} = \langle 0,\ 0,1 \rangle \cdot \frac{\langle x,\ y,\ z \rangle}{2} = \frac{1}{2}z$$

Use spherical coordinates: $x = 2\sin\phi\cos\theta,\ y = 2\sin\phi\sin\theta,\ z = 2\cos\phi,\ ds = 4\sin\phi\ d\phi\ d\theta$

$$\iint_S (\nabla \times \mathbf{F}) \cdot \mathbf{n}\ dS = \iint_s \frac{1}{2}zds$$

$$= \int_0^{2\pi}\int_0^{\pi/2}(\cos\phi)(4\sin\phi\ d\phi\ d\theta)$$

$$= \int_0^{2\pi}\left[2\sin^2\theta\right]_{\theta=0}^{\theta=\pi/2} d\theta$$

$$= \int_0^{2\pi}2\ d\theta$$

$$= 4\pi$$

9. ∂S is the triangle with vertices at $(0, 0, 0)$, $(2, 0, 0)$, and $(0, 2, 0)$.

C_1 : From $(0, 0, 0)$ to $(2, 0, 0)$

$x = 2t,\ y = 0,\ z = 0,\ 0 \le t \le 1$

$\int_{C_1} \mathbf{F}(x,\ y,\ z) \cdot d\mathbf{r}$

$= \int_{C_1} (zy^4 - y^2)dx + (y - x^3)dy + z^2\ dz$

$= \int_0^1 [(0)(2) + (-8t^3)(0) + (0)(0)]dt$

$= 0$

C_2 : From, $(2, 0, 0)$ to $(0, 2, 0)$

$x = 2 - 2t,\ y = 2t,\ z = 0,\ 0 \le t \le 1$

$\int_{C_3} \mathbf{F}(x,\ y,\ z) \cdot d\mathbf{r}$

$= \int_{C_3} (zy^4 - y^2)dx + (y - x^3)dy + z^2dz$

$= \int_0^1 [(2 - 2t)^2(0) + (2 - 2t)(-2) + (0)(0)]dt$

$= \int_0^1 (4t - 4)dt$

$= (2t^2 - 4t)\Big|_{t=0}^{t=1}$

$= -2$

$\iint_S (\nabla \times \mathbf{F}) \cdot \mathbf{n}\ dS$

$= \int_{C_1} \mathbf{F} \cdot d\mathbf{r} + \int_{C_2} \mathbf{F} \cdot d\mathbf{r} + \int_{C_3} \mathbf{F} \cdot d\mathbf{r}$

$= 0 + \dfrac{2}{3} - 2$

$= -\dfrac{4}{3}$

$\int_{C_2} \mathbf{F}(x,\ y,\ z) \cdot d\mathbf{r}$

$= \int_{C_2} (zy^4 - y^2)dx + (y - x^3)dy + z^2dz$

$= \int_0^1 [(-4t^2)(-2) + (2t - (2 - 2t)^3)(2) + (0)(0)]dt$

$= \int_0^1 (16t^3 - 40t^2 + 52t - 16)dt$

$= \left(4t^4 - \dfrac{40}{3}t^3 + 26t^2 - 16t\right)\Big|_{t=0}^{t=1}$

$= \dfrac{2}{3}$

C_3 : From $(0, 2, 0)$ to $(0, 0, 0)$

$x = 0,\ y = 2 - 2t,\ z = 0,\ 0 \le t \le 1$

11. ∂S is the circle $x^2 + y^2 = 1$ in the xy-plane.

$x = \cos t,\ y = \sin t,\ z = 0,\ 0 \le t \le 2\pi$

$\iint_S (\nabla \times \mathbf{F}) \cdot dS = \int_{\partial S} \mathbf{F}(x,\ y,\ z) \cdot d\mathbf{r}$

$= \int_{\partial S} zx^2 dx + (ze^{xy^2} - x)dy + x\sin y^2 dz$

$= \int_0^{2\pi} [(0)(-\sin t) + (-\cos t)(\cos t) + (\cos t)(\sin(\sin^2 t))(0)]dt$

$= \int_0^{2\pi} (-\cos^2 t)dt$

$= \int_0^{2\pi} \left(-\dfrac{1}{2} - \dfrac{1}{2}\cos 2t\right)dt$

$= \left(-\dfrac{1}{2}t - \dfrac{1}{4}\sin 2t\right)\Big|_{t=0}^{t=2\pi}$

$= -\pi$

13. ∂S is a triangle with vertices at $(0, 0, 0)$, $(0, 0, 1)$ and $(2, 0, 0)$. Note that the vertices are chosen in this order so that the curve is positively oriented with respect to the unit vectors pointing toward the right.

C_1 : From $(0, 0, 0)$ to $(0, 0, 1)$

$x = 0$, $y = 0$, $z = t$, $0 \leq t \leq 1$

$$\int_{C_1} \mathbf{F}(x,\ y,\ z) \cdot d\mathbf{r} = \int_{C_1} (zy^4 - y^2)dx + (y - x^3)dy + z^2dz$$

$$= \int_0^1 [(0)(0) + (0)(0) + (t^2)(1)]dt$$

$$= \int_0^1 t^2 dt$$

$$= \frac{1}{3}t^3 \Big|_{t=0}^{t=1}$$

$$= \frac{1}{3}$$

C_2 : From $(0, 0, 1)$ to $(2, 0, 0)$

$x = 2t$, $y = 0$, $z = 1 - t$, $0 \leq t \leq 1$

$$\int_{C_2} \mathbf{F}(x,\ y,\ z) \cdot d\mathbf{r} = \int_{C_2} (zy^4 - y^2)dx + (y - x^3)dy + z^2dz$$

$$= \int_0^1 [(0)(2) + (-8t^3)(0) + (1 - t)^2(-1)]dt$$

$$= \int_0^1 (-t^2 + 2t - 1)dt$$

$$= \left(-\frac{1}{3}t^3 + t^2 - t\right)\Big|_{t=0}^{t=1}$$

$$= -\frac{1}{3}$$

C_3 : From $(2, 0, 0)$ to $(0, 0, 0)$

$x = 2 - 2t$, $y = 0$, $z = 0$

$$\int_{C_3} \mathbf{F}(x,\ y,\ z) \cdot d\mathbf{r} = \int_{C_3} (zy^4 - y^2)dx + (y - x^3)dy + z^2dz$$

$$= \int_0^1 [(0)(-2) - (2 - 2t)^3(0) + (0)(0)]dt$$

$$= \int_0^1 0\ dt$$

$$= 0$$

$$\iint_S (\nabla \times \mathbf{F}) \cdot \mathbf{n}\ dS = \int_{C_1} \mathbf{F} \cdot d\mathbf{r} + \int_{C_2} \mathbf{F} \cdot d\mathbf{r} + \int_{C_3} \mathbf{F} \cdot d\mathbf{r}$$

$$= \frac{1}{3} - \frac{1}{3} + 0$$

$$= 0$$

15. ∂S is the square with vertices at $(0, 0, 0)$, $(1, 0, 0)$, $(1, 1, 0)$, and $(0, 1, 0)$.

$C_1 : x = t,\ y = 0,\ z = 0,\ 0 \le t \le t$

$$\int_{C_1} \mathbf{F} \cdot d\mathbf{r} = \int_{C_1} xyz\,dx + (4x^2 y^3 - z)dy + (8\cos xz^2)dz$$

$$= \int_0^1 [(0)(1) + (0)(0) + (8)(0)]dt$$

$$= 0$$

$C_2 : x = 1,\ y = t,\ z = 0,\ 0 \le t \le 1$

$$\int_{C_2} \mathbf{F} \cdot d\mathbf{r} = \int_{C_2} xyz\,dx + (4x^2 y^3 - z)dy + (8\cos xz^2)dz$$

$$= \int_0^1 [(0)(0) + (4t^3)(1) + (1)(0)]dt$$

$$= \int_0^1 4t^3\,dt$$

$$= t^4 \Big|_{t=0}^{t=1}$$

$$= 1$$

$C_3 : x = 1 - t,\ y = 1,\ z = 0,\ 0 \le t \le 1$

$$\int_{C_3} \mathbf{F} \cdot d\mathbf{r} = \int_{C_3} xyz\,dx + (4x^2 y^3 - z)dy + (8\cos xz^2)dz$$

$$= \int_0^1 [(0)(-1) + 4(1-t)^2(0) + (1)(0)]dt$$

$$= 0$$

$C_4 : x = 0,\ y = 1 - t,\ z = 0,\ 0 \le t \le 1$

$$\int_{C_4} \mathbf{F} \cdot d\mathbf{r} = \int_{C_4} xyz\,dx + (4x^2 y^3 - z)dy + (8\cos xz^2)dz$$

$$= \int_0^1 [(0)(0) + (0)(-1) + (1)(0)]dt$$

$$= 0$$

$$\iint_S (\nabla \times \mathbf{F}) \cdot \mathbf{n}\,dS = \int_{C_1} \mathbf{F} \cdot d\mathbf{r} + \int_{C_2} \mathbf{F} \cdot d\mathbf{r} + \int_{C_3} \mathbf{F} \cdot d\mathbf{r} + \int_{C_4} \mathbf{F} \cdot d\mathbf{r}$$

$$= 0 + 1 + 0 + 0$$

$$= 1$$

17. ∂S is the circle $x^2 + y^2 = 1$ in the plane $z = 1$. Since the exterior unit vectors point downward, the curve is oriented clockwise. $x = -\cos t,\ y = \sin t,\ z = 1,\ 0 \le t \le 2\pi$

$$\iint_S (\nabla \times \mathbf{F}) \cdot \mathbf{n}\,dS = \int_{\partial S} \mathbf{F} \cdot d\mathbf{r}$$

$$= \int_{\partial S} (x^2 + y^2)dx + ze^{x^2+y^2}dy + e^{x^2+z^2}dz$$

$$= \int_0^{2\pi} [(1)(\sin t) + (e)(\cos t) + (e^{1+\cos^2 t})(0)]dt$$

$$= \int_0^{2\pi} (\sin t + e\cos t)dt$$

$$= (-\cos t + e\sin t)\Big|_{t=0}^{t=2\pi}$$

$$= 0$$

19. ∂S is the circle $x^2 + z^2 = 4$ in the xz-plane.

$x = 2\cos t,\ y = 0,\ z = -2\sin t,\ 0 \le t \le 2\pi$

Note the negative sign in the definition of z; this gives the curve a positive orientation with respect to the exterior (rightward) normal vectors of the surface S.

$$\iint_S (\nabla \times \mathbf{F}) \cdot \mathbf{n}\ dS = \int_{\partial S} \mathbf{F} \cdot d\mathbf{r}$$

$$= \int_{\partial S} yx^2 z\ dx + x^2 \cos y\ dy + x\ dz$$

$$= \int_0^{2\pi} [(0)(-2\sin t) + (4\cos^2 t)(0) + (2\cos t)(-2\cos t)]dt$$

$$= \int_0^{2\pi} (-4\cos^2 t)\ dt$$

$$= \int_0^{2\pi} (-2 - 2\cos 2t)dt$$

$$= (-2t - \sin 2t)\big|_{t=0}^{t=2\pi}$$

$$= -4\pi$$

21. Note that C is the circle $x^2 + y^2 = 4$ in the xy–plane. For simplicity, let S be the disk $x^2 + y^2 \le 4$ in the xy-plane; it is not necessary to use the paraboloid.

$$\nabla \times \mathbf{F} = \begin{vmatrix} \mathbf{i} & \mathbf{j} & \mathbf{k} \\ \frac{\partial}{\partial x} & \frac{\partial}{\partial y} & \frac{\partial}{\partial z} \\ e^x x^2 - y & \sqrt{1+y^2} & z^3 \end{vmatrix}$$

$$= \left(\frac{\partial}{\partial y}(z^3) - \frac{\partial}{\partial z}\left(\sqrt{1+y^2}\right) \right)\mathbf{i} - \left(\frac{\partial}{\partial x}(z^3) - \frac{\partial}{\partial z}(e^x x^2 - y) \right)\mathbf{j} + \left(\frac{\partial}{\partial x}\left(\sqrt{1+y^2}\right) - \frac{\partial}{\partial y}(e^x x^2 - y) \right)\mathbf{k}$$

$$= (0-0)\mathbf{i} - (0-0)\mathbf{j} + (0+1)\mathbf{k}$$

$$= \langle 0, 0, 1 \rangle$$

$\mathbf{n} = \langle 0, 0, 1 \rangle$

$dS = dA$

$$\int_C \mathbf{F} \cdot d\mathbf{r} = \iint_S (\nabla \times \mathbf{F}) \cdot \mathbf{n}\ dS$$

$$= \iint_S \langle 0, 0, 1 \rangle \cdot \langle 0, 0, 1 \rangle\ dA$$

$$= \iint_S dA$$

$$= (\text{area of circle radius 2})$$

$$= 4\pi$$

23. $\nabla \times \mathbf{F} = \begin{vmatrix} \mathbf{i} & \mathbf{j} & \mathbf{k} \\ \frac{\partial}{\partial x} & \frac{\partial}{\partial y} & \frac{\partial}{\partial z} \\ 2x^2 & 4y^2 & e^{8z^2} \end{vmatrix}$

$\quad = \left(\frac{\partial}{\partial y}(e^{8z^2}) - \frac{\partial}{\partial z}\left(4y^2\right) \right)\mathbf{i} - \left(\frac{\partial}{\partial x}(e^{8z^2}) - \frac{\partial}{\partial z}(2x^2) \right)\mathbf{j} + \left(\frac{\partial}{\partial x}\left(4y^2\right) - \frac{\partial}{\partial y}(2x^2) \right)\mathbf{k}$

$\quad = (0-0)\mathbf{i} - (0-0)\mathbf{j} + (0-0)\mathbf{k}$

$\quad = \langle 0, 0, 0 \rangle$

$\displaystyle\int_C \mathbf{F} \cdot d\mathbf{r} = \iint_S (\nabla \cdot \mathbf{F}) \cdot \mathbf{n}\, dS$

$\qquad\qquad = \iint_S \mathbf{0} \cdot \mathbf{n}\, dS$

$\qquad\qquad = 0$

25. $\nabla \times \mathbf{F} = \begin{vmatrix} \mathbf{i} & \mathbf{j} & \mathbf{k} \\ \frac{\partial}{\partial x} & \frac{\partial}{\partial y} & \frac{\partial}{\partial z} \\ x^2 + 2xy^3z & 3x^2y^2z - y & x^2y^3 \end{vmatrix}$

$\quad = \left(\frac{\partial}{\partial y}(x^2y^3) - \frac{\partial}{\partial z}\left(3x^2y^2z - y\right) \right)\mathbf{i} - \left(\frac{\partial}{\partial x}(x^2y^3) - \frac{\partial}{\partial z}(x^2 + 2xy^3z) \right)\mathbf{j}$

$\qquad + \left(\frac{\partial}{\partial x}\left(3x^2y^2z - y\right) - \frac{\partial}{\partial y}(x^2 + 2xy^3z) \right)\mathbf{k}$

$\quad = (3x^2y^2 - 3x^2y^2)\mathbf{i} - (2xy^3 - 2xy^3)\mathbf{j} + (6xy^2z - 6xy^2z)\mathbf{k}$

$\quad = \langle 0, 0, 0 \rangle$

$\displaystyle\int_C \mathbf{F} \cdot d\mathbf{r} = \iint_S (\nabla \cdot \mathbf{F}) \cdot \mathbf{n}\, dS$

$\qquad\qquad = \iint_S \mathbf{0} \cdot \mathbf{n}\, dS$

$\qquad\qquad = 0$

Chapter 14 Review

1. See Example 1.1.

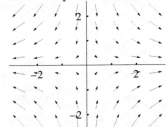

3. See Example 1.3.

$\mathbf{F}_1(x, y) = \langle \sin x, y \rangle$: Graph D, because the vectors point upward when $y > 0$ and downward when $y < 0$.

$\mathbf{F}_2(x, y) = \langle \sin y, x \rangle$ Graph C, because the vectors point downward when $x < 0$, upward when $x > 0$, to the left when $-\pi < y < 0$, and to the right when $0 < y < \pi$.

$\mathbf{F}_3(x, y) = \langle y^2, 2x \rangle$: Graph B, because the vectors point downward when $x < 0$ and upward when $x > 0$, and all vectors point to the right.

$\mathbf{F}_4(x, y) = \langle 3, x^2 \rangle$: Graph A, because all vectors point to the right and upward.

5. See Example 1.9.

If $\nabla f(x, y) = \left\langle y - 2xy^2, x - 2yx^2 + 1 \right\rangle$, then

$\dfrac{\partial f}{\partial x} = y - 2xy^2$ and $\dfrac{\partial f}{\partial y} = x - 2xy^2 + 1$.

$f(x, y) = \int (y - 2xy^2) dx = xy - x^2 y^2 + g(y)$

$\dfrac{\partial f}{\partial y} = x - 2x^2 y + g'(y) = x - 2yx^2 + 1$

$$g'(y) = 1$$
$$y = \int 1 dy = y + c$$

The vector field is conservative,

$f(x, y) = xy - x^2 y^2 + y + c$.

7. See Example 1.9.

If $\nabla f(x, y) = \left\langle 2xy - 1, x^2 + 2xy \right\rangle$, then

$\dfrac{\partial f}{\partial x} = 2xy - 1$ and $\dfrac{\partial f}{\partial y} = x^2 + 2xy$.

$f(x, y) = \int (2xy - 1) dx = x^2 y - x + g(y)$

$\dfrac{\partial f}{\partial y} = x^2 + g'(y) = x^2 + 2xy$

$$g'(y) = 2xy$$

Since this is not possible, the vector field is not conservative.

9. $\dfrac{dy}{dx} = \dfrac{2x/y}{y} = \dfrac{2x}{y^2}$

$\int y^2 dy = \int 2x\, dx$

$\dfrac{1}{3} y^3 = x^2 + c$

Since c (or $3c$) is an arbitrary constant, we may write $y^3 = 3x^2 + c$.

13. See Theorem 2.4.

$x = 2 + 2t, y = 3; 0 \le t \le 1$

$dx = 2\, dt$

$\int_C 3y\, dx = \int_0^1 3(3)(2)\, dt = 18$

15. See Examples 2.2 and 2.3.

$x = 3\cos t, y = -3\sin t; 0 \le t \le 2\pi$

$ds = \sqrt{[x'(t)]^2 + [y'(t)]^2}\, dt$

$\quad = \sqrt{(-3\sin t)^2 + (-3\cos t)^2}\, dt$

$\quad = 3\, dt$

$\int_C \sqrt{x^2 + y^2}\, ds = \int_0^{2\pi} \sqrt{(3\cos t)^2 + (3\sin t)^2}\, (3) dt$

$\quad = \int_0^{2\pi} 3 \cdot 3\, dt$

$\quad = 18\pi$

Alternative solution:

On C, note that $\sqrt{x^2 + y^2} = \sqrt{9} = 3$.

$\int_C \sqrt{x^2 + y^2}\, ds = 3(\text{circumference of circle})$

$\quad = 3(6\pi)$

$\quad = 18\pi$

17. See Theorem 2.4.

Note that C is a closed curve and the integral is of the form $\int_C f(x)\, dx$. Using the technique of Exercise 72 in section 14.2, the line integral is 0.

19. See Example 2.8.

$x = 2\cos t, \; y = 2\sin t; 0 \le t \le 2\pi$

$$\int_C \mathbf{F} \cdot d\mathbf{r} = \int_C x\, dx - y\, dy$$

$$= \int_0^{2\pi} \left[(2\cos t)(-2\sin t) - (2\sin t)(2\cos t) \right] dt$$

$$= \int_0^{2\pi} (-4\sin t \cos t)\, dt$$

$$= (-2\sin^2 t) \Big|_0^{2\pi}$$

$$= 0$$

21. See Example 2.8.

$C_1 : x = 2\cos t, \; y = 2\sin t; 0 \le t \le \dfrac{\pi}{2}$

$$\int_{C_1} \mathbf{F} \cdot d\mathbf{r} = \int_{C_1} 2\, dx + 3x\, dy$$

$$= \int_0^{\pi/2} [2(-2\sin t) + 3(2\cos t)(2\cos t)]\, dt$$

$$= \int_0^{\pi/2} (-4\sin t + 12\cos^2 t)\, dt$$

$$= \int_0^{\pi/2} (-4\sin t + 6\cos 2t + 6)\, dt$$

$$= (4\cos t + 3\sin 2t + 6t) \Big|_0^{\pi/2}$$

$$= (0 + 0 + 3\pi) - (4 + 0 + 0)$$

$$= 3\pi - 4$$

$C_2 : x = 0, \; y = 2 - t; 0 \le t \le 2$

$$\int_{C_2} \mathbf{F} \cdot d\mathbf{r} = \int_{C_2} 2\, dx + 3x\, dy$$

$$= \int_0^2 [2(0) + 0(-1)]\, dt$$

$$= 0$$

Total: $\displaystyle\int_C \mathbf{F} \cdot d\mathbf{r} = 3\pi - 4 + 0 = 3\pi - 4$

23. See Example 2.9.

The motion is perpendicular to the direction of the force field, so the work done by the force field is zero.

25. See Example 2.1.

$x = \cos 3t, \; y = \sin 3t, \; z = 4t; 0 \le t \le 2\pi$

$$ds = \sqrt{[x'(t)]^2 + [y'(t)]^2 + [z'(t)]^2}\; dt$$

$$= \sqrt{(-3\sin t)^2 + (3\cos 3t)^2 + 4^2}\; dt$$

$$= \sqrt{3^2 + 4^2}\; dt$$

$$= 5\, dt$$

$$m = \int_C \rho\, ds$$

$$= \int_0^{2\pi} 4 \cdot 5\, dt$$

$$= 40\pi$$

27. See Example 3.1.

The integral is $\int_C \mathbf{F} \cdot d\mathbf{r}$ where $\mathbf{F}(x, y) = \langle 3x^2 y - x, x^3 \rangle$. A potential function is $f(x, y) = x^3 y - \frac{1}{2}x^2$, so the line integral is independent of path.

$$\int_C \mathbf{F} \cdot d\mathbf{r} = f(x, y)\big|_{(2, -1)}^{(4, 1)}$$
$$= \left[(4)^3(1) - \frac{1}{2}(4)^2\right] - \left[(2)^3(-1) - \frac{1}{2}(2)^2\right]$$
$$= 66$$

29. See Example 3.1.

A potential function is $f(x, y) = x^2 y - y\cos x + e^{x+y}$, so the line integral is independent of path.

$$\int_C \mathbf{F} \cdot d\mathbf{r} = f(x, y)\big|_{(0, 3)}^{(3, 0)}$$
$$= \left[(3)^2(0) - 0\cos 3 + e^{3+0}\right] - \left[(0)^2(3) - 3\cos 0 + e^{0+3}\right]$$
$$= 3$$

31. See Examples 3.1 and 3.3.

A potential function is $f(x, y, z) = x^2 y - \frac{1}{2}y^2 + z^2$, so the line integral is independent of path.

$$\int_C \mathbf{F} \cdot d\mathbf{r} = f(x, y, z)\big|_{(1, 3, 2)}^{(2, 1, -3)}$$
$$= \left[(2)^2(1) - \frac{1}{2}(1)^2 + (-3)^2\right] - \left[(1)^2(3) - \frac{1}{2}(3)^2 + (2)^2\right]$$
$$= 10$$

33. See Example 3.2.
The graph appears to represent the vector field
$\mathbf{F} = \langle x, 1 \rangle$.

$$M_y = \frac{\partial}{\partial y}(x) = 0$$
$$N_x = \frac{\partial}{\partial x}(1) = 0$$

Since $M_y = N_x$, the vector field is conservative.

(Note: Even if the graph does not represent
$\mathbf{F} = \langle x, 1 \rangle$, we can see that $M_y - N_x = 0$.)

35. See Examples 4.1, 4.2, and 4.4.
Let $M(x, y) = x^3 - y$ and $N(x, y) = x + y^3$.
Note that the curve has positive orientation.

$$\oint_C \mathbf{F} \cdot d\mathbf{r} = \oint_C M\,dx + N\,dy$$
$$= \iint_R \left(\frac{\partial N}{\partial x} - \frac{\partial M}{\partial y}\right) dA$$
$$= \int_0^1 \int_{x^2}^x [1 - (-1)]\,dy\,dx$$
$$= \int_0^1 (2x - 2x^2)\,dx$$
$$= \left(x^2 - \frac{2}{3}x^3\right)\bigg|_{x=0}^{x=1}$$
$$= \frac{1}{3}$$

37. See Examples 4.1, 4.2, and 4.4.

Let $M(x, y) = \tan x^2$ and $N(x, y) = x^2$.

Note that the curve has negative orientation.

$$\oint_C M\,dx + N\,dy$$

$$= -\iint_R \left(\frac{\partial N}{\partial x} - \frac{\partial M}{\partial y} \right) dA$$

$$= -\int_0^1 \int_y^{2-y} 2x\,dx\,dy$$

$$= -\int_0^1 \left[x^2 \right]_{x=y}^{x=2-y} dy$$

$$= -\int_0^1 (4 - 4y)\,dy$$

$$= \left(2y^2 - 4y \right)\Big|_{y=0}^{y=1}$$

$$= -2$$

39. See Examples 4.1, 4.2, and 4.4.

Since the curve lies in the plane $x = 0$, we may use

$\mathbf{F} = \left\langle 0, 4y^3 - z, z^2 \right\rangle$. Then

$$\oint_C \mathbf{F} \cdot d\mathbf{r} = \oint_C 0\,dx + (4y^3 - z)\,dy + z^2 dz$$

$$= \oint_C (4y^3 - z)\,dy + z^2\,dz.$$

Thus, we can reduce the problem to two dimensions (in terms of y and z):

Evaluate $\oint_C (4y^3 - z)\,dy + z^2\,dz$ where C is

formed by $z = y^2$ and $z = 4$ oriented positively in the zy plane.

Let $M(y, z) = 4y^3 - z$ and $N(y, z) = z^2$.

Since $z = y^2$ and $z = 4$ intersect at

$(y, z) = (\pm 2, 0)$, the region extends from $y = -2$ to $y = 2$.

$$\oint_C M\,dy + N\,dz = \iint_R \left(\frac{\partial N}{\partial y} - \frac{\partial M}{\partial z} \right) dA$$

$$= \int_{-2}^{2} \int_{y^2}^{4} (1)\,dz\,dy$$

$$= \int_{-2}^{2} (4 - y^2)\,dy$$

$$= \left(4y - \frac{1}{3}y^3 \right)\Big|_{y=-2}^{y=2}$$

$$= \frac{32}{3}$$

41. See Example 4.3.

$C = \{(x, y) \mid x = 3\cos t, y = 2\sin t, 0 \le t \le 2\pi\}$

$$A = \frac{1}{2} \oint_C x\,dy - y\,dx$$

$$= \frac{1}{2} \int_0^{2\pi} [(3\cos t)(2\cos t) - (2\sin t)(-3\sin t)]\,dt$$

$$= \frac{1}{2} \int_0^{2\pi} 6(\cos^2 t + \sin^2 t)\,dt$$

$$= 6\pi$$

43. $\mathbf{F}(x, y, z) = (x^3, -y^3, 0)$

$\operatorname{curl} \mathbf{F} = \nabla \times \mathbf{F}$

$$= \begin{vmatrix} \mathbf{i} & \mathbf{j} & \mathbf{k} \\ \frac{\partial}{\partial x} & \frac{\partial}{\partial y} & \frac{\partial}{\partial z} \\ x^3 & -y^3 & 0 \end{vmatrix}$$

$$= \left(\frac{\partial(0)}{\partial y} - \frac{\partial(-y^3)}{\partial z} \right)\mathbf{i} - \left(\frac{\partial(0)}{\partial x} - \frac{\partial(x^3)}{\partial z} \right)\mathbf{j} + \left(\frac{\partial(-y^3)}{\partial x} - \frac{\partial(x^3)}{\partial y} \right)\mathbf{k}$$

$$= (0 - 0)\mathbf{i} - (0 - 0)\mathbf{j} + (0 - 0)\mathbf{k}$$

$$= \langle 0, 0, 0 \rangle$$

$\operatorname{div} \mathbf{F} = \nabla \cdot \mathbf{F}$

$$= \frac{\partial(x^3)}{\partial x} + \frac{\partial(-y^3)}{\partial y} + \frac{\partial(0)}{\partial z}$$

$$= 3x^2 - 3y^2 + 0$$

$$= 3x^2 - 3y^2$$

45. $\mathbf{F}(x, y, z) = \left\langle 2x, 2yz^2, 2y^2z \right\rangle$

curl $\mathbf{F} = \nabla \times \mathbf{F}$

$$= \begin{vmatrix} \mathbf{i} & \mathbf{j} & \mathbf{k} \\ \frac{\partial}{\partial x} & \frac{\partial}{\partial y} & \frac{\partial}{\partial z} \\ 2x & 2yz^2 & 2y^2z \end{vmatrix}$$

$$= \left(\frac{\partial(2y^2z)}{\partial y} - \frac{\partial(2yz^2)}{\partial z} \right)\mathbf{i} - \left(\frac{\partial(2y^2z)}{\partial x} - \frac{\partial(2x)}{\partial z} \right)\mathbf{j} + \left(\frac{\partial(2yz^2)}{\partial x} - \frac{\partial(2x)}{\partial y} \right)\mathbf{k}$$

$$= (4yz - 4yz)\mathbf{i} - (0 - 0)\mathbf{j} + (0 - 0)\mathbf{k}$$

$$= \left\langle 0, 0, 0 \right\rangle$$

div $\mathbf{F} = \nabla \cdot \mathbf{F}$

$$= \frac{\partial(2x)}{\partial x} + \frac{\partial(2yz^2)}{\partial y} + \frac{\partial(2y^2z)}{\partial z}$$

$$= 2 + 2z^2 + 2y^2$$

47. $\mathbf{F}(x, y, z) = (2x - y^2, z^2 - 2xy, xy^2)$

curl $\mathbf{F} = \nabla \times \mathbf{F}$

$$= \begin{vmatrix} \mathbf{i} & \mathbf{j} & \mathbf{k} \\ \frac{\partial}{\partial x} & \frac{\partial}{\partial y} & \frac{\partial}{\partial z} \\ 2x - y^2 & z^2 - 2xy & xy^2 \end{vmatrix}$$

$$= \left(\frac{\partial(xy^2)}{\partial y} - \frac{\partial(z^2 - 2xy)}{\partial z} \right)\mathbf{i} - \left(\frac{\partial(xy^2)}{\partial x} - \frac{\partial(2xy - y^2)}{\partial z} \right)\mathbf{j} + \left(\frac{\partial(z - 2xy)}{\partial x} - \frac{\partial(2x - y^2)}{\partial y} \right)\mathbf{k}$$

$$= (2xy - 2z)\mathbf{i} - (y^2 - 0)\mathbf{j} + (-2y + 2y)\mathbf{k}$$

$$= \left\langle 2xy - 2z, -y^2, 0 \right\rangle$$

Since curl $\mathbf{F} \neq \mathbf{0}$, $\mathbf{F}$ is not conservative.

div $\mathbf{F} = \nabla \cdot \mathbf{F}$

$$= \frac{\partial(2x - y^2)}{\partial x} + \frac{\partial(z^2 - 2xy)}{\partial y} + \frac{\partial(xy^2)}{\partial z}$$

$$= 2 - 2x + 0$$

$$= 2 - 2x$$

Since div $\mathbf{F} \neq 0$, $\mathbf{F}$ is not incompressible.

49. $\mathbf{F}(x, y, z) = \langle 4x - y, 3 - x, 2 - 4z \rangle$

curl $\mathbf{F} = \nabla \times \mathbf{F}$

$$= \begin{vmatrix} \mathbf{i} & \mathbf{j} & \mathbf{k} \\ \dfrac{\partial}{\partial x} & \dfrac{\partial}{\partial y} & \dfrac{\partial}{\partial z} \\ 4x - y & 3 - x & 2 - 4z \end{vmatrix}$$

$$= \left(\frac{\partial(2 - 4z)}{\partial y} - \frac{\partial(3 - x)}{\partial z} \right) \mathbf{i} - \left(\frac{\partial(2 - 4z)}{\partial x} - \frac{\partial(4x - y)}{\partial z} \right) \mathbf{j} + \left(\frac{\partial(3 - x)}{\partial x} - \frac{\partial(4x - y)}{\partial y} \right) \mathbf{k}$$

$$= (0 - 0)\mathbf{i} - (0 - 0)\mathbf{j} + (-1 + 1)\mathbf{k}$$

$$= \langle 0, 0, 0 \rangle$$

Since the components of $\mathbf{F}$ have continuous partial derivatives throughout $\mathbf{R}^3$ and curl $\mathbf{F} = \mathbf{0}$, $\mathbf{F}$ is conservative.

div $\mathbf{F} = \nabla \cdot \mathbf{F}$

$$= \frac{\partial(4x - y)}{\partial x} + \frac{\partial(3 - x)}{\partial y} + \frac{\partial(2 - 4z)}{\partial z}$$

$$= 4 + 0 - 4$$

$$= 0$$

Since div $\mathbf{F} = 0$, $\mathbf{F}$ is incompressible.

51. The divergence is positive, because if we draw a box around P, the outflow (at the top of the box) is greater than the inflow (at the bottom of the box).

53. See Example 6.3.

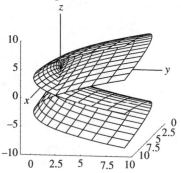

55. See Examples 6.3 and 6.4.

  a.   B

  b.   C

  c.   A

57. See Examples 6.1, 6.2, and 6.6.

$$z = f(x, y) = x^2 + y^2$$

$$\mathbf{n} = \langle f_x, f_y, -1 \rangle = \langle 2x, 2y, -1 \rangle$$

$$\|\mathbf{n}\| = \sqrt{(2x)^2 + (2y)^2 + (-1)^2} = \sqrt{4x^2 + 4y^2 + 1}$$

$$\iint_S dS = \iint_R \|\mathbf{n}\| \, dA$$

$$= \iint_R \sqrt{4x^2 + 4y^2 + 1} \, dA$$

$$= \int_0^{2\pi} \int_1^2 \sqrt{4r^2 + 1} \, r \, dr \, d\theta$$

$$= \int_0^{2\pi} \left[\frac{1}{12}(4r^2 + 1)^{3/2} \right]_{r=1}^{r=2} d\theta$$

$$= \int_0^{2\pi} \frac{1}{12}(17^{3/2} - 5^{3/2}) \, d\theta$$

$$= \frac{\pi}{6}(17^{3/2} - 5^{3/2})$$

59. See Examples 6.1, 6.2, and 6.6.

$z = f(x, y) = 12 - 3x - 2y$

$\mathbf{n} = \langle f_x, f_y, -1 \rangle = \langle -3, -2, -1 \rangle$

$\|\mathbf{n}\| = \sqrt{(-3)^2 + (-2)^2 + (-1)^2} = \sqrt{14}$

$\displaystyle \iint_S (x - y)\,dS$

$\displaystyle = \iint_S (x - y)\sqrt{14}\,dA$

$\displaystyle = \sqrt{14} \int_0^4 \int_0^{-\frac{3}{2}x+6} (x - y)\,dy\,dx$

$\displaystyle = \sqrt{14} \int_0^4 \left[xy - \frac{1}{2}y^2 \right]_{y=0}^{y=-\frac{3}{2}x+6} dx$

$\displaystyle = \sqrt{14} \int_0^4 \left[(x)\left(-\frac{3}{2}x + 6\right) - \frac{1}{2}\left(-\frac{3}{2}x + 6\right)^2 \right] dx$

$\displaystyle = \sqrt{14} \int_0^4 \left(-\frac{21}{8}x^2 + 15x - 18 \right) dx$

$\displaystyle = \sqrt{14} \left(-\frac{7}{8}x^3 + \frac{15}{2}x^2 - 18x \right) \Big|_0^4$

$= -8\sqrt{14}$

61. See Examples 6.1, 6.2, and 6.6.

$z = f(x, y) = 4 - \dfrac{4}{3}x - \dfrac{1}{3}y.$

$\mathbf{n} = \langle f_x, f_y, -1 \rangle = \left\langle -\dfrac{4}{3}, -\dfrac{1}{3}, -1 \right\rangle$

$\|\mathbf{n}\| = \sqrt{\left(-\dfrac{4}{3}\right)^2 + \left(-\dfrac{1}{3}\right)^2 + (-1)^2} = \dfrac{1}{3}\sqrt{26}$

$\displaystyle \iint_S (4x + y + 3z)\,dS = \iint_S 12\,dS$

$\displaystyle = \iint_R (12)\left(\dfrac{1}{3}\sqrt{26}\right) dA$

$= (4\sqrt{26})(\text{area of circle of}$

$\text{radius } 1)$

$= 4\pi\sqrt{26}$

63. See Examples 6.1, 6.2, and 6.6. Use the formula from Exercise 58 in Section 14.6.

$y = f(x, z) = \sqrt{x^2 + z^2}$

$\|\mathbf{n}\| = \sqrt{(f_x)^2 + (f_z)^2 + 1}$

$\displaystyle = \sqrt{\left(\dfrac{x}{\sqrt{x^2 + z^2}}\right)^2 + \left(\dfrac{z}{\sqrt{x^2 + z^2}}\right)^2 + 1}$

$= \sqrt{2}$

$\displaystyle \iint_S yz\,dS = \iint_R (\sqrt{x^2 + z^2})(z)(\sqrt{2})\,dA$

$= 0 \ \ (\text{by symmetry})$

65. See Examples 6.1, 6.2, and 6.6 for techniques of surface integration. See Exercises 47–56 in Section 14.2 for formulas related to mass and center of mass.

$z = x^2 + y^2$

$\mathbf{n} = \langle 2x, 2y, -1 \rangle$

$dS = \sqrt{4x^2 + 4y^2 + 1}\,dA$

$\displaystyle m = \iint_S \rho\,dS$

$\displaystyle = \iint_R (2)(\sqrt{4x^2 + 4y^2 + 1})\,dA$

$\displaystyle = 2\int_0^{2\pi} \int_0^2 \sqrt{4r^2 + 1}\,r\,dr\,d\theta$

$\displaystyle = 2\int_0^{2\pi} \left[\dfrac{1}{12}(4r^2 + 1)^{3/2} \right]_{r=0}^{r=2} d\theta$

$\displaystyle = \int_0^{2\pi} \dfrac{1}{6}(17^{3/2} - 1)\,d\theta$

$= \dfrac{\pi}{3}(17\sqrt{17} - 1)$

≈ 72.35

$\displaystyle \bar{x} = \dfrac{1}{m} \iint_S x\rho\,dS$

$\displaystyle = \dfrac{1}{m} \iint_R (2x)\left(\sqrt{4x^2 + 4y^2 + 1}\right) dA$

$= 0 \ (\text{by symmetry})$

$\displaystyle \bar{y} = \dfrac{1}{m} \iint_S y\rho\,dS$

$\displaystyle = \dfrac{1}{m} \iint_R (2y)\left(\sqrt{4x^2 + 4y^2 + 1}\right) dA$

$= 0 \ (\text{by symmetry})$

$\displaystyle \bar{z} = \dfrac{1}{m} \iint_S z\rho\,dS$

$\displaystyle = \dfrac{1}{m} \iint_R 2(x^2 + y^2)\left(\sqrt{4x^2 + 4y^2 + 1}\right) dA$

$\displaystyle = \dfrac{2}{m} \int_0^{2\pi} \int_0^2 r^2 \sqrt{4r^2 + 1}\,r\,dr\,d\theta$

$= \dfrac{2}{m} \cdot \dfrac{\pi}{60}(1 + 391\sqrt{17})$

$= \dfrac{1 + 39\sqrt{17}}{10(17\sqrt{17} - 1)}$

≈ 2.33

Note : See the solution to Exercises 35 in Section 14.6 for the steps to calculate the integral for $\bar{z}$.

67. See Example 7.1.

$$\nabla \cdot \mathbf{F}(x, y, z) = \nabla \cdot \left\langle y^2 z,\ y^2 - \sin z,\ 4y^2 \right\rangle$$
$$= 0 + 2y + 0$$
$$= 2y$$

$$\iint_{2Q} \mathbf{F} \cdot \mathbf{n}\, dS = \iiint_{Q} \nabla \cdot \mathbf{F}(x, y, z)\, dV$$

$$= \int_0^2 \int_0^{4-2y} \int_0^{4-2y-x} 2y\, dz\, dx\, dy$$

$$= \int_0^2 \int_0^{4-2y} [2yz]_{z=0}^{z=4-2y-x}\, dx\, dy$$

$$= \int_0^2 \int_0^{4-2y} 2y(4 - 2y - x)\, dx\, dy$$

$$= \int_0^2 \left[2y(4 - 2y)x - yx^2 \right]_{x=0}^{x=4-2y}\, dy$$

$$= \int_0^2 [2y(4 - 2y)^2 - y(4 - 2y)^2]\, dy$$

$$= \int_0^2 (4y^3 - 16y^2 + 16)\, dy$$

$$= \left(y^4 - \frac{16}{3}y^3 + 16y \right)\Bigg|_{y=0}^{y=2} = \frac{16}{3}$$

69. See Example 7.1.

$$\nabla \cdot \mathbf{F}(x, y, z) = \nabla \cdot \left\langle 2xy,\ z^3 + 7yx,\ 4xy^2 \right\rangle$$
$$= 2y + 7x + 0$$
$$= 7x + 2y$$

$$\iint_{\partial Q} \mathbf{F} \cdot \mathbf{n}\, dS = \iiint_{Q} \nabla \cdot \mathbf{F}(x, y, z)\, dV$$

$$= \int_{-1}^{1} \int_0^{1-y^2} \int_0^{4-z} (7x + 2y)\, dx\, dz\, dy$$

$$= \int_{-1}^{1} \int_0^{1-y^2} \left[\frac{7}{2}x^2 + 2yx \right]_{x=0}^{x=4-z}\, dz\, dy$$

$$= \int_{-1}^{1} \int_0^{1-y^2} \left[\frac{7}{2}(4 - z)^2 + 2y(4 - z) \right] dz\, dy$$

$$= \int_{-1}^{1} \int_0^{1-y^2} \left[\frac{7}{2}z^2 - (28 + 2y)z + (56 + 8y) \right] dz\, dy$$

$$= \int_{-1}^{1} \left[\frac{7}{6}z^3 - (14 + y)z^2 + (56 + 8y)z \right]_{z=0}^{z=1-y^2}\, dy$$

$$= \int_{-1}^{1} \left[\frac{7}{6}(1 - y^2)^3 - (14 + y)(1 - y^2)^2 + (56 + 8y)(1 - y^2) \right] dy$$

$$= \int_{-1}^{1} \left(-\frac{7}{6}y^6 - y^5 - \frac{21}{2}y^4 - 6y^3 - \frac{63}{2}y^2 + 7y + \frac{259}{6} \right) dy$$

$$= \int_{-1}^{1} \left(-\frac{7}{6}y^6 - \frac{21}{2}y^4 - \frac{63}{2}y^2 + \frac{259}{6} \right) dy$$

$$= \left(-\frac{1}{6}y^7 - \frac{21}{10}y^5 - \frac{21}{2}y^3 + \frac{259}{6}y \right)\Bigg|_{y=-1}^{y=1}$$

$$= \frac{304}{5}$$

Note: We used symmetry to drop the odd powers of y.

71. $\nabla \cdot \mathbf{F}(x, y, z) = \nabla \cdot \left\langle xz, yz, x^2 - z \right\rangle$

$\qquad\qquad\qquad = z + z - 1$

$\qquad\qquad\qquad = 2z - 1$

$\iint_{\partial Q} \mathbf{F} \cdot \mathbf{n}\, dS = \iiint_{Q} \nabla \cdot \mathbf{F}(x, y, z)\, dV$

$\qquad\qquad\quad = \int_{0}^{2\pi} \int_{0}^{2} \int_{0}^{r} (2z - 1)\, r\, dz\, dr\, d\theta$

$\qquad\qquad\quad = \int_{0}^{2\pi} \int_{0}^{2} [(z^2 - z)r]_{z=0}^{z=r}\, dr\, d\theta$

$\qquad\qquad\quad = \int_{0}^{2\pi} \int_{0}^{2} (r^3 - r^2)\, dr\, d\theta$

$\qquad\qquad\quad = \int_{0}^{2\pi} \left[\dfrac{1}{4}r^4 - \dfrac{1}{3}r^3 \right]_{r=0}^{r=2} d\theta$

$\qquad\qquad\quad = \int_{0}^{2\pi} \dfrac{4}{3}\, d\theta$

$\qquad\qquad\quad = \dfrac{8}{3}\pi$

73.   See example 8.2.

∂S is the triangle with vertices at $(0, 0, 0)$, $(0, 2, 0)$, and $(0, 0, 1)$.

C_1 : From $(0, 0, 0)$ to $(0, 2, 0)$

$x = 0, y = 2t, z = 0, 0 \leq t \leq 1$

$\int_{C_1} \mathbf{F}(x, y, z) \cdot d\mathbf{r} = \int_{C_1} (zy^4 - y^2)dx + (y - x^3)dy + z^2 dz$

$\qquad\qquad\qquad = \int_{0}^{1} [(-4t^2)(0) + (2t)(2) + (0)^2(0)]dt$

$\qquad\qquad\qquad = \int_{0}^{1} 4t\, dt$

$\qquad\qquad\qquad = 2t^2 \Big|_{t=0}^{t=1}$

$\qquad\qquad\qquad = 2$

C_2 : From $(0, 2, 0)$ to $(0, 0, 1)$

$x = 0, y = 2 - 2t, z = t, 0 \leq t \leq 1$

$\int_{C_1} \mathbf{F}(x, y, z) \cdot d\mathbf{r} = \int_{C_1} (zy^4 - y^2)dx + (y - x^3)dy + z^2 dz$

$\qquad\qquad\qquad = \int_{0}^{1} [(t(2 - 2t)^4 - (2 - 2t)^2(0) + (2 - 2t)(-2) + (t^2)(1)]dt$

$\qquad\qquad\qquad = \int_{0}^{1} (t^2 + 4t - 4)dt$

$\qquad\qquad\qquad = \dfrac{1}{3}t^3 + 2t^2 - 4t$

$\qquad\qquad\qquad = -\dfrac{5}{3}$

C_3 : From $(0, 0, 1)$ to $(0, 0, 0)$

$x = 0, y = 0, z = 1 - t, 0 \le t \le 1$

$$\int_{C_3} \mathbf{F}(x, y, z) \cdot d\mathbf{r} = \int_{C_3} (zy^4 - y^2)dx + (y - x^3)dy + z^2 dz$$

$$= \int_0^1 [(0)(0) + (0)(0) + (1-t)^2(-1)]dt$$

$$= \int_0^1 (-t^2 + 2t - 1)dt$$

$$= \left(-\frac{1}{3}t^3 + t^2 - t \right) \Big]_{t=0}^{t=1}$$

$$= -\frac{1}{3}$$

$$\iint_S (\nabla \times \mathbf{F}) \cdot \mathbf{n}\, dS = \int_{C_1} \mathbf{F} \cdot d\mathbf{r} + \int_{C_2} \mathbf{F} \cdot d\mathbf{r} + \int_{C_3} \mathbf{F} \cdot d\mathbf{r}$$

$$= 2 - \frac{5}{3} - \frac{1}{3} = 0$$

75. See Examples 8.1 and 8.2.

The boundary curve ∂S is somewhat difficult to parameterize, so evaluate directly.

$$\nabla \times \mathbf{F} = \begin{vmatrix} \mathbf{i} & \mathbf{j} & \mathbf{k} \\ \frac{\partial}{\partial x} & \frac{\partial}{\partial y} & \frac{\partial}{\partial z} \\ 4x^2 & 2ye^{2y} & \sqrt{z^2+1} \end{vmatrix}$$

$$= \left(\frac{\partial}{\partial y}\left(\sqrt{z^2+1}\right) - \frac{\partial}{\partial z}(2ye^{2y}) \right)\mathbf{i} - \left(\frac{\partial}{\partial x}\left(\sqrt{z^2+1}\right) - \frac{\partial}{\partial z}(4x^2) \right)\mathbf{j} + \left(\frac{\partial}{\partial x}(2ye^{2y}) - \frac{\partial}{\partial y}(4x^2) \right)\mathbf{k}$$

$$= (0 - 0)\mathbf{i} + (0 - 0)\mathbf{j} + (0 - 0)\mathbf{k}$$

$$= \langle 0, 0, 0 \rangle$$

$$\iint_S (\nabla \times \mathbf{F}) \cdot \mathbf{n}\, dS = \iint_S \mathbf{0} \cdot \mathbf{n}\, dS = 0$$

77. See Example 8.1.

$$\nabla \times \mathbf{F} = \begin{vmatrix} \mathbf{i} & \mathbf{j} & \mathbf{k} \\ \frac{\partial}{\partial x} & \frac{\partial}{\partial y} & \frac{\partial}{\partial z} \\ 2xy(\cos z) & y^2 + x^2(\cos z) & z - x^2 y(\sin z) \end{vmatrix}$$

$$= \left(\frac{\partial}{\partial y}(z - x^2 y(\sin z)) - \frac{\partial}{\partial z}(y^2 + x^2(\cos z)) \right)\mathbf{i} - \left(\frac{\partial}{\partial x}(z - x^2 y(\sin z)) - \frac{\partial}{\partial z}(2xy(\cos z)) \right)\mathbf{j}$$

$$+ \left(\frac{\partial}{\partial x}(y^2 + x^2(\cos z)) - \frac{\partial}{\partial y}(2xy(\cos z)) \right)\mathbf{k}$$

$$= (-x^2(\sin z) + x^2(\sin z))\mathbf{i} - (-2xy(\sin z) + 2xy(\sin z))\mathbf{j} + (2x(\cos z) - 2x(\cos z))\mathbf{k}$$

$$= \langle 0, 0, 0 \rangle$$

$$\int_C \mathbf{F} \cdot d\mathbf{r} = \iint_S (\nabla \times \mathbf{F}) \cdot \mathbf{n}\, dS$$

$$= \iint_S \mathbf{0} \cdot \mathbf{n}\, dS$$

$$= 0$$